Fundamentals of Plastics Mould Design

About the Authors

Prof.(Dr.) Sanjay K Nayak is holding a D.Sc. degree in Polymer composites and nanocomposites. He has more than 25 years of teaching and industrial experience and has published more than 200 research papers in reputed journals. He has been author and editor for 10 books in the area of Advanced Materials and Polymer Science & Engineering.

Pratap Chandra Padhi has graduated in mechanical engineering from Institute of Engineers (India) and has 20 years of teaching experience in the field of plastics moulds and die design, tooling, CAD / CAM / CAE engineering. He has Technical expertise in the state of art CNC – EDM, wire cut EDM, Milling and Lathe machines.

Y. Hidayathullah has more than eight years of teaching experience in the filed of CAD/CAM/CAE. He has good knowledge on CAD/ CAM/CAE softwares such as Unigraphics, CATIA, Pro/Engineer, Mold flow and AutoCAD. He has been involved in various projects related to plastics Products and Mould Design and Injection moulded part validation.

Fundamentals of Plastics Mould design

Prof. (Dr.) Sanjay K Nayak
Pratap Chandra Padhi
Y. Hidayathullah

Central Institute of Plastics Engineering & Technology (CIPET)
T.V.K. Industrial estate
Guindy, Chennai
Tamilnadu

Tata McGraw Hill Education Private Limited
NEW DELHI

McGraw-Hill Offices

New Delhi New York St Louis San Francisco Auckland Bogotá Caracas
Kuala Lumpur Lisbon London Madrid Mexico City Milan Montreal
San Juan Santiago Singapore Sydney Tokyo Toronto

Tata McGraw-Hill

Published by Tata McGraw Hill Education Private Limited,
7 West Patel Nagar, New Delhi 110 008

Fundamentals of Plastics Mould design

ISBN (13 digits): 978-1-25-909756-0
ISBN (10 digits): 1-25-909756-0

Vice President and Managing Director—McGraw-Hill Education: *Ajay Shukla*

Head—Higher Education Publishing and Marketing: *Vibha Mahajan*

Publishing Manager—SEM and Tech Ed.: *Shalini Jha*
Sr Editorial Researcher Acquisitions: *Vamsi Deepak Sankar*
Executive—Editorial Services: *Sohini Mukherjee*
Sr Production Manager: *Satinder Singh Baveja*

Marketing Manager—Higher Education: *Vijay Sarathi*
Graphic Designer—Cover: *Meenu Raghav*

General Manager—Production: *Rajender P Ghansela*
Asst General Manager—Production: *Reji Kumar*

Typeset at BeSpoke Integrated Solutions, Puducherry 605 008, India and printed at Rajkamal Electric Press, Plot No. 2, Phase-IV, Kundli, Haryana.

Cover: Rajkamal Electric Press

The McGraw-Hill Companies

Contents

Preface

ABOUT THE BOOK

We feel great pleasure to bringing out the book, *Fundamentals of Plastics Mould Design*, for the benefit of the students' community. This book has been written explicitly to meet the requirements of B.E./B.Tech./M.E./M.Tech. courses of Plastics/Polymer Technology branches and also with the perspective of enlightening students about the plastics' mould design. *Fundamentals of Plastics Mould Design* also meets the requirements of Diploma/ Post-diploma / Postgraduate diploma courses of Plastics Mould Technology and Plastics Technology/ Plastics Mould Design/Plastics Processing and Testing, respectively.

All the mould designers should know about Plastics Product Design, Mould Design, Plastic materials, its processing and testing. This book covers the design aspects of the plastics such as product design and mould design. Being a book on mould design, *Fundamentals of Plastics Mould Design* covers the design aspects of Injection Mould, Compression Mould, Transfer Mould, Blow Mould and Extrusion.

In addition to the fundamental design concepts such as Basic Mould Construction, Ejection Systems, Feed Systems and Cooling Systems, this book also covers the calculation aspects, the recent technologies like CAD/CAM applications in the field of product and mould design, Prototype Development Methods, Reverse Engineering, Role of Mould flow software for analysing the Plastics Injection Moulded Products, etc.

SALIENT FEATURES

- Through out the book the same system of SI units are adopted
- Also provided with equivalent fps system of unit in required areas
- Easy communication between the book and the readers
- In depth discussion in all chapters
- Total of 550 Questions covering all chapters have been included
- Solved examples for design calculations

CHAPTER ORGANISATION

The book contains 8 chapters. **Chapter 1** covers plastics product design, **Chapter 2** discusses injection mould design, compression mould design, transfer mould design are dealt in **Chapter 3 and 4** respectively. **Chapter 5** covers advanced injection mould design, while **Chapter 6** covers blow mould design. **Chapters** 7 and 8 cover extrusion dies and CAD/CAM applications in mould design.

All chapters of this book consist of simplified explanation in the introduction, relevant illustrations and figures wherever required for the better understanding of students and solved examples of numerical problems. Moreover, important questions at the end of each chapter are given for the awareness of the students to face their exams confidently. The fashion of presentation remains the same for all the eight chapters, for better communication between the book and the readers.

ACKNOWLEDGEMENTS

We are thankful to our publisher, Tata McGraw-Hill for the cooperation and support towards the outcome of this book.

Special thanks to Mr. M. Pheer Mohamed, Manager-Training, CIPET, Head office, Chennai and his team for his undaunted support in bringing out the book.

Sanjay K. Nayak
Pratap Chandra Padhi
Y. Hidayathullah

PUBLISHER'S NOTE

Any constructive criticism of the book is most welcome by the readers. You can write to the authors at *cipethq@vsnl.com* by mentioning the title and the author's name along with your feedback. Report of any piracy related problems will be highly appreciated both by the author and the publisher.

CHAPTER 1

Product Design

1.1 GENERAL PRODUCT DESIGN CONCEPT

1.1.1 Introduction

Design is the universal term being used everywhere; it is the art of giving shape to the ideas and defined as the product of creative thinking. It actually gives a complete description of an object and prescription for its production. Hence, a product design deals with conversion of ideas into reality and aims at fulfilling human needs.

1.1.2 Basic Concepts of Design

Any product design should have a definite shape, proportionate size with good appearance and reveal the function of the product. Hence, the primary concept involved in design is shape, size, aesthetics and function.

Shape is defined as the geometry and topology of a feature. All the parameters and their values that fully define a feature are called **size** or size of a feature. If we combine shape and size with suitable colours for the specific purpose of the product, it has to give attractive appearance or good eye appeal, which is called **aesthetics**.

Appearance determines the whole 'character' of the product. It should reflect pride of ownership, the function served, high product quality and value and the reputation of the maker.

Analysis of product aesthetics The appearance of a product is the total visual effect produced by structure, form, material, dimension and surface finish including colour. In composite products, structure implies combined effect produced by positioning of adjacent forms in deriving the final product.

The visual appeal of objects like jewellery is very important. On the other hand, items such as screws, nails, and ferrules have sheer functional value and negligible aesthetic value. An attractive appearance alone is not only the deciding factor for a good design but the product which is designed should fulfill the functional requirements of the consumer with reasonable cost.

The concept involved in design of a product is its quality and economy. The term quality means fitness for use. If a designer wants to produce a good quality product, then the cost of the product will come into picture.

Hence, the product must be designed in such a way that without sacrificing the quality of the product, the cost of the product should be brought down to an optimum level. Accordingly, the product configuration / geometry, type of finish, accuracy, material selection, processing methods and post-moulding procedures, etc., are to be selected.

1.1.3 Essential Factors of Product Design

1. **Need:** A design must be in response to individual or social needs, which can be satisfied by the technological status of the times when the design is to be prepared.
2. **Physical Reliability:** A design should be convertible into material goods or services, i.e., it must be physically reliable and should last long.
3. **Economic Worthiness:** The goods or services, described by a design, must be useful to the consumer which equals or exceeds the sum of the total costs of making it available to him.
4. **Financial Feasibility:** The operations of designing, producing and distributing the goods must be financially supportable, i.e., a design project should be capable for being funded by suitable agencies. The method for assessment of financial feasibility could be 'net present value' which states that the present worth of cash flows in the project when added up during the useful life of the product should be greater than the initial investment for the product.
5. **Optimality:** The choice of a design concept must be optimal amongst the available alternatives. The selection of the chosen design concept must be optimal among all possible design proposals. Optimal design, in theory, strives to achieve the best or singular point derived by calculus methods. In the context of optimisation under constraints for mechanical strength, better quality, minimum weight, minimum cost, etc., are the criteria for optimisation.
6. **Design Criterion:** Optimality must be established relative to a design criterion which represents the designer's compromise among possibly conflicting value judgments, which include those of the consumer, the producer, the distributor and his own.
7. **Morphology:** Design is progression from the abstract to the concrete. The morphology of design refers to the study of the chronological horizontal structure of the design projects. It is defined by the phases shown in Fig. 1.1 and their constituent steps of the seven phases; the first three phases belong to design such as feasibility study phase, preliminary design phase, detailed design phase; and the remaining four phases belong to production, distribution, consumption and retirement.
8. **Design Process:** Design is an iterative problem-solving process. This gives a vertical structure to each design phase. The iterative nature of design is owing to feedback

from existing design and improvement with further information in the form of technological, financial and creativity inputs.

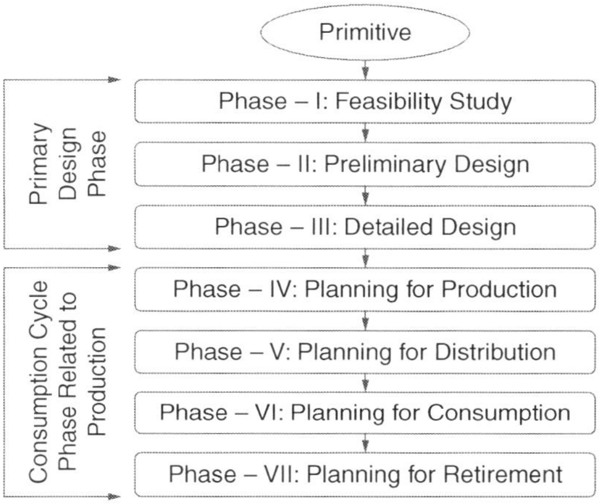

Fig. 1.1 Morphology of design process.

9. **Subproblems:** During the process of solution of the design problem, a sublayer of subproblems appears; the solution of the original problem is dependent on the solution of the subproblems. The 'Design Tree' of Fig.1.2 reveals the concept of subproblems.

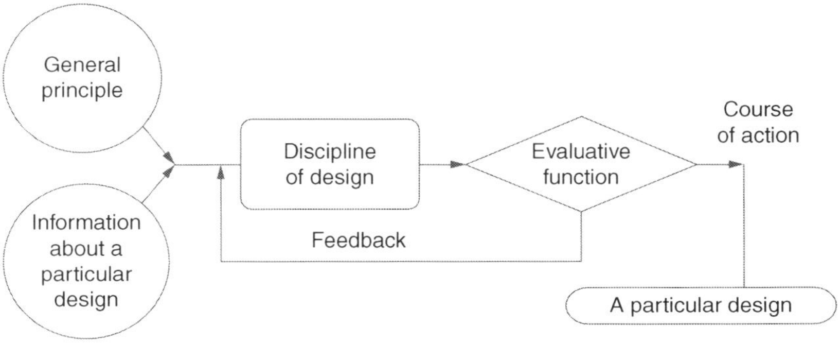

Fig. 1.2 Iterative nature of design process.

10. **Reduction of Uncertainty:** Design is derived after processing information that results in a transition for uncertainty, about the success or failure of a design towards certainty. Each step in design morphology from step (1) to step (7) enhances the level of confidence of the designer.
11. **Economic Worth of Evidence:** Information gathering and processing have a cost that must be balanced by the worth of the evidence, which affects the success or failure of the design. Authentic information should be gathered to make the design project a success. Today, information is regarded as a resource which is as valuable as money, manpower and material.

12. **Bases for Decision:** A design project is terminated when it is obvious that its failure calls for its abandonment. It is continued when confidence in an available design solution is high enough to indicate the commitment of resources necessary for the next phase.
13. **Minimum Commitment:** In the solution of a design problem at any stage of the process, commitments which will fix future design decisions must not be made beyond what is necessary to execute the immediate solution. This will allow maximum freedom in finding solutions to subproblems at the lower levels of design.
14. **Communication:** A design is a description of an object and prescription for its production. It will exist to the extent it is expressed in the available modes of communication. The best way to communicate a design is through drawings, which is the universal language of designers. Three-dimensional renderings or sectional views help explain the design to the user of the design. The present day impact of computer aided modelling and drafting has resulted in very effective communication between the designer and the user.

1.2 DESIGNING FOR PLASTICS

1.2.1 Introduction

In the present day the application of plastics is wide as compared to traditional material such as metal. Plastics offer impressive advantages over metals. They are listed below:

1. They are not subjected to corrosion.
2. Light in weight with good strength to weight ratio.
3. Cost effective.
4. Less cycle time for production.
5. Unique design freedom.
6. Good mechanical, electrical and chemical properties.
7. They are available in wide range of colours.
8. Reduced assembly time.

In addition to it, each plastic material offers some special property which serves a particular application or can be made to do so by the incorporation of suitable additives with the plastic materials.

1.2.2 Advantages of Plastics

The successful use of plastics usually derives from a combination of cost savings and improvement in performance or appearance, but often the cost saving alone is sufficient to justify the choice of plastic material. Plastics can offer the following technical advantages:

(a) **Light Weight:** All plastics have low densities, generally in the range 830 to 2500 kg/m^3. These figures can be extended upward or downward. For example, foamed materials can have densities as low as 10 kg/m^3 and filled plastics as high as 3500 kg/m^3. In comparison, the density of aluminium is about 2700 kg/m^3 and that of stainless steel about 7900 kg/m^3. It is evident that large volume of plastic material for unit weight can be obtained than with metals.

(b) **Toughness:** Some plastics are very tough, and objects made from them are difficult to destroy by mechanical treatment. Some plastics are less tough, and some others are fragile.

(c) **Resilience:** Plastics show some of the behaviour associated with rubbers in accommodating relatively large strains without fracture and in recovering their original shape and dimensions when the stress is removed.

(d) **Vibration Damping:** The quietness in use of plastics gear trains and bottle crates depends on the inherently high degradation of mechanical energy to heat.

(e) **Resistance to Fatigue:** In general plastics perform remarkably satisfactorily in situations involving dynamic stresses or strains.

(f) **Low Coefficient of Friction:** Plastics to plastics and plastics to metals combination have low coefficient of friction and can often perform unlubricated without fear of seizing.

(g) **Thermal Insulation:** Plastics are good insulators, their thermal conductivities being many orders of magnitude lower than those of metals. This low conductivity may be exploited in handles for utensils and in the design of pipes for carrying hot fluids.

(h) **Corrosion Resistance:** In general, plastics are resistant to corrosion. They are resistant to weak acids, weak bases and aqueous salt solutions, although strong oxidising acids may cause some attack, leading to discolouration and possible embrittlement. On the other hand, organic solvents, on which metals are generally inert, may cause swelling, deterioration of properties and eventual dissolution.

(i) **Colour Possibilities:** Some plastics are transparent, some are translucent and a few are opaque. Acrylics, polystyrenes, methylpentene polymers, polycarbonates and certain grades of PVC can be very transparent indeed to visible light. All plastics can be coloured by incorporating a wide range of dyes or pigments, thus avoiding the need for painting. However, subsequent painting or plating is possible with some, if required.

(j) **Manufacturing Methods:** A variety of automatic and semi-automatic techniques allows easy, economical and reproducible fabrication of articles and components. Further finishing operations are easy to carry out on most plastics.

(k) **Integrated Design:** The favourable processing characteristics and the properties offered, allow the design and manufacture of polyfunctional shapes without the need for assembly.

(l) **Price:** On a volume basis, raw materials of plastics are considerably less expensive than many metals and in spite of inflationary pressures their prices have tended to fall relatively to those of other materials.

1.2.3 Limitations of Plastic

The disadvantages frequently turn out to be not so much limitations as challenges for the designer to think of plastics as materials in their own right rather than as substitutes.

(a) Strength, surface hardness and abrasion resistance
(b) Modulus
(c) Temperature resistance
(d) UV Resistance and outdoor weathering
(e) Flammability
(f) Thermal expansion
(g) Electrostatic charges
(h) Orientation

The designer should therefore, bear this in mind and take appropriate steps to overcome the same, in order to meet the specified requirements of application.

1.2.4 The Material Selection and Its Characteristics

The technical and cost considerations would influence selection of a particular material for a particular product. Selection of materials is a highly complex process. In order to arrive at the optimum materials for a given use with some degree of efficiency and reliability, a systematic approach should be adopted to outline the proper approach to material selection and to compare plastics both with other engineering materials and among themselves in terms commonly used in material selection.

There are two criteria in the selection of materials, i.e., function and cost. How a material will perform in a given use and the cost of the material are two of the primary considerations in selecting a material.

The following criteria are required for selecting the plastic material:

1. Mechanical properties such as strength, toughness, rigidity, resistance to creep and fatigue, abrasion and wear resistance, resilience and hardness.
2. Thermal properties which include the effect of temperature (particularly that of maximum and minimum anticipated service temperature on engineering properties) thermal conductivity, expandability, etc.
3. Electrical properties such as conductive or insulating qualities.
4. Resistance to chemicals or other environments.

The structural machine parts, products or devices are subject to loading conditions or other external environments which produce internal stresses and strains. Knowledge of stress, strain or mechanical behaviour of materials is important for a number of reasons. In order to develop a safe and economical design with the minimum weight and cost, the mechanical or stress strain properties of materials are to be considered. To achieve the above requirements, part design becomes an integrative process, with many experiments and false starts along the way. Although sheer creativity is essential to good design, engineers also tend to go through a methodical series of steps in their quest for an optional design solution.

The accurate geometry, i.e., shape and size of a component made in plastics is not its functional guarantee, even if it is found fulfilling the immediate functional requirements. This is arrived at by evaluating the long-term characteristics such as

1. Creep/flexural modulus
2. Resistance to electrical/potential discharge
3. Rate of imbrittlement
4. Environmental stress cracking/crazing
5. High temperature (continuous) with standability
6. Tribological properties

1.2.5 Methodical Approach in Plastics Product Design

The design of a good plastics product requires enough information about plastics and their properties, different moulding methods, post-moulding procedures and information in key design areas. Several factors have to be considered, which may lead to a wide choice in ultimate design features of a product. To ensure proper design, close cooperation is required between the industrial designer, the engineer, the draughtsman, the tool builder, the moulder, and the raw material supplier. Preferably, each must become involved to some extent at the very beginning of the design process.

The step-wise procedure for the development of a plastic part is detailed below:

1. Define the function of the part with life requirement.
2. Define shape and size based on space and capacity.
3. Assess loading conditions, stresses and deformation.
4. Define all environmental conditions.
5. Select several materials to satisfy the above conditions and other relevant properties.
6. Do several trial designs using different materials and geometries to perform the required function.
7. Define the manufacturing process for each design.
8. Evaluate the trial designs on a cost effective basis. Determine several levels of performance and the specific costs associated with each to the extent that it could be done with available data.
9. Based on the above detailed study, select the best apparent choice and do a detailed design of the part.

10. Based on the detailed design, select the probable part design, material and process.
11. Make a model, if necessary to test the effectiveness of the part.
12. Build prototype tooling, if necessary.
13. Make prototype parts to determine if they meet the required function.
14. Redesign the part if necessary based on the prototype testing.
15. Retest.
16. Make field tests.
17. Add instructions for use.

1.2.6 Design Theory

Design concept

1. What are the end use requirements for the part or product (aesthetic, structural, mechanical)?
2. How many functional items can be designed into the part for cost effectiveness?
3. Can multiple parts be combined into one large part?

Engineering consideration

1. What are the structural requirements?
2. Are the load static, dynamic, cycling? What are the stress levels?
3. What deflection can be tolerated?
4. Is the part subject to impact loads?
5. What tolerances are required for proper functioning and assembly?
6. What kind of environment will the part see?
7. What operating temperature will it have?
8. What will its chemical exposure be?
9. What is the expected life of the product?
10. How will the product be assembled?
11. What kind of finish will be required on the parts?
12. Are agency requirements or codes involved?
13. Can the proposed product be moulded and finished economically?

Once the above questions have been considered, the next step is usually to consult data property sheets to compare material. Properties presented in these sheets are for comparative purposes and not generally for design. Seldom will a part's design conditions match the conditions used for generating the data on the poperty sheets, but the standardised tests are a valuable tool. Without standardised data properties, fair comparisons could not be made. The standardised information on mechanical strength, impact, chemical resistance, etc., must be adjusted for the end-use environments and life of the product.

After one selects the proper material for the part, calculations of wall thicknesses and part geometry are made, followed by the next design step, which is to improve the effectiveness of the design. In the case of injection moulded parts, the design should be reviewed in terms of the following questions:

1. Can a tool be built and the part moulded?
2. Are the wall thicknesses adequate for the flow of the material to fill the part?
3. Have all internal corners been 'radiused' to reduce all high localised stress points?
4. Do all changes in wall thicknesses have smooth transitions?
5. Are heavy wall sections cored out to give a uniform wall where possible?
6. Is the ratio of rib or boss thickness to adjacent wall thickness proper?
7. Is it possible to gate into the thicker wall sections and flow to the thinner sections?
8. Are weld lines going to present strength or appearance problems?
9. Have adequate draft angles been included on all surfaces?
10. Have reasonable tolerances been selected for all parts?

If you have properly evaluated the needs of the product, chosen the proper material, optimised the design for that material, and, finally, carefully considered proper manufacturing practices, you will be on the way to having a part that works.

1.2.7 Quality and Economy

In product design the quality and economy are the most important requirements from raw material selection to finished product and finally reach the customer.

To ensure the product quality the following factors must be considered:

1. Methodical approach in product design.
2. Selection of suitable processing machine.
3. Efficient and economy mould design.
4. Trouble free mould construction (minimum mould cost).
5. Adopt quality control methods.
6. Suitable correction and modification based on feedback from customers.

Economy Cost effectiveness is an important factor in product design. This can be achieved by involving effective method of product design only. Economic success essentially depends upon external influences such as market situation and competition and ecological considerations, management attention must also be directed toward the possibilities existing inside its own factory.

Product manufacturer will have to consider the following factors as most important to achieve the economy in their product:

1. Selection of raw material (least in cost, but best in product required properties like aesthetic, strength, etc.).

2. Product design concept like minimum / maximum wall thickness to meet out economic product (the optimum wall thickness as to design without loss of its strength).
3. The prototype mould are rapid prototype product before regular mould to reduce the initial mould development cost.
4. Selection of suitable processing machines to avoid wastage of raw material and time.
5. Best mould design to minimise the tooling cost and processing cost (reducing cycle time).
6. Introduce modern technology method like hot runner / flash free quality moulds using special grade of mould steels and surface treatments.
7. Engage minimum labour to minimise cost. Introduce modernised methods (like automation) if more labour is necessary.

1.2.8 Product Design Appraisal

Product design has to fulfil the essential qualities like shape, strength, aesthetic, shrinkage and tolerance, etc. These qualities can be achieved by better methodical approach in product design, good mould design concept, adopt modernised tool manufacturing methods, engaging suitable processing machineries and adopt good quality assurance methods.

Check list has to be prepared to evaluate the product in order to evaluate the product design:

1. Check over all shape.
2. Functional aspect.
3. Size with recommended tolerance.
4. Aesthetic without any processing defects.
5. Without any defect in mould design concept.
6. Free from warpage, sink mark, burn mark.
7. Ensure selection of raw material, processing machines, recommended concept of mould design and mould manufacturing methods.

With all above factors, quality products with cost effectiveness can be maintained.

1.2.9 Emphasis on Designing with Engineering Plastics

Successful manufacture of good plastic products requires a combination of sound judgments and experience. It also requires knowledge on plastic materials, their properties and applications, various moulding methods, post-moulding procedures and information of key design areas such as

1. Wall thickness
2. Parting line
3. Ribs, bosses and gussets
4. Radii and fillets

5. Taper / draft
6. Holes
7. Coring
8. Undercuts
9. Threads
10. Inserts
11. Gate size and location
12. Location of ejector pins
13. Tolerances
14. Fastening
15. Shrinkage
16. Surface finish

1.3 WALL THICKNESS

In plastic product design, selecting the correct nominal wall thickness is the most important factor. Choosing proper wall sections sometimes determine the ultimate success or demise of the product. While an inadequate wall section can lead to poor performance or structural failure, a section that is too heavy, even in just certain regions, can make the product unattractive, overweight, or too expensive. The following discussion on determining wall section thickness should help the design or production engineer to eliminate potential problems on paper (or computer screen) rather than in tool steel. In many parts, only some of the guidelines can be followed due to geometric, structural, or functional requirements, but at least the potential existence of a particular problem is known in advance and remedial action can be planned. For example, if a surface defect is discovered to be likely to appear in a visible area during moulding texture, logo or label can be planned for that region.

1.3.1 Nominal Wall Thickness

In product design, section thickness is usually governed by load requirements. But other considerations such as moulded in stress, part geometry, uniformity of appearance, resin flow or mouldability and economy are equally important. The design of wall thickness normally depends upon the selection of the material for particular application.

Just as metals plastics also have normal working ranges of wall thickness. The vast majority of injection moulded plastic parts probably range from 0.80 mm to 4.8 mm with the thickness within that range generally related to the total size of the part. That does not mean those parts cannot be moulded to be thinner or thicker or that a big part cannot be thin or a tiny part can not be thick. However, these norms can act as a starting point for the design. Tables 1.1 and 1.2 show the guidelines for wall thickness of various classes of thermoplastics and thermosetting plastics used in various applications.

Table 1.1 Suggested wall thickness for thermoplastic moulding material.

Thermoplastics materials	Minimum thickness (mm)	Average thickness (mm)	Maximum thickness (mm)
Acetal	0.4	1.6	3.2
ABS	0.8	2.3	3.2
Acrylic	0.6	2.1	6.1
Cellulosics	0.7	1.9	4.8
FEP fluoroplastic	0.3	9.0	12.7
Nylon	0.4	1.6	3.2
Polycarbonate	1.0	2.4	10.0
Polyethylene (L.D)	0.5	1.6	6.4
Polyethylene (H.D)	0.9	1.6	6.4
Ethylene vinyl acetate	0.5	1.6	3.2
Polypropylene	0.6	2.0	7.6
Polysulfone	1.0	2.6	9.5
Noryl (modified PPO)	0.8	2.0	9.5
Polystyrene	0.8	1.6	6.4
SAN	0.8	1.6	6.4
PVC- rigid	1.0	2.4	9.5
Polyurethane	0.7	12.7	38.0
Surlyn (ionomer)	0.7	1.6	19.0

Table 1.2 Suggested wall thickness for thermosetting moulding material.

Thermosetting materials	Minimum thickness (mm)	Average thickness (mm)	Maximum thickness (mm)
Alkyd - glass filled	1.0	3.2	12.7
Alkyd - mineral filled	1.0	4.8	9.5
Diallyl phthalate	1.0	4.8	9.5
Epoxy glass	0.8	3.2	25.4
Melamine-cellulose filled	0.9	2.5	4.8
Urea-cellulose filled	0.9	2.5	4.8
Phenolic-generakl purpose	1.3	3.2	25.4
Phenolic-flock filled	1.3	3.2	25.4
Phenolic-glass filled	0.8	2.4	19.0
Phenolic-fabric filled	1.6	4.8	9.5
Phenolic-mineral filled	3.2	4.8	25.4
Silicon glass	1.3	3.2	6.4
Polyester premix	1.0	1.8	25.4

The design engineer should also refer to data related to the ability of a plastic resin to flow into the mould cavity. This information, usually shown in the form of spiral flow curves, gives a relative measure of how far one can expect the plastic resin to flow from the gate (Fig. 1.3).

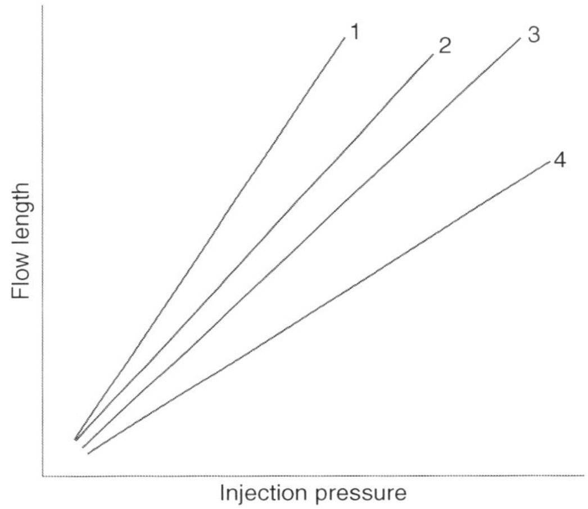

Fig. 1.3 Some typical spiral flow curves; 1. Nylon 6/6, 2. Thermoplastic polyester, PBT liquid crystal -glass reinforced, polyphenylene sulfide-glass reinforced, 3. Acetal copolymer, 4. PBT-glass reinforced.

A non-uniform wall thickness will cause more trouble than any other problem in part design. A thick section will cool last and sink away from the mould, causing a 'sink mark.' Heavy sections mean long cycle times. Fundamentals of design with any material require that the wall sections are of adequate thickness for the application of the part and be shaped for adequate strength.

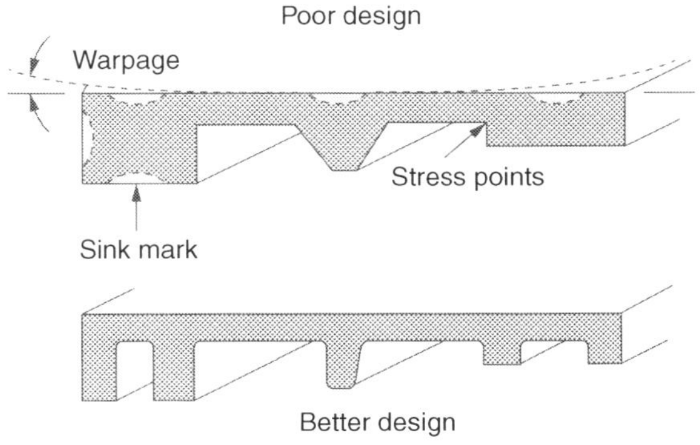

Fig. 1.4 Possibility of defects in poor design.

Wall thickness should be as uniform as possible to eliminate internal stresses, part distortion, cracking, warpage and sink marks (Fig. 1.4).

As per thumb rule, thicker the wall, the longer the part will have to stay in the mould in order to cure or cool properly. This rule is important while moulding by compression or injection. If different wall thickness in a part cannot be eliminated they should be bler.ded gradually (Fig. 1.5).

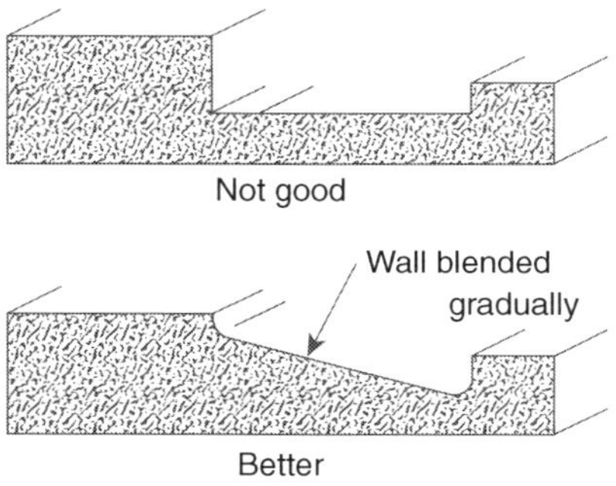

Fig. 1.5 Variable wall section on the part.

1.3.2 Variance in Wall Thickness

Wall thickness should neither be too heavy nor it should vary greatly. It should not vary more than a ratio of three to one. This is true of the fact that the part is moulded either by compression or injection methods. If parts are designed with thick and thin sections, sink marks will be evident on the thick sections. If translucent or transparent colour materials are used, a variance in depth of colour will be observed. Thick, heavy solid sections such as are found in knobs and handles should be redesigned into two individual mouldings (Fig. 1.6).

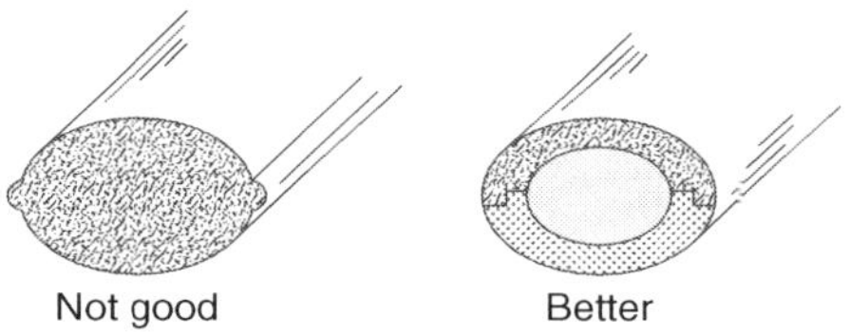

Fig. 1.6 Redesigned thick and heavy solid sections.

If a plastic part is designed with uneven wall thickness, it should be redesigned as shown in Fig. 1.7.

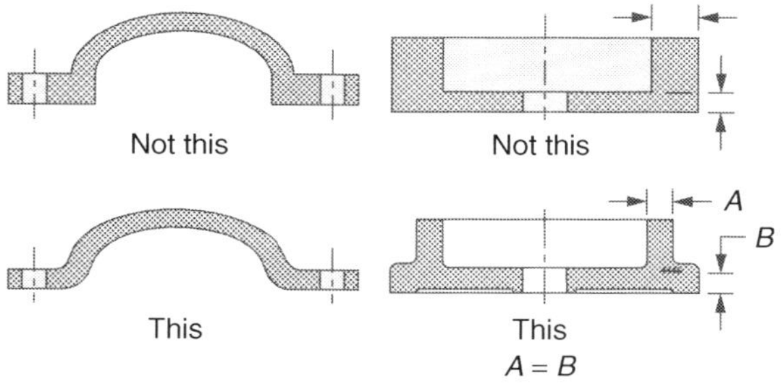

Fig. 1.7 Redesigned plastic product design for uniform wall thickness.

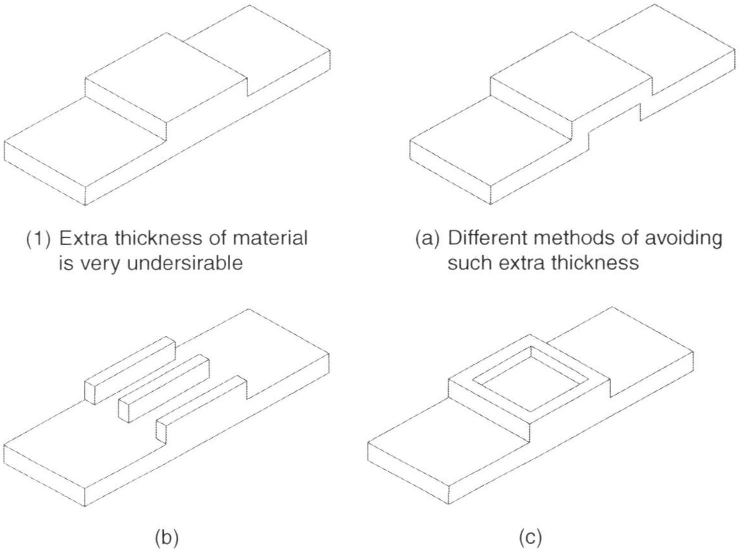

Fig.1.8 Compensating for different wall thickness.

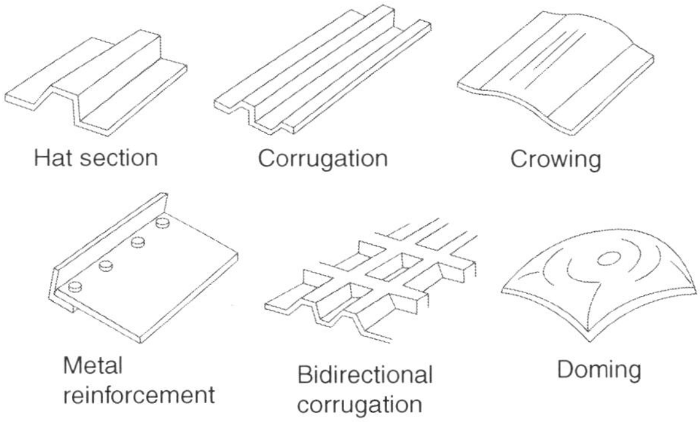

Fig.1.9 Modifications to nominal wall to improve structural response.

1.3.3 Determining the Wall Thickness

The determination of wall thickness should be the result of an analysis of the following requirements:

Functional requirements

1. Structure
2. Strength
3. Dimensional stability

4. Weight
5. Insulation

Manufacturing requirements

1. Moulding
2. Flow, setting and ejection
3. Assembly
4. Strength and precision

1.3.4 Structural Requirements of the Nominal Wall

If a part is subjected to any significant loading, the load bearing areas should be analysed for stress and deflection. If the stress or deflection is too high, the following alternatives should be considered:

1. Use ribs or contours to increase the section modulus. This is often the most economical solution and is discussed in detail under the heading 'Ribs and Bosses'.
2. Use a higher strength, higher modulus material.
3. Increase the wall section if it is not already too thick.

1.3.5 Insulation Characteristics of the Nominal Wall

Plastic materials are good insulators for electrical and heat energy. They can also serve as barriers and filters for sound and light. In general, insulating ability is directly related to the thickness of the plastic. In the case of sound transmission, change in thickness may be needed to change the resonant frequency of a plastic housing.

1.3.6 Impact Response of the Nominal Wall

The impact resistance of a particular part is directly related to its ability to absorb mechanical energy without fracture or plastic deformation. This, in turn, depends on the properties of the plastic resin and the geometry of the part. Increasing wall thickness generally improves the impact resistance of moulded parts. However, increased, wall thickness could hurt impact resistance by making the part overly stiff, unable to deflect and distribute impact. Both of these methods of absorbing impact energy should be examined when the nominal wall thickness is being selected.

1.4 PARTING LINE

The parting lines may be described as those lines made by the juncture of the male and female die and loose mould sections. It should be around the section of the part having the largest cross-sectional area. A beaded parting line helps in disguising mould misalignment. In case of peaked parting line, the flash can be cleaned without marring the surfaces. A stepped parting line will also allow clean and easy finishing of the component: the flash will be horizontal and easy to remove mechanically. In the event of bead is not permissible, a flush parting line may be necessary. The flash may be removed by adding a decorative effect made by grooving the part at the parting line. This method is possible only on round articles as shown in Fig.1.10 below.

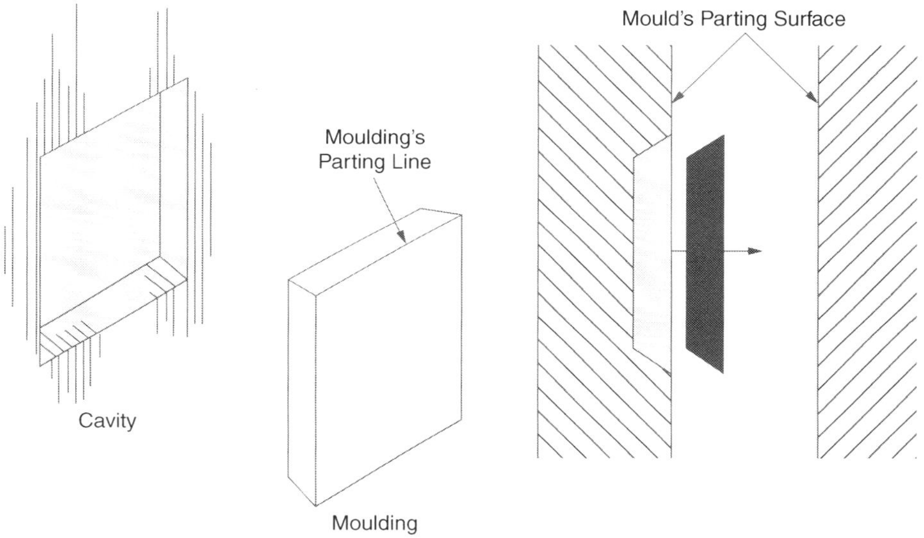

Fig. 1.10 Parting line of moulding and parting surface of mould.

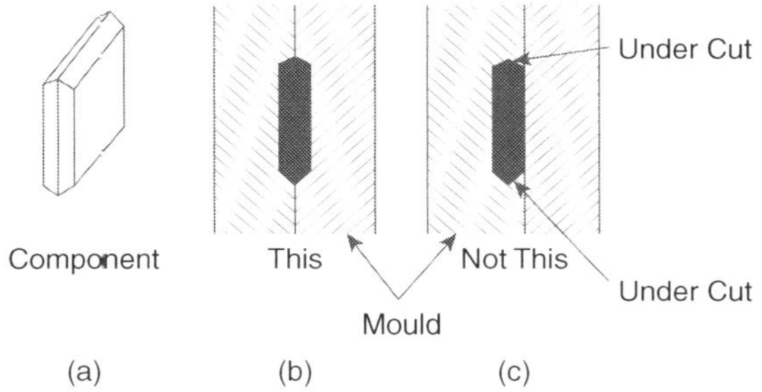

Fig. 1.11 Practicable and impracticable choice of parting surface.

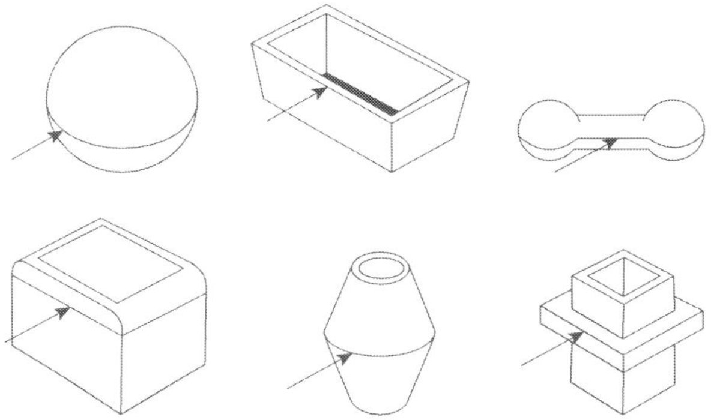

Fig. 1.12 Typical moulding which permit flat parting surface to be adopted.

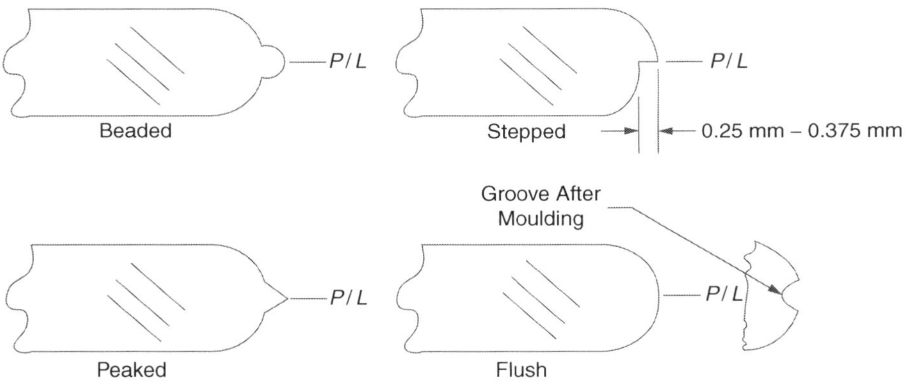

Fig. 1.13 Compression moulded parts that have parting lines on rounded sections may use any of the above designs.

Whenever possible, parting line should be located at the top of the part to facilitate finishing operations. See Fig. 1.14.

Avoid designing a part having a thin sharp wall at the parting line; it will break very easily during finishing operations.

The tool design software identifies the parting line through the draw direction. It identifies the extreme outer edge as a parting line. If there are any holes or pockets on the part it identifies the inner most edge for the inner openings.

Designing the parting line is an important factor in mould design, to the ease of mould ejection in line of draw without any hindrance of moulding cycle. The product designer should have to fulfil these conditions without sacrificing its aesthetic and functional part. The parting line may be simple flat or stepped or profile or angled.

In compression-moulded parts such as knobs, bottle caps, handles, and any other part with a rounded section where the parting line of the die must be placed, it is advisable to use one of the designs illustrated in Fig. 1.13. The standard type parting line that is flat and square at the top of a part is most often designed. If strength is needed, a slight edge may be added. If a drinking-glass or cup-type is desired, this design may be used. The parting of a die on a radius is not recommended.

The cost of finishing or removing the flash from a compression moulded part is generally a large percentage of the direct labour cost of the moulded piece. Simple straight parting lines should be designed into the part if at all possible.

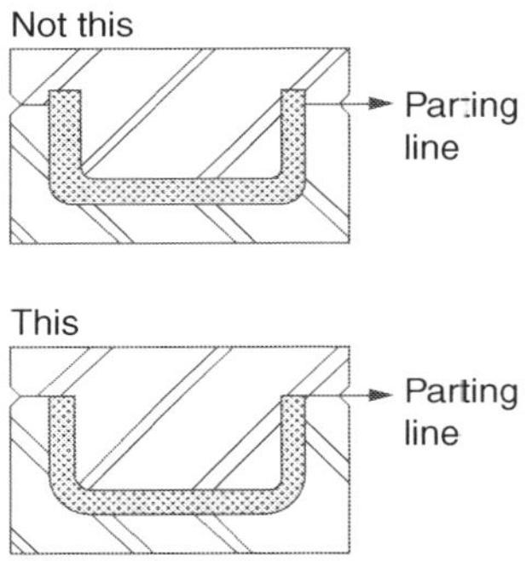

Fig. 1.14 Parting line at top of the part.

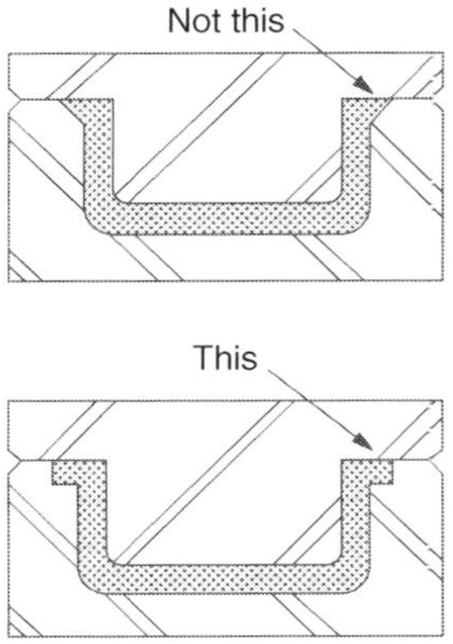

Fig. 1.15 Avoiding thin sharp wall at parting line.

1.4.1 Designed Mismatch

A mismatch is the alignment of two moulded parts (e.g., a box and a cover) usually is traceable to part warp age or differences in shrinkages or misalignment between the core and cavity part of the mould. The misalignment appears to be improved when a bead or designed mismatch is utilised, (Fig. 1.17). Different possibilities for designed mismatch are shown in Fig.1.16 varieties of products such as telephone handset, telephone base, computer mouse, etc., have been adopted this type of concept to overcome this mismatch problem.

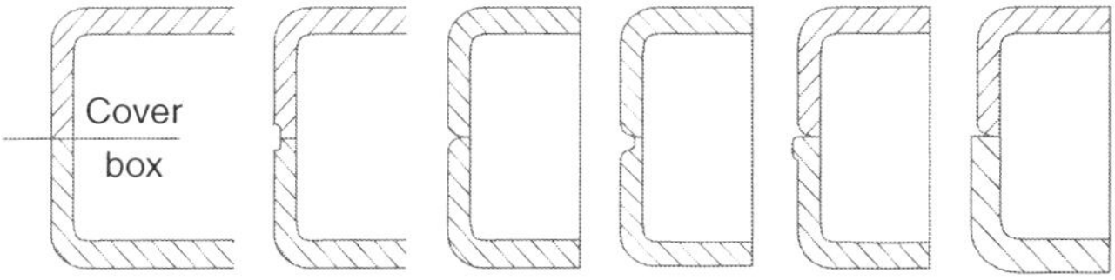

Fig. 1.16 A series of possibilities for designed mismatches.

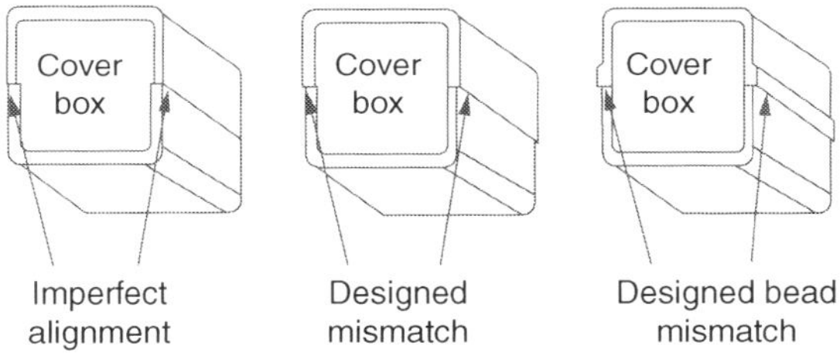

Fig. 1.17 A mismatch is the alignment of two moulded parts (box and cover) usually is traceable to part warpage or differences in shrinkages. The misalignment appears to be improved when a bead or designed mismatch is utilised.

In compression-moulded parts such as knobs, bottle caps, handles, and parts with a rounded section where the parting line of the die must be placed, it is advisable to use one of the designs illustrated in Fig. 1.16. A beaded parting line is used where the appearance is important and the adjoining surface should be protected during the finishing operation. A beaded parting line also helps in disguising mould misalignment. This is sometimes known as *placing a parting line on a ridge*. A peaked parting line is located above the surrounding surfaces of the part, and the flash can be cleaned without marring the surfaces. A stepped parting line will also allow clean and easy finishing of the moulded part; the flash will be horizontal and easy to remove mechanically. In the event that a bead is not permissible, a flash parting line may be necessary. The flash may be removed by adding a decorative effect made by grooving the part at the parting line. This method is practical only on round parts that may be spindled by turning the part rapidly on the end of a motor shaft and holding a cutting tool against it.

The cost of finishing or removing the flash from a compression moulded part is generally a large percentage of the direct labour cost of the moulded piece. Simple straight parting lines should be designed into the part if at all possible. The designer of a plastic part will have no difficulty with flash lines if they are located where they may be removed easily and where they are inexpensive and where they may be disguised or at least unobjectionable.

Parting lines for injection moulds are shown in (Fig.1.18). The standard type parting line that is flat and square at the top of a part is most often designed. If strength is needed, a slight edge may be added. If a drinking-glass or cup-type is desired, this may be used. The parting of a die on a radius is not recommended.

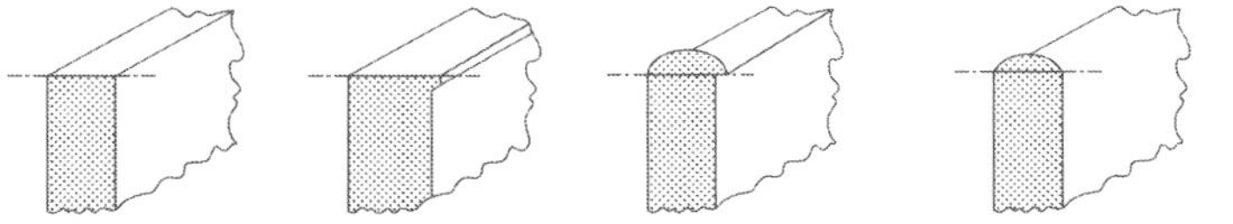

Fig. 1.18 Recommended parting line designs for injection moulding.

In order to increase the rigidity of such items as food containers moulded from the less rigid thermoplastics, e.g., polyethylene, polypropylene, EVA, polyvinyl chloride, etc., it is necessary to design the lip of the container to make it stiff.

1.4.2 Parting Line Mismatch

Repeated opening and closing of moulds will cause them to wear. Wear between the plunger and cavity pins and guide pin bushing causes the misalignment. The excessive wear at mould parting lines can create a mismatch, on a moulded part. A mismatch at a parting line of a few thousandths of an inch may appear to look many times larger. Tolerances for misalignment of cavity and plunger should total 0.15 mm for the average mould expected during the normal life of mould. When parts are first produced from the mould, the tolerance due to misalignment will be much less.

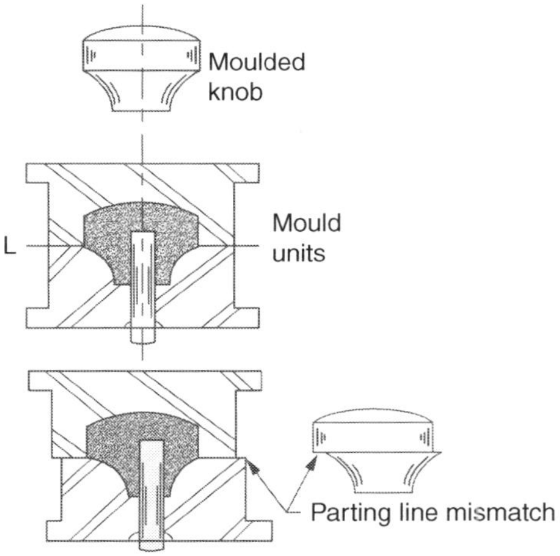

Fig. 1.19 Parting line mismatch.

1.5 RIBS, BOSSES AND GUSSETS

1.5.1 Ribs

Ribs may be defined as long protrusions on the part which may be used to decorate or strengthen the part and to prevent it from warping. The function of ribs is to increase the strength and rigidity of a moulded piece without increasing wall thickness. Proper design of ribs will usually prevent warpage during cooling, and in some cases, they facilitate smooth flow during moulding. Several features in the design of a rib must be carefully considered in order to minimise the internal stresses associated with irregularity in wall thickness.

Width, length, etc., must be analysed. For example, in some applications, thick heavy ribs can cause vacuum bubbles or sink marks at the intersection of mating surfaces and will result in appearance problems, structural discontinuity, high thermal stresses and stress concentration. To eliminate these problems, long, thin ribs should be used. It is also possible to core ribs from the underside to maintain uniform wall thickness (Fig. 1.21). A large rib placed on a part should result in a non-uniform wall thickness and cause a sink mark area. It is better to make many smaller narrow ribs instead of one large heavy rib or even better one long narrow rib (Figs. 1.22 and 1.23).

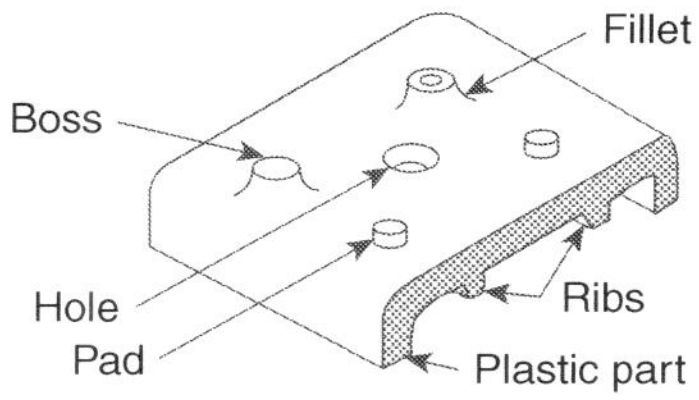

Fig. 1.20 Plastic part with ribs and bosses.

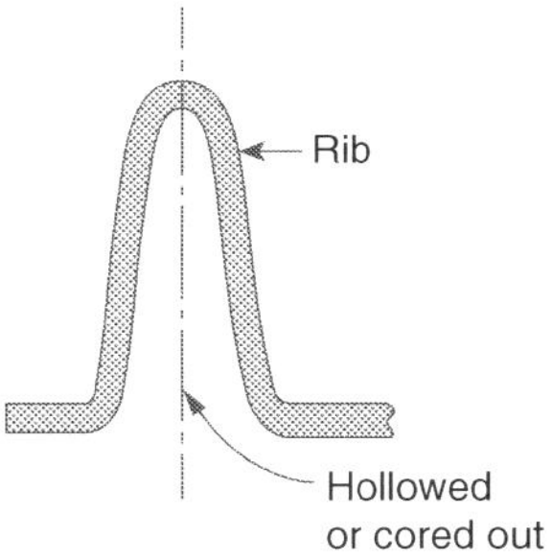

Fig. 1.21 Large rib cored out from back side.

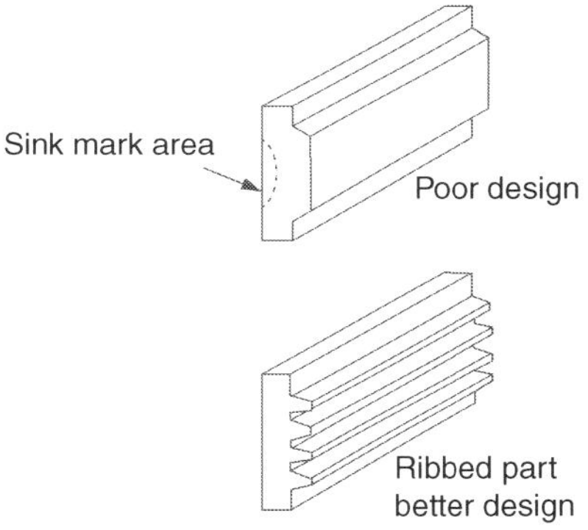

Fig. 1.22 Many small rib instead of a heavy rib.

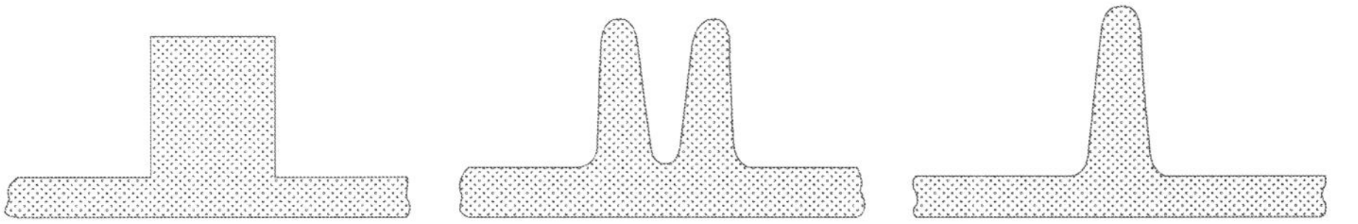

(a) Heavy rib
Not a good design

(b) Two small ribs
Better design

(c) One longer deeper rib.
Best design

Fig. 1.23 Good rib design calls for narrow ribs instead of one large heavy rib.

The use of two or more ribs is better than to increase the height of a single rib and the distance between them should be greater than the thickness of the wall to which they are attached. Sides and front of ribs should be tapered ½° to 5°, depending on length and width, to facilitate ejection. A fillet should be used where the rib joins the wall to minimise stress concentration and provide additional strength. Fig. 1.24 illustrates the proportions of ribs used in most of the thermoplastics.

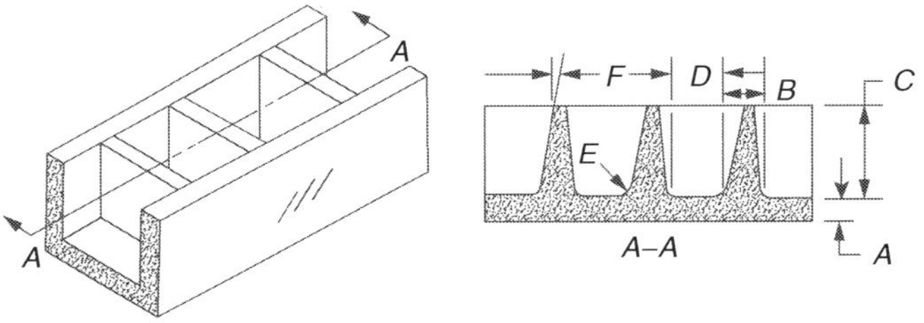

Fig. 1.24 Rib design where A = Wall thickness, $B = 0.8 \times A$, $C = 3 \times B$, $D = 2 \times B$, E = 0.125 mm radius, $F = 1\,1/2^{\circ}$ to 2°. If more strength is required, add additional ribs.

A rib should be located in the corner or side of a moulded part. This will lower the mould cost and allow easier filling of the plastic part (Fig. 1.25). Bosses and fillets will help to fill out ribs (Fig. 1.26).

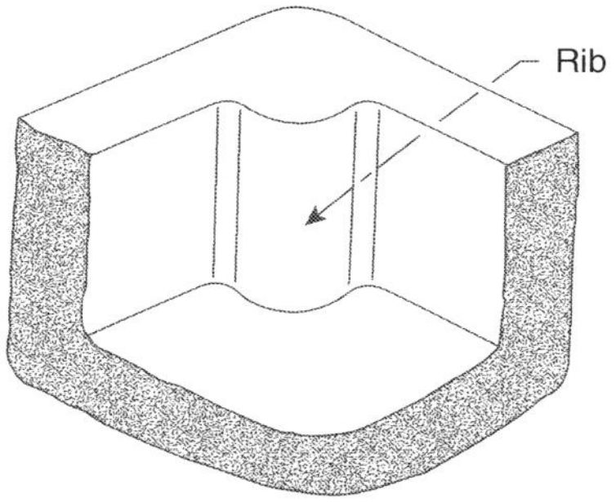

Fig. 1.25 Rib located in corner.

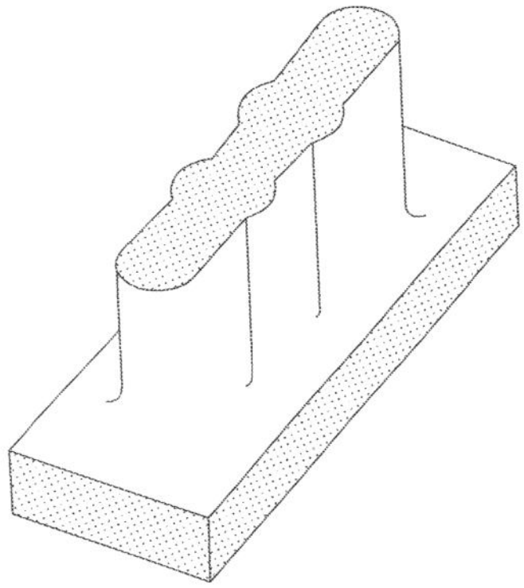

Fig. 1.26 Bosses and fillets in rib.

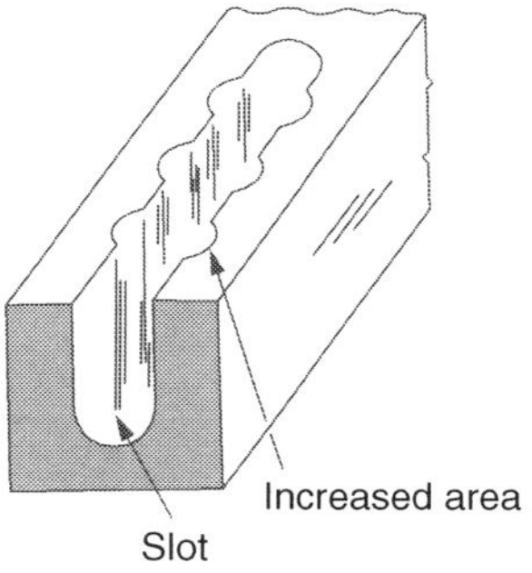

Fig. 1.27 Increasing the area of slot at intervals.

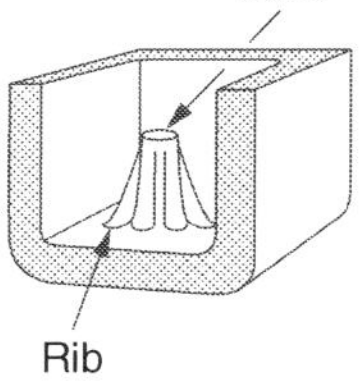

Fig. 1.28 Ribs on the side of long bosses.

Increasing the area of slots at intervals in the moulded parts will add strength to the mould (Fig. 1.27). Rib on the side of long bosses will aid in filling them (Fig. 1.28).

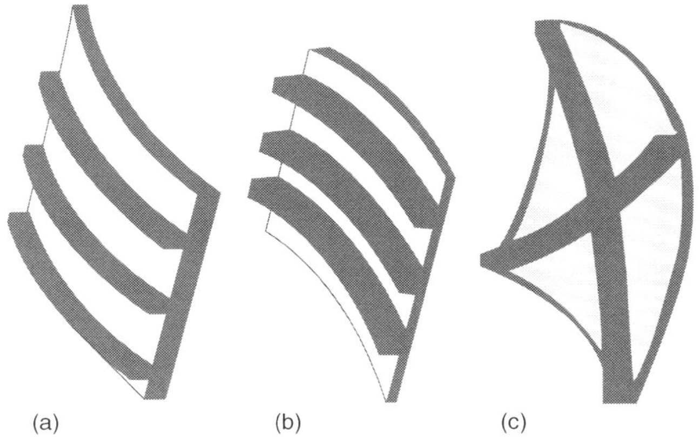

Fig. 1.29 Distortion resulting from ribs present on one side of a plate; a) Distortion from thin ribs on a thick plate, b) Distortion from thick ribs on a thin plate, c) Distortion from thick cross ribs on a thin plate.

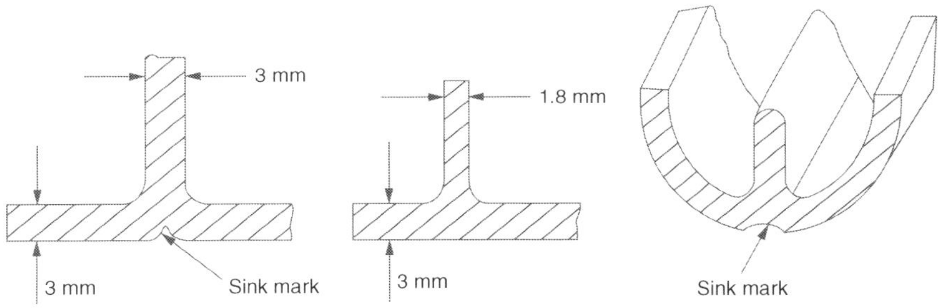

Fig. 1.30 Thickness of adjacent walls and ribs in thermoplastic parts should be about 60% of thicknesses of main bodies to reduce the possibility that sink marks will develop.

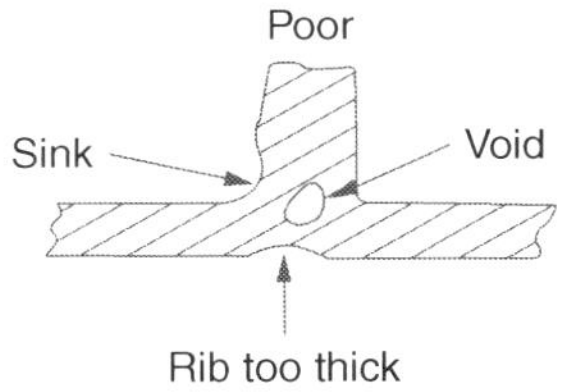

Fig. 1.31 Sink marks and voids developed due To thicker rib.

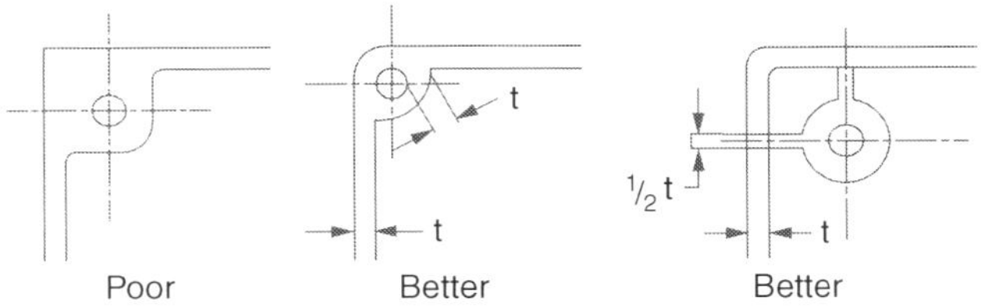

Fig.1.32 Rib design for reinforcing thermoplastic parts.

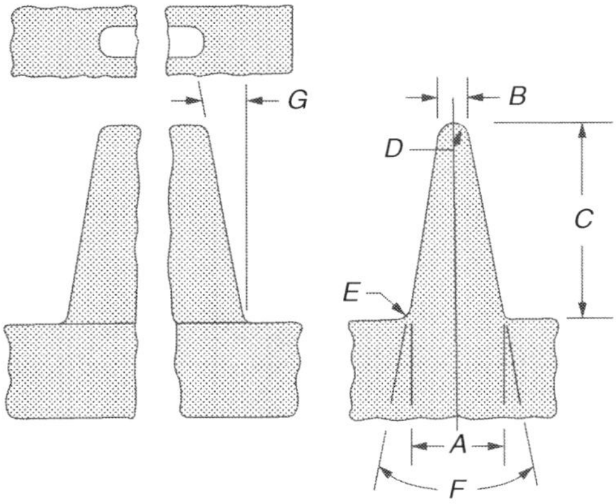

Fig. 1.33 Rib design for thermosetting materials where A = Width of rib at base, $B = A / 2$, $C = 3 \times A$, $D = A / 4$, $E = A / 4$, $F = 10^{\circ}$ and $G = 5^{\circ}$.

1.5.2 Bosses

Bosses may be defined as protruding studs on a part that assist in the assembling of a plastic part with another part. Because they are frequently the anchoring members between both part, they are subjected to strain and stresses not found in any other areas of the part. Bosses with outside diameter equal to twice the whole diameter are sufficiently strong. Also they can be used for pressed-in assembly, self-tapped screw assembly or tapped for a standard machine screw. Special attention should be paid to the design of bosses, in contrast with exterior surfaces, avoid any heavy sections to prevent voids or external sink marks. Avoid, if possible, bosses too near an outside wall, because this will cause weak sections in the mould. Sharp edges on bosses will cause more expense for the mould, hence it should be avoided. The edges of the boss should be blended with 0.5 mm radius (1/64 inch). Square or oval holes are difficult to machine in to mould steel and will result in a more expensive mould. It should be avoided. To design a square or rectangular boss, the corners should have a radius of at least 0.4 mm (0.15 in). Any attached member to the wall should not be over 80% of the wall thickness.

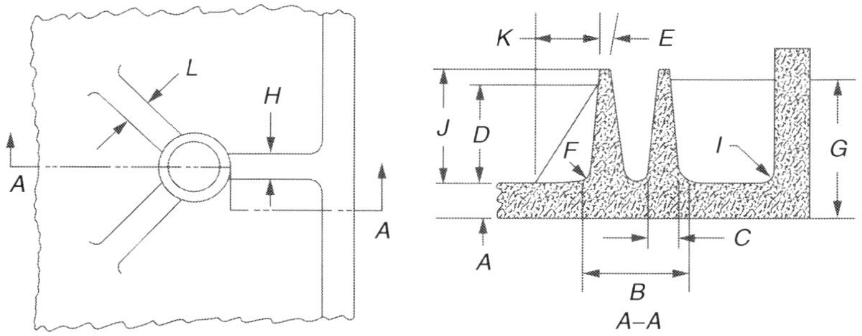

Fig. 1.34 This drawing illustrates design and proportions of a boss at a wall used for most thermoplastic materials where A = Wall thickness, B = Diameter of boss over radii, $C = 0.8 \times A$, $D = 2 \times B$, $E = 1^{\circ}$ to 2°, F = 0.125 mm (0.005″) radius, $G = D$, $H = 0.8 \times A$, $I = A / 4$, $J = 2 \times B$, $K = J$ Max. or $0.3 \times J$ Min. and $L = 0.8 \times A$.

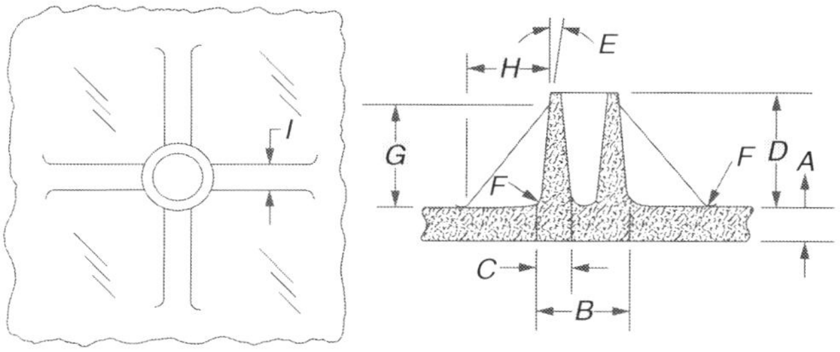

Fig. 1.35 This drawing illustrates design and proportions of a boss away from wall used in most thermoplastics materials where A = Wall thickness, B = Diameter of boss over radii, $C = 0.8 \times A$, $D = 2 \times B$, $E = 1^{\circ}$ to 2°, F = 0.125 mm (0.005″) radius, $G = 0.95 \times D$ Max., $H = G$ (Max.) and $0.3 \times G$ (Min.) and $I = 0.8 \times A$.

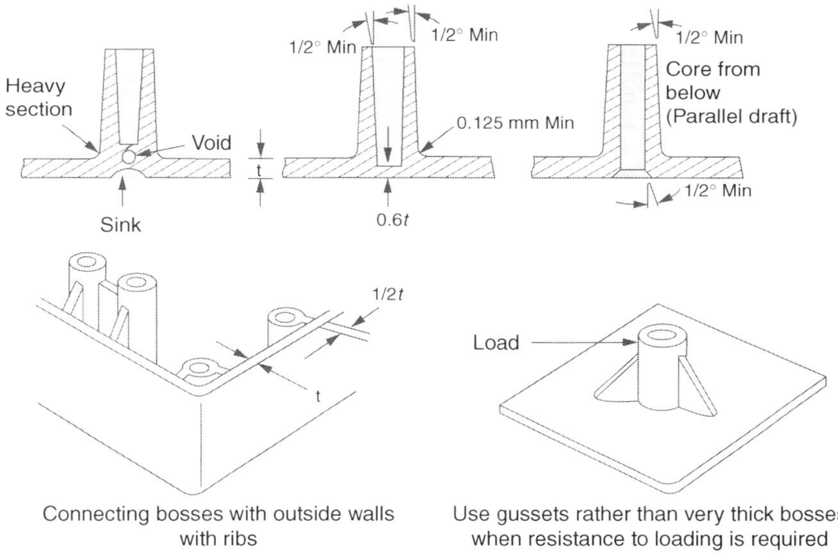

Fig. 1.36 Design of bosses.

Bosses can also be designed on the outside of the moulded part. An outside boss is used in attaching a part to an assembly. Figure 1.37 shows the recommended proportions for a boss on the outside of the moulded part.

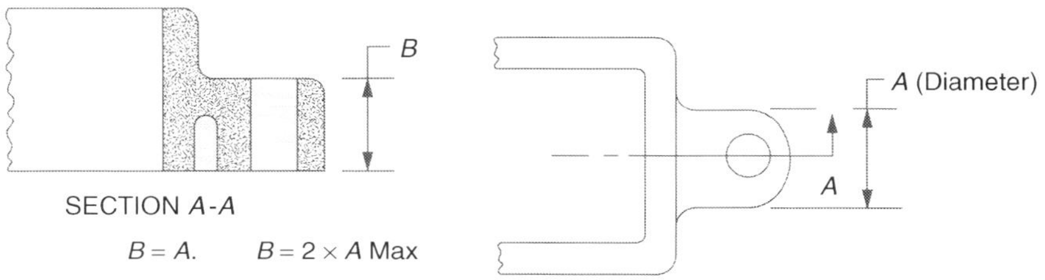

Fig. 1.37 Recommended proportions for a boss on the outside of the moulded part.

1.5.3 Gussets

Gussets are supporting arrangements for edge. All dimensions of the gussets are a function of the wall thickness (Fig. 1.38).

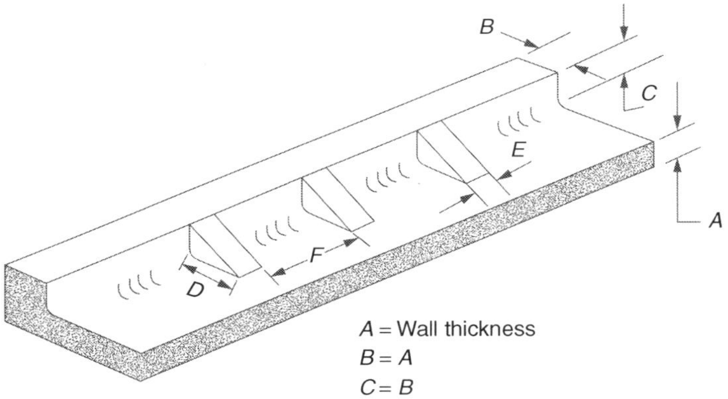

Fig. 1.38 This drawing illustrates design and proportions of gussets used in most thermoplastic materials.

1.6 RADII AND FILLETS

Fillets and radii are used at the ribs or bosses to facilitate the flow of plastic material and to eliminate sharp corners, thus, reducing stress concentration in the moulded article. The radius should be at least 0.25 mm and preferably 0.75 mm. All corners of mouldings should be filleted (rounded) to improve the flow of the material. Corner radii should be minimum ¼ of the part thickness. Figure 1.39 shows the radius/thickness ratio to avoid stress concentration.

The overall advantages of fillets and radii are:

(a) Improves flow of plastics material.

(b) Eliminates cracking and increase impact strength.

(c) Better structure with more rigidity and better stress distribution.

(d) Reduction in cycle time.

(e) Uniform density of the moulded article.

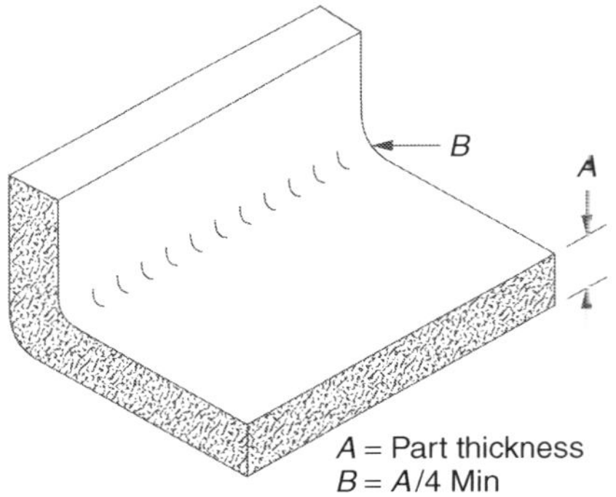

Fig. 1.39 Corner radii should be a minimum of 1/4 of the part thickness for most thermoplastic materials.

(f) Ensures more economical and long life of mould.
(g) Prevent cracking of mould parts during heat treatment.

Flow of a material at a corner presents no problem if the corner is rounded as shown in Fig. 1.40.

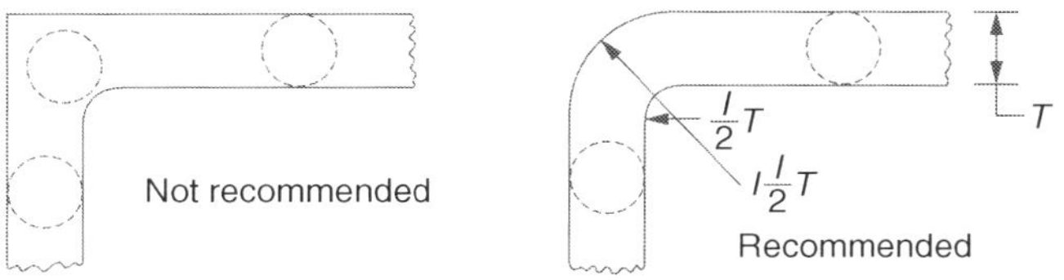

Fig. 1.40 Plastic corner flow should present no problem if the corner is rounded.

Fillets or radii are used at the base of ribs or bosses to facilitate the flow of plastic material and to eliminate sharp corners, thus, reducing stress concentrations in the moulded part. Figure 1.41 shows the stress concentration factor. All plastic parts requiring bosses should be provided with fillets at the junction of the boss with the main body of the plastic part. Radii of these fillets should be at least 0.25 mm (0.01 inch) and preferably 0.75 mm (0.03 inch). The addition of a fillet increases the strength of the mould and the moulded part. Fillets generally reduce the cost of the mould, the moulded part is more streamlined, and the corners of the moulded part are easier to keep clean of dust. All fillets should be placed at the junction of bosses and ribs with the main body of the part.

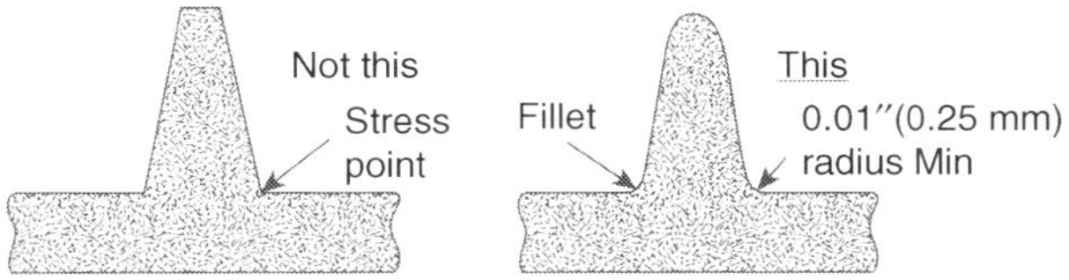

Fig. 1.41 Fillets placed at the junction of bosses and ribs with the main body of the part.

Curves and fillets in a moulded part prevent stress concentrations, add strength, and help eliminate warpage. Figure 1.42 shows redesigned part with fillet.

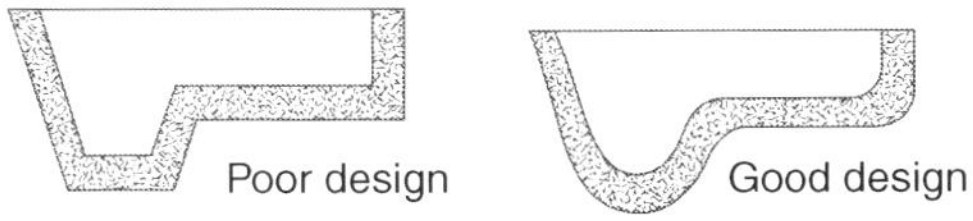

Fig. 1.42 Part with and without fillet.

Streamlining of the plastic part will help to prevent gas pockets. When the material is being moulded, it should sweep across the confined areas of the mould. Otherwise, gas pockets may develop. This results in blisters or sink marks on the surface near the pockets. In the case of thermosetting materials, gas pockets may be caused by trapped gas. Good design calls for consideration of the flow route in the part. When plastics materials flow around protruding sections they knit or weld on the other side. With thermosetting materials, knit or weld line may be a weak point due to the fact that the plastic materials has approached the last stages of polymerisation before the two streams (flow fronts) meet on the opposite side. Hence, they do not bond well. As a result, weld line will be weaker than the adjoining material.

1.7 TAPER OR DRAFT

A taper is a slight draft angle in a mould wall designed to facilitate the removal of the moulded part from the mould. Plastic parts designed to be produced by the moulding process must have taper or draft on all surfaces perpendicular to the parting lines of the mould. Draft should be provided, both inside and outside as shown in Fig. 1.44. Plastic materials tend to shrink tightly around cores. In order to remove the moulded part, adequate taper must be provided. The degree of taper will vary according to the moulding process, wall thickness and the moulding material. There are no precise calculations or formulae for taper. A minimum of ½ degree taper per side is generally adequate. Most phenolic materials can be moulded with minimum taper of ½ degree but mould for thermoplastics require taper ranging from ½ to 3 degrees. Ribs and bosses should have a taper of ½° to 5° per side.

If the taper or draft is on the inside of the part only the part may stay in the cavity as the mould opens. Knockout pins will then be required in the cavity side. If the moulded part has draft on the outside, this will cause the part to stay on the plunger as the mould opens. Knockout pins in the plunger will be required to remove the part.

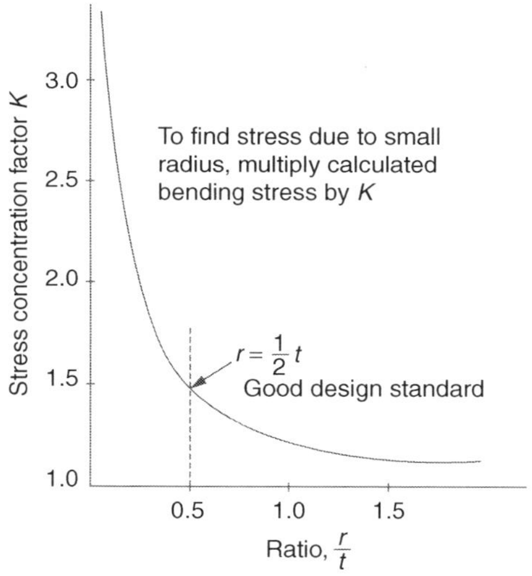

Fig. 1.43 Typical stress concentration factor.

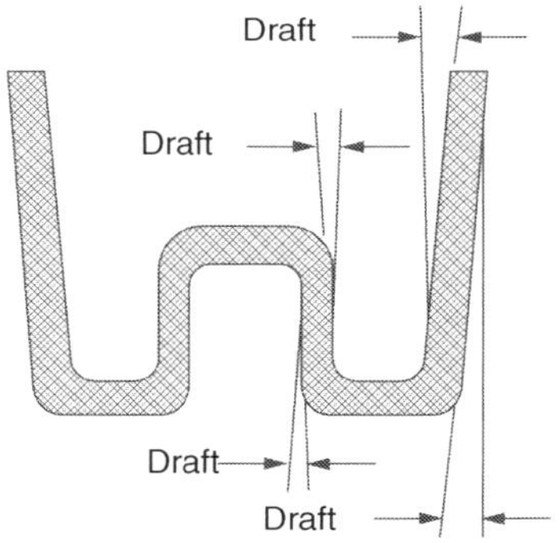

Fig. 1.44 Draft provided both inside and outside of the part.

1.8 HOLES

Moulded holes or openings in plastic parts are used for a variety of purposes. Holes are provided to allow assembly with other parts, to decorate the part and give it more eye appeal, or functional, such as ventilators or louvres.

Holes or openings can be round, square, rectangular, elliptical, etc. Moulded holes may be classified as blind, through, step, recess step and intersecting. (Fig. 1.45).

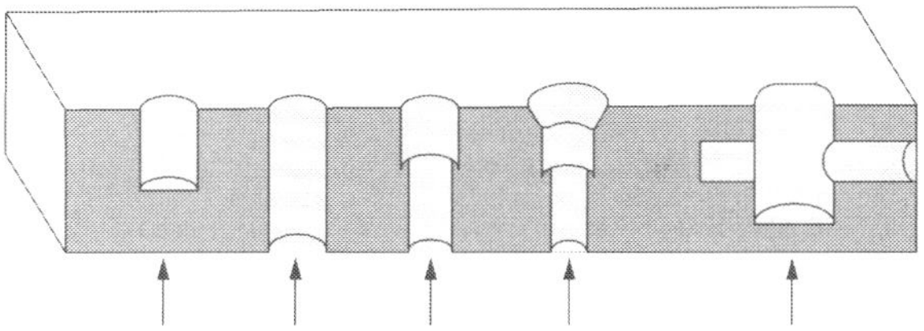

Fig. 1.45 Moulded holes classification in plastic parts.

1.8.1 Holes Parallel to the Draw

A holes parallel to the draw is a hole whose axis is parallel to the movement of the mould as it opens and closes. Holes may also be moulded at right angles to the draw and at oblique angles (Fig. 1.46). Oblique-angles holes should be avoided if possible, because it is very difficult and expensive to make a mould, to operate at oblique angles, as they require split dies and retractable core pins.

The main point is that the hole is made by a pin that is inserted into the mould. The pin is subject to breakage and wear. Holes may extend part way through the moulded piece, in which case the steel pin making the hole is supported only at one end.

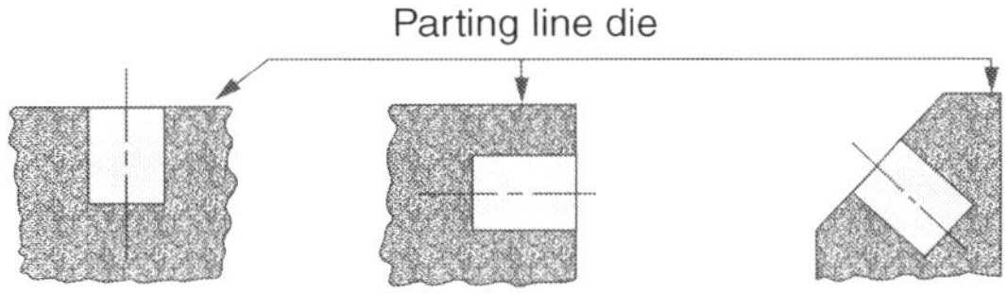

(a) Best (perpendicular) (b) Fair (parallel) (c) Difficult (oblique angle)

Fig. 1.46 Holes in plastic parts may be moulded perpendicular to the draw parallel to the draw, and at oblique angles. Oblique angles are not recommended.

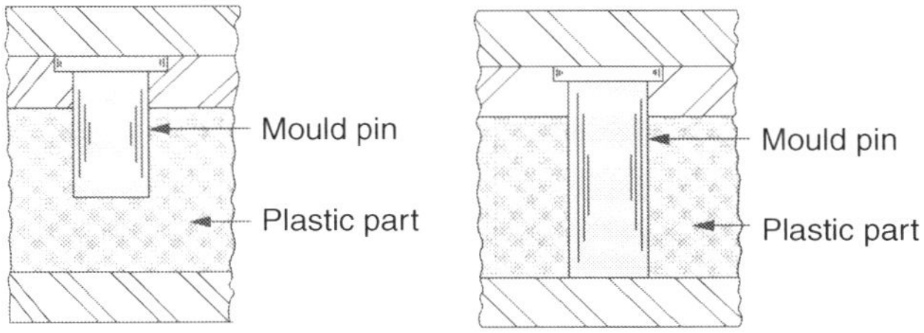

(a) Blind hole made by mould pin

(b) Through hole made by mould pin

Fig. 1.47 Moulded hole made by pin supported only at one end.

Holes may extend entirely through the piece (Fig. 1.47), in which case the pin may not be supported at both ends. The strength of the pin making the hole is influenced by the ratio of its mean diameter to its length called its '*slenderness ratio*'. Figure 1.48 lists the slenderness ratios for the average holes, using the various types of moulding methods, when these holes are moulded parallel to the draw. It is usually possible to follow the depth to diameter ratios given in Fig. 1.48 for moulded side holes. Hole moulded through a part by a pin supported at both ends is considered poor design in compression moulding. Hole moulded through a part by the pin supported at one end is most frequently used in moulding holes.

In some cases, small-diameter holes are required. If the holes are less than 1.5 mm in diameter, supported at one end and moulded by the compression method, these holes should be no longer than their diameter (Fig. 1.49). If small holes are moulded by transfer or injection, the slenderness ratio given in Fig.1.48 should apply.

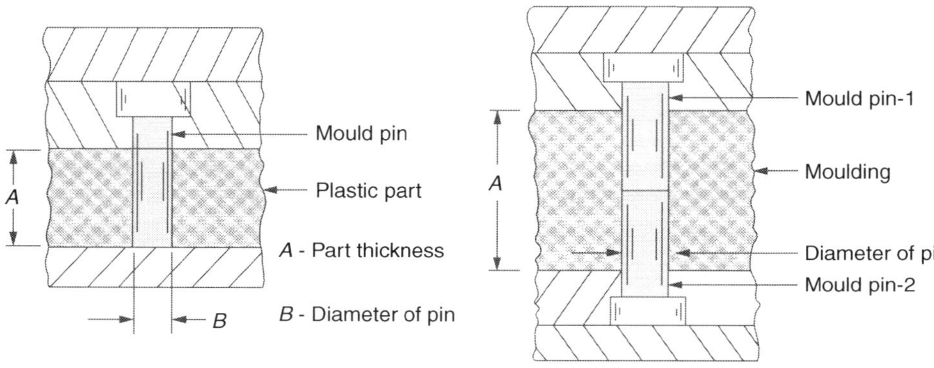

(a) One pin

Compression	$A = 2 \times B$
Transfer	$A = 6 \times B$
Injection	$A = 6 \times B$

(b) Two pins

Compression	$A = 6 \times B$
Transfer	$A = 15 \times B$
Injection	$A = 15 \times B$

Fig. 1.48 Slenderness ratio for moulded holes using various moulding methods.

If the diameter of the hole is less than 1.5 mm then the depth of hole will be its diameter. If the diameter of the hole is more than 1.5 mm then the depth of the hole will be two times of its diameter. If it is greater than 4 mm, it should be a through hole.

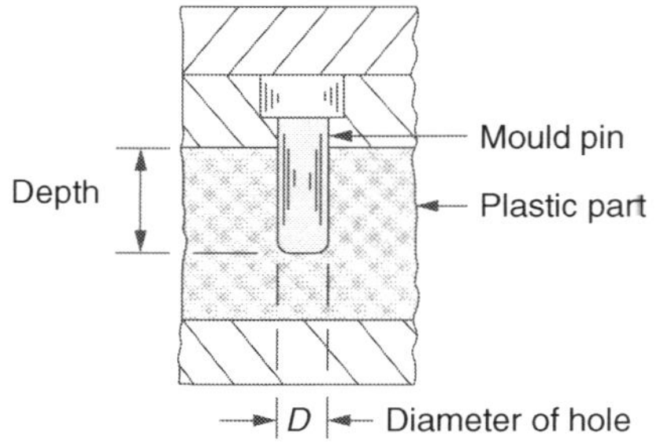

Fig. 1.49 Small diameter holes made by compression moulding should use the proportions in the above drawing.

Through holes, made with a pin supported at both ends, are not always as practical from a moulding standpoint as are holes made by butting the pins. Hole made by a butting pin is to be used in assembling the part with some other parts, namely a bolt or shaft, provision should be made for the misalignment of cavity and plunger and the possibility of the pin's bending. If either misalignment or bending occurs, the companion part may not go through the hole. Therefore, the pins used in moulding butt holes should differ in radius by at least 0.40 mm (Fig. 1.50).

In order to compensate for the misalignment of cavity and plunger sections, it is preferred that moulded hole to be made by using two pins of different diameters. Some designers may prefer to telescope the pin, thus making the hole as shown in Fig. 1.51. It must be pointed out that in the case of compression mouldings, flash may occur where the pins have been butted together. Some of the flash may remain on the pin having the recess. When the mould closes, this may cause the pins to bend and break.

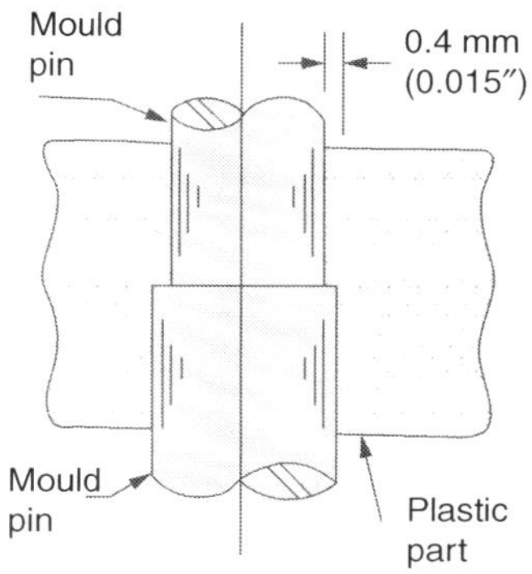

Fig. 1.50 Hole moulded by the mould pins butted together. Good design requires pins of different diameter.

Deep holes may be moulded by building the holes up in steps as shown in Fig. 1.52. The slenderness ratio, however, should be considered as resulting from the ratio of the mean diameter to the total length and should not exceed the slenderness ratios given in Fig. 1.48. Long holes of small diameter are difficult to mould. Sometimes deep holes are moulded for a short length or spotted, and then drilled after moulding.

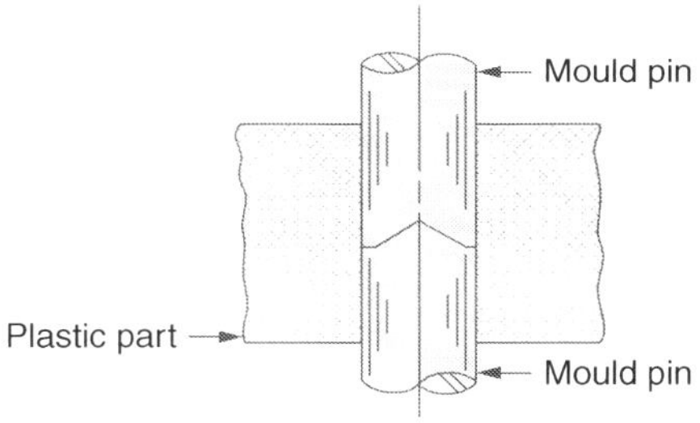

Fig. 1.51 Hole Moulded by two pins telescoping together.

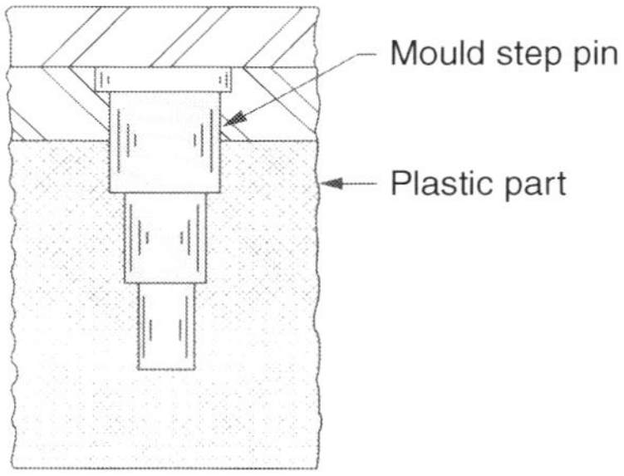

Fig. 1.52 Deeper hole can be moulded with a step pin.

1.8.2 Nearness of Holes to Each Other and Side Wall

The thermoplastic and thermosetting plastics flow and knit inside the mould, thereby setting up strains within the part, because the flow of the compound around a pin making a hole and the welding of that compound on the other side cause strain lines. Strains should be spread over as wide an area as possible so as to minimise their effect.

The wall thickness should be kept uniform. Not only will non-uniform wall thickness cause inequalities in curing time, but the part will be more susceptible to warping at very thin wall sections when these sections are joined to much thicker wall sections. Shown in Fig. 1.53 is the design of an injection moulded part with many holes. All the holes are parallel to each other or at right angles to each other. There are no oblique-angle holes with reference to the parting line.

The transfer and injection moulds require gates and if holes are moulded, knit lines will occur on the opposite side of the holes from the gate and between the hole and the side wall.

The distance between the hole and the side wall should be at least 3 mm, if maximum mechanical strength is required.

In moulding with most thermoplastic materials, the material between the wall of a hole and exterior wall of a part should be at least the thickness of the hole diameter shown in Fig. 1.53. The interior wall between holes should be at least one thickness of the holed diameter, and never less than 3 mm.

If two holes are moulded in a thermosetting material from opposite sides of the part, they should be no closer than 3 mm, if the possibility of cracking between the sharp edges of the holes is to be avoided. If thermoplastic materials are used, this distance may be as short as 1.5 mm.

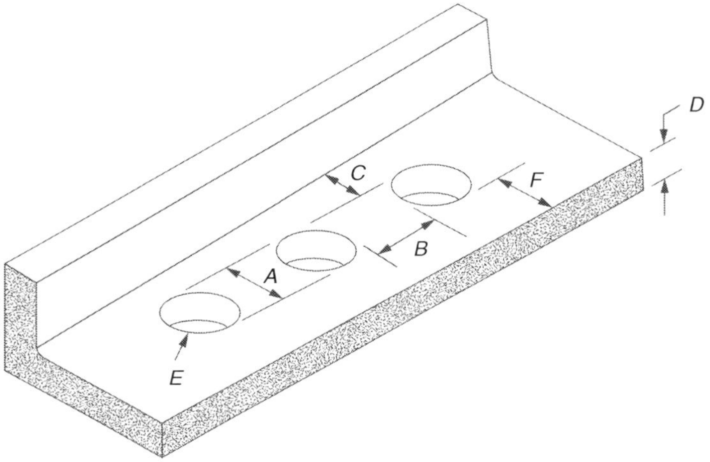

Fig. 1.53 This drawing illustrates the relationship of the distance of moulded holes from each other and from the side walls where A = Diameter of hole, $B = 1 \times A$, $C = 1 \times A$, D = Wall thickness, E = 0.125 mm radius at edges min., $F = 2 \times A$.

The bottom wall thickness, of a hole that is not moulded through, should be at least 1/6 the diameter of the hole (Fig. 1.54-A). If the wall at the bottom of the hole is less than 1/6 the diameter of the hole, the bottom will tend to bulge after moulding (Fig. 1.54-B). A better design is shown in Fig. 1.55. It will be noted that the wall thickness is uniform throughout for small blind holes, and there are no sharp corners for stress concentrations to develop.

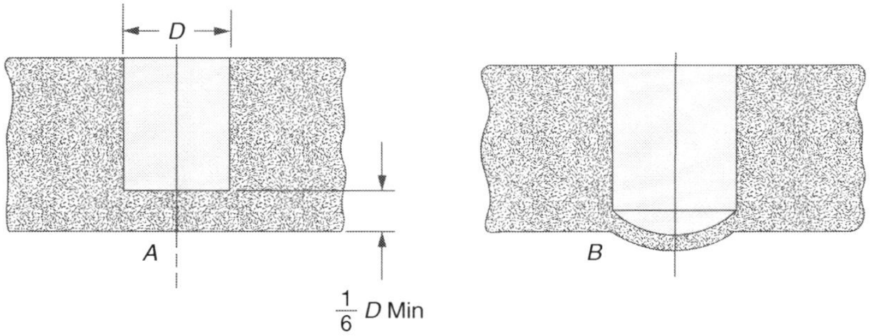

Fig. 1.54 Holes moulded only part way through should have ample thickness at the bottom.

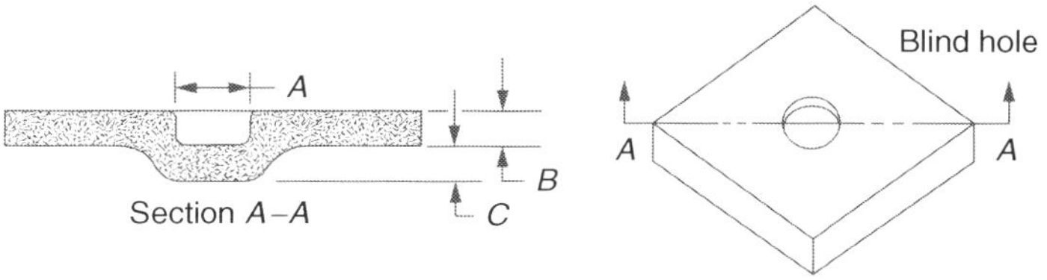

Fig. 1.55 The wall thickness should be kept uniform for small blind holes where A = Diameter of hole, $B = 1 \times A$ under 1.5 mm, $B = 2 \times A$ over 1.5 mm, $C = B$.

1.8.3 Moulding Holes not Parallel to the Draw

If the axis of a hole is not parallel to the draw, either the pin making the hole must be removed from the part, before the part is extracted from the mould or the pin must be removed with the part as it is extracted and then taken from the part. Usually, moulded holes not parallel to the draw require more complicated moulds or more labour than do holes parallel to the draw, and thus resulting in higher mould and higher parts costs. Therefore, holes entering the side of the part should be avoided if possible.

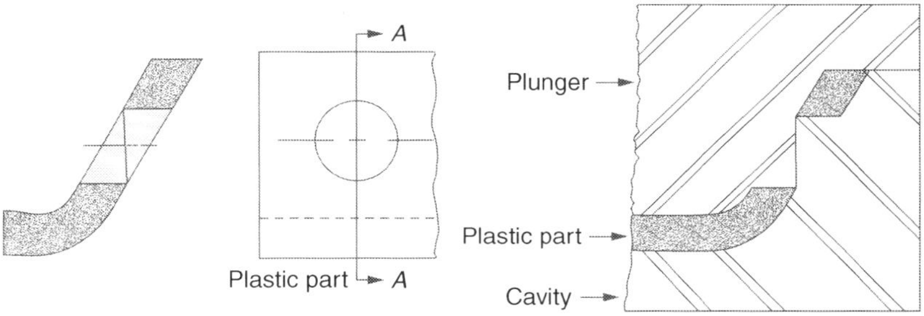

Fig. 1.56 Side holes or openings may be moulded without the use of pull pins or loose plugs in certain applications by using a 'kiss-off' type of die.

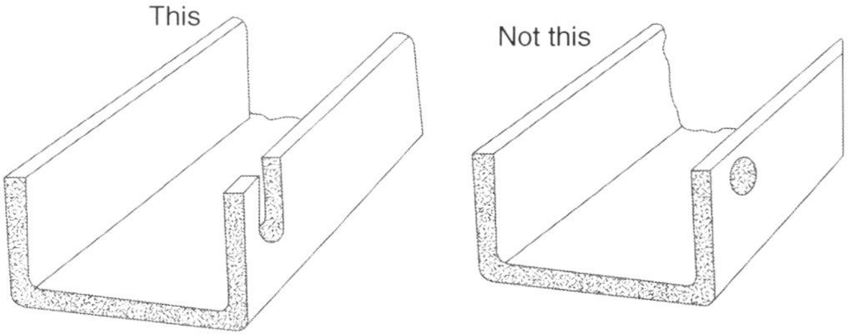

Fig. 1.57 Extending side openings to the part will lower costs.

Certain designs of side holes, generally low cost moulding (Fig. 1.56) illustrate a side hole that can be moulded without the necessity of the removal of the side pin, either before or after the part is extracted. This design may be impractical for some applications because either one or both the upper and lower surfaces of the cavity and plunger sections of the mould making the hole must slide along each other in order to obtain a straight draw of the part from the mould. Another design point that is often overlooked by engineers is the low cost of moulding that can be effected if a side hole is extended the top of the part, thus facilitating a straight draw (Fig. 1.57). Procedure is not practical in many designs, but it will lower price on parts and mould costs if it can be used. Moulded-in side holes are trouble to produce because extra provision is required to actuate the core pins from the side. Side holes may be moulded automatically by using the cam action in the mould or with hydraulic or pneumatic actuators. In compression moulding, drilling after the part is moulded usually simpler.

1.8.4 Drilled and Tapped Holes

When holes parallel to the draw are too slender to be moulded it becomes necessary to drill these holes after the part has been moulded. At times, it may be more economical to drill a side hole than to mould that hole. Good design calls for the spot locations of a hole to be drilled (Fig. 1.58). This spot acts as a guide for the drill entering the plastic. Spots should be made only for holes that are to be drilled parallel to the draw. A spot for a hole to be drilled perpendicular to the draw would be an undercut in the part and should not be used. (Fig. 1.59).

If through holes are drilled in the moulded plastic part it is a good practice to note on the drawing that the hole may chip on the edge where the drill exists. It is best to avoid designing parts so that drilling must be done on an angled surface. Drills may break the surface or 'walk' over the entering surface. Drilled holes should be so designed that the drill enters the part perpendicularly.

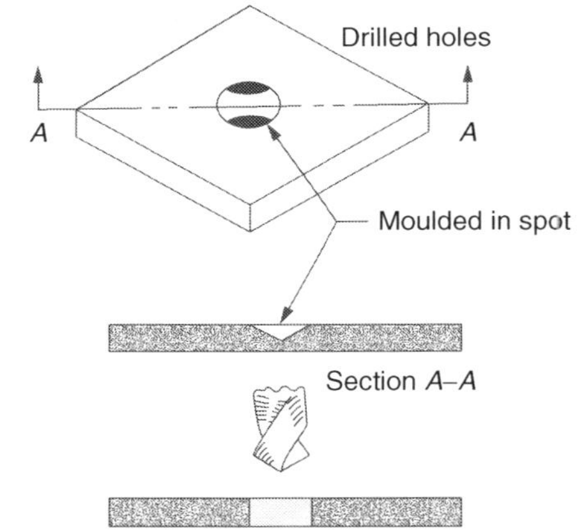

Fig. 1.58 It may be economical to drill a hole than to mould it in the plastic part. Drilled holes reduce mould costs, eliminate weld lines, and reduce mould maintenance.

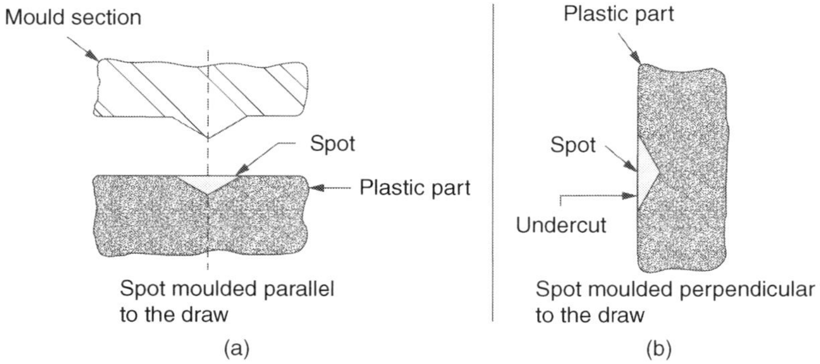

Fig. 1.59 (a) Good design calls for a spot location of a hole that is to be drilled, (b) A spot should never be located perpendicular to the draw as this will constitute an undercut.

Frequently, moulded holes are tapped. Holes to be tapped after moulding or holes for self-tapping screws or drive screws should be countersunk to allow the tap or to find its way in and to prevent chipping at the entering end.

Many moulded plastic parts have holes drilled in the part after moulding. It is generally less expensive to build a drill jig to drill holes in a moulded part than it is to incorporate elaborate retractable core in the mould. Weld lines created by moulded holes can be eliminated by moulding the holes two thirds of the way through and then drilling the remainder of the way.

An electronic instrument case injection moulded from high-impact polystyrene, holes on the side of case were drilled after moulding. This was necessary because of frequent changes in the assembly of the part, in order to make the case versatile for other uses.

A designer should not call for a perfect chamber or radius at the end of a hole, because this will call for extreme precision in the mould. If a conical-head screw is not to extend above the surface of the plastic part the hole for the screw should be designed with a 0.4 mm of vertical depth to allow for the variations in screws (Fig. 1.60).

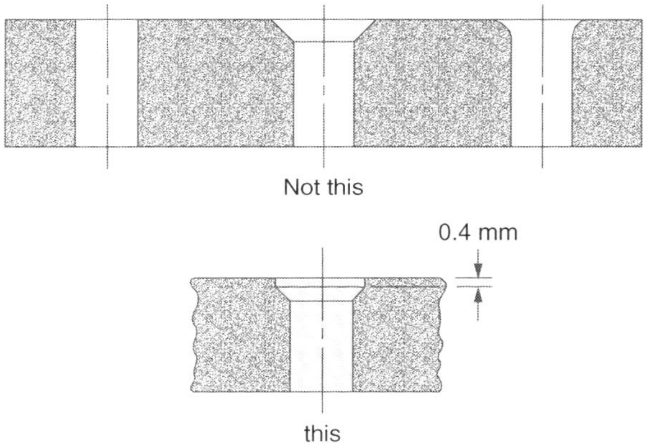

Fig. 1.60 Holes to be tapped or used with self-tapping screws should be countersunk.

Any moulded hole that requires extremely tight tolerances may necessitate size development. This means that the metal mould pin is made oversize and then gradually reduced in size by removing metal from the pin until the exact dimensions are obtained in the moulded part. (Fig. 1.61)

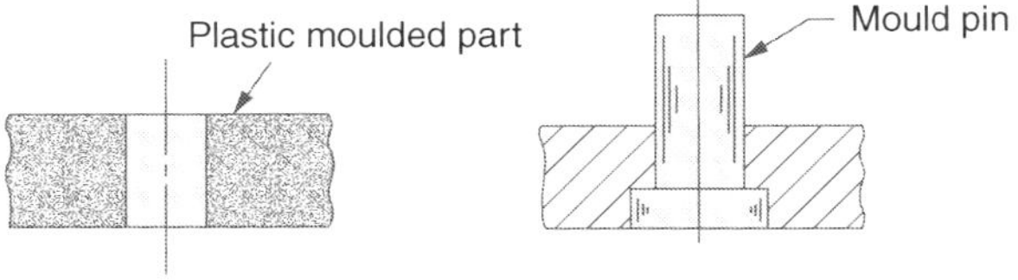

Fig. 1.61 Moulded through holes with tight tolerances may require size development of the mould pin. The mould pin is made oversize and then reduced.

Elongated holes may be used with plastic materials that have uncontrollable shrinkages such as the polyolefins shown in Fig. 1.62.

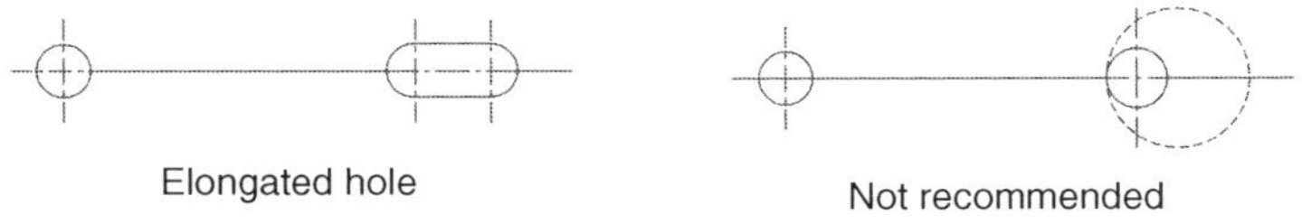

Fig. 1.62 Elongated holes may be used with plastic materials that have uncontrollable shrinkages.

Holes impractical to mould must be drilled. But they must not be so close to edges or corners because crack will result. Since a small diameter hole is difficult to drill along its intended direction to any great depth, the most practical approach in many cases is to mould it partway and drill it the remainder of the distance (Fig. 1.63).

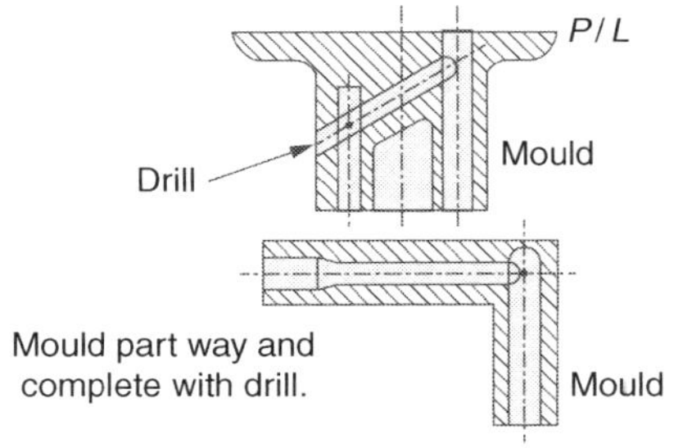

Fig. 1.63 Holes impractical to mould must be drilled.

1.9 CORING

Heavy section should be cored to provide uniform wall thickness. In the process, sink marks, blow holes and thermal cycles are reduced. Coring at right angle to the direction of moulding opening requires cams or hydraulic action which considerably increases the mould cost. Movable and loose piece cores are expensive but can be used in areas such as internal undercuts and thread moulding.

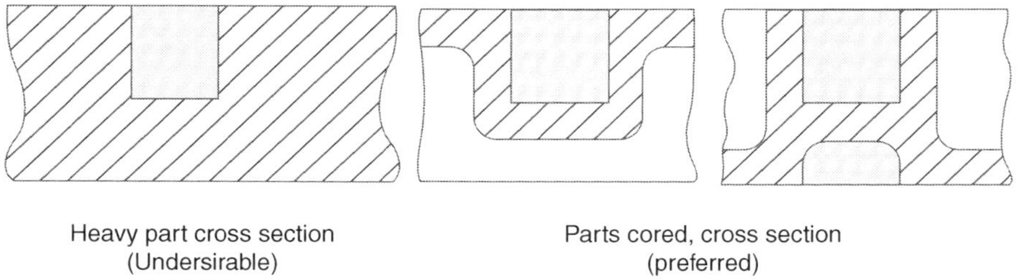

Fig. 1.64 Parts having heavy cross sections are subject to longer cycles and causes laminations or sinks, blisters, warpage and increased manufacturing costs. Core out or thin down heavy sections to preclude difficulties.

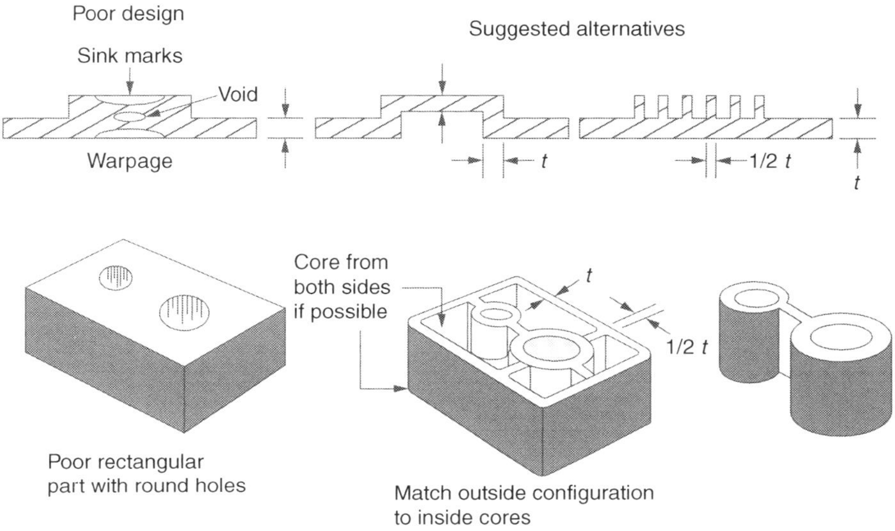

Fig. 1.65 Coring examples.

1.10 UNDERCUTS

An undercut is an indentation or projection on a moulded part, which makes ejection from the simple two part mould almost impossible. Undercuts can be classified as internal undercut, external undercut, circular undercut and an undercut on the side of the wall of a part formed by a core pin (Fig. 1.66). The undercuts in plastic parts can increase the mould cost and product cost by increasing the cycle time.

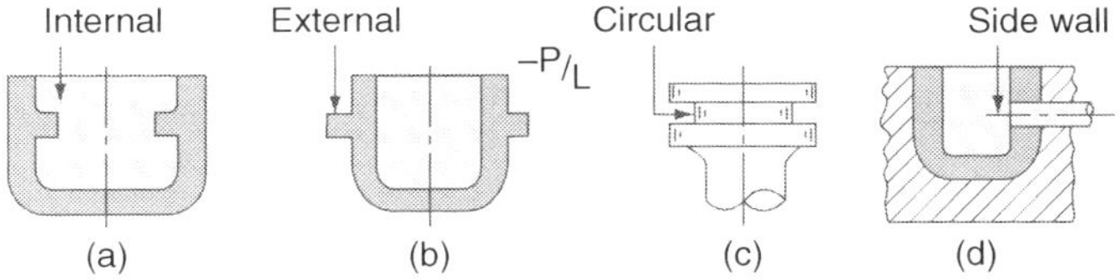

Fig. 1.66 Different types of undercuts in moulded plastic parts.

Undercuts may be moulded by means of split-mould cavity sections and by means of movable side cores that must be drawn away from the part before the part can be extracted from the mould. Because wedges or pull pins must be used to mould undercuts, flash or parting lines will be evident where the movable mould sections meet or where they meet the fixed mould sections.

Internal undercuts (Fig. 1.66(a)) are impractical and expensive and should be avoided. Whenever undercuts are encountered, it is best to design the part in two halves and assemble the two parts after they have been moulded. An internal undercut can be produced by using a removable ejector wedge as shown in Fig. 1.67. This calls for elaborate tooling.

External undercuts (Fig. 1.66(b)) are located in the outside contours of the piece. It would be impossible to withdraw a piece of such a shape from a one-piece mould cavity. In order to mould such a part, it is necessary to build the cavities of two or more loose members. After the part has been moulded, the loose members are parted, and the moulded piece is removed. It must be remembered that a parting or flash line will be visible on the moulded part.

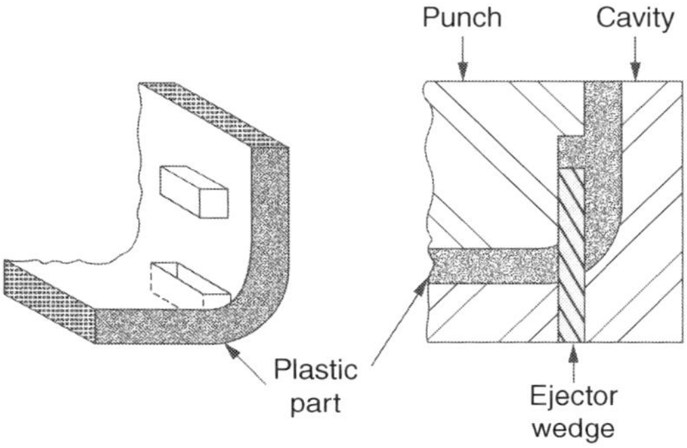

Fig. 1.67 An undercut in a moulded plastic part produced by using a removable ejector wedge.

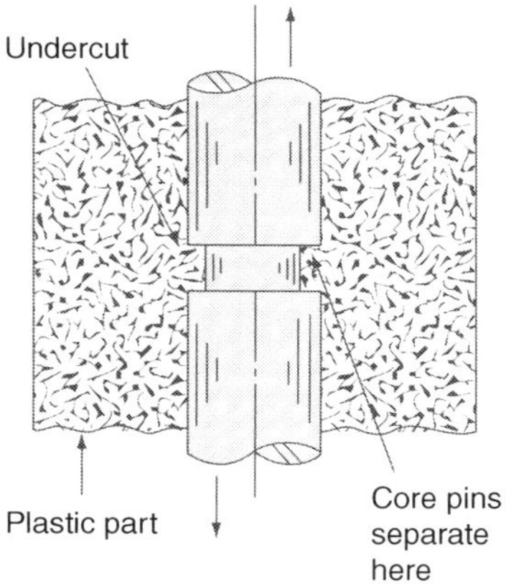

Fig. 1.68 An internal undercut can be moulded by using two separate core pins.

Circular undercuts (Fig. 1.66(c)) frequently can be made less expensive by grooving the part in a lathe with properly designed cutting tool.

Undercuts in the side wall of a moulded part (Fig. 1.66(d)) are produced by retractable core pins. Injection moulds may be designed with cam-operated side cores for automatic moulding. Side cores for transfer and compression moulds are generally operated manually, although in some cases they are run by pressure cylinder on the side of the mould.

Internal undercuts can be moulded by using two separate core pins as shown in Fig. 1.68. This is a very practical method to use, but flash will sometimes occur where the two core pins meet.

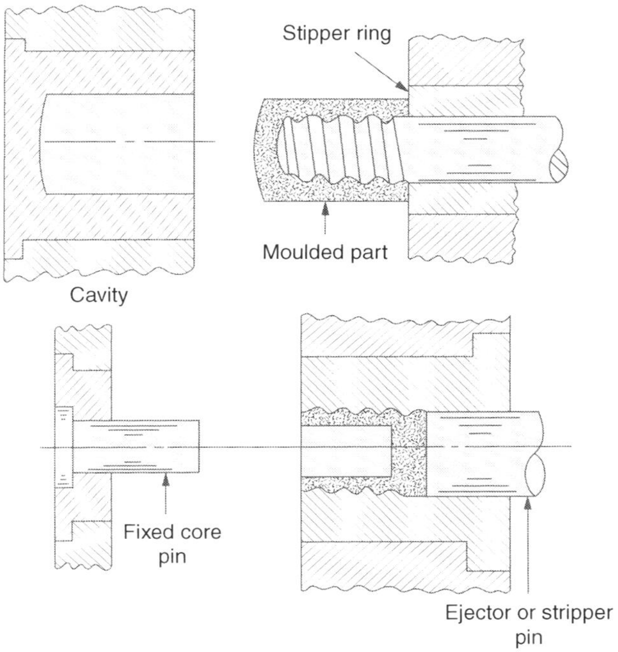

Fig. 1.69 Many thermoplastic materials can be stripped from a mould.

Some mould parts may be stripped from the mould without damage to the moulded piece (Fig. 1.69). All plastic materials, however, do not lend themselves to a stripping operation. Most flexible thermoplastic materials can tolerate a 10% strain in the mould ejection and not encounter permanent deformation.

1.11 THREADS

Threads are used in plastics for the purpose of providing a secure anchorage for a mating part. In all threads, a principle of wedge is used to lock or anchor the two parts. Thread forms in plastics part may be moulded, tapped or part of insert.

1.11.1 Thread Classes and Fits

Generally seven different types of threads are used in plastic products. They are the American standard, square, acme, buttress, bottle type, sharp '*V*', and a unified screw thread.

The fitting of threads is classified as follows :

1. Class 1 A loose fit for quick and easy assembly.
2. Class 2 A moderate or free fit for interchangeable parts.
3. Class 3 A semiprecision or medium fit.
4. Class 4 A precision or snug fit for parts assembled with tools. The parts are not interchangeable. This type of thread fit is not recommended for plastics.

Threads of classes 1 and 2 are adequate for most application in moulded plastics. The major diameter of a thread is the largest diameter of the thread of the screw or nut. The minor diameter of a thread is the smallest diameter of the thread of the screw or nut.

1.11.2 Types of Threads

The general forms of threads used in plastics are:

1. American standard
2. Square
3. Acme
4. Buttress
5. Bottle
6. Sharp '*V*'
7. Unified screw thread

Coarse threads are generally preferred in moulding. The construction of mould becomes sophisticated for getting very accurate internal thread forms. However, external threads are produced by easier methods. Secondary operation may be required for cleaning the threads (refer to Fig.1.70).

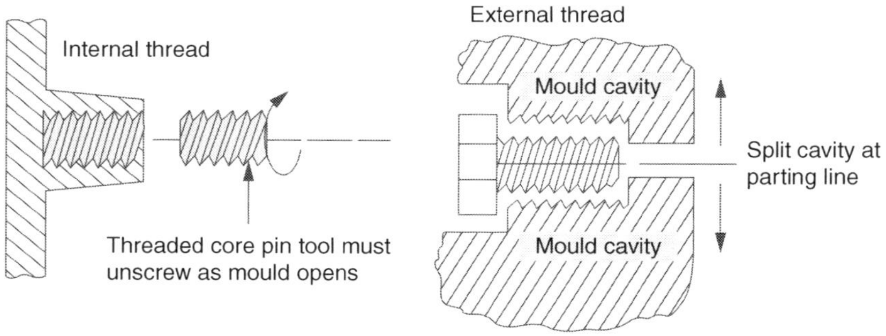

Fig. 1.70 Threads.

Different types of threads used in plastic parts are shown in Fig. 1.71.

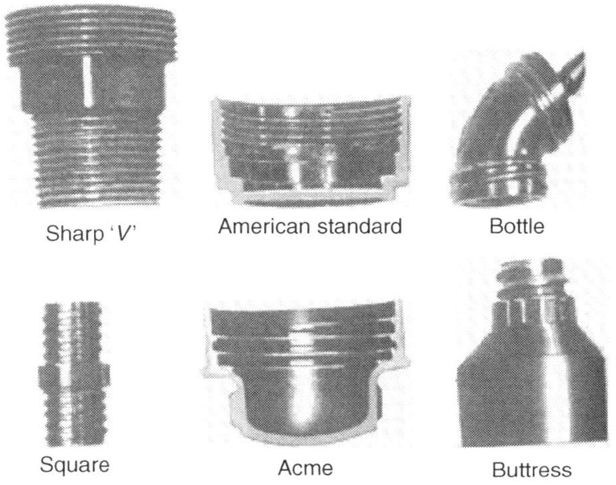

Fig. 1.71 Various types of threads used in plastics.

Profiles of the six different types of threads used in moulded plastics are shown in Fig. 1.72.

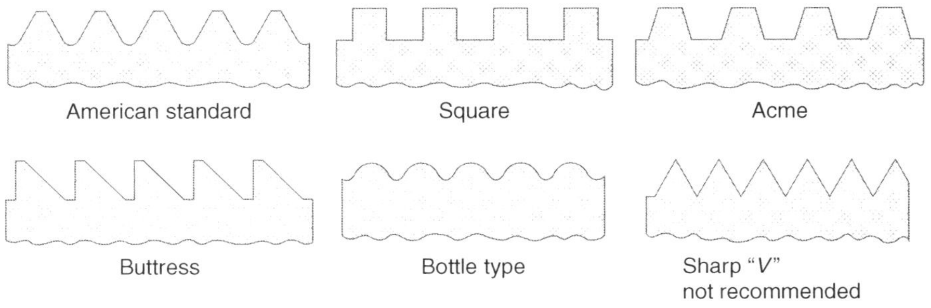

Fig. 1.72 Six different types of threads profiles used in moulded plastics.

1. **American Standard Thread:** The American standard thread is recommended for general use in plastics engineering work. It is used in threaded components where quick and easy assembly of the part is desired, and for all work where conditions do not require the use of the fine-pitch threads. This thread, in both the thermoplastics and thermosetting moulded materials, is made 75% of full depth. Tapped threads average 70% of full depth (Fig. 1.73).

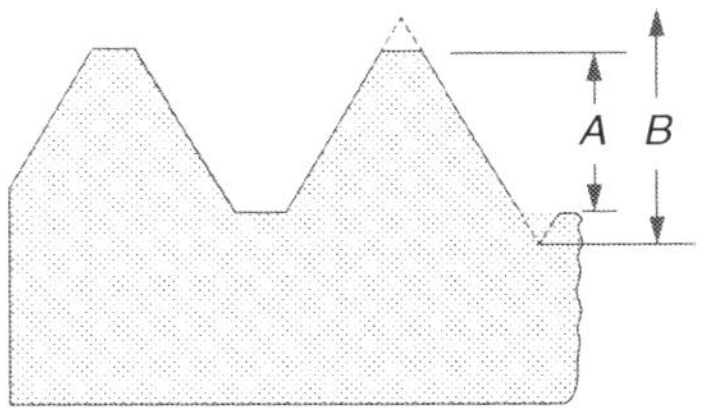

Fig. 1.73 The American standard thread is easily moulded or tapped. '*A*' is 75 per cent of '*B*' when moulded and 70 per cent when tapped.

2. **Square Thread:** A square thread is used where the highest strength is desired, e.g., in pipe fittings.
3. **Acme Thread:** The acme thread is similar to the square thread and is used in applications requiring strength. This type of thread is much easier to mould or cut than the square thread. Moulded acme threads are used in pump housing.
4. **Buttress Thread:** The buttress thread is used for transmitting power or strength in only one direction. It has the efficiency of the square thread and the strength of the '*V*' thread. It produces exceptionally high stresses in one direction only, along the threaded axis. Examples of actual applications are caps for tooth paste tubes and caulking gun cartridges.
5. **Bottle Thread:** The bottle thread is the type commonly used on glass containers. It is the accepted design standard set up by the Glass Container Association of America. All types of threads cause stress points in plastic materials, but the bottle threads results in the least. This type of thread has been developed to give the greatest ease in screwing or unscrewing mating parts. The round profile thread has been found to be very practical in plastic closures for glass containers, because of variations that occur in cast glass threads. A half thread, or a developed one-quarter-turn, round-type bottle thread is also used quite frequently.

Plastic parts incorporating round profile threads may be stripped from the mould if no undercuts other than threads exist. The plastic is actually stretched over the threads, and the method is limited to certain types of plastics (Fig.1.74). Special grades of plastics suitable for stripping are offered by the material suppliers in the ureas and phenolics, both thermosetting rigid materials. Most thermoplastic materials can be stripped, due to their more deformable nature.

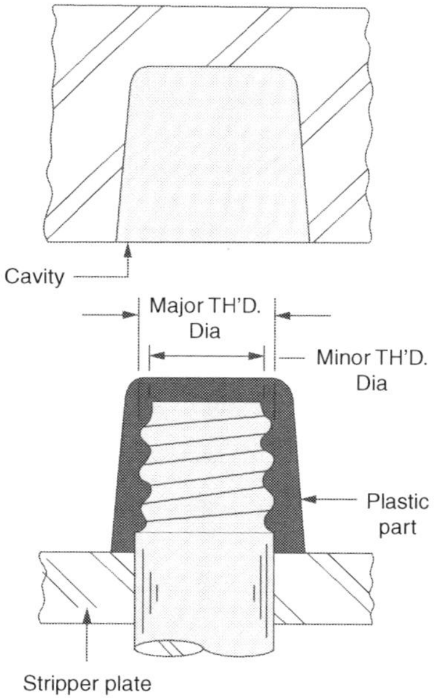

PER CENT STRAIN = MAJOR THREAD DIAMETER - MINOR THREAD DIA / MINOR THREAD DIAMETER × 100

EXAMPLE

MAJOR DIAMETER 31.80 MINOR DIAMETER 29.40

1.250 - 1.157 / 1.157 × 100 = 8% STRAIN

MATERIAL	% STRAIN AT 150°F
ABS	8
SAN	N.R
POLYSTYRENE	N.R
ACETAL	5
NYLON	9
ACRYLIC	4
POLYETHYLENE L.D.	21
POLYETHYLENE H.D.	6
POLYPROPYLENE	5
POLYALLOMER	15
POLYCARBONATE	N.R
NORYL	N.R
SURLYN	10

*N.R - NOT RECOMMENDED

Fig. 1.74 Stripping of threaded undercuts is permissible with certain types of plastic materials. A strain of 10 per cent is generally the maximum that is allowed.

6. **Sharp 'V' Threads:** Although the standard 'V' type thread is sometimes used in moulded plastic parts, it is not recommended. The sharp 'V' points, create stress points, making the plastic part notch sensitive and subject to breakage in these areas. Moulded pipe fittings that must match metal pipe fittings use this type of thread. The conventional 'V' type thread is generally used for mechanical assemblies.
7. **Unified Screw Threads:** The developed unified screw thread is used frequently (Fig. 1.75). It should be noted that the root of the thread has a radius and does not have a 'feather edge'. The tip or crest of the thread is flat and does not have a 'feather edge'.

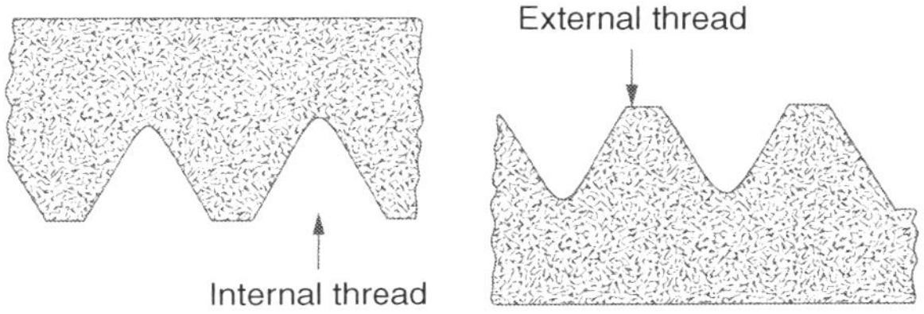

Fig. 1.75 Developed unified screw thread. This type of thread does not have feathered edges. The root of the thread has a radius and the tip or crest is flat.

1.11.3 Moulded Threads

A moulded internal thread (Fig. 1.76) to be unscrewed from threaded pin in the mould. A moulded external thread (Fig. 1.77) is unscrewed from a threaded recess in the mould, or a split mould section may be used if flash from the parting line develops between the threads. Difficult assembly with the mating part is the result. To avoid the possibility of flash developing between the threads, it is best to use a split-mould section only when moulding the thermoplastic materials by injection moulding.

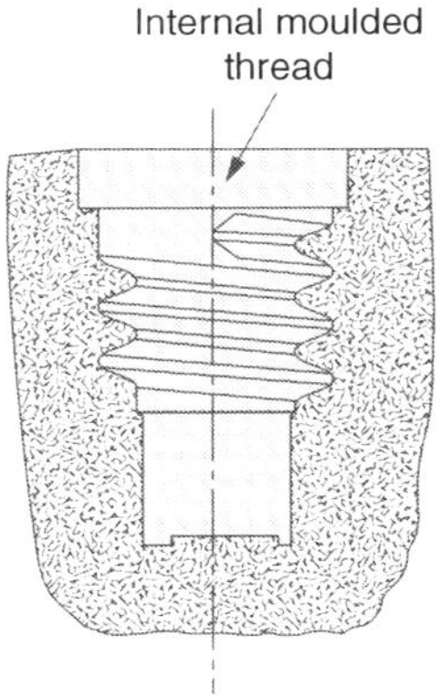

Fig. 1.76 An American standard moulded internal thread. The thread starts and stops abruptly. The bottom unthreaded portion of the hole has a diameter equal to or less than the minor diameter of the thread.

Moulded threads, particularly internal one, should be designed so that they can be removed quickly from the mould. Most of the plastics compounds will shrink in the mould. Internal threads are made by a threaded pin in the mould, and the threads will shrink quickly around the pin. It is advisable to design moulded threads no finer than 32 threads per inch and no longer than ½ in., or too much time will be taken by the press operator when unscrewing the part from the mould. If more threads per inch are required, in order to increase thread strength, it may be wise to consider double or triple threads. On a single thread, the lead and pitch are equal. On a double thread, the lead is twice the pitch, and on a triple thread, it is three times the pitch. This will lessen the time of unscrewing from the mould and will provide greater thread strength.

If an internal thread is very fine and too long, the part may shrink so tightly around the threaded metal pin that it may crack before the operator removes the part from the mould. Shrinkage will also change the pitch of a long thread and cause difficult engagement. If a part having more than 32 threads to the inch is moulded with cloth-r glass-reinforced thermosetting plastic, the filler or reinforcement may fail to enter the tip of the thread and may leave a weak tip filled only with the resin binder. If threads are required that are finer than 32 threads to the inch and more than ½ in. in length, it is advisable that they be tapped, provided they are no larger than ½ in. in diameter.

When threads are tapped in a plastic part, the cutting of the thread is done across minute weld lines in the plastic. This often results in cracking or chipping of the thread, which can be seen only on very close examination. As a rule, closer tolerances can be held with a moulded thread than with a tapped thread. This is due principally to the fact that in a moulded thread, the exact impression of the thread will be made from the mould itself.

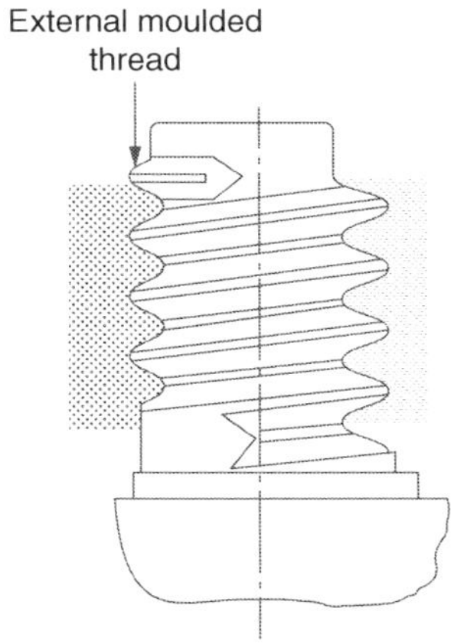

Fig. 1.77 An American standard moulded external thread. The thread may be unscrewed from the mould or a split mould may be used. The thread starts and stops abruptly. The unthreaded bottom portion of the stud should have a diameter equal or greater than the major diameter of the thread.

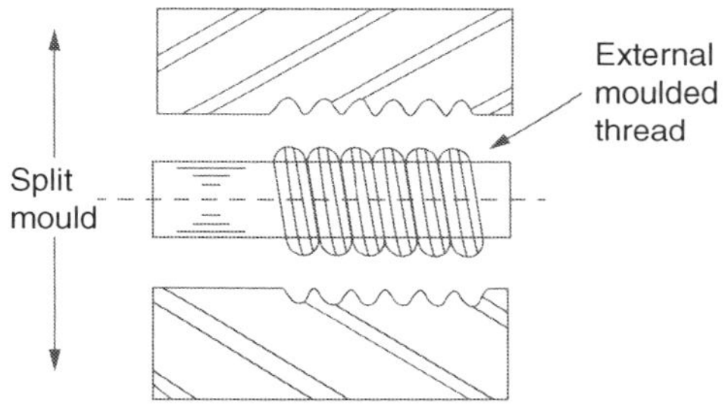

Fig. 1.78 A split thread does not mean that the thread is split, but that the mould that made the thread is split or made in two parts. A split thread generally leaves a visible parting line.

1.11.4 Moulded Thread Design

A moulded thread design is different from the design of a screw-machine thread or a tapped thread. The fundamental difference is that a moulded thread starts and stops abruptly. An internal and external thread should not feather out. A thread that feathers means a weak mould section that may break after repeated use. The thread itself will also be weak at this point. Figure 1.79(a) and (b) illustrate a thread that feathers out.

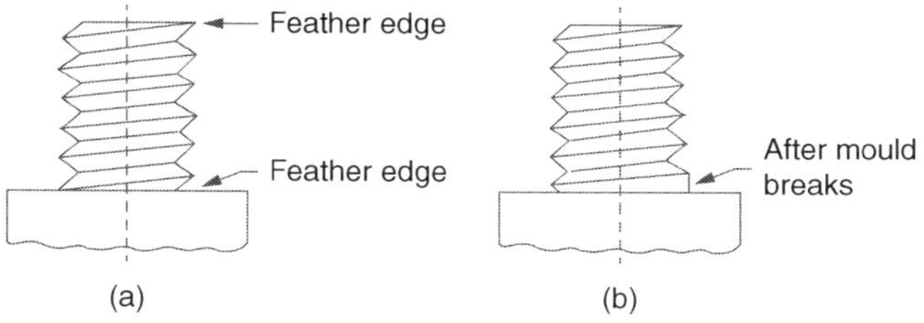

Fig. 1.79 Avoid plastic threads that feather out, as shown in (a) at the bottom thread. This will produce a weak mould section which may break, causing a heavy moulded thread as shown in (b).

The bottom unthreaded portion of a moulded male threaded stud should have a diameter equal to or greater than the major diameter of the thread. If the unthreaded portion is less than the major diameter, the threads will be stripped by removing the part from the mould. Figure 1.80 illustrates an incorrectly designed thread made by the ring section of the mould. The operator must unscrew the ring up ward. If the thread is designed as illustrated, unscrewing the ring will result in stripping the threads to diameter '*A*' on the ring.

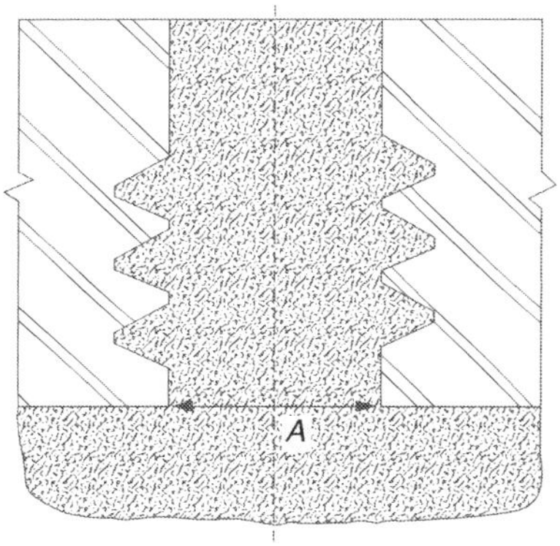

Fig. 1.80 An incorrectly designed thread moulded by a loose ring in the mould. As the operator unscrews the ring upward, the threads will strip to the diameter marked '*A*'.

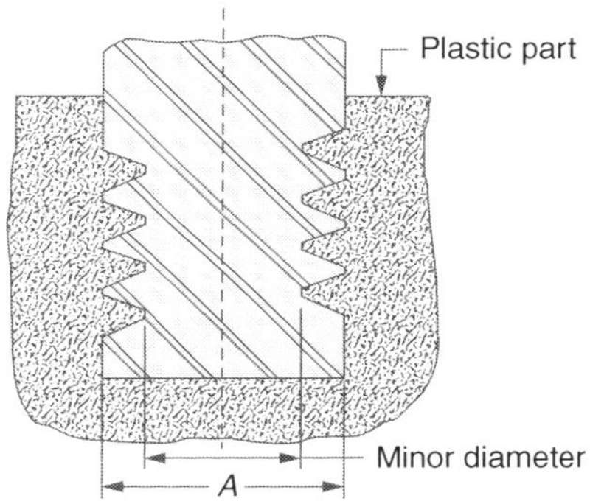

Fig. 1.81 If diameter '*A*' is larger than the minor diameter of the thread, the mould section can not be unscrewed from the part without stripping the threads.

The bottom unthreaded portion of a female threaded hole should have a diameter equal to or less than the minor diameter of the thread (Fig. 1.81). If the unthreaded portion is more than the minor diameter of the thread, the plug or pin will strip all of the threads as the screw moves upward.

If it is required that the mating part seat or meet flush with the companion plastic part, annular grooves should be provided on both male and female threads (Figs. 1.82, 1.83). Seating of a companion part is advisable if total tolerances on subassemblies are to be maintained and if maximum overall strength of the two parts is required.

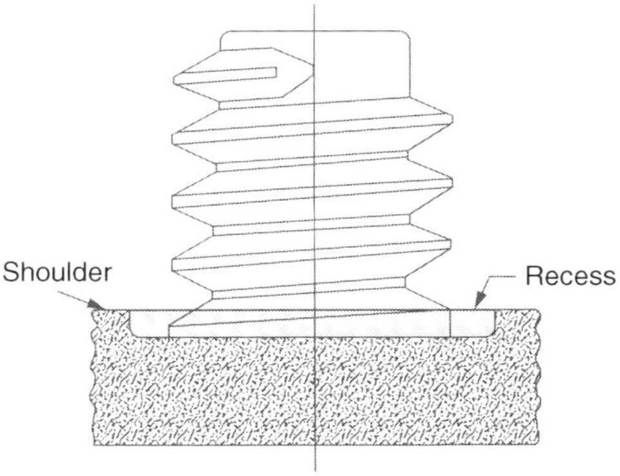

Fig. 1.82 Male thread. If the mating part is to seat properly with its companion plastic part, an annular groove should be provided at the bottom of an externally threaded stud. Also the recess at the base of the thread will allow the thread to end without having a 'feather edge'.

1.11.5 Removing Threaded Parts from the Mould

Injection moulded parts having moulded threads are often removed automatically, but when designing compression moulded parts having moulded threads, it is advisable to consider how the part will be removed from the mould. Compression moulded parts are usually removed by hand after the mould opens. There are four commonly used methods for this operation.

1. The operator may use a spanner wrench inserted in holes in the opposite side of the part.
2. He may use a screwdriver inserted in a slot provided for that purpose.
3. He may use a wrench, gripping against flat sides placed 180° apart.
4. He may unscrew by hand.

Internal and external threads are generally moulded by means of a threaded metal ring. This ring is removed with the part when the mould is opened and an identical metal ring is placed in the mould so that the mould may be reloaded at once. During the moulding cycle, the operator will unscrew the previously moulded part from the metal ring.

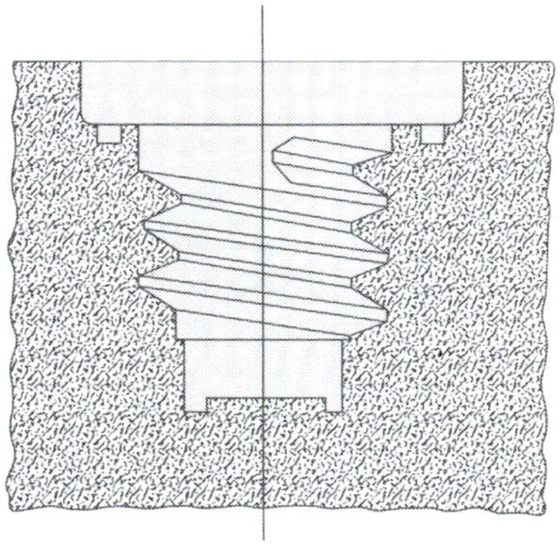

Fig. 1.83 Female thread. If the mating part is to seat properly with its companion plastic part, an annular groove should be provided.

1.11.6 Tapped Thread Design

Tapped threads are not strong, nor can they be held to as close a fit, as a moulded thread. For reasons of overall economy, self-tapping screws should be used, if the screw is not to be removed and replaced frequently. Repeated unions of the two threads eventually will damage the thread made in the plastic by the self-tapping screw.

Holes to be tapped should be slightly larger than those used in metal. The moulded or drilled hole to be tapped should be slightly countersunk or chamfered (Fig.1.84) to avoid chipping at the edge of the hole. A vertical step and chamfer (or countersink) will be better if the location of the hole to a mating part is to be encountered. A through-hole affords easier tapping than a blind hole, since chips resulting from this operation will fall through the hole. A blind hole may require frequent extraction of a tap in order to remove the chips.

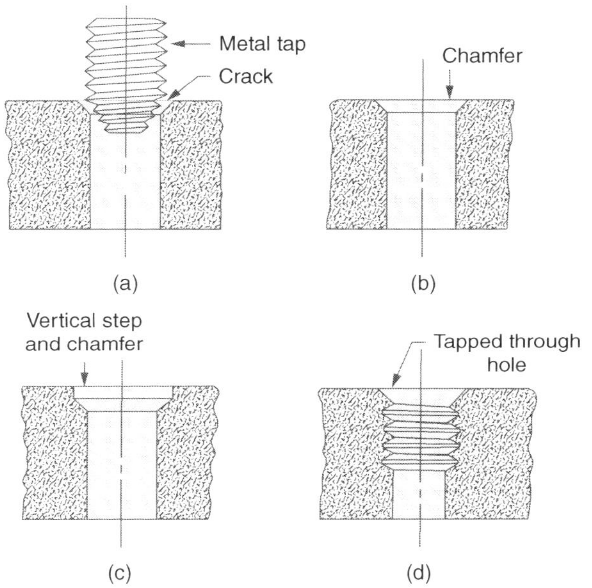

Fig. 1.84 Moulded or drilled holes should be countersunk or chamfered if they are to be tapped. This will avoid cracking at the hole entrance: (a) A plane hole will crack when tapped, (b) A countersink or chamfer will avoid cracking when tapped, (c) A vertical step and a chamfer will avoid cracking when tapped and also help in locating a mating part, (d) A through hole affords easier tapping, since chips resulting from tapping can fall through.

1.12 INSERTS

1.12.1 Materials for Inserts

Inserts are used in plastic parts to take wear and tear, to carry mechanical stresses that are beyond the limits of the plastic material, to decorate the part, to transmit electric current, and to aid in subassembly or assembly work. Inserts generally serve an important functional purpose and are made of brass, aluminium, or steel, other materials, including ceramics and plastics are used. Brass is used most frequently because it does not rust or corrode and is inexpensive and easy to machine. Inserts must be designed to ensure a secure anchorage to the plastic part, to prevent rotation as well as pulling out. Usually a medium or coarse diamond knurl provides adequate holding strength. Figure 1.85 shows a selection of metal inserts used in moulded plastic products.

Fig. 1.85 A selection of metal inserts used in moulded plastic products.

1.12.2 Types of Inserts

Various types of inserts and many methods of installation of inserts into plastic parts are used. Two types of insert attachment are moulded-in inserts and post-moulded inserts. Moulded-in inserts are placed in the mould before the mould closes. Fluid plastic (either thermoplastic or thermo-set) surrounds the insert during the moulding process and holds it in place. Post-moulded inserts are installed in the part after it has been moulded. Inserts that are installed by use of mechanical pressure alone may be used for both thermoplastics and thermosets. However, heated inserts or ultrasonic inserted inserts that permit localised melting of plastic around the insert are only available to thermoplastics.

Common uses for inserts are threads, and electrical contacts, which must be made of metal. Another outgrowth of insert moulding is encapsulation of electrical components such as coils. Here, a wound coil is placed into the mould, and an insulating material like epoxy is injected into the mould to form an insulating skin that also protects the delicate coils mechanically.

1.12.3 Factors to be considered with Moulded-In Inserts

1. The insert must provide the required mechanical strength. It should be of sufficient size to resist forces likely to be met by the part in service. Sufficient anchorage must be provided to prevent the insert from pulling out of the plastic.
2. It is not feasible to mould inserts in all plastic materials. Some plastic materials will crack around the insert after they have aged. Other materials will creep in aging, or if two inserts are rigidly located together in a mating part, one or the other of the inserts will pull out as the plastic ages and shrinks.
3. The flow of the plastic material in the moulding process should not dislodge the insert. In compression moulding, the flow of material is more violent than in transfer or injection moulding. It is not advisable, therefore, to use the compression method when moulding delicate inserts, such as thin metal stampings or phosphor bronze spring

wires, although transfer moulding may be used when moulding such inserts. It is not advisable to place fragile inserts in the path of the flow of material from the gates of transfer or injection moulds. Too heavy an impact from the flow of compound against a fragile or delicate insert will dislodge or break the insert.

4. Sufficient wall section should be allowed around the insert to prevent cracking of the plastic as it cools. Plastic materials have a higher coefficient of thermal expansion and contraction than the metals commonly used for inserts. Cracking of a moulded part around an insert will result in its becoming loose and pulling out of the part under slight strain. To help prevent cracking of plastic material around inserts, the inserts may be preheated prior to moulding. This procedure will allow a maximum expansion of the insert. Preheating inserts for thermoplastics will tend to eliminate weld marks, as well as to prevent cracking.
5. Inserts may require retapping, facing, or other expensive cleaning operations after they have been moulded in the part. If there is poor design or improper location of the insert in the part, cleaning operations to remove objectionable flash will be required. This expensive operation can be eliminated in many cases by a proper design of the insert.

1.12.4 Shapes of Male and Female Inserts

The majority of inserts used in plastic parts are made by either automatic screw machines or metal stamping machines as shown in Fig. 1.86 and gives the proper name to most of the metal inserts used in the plastic industry. Typical male and female inserts are shown in Fig. 1.87. The end of the insert to be imbedded in the plastics is chamfered or rounded, which is desirable so that the plastics will flow easily around the insert. Embedded sharp corners on inserts causes the plastics to crack at the corners. Current carrying inserts with sharp corners also bring about a concentration of electrical stresses or corona effect. Because the embedded end of a female insert is closed, compound will not flow into the threads from this end. Tolerances on the minor thread diameters of female inserts should be held to plus 0.050 mm and for precision work minus 0.001 mm. Tolerances on the major thread diameters of male inserts should be held to plus 0.05 mm and for precision work minus 0.001 mm. Close tolerances on this dimension give the insert a positive location in the moulded part and help to prevent compound from flowing into the insert during moulding.

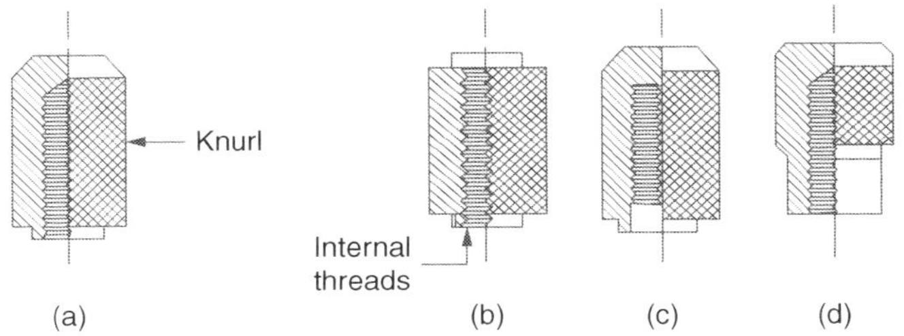

Fig. 1.86 Various designed inserts are used in moulded plastics: (a) Blind hole female insert with internal threads, (b) Open ends female insert with internal threads, (c) Blind hole female insert with internal threads and counter bore, (d) Male stud with internal thread,

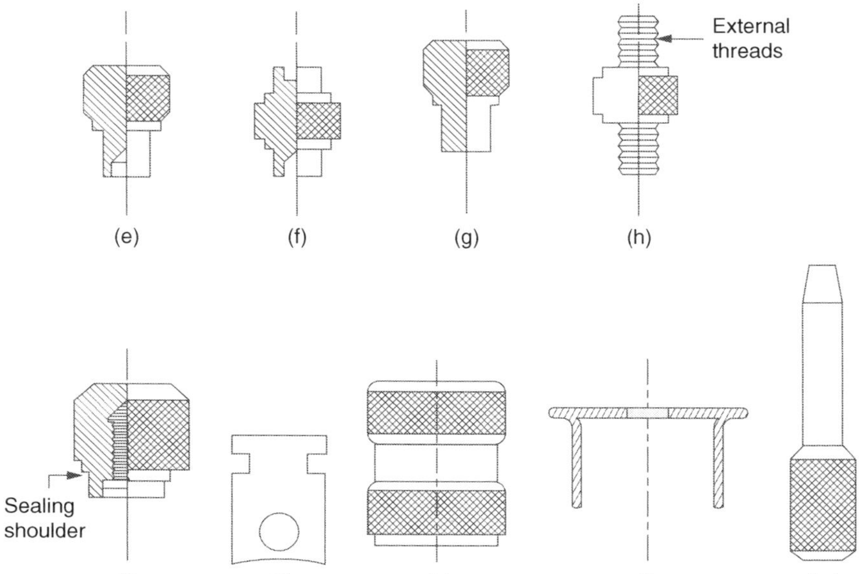

Fig. 1.86 (*Contd.*) (e) Eyelet projecting, (f) Eyelet both ends projecting, (g) Projecting rivet, (h) Double projecting insert with external threads, (i) Blind hole female insert with internal threads and double sealing shoulder, (j) Metal stamped insert, (k) Female insert with undercut, (l) Drawn eyelet, (m) Rod or pin type insert (*Courtesy: Society of Plastic Industry*).

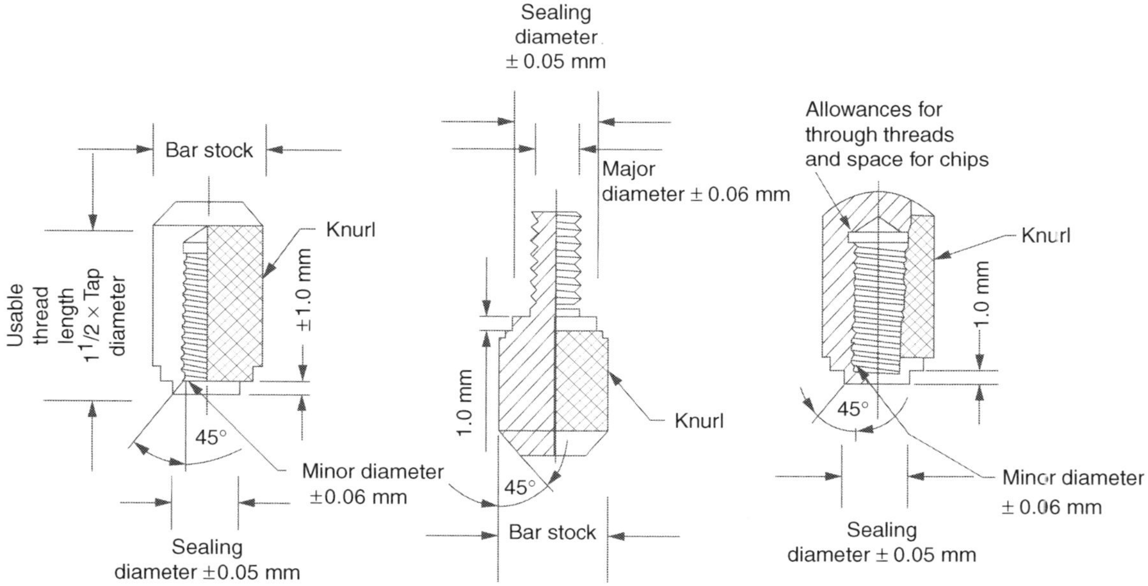

Fig. 1.87 Standard designs for male and female inserts.

Inserts should be located so as to be parallel to the movement of the mould as it opens and closes. Inserts located at right angles and oblique angles are difficult and expensive to mould (Fig. 1.88). This also holds true with moulded holes.

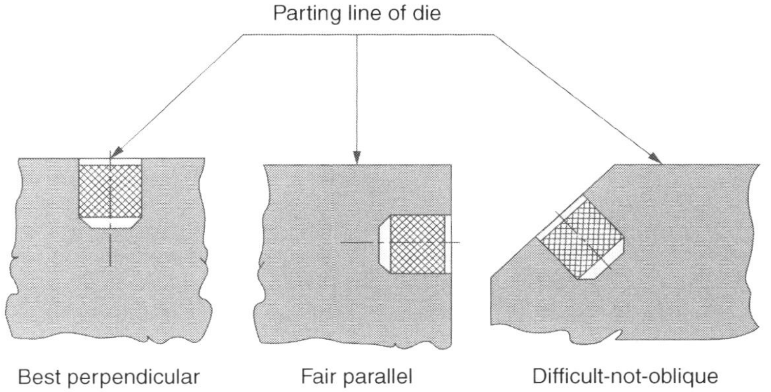

Fig. 1.88 Inserts should be located so as to be parallel to the movement of the mould as it opens and closes. Inserts located at right angles and oblique angles are difficult and expensive to mould.

Some female inserts are open at both ends and are moulded through the part (Fig. 1.89). The length of these inserts should be 0.025 mm to 0.050 mm over size; if there axes are moulded parallel to the draw. The extra length on the end of the insert aids greatly in preventing plastic compound from covering the ends of the insert and getting inside. More extra length may cause the insert to break as the mould comes together and pinches the insert. If the outside diameters of the inserts are used to hold the insert in place during the moulding operation, tight tolerances are required (Fig. 1.90).

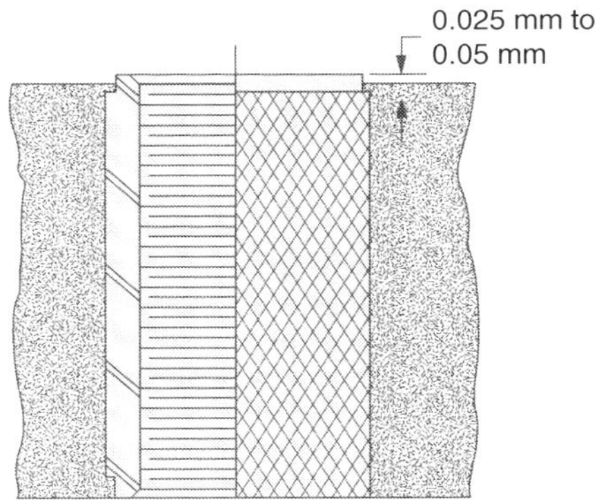

Fig. 1.89 Female inserts which are open at both ends should be designed to have an axial tolerance of plus 0.025 mm to 0.050 mm, compared with the part. This precaution will help to keep the plastic compound from entering the thread.

Female inserts moulded through the part frequently are not threaded before moulding when a thermosetting material and compression moulding are used. Plastics material may flow into the threads, necessitating a retapping operation to remove the compound. A well-constructed mould and good inserts allow little or no flash to get into the threads. A female insert located on a mould pin or insert pin requires tight tolerances in order to prevent misalignment.

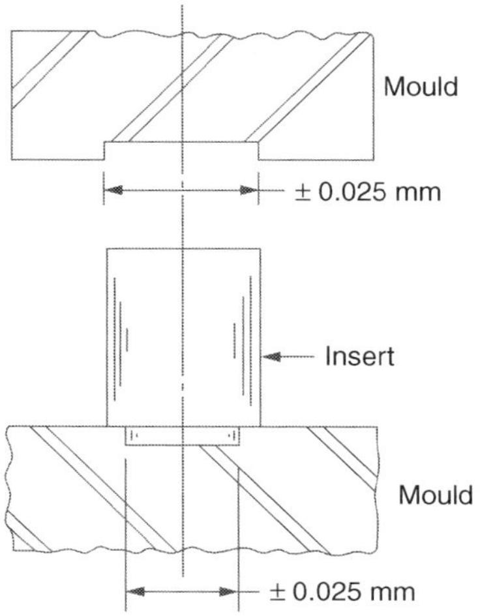

Fig. 1.90 If the outside diameters of an insert are used to hold the insert in place during the moulding operation, tight tolerances are required.

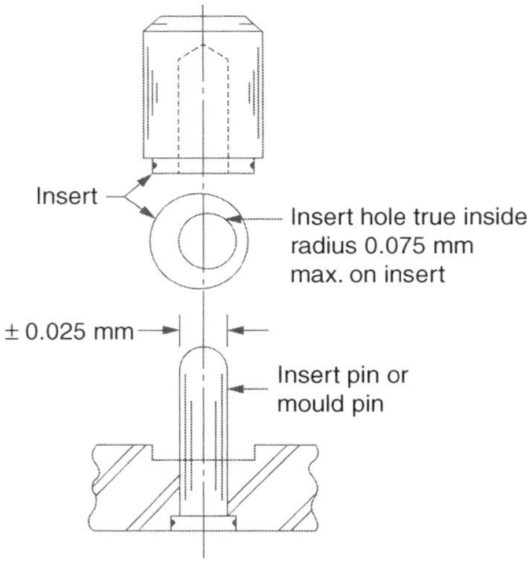

Fig. 1.91 A female insert located on a mould pin will require close tolerances in order to prevent misalignment. It will be noted in the drawing that the true inside radius of the insert must be held to 0.075 mm maximum in relation to the outside radius of the insert.

Male and female inserts should be provided with a shoulder in order to prevent plastic compound from flowing into the threads (Fig. 1.92). An adequate sealing shoulder should be allowed.

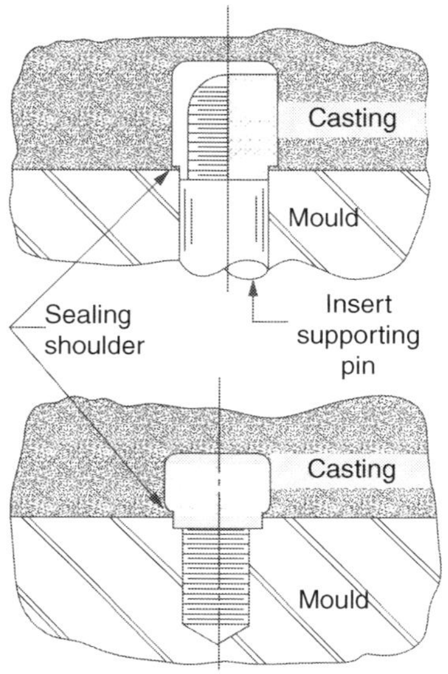

Fig. 1.92 Male and female inserts should be provided with shoulders to help prevent the plastic compound from flowing into the threads.

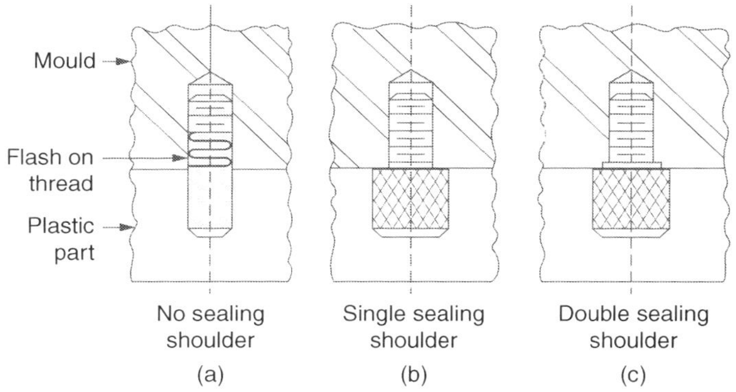

Fig. 1.93 Avoid using an insert that has not been provided with a shoulder as shown in (a). A single sealing shoulder (b) is better. A double sealing shoulder is the best (c) but it is more expensive.

Male inserts as bolts that are not provided with a shoulder (Fig. 1.93) should be avoided, because compound will flow up into the threads during moulding. A single or double-sealing shoulder is better. The necessity of chasing these threads after moulding may be eliminated on male inserts not having shoulders, if threaded lugs or bullets are used (Fig. 1.94). This procedure increases the parts cost, however, because the threaded mould section must be unscrewed from the insert after each part is moulded.

Some designs require that a portion of the insert extends above the surface of the part (Fig. 1.95). The extended portion should be round, since that portion of the insert must fit into a recess in the mould while the part is being moulded. Anything but a round recess is expensive to machine and should be avoided. Also, a round shoulder on the top of a hexagonal insert eliminates the necessity to machine a hex hole in the mould (Fig. 1.96).

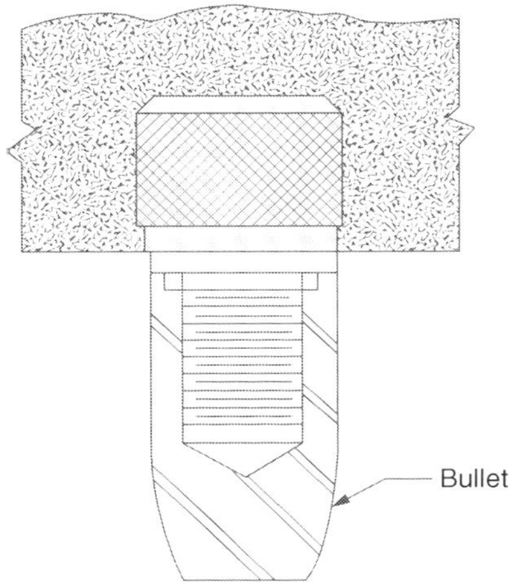

Fig. 1.94 Plastic flash may be kept out of threads on an insert by means of a threaded lug or bullet.

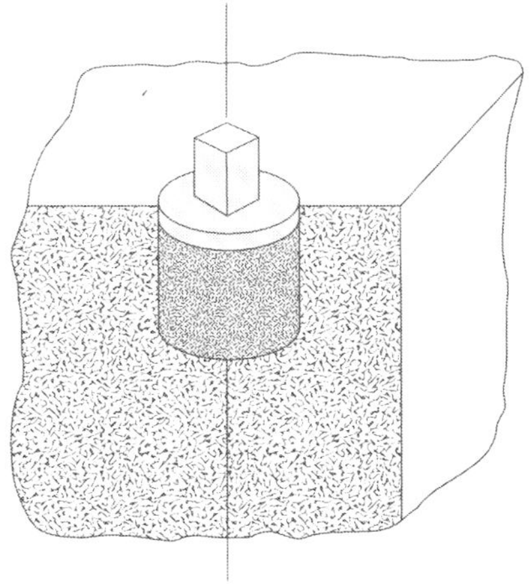

Fig. 1.95 The portion of a moulded in insert that extends above the surface of the part should be round. Square holes are difficult and expensive to machine in the mould.

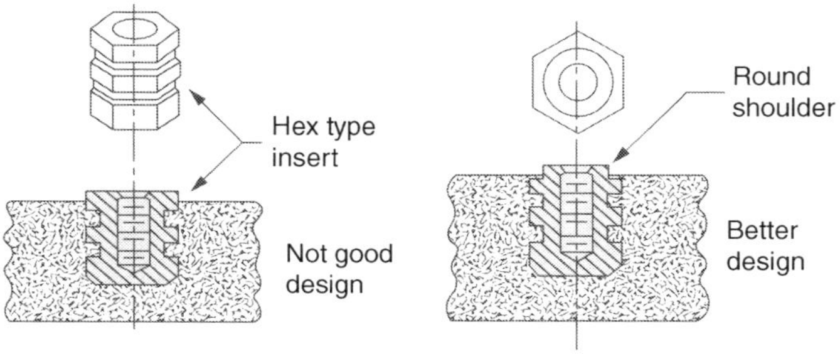

Fig. 1.96 A round shoulder on the top part of a hexagonal insert eliminates the necessity to machine a hexagonal hole in the metal mould.

Male inserts extending through the top and bottom of the part and into recesses in the mould (Fig. 1.97) are considered poor design. In cases of mould misalignment of a maximum of 0.150 mm the mould in closing may scratch one of the extended portions of the inserts.

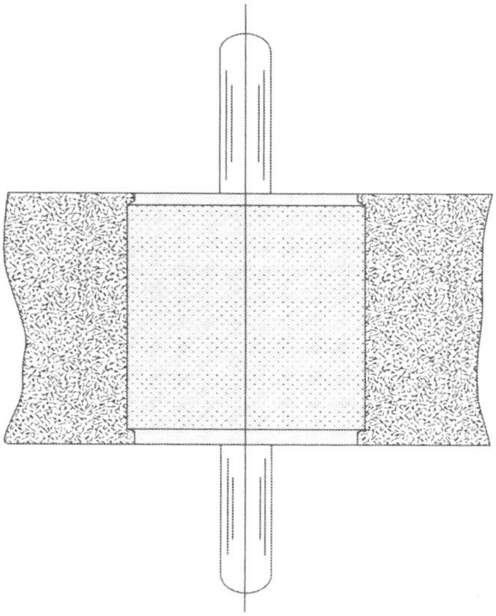

Fig. 1.97 Avoid male inserts that extend above the top and below the bottom of the part. Mould misalignment may cause damage to the insert as well as the mould, when the mould closes.

Female spun-over inserts are used to provide a permanent assembly of contact strips and washers to the moulded part (Fig. 1.98). The design features are essentially the same as those used in threaded inserts, except for the tubular projection.

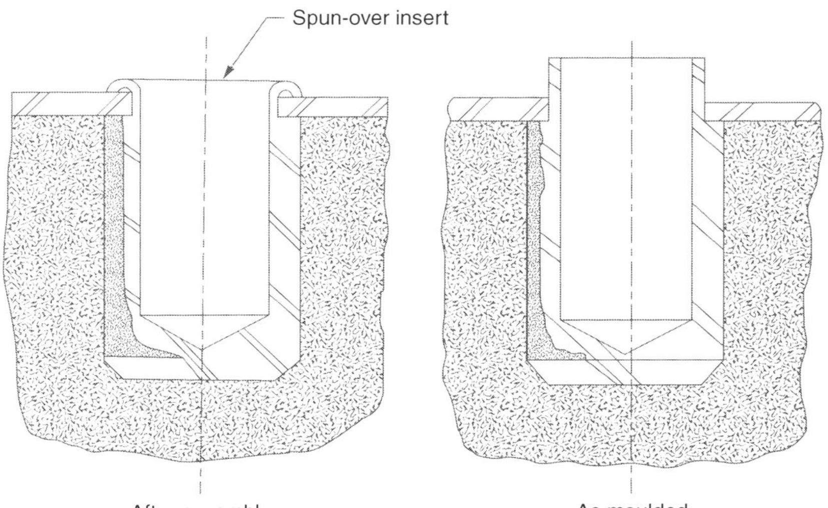

Fig. 1.98 Female spun-over inserts are used to provide a permanent assembly of contact strips and washers to the moulded part.

No insert is so well sealed in the plastic that gas, under sufficiently high pressure, or a liquid will not pass between the insert and the compound. To aid in preventing seepage around the insert, a rubber '*O*' ring may be moulded with the insert in the plastic part (Fig. 1.99). Sometimes the rubber '*O*' ring is placed around the insert and assembled after moulding (Fig. 1.100).

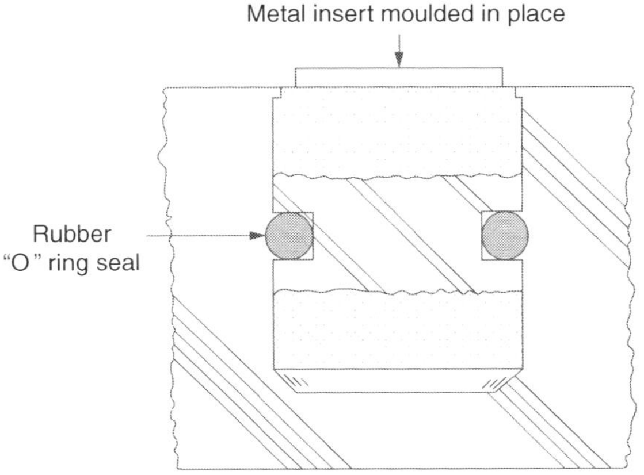

Fig. 1.99 Rubber 'O' rings are sometimes moulded with the metal insert to prevent gas or liquid seepage around the insert.

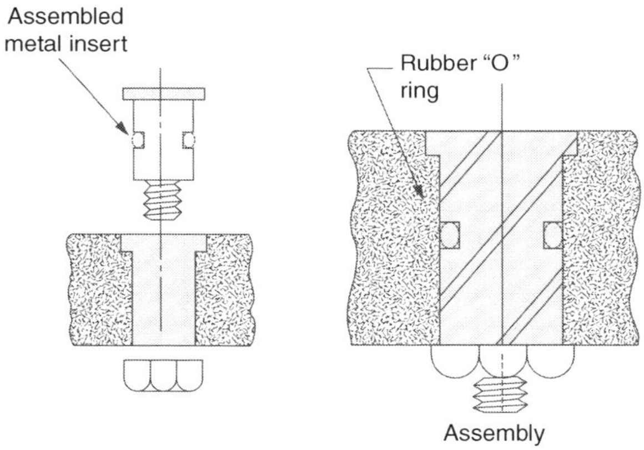

Fig. 1.100 Rubber 'O' rings are placed around inserts and assembled after moulding when it is impractical to mould in the insert.

If an insert is subjected to excessive axial strain, a firmer anchorage of the insert to the compound may be obtained by grooving (Fig. 1.101).

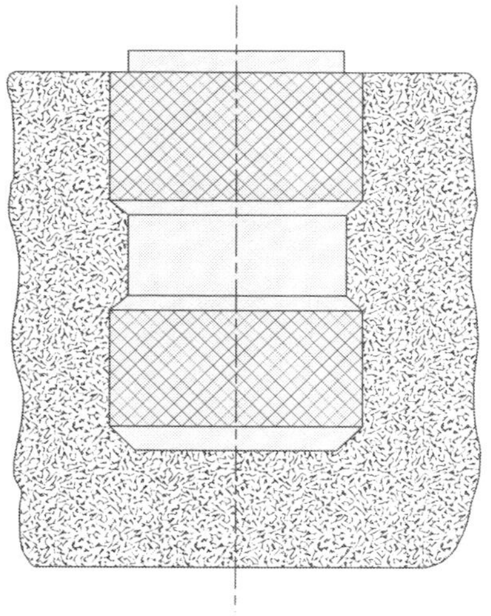

Fig. 1.101 An insert subject to excessive axial pull should be grooved in addition to being knurled to aid in providing good anchorage.

1.12.5 Effect on Mould Strength

The problems involved with the shape of inserts and their effect on mould strength are similar to those encountered with holes in the product.

If compression moulding is used, it is advisable to have the length of the embedded portion of a closed-end insert should not be more then twice its diameter, when the insert is moulded parallel to the draw (Fig. 1.102). It is best to have a through insert that is no longer than twice its diameter, when moulded parallel to the draw (Fig. 1.103).

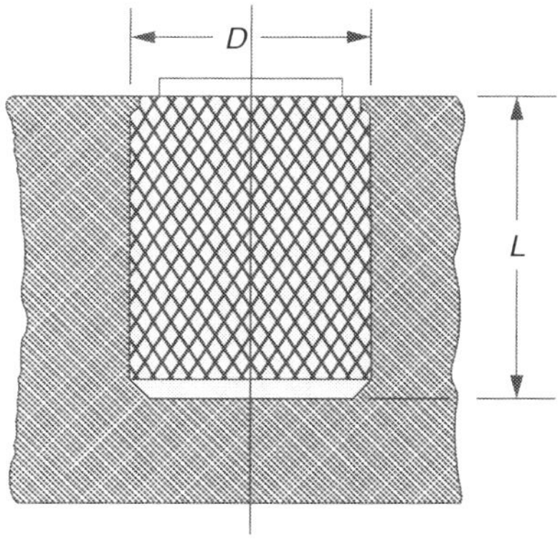

Fig. 1.102 When compression moulding is used, the length (L) of the embedded portion of an insert should be no longer than twice its diameter (D) when moulded parallel to the draw. When moulded perpendicular to the draw, the length should be no longer than the diameter.

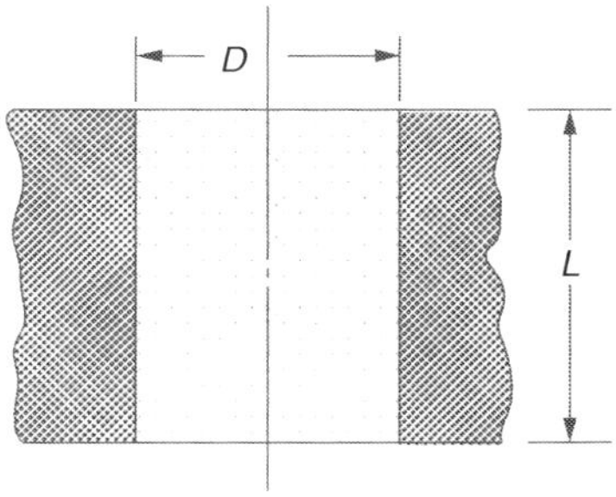

Fig. 1.103 When compression moulding is used a through insert when moulded parallel to the draw should be no longer than twice its diameter.

1.12.6 Location of Inserts in the Part

Inserts that are improperly located in a part, from a moulding stand point, may result in objectionable decorative effects, poor electrical properties, a weak part, a weak mould, or excessive finishing costs.

Because of the differences in the coefficients of expansion and contraction between metals and plastics, sink marks or concaved depressions on a part may result if the end of the insert is too close to the opposite wall (Fig. 1.104). A sink mark may be objectionable from a decorative standpoint. If letters or designs are to be hot-stamped on the surface having the sink mark, they will not be as deep on the sink mark area as on the rest of the surface. If sink marks are to be avoided, the thickness of the plastic compound at the end of the insert should be at least one-sixth the diameter of the insert.

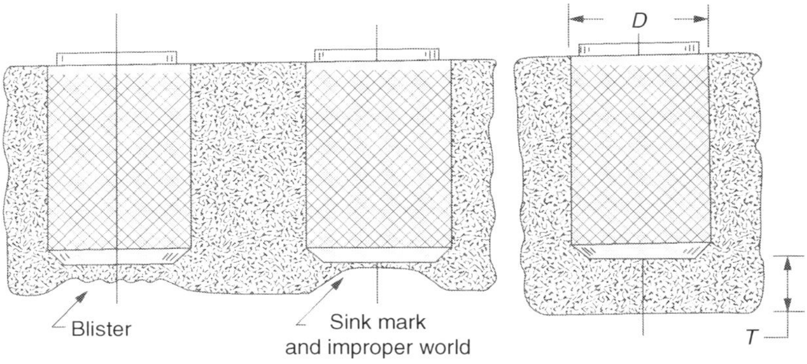

Fig. 1.104 If sink marks and blisters are to be avoided at the end of inserts, the thickness (T) of the plastic at the embedded end of the insert should be at least 1/6 the diameter (D) of the insert.

The location of male and female inserts in a moulded part may affect the mould strength. Female inserts, for example, are held in place by a steel rod or pin that has been inserted in a hole in the mould. If the hole through which the pin is inserted is less than 0.50 in. from the face of a mould section, the mould may crack at this point (Fig. 1.106).

Inserts used in bosses should extend to within one material thickness of the opposite wall, and ribs should be used to support the boss (Fig. 1.107). The end of the insert should be rounded or chamfered to avoid concentration of stresses at the sharp edges.

Moulded-in inserts are used to carry mechanical stresses that are beyond the limits of the plastic material. When heavy loads are encountered, the insert and not the plastic should carry the load (Fig.1.108).

The long male inserts present no problem as far as mould strength is concerned because the larger portion of the length is firmly held by a recess in the mould and cannot be subjected to the flow of the compound. The distance between the mould insert recess and the side of the mould should be greater than 0.5 mm (Fig. 1.109).

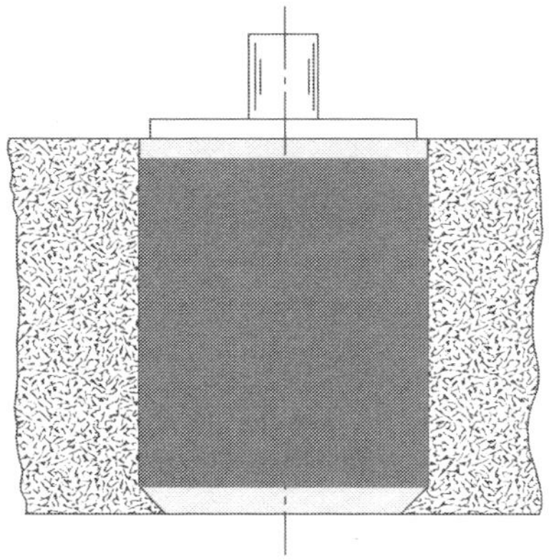

Fig. 1.105 Inserts used in making and breaking electrical circuits should extend above the plastic part.

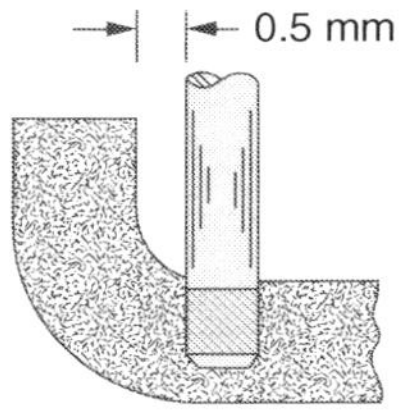

Fig. 1.106 Avoid locating inserts too near the edge of the part. If insert supporting pins are used, the thin adjoining mould section may crack. Minimum allowable thickness for the mould section is 0.50 mm.

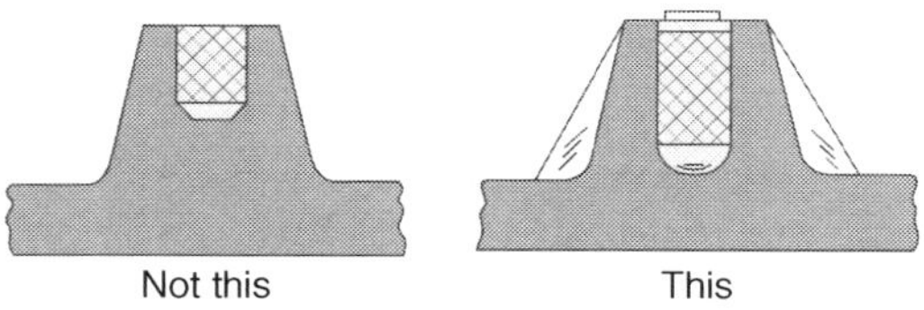

Fig. 1.107 Metal inserts in bosses should extend to within one material thickness of the opposite wall and ribs can be added for additional support to the boss.

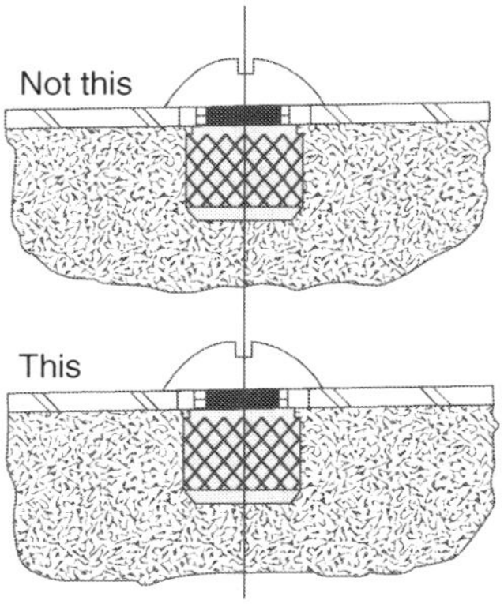

Fig. 1.108 When stresses are encountered, the insert and not the plastics should carry the load.

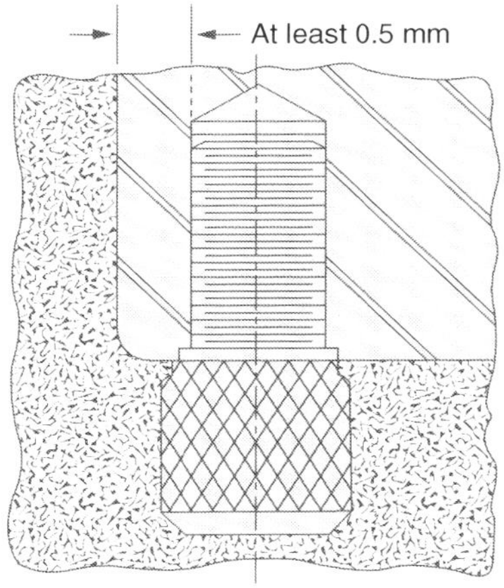

Fig. 1.109 Mould recesses for male inserts should be at least 0.5 mm from the edge of the mould if cracking of the mould is to be avoided.

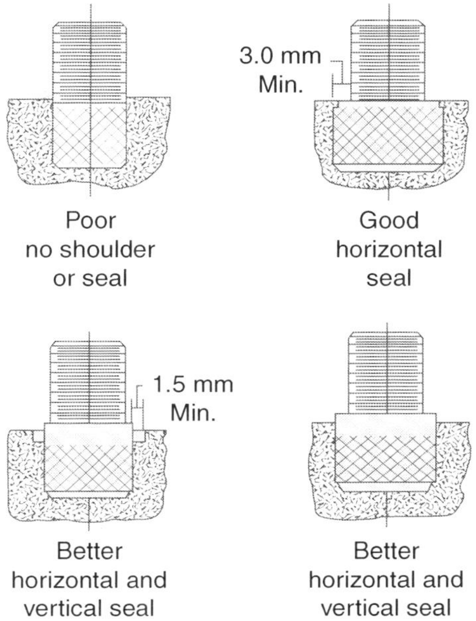

Fig. 1.110 Male inserts should be so designed that they have a shoulder in order to seal out any plastic material that might be forced around the insert during the moulding operation.

Shoulders should be located to afford a vertical as well as a horizontal seal. Providing seals in both directions is the best preventive measure against the flow of compound over the threads or over the inserts.

Figures 1.110 and 1.111 show various methods of designing inserts for effective seals. The shoulders or seals are always flush with or in contact with a portion of the mould (Fig. 1.112). Seals are located to prevent material from flowing into the threads. Female inserts generally are supported by insert pins, and male inserts are held in place by recesses in the mould.

If a female insert is used without any sealing shoulder, the minor thread diameter of the insert must be held to plus or minus 0.025 mm, in order to keep plastic material from flashing into the threads. The close tolerance makes a tight fit or seal between the threads on the insert and the locating pin in the mould.

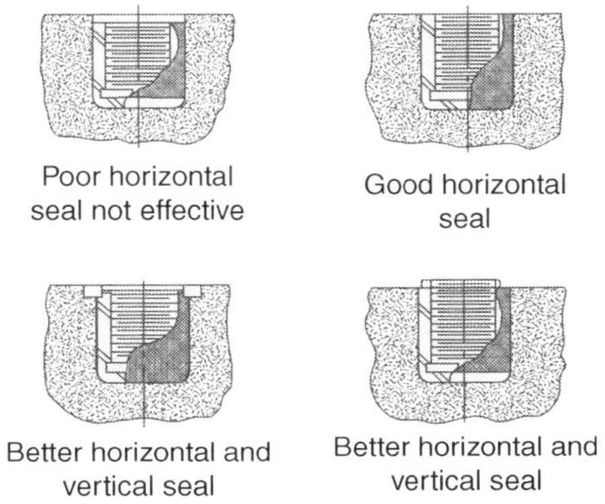

Fig. 1.111 Female inserts should be so designed that they have a shoulder in order to seal out any plastic material that might be forced into the insert during the moulding operation.

Male inserts moulded without any sealing shoulders should have a tolerance of plus or minus 0.025 mm on the major diameter, in order to help prevent material from flashing into the threads on the insert and locating pin in the mould.

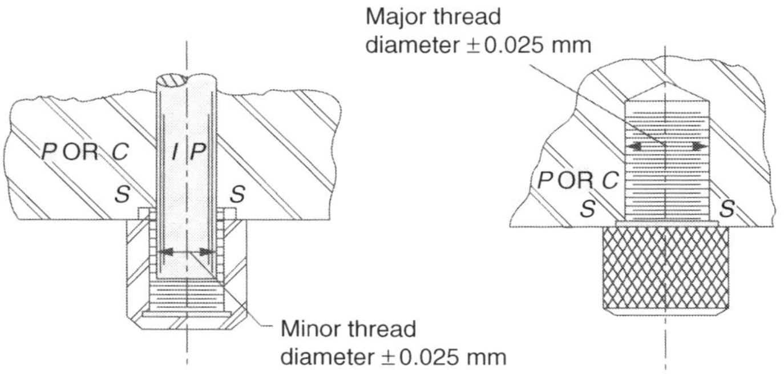

Fig. 1.112 Shoulders or seals are always flush or contact a portion of the mould. *P* or *C*, plunger or cavity; *IP*, insert pin: *I*, insert; *S*, sealing surface, shown heavy.

1.12.7 Suggested Minimum Wall Thickness for Inserts

Metal inserts moulded in either thermosetting or thermoplastic materials require a wall of compound around them of sufficient thickness to prevent cracking upon alternate heating and cooling and aging of the part. Table 1.3 gives what is considered the minimum wall thickness requirements for both the thermoplastics and thermosetting compounds. In some cases, an insert with sharp embedded corners is moulded in the plastic. Sharp corners increase the danger of cracking through the wall (Fig. 1.113).

Inserts that are moulded in opposite sides of a thermosetting part should be no closer than 3.2 mm. Cracking of the compound between the inserts may occur if they are closer. If inserts moulded in this position carry electric current, they may short circuit through the crack.

Table 1.3 Suggested minimum plastics wall thickness for inserts of various diameters.

Diameter of inserts (mm)	3.18	6.35	9.50	12.70	19.00	25.4
Plastic materials						
ABS	3.18	6.35	9.50	12.70	19.0	25.4
Acetal	1.57	3.18	4.74	6.35	9.52	12.7
Acrylics	2.36	3.18	4.74	6.35	9.52	12.7
Cellulosics	3.18	6.35	9.52	12.7	19.0	25.4
Erthylene vinyl acetate	1.0	2.15	Not Recommended			
F.E.P. (Flurocarbon)	0.63	15.24	Not Recommended			
Nylon	3.18	6.35	9.52	12.7	19.0	25.4

(Contd.)

Table 1.3 *(Contd.)*

Diameter of inserts (mm)	3.18	6.35	9.50	12.70	19.00	25.4
Polyallomers	3.18	6.35	9.52	12.7	19.0	25.4
Polycarbonate	1.57	3.18	4.74	6.35	9.52	12.7
Polyethylene (H.D.)	3.18	6.35	9.52	12.7	19.0	25.4
Polypropylene	3.18	6.35	9.50	12.7	19.0	25.4
Polystyrene	Not recommended					
Polysulfone	Not recommended					
Surlyn (ionomer)	1.57	23.6	3.18	4.74	6.35	7.92
Phenolic G.P.	23.6	4.0	4.74	5.53	7.92	8.71
Phenolic (medium impact)	2.00	3.55	4.0	5.15	7.13	7.92
Phenolic (high impact)	1.57	3.20	3.55	4.74	6.35	7.13
Urea	2.36	4.0	4.74	5.53	7.92	8.71
Melamine	3.18	4.74	5.53	7.92	8.71	9.52
Epoxy	0.50	0.76	1.0	1.27	1.52	1.78
Alkyd	3.18	4.74	4.74	7.92	8.71	9.52
Diallyl phthalate	3.18	4.74	6.35	7.92	8.71	9.52
Polyester (premix)	2.36	3.18	3.55	4.74	6.35	7.13

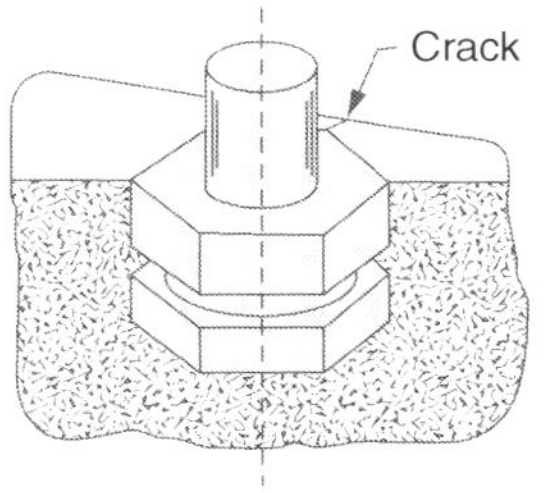

Fig. 1.113 Inserts with sharp corners should be avoided. Such inserts may cause cracking of the part.

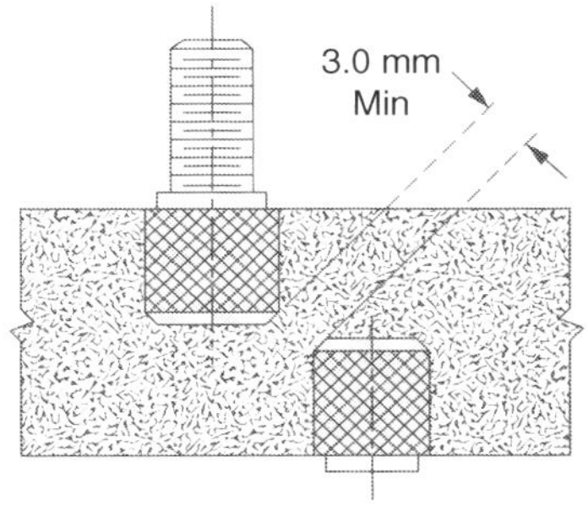

Fig. 1.114 Inserts moulded in opposite sides of a thermosetting part should be no closer than 3.0 mm.

If a boss supporting insert is required, it may be necessary to bring the boss down to a narrow fin around the insert (Fig. 1.115) is to be avoided. The same results may be accomplished by cutting the fin back, as illustrated in Fig. 1.116. Minimum compound wall thickness, as covered in Table 1.3 should be allowed.

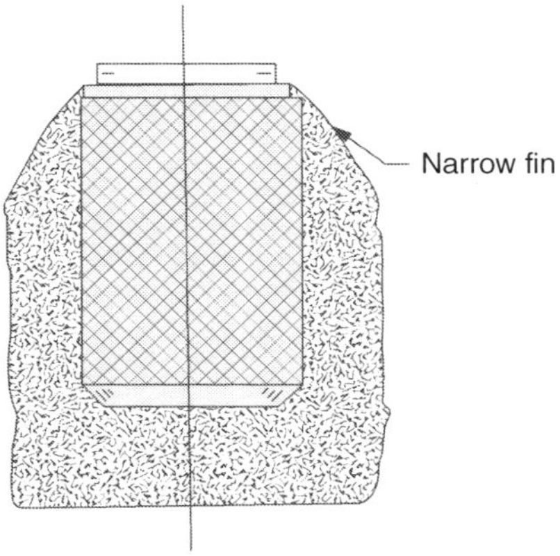

Fig. 1.116 Avoid bringing a boss down to a narrow fin around an insert. This may cause cracking of the plastic material.

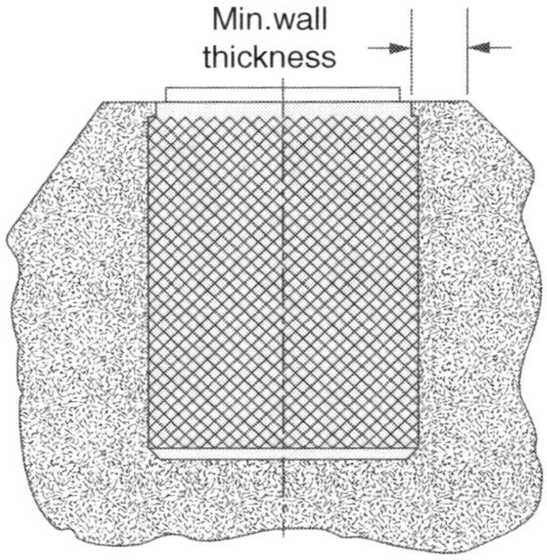

Fig. 1.117 Inserts moulded in bosses should have enough compound around them to prevent cracking. Minimum wall thickness can be obtained from Table 1.3.

A few applications may call for a plastic material to be moulded in or around the inside of a metal insert. Since the plastic material shrinks on cooling, the metal insert should be undercut to prevent it from falling off (Fig.1.118).

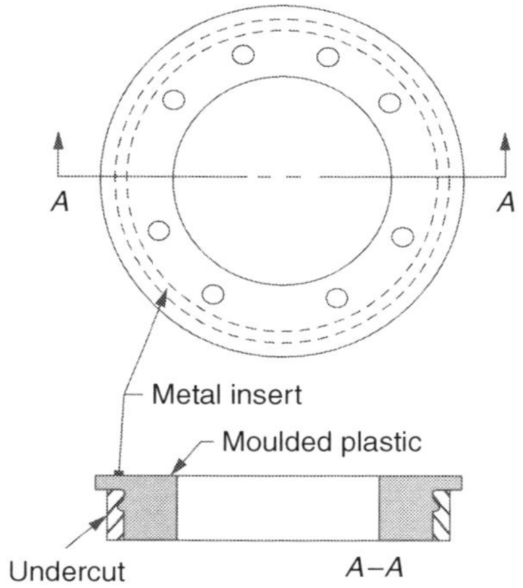

Fig. 1.118 A metal insert moulded around a plastic part should be provided with proper undercuts in order to hold the plastic and metal together.

Sometimes, inserts will be covered with a certain amount of flash after they have been moulded in the part. This is true even though the inserts have been provided with shoulders. All inserts that are to be faced after moulding should project above the surface of the part at least 0.4 mm (Fig.1.119). It may be necessary to face off an insert after moulding if a plastic material with high shrinkage values has been used (Fig.1.120).

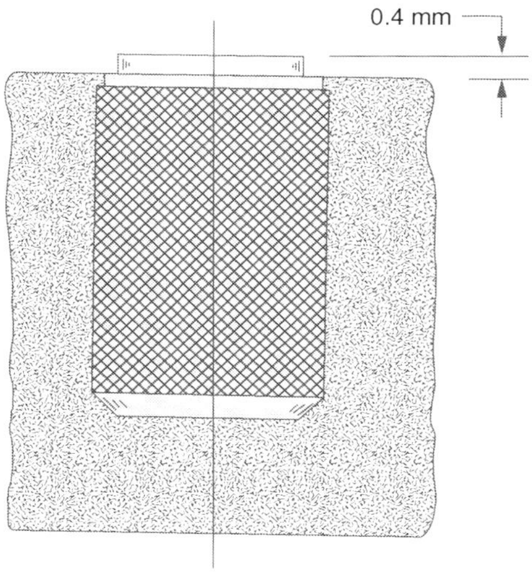

Fig. 1.119 Inserts to be faced after moulding should project above the part at least 0.4 mm.

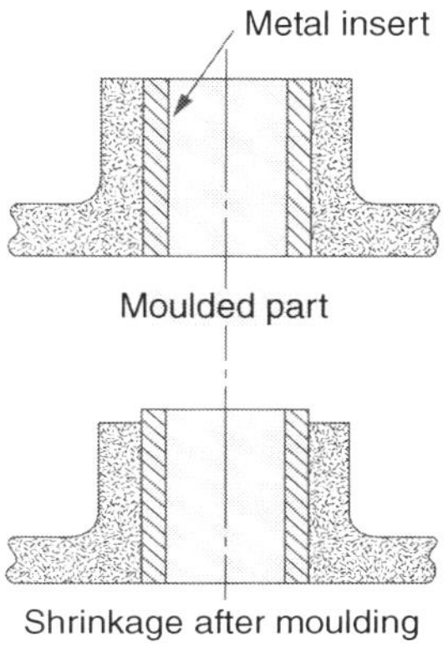

Fig. 1.120 It may be necessary to face off an insert after moulding if a plastic material with high shrinkage values has been used.

1.12.8 Pressed-in Inserts

Insert may be pressed into holes provided for them immediately after the part has been moulded. As the part cools, it shrinks around the insert, holding it securely in place. Pressed in inserts are not generally recommended for use with thermoplastic moulded parts. These parts have been cooled slightly before they are removed from the mould. Care must be taken to prevent exceeding the elastic limit of the plastic material. Pressed-in inserts are not used in more brittle plastics. Inserts that are to be pressed in after moulding require a straight knurl (Fig.1.121).

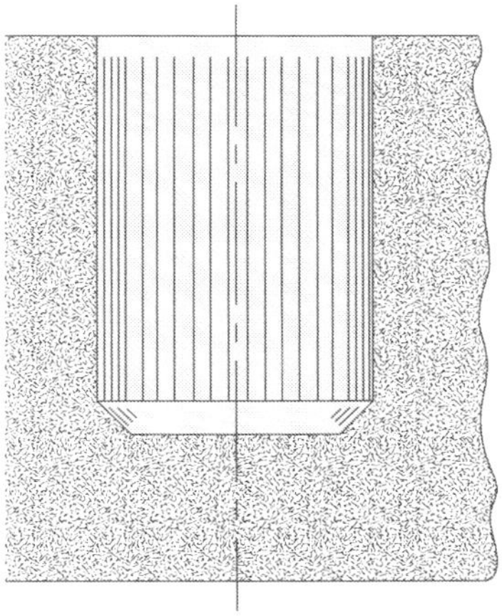

Fig. 1.121 Inserts that are pressed in after moulding should be provided with a straight knurl.

The holding qualities of these inserts are not as good as those of moulded-in inserts, because of the straight knurl feature. Sometimes it is necessary to develop the hole size for an insert that is to be pressed in place (Fig.1.122). This means that the metal mould pin is made oversize and then gradually reduced in size by removing metal part. If a diamond knurl is used, it may be necessary to use a suitable adhesive to help hold the insert in the plastic.

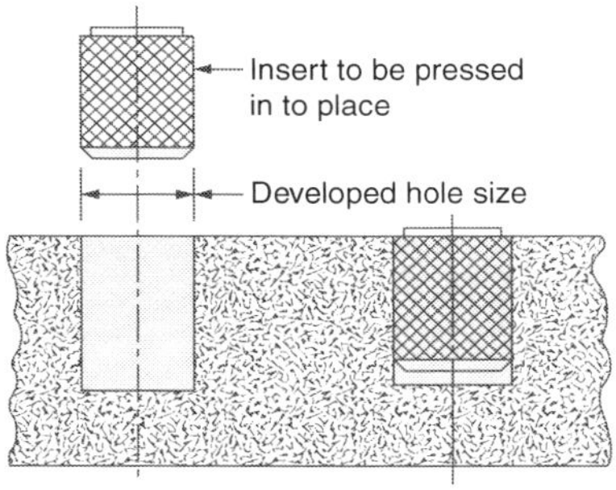

Fig. 1.122 It may be necessary to develop the hole size for an insert that is to be pressed in place. The mould pin can be made oversize and turned down until the desired press fit is obtained.

1.12.9 Metal-Stamping and Rod-Type Inserts

Metal-stamping inserts (Fig.1.123) are not advisable, as it is difficult to hold them to proper size for a close mould fit. The metal used in stamping the insert may be too soft to withstand flow pressures. Flash will flow over the insert and must be cleaned by a finishing operation. Metal-stamped inserts should be placed at the parting line or below, if scouring or pinching is to be prevented. Also, some means should be provided for anchoring it solidly in the mould and the part (Fig.1.124).

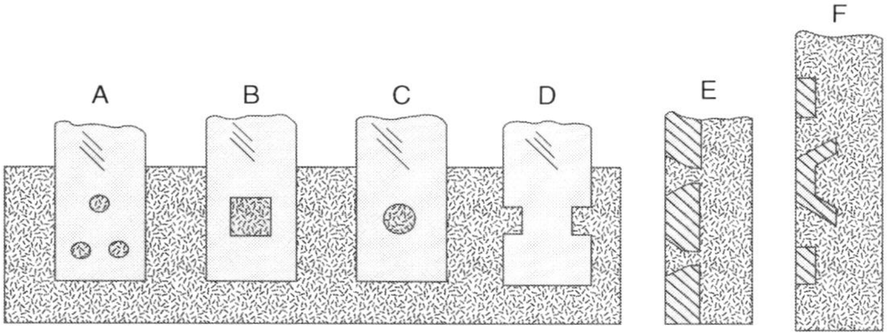

Fig. 1.123 Metal stampings or sheet metal inserts may be secured to the moulded plastic material through the use of punched holes (A, B, C), notches (D, E) and bent tabs or flanges (F).

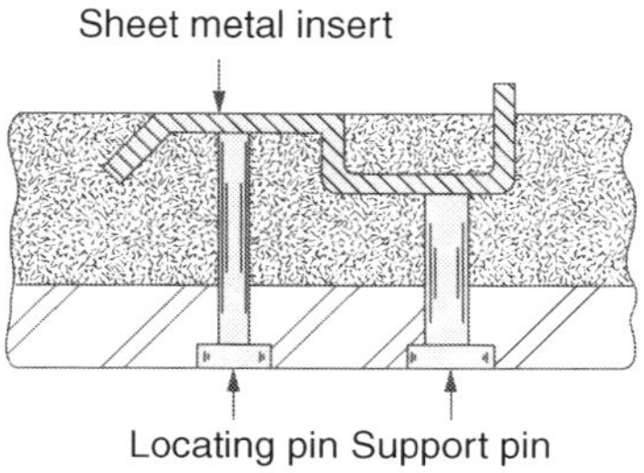

Fig. 1.124 Sheet metal inserts and metal stampings should be anchored securely by locating pins and supported with die pins.

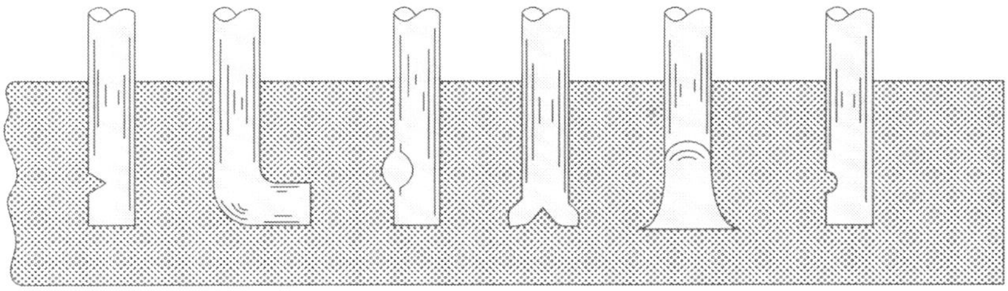

Fig. 1.125 Rod-type inserts are generally restricted to injection or transfer moulding. The rod insert may be anchored by notching, bending, swaging, grooving, etc.

Rod-type inserts (Fig.1.125) should be used with fairly soft or free flowing plastic moulding material. Injection or transfer moulding with low moulding pressures is required, or the rod insert will tend to bend readily or, in some cases, shear off. Rod-type inserts are also difficult to hold in position and will lift out of location, due to the flow of the plastic material. Figure 1.126 illustrates an automobile steering wheel that is a successful rod type insert and that has been used for many years. An injection moulded, cellulosic type plastic material is popular, because colour can be moulded-in. Other steering wheels are compression moulded from hard rubber and then painted. The same rod-type insert is used for both materials.

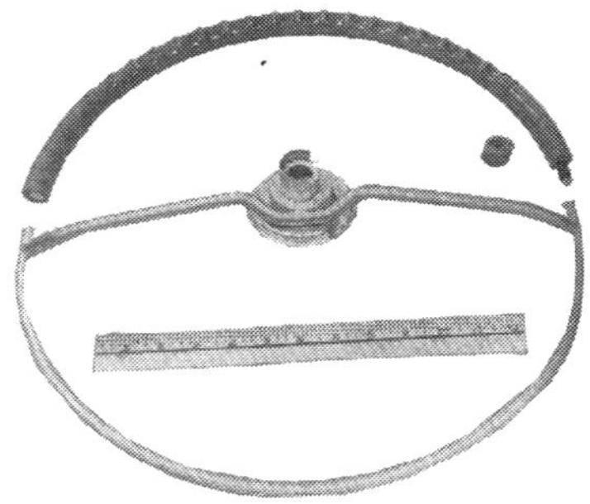

Fig. 1.126 A cross section of an automobile injection moulded steering wheel using a rod type insert.

Carriage bolts, stove bolts, machine screws, etc., are very similar to rod-type inserts and are not recommended, as they are difficult to anchor in the plastic and generally require a cleaning operation after moulding (Fig.1.127).

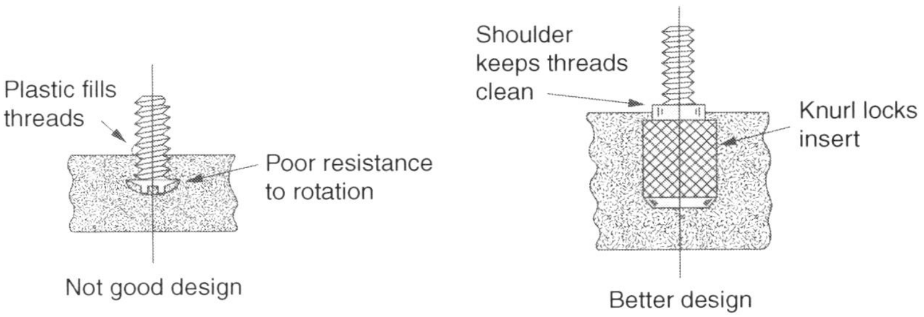

Fig. 1.127 Avoid using standard threaded bolts as they will generally require a cleaning operation after moulding.

1.12.10 Encapsulation

Many intricate electronic devices are encapsulated or moulded in diallyl phthalate moulding compounds. Figure 1.128 shows many types of metal protruding inserts that have been moulded in plastic by the transfer process. Very low transfer moulding pressures are used to prevent damage to the delicate inserts. Extreme accuracy is required in making the inserts and the moulds. It has been the development of plastic materials that can be transferred at low pressure that has made possible the embedment of delicate electronic parts.

1.12.11 Composite Parts

Composite moulding is a moulding process whereby two or more plastic materials are moulded in one operation to make one moulded part. The advantage of this type of moulding is that it utilises the different properties of the two plastic materials.

Figure 1.129 shows a nylon coupling gear that is used in conjunction with an automobile window lift motor. The whole assembly represents a composite gear made of nylon, rubber and metal. The nylon gear is injection moulded around a straight knurled metal shaft. The neoprene coupling shield is bonded to the nylon after moulding, and a metal back plate is bonded to the neoprene rubber coupling shield. The neoprene rubber seals out any water that might get into the electric motor.

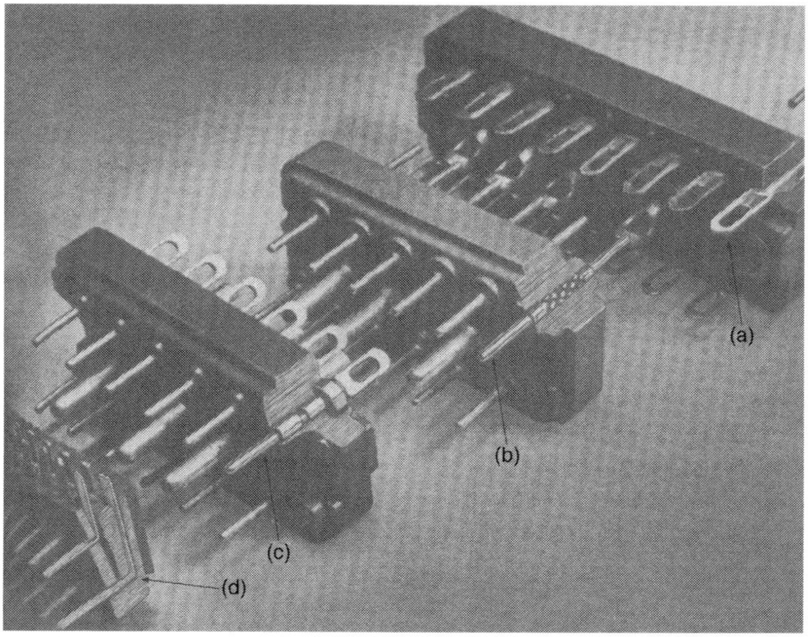

Fig. 1.128 Intricate electronic parts showing moulded in metal inserts. A section of each part has been cut away to show the insert: (a) An eyelet insert assembled after moulding, (b) A knurled wire or pin type moulded through insert, (c) An eyelet pin type moulded through insert, (d) A flat ribbon wire moulded through insert.

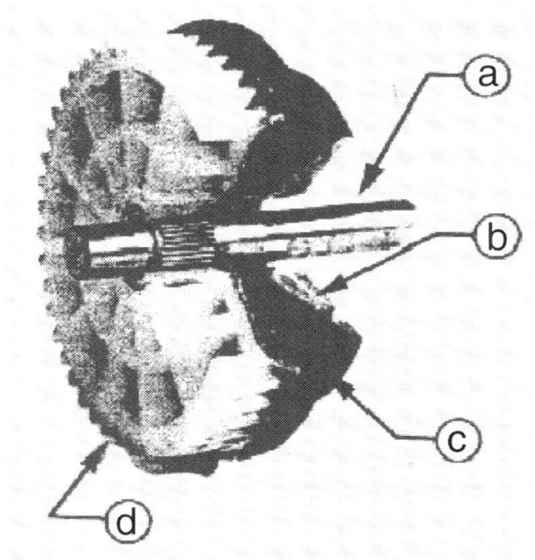

Fig. 1.129 This picture illustrates a cross section of a nylon coupling gear used in conjunction with an electric motor for an automobile window lift. Note the straight knurled metal shaft: (a) Knurled metal shaft, (b) Metal face plate, (c) Neoprene coupling shield, (d) Nylon moulded gear.

1.13 GATE SIZE AND LOCATION

The gate is the small channel in the flow path of the plastic melt that separates the mould cavity from the runner. The major consideration in gate location is the effect upon appearance or function of the moulded part. The end use of the part may place restrictions on the type of the gate, gate location and size of the gate mark that is acceptable (Fig. 1.130). The area near the gate is highly stressed due to frictional heat generated at the gate and the high velocities of the flow material. The size of the gate can be considered in terms of the gate cross- sectional area and the gate length (also called *gate land*).

A small gate is desirable so that

1. the gate freezes soon after the impression is filled and the injection plunger can be withdrawn immediately without the probability of voids being created in the moulding by suck back.
2. it allows for simple degating and in some moulds this degating can be automatic.
3. after degating, only a small witness mark remains.
4. better control of the filling of multi–impression can be achieved.
5. packing the impression with material in excess of the required to compensate for shrinkage is minimised.

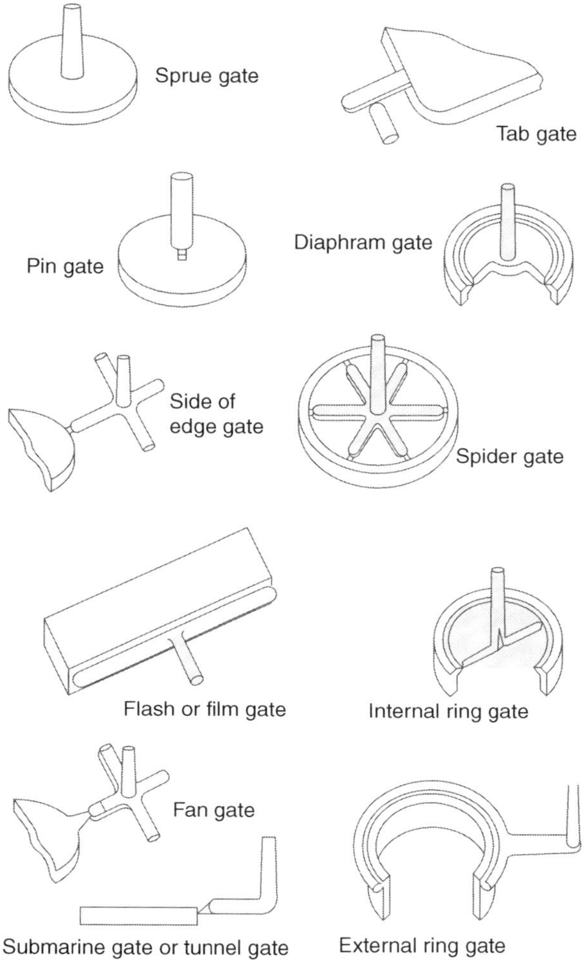

Fig. 1.130. Types of Gate.

The optimum size for a gate will depend upon the following factors:

1. The flow characteristics of the material to be moulded.
2. The wall section of the moulding.
3. The volume of material to be injected into the impression.
4. The temperature of the melt.
5. The temperature of the mould.

The position of the gate should be such that there is an even flow of melt in the impression, so that it fills uniformly and the advancing melt front spreads out and reaches the various impression extremities at the same time.

1.14 LOCATION OF EJECTOR PINS

All thermoplastic materials contract as they solidify, which means that the moulding will shrink on to the core which forms it. This shrinkage makes the moulding difficult to remove. The ejection of a part from a mould requires careful consideration. Ejection should be positive in action and also placed in a position that the ejector pins or blades marks will not be seen on the moulded part when it is in normal use. When ejector pins are placed on thin wall sections, the pin should be 0.13–0.25 mm away from the wall on the core to get maximum amount of bearing surface of the ejector pin on the piece part. Ejector pins are made so that they protrude above the parting surface about 0.05 mm.

Ejector mechanism releases the moulded plastic part from the mould. Ejector pins should be located at the thickest possible sections, preferably directly over bosses or ribs. A stripper plate instead of knockout pins may be used to remove the part from the plunger. Knockout pins should be recessed into the plastic surface approximately 0.4 mm (Fig. 1.131).

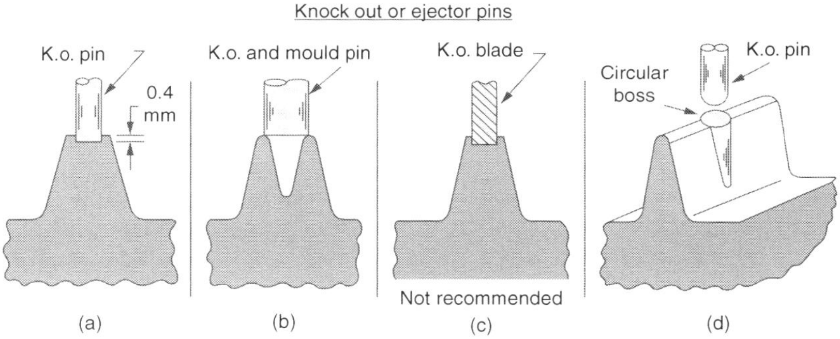

Fig. 1.131 (a) Knockout pins should be recessed into the plastic surface approximately 0.4 mm, (b) Knockout pins are sometimes used as mould pins, (c) Blade type knockout pins are not recommended. They are expensive and difficult to maintain, (d) All knockout pins should have an adequate surface to push against.

Knockout pins are sometime used as mould pins. Blade-type knockout pins are not recommended as they are expensive and difficult to maintain. All knockout pins should have an adequate surface to push against. Knockout pins are some time placed on runners. Wear between the knockout pin and the mould will cause flash to occur around the pin. Pick-up ribs are frequently used to make the moulded part remain in the cavity or on the punch.

If knockout marks are objectionable, a stripper plate may be used to remove the part from the mould. It exerts a more even pressure on the moulding than ejector pins and results in less distortion. Ejector sleeves are preferred when the moulded parts have to be stripped off of one or more cores. Ejector sleeves are subjected to severe stress and wear. The outside diameter of the sleeve should be held to 0.025 mm to 0.05 mm smaller than the hole in the cavity. Moulded parts that have thin walls and deep draws can be ejected from the moulds by using air poppet valves.

1.15 TOLERANCE

1.15.1 Definition

Dimension tolerances may be defined as follows:

Allowances are the intentional differences in dimensions to take care of fit. Tolerances are the unintentional variations that occur during manufacture.

In the case of moulded plastics, they are the differences that exists in dimensions from cavity to cavity, or the dimensions difference that exists, between parts, due to shrinkage, warpage, and other uncontrollable factors. Limits are the maximum and minimum dimensions that define the tolerances. Basic dimensions are the expressed theoretical value of a dimension from which the allowable tolerance variations are to be calculated.

The following discussion of tolerances is limited to tolerances held by a moulded part within a few hours after moulding and at normal temperature and atmospheric conditions. Holding extreme accuracy of dimensions in moulded parts is expensive. Extremely close tolerance increases the initial cost of the mould. The moulding operation costs are higher because greater care is required to maintain moulding uniformity and extra cooling or shrink fixtures may be needed after moulding. The designer should note that dimensional changes due to temperature variations alone can be three to four times as great as the specified tolerances. Also, the nature of the surrounding and processing conditions is important items to consider. Dimensional tolerances in plastic moulded articles will be considered as allowable variations, plus and minus, from a nominal or mean dimension, as used or set by the plastic industry.

1.15.2 Parameters Influencing Part Tolerance

1. **Part Design:** Part configuration (size/shape), relate shape to flow of melt in mould to meet performance requirements that should at least include tolerances.
2. **Material:** Chemical structure, molecular weight, amount and type of fillers / additives, heat history, storage handling.
3. **Mould Design:** Number of cavities, layout and size of cavities/runners/gates/cooling lines/side actions/knockout pins/etc., relate layout to maximise proper performance of melt and cooling flow patterns to meet part performance requirements, pre–engineer design to minimise wear and deformation of mould (use proper steels) layout cooling lines to meet temperature to time cooling rate of plastics (particularly crystalline types).
4. **Machine Capability:** Accuracy and repeatability of temperature / time / velocity / pressure controls of injection unit, accuracy and repeatability of clamping force, flatness and parallelism of platens, even distribution of clamping on all tie rods, repeatability of controlling pressure and temperature of oil, minimise oil temperature variation, no oil contamination (by the time you see oil contamination damage to the hydraulic system could have already occurred), machine properly leveled.

5. **Moulding Cycle:** Set up the complete moulding cycle to repeatedly meet performance requirements at the lowest cost by interrelating material / machine / mould controls.

Dimensional tolerance should be as generous as possible. Before specifying a dimensional tolerance, the designer needs to consider this demands, which will subsequently place on the mould designer and the plastics engineer - in - charge of productions. Extreme accuracy of dimensions in moulded articles is difficult and expensive to achieve. Closer the tolerance demanded, greater would be the cost of the moulds. Designer should keep this in mind while designing the product. Also the toolmaker's tolerance is very much depending on the product designer's need.

Variables to be considered when stating tolerances are:

1. Mould maker's tolerances.
2. Plastic material shrinkage variances.
3. Moulding process techniques.

Dimensional tolerances are affected by several production and tooling variables such as the number and the size of cavities and the degree of control to the moulding operations.

1.15.3 Standard Tolerances on Moulded Articles

Tolerances over various dimensions of the moulded part will vary in accordance with the method used in moulding. Figure 1.132 shows that the tolerances on the following dimensions are influenced by the mould section:

1. *'A'* in Fig. 1.132 represents dimensions established by fixed mould details.
2. *'B'* represents dimensions established by fixed mould details extending across a cut off line.
3. *'C'* represents dimensions established by fixed mould details extending across a prating line.
4. *'D'* represents a dimension established by fixed mould details extending across a cut off line and one parting line.

In order to simplify drawings and to take cognisance of the fact that dimensions across cut off lines vary from parallel dimensions that do not cross cut off lines, some manufacturers dimension drawings as illustrated in Fig. 1.133(a). All vertical dimensions of the part made by the cavity should be measured from the bottom of the part, that is the section of the part made by the bottom of the cavity. All vertical dimensions made by the plunger should be dimensioned from the bottom of the inside. Dimensions *'A'* and *'D'* will be the only ones affected by the cut off. *'A'* and *'D'* will carry the cut off tolerance applicable to the particular type of moulding used. Many engineers and draftsmen dimension their drawings from a single locating point or line, as illustrated in Fig. 1.133(b). If this is used, more than one dimension will be affected by the cut off tolerance, namely, dimensions *'E'* and *'G'* or the dimensions that are made by the cavity section of the mould. Both *'E'* and *'G'* will require cut off tolerances. The dimensions *'F'* and *'H'* which are by the plunger, will not be affected by the cut off. Figure 1.134 shows that when the cut off flash is increased, the plunger is raised, thus thickening the part at the bottom. If the designer will remember to separate the dimensions made by the plunger section, he will experience little difficulty in placing the correct tolerances on the dimensions.

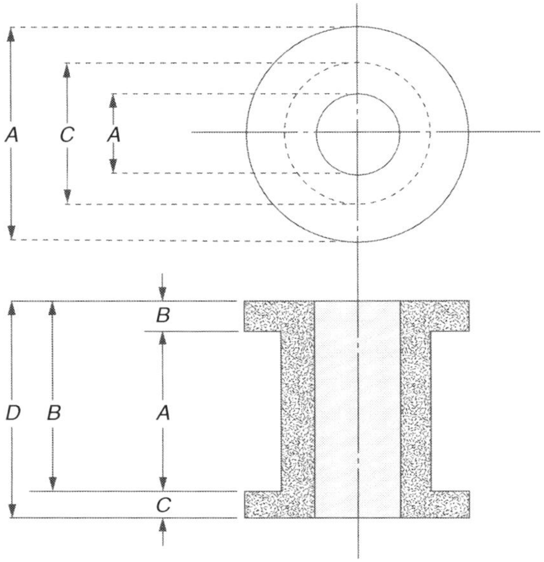

Fig.1.132 Tolerances on the dimensions of moulded parts vary, depending upon how the mould is made and how the part is moulded. It is assumed that the part shown here was made from a mould.

Figure 1.135 illustrates isometrically the dimensions affected by the parting lines. The heavy lines shown in Fig.1.135 indicate parting lines. There is a parting line on the bottom of the upper flange and a parting line through the centre of the part. Dimensions '*B*', '*C*', and '*D*' are, therefore, the only dimensions affected by the parting line. Dimensions '*E*' and '*F*', because they are made by fixed sections of the mould, do not carry additional parting line tolerances. Dimension '*A*' carries the cut off tolerance.

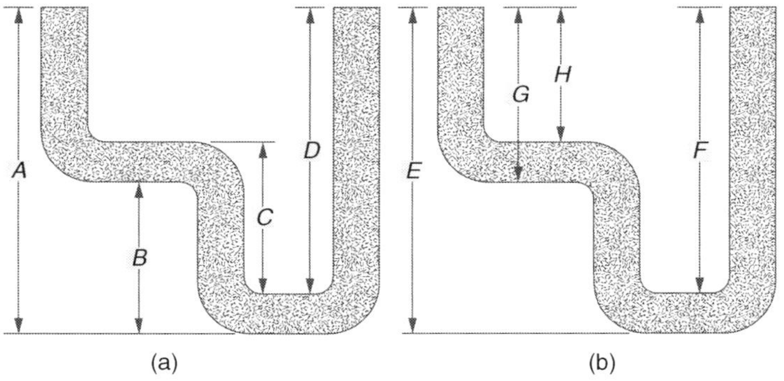

Fig.1.133 Critical dimensions should be based on locating points that are not subject to parting line variations. All dimensions in (b) refer to the parting line and hence all dimensions will vary with cut off tolerance. However, in Fig. (a), only dimensions '*A*' and '*D*' will be affected by cut off tolerance.

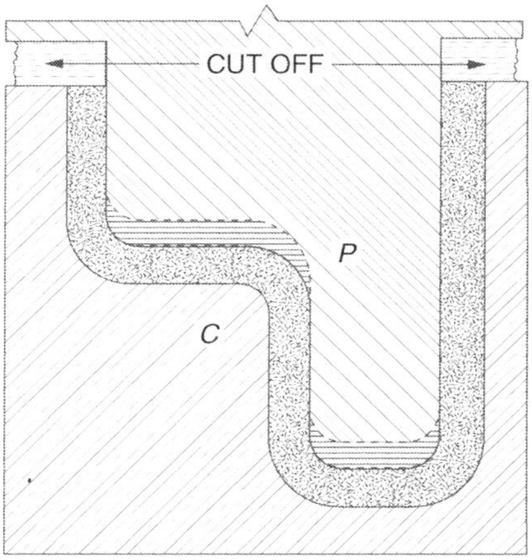

Fig. 1.134 Dimensions controlled by the plunger are not affected by the cut off flash, but dimensions controlled by the cavity are so affected. *P* is the plunger, *C* the cavity.

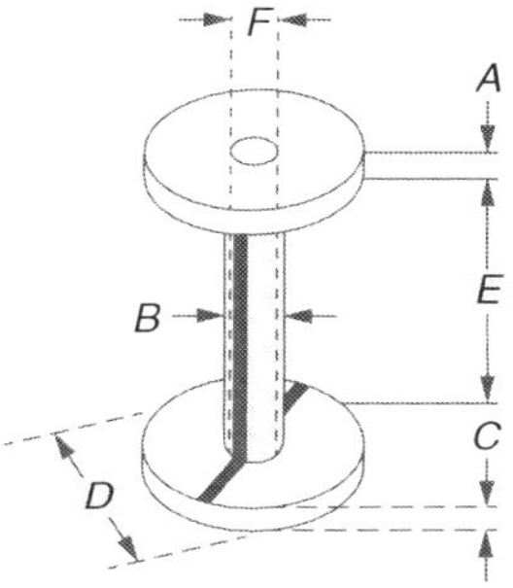

Fig. 1.135 A plastic spool will have parting lines as shown here by heavy lines.

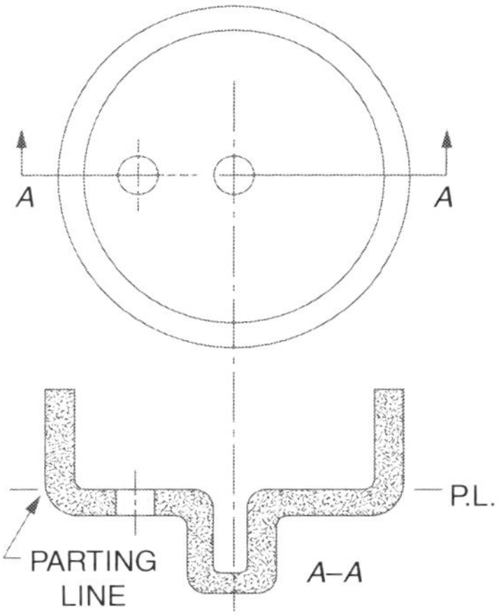

Fig. 1.136 A cross section of a round moulded part which has one moulded through-hole. The parting line of the die is indicated. Note that the parting line is placed near the bottom of the piece.

Table 1.4 on standard tolerances for acrylic material contained in this book was prepared by the custom moulders of The Society of Plastics Industry. This table is to be used only as a guide. The dimensions are based on a hypothetical moulded article with a cross section shown in the table and (Figs. 1.136 and 1.137)explains the cross section of the moulded article along with a mould that would make the plastic part. This illustration gives the reader a much clearer picture of how the dimensions are taken or derived. Note that the cavity controls some dimensions and the punch or force controls other dimensions.

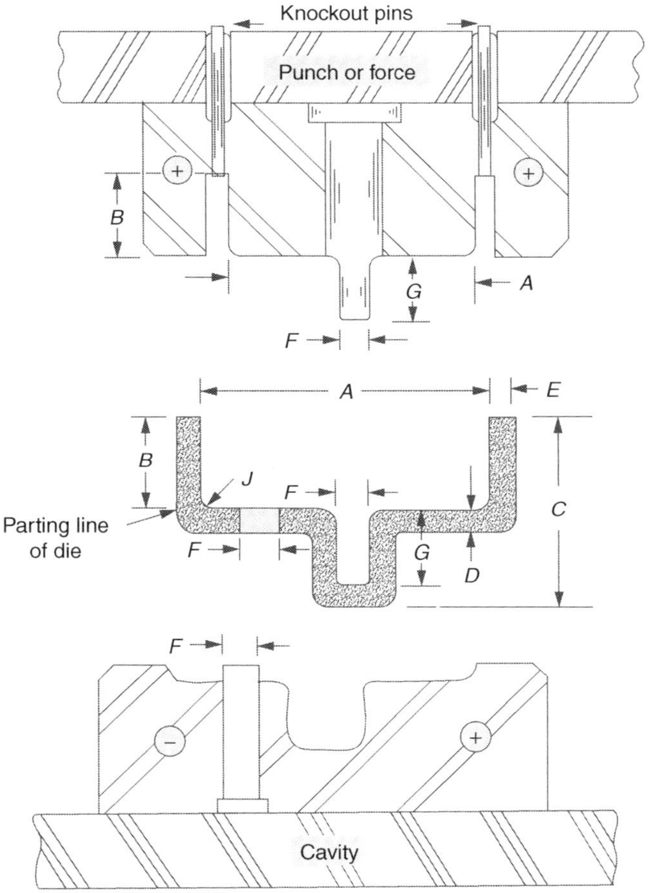

Fig. 1.137 This drawing shows each dimension on the part and the corresponding dimension of the die.

Figure 1.137 illustrates the mould dimensions and part dimensions. The following example shows the reader how to use the tables. A round ABS injection moulded part is considered in the Table 1.5. Fine tolerances represent the narrowest possible limits of variation obtainable under close supervision and control of production. Commercial tolerances will be that which can be held under average conditions of manufacture. Table 1.5 indicates dimensions taken from Table 1.4.

Table 1.4 Standard moulding tolerances for ABS moulding materials.
(Courtesy: *The Society of the Plastics Industry, Inc*).

STANDARDS AND PRACTICES OF PLASTICS CUSTOM MOULDERS

Engineering and Technical Standards
ABS

Note: The Commercial values shown below represent common production tolerances at the most economical level.
The Fine values represent closer tolerances that can be held at a greater cost.

Drawing Code	Dimensions (mm)	Plus or minus in Thousands of Millimeters [microns]
A = Diameter (see Note #1) B = Depth (see Note #3) C= Height (see Note #3)	0.000 12.50 25.00 50.00 75.00 100.00 125.00 150.00	10 20 30 40 50 60 70 80 90 100 110 120 130 140 150 160 170 180 190 200 210 220 230 240 250 260 270 280 290 300 310 320 330

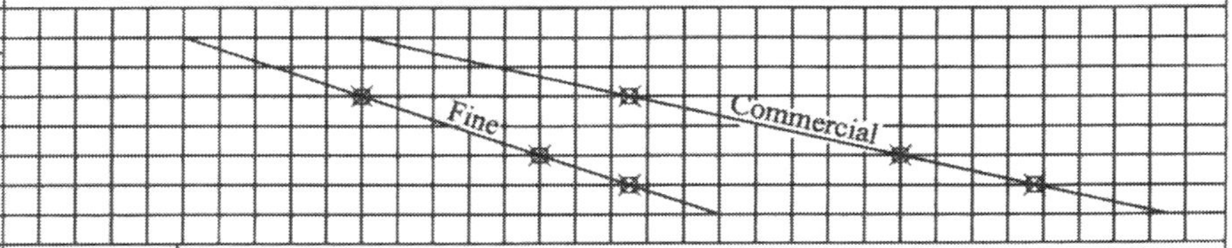

Drawing Code	Dimensions (mm)	comm. ±	Fine. ±
	150.00 to 300.00 for each additional mm add (Millimetres)	0.075	0.050
D = Bottom Wall (see Note #3)	—	0.100	0.050
E = Side Wall (see Note #4)	—	0.075	0.050
F = Hole Size Diameter (see Note #4)	0.000 - 3.175	0.050	0.025
	3.175 - 6.350	0.050	0.025
	6.350 - 12.700	0.075	0.050
	12.700 & over	0.100	0.050
G = Hole Size Depth (see Note #5)	0.000 - 6.350	0.075	0.050
	6.350 - 12.700	0.100	0.050
	12.700 - 25.400	0.125	0.075
Draft Allowance per side (see Note #5)	—	2°	1°
Flatness (see Note #4)	0.000 - 75.00	0.375	0.250
	75.00 - 150.00	0.750	0.500
Thread size (class)	Internal	1	2
	External	1	2
Concentricity (see Note # 4)	(T.I.R)	0.225	0.125
Fillets, Ribs, Corners (see Note # 6)	—	0.625	0.375
Surface Finish	(see Note # 7)	—	—
Color Stability	(see Note # 7)	—	—

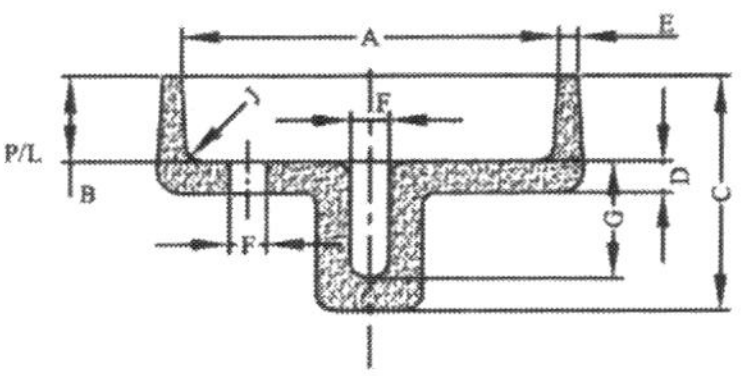

REFERENCE NOTES

1 - These tolerances do not include allowance for aging characteristics of material.

2 - Tolerances based on 3.2mm wall section.

3 - Parting line must be taken into consideration.

4 - Part design should maintain a wall thickness as nearly constant as possible. Complete uniformity in this dimension is impossible to achieve.

5 - Care must be taken that the ratio of the depth of a cored hole to its diameter does not reach a point that will result in excessive pin damage.

6 - These values should be increased whenever compatible with desired design and good moulding technique.

7 - Customer - Moulder understanding necessary prior to tooling.

Table 1.5 Tolerances for an ABS moulded part.

Dimensions of the moulded part	Tolerances (mm) ±	
	Fine	Commercial
A. 125.0 mm (made by core only)	0.175	0.290
B. 50.0 mm (made by core only)	0.100	0.175
C. 100.0 mm (made by cavity and core)	0.150	0.250
D. 2.5 mm (made by cavity and core)	0.050	0.100
E. 2.5 mm (made by core only)	0.050	0.075
F. 6.25 mm (made by mould pin in cavity or core)	0.025	0.050
G. 18.75 mm (made by core only)	0.075	0.625
J. Fillets, Ribs, Corners	0.375	0.625
Draft allowance per side	1 degree	2 degree
Flatness 125.0 mm	0.500	0.750
Concentricity (T.I.R., True Inside Radius)	0.125	0.225

1.16 FASTENING

There are four broad techniques used for joining plastics to each other and to other materials: (1) mechanical fasteners, such as rivets, pressed in inserts, self-thread cutting screws, etc., (2) mechanical means, snap-in fits, and press-in fits; (3) welding, such as spin welding, heat welding, ultrasonic welding and electronic heat sealing; and (4) adhesives, including solvents, elastomers, monomers, and epoxies. This chapter discusses in detail the mechanical fasteners and mechanical means only.

1.16.1 Screws

The self-tapping screw is perhaps the oldest type or method of fastening plastic parts. There are two types of self-tapping screws: thread-forming and thread-cutting. The self-tapping screw tends to make its own threads. This is done either by compressing and extruding, as in thread-forming screws, or displacement and cutting away of the plastic material in thread-cutting screws.

In order to select the proper type of self-tapping screw, the designer should first select the type of plastic that is to be used. If the plastic material selected is a thermoplastic, a thread-forming type of screw should be used. If the plastic material selected is a thermoset, a thread-cutting screw should be used.

1. Thread-forming Screws: The type of thread-forming screws recommended for most thermoplastic materials is shown in Fig. 1.138. The USA Standard *'B'* type screw is a blunt-point, spaced-thread screw. It is a fast driving screw with tapered threads. The *'BP'* screw is essentially the same as the type *'B,'* except that it has a 45° included angle, unthreaded cone point. The cone point helps in aligning holes in assembly. The *'U'* type screw is a multiple-thread drive screw with a blunt point. This type screw is intended for making permanent fastenings and is not recommended where removal is anticipated. The side walls should be at least as thick as the diameter of the screw. Metal threaded inserts should be considered when frequent removal of small diameter screws with high pull-out strengths is required.

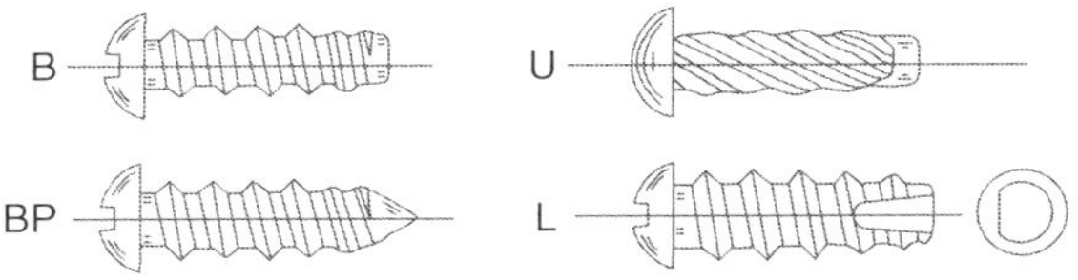

Fig. 1.138 The four most often recommended USA standard thread-forming screws used in thermoplastic materials.

A special type of screw is used with nylon. It is called type *'L'* screw and it is a combination of thread-forming and thread-cutting screw. The end of the screw has a tapered flat edge to start the thread, and then the remaining threads on the screw form the full diameter thread in the nylon.

All thermoplastic materials will attempt to return to their original shape if distorted. This is the key factor in using a thread-forming screw. As the screw is tightened, the thermoplastic material is forced out of the way by the thread engagement, but it continually tries to return to its original shape. This principle produces a secure locking and tight fit against the screw. On the other hand, if the same type of screw were driven into a thermosetting plastic material, it would set up stresses and eventually cause the part to crack.

Figure 1.139 illustrates an assembly of a thread-forming screw. The moulded or drilled hole should have a chamfer in order to guide the screw into the hole and to prevent it from any misalignment. The chamfer also helps to prevent burring or swelling of the plastic part.

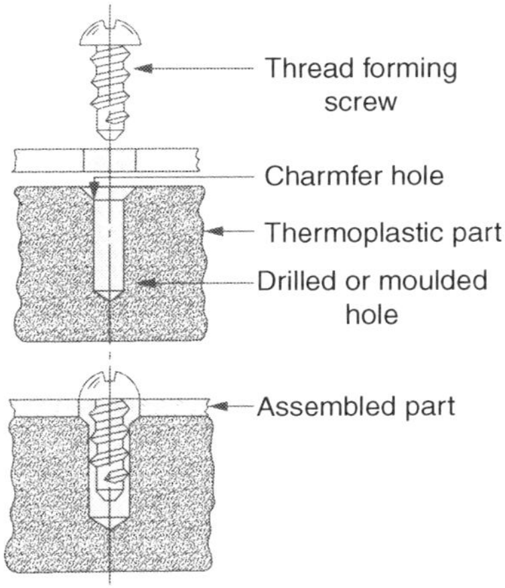

Fig. 1.139 This demonstrates a metal thread forming screw assembly in a thermoplastic material.

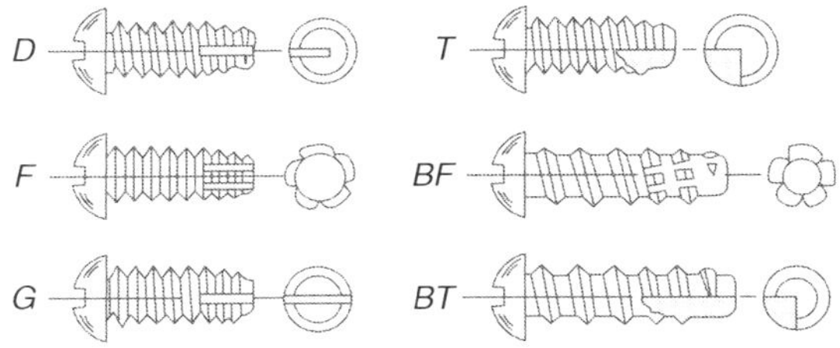

Fig. 1.140 Six types of USA standard thread-cutting screws are recommended for use in thermosetting plastic materials.

The taper on the moulded hole should carry the same taper as the thread forming screw. If the thread-forming screw is to be located in a boss, the wall thickness of the boss should be equal to one screw diameter.

2. Thread-cutting Screws: Thread-cutting screws have cutting edges and chip cavities that actually cut a thread in the rigid thermosetting plastic material. The type of thread-cutting screws recommended for most thermosetting plastic materials is shown in Fig. 1.140. The USA Standard '*D*' type screw has a blunt-point with threads of the same pitch as a standard machine screw. The flute on the end of the screw is designed to produce a cutting edge. This type of screw is very easy for a person to start to thread in a hole. The '*F*' screw is similar to a machine screw thread and has a blunt point. It has five evenly spaced cutting grooves and large chip cavities. It can be used in most thermosetting plastic materials. The '*G*' screw has a machine screw thread with a single slot that forms two cutting edges. The '*T*' screw has a blunt point with a wide flute that gives more chip clearance. The '*T*' type screw cuts easier than the '*D*' type. The '*T*' type screw is easy to start in a hole, and the threads are resizable. The '*BF*' screw is like type '*B*' (Fig. 1.138), with a blunt point, but has five evenly spaced cutting grooves and chip cavities. The recommended wall thickness for the '*BF*' screw is one and one half times the major thread diameter of the screw. This type of screw drives faster and provides good pull out strength. The '*BT*' screw is the same as the '*BF*' screw, except for a single wide flute that provides room for large chips. The '*BF*' and '*BT*' screws require lower driving torque and develop lower boss stresses than thread-forming screws.

Thread-cutting screws are recommended and used in most thermosetting plastic materials. Figure 1.141 illustrates an assembly with a thread-cutting screw. It should be noted that the moulded or drilled hole should have a chamfer in order to guide the screw into the hole and prevent it from any misalignment. If possible, the taper on the moulded hole should carry the same taper as the thread-cutting screw. If the thread-cutting screw is to be located in a boss, the wall thickness of the boss should be equal to one screw diameter. Note the extra area located at the bottom of the hole. This provides a reservoir for thread-cutting chips.

It is advisable to contact the plastic material company and the metal screw manufacturer to determine the exact type of screw sizes that will be needed in any fastening application.

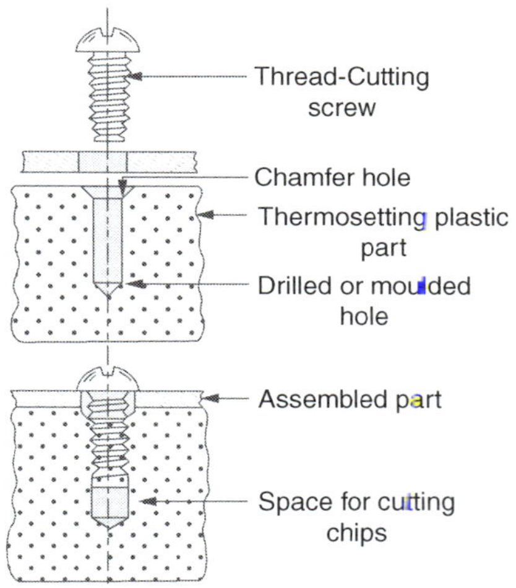

Fig. 1.141 This demonstrates a metal thread cutting screw assembly in a thermosetting plastic materials.

1.16.2 Wire-Type Screw Thread Insert

Figure 1.142 illustrates a wire-type screw thread insert. The coil wire-type insert is made from a diamond-shaped wire. The diamond-shape wire will act as an internal and external thread when it is made into a coil form. It is installed by pushing the insert into a drilled or moulded hole. This acts as a thread for the assembly screw. The thread of the insert is usually of standard size and in thread form. This same type of wire insert principle can be used in a blind hole by first taping the moulded hole and then threading in the wire coil insert. A wire-type screw thread insert assures freedom from thread wear if an assembled part must be taken apart frequently.

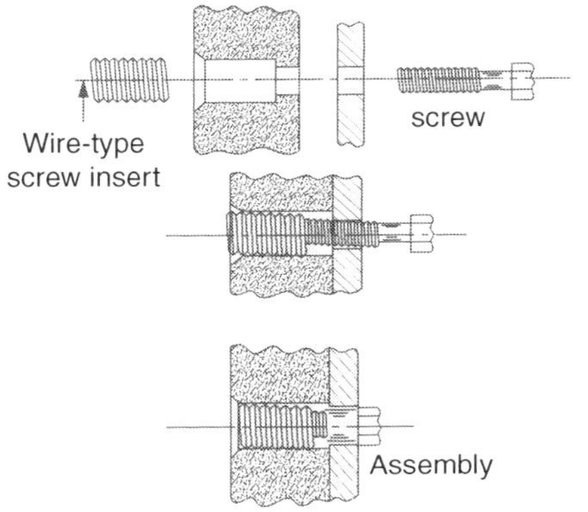

Fig. 1.142 A cross-sectional drawing of a wire-type screw thread installed in a moulded through hole.

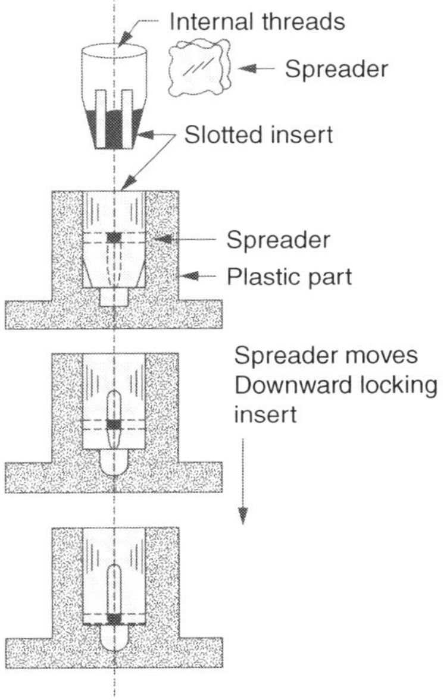

Fig. 1.143 One type of a expansion metal insert that is used after a plastic part has been moulded.

1.16.3 Expansion-Type Metal Inserts

A standard-type expansion insert is shown in Fig. 1.146. The insert is placed in a moulded or drilled hole and the tapered knurled bottom section of the insert are spread apart by the metal spreader as it is forced down the four slots in the insert. The moulded or drilled hole diameter is generally 0.050 mm greater than the insert. The moulded hole should be flat at the bottom in order to support and retain the spreader.

Figure 1.144 illustrates a cone-spread metal insert. The insert is made in one piece. It has a knurled outside surface and threads on the inside. The insert has a spreader cone attached to the closed end. As the insert is pressed down into the moulded or drilled hole, the spreader cone breaks and forces the external knurls on the insert to expand against the hole wall and lock the metal insert in place. The moulded or drilled hole diameter should be equal to the insert body diameter.

Figure 1.145 shows an expansion stud-type insert. This metal insert works on the same principle as the cone-spreader insert, except that it has a metal threaded stud above the surface of the plastic part. The cone spreader inserts are generally made out of the brass.

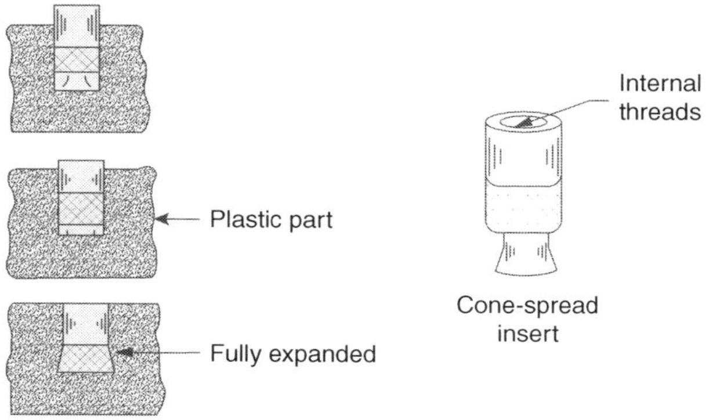

Fig. 1.144 A cone-spread metal insert. This type of metal insert is used in plastic parts that have moulded or drilled holes.

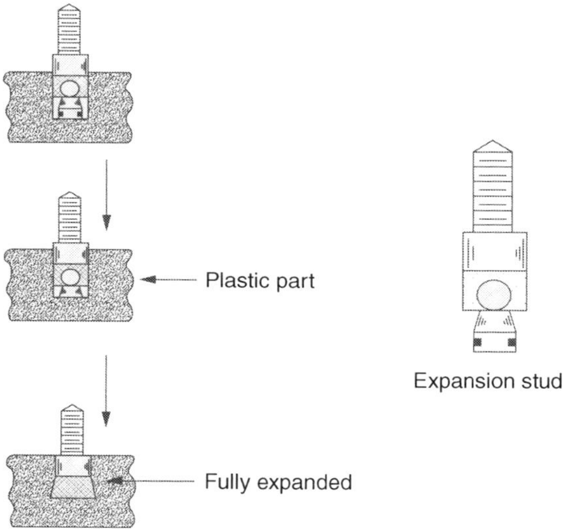

Fig. 1.145 An expansion stud-type metal insert. It is used in plastic parts that have moulded or drilled holes.

Figure 1.146 illustrates a double slotted metal insert. As the insert is pressed into the hole, it is compressed until the slot is closed. The spring tension holds the insert in place. When the screw is installed, it expands the slotted portion of the insert. The insert is then locked in place. This type of insert is suitable for use in soft plastic materials.

1.16.4 Rivets

Perhaps the oldest method of fastening plastic parts together is the metal rivet. Very few fasteners can match the advantages of the tubular and split rivets. Rivets can be used manually or on automatic bench riveting equipment, but they do not carry the precision tolerances of metal screws and inserts. However, rivets are not considered to be the best type of fastener to withstand tension. Their great advantage is that they are inexpensive and easy to install. In a semitubular rivet, the proportion for the distance of the rivet from the edge should be three times the shank diameter. The proportion for the clinch allowance should be six tenths (.6) times the shank diameter.

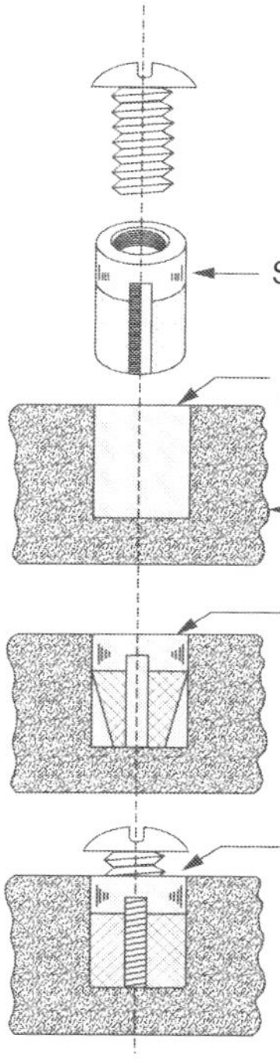

Fig. 1.146 A metal expansion-type insert. The double slotted insert is expanded when the screw is installed.

Blind Rivets A blind rivet is used when it is impossible to have access to the reverse side of the joint. Blind rivets are available both in metal and plastic and are designed for installation from one side only. Essentially, the blind rivet consists of a hollow body and a solid pin. The setting of the rivet is done by driving or pulling the solid pin through the hollow shank and flaring the shank on the blind side of the rivet and joint. This provides a positive locking action. There are a variety of proprietary designs for plastic rivets.

1.16.5 Hinges

No discussion on the subject of fastening would be complete without covering hinges, which act in effect as inserts to hold two plastic parts together. The two parts, top and bottom, of a plastic closure may be held together by a hinge. Three types of hinges that are standard in the plastic industry and have been used in the past are the Rathbun hinge, the piano hinge, and the lug and pin hinge (Fig. 1.147).

1. **Rathbun hinge:** The Rathbun hinge uses elliptical shaped steel spring clips to hold a box lid tightly closed or wide open. The steel spring clips are placed in small undercuts or slots provided in both top and bottom of the lid and box. Special tooling in the mould is required in order to mould the undercuts. This type of hinge provides a sturdy spring action and holds the cover of a box tightly in place.
2. **Piano Hinge:** The piano hinge is made of metal and is the same type design and construction that is used to fasten the hinged cover of a piano. This type of hinge is used in

the plastic industry on plastic containers, boxes, etc. The hinge is fastened to the plastic parts by rivets or self-tapping screws. The holes for the rivets and self-tapping screws are generally drilled after the part has been moulded.

3. **Lug and Pin Hinge:** The lug and pin hinge is the least expensive from a moulding and assembly standpoint. This hinge requires a recess in the side of the box and a moulded lug or prong on the cover. A hole is drilled through the side of the box and through the centre of the lug, and then a metal pin is driven into the hole.

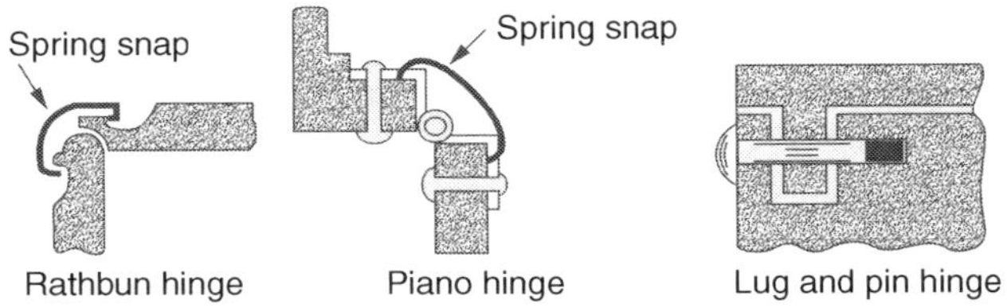

Fig. 1.147 Hinges used in plastics. The Rathbun hinge with moulded mating curved sections and moulded special undercut grooves. The piano type of spring snap hinge assembled by rivets. The lug and pin hinge with moulded mating lug and slot.

4. **Standard Pin Hinge:** The standard pin hinge will require moulded holes or drilled holes. Moulded holes are expensive and will require cams in the mould (Fig. 1.148).
5. **Inexpensive Pin Hinge:** The inexpensive pin hinge design will eliminate the drilling or cam operation that is required in the standard pin hinge (Fig. 1.148).
6. **Heat Sealed Hinge:** Heat sealed hinges are very strong and durable. One half of this hinge is moulded with a pin and the other half is moulded with two tabs. After the hinge is assembled, the lower part of the two tabs are heated and bent permanently around the two pins (Fig.1.148).

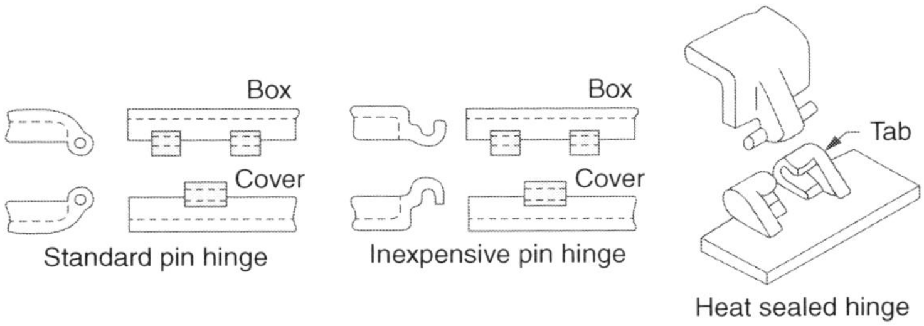

Fig. 1.148 Hinges used in plastics. The standard pin hinge will require moulded or drilled holes. The inexpensive pin hinge design will eliminate moulding or drilling of holes. The heat sealed hinge is assembled and the two tabs are heated and bent permanently around the two pins.

7. **Integral Hinges:** The integral hinges are shown in Fig.1.149 illustrates three modern-type hinges that are classified as (a) Integral moulded strap hinge, (b) Integral moulded hinge, and (c) Integral coined hinge.

(a) **Integral Moulded Strap Hinge:** The flexible strap hinge is generally moulded out of the polyolefin's and is approximately 6.35 mm wide and 1.0 mm in. thick. The straps can be spaced any distance apart.

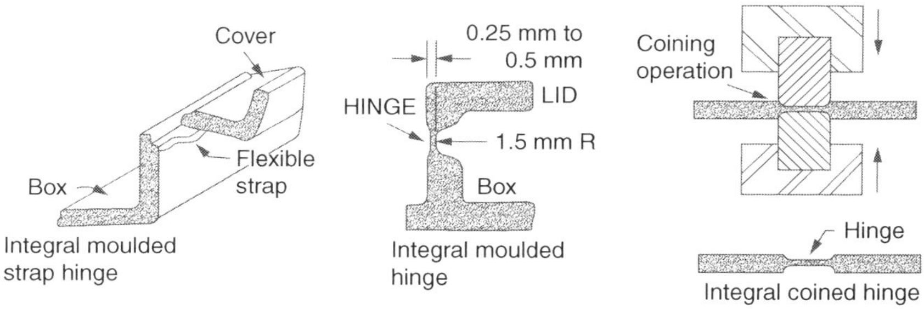

Fig. 1.149 Three types of integral plastic hinges. The integral moulded strap hinge and the integral moulded hinge are made from the polyolefins. The integral coined hinge has been made from nylon, acetal, and the polyolefins.

(b) **Integral Moulded Hinge:** The integral moulded hinge is the most popular and practical hinge used today. It is generally moulded out of polypropylene and can be flexed many hundreds of thousands of times without failure. In order to orient the molecular chains of the materials across the hinge for increased strength and life, the part should be opened and closed a few times right after moulding. The gate in the mould must be placed so that the flow of plastic is straight across the hinge and not lengthwise. The gate should also be located in the heavier half of the part so that the flow is across the hinge to the lighter half of the piece. Hinge thickness is usually 0.25 to 0.50 mm.

The integral hinge can be extruded by the extrusion process; however, the hinge has poor flex life as compared to the standard injection moulded hinge. This is because the hinge is formed in the direction of plastic polymer flow, and as a result it cannot be sufficiently oriented when flexed.

(c) **Integral Coined Hinge:** A coined hinge is made by placing the part to be coined between two coining bars and applying enough pressure until the desired hinge thickness is reached. Heat is sometimes applied to the coining bars. The pressure is released and the hinge is removed. The hinge thickness ranges from 0.25 mm to 0.40 mm. The flex life of the coined hinge is not as good as the integral moulded hinge, but the ability to resist tearing is greater. Coined hinges are made from nylon, acetal, polypropylene, and polyethylene materials.

8. **Ball Grip Hinge:** Figure 1.150 shows a ball grip hinge design. This type of hinge is used on small boxes and is an accepted standard in the box industry. The two balls are approximately 3.2 mm in diameter and are moulded into one half of the box. The balls snap into the two depressions on the other half of the box or container. Although the two depressions are undercuts in the mould, the parts are stripped out. The depth of the depression is approximately 0.45 mm.

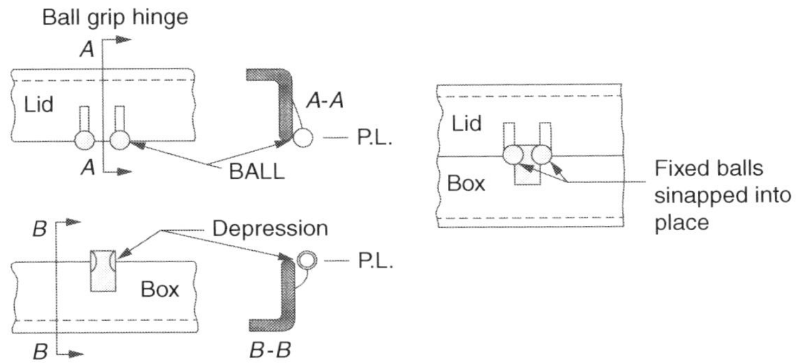

Fig. 1.150 The ball grip hinge design. This type of hinge is used on small boxes and containers.

1.16.6 Clasps

Figure 1.151 displays many types of clasps that are used on plastic containers or boxes. It will be noted that most of the clasps work on the friction hook principle.

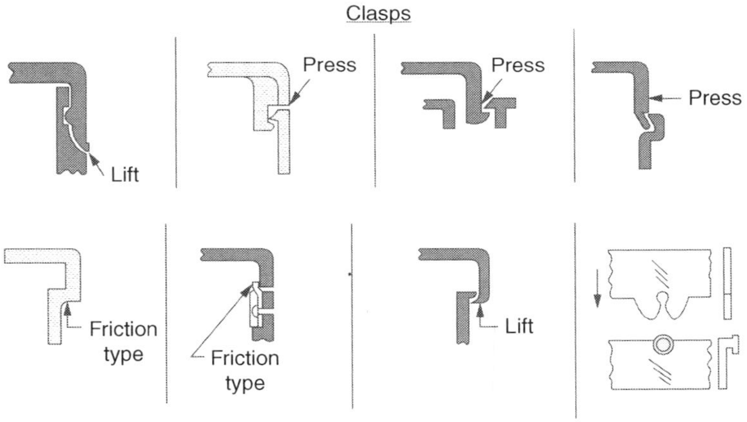

Fig. 1.151 Types of clasps used on plastic containers.

1.16.7 Snap Fit

There are two basic methods of mechanical means of fastening. They are snap-in fittings or snap-on fittings, and press or interference fits. Plastic parts can be joined together by using a snap-on fit or a snap-in fit.

1. **Snap-in and Snap-on Fittings:** Figure 1.152 illustrates a snap-in fit and a snap-on fit. The strength of snap fit comes from mechanical interlocking as well as from friction. Snap-in fittings require a tough, stiff, plastic engineering material. The two engineering materials recommended are ABS, nylon, acetal and polycarbonate. The softer materials such as polyethylene, polypropylene, flurocarbons, flexible vinyls, etc., are not recommended for these two types of fittings. In the snap-in fit, the part design should provide sufficient flexibility for the prongs to flex and snap back when they are inserted into the joining part. The

strength of the part is limited by the amount of undercut that can be produced by injection moulding. Snap-on fits are used with the more flexible materials in food containers, etc.

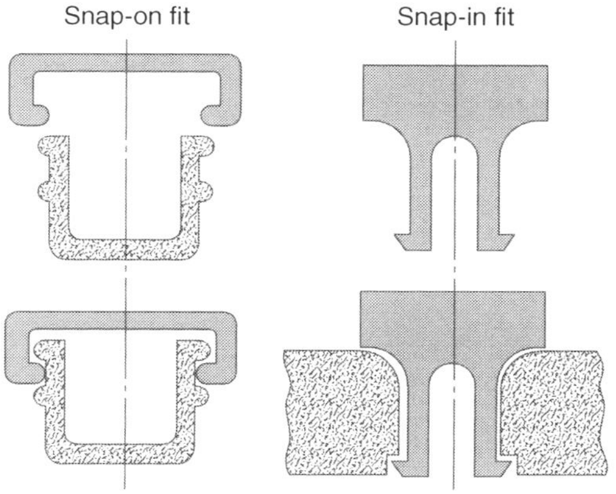

Fig. 1.152 Snap-on and snap-in fittings.

2. **Press or Interference Fit:** The press or interference fit is the forcing of a slightly oversize part into a standard hole or opening. This is generally used in pressing metal inserts into a plastic part. It is limited to the stiff and tough plastic engineering materials. The designer should be aware of the fact that the joint may loosen in time, due to creep in the plastic material.

1.16.8 Staking

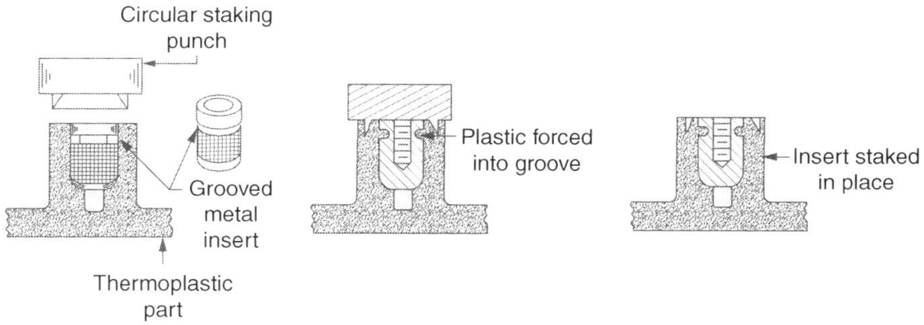

Fig. 1.153 This illustrates a method of staking a metal insert in a plastic part.

In this process, a metal insert is held in place by cold flowing a thermoplastic material into an external undercut on the insert (Fig.1.153). The metal insert should be knurled and slightly oversize to help it from twisting loose when in use. An interference of approximately 0.25 mm between the outer diameter of the knurl and the moulded hole is generally used. The metal insert is pressed into the moulded hole and a circular staking punch comes in contact with the plastic and forces the plastic material into the groove.

This is sometimes referred to as swaging or peening and is accomplished by compressively loading the end of a plastic rivet or stud, while holding and containing the rivet or stud body. A head is formed at the end of the rivet or stud by cold flow of the plastic when the compressive stress exceeds the yield point. All thermoplastic materials cannot be cold headed. Examples of cold-headed devices are snap fasteners in children's clothing and electrical connectors.

1.16.9 Heat Staking

This process also known as hot heading. In this process, a heated metal staking tool is pressed down and over a thermoplastic stud (Fig.1.154). The thermoplastic stud melts, taking the form of the staking tool. This type of staking is not as fast as ultrasonic staking. The amorphous type of thermoplastic materials is much easier to heat stake than the crystalline materials.

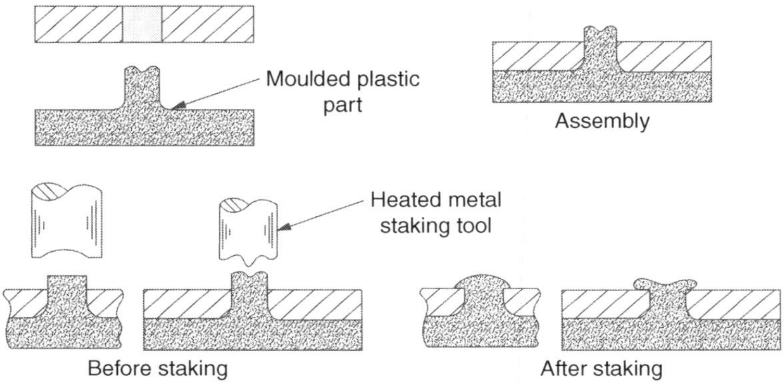

Fig. 1.154 The process of heat staking a thermoplastic assembly part.

1.17 SHRINKAGE

Shrinkage is the contraction in dimensions of the product after it is moulded. All the plastic materials have tendency to shrink during solidification. The shrinkage value differs from material to material and it is very difficult to predict. So the use of correct information is very important, not only for having the desired proportions of a product, but also for fitting with other parts and for functional purposes. The shrinkage data are usually given in a range of two values. The lower value is intended to apply to thin parts, whereas the higher value for thick parts.

The choice of shrinkage for a selected material and a specific design is the responsibility of the mould designer, moulder and product designer. In cases where very close tolerance are involved, preparing a prototype of the part may be necessary to establish critical dimensions. If this test is not practical, it may be necessary to test a mould during various stages of cavity and core fabrication with allowances for correction in order to determine the exact shrinkage needed.

Considering the factors that can contribute to variations in shrinkage, it will be fully appreciated how significant it is to select the appropriate numbers.

The thermosetting compression moulded parts will have a higher shrinkage

1. When cavity pressure is on the low side.
2. When mould temperature is on the high side.
3. When cures are shorter.
4. When parts are thicker (over 4.5 mm).
5. When a material is soft flowing (highly plasticised).
6. When a material is pre-heated at relatively low heat.
7. When a high moisture content is present in the raw material.

Transfer moulded and injection moulded thermosetting resin will have higher shrinkages in comparison with compression moulded parts of the same design and material. Some of those higher shrinkages are due to the imparted directional flow and others are due to a tendency to use small gates that do not permit the application of higher pressures to the cavity.

1.17.1 Factors Affecting Shrinkage

The shrinkage of injection moulded thermoplastics will be affected as follows:

1. Higher cavity pressure will cause lower shrinkage
2. Thicker parts (3 mm or more) will shrink more than thinner ones.
3. Mould temperature 27°C or less will bring about lower shrinkage, whereas temperature 40°C and above or more will produce higher shrinkage.
4. A melt temperature of the material at the lower end of the recommended range will produce a lower shrinkage, but the upper end of the range will produce a higher shrinkage.
5. Longer cycle time, above the required solidification point, will bring about lower shrinkage.
6. Openings in a part will bring out lower and varying shrinkage than the part without opening.
7. Larger gates permit higher pressure build - up in the cavity and will cause lower shrinkage.
8. For crystalline and semicrystalline materials, the shrinkage value will be higher in flow direction and lower in perpendicular direction. But in a symmetrical part, when centre gated, the shrinkage will average out and be reasonably uniform.
9. Glass reinforced or otherwise filled thermoplastics have considerably lower shrinkage than the basic polymer.

Most thermoplastics attain their full shrinkage after 24 hours, but there are some, which may take weeks time to stabilise their dimensions fully. The manufacture of the material usually indicates whether there is a delayed shrinkage effect present. The data on shrinkage have to be approached with much care if one to avoid dimensional problems with the plastic product.

1.17.2 Basics of Shrinkage

Shrinkage is caused by volumetric change in the polymer as it cools from molten to solid. Shrinking is not a single event but occurs over a period of time. Most of the contraction (70 – 90%) occurs in the mould but it can continue for 24 to 48 hours until the part stabilises naturally. The cause of part shrinkage is the internal stress. There are two types of stresses. Molecular orientation stress within each molecules and stress between molecules.

1.17.3 Shrinkage in Amorphous and Crystalline Material

Shrinkage rate depends greatly on whether the polymer is crystalline or amorphous and on how much reinforcement or filler it contains. Amorphous plastics shrink less than crystalline. In amorphous plastics, shrinkage will be uniform in all directions. Amorphous plastic requires less heat (mould and melt temperature) than crystalline plastics. In crystalline plastics shrinkage is typically greater in one direction. Crystalline plastics shrinkage will be more compared to amorphous plastics. Crystalline plastic requires more heat (mould and melt temperatures) than amorphous plastics. Stress and low plastic rigidity cause additional shrinkage. Thin wall parts typically shrink less than thick wall parts because, higher plastic pressures are required to fill and pack it. Molecular orientation is affected by plastic fill rate and plastic cooling rate. Polymer molecules are oriented in flow directions and the shrinkage is less. Polymer molecules are non-oriented in perpendicular to the flow direction and the shrinkage will be more.

Whether the shrinkage is isotropic or anisotropic is another important factor. Amorphous materials generally shrink isotropically, so they have the same mould shrinkage rate in both the flow and cross flow directions. When glass fibre is added to amorphous materials, however, the degree of anisotropic shrinkage typically increases in the material because the glass fibres align in the direction of materials flow.

1.17.4 Mould Design Consideration

The main considerations in mould design affecting part shrinkage are adequate cooling, proper gate size and location, and structural rigidity. Of the three, cooling conditions are most crucial, especially for crystalline resins. The cooling system must be adequate for the heat load. Slow cooling, increases shrinkages by giving resin molecules more time to reach a relaxed state. As in crystalline resins, longer cooling time leads to a higher level of crystallinity, which in turn accentuates shrinkage.

For the same reason, the cooling line layout should be engineered to make all melt containing surfaces in the mould run close to the same temperature as possible. Hot spots can produce local shrinkage, which is an open invitation to localised stress and warpage. Small cores and core pins can be particularly troublesome because they tend to accumulate heat. Hot core pins promote shrinkage in hole diameters.

Gate location and size are other factors to consider. As explained above, if other considerations permit, gate placement to produce uniform flow length is highly desirable; it aids uniform shrinkage by promoting the most uniform melt condition during filling and pressure profile during packing (assuming no severe pinch point in the flow path). Edge gating of long thin parts run the risk of differential shrinkage along its length and warpage. In such causes, multiple gates may be needed.

Small gates promote shrinkage because they can freeze off prematurely and thus terminate part packing too soon. Low material density encourages shrinkage. In general, the effect of undersized gates increased with wall thickness and decreases with lower mould temperature. Under the most adverse combinations, the shrink rate increased by over 50%.

1.17.5 Warpage vs Shrinkage

Warpage, an equally troublesome tolerance problem, is usually defined as the result of uneven shrinkage in different areas of the part. Warpage is much more closely associated with part geometry, such as thickness changes and with long, thin sections generally, than with the size of the shrink rate of the polymer.

Basically, warpage is a reflection of uneven stress distribution on the part, which can be produced by the combined effects of processing conditions, mould design and part geometry. Many of the part, and mould designing and processing conditions that influence mould shrinkage also affect the potential for warpage.

Generally, shrink rates vary substantially with wall thickness and part geometry. Since shrinkage is so heavily affected by the specifications of the product, it makes sense to co-ordinate the three functions in part development – part design, mould design and moulding parameters – to prevent or at least minimise surprises when critical dimensions of the first moulded part are checked.

1.17.6 Part Design Variables

Part configuration influences the amount of shrinkage and where it occurs. One of the most significant factors is wall thickness. Doubling the wall thickness, from 1.0 mm to 2.0 mm, for example, will increase shrinkage by about 50%.

Non-uniform wall thickness will produce different amounts of shrinkage along with a greater potential for warpage. If part thickness must vary, the transition should be as gradual as possible. Sharp corners produce localised shear heating and therefore slower cooling, higher stress (and tendency to low) and higher crystallinity around those points. On edges and corners, particularly in hollow shapes such as boxes, make the outside radius larger than the inside radius in order to maintain a constant cross-sectional thickness between the edge and adjacent walls.

Draft angles on deep-draw parts can be reduced for resin having large shrink rates. The high shrinkage helps free the part from the cavity, reducing drag as the core extracts the part for ejection.

1.17.7 Shrinkage vs Tolerance

Several factors that affect tight-tolerance capability are material, general design, the types of moulding machine and process variables, and tooling. The ability to hold tight tolerances in moulded parts can be improved if the role and importance of each of these factors are thoroughly understood.

1.17.8 Material

The most critical element involved in holding tight-tolerances is materials shrink rate. The materials which shrink less are easier to hold to specified tolerances. Consequently, most materials used for tight-tolerance applications have high reinforcement contents. While glass fibre is the typical reinforcement used, materials are also available with mineral, mineral/glass-fibre combinations, glass flaked and spheres and so on.

Whether the shrinkage is isotropic or anisotropic is another important factor. Amorphous materials generally shrink isotropically, so they have the same mould shrinkage rate in both the flow and cross flow directions. When glass fibre is added to amorphous materials, however, the degree of anisotropic shrinkage typically increases in the material because the glass fibres align in the direction of materials flow.

The addition of mineral to amorphous materials, on the other hand, generally reduces mould shrinkage without increasing the degree of anisotropy, but does not impart equivalent mechanical properties to the material. Meanwhile, glass flake reduces the degree of anisotropy relative to the glass fibre, as well as provides marginally more mechanical properties than the mineral or the glass fibre, but not without incurring additional costs. Finally, combinations of these reinforcements can improve mechanical properties, reduce the mould shrinkage and minimise degree of anisotropic shrinkage. It is much more difficult to predict and hold tight tolerances where filled crystalline materials are concerned because they tend to have higher shrinkage rates than equivalently filled amorphous materials.

The higher shrinkage rate, coupled with the different between flow and cross flow shrinkage (anisotropy), is the leading cause of part warpage for filled crystalline materials.

Although filled amorphous plastics also shrink anisotropically, the actual shrinkage and the difference between the flow and cross flow shrinkage are not as severe as in crystalline plastics. Therefore, filled amorphous materials are more effective in tight-tolerance applications.

1.17.9 ASTM Shrinkage Values

The mould shrinkage rates listed on property data sheets can sometimes mislead designers. Generally taken from ASTM, these mould shrinkage values are derived from a flammability bar for shrinkage in the flow direction.

While these values are useful, wall thickness, flow length, and packing pressure also affect materials mould shrinkage. Therefore, while the ASTM flow-direction mould shrinkage is applicable for the given wall thickness and approximate flow length, it must be understood that mould shrinkage typically increases as either flow length or wall thickness increases. Mould shrinkage also increases with a decrease in packing pressure.

ASTM mould shrinkage in the cross flow direction is even more misleading. A measurement is made across the middle of an edge-gated disk and defined as 'cross flow' shrinkage. However, the fibre orientation in the disk is actually pure cross flow only in a small portion of the sample, while the rest of the sample varies between pure flow and pure cross flow. In a pure cross flow situation, the actual shrinkage can be much greater than the ASTM value. Therefore, the ASTM cross flow shrinkage is applicable only if the flow approximates the radial flow pattern.

The use of regrind in tight-tolerance parts also must be closely monitored. Although the mechanical properties of a material will not vary significantly at regrind levels up to 20%, the regrind process will invariable shorten the length of the glass fibres in the materials, as well as reduce the melt viscosity of the resin which may affect later the shrinkage characteristics of the material to the point where part dimensions are affected. If regrind must be used, low levels should be added initially, and increased only after each previous regrind level has proven to be satisfactory in the moulded parts.

1.17.10 General Design

Tolerance capability is extremely design dependent. For simple geometrics, material advances have measurably increased performance. However, this only means that a simple geometry can be repeated within this tolerances - not that the initial design can be moulded within these tolerances. As the specific dimension increases, the tolerance capability becomes even more dependent on part design.

In standard tolerance applications, the rules for plastic part design are commonly stretched or broken without significantly sacrificing end use performance. Designs for tight tolerance parts, however, cannot violate these rules with impurity, as each violation reduces the capability of the design. Maintaining constant nominal wall thickness improves the flow patterns in the part and reduces the risk of warping. Generous radii are necessary to reduce stress concentration at all corners and to prevent poor fill patterns. The design must also include a minimum of ½° draft per side and solids eject areas to permit easy removal of the part from the mould. Difficulty in ejecting a part has known to render a good tight tolerance design ineffective.

The dimensions of small holes and bosses are relatively easy to hold to tight-tolerances. Understanding this phenomenon has led to the use of edge details to further help to hold tolerances. In some areas of the mould, part shrinkage is uncontrolled - a function only of material shrinkage. However, in other areas, the part shrinks around the core, or around a core pin. In these cases, the tool steel provides additional control over the part shrinkage, especially with proper cooling control.

Extending the nominal wall or adding an edge rib to the design forces the material to cool around a core, and increases a designer's tolerance capability. Since this often requires the use of inserted side actions, however, it is normally employed only in the direction of draw, or where a side action already exists for the formation of other features.

1.17.11 Types of Moulding Machine and Process Variables

The success of the part and mould design is determined by what happens in the moulding process. Basically, part shrinkage is affected by a combination of temperature, pressure and time in the mould. Mould parameters that affect orientation and the degree of crystallisation in crystalline resins influence the amount of shrinkage.

The optimum combination consists of a melt injected at a rate and temperature in a mould hot enough to fill the mould with minimum viscous shear. Packing pressure should be high enough and gate freeze delayed long enough to fully pack the part. Pressure drop between the gate and end of flow -largely a function of mould design in relation to part size and geometry should be minimised to ensure proper fill and uniform pack - out density. High material density inhibits shrinkage. The cooling rate must provide enough time for the stressed layers of resin to relax, but still fast enough to prevent a higher level of crystallinity in crystalline resins than the strength requirements of the part. The higher the crystallinity, the higher the shrinkage. Any process parameter that increases cooling time will enhance shrinkage.

Logically these parameters include raising the temperature of the mould and melt and conditions that reduce cooling efficiency. Bear in mind, though, that shrinkage (and warpage) cannot be permanently foiled by fast ejection. The part will continue to change size and perhaps shape until enough stress is released to stabilise it, which may take from a few hours to two days depending on its demoulded condition.

The injection moulding machine is another important variable in holding tight tolerance plastic parts. The type of machine used can affect tolerance capability. Toggle injection moulding machines often have tighter specifications for parallel precision of die-plate alignment than do hydraulic machines. However, over time, there may be more wear on the toggle machine, thus negating this advantage for high volume parts.

The use of closed loop process controls, in conjunction with the injection moulding process, is also highly recommended. Variation in packing pressure, melt temperature, and material mixing can have a substantial impact on material shrinkage in the part. Closed loop process controls allow the moulder to make adjustments to maintain moulding parameters as constants.

Required press size for tight-tolerance injection moulded parts depends on the material, part weight, part complexity and other processing and design concerns. However, initial estimates can be based on the anticipated projected surface area of the part.

The highly filled amorphous resins used in tight-tolerance applications typically will need 4.5 to 5.5 tons of clamping force per square inch of projected surface area.

1.17.12 Tooling

In selection steel for tight – tolerance part, the reinforcement content of the resins being used must be taken in to account.

For example, the pressure of glass and mica in the material can cause abrasion and wear on the mould, changing the dimensions over time and reducing tolerance capability.

P-20, the steel of choice for many moulds, offers minimal resistance to the resulting abrasion and wear when these materials are used and would not generally be recommended for such applications, with possible exception of very large parts. Steel with higher carbon content for overall hardness and higher chromium content for better wear resistance is suggested for tight-tolerance design. Likewise, air-hardened steel is preferred over oil – hardened steel because of improved dimensional stability during heat treatment.

Table 1.6 Shrinkage values for thermoplastics and thermosets.

Material	Mould linear shrinkage%	Material	Mould linear shrinkage%
Polyethylene and Ethylene Copolymers		**Acrylics**	
1. Low density	1.5–5.0	1. Moulding	0.1–0.4 (C)
2. Medium density	1.5–5.0	2. Impactacrylic moulding compound	0.2–0.8 (I)
3. High density	2.0–5.0	3. MMA alphamethyle-styrene copolymer	0.20–0.6
4. Ethylene ethyl-acrylate copolymer	1.5–3.5	4. Acrylic multipolymer	0.6–1.0
5. Ethylene vinyl acetate copolymer	0.7–1.1	**Allylic Resins and Monomers**	
Polypropylenes		1. DAP Moulding compound glass filled	0.20–0.06
1. Unmodified	1.0–2.5	2. DAL Moulding Compounds a. Mineral filled b. Synthetic fibre filled	 0.5–0.7 0.9–1.1
2. Copolymer	0.9–2.0	**Compounded with Butadiene Acrylonitrile Copolymers**	
3. Inert filled	0.5–1.5	1. Wood flour and cotton flock filled	0.4–0.9
4. Glass reinforced	0.4–0.8	2. Asbestos filled	0.4–0.7
5. Impact(rubber modified)	1.0–2.5	3. Rag filled	0.2–0.4
Vinyl Polymers and Copolymers		4. Metal filled (iron, or lead)	0.3–0.4
1. Rigid	0.1–0.5	**Cellulosic Moulding Compounds**	
2. Flexible unfilled	0.1–0.5 (Varies with Plasticiser)	1. Ethyl cellulose moulding compounds and sheet	0.5–0.9
3. Flexible filled	0.8–3.5	2. Cellulose acetate moulding	0.3–1.0 (C) 0.3–0.8 (I)
4. Vinyl butyral moulding	0.5–2.5	3. Cellulose propionate moulding compound	0.3–0.9 (C) 0.3–0.8 (I)
5. Vinyl formal moulding compound	0.15–0.30	4. Cellulose acetate butyrate moulding compound	0.3–0.9 (C) 0.3–0.6 (I)

(Contd.)

Table 1.6 (*Contd.*)			
6. Chlorinated polyvinyl chloride compound	0.3–0.7	Chlorinated polyether	0.4–0.6
7. ABS modified vinyl chloride Rigid	0.4–1.5	Ionomers	0.3–2.0
8. PP modified vinyl chloride rigid	0.2–0.5	**Acrylonitrile-Butadiene - Styrene**	
Polystyrene		1. Extrusion grade	0.4–0.5
1. Unfilled free flowing general purpose heat resistant	0.1–0.6 (C) 0.2–0.6 (I)	2. High impact	0.5–0.8
2. Impact Resistant, high impact medium impact	0.2–0.6	3. High heat resistant	0.5–0.8
3. Special Heat and Chemical Resistant Type	0.1–0.8	4. Medium impact	0.5–0.8w
4. 20-30% glass filled	0.1–0.2	5. Self-extinguishing	0.6–0.8
5. Styrene Acrylonitrile	0.2–0.7	6. 20–40% glass filled	0.1–0.2
6. 20-30% glass filled	0.1–0.2	7. Clear	0.6–0.8
Styrene-Butadiene			
Thermoplastic	0.1–0.5		
Nylons		**Polyphenylene Sulphides**	
1. Unmodified	0.8–1.5	1. Unfilled	1.0
2. Nucleated	0.7–1.2	2. Polysulfone	0.7
3. 30–35% glass reinforced	0.5	3. Polymethyl pentene polymer	1.5–3.0
Type - 6		**Phenol Formaldehyde and Phenolfurfuralmoulding Compounds**	
1. Unmodified	06–1.4	1. No filler	1.0–1.2
2. 30 - 35% glass reinforced	0.4	2. Wood flour and cotton flock filled	0.4–0.9
Type - 6/6		3. Asbestos filled	0.2–0.9
Copolymer	0.6–1.5	4. Mica filled	0.2–0.6
Type 6/12		5. Glass fibre filled	0.2–0.4
1. Unmodified	1.1	6. Macerated fabric and cord filled	0.2–0.9
2. 30–35% glass reinforced	0.3	7. Pulp performed and moulding board	0.18–0.8

(*Contd.*)

Table 1.6 *(Contd.)*

Type 6/10		**Melamine-Formaldehyde Moulding Compounds**	
1. Unmodified	1.2	1. No filler	1.1–1.2
2. 30-35% glass reinforced	0.4	2. Alpha cellulose filled	0.5–1.5
Type - 11		3. Cellulose filled	0.6–0.8
Unmodified	1.2	4. Flock filled	0.6–0.7
Type - 12		5. Asbestos filled	0.5–0.7
1. Unmodified	0.3–1.5	6. Macerated fabric filled	0.3–0.4
2. 30-35% glass reinforced	0.3	7. Macerated fabric filled (phenolic modified)	0.3–0.6
Acetals		8. Glass fibre filled (Including nodular)	0.1–0.4
1. Homopolymer	2.0–2.5	9. Melamine phenol moulding compounds	0.4–1.0
2. Copolymer	2.0 (Avg.)	UREA formaldehyde moulding compound	0.6–1.4
3. 20% Glass filled	1.3–2.8	**Epoxy Resins**	
4. 25% glass filled Copolymer	0.2–0.6	1. Glass Fibre filled	0.1–0.5
5. TFE-fibre reinforced	2.0–2.5	2. Mineral filled	0.2–0.8
Fluoroplastics		3. Low density.	0.6–1.0
1. Polycholorotrifluoroethyl-ene	1.0–1.5		
2. FET fluoroplastic	3.0–6.0		
3. Poly vinylidene fluoride	3.0		

1.18 SURFACE FINISH

Decorative textures are used on mould surfaces to improve the aesthetic look of parts. Such textured surface are often helpful in hiding some moulding defects such as weld line, flow lines, sink marks, etc. Electroplating of metal on plastics is also used to increase the decorative effect. The necessary radii are provided so that, while plating the concentrated effect is uniform throughout the surface. Raised letters on the components are easier and cheaper to produce than depressed letters because the lettering is machined inside the mould cavity (Refer to Fig.1.155).

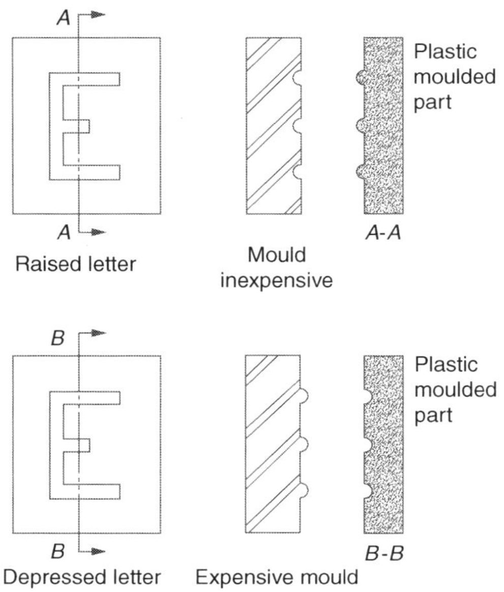

Fig. 1.155 Raised letters on the moulded plastic part are depressed letters on the mould. This allows engraving the mould which is a cheap procedure. Depressed letters on the moulded plastic part are raised letters on the mould. This means cutting away the mould around the letter which is an expensive operation. If a hob is used in making the mould, the reverse is true.

1.19 DESIGNING WITH PLASTICS FOR LOAD BEARING APPLICATIONS

The excellent mechanical, frictional, and chemical-resistance properties of engineering plastics encourage their direct substitution for metals in applications such as sleeve bearings, gears, and piping. Some of these plastics applications have been in wide use over a long time to develop adequate technology for design purposes. In this topic, design of some of these mechanical components like bearings, gears and sandwich laminates will be described.

1.19.1 Plastics Bearings

Some engineering plastics are sufficiently low in coefficient of sliding friction against metals that additional fluid lubricants are not needed and a simple concentric sleeve suffices as a journal bearing. In other applications, loads and speeds may be such that the plastic must be reinforced with fibres and/or filled with an internal solid or liquid lubricant. Compared with porous, oil-filled bronze bearings, some plastics have lower wear rates, lower coefficients of friction, better resistance to chemicals, and lower noise generation. A major advantage is the possibility of combining functions of several parts, including the bearing function, in one moulded shape. Many different plastics are in use for bearings; some of the main resins for sleeve bearings are nylons 6 and 6/6, acetal copolymers and homopolymers, polycarbonates, polyesters, phenolics, polyimides, and PTFE. Among commonly used fillers are graphite, molybdenum disulfide, PTFE, silicone oil, and fibreglass.

The chief rating factors for plastic or metal sleeve bearings are the following: maximum static pressure P, maximum relative sliding speed V, the PV product (discussed in detail below), maximum and minimum bearing temperatures, and permissible chemical environments. The use of manufacture's data for bearing design is best presented by means of a numerical example. The bearing material chosen for illustration is a crystallisable polyester or CPET sold by ERTA, Inc., of Exton, PA, under the trade name ERTALYTE.

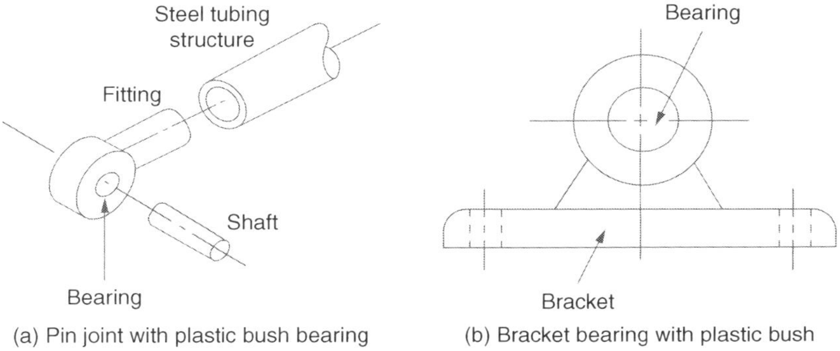

Fig. 1.156 Bearing design with plastics.

Table 1.7 shows the wear and friction data for acetal, nylon, and polysulphone. Wear factors of 240 or less are considered very good for most design purposes. The pressure at the bearing is the load divided by the projected area of the bearing (length × diameter).

Table 1.7 Friction coefficients and wear factors of plastics (*).

Bearing material	Shaft material	Wear factor $= \frac{\text{radial wear}}{P \times V} = \frac{(\text{cm/hr} \times 10^{-8})}{\text{kg/cm}^2 \times \text{mpm}}$	Coefficient of friction	
			Static	Dynamic
Acetal	Acetal	10,200	0.19	0.15
Acetal	Steel	78	0.14	0.21
Acetal (20% TEF)*	Steel	21	0.07	0.15
Nylon 6/6	Nylon 6/6	1380	0.12	0.21
Nylon 6/6	Steel	240	0.20	0.26
Nylon (20% TEF)	Steel	144	0.10	0.18

(Contd.)

Table 1.7 *(Contd.)*

Nylon (30% glass) (15% TEF)	Steel	192	0.19	0.26
Polysulphone (30% glass) (15% TEF)	Steel	66	0.12	0.10
Polycarbonate (30% glass) *25% TEF)	Steel	36		

Courtesy: * *Theberge, J.E., B. Arkles, and P.J. Cloud, How plastics wear against plastics, Jour. Machine Design, Oct. 31, 1974, pp. 60–61*

Operating conditions and formulae for Table 1.7 are tabulated as follows:
The wear factors in the above table were measured at

$$P = 2.8 \text{ kg/cm}^2,\ V = 150 \text{ mpm, TEF – fluorocarbon}$$

All fillers are percentage by weight.

Nylon and fluorocarbon bearings have a tendency to creep under moderate pressures. The contact pressure can be decreased by increasing the length of the bearing. The product of pressure and velocity, *PV*, is the power rating of the bearing per unit area. The product of *PV* and the coefficient of friction gives the energy dissipation (or the rate of heat generation). The wear rate is the product of the wear factor and *PV*. Increasing the material hardness by glass filler or metal backing can reduce wear. Glass reinforcement is an important feature of most of the bearings because the hardness of the resin alone is insufficient to protect a bush from wear.

Friction and wear Plastic bearings are sensitive to temperature. When the temperature rises above a certain point, wear becomes extremely fast. Because of the slow heat conduction through plastics, one of the methods used in removing heat is by a fluid such as oil or water.

The coefficient of friction of ERTALYTE is about 0.20 compared with oil-filled porous bronze at 0.30. One advantage of plastics, such as ERTALYTE, is that their static and dynamic friction coefficients tend to be about equal. In some plastics the static coefficient is lower than the dynamic coefficient. This suggests that there is little or no stick-slip action with plastic-metal combinations; hence, there is easier starting and less wear. Table1.8 summarises friction and wear properties of some commonly used plastic bearing materials. Friction and wear are given for plastic bearings running on hardened steel shafts. The wear factor *K* in Table 1.8 is defined by the following equation:

$$W = KFVt$$

Table 1.8 Friction and wear properties of plastic sleeve-bearing materials.

Material and filler	Maximum *PV* unlubed/lubed (KN.m/m^2/S	*K*	Dynamic coefficient of friction	Maximum temperature (°C)
PA/PTFE + glass	571400	16	0.26	93
PTFE/glass	1514260	7	0.09	260
CPET (ERTALYTE)	142850/399980	10	0.25–0.08	93
PA/Oil (ERTALON LFX)	199990/399980	2	0.20–0.08	93
Acetal (celcon)	114280/399980	65	0.25–0.05	105
Bronze/oil	1428500	100	0.20	66
Phenolic / PTFE	428550	30	0.26	83
Polyimide (DuPont SP-1)	8571000	40	0.29	288
Polyimide/graphite (DuPont SP-21)	2857000	40	0.24	288*

* Intermittent contact to 393°C

where t = time in hours, V = sliding speed in m/min, F = load in N, and W = total material removal in kilograms. There is a different K factor for every different bearing-shaft material combination, speed, and temperature.

There is an as yet unexplained inverse correlation between shaft hardness and wear rate for plastic bearings. Relative wear decreases by as much as five times between a shaft with a 20 RC and one with a 50 RC hardness for PTFE bearings. As might be expected, the smoother the shaft, the lower the wear rate. *PV* ratings for plastic against plastic are reduced due to higher temperature build ups. In addition, most plastic-metal combinations work better with lubrication; even an initial grease application alone helps. Lubricants remove wear debris, reduce friction coefficients, and help cool the bearing.

1.19.2 Plastic Gears

The main advantages of plastic gears are low cost for moulding or casting, low weight, reduced noise even without lubrication, low friction, possibility of parts consolidation, and environmental resistance. Commonly used plastic gear materials are nylons, acetals, and phenolics. For gears in toys, any rigid commodity thermoplastic will suffice. However, no plastics or plastics or plastic composites are known that can economically replace hardened steels or ductile cast irons for highly loaded gears.

The design of plastic gears for strength and durability has not yet advanced to the level of detailed consensus found in current standards for metallic gears, e.g., the A.G.M.A. gear standards used in the U.S. The following numerical example indicates current technology for plastic gears using data for Celcon, and acetal copolymer, published by the manufacture - Hoechst Celanese. The approach of other plastic gear material suppliers and fabricators may vary somewhat from that described here. Familiarity with basic terminology for gearing and gear stresses, as presented in machine design courses, is assumed. The U.S. conventional unit system is used. Subscripts p and g pertain to the pinion and gear, respectively.

Example 1.1

Given:

Pressure angle $\phi = 20^{o}$

Diametral pitch $P_d = 32$

Gear ratio $m = 5$

Centre distance $C = 1.5$ in

Face width $f = 0.300$ in

Pinion speed $n_p = 1500$ rpm

Horsepower $P = 0.1$

Steady load, 3 h/day

Continuously lubricated

Steel pinion

Celcon M90 gear

Operating temperature 100°F

Minimum life 10^7 cycles

Check the feasibility of this design for bending and contact (compressive fatigue) stress.

1. From the given data, pitch diameters of the pinion and gear, numbers of teeth, and tangential velocity are calculated:

$$D_p = 2\,C/(1 + m) = 2 \times 1.5/(1+5) = 0.50 \text{ in}$$

therefore,

$$D_g = mD_p = 5 \times 0.50 = 2.50 \text{ in}$$
$$V = \pi D_p\, n_p/12 = 196 \text{ fpm}$$

Since V is less than 2750 fpm, no correction for dynamic effects is necessary. With plastic gears, low elastic modulus and high damping tend to give low dynamic loads. In metal gearing, dynamic correction effects are significant at much lower V values. In high modulus metal spur gears, the number of teeth in contact averages 1.2 or 1.3, but in plastic-metal gear combinations it is closer to 2 teeth in contact. The number of teeth in the Celcon gear is

$$N_g = D_g \times P_d = 2.50 \times 32 = 80$$

2. The bending stress formula for the Celcon gear is adapted from the Lewis equation. It gives predicted unit load for failure L_f in terms of various factors that have been empirically determined by Hoechst Celanese for Celcon.

$$L_{fg} = L_u K_t K_L K_m K_s K_g / C_s \qquad (1)$$

where Lu is the reference tangential load per unit face width per Fig.1.157 in this case $L_u = 182$ lb/in. Applied tangential load F_t is given by

$$F_t = 33000\, P/V = (33000 \times 0.1)/196 = 16.8 \text{ lb}$$

therefore, $L_t = F_t/f = 16.8/0.300 = 56$ lb/in = applied unit load to be compared with the unit load at failure L_{fg}. The correction factors in E_q.: 1 are found as follows:

K_t = temperature factor = σ_y at ambient temp./ σ_y at 73F

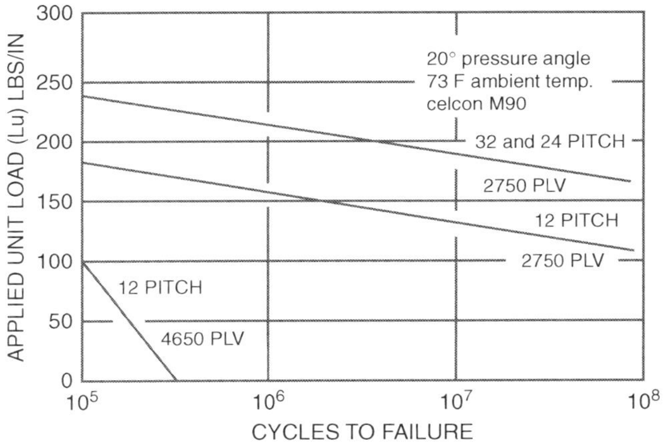

Fig. 1.157 Allowable unit load versus revolutions to failure by flexural fatigue (fluctuating tooth bending stress). For unfilled, acetal copolymer Celcon gear with steel pinion and initial grease lubrication. For Ex. 19.2, Lu = 182 lb/in at 10^7 cycles for a 32-pitch gear. (*Hoechst Celanese Corp*).

$$Kt = 7000/8800 = 0.80 \text{ at } 100^\circ\text{F}$$

(See Table 1.8. Kt = 1.0 for metal gears)

K_L = lubrication factor = 1.0 for initially lubricated pair (this case) = 1.5–3.0 for continuous lubrication
K_m = mating material factor = 1.0 for steel pinion, 32 μ in rms surface finish or better with friction coefficient of 0.15
K_m = 0.75 for acetal-steel pair with friction coefficient of 0.35
K_s = tooth type factor = Y_{des} for design gear/Y_{ref} for reference gear
where Y = Lewis form factor: Y_{ref} = 0.52, 0.64, or 0.69 for 12, 24, or 32 diametral pitch gears, respectively. Y_{des} is given in Table 1.10 for full-depth 20° pressure angle gears. In this example, with 80 teeth in the acetal gear, Y_{des} = 0.739 and K_s = 0.739/0.69 = 1.07.
K_g = gear type factor = 1.0 for spur, internal, and helical gears.
C_s = service factor = 0.80 (Table 1.10)
Combining these factors gives the predicted load to failure in 107 cycles from Eq. 2.2:

$$L_f = (182 \times 0.80 \times 1.0 \times 1.0 \times 1.07)/0.80 = 195 \text{ lb/in}$$

which is 348% greater than the applied unit load of 56 lb/in.

3. Plastic and metal gears generally fail because of wear or pitting associated with compressive fatigue. The contact stress in this example will now be checked against the allowable contact stress for Celcon. The allowable contact stress Sc for unlubricated pairs is proportional to the reference value Sc in Fig.1.158 multiplied by the factor C_k, which is given by

$$C_k = \{0.70/[(1/E_1 + 1/E_2) \cos\phi \sin\phi]\}\, 0.5$$

E_1 and E_2 are Young's moduli for the mating materials, and f is the pressure angle. E_1=29×106 psi for the steel pinion. E_2=315,000 psi for the Celcon gear from Table 1.11 at 100F. Equation 2.3 gives $C_k = 824$. Figure 1.158 gives *Sc* - 2800 at 107 cycles. Therefore, the allowable contact stress is

$$\text{Sac} = Sc\ C_k/639 = 2800 \times 824/639 = 3610 \text{ psi}$$

The applied contact stress is given by

$$\sigma_c = \{[L_t\ C_k\ (m+1)]/D_p\ m\}0.5 = 9550 \text{ psi}$$

Since the applied σ_c is considerably larger than the allowable σ_c, this design is not feasible.

The simplest fix is to increase gear face width from 0.300 to at least 2.10 in. The problem then is that the pinion face width is about four times its pitch diameter, which will require careful shaft alignment. A better solution might be to use a Celcon–Celcon pair. The advantage is the much greater accommodation of plastic-plastic gear pairs to tooth misalignment, which allows design of gears with higher face width/pitch diameter ratios.

Table 1.9 Tensile yield strength versus temperature for Celcon acetal copolymer*.

Temperature °C	–18	10	38	66	93	100
Yield strength (MPa)	83	67	50	38	29	28

* Celcon M90, unfilled virgin resin (*Hoechst Celanese Corp.*).

Table 1.10 Lewis form factor *Y* for 20° pressure angle, full depth gear teeth.

Number of teeth	12	14	16	18	20	24	30	38	50	75	100	300
Y	0.415	0.468	0.500	0.520	0.544	0.571	0.605	0.650	0.696	0.734	0.758	0.802

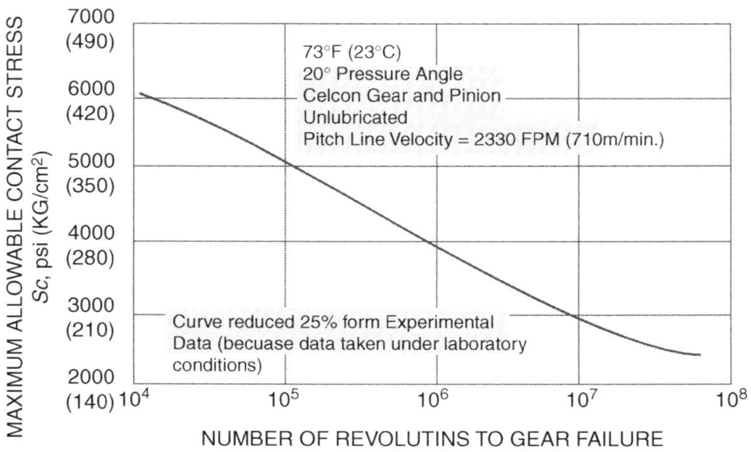

Fig. 1.158 Allowable surface contact stress versus number of gear revolutions to failure by surface wear for unfilled Celcon (*Hoechst Celanese Corp.*).

Table 1.11 Service factor Cs for Celcon gears mating with steel pinions, initial gear grease lubrication.

Type of load	24 hrs/day	8–10 hrs/day	3 hrs/day	1/2 hrs/day
Steady	1.25	1.00	0.80	0.50
Light shock	1.50	1.25	1.00	0.80
Medium shock	1.75	1.50	1.25	1.00
Heavy shock	2.00	1.75	1.50	1.25

4. There are many more details to be considered in gear design for plastics. Designers are advised to consult closely with resin suppliers on design of these components to minimise trial-and-error development procedures (Fig. 1.158).

Table 1.12 Effect of ambient temperature on flexural modulus *E* for Celcon*.

Ambient temperature (°C)	23	38	66	93	121
E (GPa)	2.8	2.2	1.4	0.90	0.51

Moulded plastic gear manufacturing technique Making an accurate plastic gear is more difficult than making a similar machined gear because, a plastic gear is made by moulding process. The engineer has to consider errors in the machining of the mould in the moulding of the product, and the shrinkage of the particular material after moulding.

Design calculation Production of moulded plastic gear involves the following steps:

1. Cut a gear in a brass blank by hobbing.
2. With the brass gear as the tool, use EDM to create mould cavity in the die.
3. Injection of molten plastic into the mould to obtain plastic gear.

The final moulded plastic part will shrink and become smaller than the mould cavity.

Design calculations to combat shrinkage of plastic in the mould: The diameter of the mould cavity should be the final diameter of the piece multiplied by $(1 + s)$, where s is the fractional shrinkage. In order to have the same number of teeth on a larger diameter, the pitch on the mould has to be non-standard. The module of female gear in the mould has to be increased by the factor $(1 + s)$. The pressure angle in the mould should be smaller than the standard angle, in order to get a moulded piece with the standard pressure angle. As an illustration, the shrinkage (S) in acetal is 0.023. In order to get a 50 mm diameter gear of module $m = 1.04$ and 20° pressure angle, the mould should be made with 51.15 mm diameter, with a module of 1.065 and a pressure angle of 15.99°.

1.19.3 Sandwich Laminated Plastics

Laminated plastics are one form of 'reinforced plastics'. The term 'reinforced plastics' is used extensively, since it includes moulded parts, in which the reinforcing is not usually in the laminated form, and both thermoplastic and thermosetting resins. Laminated plastics are piles of sheet material (bases), usually impregnated with a thermosetting resin (binder), and bonded together by means of heat and pressure to form sheets, tubes, rods, or moulded shapes.

The principal resins used in laminated thermosetting products are shown in Table 1.13. Phenolics are low cost and have good electrical and physical properties. Another class of thermosetting resins include the melamines. Melamines are more costly but offer flame resistance and have excellent electrical properties. The polyesters are low in cost with average properties. The epoxies are high in chemical resistance and are extremely moisture resistant. Silicones are used primarily for their retention of mechanical and electrical properties, even at very high temperatures, when their higher costs can be justified.

Table 1.13 Thermosetting resins used in laminated products.

	General properties of thermosetting resin types				
Property	**Phenolics**	**Melamine**	**Polyesters**	**Epoxy**	**Silicones**
Specific gravity	1.3	1.48	1.3	1.25	1.3
Cost of price	Low	Medium	Low	Medium-high	High
Advantages	Good all round properties	Good electrical properties	Many types and properties	Shrinkage nil	Heat resistance
Heat resistance	Excellent	Excellent	Good	Fair	Excellent
Physical properties	Good	Good	Good	Good	Fair
Electrical properties	Excellent	Excellent	Good	Excellent	Excellent
Water resistance	Good	Fair	Good	Excellent	Good
Machining qualities	Fair to good	Fair to good	Fair to good	Good	Good
Moulding pressures	Low to high	High	Low	Low to medium	Low to high
Moulding qualities	Excellent	Good	Excellent	Fair	Good

Laminating process Basically, the laminating process is carried out by impregnating base sheet stock with the liquid thermosetting resin. If a flat laminate is to be made, the correct number of sheets (stacked one upon the other) are simultaneously subjected to heat and pressure between two polished plates in a laminating press. Heat and pressure are applied to the layers, causing the resin to flow and harden the laminate into one solid mass. The equipment for this consists of a platen press that is capable of squeezing the laminate together with sufficient force, and some means of applying heat, usually through the platens of the press. Usually a multiple platen is used so that more than one sheet can be laminated at one time.

The stock to be laminated is placed between highly polished plates that are then placed between press platens. Heating is accomplished by passing steam through cores in the platen or by using electricity to heat the press platens. The term 'high-pressure laminate' is normally confined to those laminates moulded and cured in their final form at pressures no lower than 70 kg/cm^3 or 7 MPa and more commonly in the range of 84 to 140 kg/cm^2 or 8.4 to 14 MPa. If the pressure is under 70 kg/cm^2 or 7 MPa, the product is called low-pressure laminates. Those laminates made with little or no pressure, such as hand lay-ups, are sometimes called contact pressure laminates. The term laminate is sometimes used to include composites of resins and fibres that are not in distinct layers, such as filament wound structures and spray-ups.

After the compressed sheets have been cured into a solid state by the heat of the platens and the platens have been cooled, the press is opened, and the sheets are removed. Cooling the sheets before removing them from the press helps to prevent warpage. After the sheets are trimmed at the edges, they are ready for fabricating.

Thermosetting rods and tubes are treated differently from laminated sheets. Solid laminated rods are made by winding the impregnated filler web on a very thin mandrel, which is withdrawn before moulding. The centre channel is filled-up when pressure is applied in the metal mould. The mould, as it comes together, closes the centre hole and flash develops at the lands of the mould. The flash is removed by centreless grinding. In making a tube, the mandrel is left in the tube, and the tube is moulded in the same manner as the rod.

Articles with irregular shapes, such as gears and bearings, are often formed by cutting an uncured impregnated sheet to a pattern and then stacking and moulding. Places where small parts are required, as for spacers, cams, contact arms and levers, it is economical to saw them or punch them from stock sheets, rods, or tubes.

Classification of laminates The high-pressure laminated grades of sheet stock, rods, and tubes are classified by the National Electrical Manufacturers Association, commonly referred to as NEMA. In as much as laminates have their chief use in the electrical field, it is essential that the electrical properties must be classified along with the physical. The following list of laminates explains some of their characteristics and applications:

NEMA Grade

X A laminate made with high-strength kraft paper with phenolic resin as the binder. It is used in mechanical parts where electrical properties are of secondary importance. It is used in household appliances, insulating washers, and coil forms.

P A paper-base laminate made with plasticised phenolic resin. This laminate is a punching stock intended for general hot punching operations 120 to 140°C.

PC A paper-base laminate made with phenolic resin. It is used for cold punching electrical and mechanical parts.

XX A paper-base phenolic laminate. The electrical and mechanical properties make it suitable for usual electrical applications, except where low losses or high humidity are involved. It is used in instrument panels and machined washers, barriers, relays and switch bases.

XXP A paper-base phenolic laminate. This laminate is a general-purpose hot punching stock. It is used in terminal boards, insulating washers, and switch parts.

XXX A paper-base phenolic laminate. It is a low-cost electrical grade for high-voltage and radio-frequency uses. It has good dimensional stability under humid conditions. This type of laminate is used for jack spacers, coil forms, radio and TV parts, and high-voltage switchgear.

XXXP A paper–base laminate with a plasticised phenolic resin. It is used in hot punching applications. It has high insulation resistance and low dielectric losses at high frequencies. It is used in terminal boards and radio and TV panels.

C A strong cotton fabric laminate bonded together with phenolic resin. Items fabricated from this type laminate include gears, cams, pinions, bearings and structural parts. It is not to be used for electrical parts.

CE A medium-weave, canvas-base phenolic laminate. It is used in panel boards, electrode supports, switches, small gears, and small bearings.

L A laminate made from fine-weave cotton fabric and phenolic resin. It is used for fine machined parts such as pinions, gears, breaker arms, and communication equipment.

LE A laminate made of cotton-linen base with phenolic resin as the binder. It is used for terminal blocks and strips, radio parts, and ball-bearing retainer rings.

N1 A nylon-cloth-base laminate with phenolic resin as the binder. It has good impact strength and good machinability. It is used in high-voltage applications. It can be post formed.

MC A laminate made of a cotton canvas base with melamine resin as the binder. It is used in applications requiring good resistance to caustics such as tank linings.

GP A laminate made from fibreglass mat and polyester resin. It is extremely tough mechanically and has good arc resistance. It is used in transformers as supports, wedges, and spacers. It can be punched cold.

A A laminate made from asbestos paper and phenolic resin. It has good dimensional stability under humid conditions and has good flame resistant properties. It is used for heat controls for household, ovens, electrical ranges, and furnace parts.

AA A laminate made from asbestos fabric and phenolic resin. It is used in clutches, in machine tools, rotor vanes, and insulation gaskets.

AAA A laminate made from asbestos mat and phenolic resin. It is heat-resistant and has excellent physical properties.

G-1 A laminate made from a staple-fibre-type glass cloth and phenolic resin. It is the weakest of the glass-base grades, but it has good dimensional stability.

G-3 A laminate made from a continuous-filament-type glass cloth and phenolic resin. It is used in armature slot wedges and structural parts requiring good electrical properties.

G-5 A laminate made from a continuous-filament-type glass fabric with melamine resin as the binder. It has excellent electrical properties under dry conditions. It is used in switch parts in electrical and communications apparatus.

G-6 A laminate made from a staple-fibre-type glass cloth with silicone resin as the binder. It has high heat and is arc resistance.

G-7 A laminate made from a continuous-filament-type glass fabric with silicone resin as the binder. It is used in motor slot wedges, slot liners, and high-frequency radio and radar insulators.

G-8 A laminate made from fibreglass mat and melamine resin as the binder. It has high mechanical strength and good electrical properties.

G-10 A laminate made from a continuous-filament-type glass cloth with epoxy resin as the binder. It is used for printed circuits and electronic appliances.

ES A paper–base laminate with melamine as the binder. It is made with a black or gray surface and a white opaque core. It is used mostly in engraving nameplates, etc.

Design consideration of laminates

1. Punching of Laminates: The term 'punching' is used to describe the production methods of making laminated parts, by blanking, piercing, and shaving, or combination of these operations. Holes of nearly every geometrical shape have been punched in laminated sheets (Fig. 1.159). Certain basic principles and limitations of design, however, must be followed if the part is to be produced successfully.

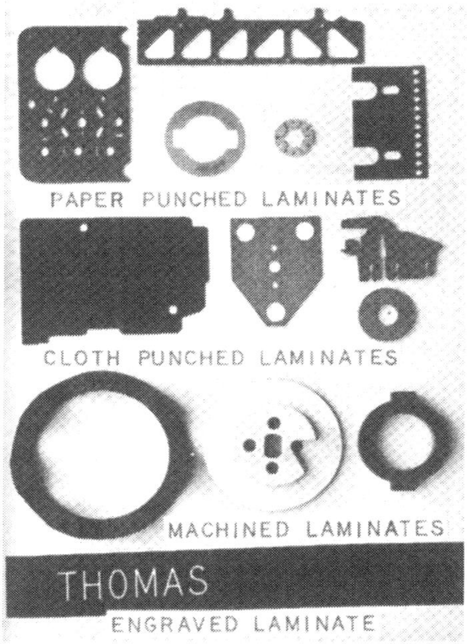

Fig. 1.159 Plastic laminates can be punched, machined and engraved.

Punchable laminates can be cold punched in maxium thicknesses ranging from 0.8 mm to 3.2 mm and hot punched in maximum thicknesses ranging from 2.4 to 3.2 mm. All laminates do not punch alike, and the only test of whether a laminate is a punchable stock is actually to punch parts from the laminate in question. Punch-grade laminates can be divided into hot and cold punching groups. The heating range of laminates can be from 38°C to 140°C. The heating may be accomplished by plate, heated by steam or electricity, ovens, hot liquid baths, infra-red lamps, or dielectric heating equipment. The laminated material should be heated as rapidly and uniformly as possible and should not be kept hot longer than necessary.

Punching is the forcing of a hole in a sheet, and blanking is the knocking (punching out) of a piece from a sheet. Figure 1.160(a) illustrates a pierce, or punch, and black die and a typical part produced by this die. Note that the part first is punched and then is blanked. Because of the resilience (yield) of laminates, punched holes tend to be smaller than punches, and blanks tend to be larger than blanking dies. With hot punching grades, thermal contraction results in holes and blanks smaller than the corresponding punches or dies. The die designer must rely on his knowledge of the characteristics of a certain type of laminate to be used and design the die accordingly.

Parts having thicknesses greater than the maximum allowed for punching may be shaved without damage to the piece or die. The part to be shaved is first rough cut to shape by blanking, fly cutting, or band sawing. The rough cut part is then placed in a shaving die. Figure 1.160(b) illustrates a shaving die. The die is mounted in the die shoe under the punching die and shaves approximately 0.38 mm to 0.50 mm from the edge of the blanked part. The shaving die finishes the rough blanked part to size and produces a very smooth clean surface.

The shaving tool is made by machining the tool to the shape of the part. The shaped tool is then ground on a 45° angle from the edge. This leaves a knife edge at the contour end of the shaving die.

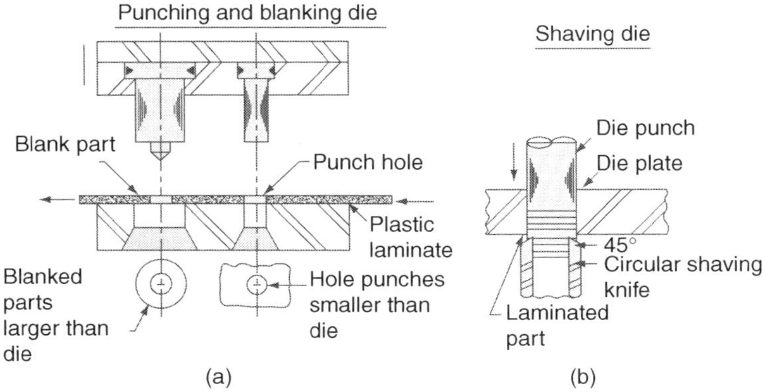

Fig. 1.160 (a)The design of a punching and blanking die, (b) A shaving die.

The diameter or diagonal of a hole to be punched should be no less than the thickness of the stock (Fig. 1.161(a)). Holes in laminate parts should be no closer to each other or to the edge of the part than the thickness of the material (Fig. 1.161(b)). Square or rectangular holes should have their corners even farther apart than the thickness of the stock, in order to prevent cracking between the holes (Fig. 1.161(c)). If it is necessary to place holes closer together and closer to the edge than the limits specified, the holes may be drilled.

Tolerances between holes and on the diameters of the holes vary with the thickness of the stock being punched. Table 1.14 gives what is considered to be good shop practice on tolerances for punched holes and slots.

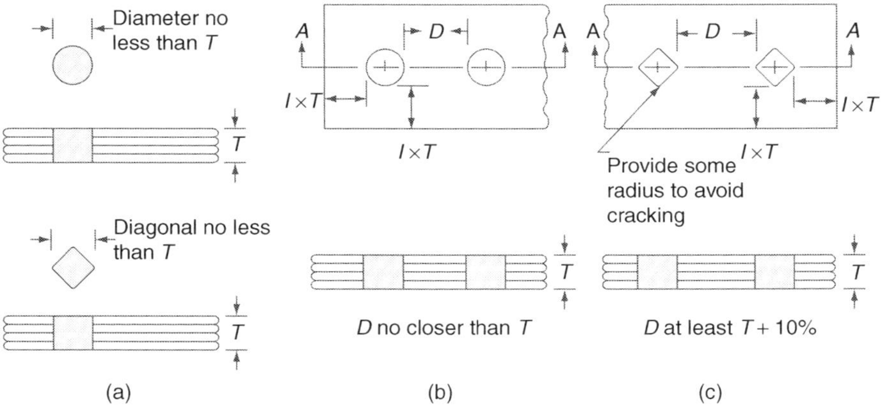

Fig. 1.161 Punched holes in laminated sheet stock have limited sizes compared to the thickness of the laminate as shown in (a), (b), and (c).

Punching beyond the maximum allowable thickness not only wears the die rapidly, but also results in a poor piece (Fig. 1.162(a)). The face sheet on both sides of the laminate have been pinched or pulled together at the edge, and this has resulted in squeezed-out laminations in the core. With some designs, it may be desirable to punch a part with an offset wall section or a hole part way through (Fig. 1.162(b)). When laminated sheet stock is used, the part is made by forcing the punch only part way through the stock. Such a technique can be used with any good grade of punching stock, but it must be remembered that the laminations will be broken, and that this will encourage water absorption and arcing through the lamination, if the part is subjected to electric current. The part should be considered only as a simple insulator and a dust cap. Erratic shaped holes should be avoided, unless the laminate stock is thin enough to allow the holes to be punched (Fig. 1.162(c)). Irregular holes in thick laminates require special tools and additional operations.

Table 1.14 Standard tolerances for punched holes and slots maximum punching tolerances on sheet stock.

Material thickness in inches	Distance between holes Under 50 mm	50–75 mm	75–100 mm	100–125 mm	Size of slots or diameter of holes	Overall dimension
Under 1.6 mm	0.076	0.10	0.12	0.15	0.38	0.20
1.6–2.4 mm	0.12	0.15	0.18	0.20	0.076	0.25
2.4–3.2 mm	0.15	0.18	0.20	0.22	0.12	0.38

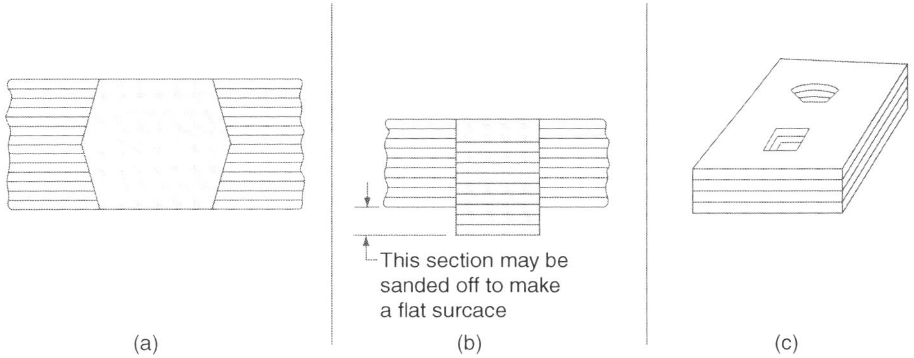

Fig. 1.162 Punched holes in laminated stock, (a) A hole punched beyond the allowable thickness, (b) A hole punched part way through, (c) Erratic shaped holes should be avoided.

2. **Dies Used for Punching:** Dies used for punching laminates are of the same general types as those used for punching metals, except that the clearances should be less than those normally employed in punching metals. Progressive dies are usually preferred, because they are more economical to make and permit higher production rates. For clean punching of laminated parts it is recommended that close tolerances between the punch and die be held. For standard tolerances, the die hole should be no more than 0.10 mm larger than the punch, giving 0.050 mm clearance all around. Deviation from standard tolerance in the die construction can be permitted if several thousandths of an inch tolerance on the size of the hole is allowed.

3. **Milling Laminates:** The milling of plastic laminated materials is very similar to that of milling brass. Because of the laminated structure of the plastic laminate, climb or down milling is always used to prevent any tendency towards delamination (Fig. 1.163). Cutters should have a negative rake of about 10°. The width of a ridge between two milled slots should be greater than the depth of the slots. This will keep the ridge from breaking (Fig. 1.163(a)).

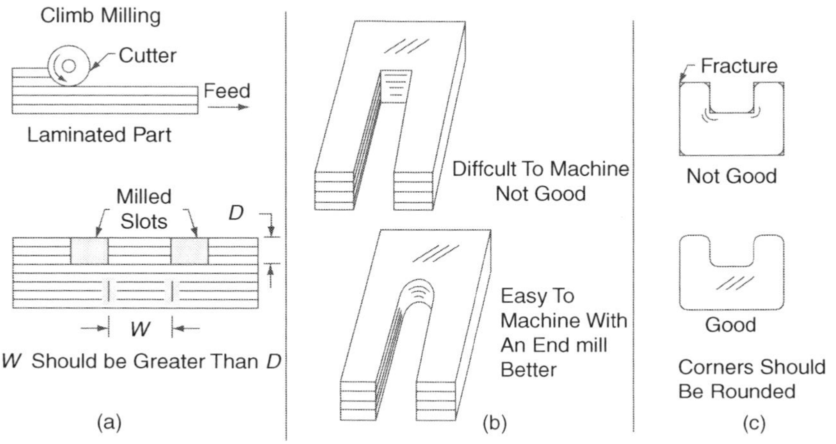

Fig. 1.163 Machined laminates: (a) Climb milling is recommended for cutting slots, (b) Slot ends should be round instead of square, (c) Corners on punched laminated parts should be rounded to prevent fractures.

Slot ends should be designed to be round instead of square (Fig. 1.163(b)). This makes a stronger part and will permit the use of an end mill, to make the rounded end, instead of hand filing. Corners on machined or punched laminated parts should be rounded to prevent delamination and fractures (Fig.1.163(c)).

Laminated plastics can be machined on standard wood or metal working equipment. On glass-base laminates, the cutting tools dull rapidly. Diamond or tungsten-carbide tools give a more satisfactory working life.

4. **Threads Machined in Laminates:** Threads are generally always machined and not moulded in plastic laminated materials. Machined threads do not start abruptly, nor do they come to an abrupt end, as do moulded threads. Threads start and stop with a feather edge. The feather edge in laminated materials is weak. If the axis of a thread is perpendicular to the laminations (Fig.1.164(a)), the thread will have maximum strength. Assume that a thread is cut on the outside diameter of a tube or rod (Fig.1.164(b)). If the mating part exerts strain in a direction parallel to the axis of the tube or rod, the threads of the laminated part will tend to strip or delaminate.

When drilling or tapping parallel to the laminations, always clamp the workpiece between two supports to prevent splitting. This is not necessary when working perpendicularly to the lamination, although a back up plate to prevent chipping makes a cleaner hole. A blind tapped hole should have a clearance at the bottom in order to prevent stripping of the thread or delamination (Fig.1.164(c)). Through tapped holes should be reamed before tapping as the edge of the cut thread will be much smoother. All threads cut into laminated materials should be specified as U.S. Standard with thread tolerance no closer than Class 2 fit.

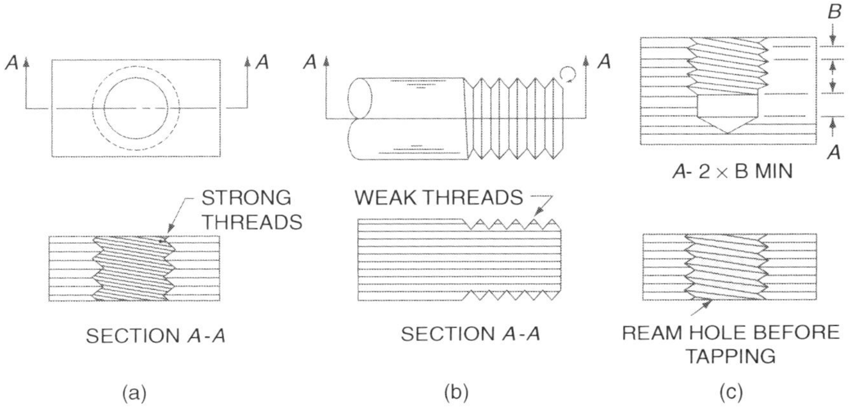

Fig. 1.164 Threads machined in plastic laminates: (a) Cut threads with axes perpendicular to the laminations are strong threads, (b) Cut threads with axes parallel to the laminations of the rod or tube are weak, (c) Blind tapped holes should have clearance at the bottom in order to prevent stripping of the thread or delamination. Through tapped holes should be reamed before tapping.

1.20 TOOLING ASPECTS ON PRODUCT DESIGN

1.20.1 Introduction

While designing any plastics product, the product designer should think of the simplicity of the product design. That means the configuration / profile of the product should be as simple as possible. The geometrical shape of the product should simplify the tooling operations. Unless it is purposely required for the particular application the designer should not complicate the profile of the product. The intricacy of the product design unnecessarily increases the tool manufacturing cost which reflects in increase of product cost. Hence, the product designer must be very careful while selecting the geometrical shape of the product for the particular application. While fixing tolerance for the plastic product, the designer should aware of the functional dimensional of the product. For example, a box with lid should have a tight fit means that the matching area dimensions are critical one; there we have to provide closer tolerance for the dimensions say +/−0.05 or +/−0.01 mm. In other area we may provide wider tolerance say +/−0.1 mm or as applicable.

1.20.2 Use of Special Machines

Now-a-days sophisticated special machines such as CNC MILLING, CNC LATHE, CNC EDM, CNC WIRE CUT EDM, CNC GRINDING etc are used for making moulds for very precised components. The machine hour rates (MHR) for special machines are very high when compared to other conventional machines such as lathe, milling, shaping, planning, EDM, cylindrical grinding, surface grinding, drilling machines etc. These conventional machines cannot maintain very closer tolerance to the workpiece which are possible in special machines..

1.20.3 Tooling Cost vs Product Cost

The product designer should study / analyse the application of the product thoroughly before deciding the cross-sectional shape of the product. Because circular cross section will always easy to machine than other cross sections such as square, rectangle, elliptical, polygon, irregular shapes, etc. Hence, the tool manufacturing cost will be higher for the products other than circular shape. Moreover machine hour rate are very high for CNC machines. Unless it is required, the product design should not complicate the cross-sectional shape of the product, which results in increasing tool cost. The product which needs closer tolerance and other than circular cross section may require CNC machines / special machines will leads to high tooling cost. The case study of preferred product design for various applications of the plastics products are shown in Figs. 1.165, 1.166, 1.167, 1.168 and 1.169.

This shows how the tooling cost is reduced in enormous proportion by way of changing cross section of the product which ultimately reduces the product cost, to a minimum. Hence, the product designer must carefully think of the cross-sectional shape and types of fits and tolerance required for the product for the particular application accordingly that have to be implemented in the product design.

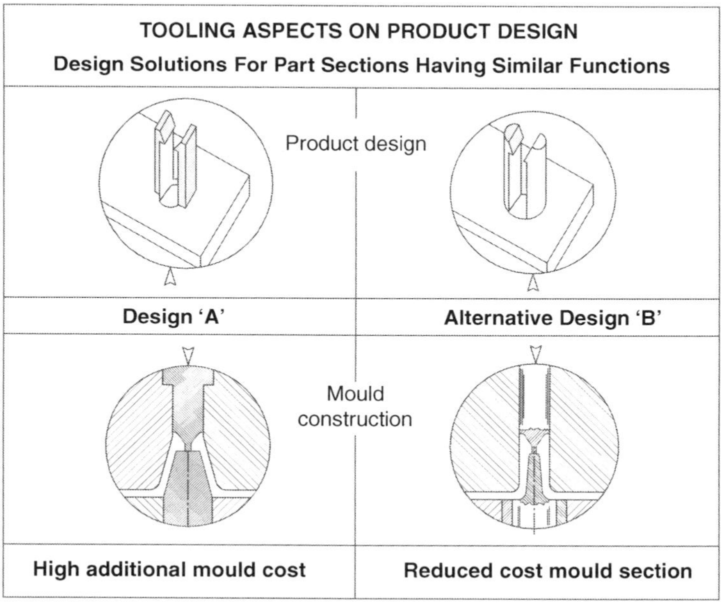

* 75% design 'A', less mould parts, less machine cost, less labour time

Fig. 1.165 Tooling aspects on product design.

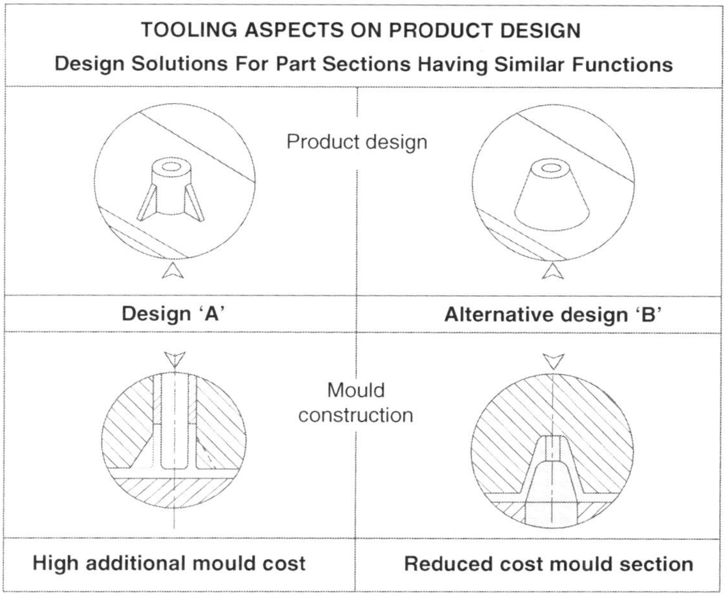

* 80% design 'A', less labour time, less machine cost, faster cycle

Fig. 1.166 Tooling aspects on product design.

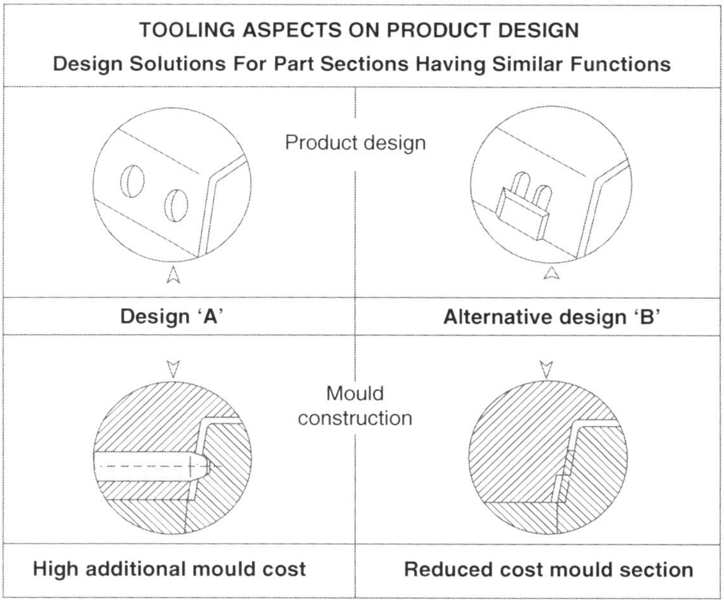

* 50% design 'A', less mould parts, less labour time

Fig. 1.167 Tooling aspects on product design.

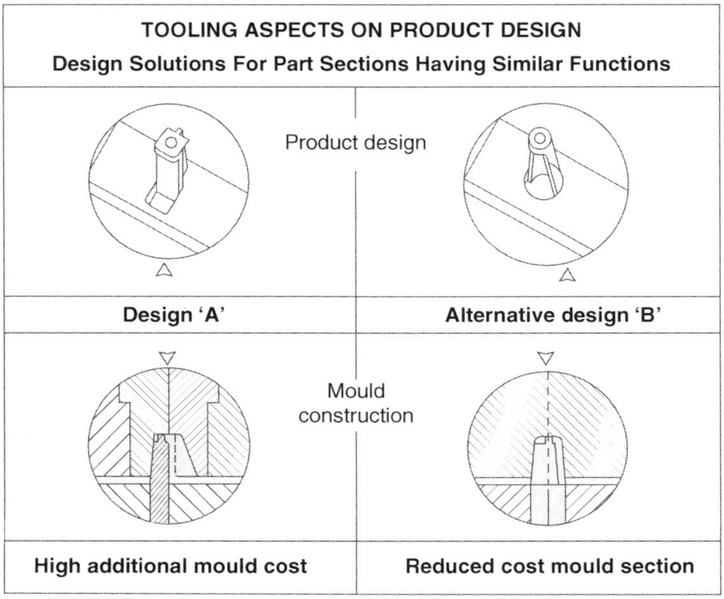

* 60% design 'A', less labour cost, less machine cost, faster cycle, better cooling

Fig. 1.168 Tooling aspects on product design.

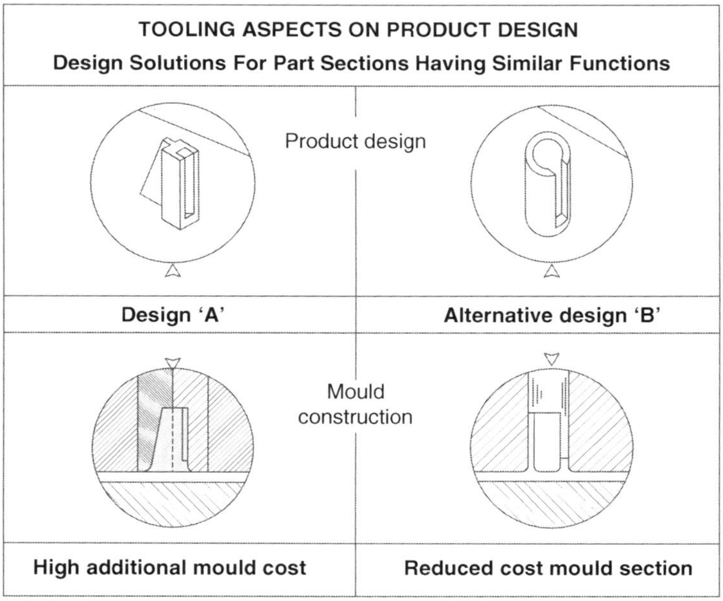

* 50% design 'A', less mould part, less labour time, less machine cost, better cooling

Fig. 1.169 Tooling aspects on product design.

1.20.4 Mould Material

In selecting steel for tight-tolerance part, the reinforcement content of the resins being used must be taken into account. For example, the pressure of glass and mica in the material can cause abrasion and wear on the mould, changing the dimensions over time and reducing the tolerance capability.

H13 or P–20, the steel of choice for many moulds, offers minimal resistance to the resulting abrasion and wear when these materials are used would not generally be recommended for such applications, with possible exception of very large parts. A steel with higher carbon content for overall hardness and higher chromium content for better wear resistance is suggested for tight-tolerance designs. Likewise, air-hardened steel is preferred over oil-hardened steel because of improved dimensional stability during heat treatment.

There are several other variables impacting tight-tolerance control that must be considered.

1.20.5 Cooling

Cooling is crucial for tolerance control. In order to allow the part to shrink evenly, an even temperature must be maintained across the mould. Otherwise, uneven shrinkage will result, leading to dimensional tolerance variations.

1.20.6 Runner System

These are other variables in the tolerance capability of a process. Three plate moulds with naturally balanced runner systems are preferred for tight-tolerance designs. Artificial balancing of these runner systems through the use of mould-filling analysis and variations in runner size can be effective; however, this introduces another variable to overall tolerance capability.

The same is true of runner less (hot-runner) systems. While they can be effective, using them may lead to reduced tolerance capability because the heat required to keep a runner hot will also heat a portion of the mould. As a result, additional cooling will be necessary, thus introducing another variable into the tight-tolerance equation. If a runner less system is chosen, an adequate cooler plate must be used to control the mould temperature properly.

1.20.7 Venting

While venting is not as critical factor as cooling and runner system, a mould should have as much venting as possible. Too little venting can trap air in the mould and lead to variations in the melt temperature and the cavity pressure as the part is filled; either condition can affect tolerance capability. One method of venting injection moulds for thermoplastics materials is shown below:

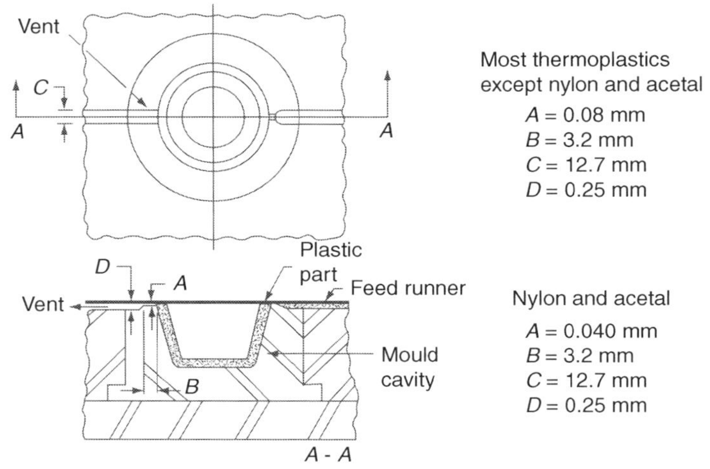

Fig. 1.170 This drawing illustrates one method of venting injection moulds for thermoplastic materials.

1.20.8 Core Pins, Slides and Side Actions

These are common variables in tight-tolerance designs. Whenever possible, these components should go through the part and lock in to a rest in the other mould half. Otherwise, deflection may result over time, merely by repeated impacts of viscous plastic flow fronts during processing.

1.21 PROCESSING VARIABLE Vs PRODUCT DESIGN

1.21.1 Effect of Temperature, Pressure and Cooling Time

The success of the part and mould design is determined by what happens in the moulding process. Basically, part shrinkage is affected by a combination of temperature, pressure and time in the mould. Mould parameters that affect orientation and the degree of crystalline resins influence the amount of shrinkage.

The optimum combinations consist of a melt injected at a rate and temperature in a mould hot enough to fill the mould with minimum viscous shear. Packing pressure should be high enough and gate freeze delayed long enough to fully pack the part. Pressure drop between the gate and end of flow – largely a function of mould design in relation to part. High material density inhibits shrinkage. The cooling rate must provide enough time for the stressed layers of resin to relax, but still fast enough to prevent a high level of crystallinity in crystalline resins than the strength requirements of the part dictate. The higher the crystallinity, the higher the shrinkage. Any process parameter that increases cooling time will enhance shrinkage.

Logically these parameters include raising the temperature of the mould and melt and conditions that reduce cooling efficiency. Bear in mind, though, that shrinkage (and warpage) cannot be permanently foiled by fast ejection. The part will continue to change size and perhaps shape until enough stress is released to stabilise it, which may take from a few hours to two days depending on its demoulded condition.

1.21.2 Design for Flow and Shape

In moulding with any plastic material, parts should be designed with ample curves, expect at the parting lines of the mould cavity section. If the material, as it is being moulded, does not sweep across the confined areas of the mould, gas pockets or voids may develop. This results in blisters or sinks marks on the surface nearest the pocket. With thermosetting materials, these gas pockets may be caused by trapped gases produced during chemical cross-linking of the material (Fig.1.171). With thermoplastics, these voids may be the result of 'case hardening' of the melt as it momentarily hesitates and cools in the corner. If possible, avoid thick and thin or uneven wall thicknesses in compression moulded phenolic moulded parts as gas pockets may form.

The strength of a plastic part depends largely on good design. When plastic materials flow around protruding sections of a mould, they knit or weld on the other side. Good design calls for consideration of this flow route in the part. With the thermosetting materials, this knit or weld line may be a weak point, due to the fact that the plastic material has approached the last stages of polymerisation before the two streams meet or weld on the opposite side. Thus, they do not bond well. The thermoplastic materials tend to cool as they fill the mould cavity, with the result that the weld or knit line will be weaker than the adjoining material. In using the thermosetting materials, a case or cover with open louvers perpendicular to the draw and made by the side of the mould should be avoided. Long louvers or open slots should be moulded in the bottom of the mould so as to avoid welding as much as possible. If louvers or slots have to be made by the side of the mould, they should run up and down the face (Fig. 1.172). When compression moulding parts of high impact materials, thin sections some times loose strength, because fibres do not flow into the narrow moulded paces (Fig. 1.173). Sharp corners, straight sides and improper venting impade plastics flow during moulding, resulting in strains and possible cracks. Optimum flow require maximum radii at corners reasonable inside and outside tapers (1° per side under 25mm, $1/2^\circ$ per side upto 50 mm) and proper vents, mould parting line, pockets and blind holes (Fig.1.173) when holes are too near or near a corner, material may not weld properly around mould pins and the flow of plastic can bend mould pins for blind holes when length exceeds diameter by 2½ times and when holes are to be long with small diameters even if these are supported (anchored) at both ends.

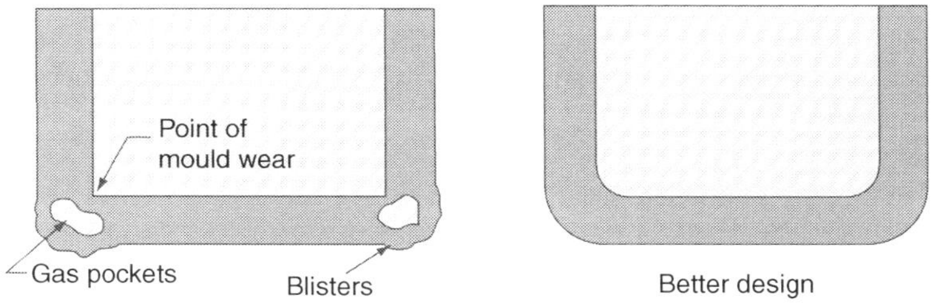

Fig. 1.171 Streamlining of the plastic part with help to prevent gas pockets.

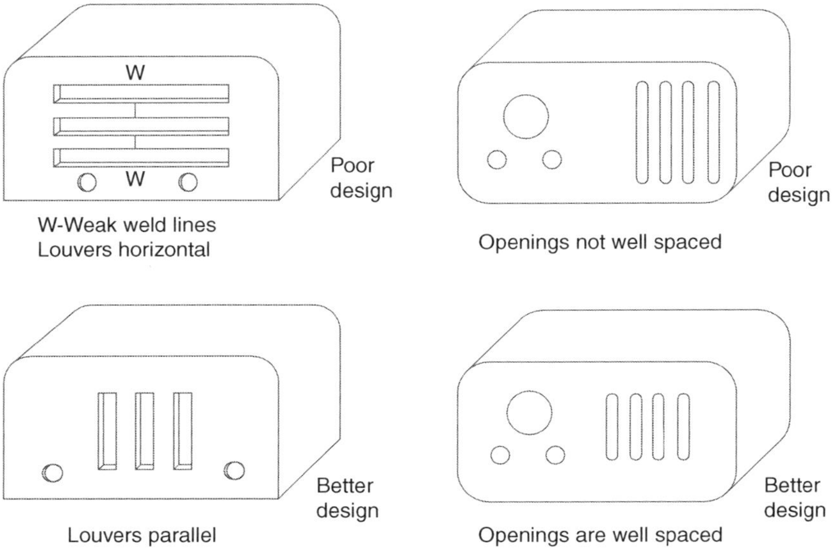

Fig. 1.172 In compression moulding, louvers should be parallel to the draw when made by the side of the mould. This will prevent weld or knit lines between the louvers.

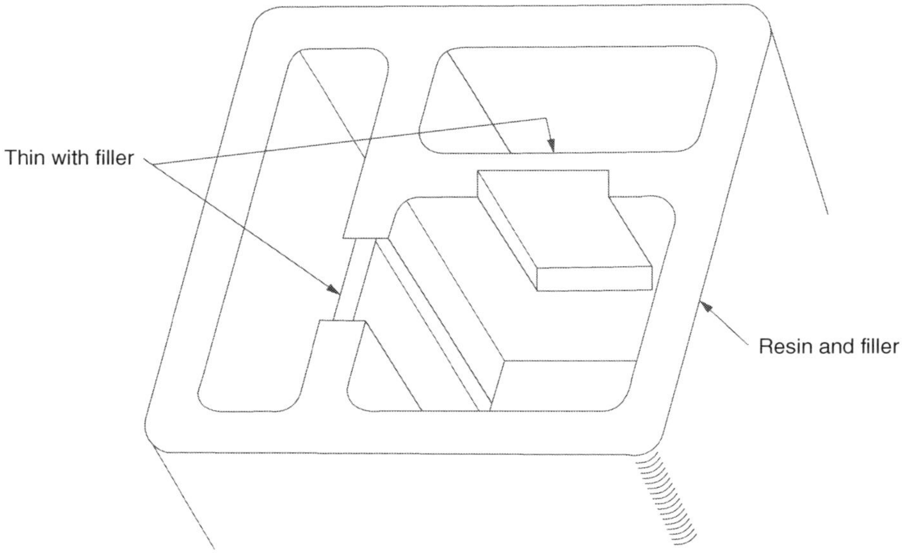

Fig. 1.173 When compression moulding parts of high-impact materials, thin sections sometimes loss strength because fibres do not flow into the narrow moulded spaces.

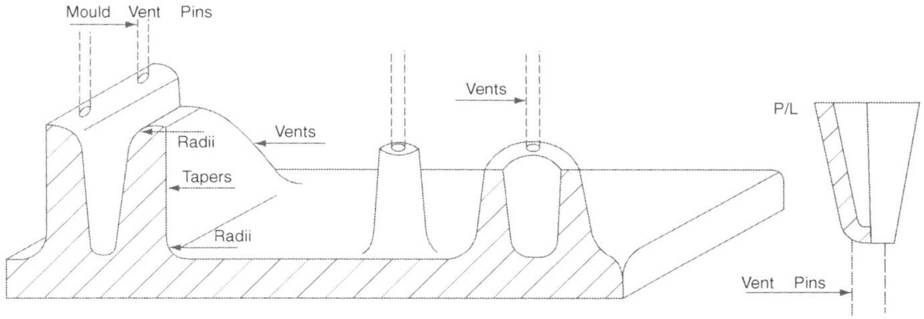

Fig.1.174 Sharp corners, straight sides and improper venting impede plastics flow during moulding, resulting in strains and possible cracks. Optimum flow requires maximum radii at corners, reasonable inside and outside tapers (1° per side under 25 mm ½° per side up to 50 mm) and proper vents at mould parting lines, packets and blind holes.

1.21.3 Flow Length vs Wall Thickness

When the polymer melt is entered from the gate point to the impression, the maximum distance at which it can reach to the impression with respect to the wall thickness of the product for certain thermoplastic materials is listed in Table 1.15 for reference. The material supplier may provide this detail or it may be obtained from the standard materials data book. Based on this flow length vs wall thickness (L / T) ratio, the designer should choose the gate position in the product for the optimum filling the cavity without any short.

Table 1.15 Approx. max. flow path – to thickness ratio of thermoplastics.

A B S	: 175 : 1
ACETAL	: 140 : 1
ACRYLIC	: 130–150 : 1
NYLON	: 150 : 1
POLY CARBONATE	: 100 : 1
POLYETHEYLENE (L D)	: 275–300 : 1
POLYETHEYLENE (H D)	: 225–250 : 1
POLYPROPYLENE	: 250–275 : 1
POLYSTYRENE	: 200–250 : 1
PVC – RIGID	: 100 : 1

1.22 MECHANICAL PROPERTIES

Mechanical properties are crucial since virtually all end-use applications involve some degree of mechanical loading. Material selection for a variety of applications is often based on mechanical properties such as tensile strength, modulus, elongation, and impact strength. These values are normally available in the marketing data sheets provided by material suppliers.

In partical applications, materials are seldom, if subjected to a single, steady deformation without the presence of other adverse factors such as the environment and temperature. Since published values of mechanical properties are generated from tests conducted in laboratories under standard test conditions, the danger of selecting and specifying a material from these values is obvious. A thorough understanding of mechanical properties and tests employed to determine such properties, as well as the effect of adverse (or beneficial) conditions on mechanical properties over long periods of time is extremely important. Some important basic definitions of various mechanical properties are as follows:

1.22.1 Stress

Consider a three-dimensional body with a balanced system of forces acting on it, F_1–F_5 in Fig. 1.175, such that the body is at rest. A body subjected to external forces develops internal forces to transfer and distribute the external load. Imagine that the body as shown in Fig. 1.175 is cut at an arbitrary cross section and one part is removed. To keep the body at rest, a system of forces must be acting on the cut surface to balance the external forces. These same systems of forces exist within the uncut body and are called *stresses*. Stresses must be described with both a magnitude and a direction. Consider an arbitrary point P on the cut surface as shown in Fig. 1.175, where the stress S, is as indicated. For analysis, it is more convenient to resolve the stress, S, into two stress components. One acts perpendicular to the surface and is called a *normal* or *direct stress*, σ. The second stress acts parallel to the surface and is called a *shear stress*, τ.

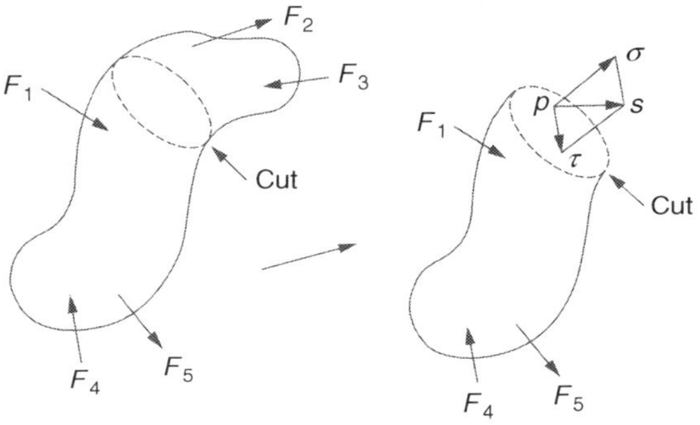

Fig.1.175 Internal forces and stresses in a body.

1.22.2 Normal Stress

A basic understanding of load, deflection, and stress starts with a simple tension test, shown in Fig.1.176. Direct stress is the ratio of applied load to the original cross sectional area, expressed in kg/cm². In the International System of units (*SI*), stress is expressed as Newtons per square metre, or Pascals (*Pa*).

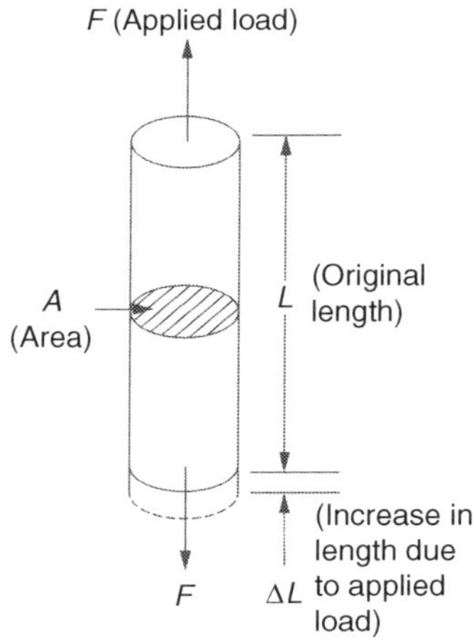

Fig.1.176 Simple Tension Load.

$$\text{Stress} = \text{Load}/\text{Area}$$

(or)

$$\sigma = F/A$$

If the load is applied as shown, the member is said to be in tension. If the load is reversed, the member is in compression.

1.22.3 Normal Strain

If a bar is subjected to a direct load, and thus a stress, the bar changes in length. If the bar has an original length L and changes in length by an amount ΔL, the strain produced is defined as

Strain = Change in length / Original length

(or)

$$\varepsilon = \Delta L / L$$

Strain is a measure of the deformation of the material and is dimensionless, i.e., it has no units. It is simply a ratio of two quantities with the same units.

In general, the extensions of materials under load are very small. With most metals, it is convenient to measure and express strain in the form of cm/cm. The symbol $\mu\varepsilon$, called microstrain, expresses this. With plastics, which generally undergo greater deformation than metals under the same loading, strain is normally expressed as 10^{-5} cm/cm. Another common way to express strain is as per cent strain. The equivalence of the three is as follows:

$$1000\ \mu\varepsilon = 1000 \times 10^{-6} = 0.001 = 0.1\ \%\ \text{strain}$$

$$10000\ \mu\varepsilon = 10000 \times 10^{-6} = 0.01 = 1.0\ \%\ \text{strain}$$

Figure 1.177 illustrates a typical tensile testing arrangement with a common test specimen. The results obtained from this testing may be plotted in the form shown in Fig. 1.178. This is a stress – strain curve, which characterises the mechanical behaviour of a material in tension.

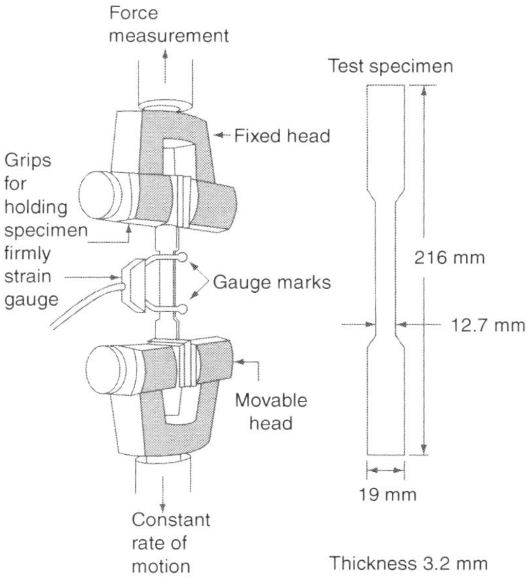

Fig. 1.177 Typical test setup and specimen.

1.22.4 Modulus of Elasticity

Most materials, including metals and plastics, have a deformation that is proportional to the imposed loads over at least a range of loads. Since stress is proportional to load and strain is proportional to deformation, this also implies that stress is proportional to strain. Hooke's law is the statement of that proportionality.

$$\text{Stress/Strain} = \text{Constant}$$

The constant E is called the *modulus of elasticity*, Young's modulus, or in the plastics industry, tensile modulus. In terms of the bar in (Fig.1.178) the tensile modulus is given by

$$E = \frac{F / A}{\Delta L / L} = \frac{FL}{A\Delta L}$$

The modulus is, therefore, the slope of the initial portion of the stress–strain curve. It must be noted that an elastic material does not necessarily obey Hooke's law. A material may return to its original shape without the stress being proportional to the strain. However, if a material obeys Hooke's law, it is elastic. In many plastic materials, the straight region of the stress– strain curve is so difficult to locate that a straight line tangent to the initial portion of the curve must be constructed to obtain a 'modulus'. A modulus obtained in this manner is called the initial modulus. For some plastic materials, the initial modulus can be misleading due to the material's non-linear elasticity. For this reason, some suppliers provide the 1 % secant modulus as a better representation of the material's behaviour. The designer is cautioned that the marketing data sheets do not always clarify whether the supplier is providing Young's modulus, an initial modulus, or a secant modulus. Thus, the designer is reminded of the warning at the beginning of this chapter on placing too much emphasis on the published data.

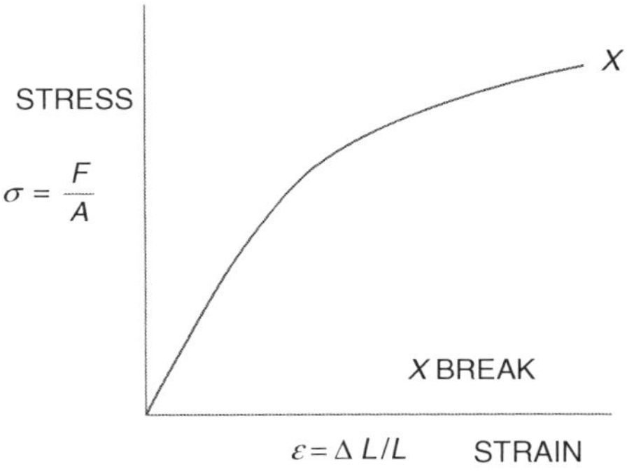

Fig.1.178 Plot of results of tensile test (stress-strain curve).

For metals, Young's modulus is normally expressed in terms of 10^6 Mpa or Gpa. For plastics, the tensile modulus is often expressed as 10^5 MPa.

A number of stress – strain curves are shown in Fig.1.179. The explanation of points A through F on the curves is provided below.

1.22.5 Proportional Limit

With most materials, some point exists on the stress–strain curve where the slope being to change and the linearity ends. The proportional limit is the greatest stress at which a material is capable of sustaining the applied load without deviating from the proportionality of stress to strain. This limit is expressed in *Pa* and is shown as point *A* in Fig.1.179. Note that some materials maintain this proportionality for large measures of stress and strain while others show little or no proportionality as previously discussed.

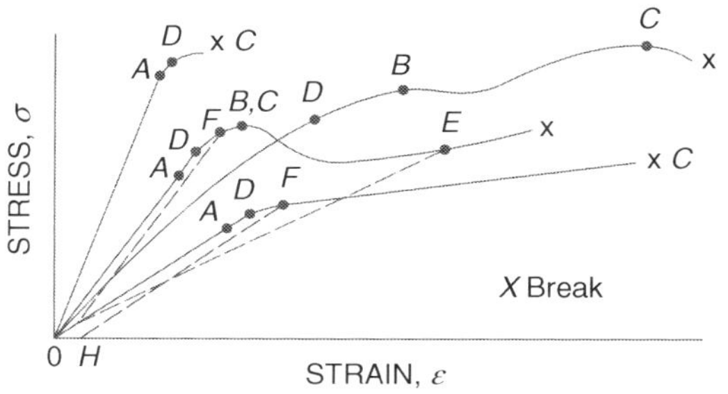

Fig.1.179 Typical stress-strain curves.

1.22.6 Yield Point

Yield point is the first point on the stress–strain curve where an increase in strain occurs without an increase in stress. This is shown as point B in Fig.1.179. The slope of the curve is zero at this point.

1.22.7 Ultimate Strength

The ultimate strength is the maximum stress a material withstands when subjected to an applied load. This is expressed in Pa and is denoted by Point *C* in Fig.1.179.

1.22.8 Elastic Limit

Many materials may be loaded beyond their proportional limit and still return to zero strain when the load is removed. Other materials, particularly some plastics, have no proportional limit in that no region exists where the stress is proportional to strain (the material obeys Hooke's law). However, these materials may also sustain significant loads and still return to zero strain when the load is removed. In either case, the point on the stress–strain curve, *D* in (Fig.1.179) represents the point beyond which the material is permanently deformed if the load is removed. This point is called the *elastic limit*.

1.22.9 Secant Modulus

The secant modulus is the ratio of stress to corresponding strain at any point on the stress–strain curve. For instance, in Fig.1.179, the secant modulus at point *E* is the slope of the line *OE*.

1.22.10 Yield Strength

Some materials do not exhibit a yield point. For such materials, it is desirable to establish yield strength by picking a stress level beyond the elastic limit. Although developed for materials that do not exhibit a yield point, this value is often used for plastics that have a very high strain at the yield point to provide more realistic yield strength. This is shown as point *F* on the curves on Fig.1.179. The yield strength is generally established by constructing a line parallel to *OA* at a specified offset strain, point *H*. The stress where the line intersects the stress – strain curve at point *F* is the yield strength at *H* offset. For instance, if point *H* were at 2% strain, then point *F* would be termed the 'yield strength at a 2% strain offset'.

1.22.11 Poisson's Ratio

Under the action of a tensile load, the bar shown in Fig.1.180 increases in length by an amount ΔL, giving a longitudinal strain in the bar of

$$\varepsilon = \Delta L \,/\, L$$

The bar also exhibits a reduction in dimensions laterally, i.e., its breadth and depth both decrease. The associated lateral strains are opposite in sign (contracting vs. stretching) to the longitudinal strain, and are given by

$$\varepsilon_{\text{lateral}} = -\Delta b/b = -\Delta d/d$$

Provided the material deformation is within the elastic range, the ratio of the lateral to longitudinal strains is always constant. This ratio is called *Poisson's ratio*, and is designated by the Greek letter v.

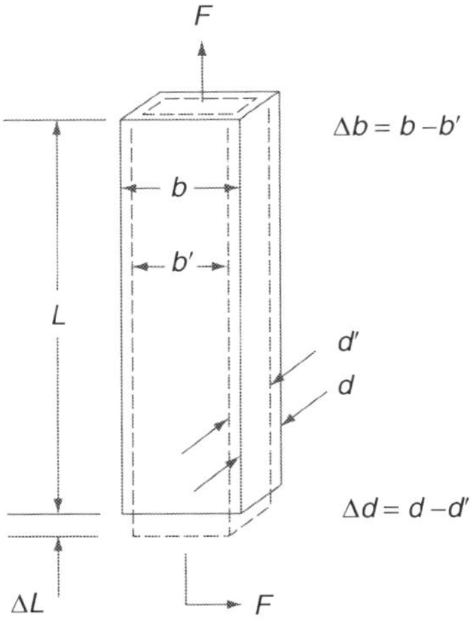

Fig.1.180 Loaded tensile bar showing dimensional change in length and width.

$$V = \frac{\text{Lateral strain}}{\text{Longitudinal strain}} = \frac{\Delta d/d}{\Delta L/L}$$

For most engineering materials, the values of v lie between 0.20 and 0.40; however, a default value of 0.35 is usually sufficient. Classically, v is between zero (no lateral contraction) and 0.5 (constant volume deformation). Table 1.16 shows typical values of v for various structural materials. Poisson's ratio is a necessary constant for the stress and deflection analysis of plastic structures such as plates, shells, and rotating discs.

Table 1.16 Typical values of Poisson's ratio.

Material	Range of Poisson's ratio
Aluminium	0.33
Carbon Steel	0.29
Rubber	0.50
Rigid thermoplastics neat	0.20–0.40
Filled or reinforced	0.10–0.40
Structural foam	0.30–0.40
Rigid thermosets neat	0.20–0.40
Filled or reinforced	0.20–0.40

1.22.12 Shear Stress

A block of material shown in Fig. 1.181 is subjected to a set of equal and opposite shearing forces, Q. If the material is imagined as an infinite number of thin layers as shown in

Fig. 1.181(b), a tendency exists for one layer of the material to slide over another to produce a shear form of deformation, or failure if the force is great enough.

The shear stress, τ, is defined as

$$\tau = \frac{\text{Shear load}}{\text{Area resisting Shear}} = \frac{Q}{A}$$

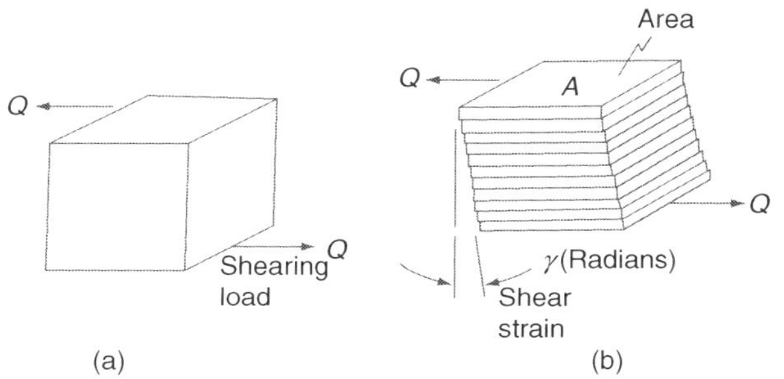

Fig. 1.181 Shear stress.

The shear stress is always tangential to the area on which it acts. The shearing strain is the angle of deformation, γ, and is measured in radians.

1.22.13 Shear Modulus

For materials that behave according to Hooke's law, the shear strain is proportional to the shear stress producing it. Thus,

$$\frac{\text{Shear stress}}{\text{Shear strain}} = \frac{\tau}{\gamma} = \text{Constant} = G$$

The constant G is called the *shear modulus* or *modulus of rigidity*, and is directly comparable to the modulus of elasticity used in direct stress applications.

1.22.14 Relating Material Constant

It was previously noted that only two material constants are required to characterise a material if one assumes the material is linearly elastic, homogeneous, and isotropic. However, three material constants have been introduced tensile modulus, E; Poisson's ratio, v; and shear modulus G. Therefore, an equation relating these three constants is needed. On the basic of elasticity principles that are beyond the scope of this manual, the following equation may be developed:

$$E/G = 2\,(1 + v)$$

This holds true for most metals and is generally applied to injection-mouldable thermoplastics. However, the designer is reminded of the inherently non-linear, anisotropic nature of most plastics, particularly fibre-reinforced and liquid crystalline materials.

1.22.15 Direct Shear

Figure 1.182 shows a typical shear strength setup used for plastics. Data obtained by this method is often reported in marketing data sheets as the shear strength of the material. In strength of the materials literature, this type of test is called *direct shear*. The reader is cautioned to use the 'shear strength' reported by this method only in similar direct shear situations. This is not a pure shear test. The test cannot be used to develop shear stress-strain curves or to determine the shear modulus because a considerable portion of the load is transferred by bending and/or compressing rather than by pure shear. In addition, the results can be depend on the susceptibility of the material to the sharpness of the load faces. When plastics are analysed in a pure shear situation or when the maximum shear stress is calculated in a complex stress environment, the use of a shear strength equal to half the tensile strength, or the above reported shear strength, is recommended, whichever is less.

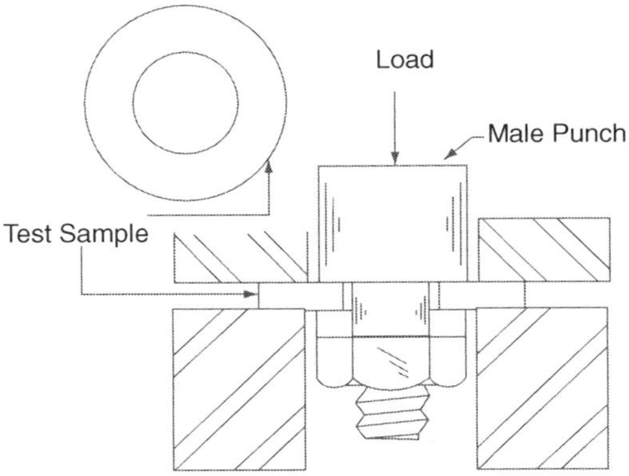

Fig.1.182 Direct shear stress test used in plastics industry.

1.22.16 True Stress

Though infrequently used, the terms 'true stress and strain' are worth noting. In Fig.1.183, the normal stress is calculated on the basis of an increasing load, F, acting over a constant area, *A*. This form of the direct stress, discussed previously, is often called 'engineering stress.' With most materials, however, a 'necking down' occurs in a critical area where failure will eventually result. If the smaller cross section, *A*, were used in place of *A*, then the calculated stress would be referred to as '*true stress*'. In addition, the direct strain discussed previously, i.e., total change in length over original length, is often called 'engineering *strain*'. The true strain would be the instantaneous deformation over the instantaneous length.

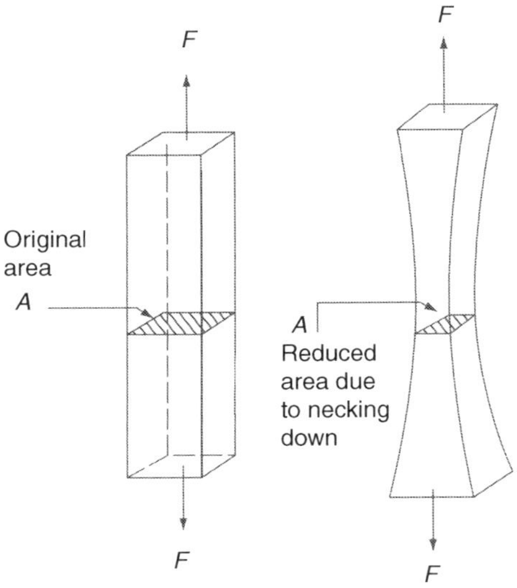

Fig.1.183 True stress.

Therefore, the shapes of a true stress-strain curve. Almost universally, however, modulus values and stress-strain curves are based on engineering stress and strain.

1.22.17 Other Measures of Strength and Modulus

For many engineering materials that are treated as linearly elastic, homogenous, and isotropic, the tensile and compression properties are considered identical. This eliminates the need to measure properties in compression. Further, if tension and compression properties are identical, there is no need to measure the properties in bending (under standard beam bending theory). However, in concession to the non-linear, anisotropic nature of most plastics, these properties, particularly flexural properties, are often reported on marketing data sheets.

1.22.18 Compression Strength and Modulus

Because of the relative simplicity of testing in tension, the elastic modulus of a material is usually measured and reported as a tension value. A material can also be loaded in compression. However, for design, the stress-strain curve for compression loading is often required.

With most elastic materials at low stress levels, the tensile and compressive stress-strain curves are nearly equivalent, as depicted by the curve in Fig.1.184. However, at higher stress levels, the compressive strain is less than the tensile strain. Unlike tensile loading, which usually results in a clear failure, stressing in compression produces a slow, indefinite yielding; this seldom leads to a failure. Therefore, the compressive strength is customarily expressed as the stress in Pa, required to deform a standard plastic specimen to a certain strain.

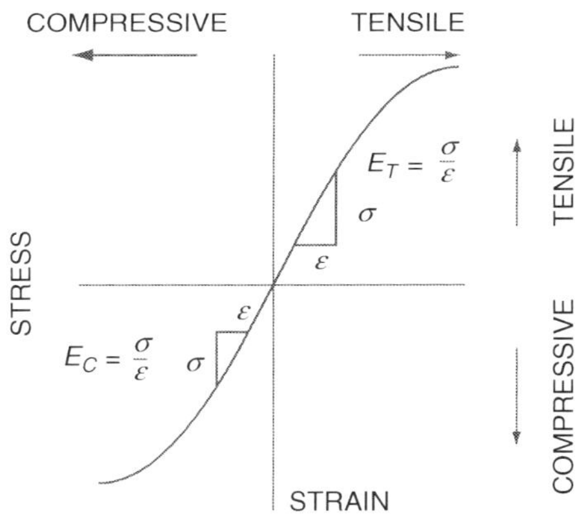

Fig. 1.184 Tensile and compressive modulus.

Compression modulus is not always reported, since defining a stress at a strain is equivalent to reporting a secant modulus. However, if a compression modulus is reported, it is generally an initial modulus.

1.22.19 Bending Strength and Modulus

If a piece of plastic or metal, most conveniently a rectangular cross section, is bent between one's figures, it is readily apparent that one surface of the material stretches in tension while the opposite surface compresses (Fig.1.185). It follows that there is a line or region of zero stress between the two surfaces, called the *neutral axis*. In simple beam bending theory, the following assumptions are made:

1. The beam is initially strong, unstressed, and symmetric.
2. The material of the beam is linearly elastic, homogeneous, and isotropic.

3. The proportional limit is not exceeded.
4. Young's modulus for the material is the same in tension and compression.
5. All deflections are small, so that planar cross sections remain planar before and after bending.

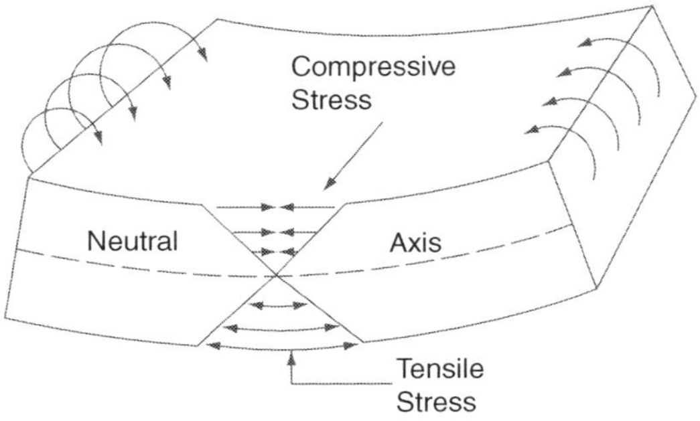

Fig. 1.185 Beam in bending.

Bending properties can be measured as shown Fig. 1.186. Using classical beam formulas and section properties, the following relationships can be derived:

Bending Stress $\sigma = 3FL/2bh^2$

Bending or Flexural Modulus $E = FL^3/4bh^3Y$

where Y is the deflection at the load point.

With these relationships, the flexural strength and flexural modulus (of elasticity) can be determined in a testing laboratory. Again, the reported flexural modulus is usually the initial modulus from the load deflection curve. Since most plastic parts must be analysed in bending, flexural values should lead to more accurate results than if the corresponding tensile values are used.

1.22.20 Rate Dependence of Mechanical Properties

The tensile and flexural data reported in marketing data sheets are measured at a particular displacement rate. Unfortunately, this rate is rarely consistent with the end-use loading environment. The same plastic material, under differing rate and/or environmental conditions, can produce different stress-strain curves. The designer must be aware of the loading rate in a particular application and request the appropriate data. Often the data are not available. Therefore, the importance of end-use testing must be kept in mind.

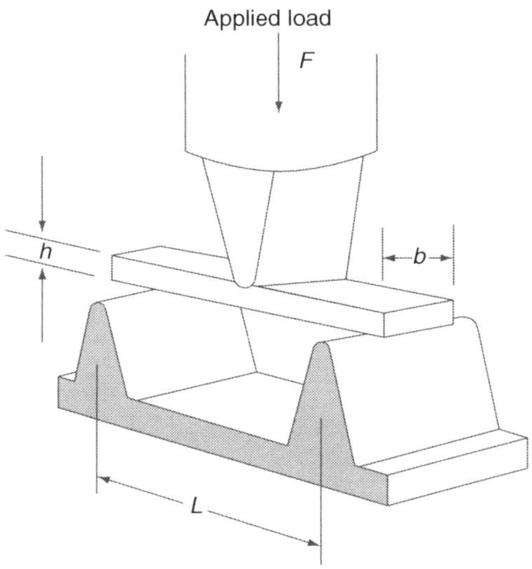

Fig. 1.186 A simple bending fixture.

1.22.21 Time Related Mechanical Properties

In the previous discussion, the given mechanical properties involved a gradually applied, short-term load. In this section, the effects of time-related loading, both long term and very short term, are considered. With high-performance thermoplastic materials, creep, impact, fatigue, and related issues are important considerations. Unfortunately, laboratory test data are not always directly applicable in estimating the structural response of actual parts. The laboratory method of applying the time - dependent load is rarely consistent with the end-use environment. Furthermore, in the actual end use, other factors are usually involved which are covered by the laboratory test conditions.

Creep properties When a part or structure is subjected to a given load, there is a corresponding predictable deformation. If the deformation continues to increase without any increase in load or stress, the material is said to be experiencing cold flow or creep. Thus, creep can be defined as increasing strain over time in the presence of a constant stress. The rate of creep for any given material depends on applied stress, temperature, and time.

In parts that are to be subjected to loads for extended periods of time where the maximum deflection is critical, the creep behaviour of a material is very important. Test samples may be loaded in tension, compression, or flexure in a constant-temperature environment. With the load constant, the deflection or strain is noted at regular intervals of hours, days, weeks, and/or months. Generally, results are obtained at four or more stress levels.

Stress-strain-time data are usually presented as creep curves of strain vs. log time. Sets of such creep curves, illustrated in Fig.1.187 can be produced by smoothing and interpolating

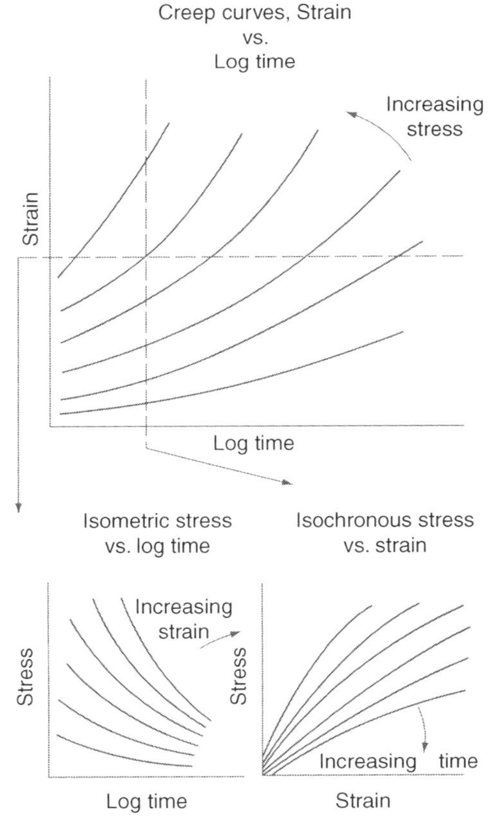

Fig.1.187 Typical presentation of creep data.

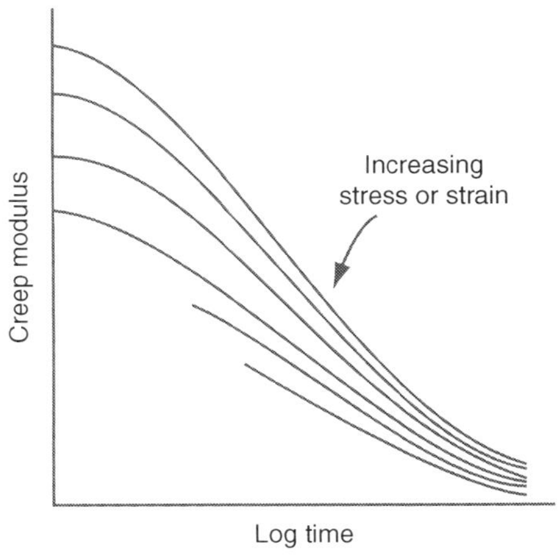

Fig.1.188 Creep modulus.

on a computer. These same data may be presented other ways to facilitate the selections may be taken through the creep curves at constant times, yielding isochronous stress-strain curves; or at constant strain, giving isometric stress vs. log time curves. These derivations are shown schematically in Fig.1.187.

In general, crystalline materials have lower creep rates than amorphous resins. Glass reinforcement generally improves the creep resistance of a plastic material.

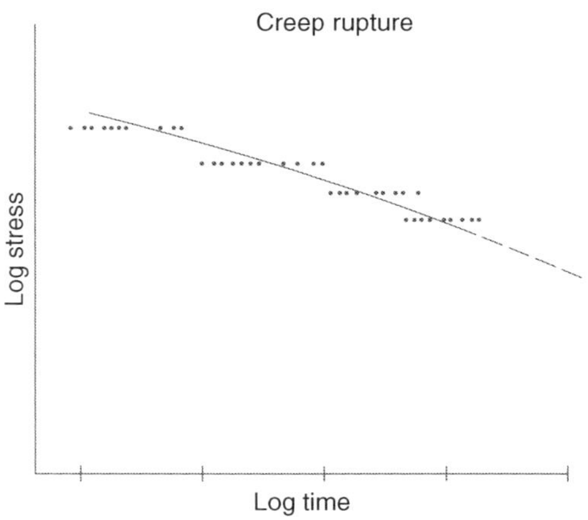

Fig.1.189 (a) Creep rupture data - a curve showing one cycle log time projection.

Apparent or Creep Modulus If the deflection of a part subject to continuous loading is calculated by using the modulus of elasticity, E, the result is likely to be inaccurate since the effects of creep have not been considered. However, if the stress level and temperature are known and the creep curves are available at the temperature, an apparent or creep modulus, E_{app} can be calculated by using the creep curves as follows:

$$E_{app} = \sigma / \varepsilon_c$$

where σ is the calculated stress level.

ε_c is the strain from the creep curves at the expected time and temperature.

This value, E_{app}, can then be used instead of E to predict the maximum deflection.

Curves of creep modulus or log creep modulus vs. log time at either constant stress or strain are often derived from the creep data and plotted as in Fig. 1.189(a). Data may also be provided as tables at constant stress and temperature at various times. Many material suppliers provide creep data in the form of creep modulus rather than by the various curves of Fig.1.188.

Creep Rupture Failure may occur in creep when a part exceeds allowable deformation or ruptures. In creep rupture failures, the fracture may be brittle or ductile with some degree of necking. Creep rupture data are obtained in the same manner as the creep data expect that higher stresses are used and time is measured to failure. The strains are sometimes recorded, but are not necessary for creep rupture. The results are generally plotted as log stress vs. log time to failure (Fig.1.189(a)).

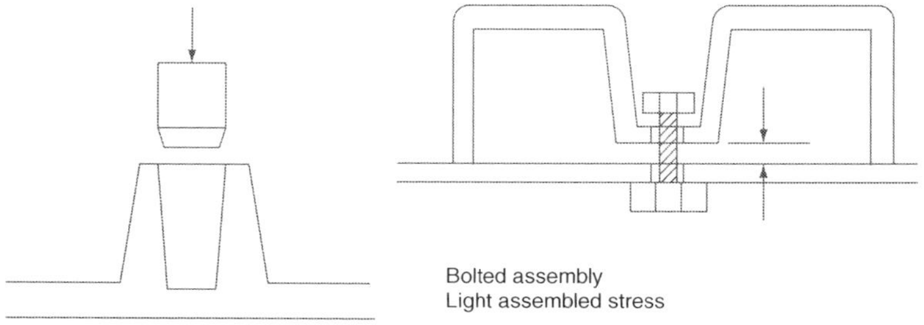

Fig. 1.189 (b) Examples of constant strain loads.

Stress Relaxation In many cases when plastic parts are assembled, they are placed into a permanently deflected condition. Examples are press-fits, bolted assemblies, and some plastic springs. In time, with the strain constant, the stress level decreases due to the same internal molecular movement that produces the creep phenomenon. This gradual decay in stress at constant strain is known as *stress relaxation*. This becomes important in applications such as bolt preload and springs, where there is a concern for the retention of load.

The relaxation can be measured by applying a fixed strain to a sample and measuring the load with time. The resulting data can then be presented as a series of curves very similar to the isometric stress curves in Fig.1.188. In addition, a relaxation modulus, similar to the creep modulus, can be derived from the relaxation data. Generally, relaxation data are not as available as creep data. However, the decrease in load due to stress relaxation can be approximated by using the creep modulus, E_{app}, calculated from the creep curves as in Fig. 1.189(a).

Plastic parts often fail due to excessive fixed strains imposed on them for extended periods of time. An example would be the splitting of a plastic tube press fit over a steel shaft. Unfortunately, there is no 'relaxation rupture' corollary to creep rupture. For purposes of initial design concept development, a strain limit of 20% of the strain at the yield or yield strength is suggested for high- elongation plastics. Likewise, 20% of the elongation at break is suggested for low-elongation, brittle materials that do not have a yield point. However, this is only a guideline for initial design. Prototype parts should be thoroughly tested at end-use conditions to confirm the design. In addition, data that suggest a higher or lower limit may be available on the specific material of interest.

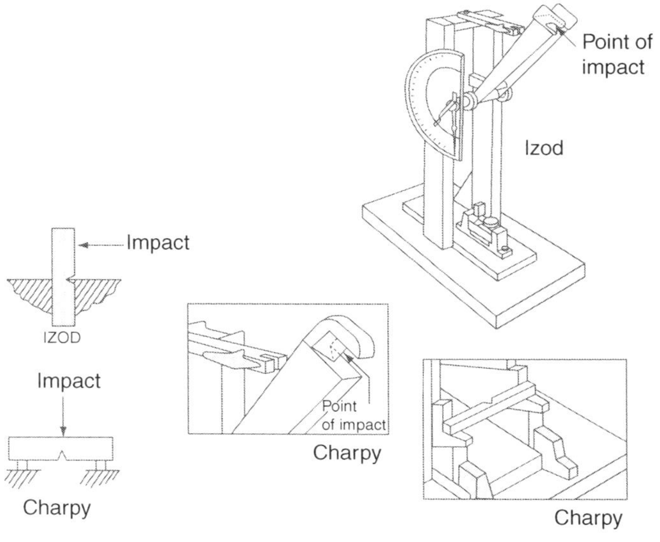

Fig.1.190 Izod and charpy impact tests.

1.22.22 Impact Loading

Whenever a part is loaded rapidly, the part is subjected to impact loading. Any moving body has kinetic energy. When the motion is stopped due to a collision, energy must be dissipated. The ability of a plastic part to absorb energy is determined by its shape, size, thickness, and the type of material. Unfortunately, the impact testing methods presently available do not provide the designer with information that can be used analytically. However, the tests are useful for comparing the relative notch sensitivity of materials or the relative impact resistance. This can be very useful in choosing a series of materials to be evaluated in an application or in rank ordering materials within a series.

Izod Impact Probably one of the most widely used tests for impact strength is the notched izod impact test. In this test a pendulum arm swung from certain height is made to impact a notched, cantilevered beam (*see* Fig.1.190). Manufacturing the test specimen, the pendulum continues to travel in the same direction, but with less energy due to the impact with the specimen. This loss of energy, measured in J/m of beam thickness, is known as the izod impact strength. This test can also be run with an unnotched specimen or with the notch reversed, in which case it is reported as unnotched or a reversed notch izod impact strength, respectively.

Charpy Impact Charpy impact is less common in the United States, but is widely used in Europe. The test is essentially identical to the izod test expect that the test specimen is a simply supported beam that is impacted midway between the supports. Like the izod test, the specimen can be notched or unnotched, and the results are reported in J/m of beam thickness.

Tensile Impact This test uses a swinging pendulum similar to that used in the izod impact test, except the sample specimen is a tensile bar that is mounted as shown in Fig. 1.191 to measure energy required to fracture it due to tensile impact loading.

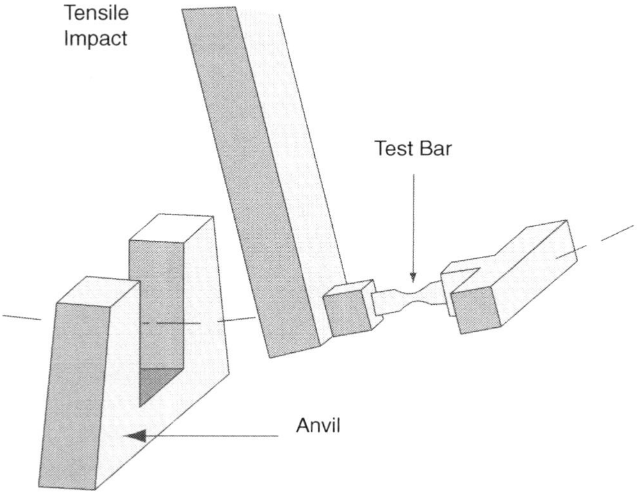

Fig. 1.191 Tensile impact.

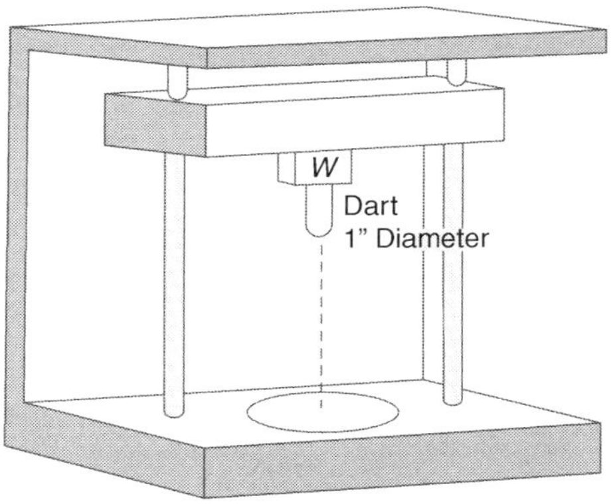

Fig.1.192 A typical dart impact apparatus.

Falling Dart Impact Test In this test, a weight is dropped onto a flat disk of the material being tested. The leading edge of the dart where it impacts the specimen has a specific diameter. Figure 1.192 is one example of a falling dart apparatus. This test is valuable for ranking materials since it tends to better represent the impact on actual parts in certain applications.

1.22.23 Fatigue Endurance

Generally, a material is subjected to fatigue when it is stressed repeatedly or in some defined cyclic manner. Examples are a snap-action plastic latch that is constantly opened and closed, a reciprocating mechanical part on a machine, a gear tooth, a bearing, any structural component subjected to vibration, and any part that is to be subjected to repeated impacts. Such cyclic loading can cause mechanical deterioration and progressive fracture of the material, leading to ultimate failure.

Typical fatigue tests are conducted on a machine which subjects a cantilever beam to reverse flexural loading cycles at different maximum stress levels. The number of cycles to failure is recorded for each stress level. The data are generally presented in a plot of log stress vs. log cycles, called an S-N curve. The cycle rate loading profile and environmental temperature should be reported with the curve. Figure 1.193 illustrated a typical S-N curve.

With thermoplastic materials there is the added complication of the heat build up, that can sometimes contribute to the actual failure. Figure 1.193 also illustrates the possible thermal failure that can in a fatigue test. Thermal failure is attributed to heat build up resulting from the frequency of the cyclic stress. Significant differences in the S-N curve can also be produced by testing at different frequencies, different mean stresses, different wave forms, and different test methods, i.e., tension rather than bending.

Although fatigue test data give some indication of the relative ability of plastic materials to survive fatigue, the designer must be aware of the above variables. The tests are run on specially prepared samples in a test environment, which ever resembles the actual loading and environment of the actual injection-moulded parts under end-use operating conditions to determine the true fatigue endurance of any part on which significant cyclic loading occurs.

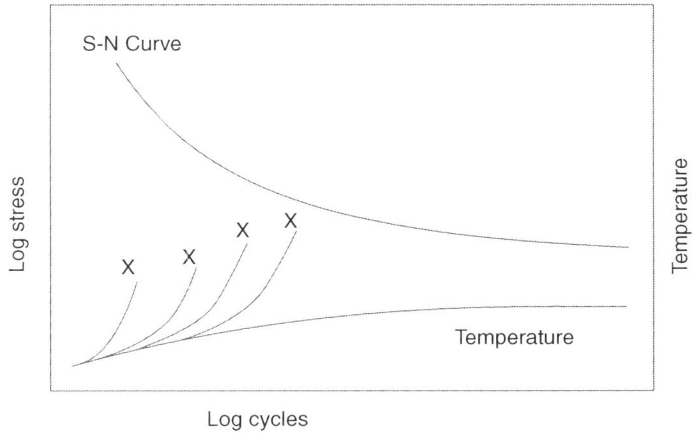

Fig.1.193 Typical S-N curve is shown along with thermal effects which sometimes occur when plastics are fatigue tested.

1.23 PRODUCT DESIGN FOR COMPOSITES

1.23.1 Introduction

The plastics industries are still in the developmental stages only even though there has been remarkable growth of composites in the past few decades. Much of the research and development work has been accomplished for aerospace and military applications, although some civilian applications during the past two decades have resulted from the initial R&D work. An analogy of the composite industry may be made with the iron age. We are in a period of rapid technological growth and transition. Many improvements in the polymer matrices are made possible by alloying and the addition of numerous reinforcing agents. Pure iron is soft, easily bent, and has little strength, but with the addition of carbon and selected alloys, iron is transformed into a very durable, useful material. New designs, applications, and industries evolved from the development of composites. Like steel, composites are having and will continue to have a pervasive, dramatic impact on civilisation.

There are a number of trends that have changed the way we view composite materials. Most of today's applications require high and often sophisticated performance. Parts are being designed to use the merits of composite properties; they are no longer only substitutes for other materials. The major drawback to more rapid adoption of composites is a lack of design technology. The design considerations for composites are more complex than those of homopolymers or metals. The viscoelasticity of a polymer matrix cannot be compared with the elasticity of structural metals. An organic composite varies with time under load, rate of loading, small changes in temperature, matrix composition, material form, reinforcement configuration, and fabrication method. Isotropic materials have a well-defined elastic and plastic stress-strain behaviour. Composites may be made isotropic, quasi-isotropic, or anisotropic depending on design requirements.

1.23.2 Advantages of Composites

The principal advantages that may be gained from the use of organic matrix composites in design are the following:

(a) Low energy costs per volume for manufacture and long term energy savings from lighter components and structures are important economic considerations.

(b) Parts may be designed to be anisotropic to exploit directional properties with specific strength and stiffness.

(c) Labour costs are reduced by automation.

(d) Fabrication processes allow the rapid manufacture of large, integrated, high-performance components.

(e) Many designs provide excellent fatigue resistance.

(f) Military and civilian applications have shown that many critical component applications can withstand a high degree of damage tolerance.

(g) Corrosion resistance in a number of hostile environments is an asset for many applications.

(h) Composites can be made with varying degrees of electrical and thermal conductivities.

(i) The variety of materials, additives, blends, alloys, and processing techniques allows for greater flexibility in the design of most components or structures.

(j) Many designs result in low scrap and post-processing operations.

1.23.3 Classification of Composites

A composite may be classified as an article containing or made up of two or more different substances. In the plastic industry, composites apply generally to structures of reinforced elements. There are five general classifications of composite materials: (i) laminar; (ii) fibre; (iii) flake; (iv) separate particles; and (v) filled. This is illustrated in Fig. 1.194.

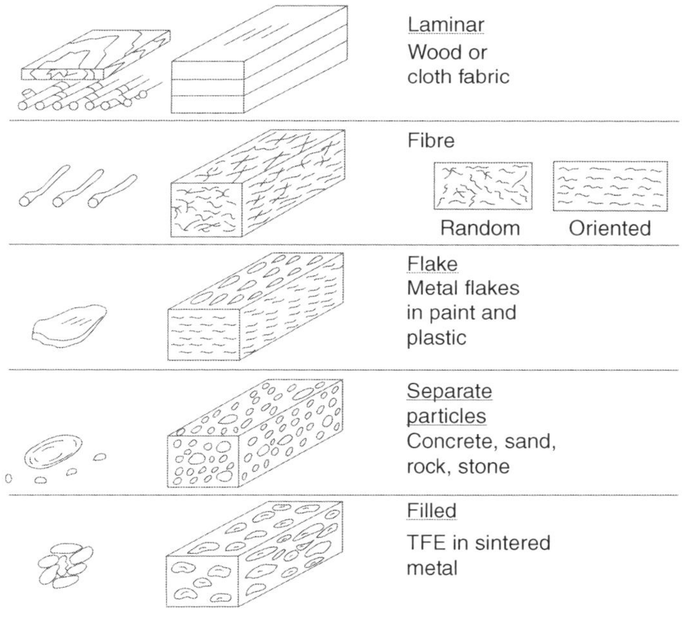

Fig. 1.194 The five general classes of composites.

1. **Laminar Composites:** This is the first composite that was made by the man. It is made of layers of different materials bonded or fastened together with different plastic resins. In plywoods, the layers are generally of the same type of wood, but the orientation of the layers differs. Plywood is made with the grain of alternating plies at right angles. The wood layers are bonded together with an adhesive (generally plastic) to make a solid, rigid article. Plywood is made from many species of hardwoods and softwoods,

ranging chiefly from 6 mm to 22 mm in thickness, and with plastic adhesives that differ in moisture resistance. Other composite materials made of wood consist of resin-impregnated woods, softboard, hardboard, and particle board.

Laminar composites in the plastic field consist of layers of resin-impregnated fabrics, paper, glass cloth, etc., which possess high strength to-weight ratios. If highly directional materials like wood, woven fabrics, or bundles or layers of continuous filaments are employed, the resulting laminate is highly directional in its mechanical properties.

2. **Fibre Composites:** Practically every type of plastic, rubber, elastomer, and ceramic has been reinforced with fibres. In selecting a reinforced fibre for a composite, the following should be considered: (i) fibre orientation; (ii) length, shape, and composition of the fibres; (iii) mechanical properties of the matrix; and (iv) integrity of the bond between fibres and matrix.

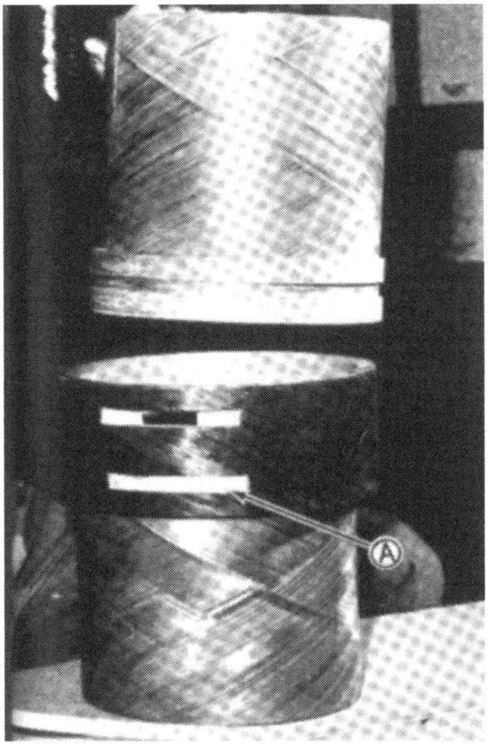

Fig. 1.195 A fibreglass epoxy filament wound water well casing. The double-keyed coupling can be quickly joined without adhesives. A flexible key strip (A) slides into the conduit formed by matching grooves in the pipe end and the coupling (Courtesy Westinghouse).

Fibres are arranged in a random, unidirectional, and bidirectional pattern. Continuous filaments are generally used in the filament winding process. (Figure 1.195) illustrates a filament wound fibreglass epoxy water well casing. The casing uses a double-keyed coupling that requires no adhesives, or screws for attachment. The slot for the keyed coupling is parallel to the helically wound filaments, which also contributes to the retention of pipe tensile strength by leaving a maximum number of uncut filaments. The key strips are made from acetal resin, aluminium or copper and copper alloys. The aluminium pigments reflect and produce brilliant blue white highlights. The copper based pigments produce a bronze or gold colour.

Numerous problems have been encountered in using metallic pigments in plastic materials. In injection moulding, the orientation of the flake in the plastic material is upset during the moulding process as flow and knit lines, producing an unsightly appearance. This condition can be improved by using a large fan type gate.

3. **Flake Composite:** A very pleasing decorative effect can be obtained by using small metallic flakes in paints and plastics. The flakes produce a brilliant metallic highlight.

The highly reflective flake pigments are particles of either aluminium or copper and copper alloys. The aluminium pigments reflect and produce brilliant blue-white highlights. The copper-based pigments produce a bronze or gold colour.

Numerous problems have been encountered in using metallic pigments in plastic materials. In injection moulding, the orientation of the flake in the plastic material is upset during the moulding process at flow and knit lines, producing an unsightly appearance. This condition can be improved by using a large fan-type gate.

4. **Separate Particle Composite:** The oldest most widely used particle composite is concrete. The gravel and sand are the particles and cement is the matrix. The cement and water form a paste that hardens by chemical reaction into a strong stone-like mass. No more cement paste is used than is necessary to coat all the aggregate surfaces and fill all the voids. The quality of the paste formed by the cement and water largely determines the character of the concrete.

   In the plastic industry, decorative panels for store fronts are made from sand and polished pebbles with polyester resin as the matrix. The sand is thoroughly mixed with the catalyzed polyester resin and cast into a slab mould. The polished stone pebbles are partially imbedded on the surface of the sand resin mixture. After the polyester resin composite mixture has polymerised and become, hard, it is removed from the mould. The cast slab is coated with a silicone resin to protect the composite from the outside elements. This type of slab composite is used only for a decorative item for building fronts.

5. **Filled Composites:** A filled composite is an open matrix or sketal structure filled with another material. It might be described as similar to filling the voids in a sponge with a solid material such as plaster. A filled honeycomb is another example.

   Materials that are porous and can be filled are metal castings, powder metal parts, ceramics, graphite, and foams. Powdered metal parts can be impregnated with a PTFF fluoroplastic to provide a good combination of bearing properties. The PTFE acts as a lubricating medium. Porous castings of aluminium can be sealed with liquid plastic resin such as polyester or epoxy. The liquid resin is forced into the porous metal by air pressure or a vacuum. The liquid resin fills all of the voids and is polymerised to a 100% solid. Almost all of the known plastics today can be filled, reinforced, or both filled and reinforced with other materials. A filler is a material added to a plastic resin in order to obtain desirable mechanical, chemical, or electrical properties not possessed by the plastic resin itself. Reinforcing fillers or fibres are added to plastic resins to enhance mechanical strengths. Table 1.17 shows the various fillers and reinforcements that are used with plastic resins. A great number of fillers and reinforced materials are used in the plastic industry. The following are the principal fillers and reinforcements:

   (a) **Asbestos:** Asbestos is a naturally occurring fibrous hydrated magnesium silicate. This material is used as a reinforcement in thermosetting resins and laminates. A finer grade is used as a filler in polyethylene, polypropylene, nylon, and vinyl. In the vinyl field, it is used mainly in vinyl asbestos floor tiles. Asbestos increases the heat resistance and flame resistance of the material and decreases the water absorption and shrinkage. Asbestos lowers the flexural and tensile strength and raises the specific gravity. Asbestos is exceptionally good in filling out in a mould, but has very poor colour properties.

(b) **Mica:** The advantages of this material as a filler is its excellent electrical properties. Mica is used in finely powdered form and does not wet readily in plastic resins. The principal use of mica is to make phenolic compounds for electrical applications requiring low-loss characteristics. Mica reduces the mechanical strength, and the specific gravity is increased.

Asbestos and mica are added to thermosetting resins to impart higher heat resistance and better electrical properties, rather than to reinforce the plastic resin. Fibreglass performs the same functions and reinforces as well.

(c) **Glass fibre Reinforcements:** This family or reinforcing material is used in both thermoplastic and thermosetting resins. It is used as a continuous filament, chopped into short lengths, woven into cloth, or made into a mat. Glass fibres are used in plastic resins to increase rigidity, toughness, dimensional stability, load-bearing properties and heat resistance.

(d) **Calcium Silicate:** This is a naturally occurring white powder that is used as a reinforcing agent in rubber and as filler in paints and plastic resins. This filler is sometimes referred to as Wollastonite. Calcium silicate is used as filler in epoxies, phenolics, PVC and polyolefins. It imparts all around good physical properties and provides a smooth white surface.

(e) **Diatomaceous Silica:** This material is essentially amorphous silica of organic origin, i.e., the fossil remains of microscopic plants known as *diatoms*. This material is very fine in texture and very light in weight. It has a high absorptive capacity, is chemically inert, and is heat stable at temperatures above those used for plastic resins. It is available in several grades of different particle size, distribution, and fineness. It mixes and bonds well with nearly all plastic resins, and gives excellent and uniform surface finish and fine gloss.

(f) **Talc:** This filler is magnesium silicate. The structure is closely related to that of mica. It is a soft friable mineral of fine colloidal particles with a soapy feel. Talc imparts greater resistance to mechanical stresses, arising from temperature differences, and may also prevent crazing. Talc is used for cosmetics, paper coatings, and as a filler for paints and plastics.

(g) **Clay:** This filler is known as kaolinite, china clay, or aluminium silicate. Clay helps to maintain uniformity of the fibreglass resin mixture in polyester resins. When this filler is used in plastic resins, it helps to control shrinkage, increase resistance to water, and improves mouldability and surface gloss. This filler is used in epoxy potting compounds, reinforced plastics, and PVC resins.

(h) **Calcium Carbonate:** Calcium carbonate is ground limestone. There are two categories of calcium carbonates, natural ground limestone and precipitated grades of calcium limestone. The precipitated grade of calcium carbonate is used mostly in plastic resins. This filler is used where a smooth surface is desired. It is used mostly in PVC compounds and in polyester resins.

(i) **Silica:** Silica (silicon dioxide) is a family of materials. Natural silica occurs chiefly in the form of quartz, sand, flint, opal, agate, etc. In the powdered form, natural silica is used as filler in phenolic resins for high-heat applications. Synthetic silica is made by treating silicon compounds with a flame or hot fumes, producing amorphous colloidal particles. It has the appearance of a white fluffy powder. Synthetic silica is used in polyester resins to increase the viscosity and impart thixotropy, without materially raising the percentage of solids in the compound. It is used in epoxy resins as a flatting agent to prevent the resin from running when it is placed on a slanted or curved surface. Paints and varnishes made from alkyd, epoxy, acrylic and urethane use synthetic silicas to improve thixotropy and flatting.

(j) **Molybdenum Disulfide:** This is a white, shiny crystalline material used as a lubricant in many plastic resins. It lowers the coefficient of friction.

(k) **Carbon Black:** Carbon black is used as a filler and pigment in several plastic resins. In polyethylene, carbon black acts as a cross-linking agent. It is used in colouring black phenolic material and is restricted to those applications where the moulded article must not bleed in contact with solvents. An example would be phenolic moulded bottle caps. Carbon black helps in weather resistance, produces greater mouldability, promotes electrical conductivity, and improves lubricity. It helps most plastic materials in the resistance to ultraviolet light rays.

(l) **Graphite:** This material is a soft, greasy form of carbon. It is used as lubricating filler for plastic resins. Graphite in the whisker form is used as reinforcement.

(m) **Wood Flour:** Wood flour is derived from resin free softwoods such as Douglas fir, pine, and spruce. Wood flour is finely pulverised dried wood and is used mostly as a filler in thermosetting moulding compounds. The wood flour enhances impact strength, cuts cost, and provides greater shrinkage control in moulding. It gives excellent mouldability and fair to good physical properties. It does absorb moisture, as do other cellulosic fillers.

(n) **Cotton Flock:** This filler is finely divided cotton fibres. It is used to give higher impact strength to thermosetting plastic moulding compounds. White bleached flock is used in melamine and urea resins. Fairly dark flock is used in phenolic brown or black materials. The water absorption and electrical properties are fair.

(o) **Macerated Fabrics:** This type of filler is produced by cutting various types of clean cotton cloth into small pieces. Chopped cotton cloth is used in phenolic moulding materials to raise the impact strength and lower the mould shrinkage. It has a bulk factor four times greater than wood-flour-filled material and cannot be readily preformed. The water absorption is high, and the electrical properties are poor. Chopped cotton cord is used as a filler and yields the highest impact characteristics for this type of moulding compound.

(p) **Shell Flour:** Shell flour is obtained by grinding walnut, pecan, peanut, or coconut shells. It is primarily used in thermosetting moulding compounds. These fillers lack the fibrous structure of wood flour and consequently have lower physical properties in the moulded plastic part. They do provide a high luster and gloss to the moulded plastic part.

(q) **Synthetic Organic Fibres:** Nylon in the form of flock, filament, and chopped fabric is used as a filler. Chopped fibres and filaments of polyester and polyacrylonitrile are used to reinforce DAP resins.

(r) **Miscellaneous Fillers and Chemicals:** There are many chemical additives that are used with plastic resin. Some of these additives are: plasticisers; pigments; oxidation stabilisers; antistatic compounds; internal lubricants; ultraviolet light stabilisers; and flame retardant compounds.

Table 1.17 Basic types of fillers and reinforcements used in plastic resins (*Courtesy : Materials in Design Engineering*).

	Asbestos	Glass fibres	Glass cloth	Wood flour	Cotton flock	Chopped cotton cloth	Mineral fillers	Carbon block	Graphite	Molybdenum disulfide	Fluoro-carbon powder	Talc	Metal powders or fibres	Ceramic fibres
Thermoplastics														
ABS		x						x						
Acetals		x						x	x	x	x			
Acrylics			x											
Cellulosics									x		x			
Chlorinated polyether		x							x					
Fluorocarbons										x				
PTFE		x							x	x				
PFEP		x	x											
PTFCE		x												
PVF_2								x						
Nylons	x	x						x	x	x				
Polycarbonates		x						x			x			
Polyethylene		x						x				x		
Polyimides			x				x							
PPO		x						x	x					
PPO modified		x					x	x						
Polypropylene	x	x		x				x				x		
Polystyrene		x								x	x			
Polysulfones		x						x						
Polyurethanes		x									x			

(Contd.)

Table 1.17 *(Contd.)*

	Asbestos	Glass fibres	Glass cloth	Wood flour	Cotton flock	Chopped cotton cloth	Mineral fillers	Carbon block	Graphite	Molybdenum disulfide	Fluoro-carbon powder	Talc	Metal powders or fibres	Ceramic fibres
Vinyls Flexible							x	x						
Rigid		x					x							
ABS/polycarbonate Alloy														
ABS / PVC alloy														
PVC / acrylic alloy														
Thermosets														
Alkyds		x					x							
DAP	x	x	x				x							
Epoxies	x	x	x				x	x			x		x	x
Melamines	x	x	x	x	x	x					x			
Phenolics	x	x	x	x	x	x	x			x	x			
Polyesters	x	x	x				x					x	x	
Ureas				x	x									
Urethanes		x												

1.23.4 Basic Design Practise

The major emphasis in composite design and processing has been on automation, energy reduction, lowering scrap rate, increasing productivity, lowering labour costs, and improving reliability and processing techniques. Many of these concepts are being met by changes in process technology. It has been estimated that flexible manufacturing systems will account for more than half of all U.S. production by 1990. Increased utilisation of computers will result in continued gains in productivity, product quality, and cost improvement. Programmable robots should be used to relieve monotony, again repetitiveness, and protect workers from high noise and hazardous processes.

Great strides have been made with the introduction and use of CAD/CAM/CAE systems. The designer is able to use a computer in the design, engineering, and manufacturing of composite products. The entire production cycle may be studied using modelling and simulation. Some of the high-cost elements such as material handling, tooling, tool maintenance, raw material costs, and scrap losses may be evaluated and reviewed before production begins.

The data base may contain valuable information on the micromechanics and macro mechanics of the matrix material selected. The detailed effects of temperature, stress, strain, time, and environmental exposure on composite stiffness, strength, and fracture toughness can be generated. The system may alert the user that a design is outside the parameters of the material or process selected. Interaction with the computer model and personnel will allow for accurate computations and design requirements. Productive models and computer-aided design and manufacture can play an active role in achieving polymer composite reliability.

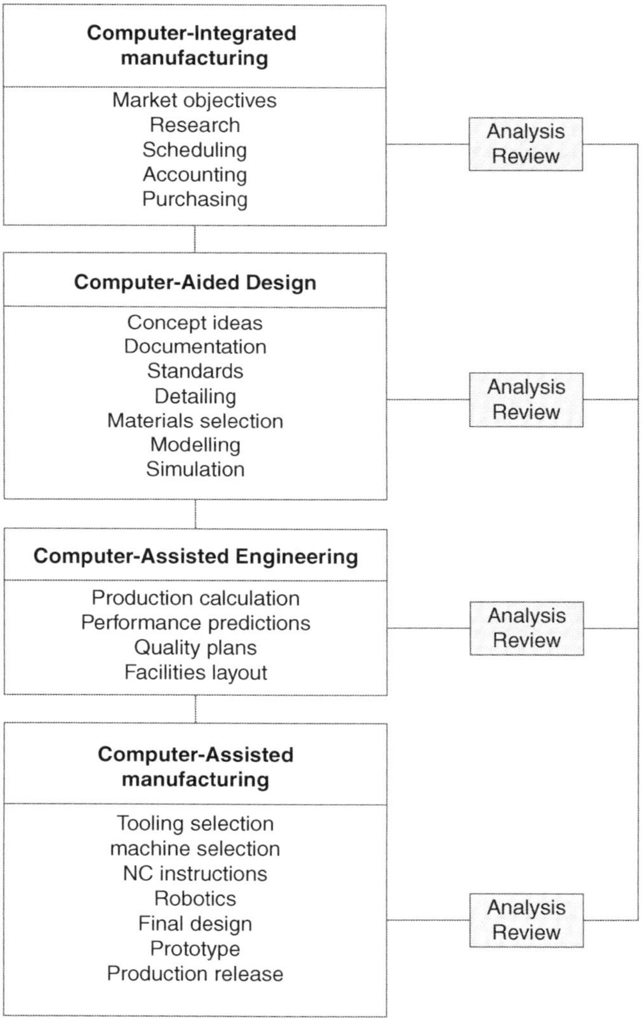

Fig. 1.196 CIM, CAD, CAE and CAM in the design production cycle.

The inter-relationship of the complex design process is shown in Fig. 1.196. This coherent design process may vary, with many functions overlapping. The order and sequence of events may vary at any phase in the development. Computer-Integrated Manufacturing (CIM) is a concept where all manufacturing processes are integrated, allowing designers, engineers, technicians, accountants, and others access to the same data base. The primary objective of CIM is to develop a cohesive digital data base that integrates the functions of manufacturing, design, and business operations. As a result, less human intervention is required on the shop floor. Part of this integration implies there would be benefits from flexible automation or manufacturing and savings in terms of reduced down time, labour costs, just-in-time (JIT) manufacturing or zero-inventory, quick changeover for batch manufacturing, and allowance for design changes.

1. Computer-aided design (CAD) is that part of the system that aids or assists in the creation, modification, and display of a design.
2. Computer-aided engineering (CAE) is that part of the system that analyzes the design and calculates the performance predictions.
3. Computer-aided manufacturing (CAM) is that part of the system that generates the manufacturing oriented data. CAM may involve production programming, robotic interfacing, quality control, and plant operations.

It is important to note that the design production cycle is an open-loop system with data being shared. A review and analysis connection is shown between each phase. This review procedure is conducted to confirm that all variables and constraints are considered in meeting the product objectives and requirements. The final design review is recommended prior to commitment to manufacture.

The design of composite parts involves three major considerations: (i) material, (ii) production, and (iii) design.

1.23.5 Material Considerations

In the past, the design was changed to meet the material limitations or properties. Today, materials must be selected with the right properties to meet the design, economic, and service conditions. This is not always an easy task. It must be remembered that most matrix compounds are manufactured with a wide range of formulations and are available in a number of forms. Some are reinforced with particles or short fibres and sold in pellet form for injection moulding, while others are available as pre-impregnated (resin) fabrics, fibres, tapes, or compounds. Material form is usually associated with a specific production technique.

It cannot be assumed that the information obtained from data sheets or the manufacturer is adequate or predictive of matrix performance. Much of the data is based on laboratory-controlled evaluation. It is sometimes difficult to compare proprietary data from several different suppliers. Generally, values are derived from statistical models requiring the evaluation of numerous laboratory test coupons.

The following three requirements may be used as systematic screening methods of material selection for a specific design: (i) functional property factors, (ii) processing parameters, and (iii) economics. This does not imply that there is one best method to screen materials or that it is a step-by-step procedure. During analysis and selection, new information, technologies, or problems may require repetition of earlier steps.

Systematic methods are used with the aid of the computer. Failure analysis, cost-versus-property indices, and weighted property indices are familiar examples. In weighted property indices, each parameter is assigned a value depending on importance. The performance of materials may then be compared. Computer models can predict and anticipate most of the ways a material can fail. This method is called *failure analysis*. The complexity of composites and the combination of materials and processes compound the difficulty of failure analysis.

1.23.6 Functional Property Factors

One of the first requirements is to list (in quantitative terms) the functional property factors that the part is expected to tolerate. Examples of these properties are tensile strength, creep, thermal expansion, permeability, and impact strength. The following product property parameters are given as an example:

1. Relative density to be less than 2.6 g/cm^3.
2. Withstand repeated deflection of not more than 0.20 mm at 43°C.
3. Have a tensile strength of more than 1,400 kg/cm^2.
4. Have a linear expansion of less than 10×10^{-6} cm/cm/°C.
5. Have a paintable, smooth surface.
6. Be able to withstand operating temperatures of more than 70°C.
7. Withstand repeated exposure in a hostile environment of petroleum fuel and saltwater.
8. Be able to withstand an impact of more than 0.02 J/m.
9. Withstand a service life of more than 10 years.
10. Meet UL standard 94 flammability of HB of not more than 38mm /min in 3.0 – 3.1 mm thick.

One question that should be asked is: "Do analogous applications exist?" A fully documented list of performance requirements may be available for a similar product.

The consumer is probably most aware of a product's appearance, utility, and reliability and is more likely to be interested in service life, wear resistance, ease of operation, ease of repair, and cost.

The composite designer must be concerned with structural and environmental questions at this phase. Everyone involved in the design operation must know where and how the matrix and finished composite component will be used.

The glass transition, melting, and crystallisation of most matrices are reversible phenomena, while thermal degradation and cross linking are not. There are limitations on the maximum service temperature to which organic matrix composites may be exposed.

It is the matrix that degrades and results in composite failure. Boron, graphite, carbon, and most ceramic and metallic fibres have high temperature resistance. Polyimide-boron fibre composites have high temperature resistance and high strength. Polyimide-graphite fibre composites can compete with metals in strength and achieve a significant weight saving at service temperatures up to 316°C. Epoxies in many epoxy-graphite fibre composites do not resist fire well. When exposed to excessive heat the epoxy matrix is damaged and releases the adhesion to the graphite fibres.

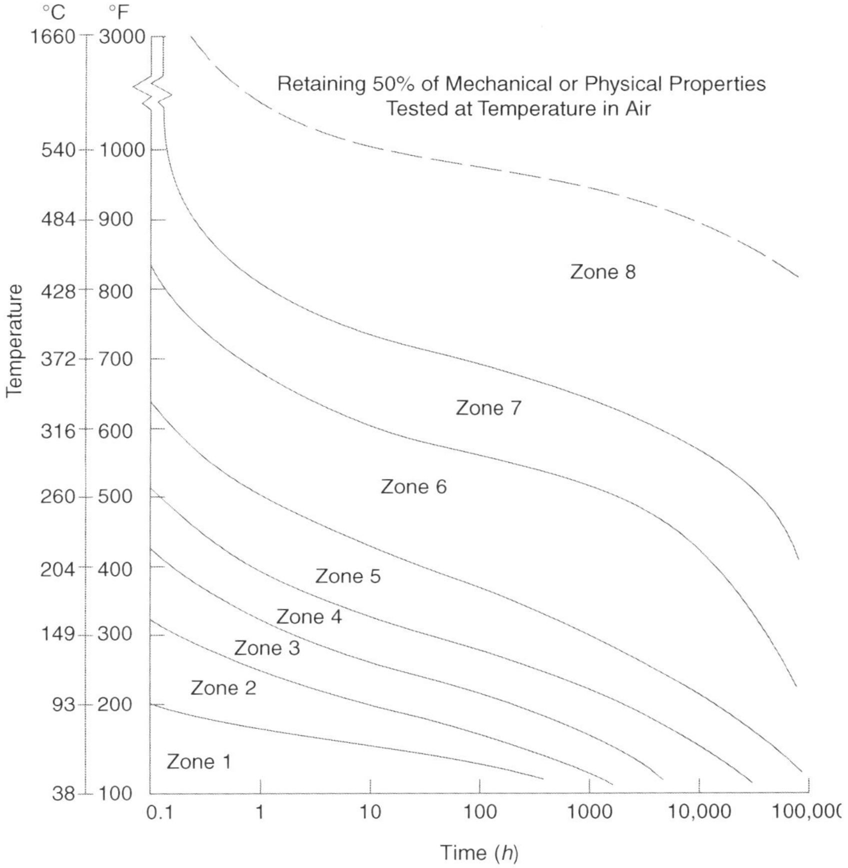

Fig. 1.197 How plastics perform on the basis of temperature and time. *(D.V. Rosato. Plastics World. p.30 (Mar.1968).*

If the polymer matrix is to be exposed to intense heat or flame, only those materials that will not melt or seriously deteriorate should be evaluated. Boron powder may be added to the matrix to stabilise the char that forms in thermal oxidation. Other flame-resistant additives or an ablative matrix may be selected. The cryogenic temperatures of space or other environments are opposite thermal extremes. Radiation can cause cross-linking in some thermoplastic materials with a resulting change in properties. Weathering affects colour, finish, transparency and other properties.

If the component is for under-the-hood applications in automobiles, it must also be able to resist a chemical environment. Petroleum liquids, fumes, and salt solutions from the roadway may attack the matrix. Some polymers simply swell while others dissolve.

The laminar composites used for kitchen countertops and many furniture surfaces must also withstand chemical attack. Products requiring selective permeability and chemical resistance may select multilayered barrier films and packages.

Moisture may be absorbed into the matrix, resulting in deterioration and weakening of the reinforcement bond. Epoxies and polyimides are subject to moisture absorption which results in a loss of strength at elevated temperatures. Any exposure of reinforcement fibres greatly accelerates this potential problem. Allowance must be made for humidity and attraction of dust and dirt because electrical arcing over the surface may result. Surfaces or machined edges and holes may require a protective coating to prevent moisture infiltration or wicking. Carbon, graphite, and metal or metal-coated fillers and reinforcements may be beneficial in electromagnetic interference (EMI) shielding applications.

All the candidate materials for consideration can be subjected to the design parameters entered into the computer. This should yield a list of the possible matrix candidates which includes numerous matrix alloys and blends. Proprietary compounds or special mixtures of additives may be recommended. Do not be tempted to enter the cost of materials into your consideration at this phase. Materials with marginal performance properties and low cost should still be considered. Redesign may compensate or bring the part into line with specified requirements. Expensive materials can remain as possible candidates depending on the processing parameters and the economics of assembly and finishing.

At this point we should narrow the number of material choices to only a few possibilities (preferably less than six).

1.23.7 Processing Parameters

The next phase in the selection and systematic screening of a candidate matrix would include processing parameters. This task must consider many variables, including (i) shape of the product, (ii) tooling, (iii) number to be produced (quantity), (iv) rate of production, (v) capital investment for new equipment or technologies, (vi) type and/or orientation of reinforcement, (vii) required performance reliability and quality and (viii) rheological (viscosity, flow) considerations of the matrix.

There are a number of processing techniques used by industry to produce composite components and structures. Most of these processing techniques are listed in Table 1.18.

Table 1.18 Principle of composite processing techniques.

Process	Remarks
Autoclave	Modification of vacuum and pressure bag; low production; low void content; dense parts; limited to autoclave size; wet
Blow moulding	Mainly closed-mould process; multilayers; short fibres or particulates; high volume; small to medium-sized products
Casting (simple)	Open-mould process; low production; little control of reinforcement orientation; monomers or polymers
Compression moulding	Closed-mould process; preforms available; some control on orientation; dense products
Expanding	Mainly closed-mould process; low to high production rates; limited control of reinforcement orientation; small to large products; monomers or polymers

(Contd.)

Table 1.18 *(Contd.)*

Process	Remarks
Extrusion	Closed-mould process; continuous lengths; multiple layers; continuous long fibres possible; preforms possible; some control on orientation
Filament winding	Mainly open-mould process; low production; control of orientation; wet
Hand lay-up	Open-mould process; low production; control of reinforcement orientation; large parts; wet
Injection moulding	Closed-mould process; short fibres or particulates; little control on reinforcement orientation; small to medium-sized products
Laminating (continuous)	Mainly closed-mould process; medium to high production; control of reinforcement orientation; small to large products; continuous; monomers or polymers
Matched-die moulding	Closed-mould process; low to high production; some control of orientation; preforms; medium-sized products; monomers or polymers
Mechanical forming	Mainly closed-mould process; medium to high production; preforms; some control of reinforcement orientation
Pressure-bag moulding	Open-mould process; low production; control of reinforcement orientation; performs; wet
Pulforming	Mainly closed-mould process; continuous; some control of reinforcement orientation; preforms; monomers or polymers
Pultrusion	Mainly closed-mould process; continuous; some control of orientation; preforms; monomers or polymers
Reaction injection moulding	Closed-mould process; small to medium sized products; medium production; monomers or polymers
Rotational casting (centrifugal)	Open-mould process; low production; little control of reinforcement orientation; small to large products; powders and wet
Spray-up	Open-mould process; low production; little control of reinforcement orientation; preforms; wet
Thermoforming	Mainly open-mould process; preforms; medium to high production; little control of reinforcement orientation; mostly short fibres or particulates
Transfer moulding	Closed-mould process; high production; dense parts; little control of reinforcement orientation; small to medium products; short fibres or particulates; monomers or polymers
Vacuum-bag moulding	Open-mould process; low production; control of reinforcement orientation; preforms; wet

1. **Shape of Product:** The physical shape of the product may narrow the matrix choice to only a few. For example, it may not be practical to select polyolefin composites for a marine hull, but they may be considered for septic tanks. Epoxies and polyesters are considered the workhorses of the composites industry. They are used in the production of boat hulls, aircraft sections, helicopter blades, drive trains, and other structural parts. Some composite materials are more easily produced into small intricate, and complex shapes than others.

2. **Tooling:** In conjunction with shape, tooling must be considered. It must be ascertained if it is practical to shape the matrix on conventional tooling. Polyamide with short fibres or particulates is successfully moulded into small parts in automobiles, power tool casings, and components. Heavily filled and reinforced composite materials may require special precautions and preparations in tooling.
3. **Quantity:** If a large quantity of parts are to be produced, some materials may not be practical. Some are not easily processed, or a great deal of reinforcement may be needed to obtain the required mechanical properties. These additives may result in moulding problems. In addition, the increased viscosity may require a different material or processing technique to be selected.
4. **Rate of Production:** The rate of production may require that only fully polymerised thermoplastic materials be used; for example, perhaps only those materials that lend themselves to full automation can be selected. This would eliminate most wet-type processes. Monomeric materials are often used where larger shapes or lower production volume is needed.
5. **Capital Investment:** Shape, tooling, quantity, and production rate criteria may determine the selection of a material that requires additional capital investment for new equipment, such as jigs, fixtures, or curing racks. The physical space of producing large composite structures in a controlled atmosphere such as an autoclave may result in consideration of different materials and/or processes. Many moulders specialise in one or two moulding techniques which limits the potential materials selection.
6. **Reinforcements:** The type and orientation of reinforcement may not be available in the desired matrix formulation. Fibre- and particulate-reinforced materials are available in a number of forms, from pellets to preforms. If the product requires specified fibre orientation, a monomeric preform or impregnating system may be needed. The use of LRIM over RIM processes permits heavy reinforcements and mass production. The surface condition of the reinforcement and matrix adhesion can be critical to the production of composites of high quality. Carbon-reinforced composites are brittle and have no yield behaviour. In addition, the matrix expands more than other materials; consequently, there may be thermally induced stresses. Compatibility must be considered any time several different materials are combined. Thermal expansion and galvanic corrosion and electrolysis may present a problem with metallic reinforcements or assemblies.
7. **Performance Reliability and Quality:** A matrix candidate with minimal performance characteristics may not be the best choice if reliability and high quality are important. Reliability of a proposed composite material is difficult to measure because it is dependent not only on material properties but on design and processing. The moisture sensitivity of some E-glass composites may be overcome by the proper choice of resin system and ratio of curing agent. Post-processing surface treatments may be required. Heavily reinforced or cellular composites sometimes result in characteristic flow marks, swirls, and weld lines. Matrix, design, and processing greatly influence the appearance of the product. Moisture and corrosion of the composites can result in notable decrease in the strength and thermal resistance of the matrix. Moisture acts as a plasticiser in the epoxy and polyimide polymers.

8. **Rheological:** Closely associated with the material form and production process are the rheological considerations of the matrix. Reinforced fluoroplastics may have all the desired performance properties but some are not melt processible. Heavily filled, viscous materials may not be easily forced into the mould cavity, resulting in poor surface quality. Mould temperatures, melt temperature, and flow speed (injection) all have a significant effect on the gloss of moulded parts. Rapid filling of a mould cavity also minimises fibre orientation and enhances weldline integrity. Thixotropic additives may greatly aid processing because the matrix is gel-like at rest but fluid when agitated. The properties of a matrix affect the wetting, reaction, compatibility, and stress transfer to the reinforcements.

1.23.8 Economics

The final phase in material selection is to consider the economics. For many, cost may be the most important single factor in selecting suitable materials for the composite product. Some matrix materials or reinforcements with the most desirable properties may be too expensive to market. For composite products to compete with other materials in a variety of applications, companies must be able to sell consumer products at a profit. Better education, planning, design, and the use of newer technological operations, including automatic fabrication of composite parts, should improve productivity, quality, reliability, and profitability. Many wet or open moulding operations are labour intensive and cannot hope to compete with companies that have increased production of composite parts in automated facilities. The CAD/CAM/CAE/CIM systems are essential for cost-effective, large-scale production. In addition to the development and production of superior composite parts, these systems may reduce materials handling and inventory and maximise utilisation of equipment and personnel.

There are a number of methods to enhance the economic attractiveness of the matrix. It is possible to change or optimise the micromechanics of the matrix material. Additives may increase the desired property. Remember that a change in one property may result in a change in several others. For example, the addition of glass fillers may increase the chemical and thermal resistances but reduce toughness and rheological properties. Diluting expensive matrices with fillers or other additives may provide a cost advantage and improve or maintain performance properties.

Cost is often based on the production method and the number of items to be produced. For example, a one-piece seamless gasoline tank may be rotationally cast or blow moulded. The latter process uses more costly equipment but can produce the tanks more quickly, thus reducing costs. A large storage tank may be produced at less cost by rotational casting than by blow moulding. To amortise the tooling cost, volume sales are needed. Some processing operations may require a special atmosphere or protection for personnel. One material may be more costly because it is more difficult to machine, fabricate, or finish. A comparison of processing and economic factors is shown in Table 1.19. It should be obvious that equipment and tooling costs will vary depending on part size, performance needs, and complexity of design.

Composite materials are expensive when compared with other materials on a per volume or weight (mass) basis. On a cost per kilogram basis, the matrix may cost ten times more than steel; yet on a volume basis, some are lower in cost than steel.

Table 1.19 Economic factors associated with different processes.

Production method	Economic minimum	Production rate	Equipment cost	Tooling cost
Autoclave	100–1,000	Low	High	Low
Bag moulding	100–1,000	Low	Low	Low
Blow moulding	1,000–10,000	High	Low	Low
Casting processes	100–1,000	Low-high	Low	Low
Compression moulding	1,000–10,000	High	Low-high	Low-high
ERM	100–1,000	High	Low-high	Low
Expanding processes	1,000–10,000	High	Low-high	Low-high
Extrusion (metres)	1,000–10,000	High	High	High
Filament winding	100–10,000	Low-high	Low-high	Low-high
Injection moulding	10,000–1,00,000	High	High	High
Laminating	1,000–10,000	Low-high	Low-high	High
Lay-up	100–1,000	Low	Low	Low
Matched die	1,000–10,000	High	High	High
Press moulding	100–1,000	High	Low-high	Low
Pultrusion (metres)	1,000–10,000	High	High	High
RIM, LRIM, RRIM	1,000–10,000	High	High	Low-high
Rotational casting	100–1,000	Low-high	Low-high	Low
RTM	1,000–10,000	High	Low-high	Low-high
Thermoforming	100–1,000	High	Low	Low
Transfer moulding	1,000–10,000	High	High	High

Apparent density and bulk factors are sometimes used to compare the costs of different matrices.

Apparent density (bulk density) is the weight (mass) per unit volume.

$$\text{Apparent density} = W / V$$

where

V = volume (in cm^3) occupied by the material in the graduated cylinder, HA
H = height (in cm) of the material in the cylinder
A = cross-sectional area (in cm^2) of the measuring cylinder
W = mass (in grams) of the material in the cylinder

Bulk factor is the ratio of the volume of loose moulding powder to the volume of the same mass of matrix after moulding.

$$\text{Bulk factor} = D1 / D2$$

where

$D1$ = average density of the moulded or formed specimen

$D2$ = average apparent density of the matrix material prior to forming

Composite parts may provide long-term benefits. Cost-effectiveness calculations must consider that many composite designs and processing techniques will result in a net savings by reducing the number of parts and the need to fabricate or assemble many components. One-piece hulls, fuselage, or floor pan for an automobile may greatly reduce multiple tooling and assembly of components. Reduced weight can more than offset the higher-cost material by decreasing fuel costs. It has been estimated that for every 50 kg removed from a 1400 kg automobile, there will be a fuel savings of 0.1 km / lit. These composite parts may also reduce corrosion, dampen sound vibrations, reduce thermal transmission, and improve fatigue properties. For example, a composite transportation vehicle may have an increased service life.

Since composites are mostly petroleum derived. It is apparent that competition for raw materials will continue. Any material selection must consider the availability of the material resources. For example, matrix and reinforcing materials may become scarce if one company or country withholds the materials from the market.

A principal advantage of using organic matrix composites in many transportation designs is the lower energy cost. This represents lower energy cost for a given volume of fabricated composite and lower energy cost for fuel during the lifetime of the composite part. For example, a composite automobile body may reduce the total energy cost over a steel body by more than 40%. The energy requirement of selected polymers and metals is shown in Fig. 1.198.

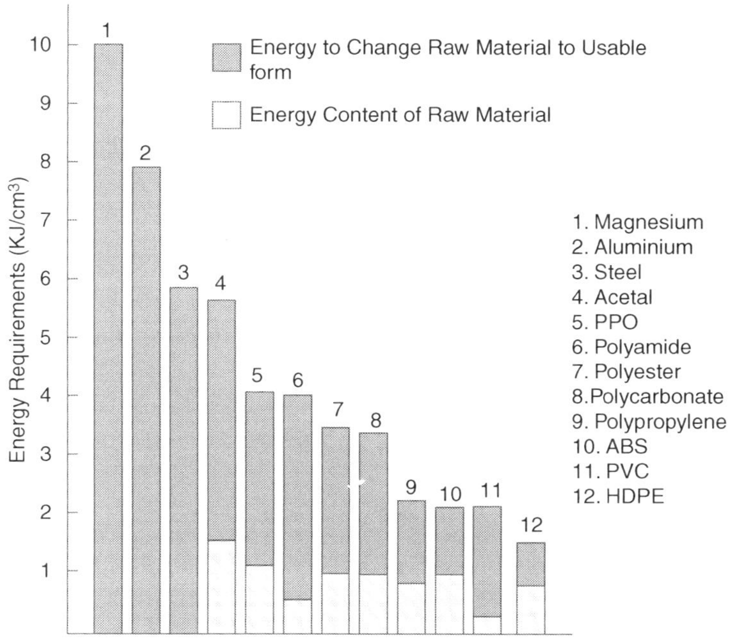

Fig. 1.198 Energy requirements of selected polymers and metals.

An increasingly important factor in materials selection is disposability and recyclability because of concern about the environment. Many composites do not degrade easily when compared with metals. Only a small portion of the materials used by the composite industry is economically recyclable. Homopolymers are more easily reused than heavily reinforced and filled polymers. In some operations such as injection moulding and extrusion, a percentage of regrind may be used as filler. Although it is difficult to visualise how a composite boat hull or automobile fender could be recycled into the useful products, this does not mean that researches could not provide an economical method and a useful product. One obvious method of disposal would be to incinerate these organic materials and capture the energy that they contain. High-technology incineration of municipal waste, including composite parts, could provide a safe and relatively pollution-free energy source for generating electrical power.

1.23.9 Product Considerations

The part shape, size, matrix formulation and matrix form often limit the means of production to one or two possibilities. Although selecting the method of production appears relatively straightforward, a number of parameters must be considered. The feasibility of making the tooling, the capacity of the moulding equipment, and the material exhibit, a close relationship. Details such as forming pressure, temperature, surface quality, post-curing cycles and production rates all come together in selecting the processing method.

As a rule, short-fibre and particulate reinforced materials are moulded by high-production methods such as injection moulding. These moulded parts are not as strong as parts moulded by processes that allow controlled filament orientation. Continuous fibre methods produce highly aligned fibre arrangement yielding high-quality, nearly ideal composites. Pultrusion, filament winding and laminating are typical processing methods.

Large composite structures such as pressure vessels, helicopter blades, fuselage structures, spacecraft parts and storage tanks may require considerable investment in equipment and tooling. Large (> 9000 Lit) multilayered, isotropic, reinforced storage tanks have been blow moulded and rotationally cast.

Any post fabrication or finishing operations must also be considered as part of the production operation. If the composite product requires surface treatment such as painting, shielding, or other coating operations, alternate production techniques may be required. Trimming, cutting, boring holes or other fabricating or assembly techniques may slow production lower performance properties and increase costs.

1.23.10 Product Design Considerations

Designing composite components or structures is an extremely complex activity. Many are designed to be anisotropic to fully benefit from the directional properties of the reinforcing additives. Composites with short fibres or particulars may be relatively isotropic or designed for anisotropic orientation of properties. Some crystalline polymers are processed with a directional molecular orientation.

In planning the preliminary ideas for a composite design there are a number of features to kept in mind: (i) overall design parameters that have been produced from design studies,

(ii) overall design conditions that the part must meet, (iii) tooling parameters and (iv) design analysis.

The advantages of composite materials are becoming more familiar to a wider range of designers and consumers. Composites have allowed designers to develop components and structures not possible with isotropic materials. One of the earliest steps in product design is to review the data and design studies that have been conducted. These may provide valuable information and help in the preliminary design of a new product.

Many problems may be overcome if the designer is well informed. Composites simply do not perform like isotropic and homogenous materials which have well-defined elastic and stress-strain properties as shown in Fig.1.199. The characteristic properties of a composite are derived from its constituents, from its processing, and from its microstructure. Experience plays an important part in the understanding of composite materials and processing.

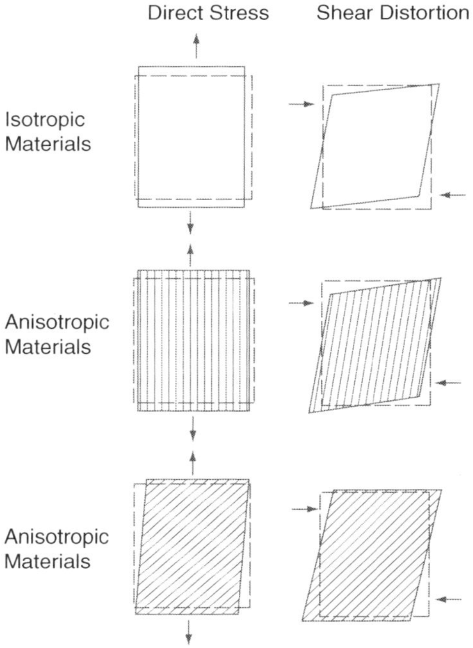

Fig. 1.199 Tensile stress causes lengthening and contraction. Shear stress cause angular distortion.

In considering the overall design conditions, the intended application or function, enviornment, reliability requirements, and specifications must be reviewed. It is helpful to list the anticipated conditions, use, and performance requirements that the part is likely to be expected to withstand. During the preliminary design review, it may be necessary to consider the performance benefits of weight, cost, and property trade offs. Deliberate strength anisotropy may result in increasing glass content or wall thickness. This generally results in weakening at right angles to the fibres. The reduction of weight (mass) may be desirable but requires more expensive materials and processing. Care must be taken during this stage that generalised views and data do not mislead the designer. The complexity of designs and sensitivity to material prices and innovative production techniques may alter many decisions.

Part of the preliminary design must include tooling parameters. Next to materials selection, tooling and processing have a marked effect on the properties and quality of composites. Properly designed tooling may minimise stress concentrations and maximise desirable properties. Tolerance specifications may demand that closed moulding methods be used. It may be necessary to develop the composite design for an appropriate method of production. Since in industry a major emphasis is placed on design to cost, the composite must be designed for rapid production technologies.

For composite materials to gain increased use, they must be cost effective with other material choices. There have been significant savings in assembly and tooling costs in the automobile industry by new designs and integration of parts. Integrated composite structures are presently made into bath fixtures, aerospace panels, and aircraft and automotive parts.

1.23.11 Design Analysis

Design analysis is most complicated for polymeric composites. There is a large data base of properties needed for the design and analysis of the part over a variety of conditions. The amount of calculations and data require that a computer be used. Finite-element analysis is a method to analyze the stress response in structural elements. The part to be studied is divided into elements which are jointed by node points. This network of nodes is used to show how stresses are transmitted. By using the computer model it is possible to show the stress response of the part with a specific geometry. CAD systems greatly reduce the model construction time required. Finite-element analysis is only one step in the total product development process.

1.23.12 Safety Factor

A designer must produce a plan that will satisfy cost, functional, and reliability requirements. It is more important to be able to predict property or part failure. A safety factor (sometimes called design factor) is defined as the ratio of the ultimate strength of the material to the allowable working stress:

$$SF = \text{Ultimate strength} / \text{Allowable working stress}$$

Allowable safety factors depend on a number of variables, many of which are specified by codes or recognised authorities. The safety factor value should be based on (i) accurate, reliable load estimates; (ii) analysis and stress determination; (iii) expected adverse environmental conditions; (iv) quality of the processing technique to produce reliable parts; (v) the nature and inhomogeneity of the loads; and (vi) criticality of application.

High safety factors have been used in designing polymeric composites for years. This over design has been a result of inexperience. While metals have a well-established and demonstrated performance level, composites lack this homogeneity. All values of composites are a function of many variable and testing methods.

The safety factor for many composites is 4.0. For critical structural composites, this factor may be as high as 10.0. Obviously, a composite helicopter rotor blade is a more critical application than the drive link in an automobile.

With accurate, reliable data some designs are using a safety factor of 1.5. This means that there must be a high degree of confidence in the method of fabrication, quality of materials, available methods of testing, and knowledge of loads. Weight savings is a common goal in most composite designs. If the safety factor and weight are to remain low, proper design and processing are key elements. Many of the aerospace processing techniques are extremely labour intensive to asure that design and processing defects are minimal.

The ultimate test is service. Computer modelling and other aids are simply not sufficient for some designs. It may be necessary to test functional models and prototypes to optimise the design and determine performance. Even if the safety factor is not a major concern, the quality of parts may warrant a prototype. A prototype mould is invaluable in providing answers to questions on the moulding needs of the part. This information can then be incorporated into the production mould.

1.23.13 Design of Simple Structural Elements

There are numerous sources of information concerning basic design practices: Here only two broad categories of guide lines are discussed: (i) product design and (ii) mould or tooling design. This does not imply that they are to be considered as separate concepts. The very fact that a part is to be moulded, not cast or machined, should illustrate the interdependence of design considerations.

General product and mould design guidelines must be further divided into the three major classifications of composites: (i) fibrous, (ii) laminar and (iii) particulate.

As a rule, we know that components benefit in several performance areas from fibrous reinforcements. This can be seen in selected properties of unreinforced versus (30%) glass-reinforced polymers in Table 1.20. Theremosetting compounds have similar improvements.

The following generalised guidelines may be applied to most thermosetting and thermoplastic materials moulded by injection, compression, and transfer moulding techniques. Some may also apply to other processing techniques.

Standard stress and deflection formulas may be calculated under various loading conditions at room temperature. Figure 1.200 shows the beam dimensions and calculation of the moment of inertia (I).

W = load (kg)

L = length of beam between supports (cm.)

Table 1.20 Selected properties of reinforced versus unreinforced thermoplastic polymers.

Polymer	Mould shrinkage (mm)	Tensile strength (10^3 kg/cm^2)	Thermal expansion (10^{-5} cm/cm/ °C)	Deflection temperature at 210 kg/cm^2 (°C)
Acetal	0.08 (0.50)	1.37 (0.62)	4.0 (8.1)	162°C (110°C)
Polyamide 6/6	0.10 (0.40)	1.54 (0.83)	3.24 (8.1)	252 °C (77 °C)
Polycarbonate	0.02 (0.15)	1.30 (0.63)	2.34 (7.0)	149 °C (129 °C)
Polyester (PBT)	0.07 (0.50)	1.37 (0.60)	2.16 (9.5)	221 °C (54 °C)
Polyetherimide	0.05 (0.15)	2.0 (1.07)	2.0 (5.6)	216 °C (200 °C)
Polyetheretherketone	0.07 (0.27)	1.76 (1.02)	3.24 (9.0)	316 °C (182 °C)
Polyether sulfone	0.07 (0.18)	1.33 (0.84)	3.24 (5.6)	213 °C (204 °C)

Values in parenthesis are for unreinforced thermoplastic polymers.

c = distance from the outermost point in tension to the neutral axis (cm)
b = beam width (cm)
d = beam height (cm)
E = modulus of material example (kg/cm^2)
S_{max} = maximum stress (kg/cm^2)
Y_{max} = maximum deflection (cm)
C = cyclic stress of material example
I = moment of inertia (cm^4)
Z = section modulus (cm^3)
M = load × distance to support (kg × cm)

Example: Moment of Inertia

$$I = \frac{bd^3}{12} \frac{(0.64)(1.27)^3}{12} = 0.1092 \text{ cm}^4$$

Example: Steady Load (Creep)

$$E = 1190 \text{ kg/cm}^2$$

$$W_{max} = \frac{4S_{max}I}{LC} = \frac{(4)(1190)(0.1092)}{(7.5)(0.64)}$$

$$= 108.28 \text{ kg}$$

Example: Cyclic Load (10^7 Cycles)

$$C = 320 \text{ kg/cm}^2$$

$$W_{max} = \frac{4S_{max}I}{LC} \frac{(4)(320)(0.1092)}{(7.5)(0.64)}$$

$$= 29.11 \text{ kg}$$

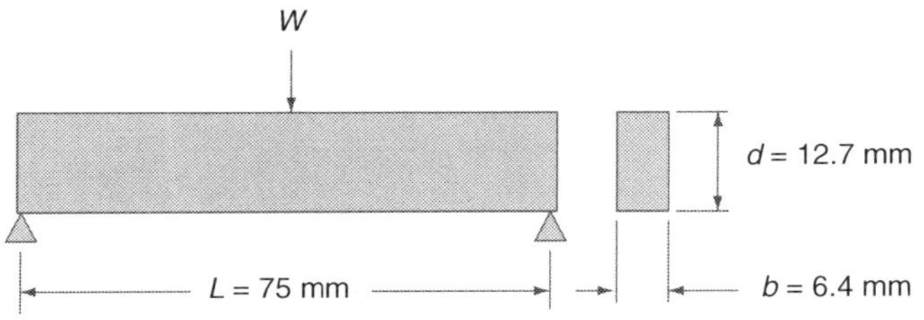

Fig. 1.200 Beam used in examples.

Example : Short-Term Loading

$$L/2 \text{ and } M = \frac{WL}{4}$$

$$S_{max} = \frac{WLc}{4I}$$

$$E = 1256 \text{ kg/cm}^2$$

$$W_{max} = \frac{4S_{max}I}{LC} \frac{(4)(1256)(0.1092)}{(7.5)(0.64)}$$

Short-term working stress = 114.3 kg

$$Y_{max} = \frac{WL^3}{48EI} \text{ at } L/2$$

$$E = 1256 \text{ kg/cm}^2$$

$$= \frac{(114.3)(7.5)}{(48)(1256)(0.1092)}$$

$$= 0.130 \text{ cm}$$

Examples of load considerations Some additional examples of load considerations are shown in Figs. 1.202, 1.203, 1.204.

For a given composite, creep strain is directly related to the applied load. Creep resistance is particularly important for extended service. In a thermoplastic matrix composite, creep rate is inversely related to the amount of fibre it contains although not proportionally to stress. Flexural creep is shown in Fig.1.201.

Integrated part designs are most efficient and generally reduce overall part cost. If the designer can combine several functions and components into a single moulded part, such as grills, ribs, or brackets, assembly and additional tooling costs can be reduced.

Design efficiency may require a decrease or an increase in wall section. Composite designs may have reduced wall thickness up to 50% over unreinforced materials. However, reduction in wall thickness may not always be the proper choice for decreasing costs. The use of ribs, contours, corrugations, and other geometric factors may be a better alternative. When a composite design requires a varying wall thickness, gradual transition is recommended to eliminate distortion and reduce internal stresses. Part geometry is directly related to how the matrix will fill the mould. This affects appearance, cycle time, flatness, dimensional stability, and other performance properties of the part. Wall transition is shown in Fig. 1.205. Fibre reinforced polymers shrink more along the axis traverse to flow than along the axis of material flow. Because some fibres are broken during the moulding process, careful mould design is important to ensure some degree of control over fibre length and orientation.

To facilitate removal of the part from the tooling, a small draft angle may be required. Draft angles vary from 0.5° to 3° depending on complexity of design, depth of draw, and texture of the mould surface. This is especially true with cores. With textured designs, the draft angles should be at least 1° per side (inside and outside) for every 25 mm of depth. Typical shrinkage of fibre-reinforced polymers is about one-third to one-half that of non–reinforced polymers.

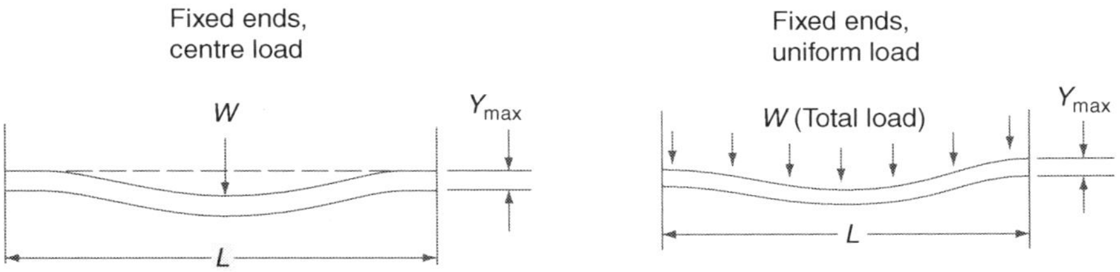

Fixed ends, centre load:

$$S_{max} = \frac{WL}{8L} \quad \text{(At supports)}$$

$$Y_{max} = \frac{WL^3}{192EI} \quad \text{(At load)}$$

Fixed ends, uniform load:

$$S_{max} = \frac{WL}{12Z} \quad \text{(At supports)}$$

$$Y_{max} = \frac{WL^3}{384EI} \quad \text{(At centre)}$$

Fig. 1.201 Examples of load considerations.

Warpage is somewhat proportional to the amount of shrinkage of the matrix. Residual stresses are developed as a result of forcing the material to conform to a mould shape (molecular and fibrous orientation). During cooling or curing, matrix shrinkage also locks in stresses. These stresses make it difficult to produce an absolutely flat surface in moulded parts. Straight side walls or surfaces may be designed with a slight dome of 0.05 mm or more to improve the surface appearance, strengthen the part, and resist warpage.

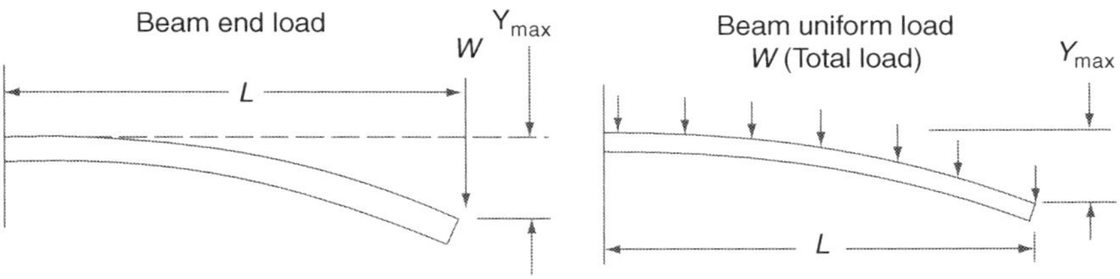

Beam end load:

$$S_{max} = \frac{WL}{Z} \quad \text{(At supports)}$$

$$Y_{max} = \frac{WL^3}{3EI} \quad \text{(At load)}$$

Beam uniform load:

$$S_{max} = \frac{WL}{2Z} \quad \text{(At supports)}$$

$$Y_{max} = \frac{WL^3}{8EI} \quad \text{(At free end)}$$

Fig. 1.202 Examples of load considerations.

Coring is an effective way to reduce heavy sections, when the heaviness is not needed for strength. Every effort should be made to keep all coring in the direction of pull or parallel to movement of the mould when it opens. Blind cores should be avoided. As a rule, cores less than 5 mm in diameter should be no greater than twice the diameter. Cored-through holes should not exceed six times the diameter. Coring recommendations are shown in Fig. 1.206.

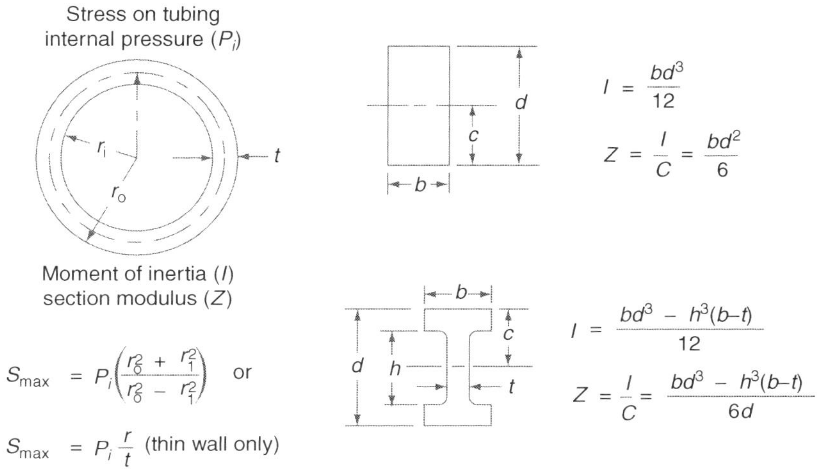

Fig. 1.203 Examples of load considerations.

Ribs are used to reduce wall thickness yet support the desired loads of the part. Thick, heavy ribs may cause vacuum bubbles or sink marks at the intersection of surfaces, therefore, thin ribs are preferred. In general, rib size should have a width at the base equal to one-half the thickness of the adjacent wall. They should be no higher than three times the wall thickness. A taper should be used. Some sink marks may be eliminated by locating a sprue or gate close to the ribbed section. Rib design recommendations are shown in Fig.1.207. General practice is to gate into the thickest section of the part to minimise sinks or voids. Fibres may be damaged if gated against a cavity wall or core pin. Round or rectangular gate size should be equal to the full width of the cavity wall. Full round runners of more than 6 mm in diameter or the wall thickness of the part are recommended.

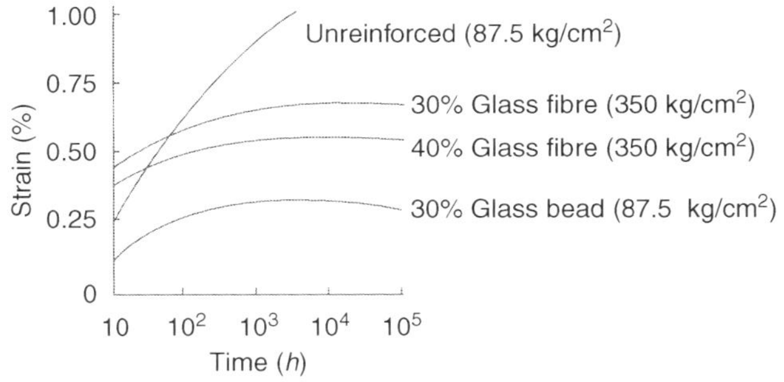

Fig. 1.204 Creep (23°C) of reinforced polyamide 6/6.

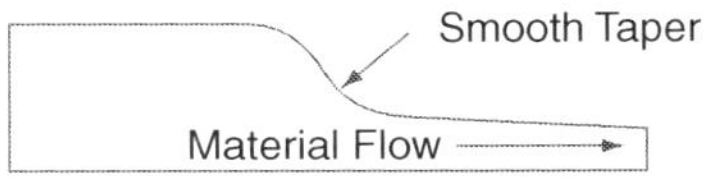

Fig. 1.205 Gradual blending between different wall thicknesses.

Flash forming around wedges, knockout pins, plugs, or name plates should be easy to clean. Bosses are commonly used to facilitate mechanical assembly and add support. Generally, the outside boss diameter should be equal to twice the inside diameter of the hole. They should be no higher than twice their diameter. Several boss designs are illustrated in Fig. 1.208.

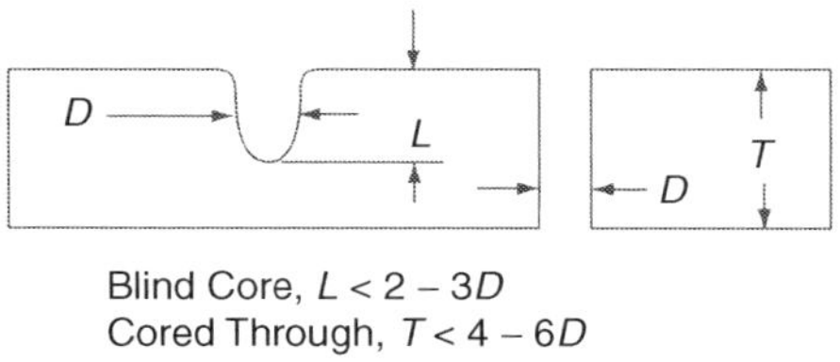

Fig. 1.206 General coring recommendations for fibrous-reinforced parts.

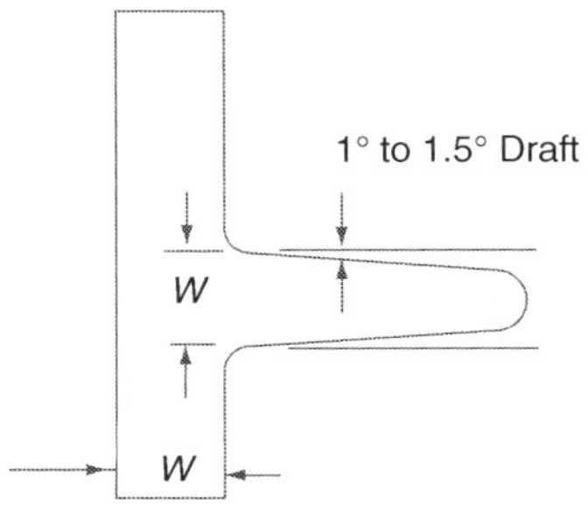

Fig. 1.207 General recommendations for rib size and design.

Undercuts require collapsing or removable cores and should be avoided to minimise tooling costs.

Sharp corners should also be avoided. Sharp corners result in stress concentrations, while liberal fillets and radii on internal corners reduce stress concentrations. A radius equal to half the adjacent wall thickness is the recommended minimum.

Both internal and external threads may be moulded into fibrous composite parts. Metallic inserts moulded into the part have greater strength. Generally, the ratio of wall thickness around the insert to the outer diameter of the insert should be slightly greater than one. This will allow sufficient material for strength. Any time two different materials are used (metal / polymer) allowance for differential expansion must be considered.

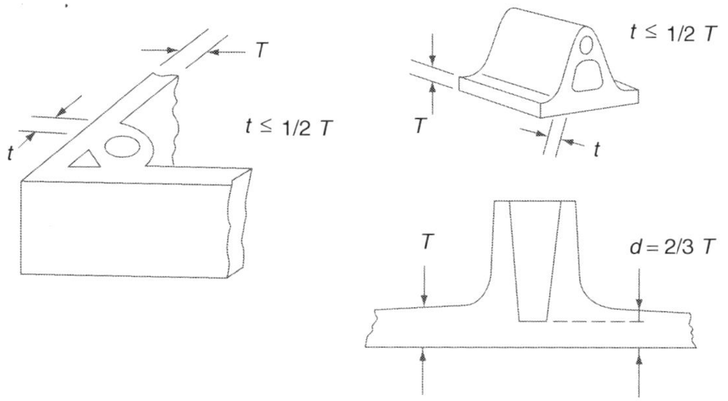

Fig. 1.208 Boss design.

There are a number of design variables when holes are required in the part. Fibres tend to orient with a resulting weld line around holes. This presents a potentially weak point. Proper gating and rapid filling are important to avoid this weld line.

Because of the high viscosity of most fibrous matrices, the mould cavity should be vented wherever air may become entrapped. Vents of 0.1 mm long × 5 mm wide are sometimes used.

Louvers or grillwork should be oriented in the direction of material flow. If this is not possible, a runner should be used across the middle of the grillwork to allow easy filling of the grill from the centre toward the sides.

The assembly methods vary greatly. There are a variety of mechanical joining techniques available for fibrous-reinforced composites, varying from snap fits to mechanical fasteners such as screws and rivets.

The assembly methods for ultrasonic, solvent, and adhesive bonding are well established. Remember, assembly is associated with stress, both chemical and mechanical. Care should be taken to minimise assembly stresses. Washers are used under bolt and screw heads to help distribute the compression forces caused by torque assembly. Self-tapping screws may be used with most thermoplastic and thermoset materials.

A snap-fit assembly, an economical and simple method of joining many composite parts, is generally used less than ten times and little or no stress is left on the flexing finger after being snapped into place. As shown in Fig. 1.209 graphite fibre composites are not suitable for snap-fit assembly. No finger beam should be expected to exceed the recommended dynamic strain limit.

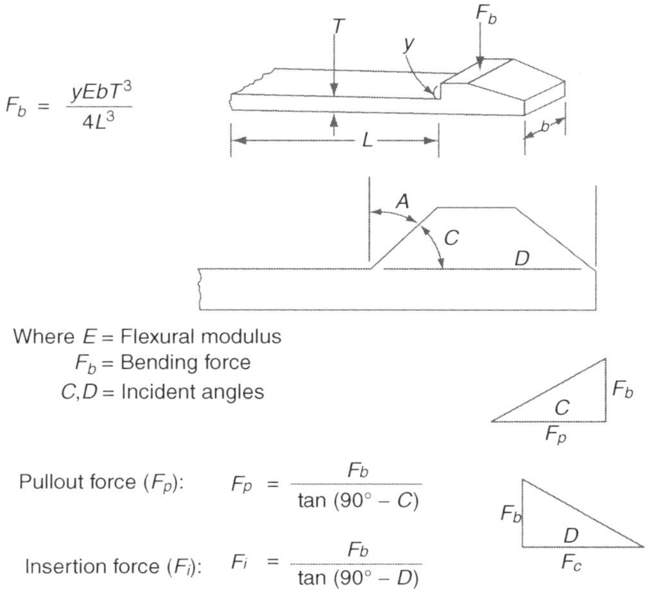

Fig. 1.209 Snap-fit assembly forces.

Figure 1.210 shows how to calculate straight – and tapered-finger beam strain. The difference between a snap and press fit is the undercut.

In blow moulding and thermoforming processes, the material is stretched at temperatures below the melt temperature. It should be apparent that these processes result in stretch orientation of the material and produce frozen-in stresses. The susceptibility of the part to damage correlates to the extent of stretching and orientation. Designs for these processes should attempt to minimise the amount of stretching.

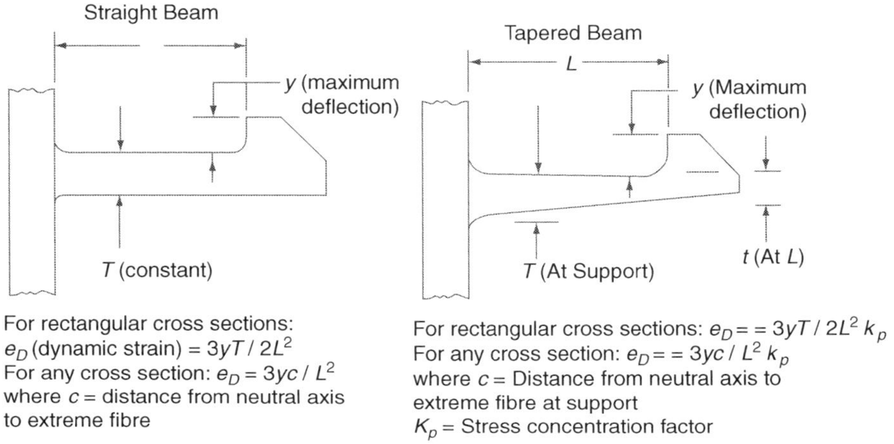

Fig. 1.210 Straight and tapered finger beam calculations.

Part design for structural foamed composites are similiar to injection moulding. The cellular core and solid skin of structural foam parts are made by both low- and high-pressure processes. In low-pressure methods (70 kg/cm^2), there is very little moulded-in increased strength properties. Because of the cellular core, bosses should have interconnecting ribs or gussets to distribute loads. Ribs should be much heavier than those in solid configurations. In general, ribs should have a width at the base equal to the thickness of the adjacent wall. To avoid sink marks and warpage problems, a uniform wall thickness should be maintained when possible.

1.23.14 Product Applications

The potential and future of polymer composite is very high. Most composites have a good balance of physical properties, excellent strength-to-weight ratios, corrosion resistance, good electrical properties, and low tooling cost. It is lower material costs, ease of fabrication, integrated part designs, faster moulding cycles, improved finishing, product reliability, environmental impact, and energy savings that will accelerate the growing consumer acceptance of composites.

Aerospace and military research has changed our concept of design and traditional construction techniques. Wood and fabric were major materials in aircraft and other transportation systems at the time. The largest growth potential for composite materials is in the areas of transportation and construction.

Automobiles, trucks, buses, vans, rapid transit vehicles, aircraft, boats, ships, aerospace vehicles, and trains will benefit from composite design . The spin-off from space and military program research has allowed the civilian markets to make impressive gains. Present production applications include drive shafts, side rails, doors, cross members, oil pans, suspension arms, leaf springs, wheels, quarter panels, trunk decks, hoods, hinges, transmission support, front and rear bumper components, seat frame, and wheels. Many engine components are being tested in present racing engines (Fig. 1.211).

Composite materials, in addition to providing weight savings over the life of a vehicle, are also energy efficient. It has been estimated that for every 50 kg reduced from a 1150 kg automobile, there will be a fuel savings of 0.1 km/lit. The energy resource requirements per kg of most composite materials are only 1/4 of that of aluminium or steel. Competition for raw material supplies between energy and petrochemical uses will continue to be a concern.

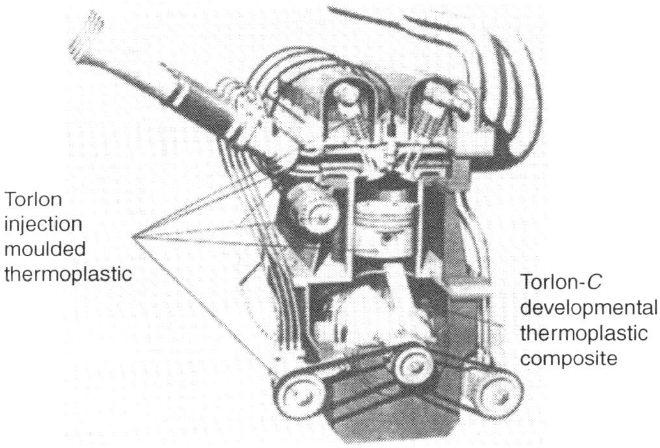

Fig. 1.211 Schematic view of an engine includes piston skirts, piston pins, piston rings, connection rods, valve stems, engine block, and cylinder heads made of carbon fibre - polyamide - imide composite (*Amoco Chemicals Company*).

New truck designs will incorporate light, strong, aerodynamic bodies and cargo containers. Reduction in wind resistance and vehicle weight will allow room for more cargo with less fuel consumption. Many of these concepts will be used for buses, trains, and mass transportation systems.

The search for higher strength-to-weight ratios and design flexibility has widened in aircraft and aerospace designs. Glass fibre composites have a specific strength five times that of aluminium, while graphite and boron in epoxy matrix have about a five times higher specific stiffness.

The current list of composite aircraft and aerospace applications is impressive. The all-composite horizontal stabiliser for the B-1 bomber, vertical and horizontal stabilisers on the F-16 fighter, wing skins on the F-18, rudder and flaps for the A-4, helicopter rotor blades, rudders, elevators ailerons, and spoilers on the B767 and B757, and payload doors of the Space Shuttle Orbiter are the most familiar applications. It has been estimated that 20% of the total weight of the AV-88 (barrier) will be composed of polymer composites. Even greater projected use of composites are in the future for AFT, V/STOL, HOTOL, stealth bombers, and Mach 20 transports. It has been suggested that up to 65% of the structure weight may be composed of polymer composites.

Glass fibre composite pleasure boats have been used for years. The 46 metre, Wilton in the British navy is the world's largest composite ship. It is likely that continued production of small ships in the minesweeper, hydrofoil, and surface effect ship class will continue to grow. Corrosion resistance, light weight, and low maintenance make marine use in hull structures, fairwaters, sonar domes, antennas, floats, buoys, masts, spars, deckhouses and all kinds of tanks an attractive application. Carbon composites are very radar absorptive and are likely to find additional structural applications in both aircraft and ships.

The market penetration in sporting goods is expected to grow from 40 to 50% in the next five years, replacing wood and metal, Graphite, boron, and Kevlar in an epoxy matrix have found acceptance in golf carts, surf boards, hang-glider frames, javelins, hockey sticks, sailplanes, sailboats, ski poles, safe playground equipment, golf shafts, fishing rods, snow and water skis, bows, arrows, tennis rackets, pole-vaulting poles, skateboards, bats, helmets bicycle frames, canoes, catamarans, oars, paddles and other items.

With changes in building codes, craft unions, consumer acceptance and acceptable designs, construction has potential for considerable growth in composite usage. Not everyone can afford a custom-built home made by traditional methods. Mass-produced, modular manufactured composite homes may be in our future. The growing mobile-home, apartment, hotel, and motel market is a natural for modular composite construction.

There are a number of miscellaneous applications, including soundboards for guitars and violins, shells for musical drums, and lightweight armored products for personnel, vehicle, and equipment protection. The appliance and furniture industries will continue to initiate new designs using the merits of composites. Housings, frames, bases, tanks, and fans will be used in the manufacture of computers, vacuum cleaners, air conditioners, refrigerators, and other appliances. Chairs, lamps, tables, and other modern furniture designs are destined to be made of composite materials. Biocompatible implants, prostheses, electrical circuit boards, hammer handles, ladder rails, highway signs, wheel chairs and numerous pipes and ducting for the food and chemical processing industries are present applications.

1.24 MODERN APPROACHES TO PRODUCT DESIGN

Several modern approaches have been made with the help of computer in plastic product design, manufacturing and engineering. Some of the modern approaches which have become very popular are:

1. Prototype Development Methods
2. Reverse Engineering Approach
3. Concurrent Engineering Approach
4. Quality Function Deployment

1.24.1 Prototype Development Methods

Most of the investors do not believe that the product developers have really considered all of the aspects, such as ramifications, variations, loads, permutations, and abuse that a new design will engender. They want to see what the thing looks like and how it will perform. These same investors will not commit funds to provide parts when they are expected to be manufactured, until they are satisfied that the risk is acceptable. These people may be senior management, marketing and sales, or customers. Since they pay the bills, product development must do something to satisfy them. Construction of prototype parts and the products is the necessary answer to the problem. Construction of prototypes implies that (i) the full production system is not completed; (ii) time to prepare prototype tooling is short; and (iii) that the cost of the prototype parts and assemblies will be higher than expected for production. However, because the capital investment is low, the risk of prototype construction is also low.

Rapid Prototyping (RP) is the automated fabrication technologies of seamless and rapidly creating accurate representative physical models of mechanical parts directly from three-Dimensional Computer Aided Design (CAD) data without the use of tooling and with minimal human intervention.

RP uses state of the art laser technology, positioning systems, materials and computer technologies in the various processes. There are many RP processes that are widely used, each one using different methods and materials to produce the final part.

The benefits of rapid prototyping, basic steps involved in it, its limitations, prototyping materials, its applications, future developments in it and its various types are explained in detail in Chapter 8, CAD/CAM applications in mould design.

1. **CAD Solid Model:** Engineering drawing programs, or computer aided design (CAD), can provide very good visual and dimensional replication of the parts on the computer screen. A solids rendition of the part can be shaded from different angles, to show highlights and irregular surfaces. Perspective are also available, so that a true picture of the part is obtained. Some colouring capabilities are also available. Often, a CAD model will suffice for people who want to see a prototype part (Fig.1.212).

Fig. 1.212 CAD Solid Model.

Advantages of CAD

(a) **The Process is Fast:** In the hands of the typical operator, a visual representation of the part is no more than an adjunct to the drafting process. A few commands, a few buttons to push, and a solids rendition of the part appears on the screen.

(b) **Least Expensive:** Since the rendition is a by-product of the drafting process, the cost is part of the design process.

(c) **Checks Form and Fit:** Many of the current high-end drawing programs allows two or more parts to be brought on to the screen at the same time. Fit between parts can be checked either visually or numerically.

Disadvantages of CAD

(a) **Cannot Touch or Feel:** For those who want to touch a part or assembly, CAD will not be acceptable.

(b) **Cannot Check Part Functions:** Although form and fit can be checked with a high-end CAD program, functionality cannot. That is, movement of the components of assemblies relative to each other, with restraints of motion from stops or walls, is beyond the capabilities of these programs.

(c) **Cannot Show Parts to the Customer:** Since there are no parts, only the computer version of the part can be shown to the customer. This may not be sufficient to the people who wish to see the actual part.

Details about the computer aided design, and its modelling types, creation, editing and support tools, its advantages are explained in detail in Chapter 8, CAD/CAM applications in mould design.

2. **Computer Aided Engineering (CAE) Simulation:** CAE is differentiated from CAD and implies the numerical simulation of a variety of actions on the part. CAE provide the responds of the parts to loads, heat or cold soaking, moulding, and the functional relationship between parts. CAE programs divide the part into smaller entities, called mesh elements, and provide calculations for the various functions from one mode of the mesh to the next. The integration of the effects over the whole part is the end point of the program (Fig.1.213).

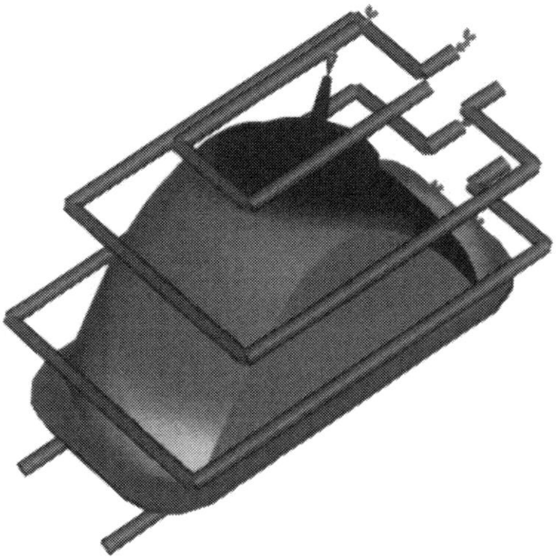

Fig. 1.213 CAE Simulation.

Common CAE analyses include

(a) Thermal stress

(b) Structural analysis

(c) Crush simulation

(d) Process simulation, such as for injection moulding

Advantages of CAE

(a) **Fast:** Although not a part of development of the part drawing, CAE is a natural next step. Since it is done on a computer, the CAD drawing is most often used as the starting for generating the mesh. Actual execution of the analysis program is depended on the number of nodes in the mesh, the particular functions that are being modeled, and the power of the computer. Typical times are measured in hours to a few days.

(b) **Low expenses:** The expense of the simulation is related to the resident expertise available and how many hours are needed for: (i) mesh preparation, (ii) analysis time, and (iii) interpretation of the results. The hours that the expert spends on the project will be the significant cost of the project.

(c) **Checks functionality:** Motions between parts can be checked for assemblies as well as distortion caused by various loading regimens. Thermal expansion and contraction from heat or cold soaking can be checked. Processing of the parts can be simulated.

(d) **No tool cost:** With good expertise in-house, the cost of the analysis program will be amortised over a number of analyses.

Disadvantages of CAE

(a) **Cannot show the effect of wear or cumulative loading:** No programs are available that will simulate the wear rate of one part against another or the effects of repetitions of loading.

(b) **Depends on the skill of the user:** Except for simple analysis, the user should be familiar with the functions that are being simulated. Detailed explanation is required, have often been called on to explain why the simulation produced results that were far from reality. The most typical solution to the discrepancy is that the user did not provide adequate inputs to the simulation program.

(c) **Depends on accurate characterisation of the materials:** Materials testing is necessary before starting the simulation, whether the user is looking for simulation of processing, stress, or temperature. In every case, the constants that are used to characterise the material for that kind of simulation are critical to the accuracy of the results. Generic data, as for a class of materials, are not adequate. Data that are averaged for a family should be used only in the absence of all other data. The best and most accurate simulation will be from characterising the exact component that will be used and that has been prepared in the same way that the final product is expected to be prepared.

(d) **Depends on the accuracy of the models used to simulate functions:** Model is the term used to describe the numerical processes and equations that are used as the basis of the simulation. For example, it is recognised that the simulation of plastics flow will not be accurately described by assuming that the viscosity is constant. Several models of the viscosity behaviour are available that allow for the change in viscosity from both temperature and shear rate changes.

Details about the computer aided engineering, finite element method and steps involved in it, its advantages, disadvantages, types of mesh and elements used for it, factors considered while taking finite element analysis decision, etc., are discussed in detail in Chapter 8, CAD/CAM applications in mould design.

3. **Machining:** Extruded or cast stock shapes can be machined to provide the prototype parts. Most thermoplastic materials are available in this form. The restrictions are the ingenuity of the machinist in making the necessary cuts in the stock to get the desired shape. With some materials, it is possible to bond parts together to provide a finished assembly.

Advantages of machining

(a) Form, fit, and function can be checked with the material of choice. This is especially important when wear and fatigue are involved. However, machining marks can also be points of stress concentration.

(b) The parts will be the same size as the production parts, and tolerance variations can be machined to check fits at high and low tolerance ranges.

(c) Typically, amorphous plastics can be painted to provide a colour that matches the end use.

(d) Tool cost is low to non–existent. At most, a few jigs and fixtures may be constructed to hold parts while they are being bonded.

Disadvantages of machining

(a) Appearance may not match production needs, since machining marks may not be appropriate to the use of the parts.

(b) Although the base material selection is the same as the material that will be used in the end product, the bar stock will not have properties identical to those of the moulded product.

(c) Surfaces will have a texture that is different from that of the moulded part. The machining process will leave a texture of its own. Although polished surfaces can be provided, specific textures cannot be included.

(d) The molecular structure will not match that of the moulded parts. This effect is especially evident for the semicrystalline materials. The crystallisation process from extrusion will result in a different orientation of the crystallites from that expected from moulding.

(e) Molecular and fibre orientation will not match that in the moulded product because the orientation processes of making rod stock are not the same as those encountered in injection moulding. In the later process, the orientation resulting from flow through the cavity is frozen into the skin layer. Interior layers have a more random orientation. When machining an extrusion, the skin layer, where any orientation may occur, is removed or at least interrupted.

(f) The very high part cost of machining will limit the number of parts that can be produced. Thus, machining intricate parts with many features is not considered appropriate for a large number of parts.

(g) For similar reasons, machining is typically not appropriate for very large parts. A combination of thermoforming, machining, and bonding may be more appropriate for large parts.

4. **Casting:** This process starts with liquid resins that can be catalyzed and polymerised. Either a flexible mould is cast from a machined model of the part or a cavity is machined. Since the pressures are from gravity, the cavities can be lightweight or even flexible. In either case, multiple parts can be made from the models. For large parts, fibre reinforcements can be added in a process akin to fibre-reinforced plastics production.

Advantages of casting

(a) Form, fit, and appearance can be checked. If the gravity process is enhanced with evacuation of the mould, very fine details can be cast.

(b) Low tool cost, since there is no need to build the mould to resist the high pressure of injection moulding. If the mould is cast from a flexible material, polished surfaces, fine detail, and undercut surfaces can be accurately replicated.

(c) Casting material can be selected that have electrical properties equivalent to those of the production parts.

(d) Casting material can be selected that have very low shrinkage rates. This allows parts with close tolerances.

(e) Appropriate for tens of parts.

Disadvantages of casting

(a) Material properties may not match the functional needs, since the number of plastic families that can be used in the process is limited. Wear and fatigue resistance are seldom matched by any of these materials; reinforcements are limited to woven and non–woven fabrics.

(b) Parts costs are high to pay for the relatively long cycle time per part and the high cost of the equipment. For this reason, the process is considered appropriate only for a few parts.

5. **Thermoforming:** This process is appropriate for parts that are too large to be machined from available stock shapes. The idea is that the basic shape can be formed from sheet and other features are added by being machined and then bonded in place.

Advantages of thermoforming

(a) Parts too large for available stock shapes can be fabricated with minimal tooling.

(b) The thermoforming tool can be fabricated from low-cost materials, such as wood: Tool cost will be relatively low.

(c) Form, fit, function, and appearance can be checked.

(d) Appropriate for tens of parts.

Disadvantages of thermoforming

(a) The process is limited to materials that can be thermoformed. Semicrystalline and fibre-reinforced materials should not be used. The properties may not match those expected of the material selected for production.

(b) The costs of the part will be high. Thermoforming a few pieces, without the trimming jigs and fixtures typical of the high production versions of the process will be expensive. Fabrication of the auxiliary parts will also be costly, as with the bonding process to locate and secure these parts.

(c) The joints between component of the fabricated assembly may be weaker than those in the equivalent moulded part.

(d) Tolerances will be wide, to accommodate the thermoforming process. Thickness is especially difficult to control to tight–tolerances.

(e) Depending on the materials available for thermoforming, the parts may not be functional equivalents of the injection moulded parts. Friction and fatigue life will not match those of the selected materials.

6. **Injection Moulding:** Prototyping parts by injection moulding relies on low-cost tooling approaches. The mould can be constructed by one of the following methods:

(a) Cast from metal-filled epoxy or urethane resin. The pattern is usually machined from aluminium. The metal filling in the casting material provides higher thermal conductivity and greater strength than the unfilled resin.

(b) Cast from low-temperature alloys, such as Kirk site .The pattern for the casting can be machined from wood and replicated in plaster for casting the metal cavity.

(c) Electroformed surface with low-temperature alloy backing. A pattern is machined from a non–conductive material. Copper is deposited from a chemical bath to provide a conductive surface. This is plated, usually with nickel, to form a shell that is backed up with cast plastic or low-temperature metal alloy.

(d) Aluminium plates can be machined more easily than the equivalent steel cavity. However, the polished finish on the aluminium will not replicate that for hardened steel.

(e) Mild steel can be machined more easily than the equivalent tool steel. If the cavity is fabricated, that is, made from sections, the machining can be simpler than expected for the production tooling.

Advantages of injection moulding

(a) Form, fit, and appearance can be checked, since the material will be the same as for production parts.

(b) Appearance depends on the mould surface, to reduce the costs, prototype moulds may not be polished as finely as the production tool.

(c) The low part cost makes the process appropriate for hundreds to thousands of parts. A good prototype mould can produce parts for early production.

Table 1.21 Comparison of materials for prototype injection moulds

Mould material	Advantages	Disadvantages
Cast epoxy or urethane	Low cost	Low thermal conductivity
	Fast turn around	Low thermal capacity
	Good for small, intricate parts	Metal filling may not get into sharp exterior corners
	Life expectancy is less than 50 parts	Special efforts needed to avoid bubbles in casting
		Surface finishes may not match production parts
		Must be contained in strong mould based
		Fine details will be fragile
		Prone to chips around machined ejector pin holes
		High wear rate with fibre reinforced materials limited life.
Cast low-temp alloys	Low cost	Must be contained in strong mould.
	Fast turn around	Easily damaged by moulding debris.
	Fair thermal conductivity	Tends to peen near ejector pin holes.
	Good thermal capacity	High gate wear rate with fibre reinforced materials
	Cooling lines can be machined	Difficult to get good surface polish
	Good for large parts	
	Life expectancy is about 100 parts	
Electroformed shell	Low cost	Must be contained in strong mould base.
	Fast turn around	Easily damaged by moulding debris
	Good replication of the pattern surface	Prone to flash

(Contd.)

Table 1.21 *(Contd.)*

Mould material	Advantages	Disadvantages
	Can be polished Good wear rate Life expectancy is several hundred parts	
Aluminium plates	Fast machining Good thermal conductivity Good thermal capacity Cooling lines can be machined Good for large parts Life expectancy is about 1,000 parts	Easily damaged by moulding debris Too soft to hold a good polish Wear rate is high from fibre reinforced materials
Mild steel	Good Good thermal conductivity Good thermal capacity Cooling lines can be machined Life expectancy is several thousand parts	Polished surface is easily damaged Wear rate is higher than for hardened tool steel

Disadvantages of prototype injection moulding

(a) Gate location may not match production. To accelerate the tool production, the tool design may be simplified. For example instead of three-plate mould, including a centralised gate, a two-plate mould with an edge gate may be used.

(b) Gate scars may be more prominent than for the production tool. For example, submarine gates may not be appropriate for the mould material.

(c) Filling patterns will be different from those expected from the production tool. This implies that weld line location, flow orientation, shrinkage, and residual stress patterns will also be different.

(d) Mould cooling may not match production. The simplified tool design may not pay as much attention to the mould cooling as will be expected from the production tool. In addition, the lower heat capacity or thermal conductivity of the mould material may not much that of the production tool material.

(e) Uneven cooling may result that will cause different shrinkage rates than expected from the production tool.

(f) The cooling rate may be different if the mould material thermal capacity and thermal conductivity are different from those of the production tool.

(g) Production rates will be lower than that expected from the final tool. The tool design compromises will often imply more mould open time and limited ability to run the mould in an automatic mode. For example, rather than include cam-operated side cores, loose insert may be used. These must be placed in the mould by hand between each shot.

(h) Tool cost and lead time are higher than for any of the other prototype processes. The longer time must be factored into the production schedule.

1.24.2 Reverse Engineering Approach

Reverse Engineering (RE) is a process that is used to create 3D CAD models directly from physical parts with little or no additional design documentation. It is simply the act of figuring out the real parts using software that you have.

When a detailed drawing of a component, is missing or insufficient to understand, or the part is of a contour shape which cannot be measured easily by the measuring instrument, then they are developed by the reverse engineering process. Reverse engineering process requires a coordinate measuring machine or digitiser (3-D scanner) to extract the input data in terms of point clouds from which the surface and model will be developed again. To develop the model again as per the existing one or to modify the existing one, it requires the 3-D modelling software. Hence, the knowledge of 3-D modelling software is essential for the designer. The designer should also have a sound knowledge on CAE software for analyzing the product for its functional requirements or its tool design verification.

The details about the reverse engineering process, steps involved in it, coordinate measuring machine, digitiser, etc., are explained in detail in Chapter 8.

1. **Use of 3-D Modelling Software in Product Design:** Plastics engineering is an area where CAD/CAM/CAE is being successfully used for quality improvement and for complex designs. The complexities of designs, bulk of operations and the need for quick changes have made the computer applications in plastics product and mould design and tool making as more important.

   With computer capability a person can modify designs quickly and even review them with the customer on the spot to achieve the best package to meet the customers' requirement in the shortest possible time. Knowledge of the basic design and related information has to be used by the designers regardless if CAD/CAM/CAE or traditional drawing boards are used with the proper software package will 'remind' the designer to follow a logical error free procedure.

   The type of analytical problems are encountered in the product design/mould design process generally fall into the sciences of mechanics, heat transfer and fluid mechanics and material science. These fields encompass many mathematical functions and relationships that are too time consuming to evaluate in manual or conventional designs. The ability of the computer to remember and execute these computations quickly adds new dimensions to the CAD/CAM/CAE process allowing prospective alternative designs to be evaluated and simulated.

With the high speed digital computers on desktop category itself (beyond 1 GHz) used for analysis, data processing activities and complete graphical information are helping to automate a domain once thought of creative discipline, i.e., the design process itself. The repetitive tasks like creating standard geometry, pack profiles, mould base, adding shrinkage to the component have been cut down. The computer ability to perform these tasks untiringly revolutionise the entire process we witness today.

We may find that using CAD/CAM/CAE techniques the time required in the conventional trial and error methodology is greatly reduced. Already there are enough indications that the managements of various organisations have started realising the impact of CAD/CAM/CAE in their industry.

With CAD/CAM coming a long way, there is an encouraging trend that the future is still brighter unlike the other areas like software development, which see their rise and fall in short periods of time. It is expected that new and more advanced design algorithms continue to flood the market, which makes the job of designer and manufacturer still easy.

The basic reasons for implementing a computer-aided design system (3-D modelling) are many. Some of them can be listed as follows.

(a) Provides better tools for reality in visualisation.

(b) Increases the effective skills of the designer.

(c) Improves accuracy and quality of the design.

(d) Reduces the visualisation and presentation problems.

(e) Creates a CNC manufacturing database based on design model.

We are very well aware that CAD and CAM are the divisions supporting the design and manufacturing sections. If we tend to subdivide CAD, we may state that CAD is inclusive of 3-D modelling, graphics operations and other engineering design tools. In addition, CAD tools can provide facilities like analysis of model like fits, tolerance studies, model properties like mass, moment of inertia, centre of gravity, etc. Also nowadays the CAD systems are having the facility element analysis of models also for structural analysis mainly.

2. **2-D Modelling:** Creating a drawing using only the x and y coordinates on a paper is termed to be a two-dimensional drawing. As such, 2-D draughting refers to the extension of manual drawing methods in a computer. The various views such as plan, elevation, end view, if necessary sectional view (to show inner details of the product) enlarged view or reduced view on various scales can be drawn to give complete description of the product.

The various functions used in the drafting can be grouped as:

(a) Drawing functions

(b) Editing functions

(c) Viewing Functions

(d) Drawing aid functions

(e) Utility functions
(f) Block functions
(g) Automatic dimensioning features

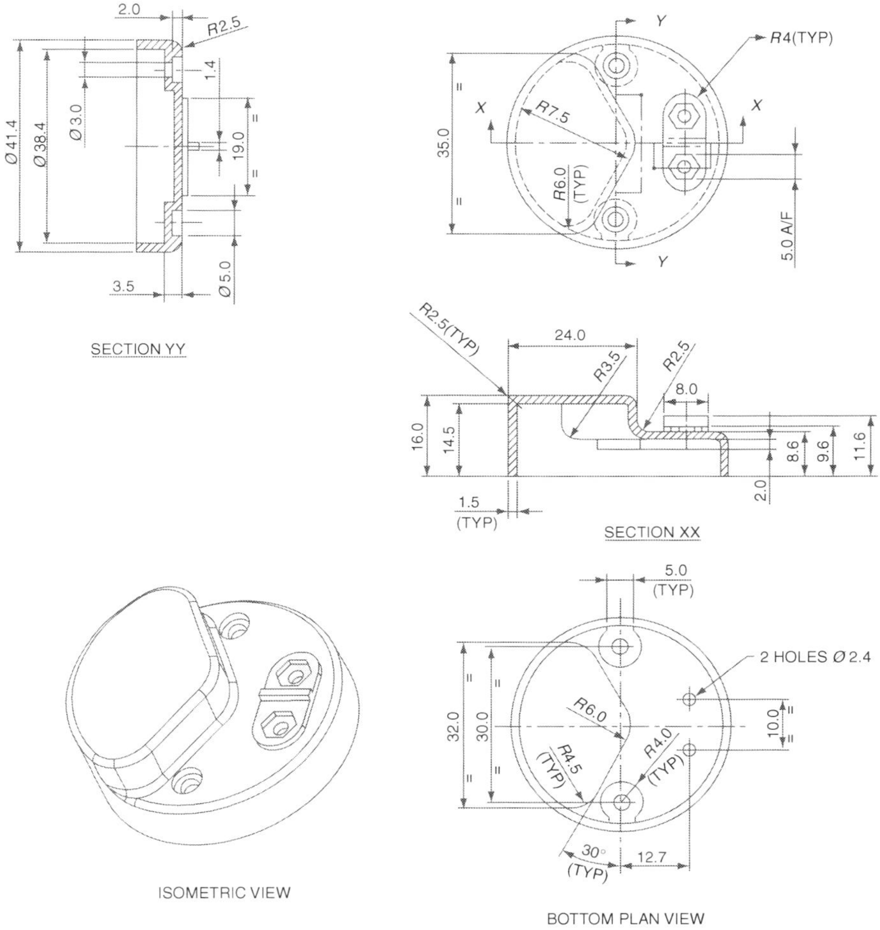

Fig. 1.214 2-D Drawing.

3. **3-D Modelling:** There are certain limitations of orthographic as a means of representing product geometry. As a consequence of these limitations, various methods have been developed for the representation of geometry with a more realistic view. These utilise the construction of a single representation of the component geometry in three-dimensional space. By using a single representation, the potential for error inherent in the use of multiple views of a component is avoided. Also, a single representation of an object is potentially far more useful as a basis for applications that

utilise the same master model to extract information for analysis and manufacture related activity.

The methods that have been developed for three-dimensional modelling involve the representation of geometry as a collection of lines and other curves, or of surfaces, or of solids in space. Of these, the solid modelling technology is the latest. It has been seen that drawings are constructed in two – dimensional coordinate system. Three-dimensional (3-D) models are constructed in 3-D space – typically in a right-handed Cartesian coordinate system. The various types of 3-D models are:

(a) Wire frame models

(b) Surface models

(c) Solid models

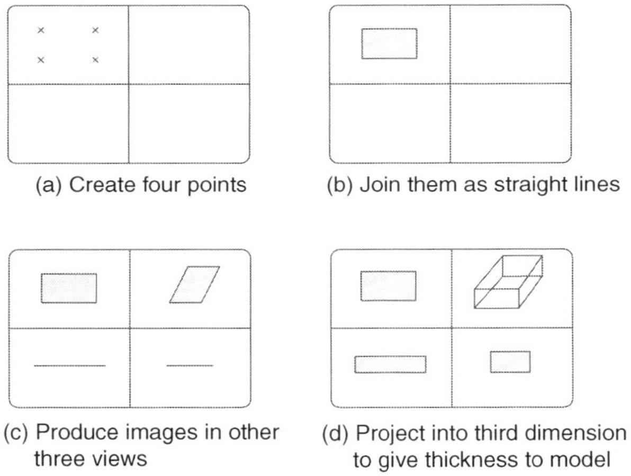

Fig. 1.215 Development of a wireframe model on a computer.

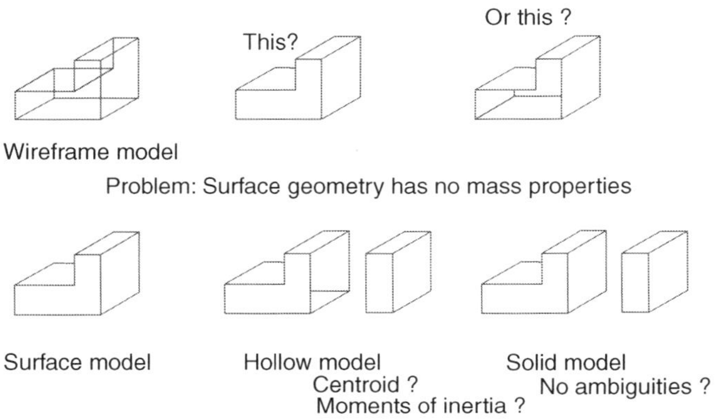

Fig. 1.216 Various methods of modelling a solid.

Some of the 3-D modelling software that are used at present are:

(a) PRO.E
(b) UNIGRAPHICS
(c) CATIA
(d) IDEAS, etc.

By using any one of the above software, we can generate any shape and size of the product with necessary colour contrast as per our requirements. If we need only modification, addition, deletion that can also be done very easily within a short spell of time we can also rotate the product to any convenient direction for better visual, easy understanding.

4. **Use of Mould Flow Software in Product Design:** The benefits of using CAE software to design and engineer components include:
   (a) Improved and consistent component quality.
   (b) Lower costs associated with the need for less prototyping, rework and lower product development cycle time.
   (c) Improved product design before the commencement of manufacturing.
   (d) Lower manufacturing cycle times.
   (e) To establish the idea of what will happen inside a mould when plastics flow inside.
   (f) To study the effects of post-moulding defects.

   The details about mouldflow software, its uses in the filed of analyzing the plastics products, steps involved in it, types of meshing used in it, various analyses done in mouldflow, etc., are explained in Chapter 8.

5. **Principles of Flow Design:** The flow analysis software relies on certain fundamental design principles to verify and optimise the flow pattern of the plastic parts. The basic design principles used are:
   (a) Flow must take place always in a single direction before it ends with the extremities of the product. This ensures a uniform filling pattern.
   (b) The pressure drop occurring in a plastic part must be constant over the entire length of the flow paths.
   (c) The maximum shear stress induced in the component part must be less the critical shear stress acceptable for the particular plastic material chosen. If the stress exceeds a critical value, then the part is likely to degrade.
   (d) The cooling must be uniform throughout the plastics part to avoid warpage.
   (e) Thin and thick sections are to be avoided to avoid the hesitation effects. Hesitations occur when the flow takes a deviation from a thicker section to thin section. Because there is no ease of flow, the material flow stops until sufficient pressure builds up.
   (f) It is common knowledge that weld lines cannot be avoided altogether in all cases of parts. The weld lines are to be positioned in the least sensitive

locations of the part so that while in service they will not affect the performance of the part.

(g) Controlled frictional heating can be used to the advantage so that the thermal properties of the plastic material can be protected to a certain extent.

(h) The feed system must be designed for thermal shut off so that any backflow can be avoided.

1.24.3 Concurrent Engineering Approach

Nowadays the fast developing industrial world has given rise to two important and novel approaches to product design. The first approach is concurrent design while the second approach is quality function deployment. At the concluding stage, we will highlight the two important approaches in the product design.

Development from concept to product requires the consideration of four basic elements. Central to this concept is the function of the product. Related to the function are the shape, material, and production techniques used to manufacture and assemble the product.

Concurrent design is the simultaneous planning of the product and the process for producing it. The role of manufacturing was to build what the designer conceived, improve the manufacture and assembly of the product. A certain industry survey showed that 60 per cent of all manufactured parts were not made exactly as represented in the drawings. The reasons varied:

(i) The drawings were incomplete.

(ii) The parts could not be made as specified.

(iii) The drawings were ambiguous.

(iv) The parts could not be assembled if manufactured as drawn.

Many of these problems have since been overcome by evolution of the design team and of the philosophy of concurrent design. The process of translating from concept to a manufacturable product is rarely accomplished now by the designer alone. Generally, a team comprising a design engineer, a manufacturing engineer and a materials engineer plays a role in supporting the chief designer. There are too many materials and manufacturing processes available for the designer to be able to make good decisions without the help of specialists.

The design team Design is both a private, individual experience and a social, group experience. The ability to recall previous designs from memory and to synthesise partial concepts together to form new ideas is unique to an individual designer. However, since most design projects are large and require knowledge in many areas, they are generally accomplished by teams of engineers with varying views and backgrounds.

We now provide a list of individuals who may fill a role on a product design team. Their inclusion on the design team will vary from product to product, and their titles will vary from company to company. Each position on the team will be described as if filled by one person, but for large design projects, there may be many persons filling that role.

1. **Product Design Engineer:** The major design responsibility is carried by the product design engineer. He must be certain about the needs for the product are clearly understood and

that engineering requirements are developed and met by the product. This usually requires both creative and analytical skills. The design engineer should bring knowledge about the design process and about specific technologies to the product.

2. **Marketing Manager or Product Marketing Manager:** In many companies, the marketing manager has the ultimate responsibility for the development of the product and represents the major link between the product and the customer. Because the product manager is accountable for the success of the product in the market, he is also often referred to as the marketing manager or the product marketing manager. The product manager also represents the interests of sales and service.
3. **Production Engineer:** It is not possible for the design engineer to have the necessary know-how about production processes. This knowledge is provided by the person who must have a grasp not only of in-house manufacturing capabilities, but also of what the industry as a whole has to offer.
4. **Design Detailer:** In many companies the design engineer is responsible for specification development, planning, conceptual design, and the early stages of product design. The project is then turned over to detailers (often called designers), who finish detailing the product and developing the manufacturing and assembly documentation.
5. **Testing Engineer:** The testing engineer aids the design engineer in developing test apparatus, performing experiments, and gathering data in the development of the product. The insights gained from the technician's hand-on experience are usually invaluable.
6. **Materials Engineer:** In some products the choice of materials is forced by availability. In others, materials may be designed to fit the needs of the product. The more a product moves away from the use of known, available materials, the more a materials specialist is needed as a member of the design team. Many suppliers actually provide design assistance as part of their services.
7. **Quality Control Specialist:** A quality control (QC) specialist has training in techniques for measuring a statistically significant sample to determine how well it meets specifications. This inspection is done on incoming raw material, incoming products from vendors, and products manufactured in-house.
8. **Industrial Designer:** Industrial designers are responsible for how a product looks and how well it interacts with human users; they are the stylists who have a background in fire arts and in human factors analysis. They often design the scope within which the engineer has to work.
9. **Assembly Engineer:** While the production engineer involved in making the components from raw materials, the assembly manager is responsible for putting the product together. Concern for the assembly process is an important aspect of product design in modern times.
10. **Supplier's Representative:** Very few products are made entirely inside one factory. Usually there will be many suppliers of both raw material and finished goods. Often it is important to have critical suppliers on the design team because the success of the product may be highly dependent on them.

The details about concurrent engineering approach, product development cycle, characteristics of concurrent engineering and its advantages are explained in Chapter 8.

1.24.4 Quality Function Deployment (QFD)

QFD is the latest approach to product design. It essentially consists of converting customer's need statement (which is usually qualitative) into technical specifications. For example, a user of automobile insists upon 'easy closure' of the door. This voice of the customer enables the design task force to derive the specifications of door closing mechanism in terms of kilograms of force required for the mechanism. QFD enables organisations to be proactive rather than reactive in QC. QFD involves (i) the customer, (ii) what is the customer requirement, (iii) how to fulfil his requirements.

The details of quality function deployment, steps involved in it and its implementation in production are explained in Chapter 8.

Questions

1. Explain the primary concepts involved in design.
2. What is the role of aesthetics in product design?
3. Write the disadvantages of evolutionary design.
4. Write two examples of design by innovation.
5. How does morphology in design influences?
6. What are the advantages of plastics over metals in product design?
7. Brief the limitation of plastics in product design.
8. Prepare a check list to evaluate the product design.
9. Give suggested wall thickness for thermoplastic and thermo-set moulding material.
10. Write short notes on wall thickness variance.
11. Why are ribs needed in product design? Explain the methodology with a neat sketch.
12. Write down the advantages of fillets and radii.
13. What is gusset in product? Explain with a neat sketch.
14. What is the significance of taper and draft in product design?
15. Why is coring needed?
16. Write the types of materials used for inserts.
17. Describe shapes of male and female inserts with a neat sketch.
18. What is pressed-in inserts in product design? Explain with neat a sketch.
19. Explain metal-stamping and rod-type inserts in product design with a neat sketch.
20. How metal inserts are replaced by plastic insert?
21. Write short notes on the following
    (a) Encapsulation (b) Composite parts

22. What are the advantages of an ideal gate size and location?
23. List out the factors on which an optimum gate size depends.
24. Describe the ideal location of ejector pins as per the product design with a neat sketch.
25. Describe wire-type screw thread insert with a neat sketch.
26. How speed nuts and clips are used for fastening?
27. What is rivet?
28. Define clasps.
29. What are the two basic methods of mechanical means of fastening? Explain with a neat sketch.
30. What are the factors affecting shrinkage?
31. Give a brief idea on bearing design with plastics.
32. Describe the manufacturing technique of plastic gear with its design calculation.
33. Explain Punching of laminates.
34. Give a brief note on product design variance pertaining to tooling aspects with a neat sketch.
35. How are composites used in product design?
36. List out the advantages of composites in product design.
37. Explain CAD solid model. Write down the advantages and disadvantages of CAD.
38. What is machining? Write down the advantages and disadvantages of machining.
39. What is casting? Write down the advantages and disadvantages of casting.
40. What is thermoforming? Write down the advantages and disadvantages of thermoforming.
41. What is injection moulding? Write down the advantages and disadvantages of injection moulding.
42. Write short notes on the following:
    a) 2-D Modelling b) 3-D Modelling
43. What are the essential factors of product design?
44. Explain methodical approach in plastics product design.
45. What are the important factors to be considered for maintaining the quality and economical product?
46. What is the significance of wall thickness in product design?
47. What is the significance of parting line in product design?
48. Why bosses are needed in product design? Explain methodology with a neat sketch.
49. What is the methodology of providing holes in a product as per product design aspect? Explain with a neat sketch.
50. Where are threads used? Classify threads and explain with a neat sketch.
51. Describe the location of inserts in the product with a neat sketch.
52. Why inserts are need for designing of a product? Explain with a neat sketch.
53. Define tolerance. Explain the parameters that influence part tolerance.

54. Explain with a neat sketch standard tolerances on moulded articles.
55. What are the techniques used for joining plastics to each other and to other materials? Explain with a neat sketch.
56. What are hinges? Explain the types of hinges with a neat sketch.
57. What is shrinkage? At what condition thermosetting compression moulded parts will have a higher shrinkage?
58. What are laminated plastics? Classify laminates and explain its process?
59. How processing variable influences product design?
60. Describe the mechanical properties such as stress, strain influences product design.
61. What is composite? Explain types of composites briefly.
62. Illustrate the steps involve before going to a product design.

    What are the modern techniques used for a product design?

64. Explain CAE. Write down its application explain its advantages and disadvantage of CAE?
65. Explain the role of 3-D modelling software in product design.
66. Describe the principles of flow design. What is the use of mould flow software in product design?
67. Describe the individuals who may fill a role on a product design team.

References

- Chitala, A. K. and Gupta R. C., *Product Design and Manufacturing*, Prentice Hall of India Pvt. Ltd., New Delhi.
- CIPET, 'Technical Manual', Chennai.
- Douglas M. Bryce, *Plastic Injection Moulding*, Society of Manufacturing Engineers, Michigan.
- Ronald D. Beck, *Plastic Product Design*, Van Nostrand Reinhold Company, New York.
- Sidney Levy, P. E. and Harry DuBois J., *Plastic Product Design Engineering Handbook*, Van Nostrand Reinhold Company, New York.
- Society of Plastic Industry Inc., *Plastic Engineering Handbook.*

CHAPTER

2

Injection Mould Design

2.1 INTRODUCTION

When identical plastic products are to be manufactured in large quantities then the injection moulding process is mostly preferred. The manufacturing process of the components is an automatic process in nature but generally tends to be operated manually. The plastic cools and takes the shape of the mould cavity. When the plastic component has solidified due to cooling, the mould is opened and the moulding is ejected.

2.1.1 Use of Mould Engineering for Thermo Plastic Material

Most of the items from electrical and home appliances are made from plastics. The technology to make plastic product is by using injection mould of thermoplastic. When thermoplastic material is heated, they change their state from solid to liquid. They will become soft, melt and flowable. When cooled again, they become solid again. Such small items but in huge quantities as bottles, sink plugs, model kits, toys,dustbins, bowls and milk crates, etc., can be made using this system.

2.1.2 Basic Concept of Injection Moulds

Injection moulding process Injection moulding is the process of forcing melted plastic into a mould cavity. Once the plastic has cooled, the part can be ejected. With this process, many parts can be made at the same time, out of the same mould. Injection moulding is often used in mass-production and prototyping. Injection moulding is a relatively simple technology to manufacture parts.

Injection moulding is a plastic-forming process used in the production of most (about 65%) of plastic parts. Other plastic-forming processes include blow moulding, pressure-forming, and thermo-forming. Injection moulding is generally used in the high-speed manufacture of low-cost, high-volume parts, like videocassette cases, plastic cups, printer parts, refrigerator parts, automotive parts, and other electronic parts like casing, gear, etc.

The process of injection moulding starts with a barrel (a hollow cylinder) hot due to heaters attached with barrel which change the status of granules into liquid plastic. The plastic is rammed by the screw at high pressure into a mould. Once the plastic fills the mould, it is allowed to cool and solidify. The finished part is then extracted automatically from the mould.

Stages in Injection Moulding Stage 1: Granulated or powdered thermoplastic material is fed from a hopper (conical shaped device to receive materials) into the injection moulding machine.

Stage 2: The injection moulding machine consists of

(a) Hollow steel barrel

(b) Rotating screw (Archimedean screw)

(c) Heaters

The barrel contains a rotating screw. The screw has its flight with some depth due to which the screw carries the plastic along the barrel to the mould. Heaters surrounding the barrel melt the plastic as it travels along the barrel.

Stage 3: Due to collection of melted plastic material at the end of barrel, the screw is forced back. Once enough plastic has been collected at the end, hydraulic ram forces the screw forward for injecting the plastic through a sprue into a mould cavity.

The mould is heated sometimes before injecting and the plastic is injected quickly to prevent it from hardening before the mould is full because the entry point of material (gate) is very small.

Stage 4: Pressure is maintained in the machine for a short time (known as dwell time) to prevent the material not to return back during hardening which prevents shrinkage and hollows, therefore, giving a better quality product. The moulding is left for some time to cool before ejecting from the mould. The moulding takes the shape of the mould cavity.

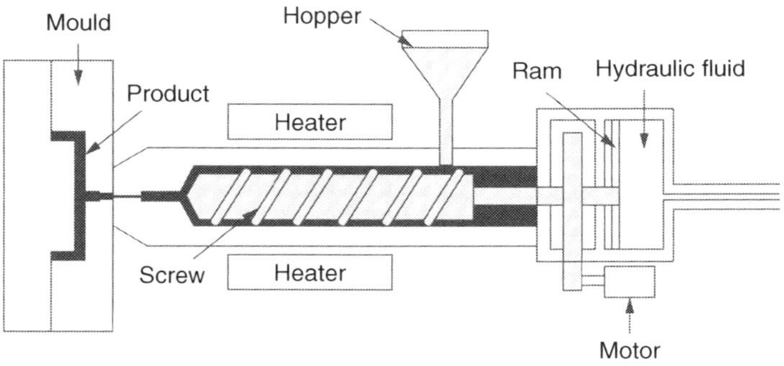

Fig. 2.1 Injection moulding machine.

As it can be seen from the above Fig. 2.1, the plastic is fed into the hopper. It is forced forward under pressure by the Archimedean screw. It is heated by the surrounding heater coils. As the plastic builds up at the front of the chamber the screw moves backward. Eventually the screw is forced forward by the hydraulic ram forcing the plastic through a sprue into the mould.

By this technique the car bonnet, crates, impeller and high precision components such as camera parts, razors and gear wheels are injection moulded.

2.1.3 Types of Injection Moulds

Classification of injection mould depends upon what is required to make by plastic parts, what is its applications, because every part has its own specific and unique design. When designing the moulds it is required to decide some of the influencing factors like geometry, number of cavities, ejection principle, plastic material and shape of parts.

Based on cavity layout The types of the moulds can be classified on the basis of number of cavities or cavity layout.

1. **Single Cavity:** When the mould is designed for single cavity, the material is injected by direct gate called sprue gate.
2. **Multicavity:** When the mould is designed for more than single cavity, the material is injected by side or submarine gate. In two-plate mould and for three-plate mould, all types of gates may be used.

Based on type of opening method and runner system

1. Standard moulds (two-plate moulds)
2. Three-plate moulds
3. Split-cavity moulds (split-follower moulds) stripper plate moulds
4. Stack moulds
5. Hot runner moulds

Based on injection moulding process Injection moulds may be divided into seven types. They are:

1. Reaction injection moulding
2. Liquid injection moulding
3. Gas assisted injection moulding
4. Co-injection moulding
5. Shot injection moulding
6. Fusible core injection moulding
7. Rapid injection moulding

Depending on material to be injected

1. Thermoplastic injection moulds
2. Elastomer moulds
3. Thermoset moulds
4. Structural foam moulds

Introduction of Different Type of Injection Moulds

Two-plate mould The standard mould is the simplest design used for making plastic articles. It is also known as two-plate mould. The mould is divided into two halves: cavity side and core side. That is why this mould is known as two-plate mould. It consists of single daylight where the mouldings and feed system get ejected during mould opening. Cavity side is the external part of the component. On the base of external surface of the component, cavity plate is designed.

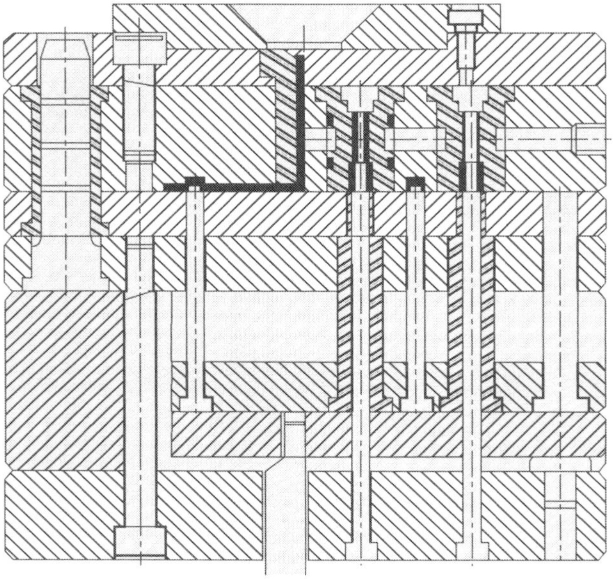

Fig. 2.2 Two-plate mould.

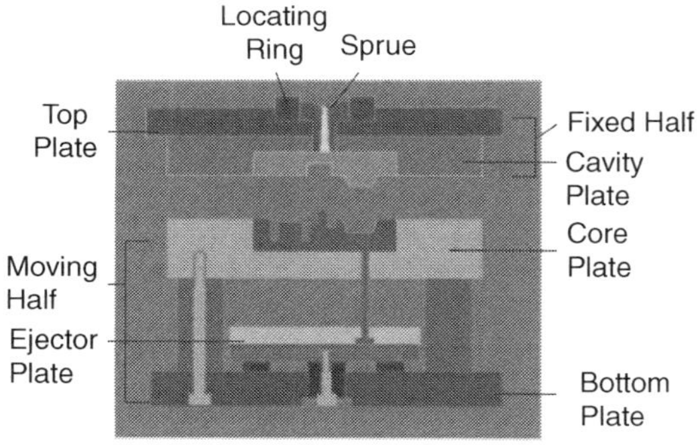

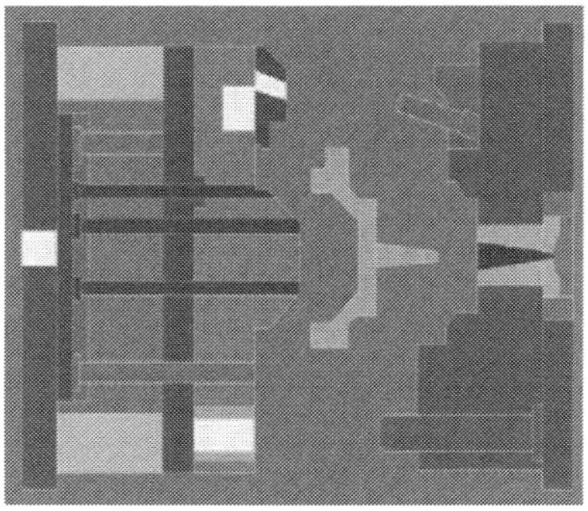

Fig. 2.3 (a) and (b) Standard mould (Two-plate mould).

In this type ejection system is designed as shown in Fig. 2.2. Standard mould has one parting line and has one opening direction. This type of mould is used in all kinds of plastic parts that do not have undercut. In this type of mould, side gate is generally preferred. Subsurface gate is also used as self degating in two-plate mould. In this type of mould, the waste (comprising runner, gate and sprue) is also ejected with the component which is to be also trimmed as secondary operation. Some other moulds as standard mould are shown in Fig. 2.3 (a) and (b).

Three-plate moulds Basically three-plate moulds have two openings/daylight and a floating plate. Floating plate is guided by guide pillar as shown in Fig. 2.4 of opened and Fig. 2.5 of closed condition of three-plate mould. Since the mould has two parting planes, the runner system can be located on one side of floating plate. There is a special plate that is attached with floating plate which is called runner ejector plate for ejecting out the runner system separately. Three plate

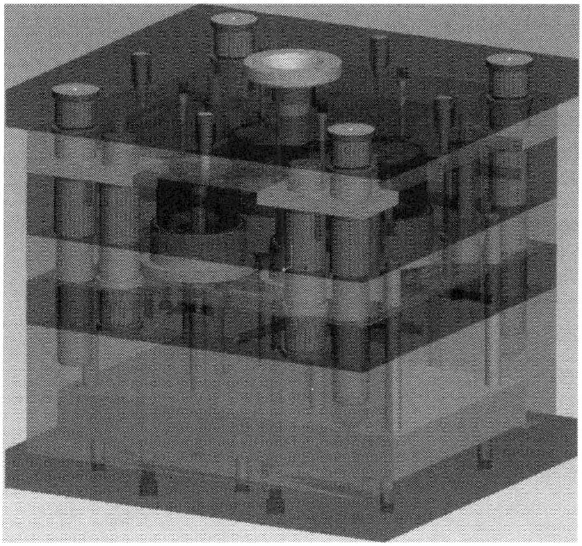

Fig. 2.4 Three-plate mould (Closed).

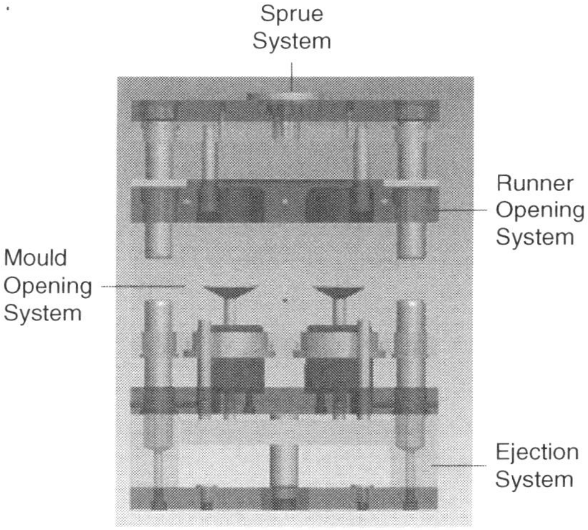

Fig. 2.5 Three-plate mould (Opened).

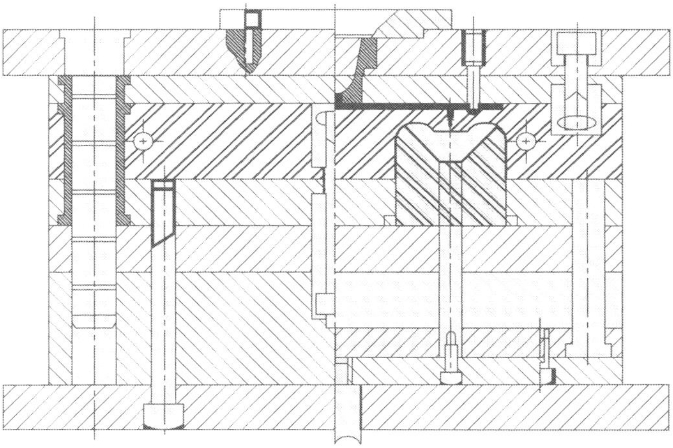

Fig. 2.6 Drawing for a small box (three-plate mould).

moulds are mostly used because of their flexibility in location of gate and are even used in multiple cavities. The 2D drawing for a small box is shown in Fig. 2.6.

Basically there are three opening types in three-plate moulds as shown in Fig. 2.7.

1. Sprue cut opening
2. Runner opening
3. Product opening

1. Sprue Cut Opening: This opening will cut sprue from nozzle, so the feeding system (runner, sprue and gate) can be easily removed manually. This opening uses link bolt and runner ejector plate to cut the sprue. The stripper plate will move after link bolts are pulled by cavity plate. The link bolts are also end mounted in cavity plate. The opening length should be between 5 mm and 10 mm. During movement of cavity plate the runner stripper plate comes forward due to the link bolt which is attached with the cavity plate and sprue.

2. Runner Opening: This opening is used to eject runner system and then the runner will be removed manually. This opening will make a space between runner, stripper plate and cavity plate. To make sure this runner opening is done safely or degating perfectly, spring is to be attached to puller bolts between runner stripper plate and cavity plate. Length of this opening is same with runner length plus 20 mm which is decided on the basis of subrunner length.

3. Product Opening: As we know, this opening is to eject the product. This opening length is about the length of the product plus space required for allowing the product to fall down plus 10 to 20 mm for ejection. This will be approximately one and half to two times of product length.

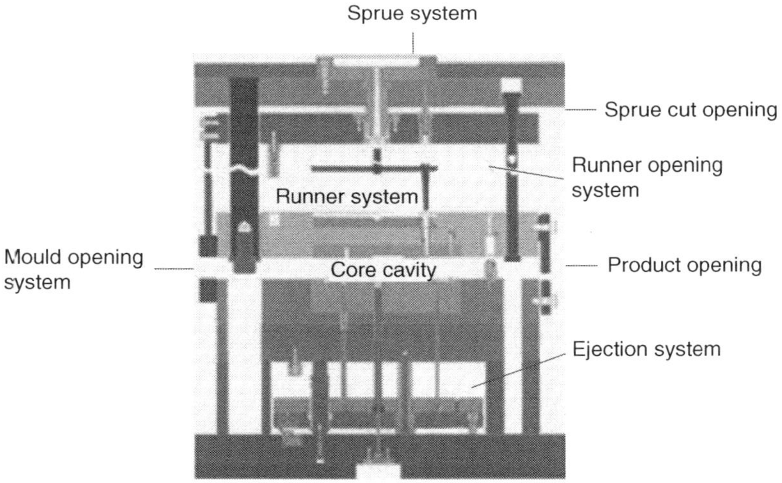

Fig. 2.7 Opening system of three-plate mould.

Opening Sequence Actually when the cavity plate is pulled, it pulls the runner stripper plate by puller bolts then open the stripper plate. It has two types, using tension link and magnet.

1. First Opening is Runner Opening: If first opening is not from runner opening side, there is a chance of degating of the article without separating runner from the mould. To get first opening from runner, some springs are always attached to make the first opening from here. Some mould designers attach spring at puller bolts or in a pin with a hole each side to attach the spring. But it is better to attach the spring in puller bolts, which is easier in process.

2. Second Opening is Product Opening: Parting Line between core-cavity plate is always easy to open. When moving platen at the injection moulding machine pulls the bottom plate, core plate and moves along with bottom plate thus creates product opening between the core and cavity.

3. Sprue Cutting Opening: After moving the platen, pull the core plate with tension link, core plate will pull the cavity plate by some mechanism until cavity plate pulls the puller bolt of runner stripper plate and make some space to cut the sprue from nozzle. Sometimes, it can be done using a simple spring by placing it in some hole with pin.

Stripper ejector mould This type of mould have special purpose ejection to make a cup shaped article without undercut. The stripper plate will make ejection easier and without a lot of marks in product as shown in Fig. 2.8.

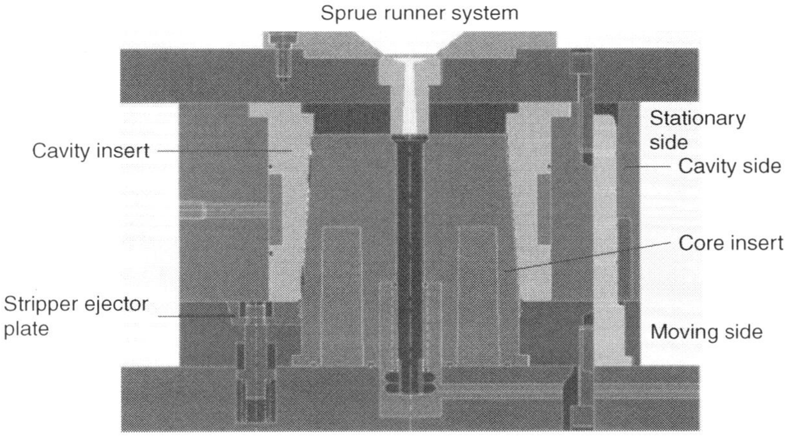

Fig. 2.8 Stripper ejection.

Split mould In this type of mould, the sliders are used for getting movement on core plate by using the movement of the moving platen. The change of movement from one to another (horizontal to vertical) is arranged by some mechanism like finger cam, dog-leg cam and cam track as per the requirement of the mould as shown in Fig. 2.9 with finger cam. This type of mould is used to make parts with undercut.

Mould with screw device This mould is special to make threading components. For releasing threads, the core can be rotated during the mould opening. Both internal and external threads can be formed by this type of mould as shown in Fig. 2.10.

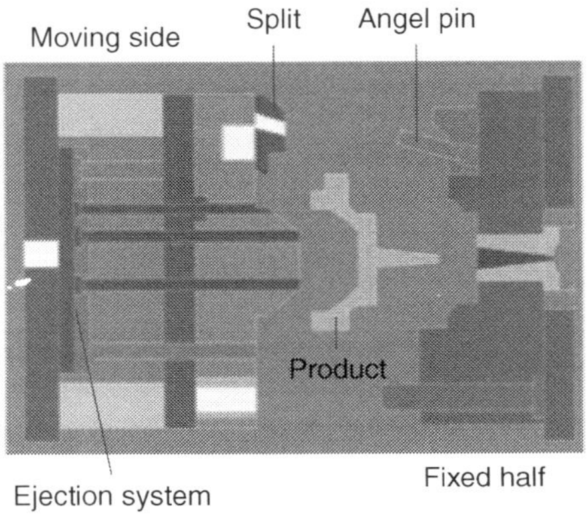

Fig. 2.9 Split mould.

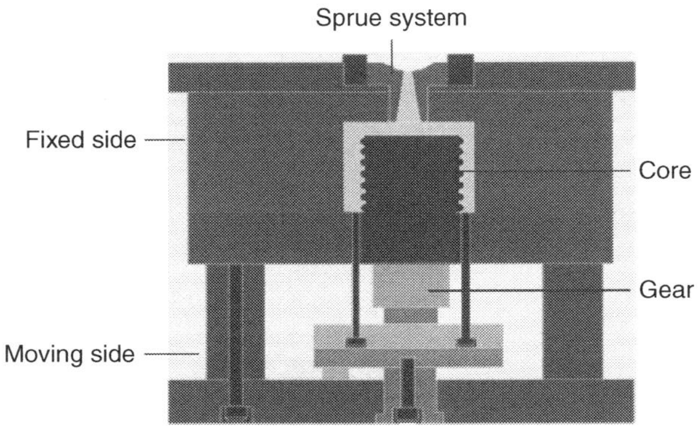

Fig. 2.10 Unscrewing mould.

From the above Fig. 2.10, it can be easily seen that it has a gear system, which will rotate with the movement of the machine platen and thread releases.

Collapsible core and cavity mould There are many applications for which a complete thread form is not required. Bottle closures consist of partial threads which can be moulded using this method. Two-segment or more segments cores may be used. Collapsible cores are like destroying and further assembling complete core by using some mechanism and movement of platen as shown in Fig. 2.11(a). In Fig. 2.11(b), the central core is withdrawn downward away from the moulding. As it moves, it forces the two small side cores to move inwards. This clears the cores from the moulded thread. Finally, the part is stripped off the tool face.

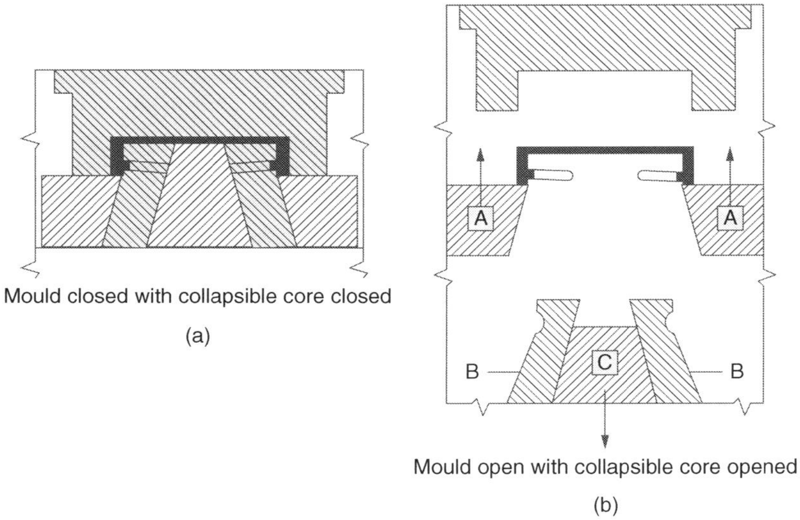

Fig. 2.11 (a) and (b) Two-segment collapsible core

Hot runner mould In a three-plate cold runner mould, the runner system must be reground and the material is reused. In case of hot runner mould, it eliminates the solidification of runner system entirely by keeping it in fluid by using heating system. The material is kept plasticised by the hot runner manifold, which is heated with heating element like electric cartridge heaters as shown in Fig. 2.12. The manifold block with cartridge heaters and the band heaters mounted on round the nozzle or on subrunner are thermostatically controlled. The plastic is maintained at liquid stage and the injection pressure is transmitted through the hot runner manifold.

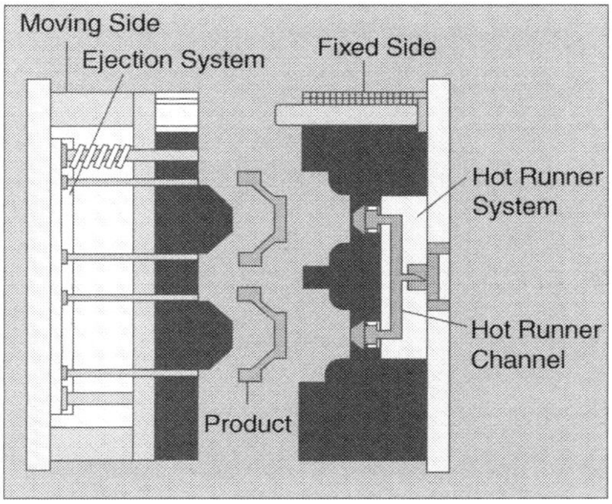

Fig. 2.12 Hot runner mould.

Stack mould The stack mould is used for moulding shallow and small parts in large quantity such as tape cassettes. In this, the cavities are located into two planes corresponding to the parting lines and are filled at the same time as shown in Fig. 2.13. For moulding the stack mould, it requires an injection moulding machine with large opening stroke and clamping force of 15% higher than a standard mould. Stack moulds were originally operated with cold runner design, which has to be moulded in each shot. Further, hot runner manifold is employed to increase the productivity.

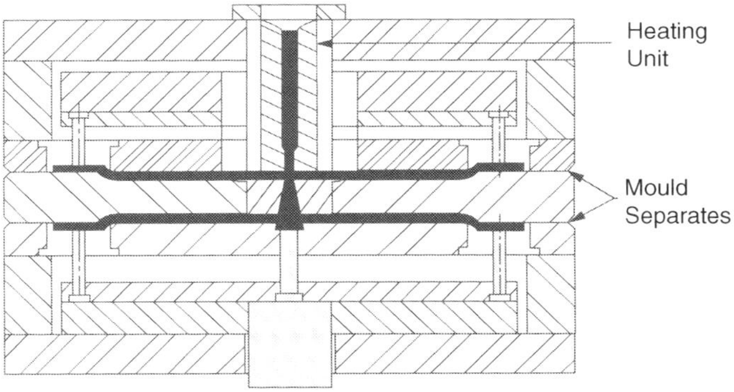

Fig. 2.13 Stack mould.

Gas-injection mould Gas-injection moulding has been developed to

1. Shorten the cycle times
2. Save the material
3. Improve the surface of article of thick-walled injection-moulded parts

General principles of gas-injection mould

For gas-injection moulding, a standard injection-moulding machine can be used, except one equipment for entering gas (normally nitrogen) in series and parallel directions with the injection of the plastics melt as shown in Fig. 2.14.

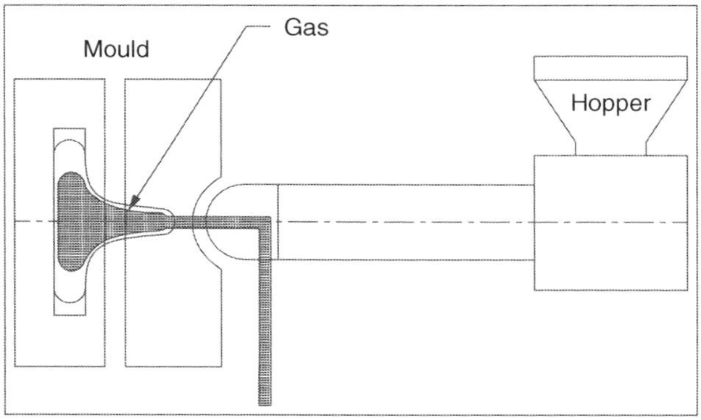

Fig. 2.14 Gas injection moulding.

Others New developments are water injection moulding, low pressure injection moulding and others.

2.1.4 Parts of Injection Mould

A detailed 3D view of three-plate mould with its parts is shown in Fig. 2.15 and 2.16.

1. **Fixed Clamping Plate or Top Plate:** It holds the fixed side of the mould to be attached with the fixed platen of the injection moulding machine. On this plate, locating ring, eye bolt and sprue bush are attached.
2. **Runner Stripper Plate:** This plate is used in three-plate moulds. The function is to cut the runner from the nozzle and pull the runner by length bolts and puller pin.
3. **Cavity Plate:** It is used to hold cavity side of product, leader pin, guide pillar, support pin, puller bolts and finger cam with the slider attached.
4. **Core Plate:** It is used to attach the core side of product, return pin, leader bush and split, if needed.
5. **Back Plate:** It is used to support cavity plate and to attach the hole for return pins and cooling channel.
6. **Spacer Block:** It is mounted between the bottom mounting plate and the movable core back plate to give space for ejection system. The required length of spacer block depends on ejection stroke required to eject the component.
7. **Ejector Retainer Plate:** It is used to hold the ejector pin, sprue puller, push back pin, and place for ejector pin and support pillar shown in Fig. 2.16.
8. **Ejector back Plate:** It pushes the ejector pins and return pins, fixed in the plate. In Fig. 2.16, number 8 is the ejector back plate.

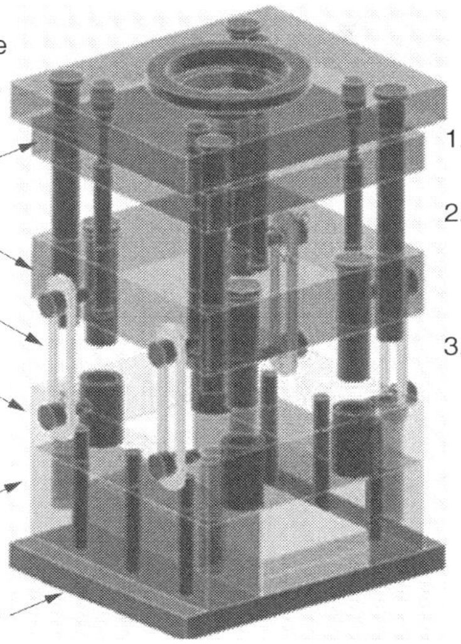

Fig. 2.15 Detailed 3D view of three-plate mould.

9. **Movable Clamping Plate:** It holds the movable side of the mould like spacer block, support plate, core plate and ejector mechanism to the movable platen of the injection machine shown in Fig. 2.16.
10. **Impression:** When the fixed and moving platens are touching, the space formed by the cut-out portion is called impression, which defines the shape of the part. It means the space between core and cavity when moulds are in closed position.
11. **Sprue:** The sprue is a tapered pathway for plastic material from machine nozzle to runner through a hole cut into a bush fixed in the centre of the fixed platen. Molten plastic flows from the sprue to fill the cavity.

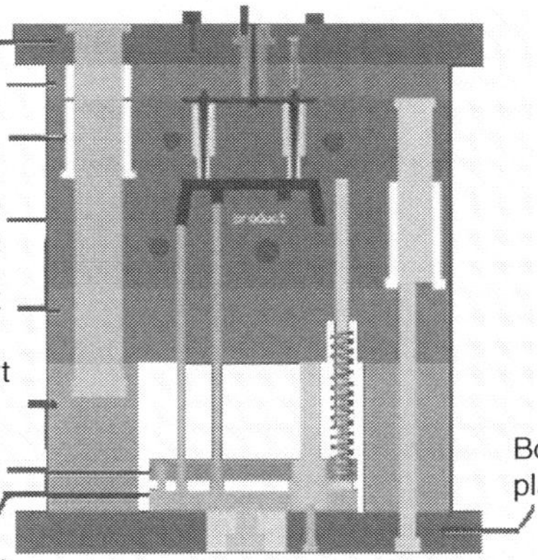

Fig. 2.16 Injection moulds and its parts.

12. **Runners:** Runners are channels or pathways machined through which the plastic melt passes from the sprue to the gate.
13. **Gates:** Gates are small openings between runners and impression, through which the plastic melt enters the cavity. They are generally small so that the finished part may be easily broken away from the sprue and runner material.

2.1.5 Steps for Quality Moulding

1. Sharp corners and sudden changes in section must be avoided. This may weaken the plastic moulding wall.
2. Large, flat surfaces should be avoided on a mould as the plastic will not come out completely flat.
3. All surfaces on the mould should be smooth and highly polished.
4. The sprue bush is to be designed very carefully. Its passages are to be mirror finished for allowing easy flow of resin. The sprue also needs to be small to allow the mould to be filled very quickly.

2.1.6 Machine Specifications

The plastics mould design depends on the specification of the injection moulding machine, which covers all the information related to the machine for the production of items.
The specifications of the injection moulding machine are given by the manufacturer along with the machine in the below mentioned tabulated format.

Table 2.1 Screw type injection moulding machine specification format.

Machine series		Machine types				
Item	**unit**	**ST30**	**ST80**	**ST130**	**ST180**	**ST200**
Screw diameter	mm					
Shot weight	g					
Shot volume	cc					
Injection pressure	Mpa					
Injection rate	cc/sec					
Screw L/D ratio						
Injection stroke	mm					
Screw speed	rpm					
Plasticising capacity	g/hr.					
Nozzle retract stroke	mm					
Hopper capacity	m^3					

ST80 means that the machine has the clamping capacity of 80 tons.

Clamping specification Clamping force, minimum mould heights (minimum day light), maximum day light, mould opening stroke, space between tie bars, platen size, ejection stroke, ejection force, etc., are also mentioned in the tabulated format.

General specifications Pump motor, dry cycle time, system pressure, oil tank capacity, heating capacity, heating zones, machine dimensions and machine weight, etc., are also mentioned.

2.1.7 Deciding Number of Impressions

Number of impressions may be calculated on the basis of machine shot capacity, plasticising capacity and clamping tonnage.

Shot capacity: It is defined as the maximum amount of plastic melt to be injected in one complete stroke of the machine screw or plunger.

If the weight of the moulding = w

Some of the material is getting wasted in the form of sprue, runner and gates, so for calculation this waste material will also be taken into consideration.

Here we may assume that the runner and gate weight on the basis of component weight. Generally it takes 10% of the component weight.

Total weight of the material = w + 0.1w (10% of the component weight) = 1.1w

If the efficiency of machine be assumed as 85%. Then,

No. of impression by shot capacity $\eta s = 0.85\ W/\ 1.1w$

Plasticising capacity: It is defined as the amount of material that the machine can bring per hour to moulding temperature.

If the weight of the moulding = w

Some of the material is getting wasted in the form of sprue, runner and gates, so for calculation this waste material will also taken into consideration.

Here we may assume the runner and gate weight on the basis of component weight. Generally it takes 10% of the component weight.

Total weight of the material = w + 0.1w (10% of the component weight) = 1.1w

Cycle time = t sec. = $t/3600$ hour

In one hour, required material = $1.1 \times 3600\ w/t$

No. of impression by plasticising capacity = $0.85Pt\ /1.1 \times 3600\ w$

Clamping force: The clamping force required to keep the mould closed during injection must exceed the force given by the product of the injection pressure in the cavity and the total projected area of all impressions and runners.

So clamping force > Injection force

> Injection pressure × Projected area

Clamping force > Injection pressure at the nozzle tip × Projected area (with runner and gate)

For n number of impression

Clamping force > Injection pressure at the nozzle tip × n (no. of projected area (with runner and gate generally 10% extra can be assumed))

As per the above,
No. of impression > 0.85 × Clamping force / (Injection pressure at the nozzle tip × 1.1 Projected area)

By the above three methods, the mould will be designed for the value which is minimum out of above three values.

2.1.8 Mould Alignment

Locating Ring Other name of locating ring is register ring which is a round circular member fitted at the top of the mould to locate the mould with the machine. Locating ring is simple part but, very important when mould base is attached to injection moulding machine. This part is used for the alignment of mould to machine. Locating ring is used in the injection mould to maintain the mould axes and the axes of machine in one line.

Types of Register Ring (Locating Ring)

1. Constant diameter type
2. Increased diameter type
3. Reduced diameter type
4. Increased depth type

Constant Diameter Type: In this type of locating ring, the diameter of ring is constant throughout the depth as shown in Fig. 2.17 which is fixed with front face of the mould by using socket-headed screw.

Reduced Diameter Type: In this type the platen hole diameter is less compared to mould recess diameter as shown in Fig. 2.18.

Increased Diameter Type: In this type the platen hole diameter is more when compared to mould recess diameter.

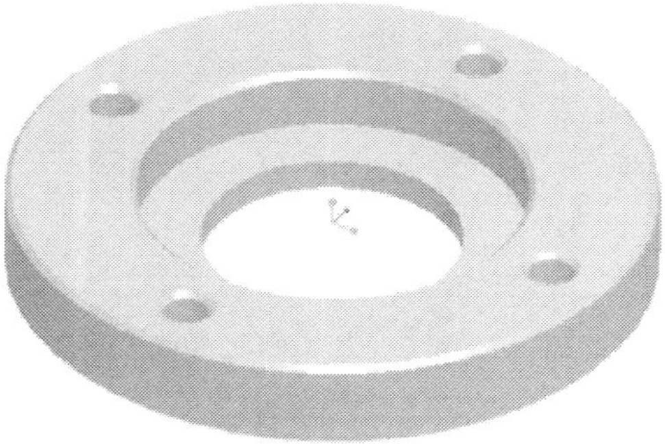

Fig. 2.17 Constant diameter.

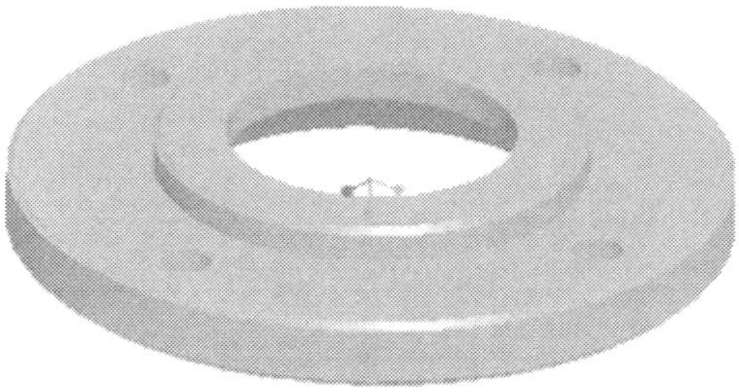

Fig. 2.18 Reduced diameter.

Increased depth type: It is identical to increased depth type except the depth of the mould fitting diameter which is increased to accommodate insulated metal sheet adjacent to the front plate.

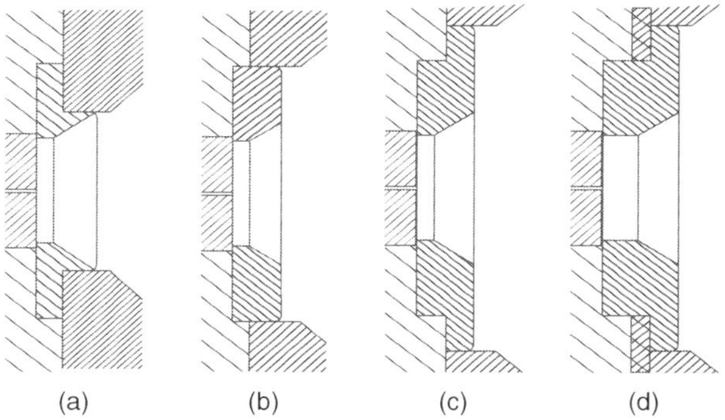

Fig. 2.19 (a) Reduced diameter type, (b) Constant diameter type (c) Increased diameter type (d) Increased depth type

Rules for Design

1. When designing a locating ring, first rule is to refer to the injection machine specification chart, and find the dimension of hole provided in the machine's fixed platen. Hole dimension can be easily obtained from the specification chart. In specification chart, the dimension is mentioned in diameter with tolerance (generally H7). If the dimension of hole is diameter 140H7 for a particular machine and the diameter of nozzle is 36 mm, so the locating ring's outer diameter must be same as Diameter 140g6 used for the machine.
2. The H7 means the tolerance in the dimension is +0.02, –0.00 and g6 means the tolerance in dimension is +0.00, –0.02 mm. Generally, the locating ring recess diameter is a standard. In some cases it is considered 90 mm or in other places it is 100 or 120 mm. So the reduced diameter, constant diameter and increased diameter locating ring is decided on this basis. It means that if in any industry, the recess diameter of the locating ring is 120 mm then
   (a) If the machine used for moulding has the platen hole size of 140 mm, the shape of the ring would be reduced locating ring.
   (b) If the machine used for moulding has the platen hole size of 120 mm, the shape of the ring would be constant locating ring.
   (c) If the machine used for moulding has the platen hole size of 90 mm, the shape of the ring would be increased locating ring.
3. The locating ring's internal diameter is decided on the basis of nozzle diameter. The internal diameter should be more than the nozzle diameter and it should not be more than the sprue bush's maximum diameter. Otherwise, there will be a chance of the sprue bush coming out of the mould. Locating ring attached with mould is shown in Figs. 2.20 and 2.21.

Selection of locating ring in mould depends on injection machine nozzle, locating hole diameter and nozzle length. Locating ring is fixed in the top mounting plate by socket-headed screws.

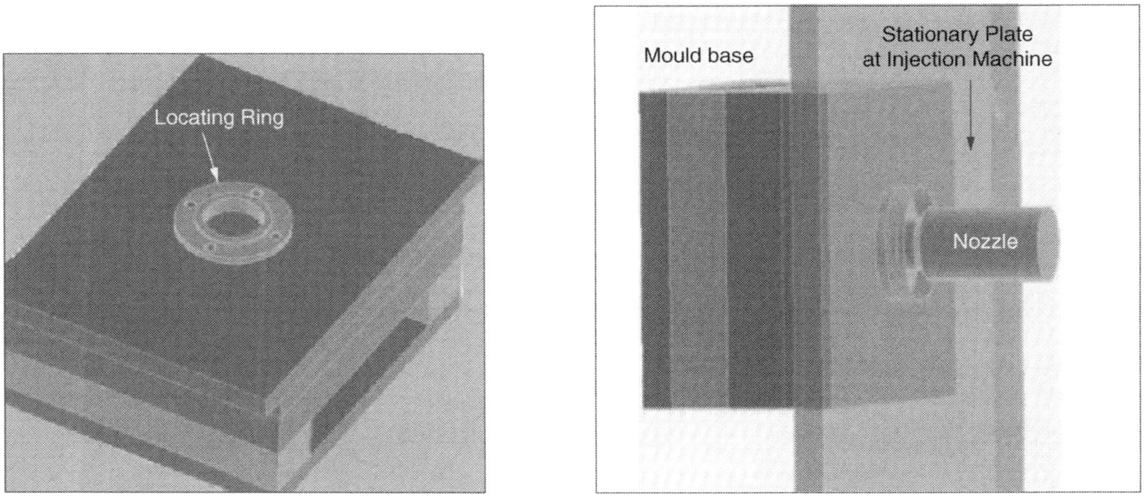

Figs. 2.20 and 2.21 Injection mould after attaching the locating ring.

Guide pillar and guide bush

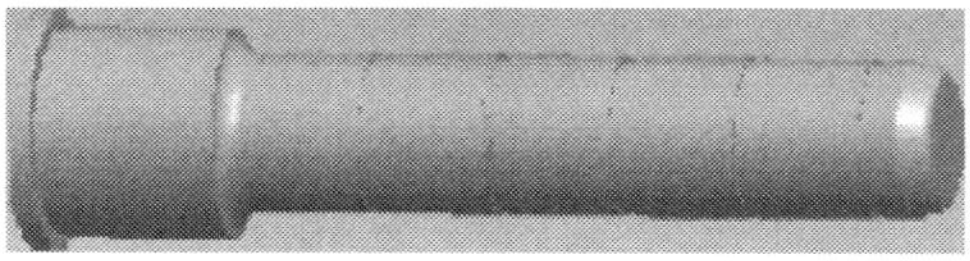

Fig. 2.22 Guide pillar 3D view.

Fig. 2.23 Guide bush 3D view.

Guide bush A guide bush is used in a mould to provide a suitable wear resisting surface for a movable guide pillar as shown in Fig. 2.23. The internal diameter of the hole of guide bush is designed as a slide fit on the adjacent guide pillar whereas external diameter is a press fit into the mould plate. A radius is made at front end of the bore to provide a lead in for the guide pillar. The rear end of the bush is often counter bored to a greater diameter than the working diameter. On each stroke the guide pillar should pass through the working diameter of the bush.

Guide Pillar A guide pillar is a male part for guiding of core and cavity with the guide bush as shown in Fig. 2.22. It is constructed in two parts. One is fitting diameter which fits with the core plate and other is sliding diameter which slides on the internal surface of guide bush which fits with the cavity plate.

Different types of guide pillars are given below:

1. Leader Pin: The alignments between two plates were achieved by incorporating shouldered pins in one half and by machining accompanying hole in the other half. It is a simple type of pin which was being used in early period when only core and cavity plate are designed in the mould. **Disadvantage:** Due to opening and closing of the mould, wear takes place in the hole and once the hole of the plate is enlarged, the perfect matching of two halves is lost. And if due to any chance the pin is bent, the removing of pin from plate will be very difficult.

2. Standard Guide Pillar and Guide Bush: As shown in Fig. 2.24, the fitting diameter is greater than the working diameter by 8 to 10 mm. The working diameter is designed as a slide fit with the adjacent hole in the guide bush. The fitting diameter is smaller than collar diameter by 10 mm; the fitting diameter is designed as a press fit with the adjacent to the hole diameter in the plate. Here the size of holes which is maintained for guide pillar and guide bush in two different plates is maintained the same and the holes are easily maintained in the plate.

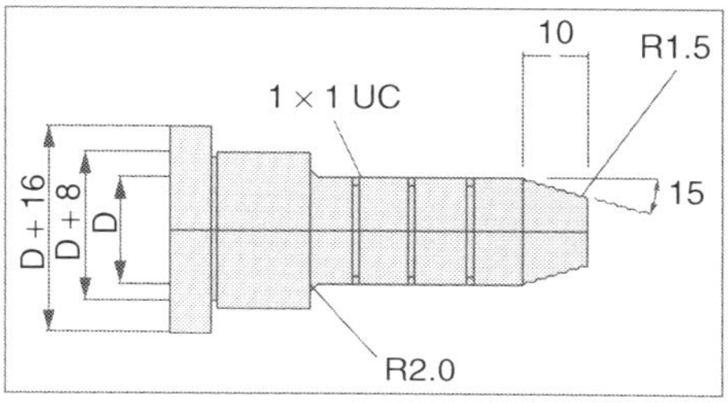

Fig. 2.24 Guide pillar.

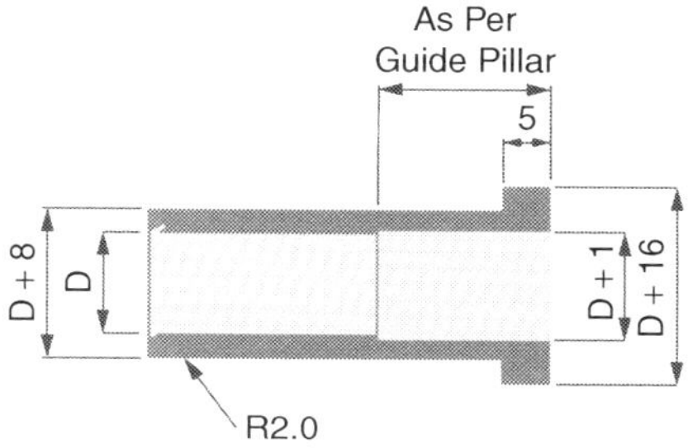

Fig. 2.25 Guide bush.

In the working diameter, 1mm undercut at two or three places are provided for lubricant between guide pillar and guide bush. As shown in Figs. 2.24 and 2.25, before starting the process or closing the mould the lubricant is given on the surface of guide pillar and with the undercut the lubricant spreads over the surface of pillar and bush. One undercut is also provided between collar and fitting diameter to insure the sharp corner. A taper portion with 15 degree angle is made at the front face of the pillar with radius of 1.5 to provide guide pillar a lead in the guide bush.

Position of guide pillar:

(a) The number of guide pillars incorporated in a mould varies from two to four. For circular moulds three numbers of guide pillars are generally preferable.

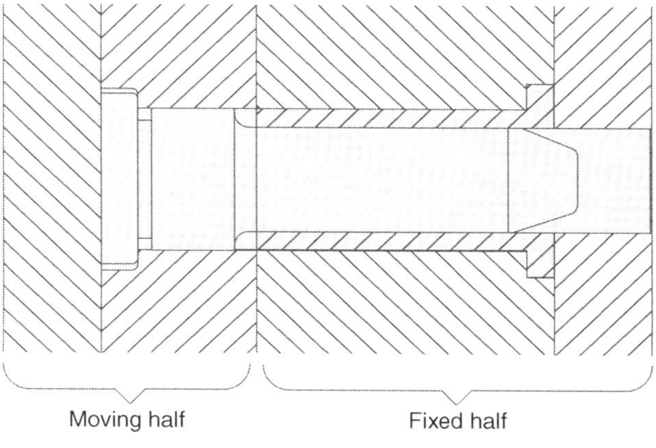

Fig. 2.26 Position of guide pillar.

(b) The guide pillars are to be incorporated generally in the moving half of the mould shown in Fig. 2.26 at the four corners of a rectangular mould. For circular mould, the pillars are positioned at 120° apart.

Strength of guide pillar The size of the working diameter of guide pillar depends on the size of the mould and whether or not a side force is likely to be exerted on it. The moulds with deep and heavy cross sectional cores exert side thrust and the guide pillars should be strong enough to absorb them without any damage.

For circular cores, side thrust $(Q) = 2/3 \times d \times h \times P_f$

where d = Maximum diameter of cores (cm); h = Height of core (cm)

P_f = Cavity pressure causing side thrust taking into consideration the effect of clamping force (kg/cm^2)

For rectangular cores, $Q = a \times h \times P_f$

where a = Maximum side of core (cm); h = Height of core (cm)

Working diameter (d) of guide pillar is $d \geq \sqrt{4Q/(N.f_s.\,\Pi)}$

where Q = Side thrust (kgf) ; N = no. of pillars ; fs = Shear stress

3. Spigotted Guide Pillar and Guide Bush: When an extra land called spigot is provided in the standard guide pillar and bush to guide for a respective plate is known as spigotted guide pillar and guide bush. It has very similar design as standard guide pillar and guide bush except a spigot is provided in both the components for giving alignment to the respective mould plate for both the members. The spigot of guide pillar is guided with an accommodating hole in the backing plate similarly the spigot of guide bush is guided with the backing plate. In addition to providing guiding to core and cavity, it additionally provides guiding to respective plates also as shown in Fig. 2.27.

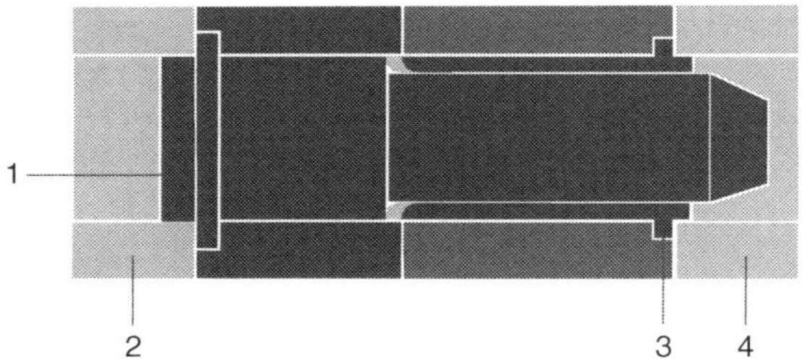

Fig. 2.27 Spigotted guide pillar and bush.

1. Spigotted guide pillar
2. Back respective mould plate
3. Spigotted guide bush
4. Front respective mould plate

Taper locator and taper location For the small moulds, the guide pillars and guide bush are able to maintain guiding for perfect matching of core and cavity within closed tolerance, irrespective of the force applied during injection. But for thin walled multicavity component mould, by guide pillar and bush, the closed perfect alignment cannot be maintained. So in design stage with the guide pillar and bush, a tapered location is used in addition to guide pillar as shown in Fig. 2.28.

The core and cavity of the mould is separately machined in the tool room and when it is matched, there may be a chance for not getting a perfect uniform impression of wall thickness. In a thin walled component, the material flows into the impression to fill the cavity and as per the fluid phenomenon. First, it fills the thicker portion of the part and apply force on core which create discrepancy in the alignment of core and cavity. Due to first filling of thickest portion, the result is that a differential force will be applied on face which is resisted by the guide pillars and the result will cause variation in wall thickness and misalignment.

For fixing the taper location as shown in Fig. 2.28, a recess is machined in the core plate into which the tapered projection part of cavity plate fits. Sometimes, taper locators are also used. For fixing the taper locator, a taper hole is machined in the cavity plate. A tapered projection part of same size is also fixed in the core plate. During matching of core and cavity plate, taper projected pin of core plate perfectly fits into the taper hole of cavity plate. Taper locator is turned from round bar and in the mould; four or more can be used.

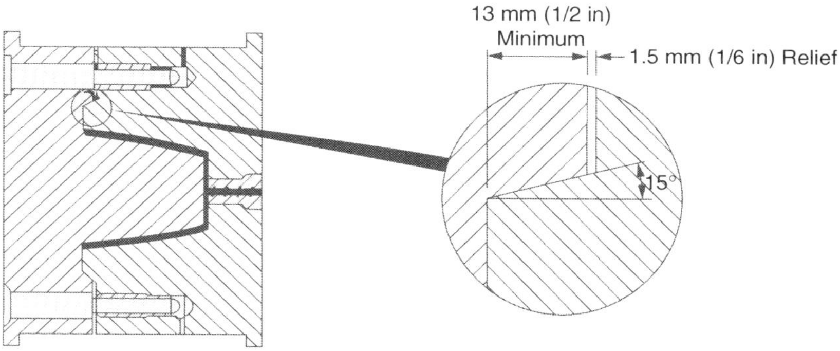

Fig. 2.28 Taper location.

In the taper locator, the disadvantage is that the applied internal forces may force the cavity to expand and hence guiding of both the halves is lost. The taper locator consists of a pair of circular matched members which are incorporated in the opposite faces of two mould halves.

The preferred method is to provide a tapered location on the cavity side fitting into a recess on the core side. This achieves the same result, as in the previous case, but overcomes disadvantage. It has also additional advantage of acting as a form of chase for the cavity, helping it to resist the expansion forces.

Support Pin for the Mould Main part of support pin may be divided into three types.

1. Support pin: It is a simple leader pin known as the main support pin.

2. Bushings: As shown in Fig. 2.29, the number of bushes that can be needed depends on the type of mould base, for example for three plate type of mould, it requires at least two bushings, one in stripper plate and another in cavity plate. Bushings prevent the direct contact of support pin with mould plate, because mould plate usually uses more soft material and support pin is made of harder material, it can cause scratch if used for long time in mass production. Bushings always use harder material with good tolerance to support pin. So the other advantage is that the mould can be opened and closed with precision and can be used in mass production.

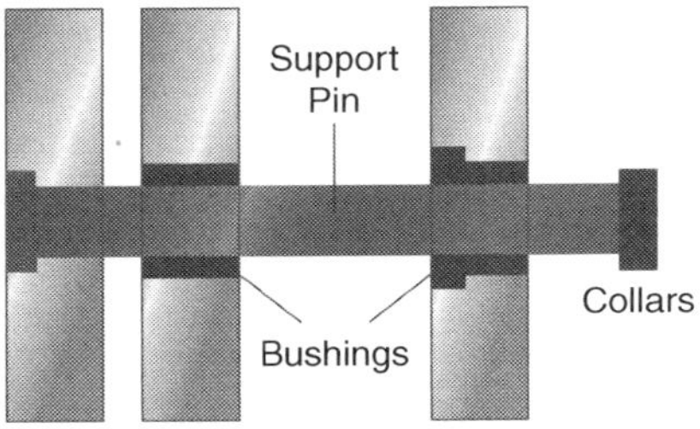

Fig. 2.29 Support pin with bush.

3. **Collar:** It is used to prevent mould plates from leader pin when puller bolt is broken. Collar gives security to the mould construction.

Ejector guide pillar and bush It is used to guide ejector plate and ejector back plate during ejection. The ejector pin is located in the ejector plate. For safe working of pin, it is required to guide the ejector pin which can be easily done by guiding the ejector plate by ejector guide pillar and guide bush. With the understanding of ejector system and ejector guide pin function, ejector guide pillar location is decided between ejector plate and ejector back plate.

To fit the ejector pillar, the bolt (some times the pillar is fixed press fit with bottom mounting plate) can be used at the bottom of ejector guide pillar, and clamped it in bottom mounting plate of mould and ejector guide bush is fixed in the ejector plate and ejector back plate as shown in Fig. 2.30. The working diameter of guide pillar is machined as slide fit to the inner surface of guide bush and the pillar slides on the inner surface of the bush.

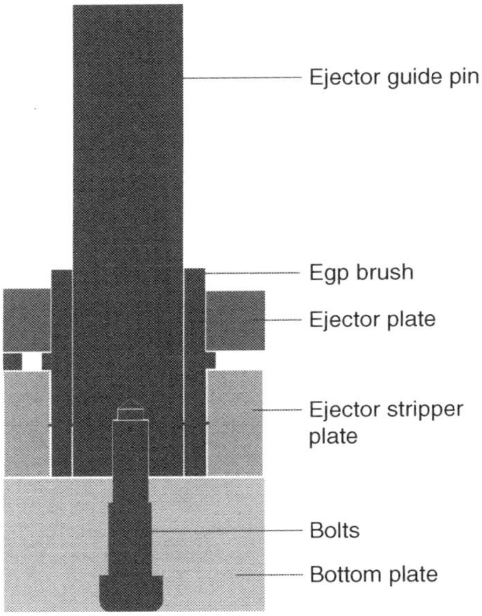

Fig. 2.30 Assembly of ejector guide pillar.

2.1.9 Mould Location

The cavity half of injection moulds is located and fixed on fixed platen of the machine by matching the platen hole and the projected part (locating ring) of the mould. As per the machine specification chart, dimension of platen hole is taken for the fabrication of locating ring. During fabrication of locating ring, the given tolerance (slide fit) is adopted. During loading of the mould, the fixed half of the mould is fixed in the fixed platen of the machine, first by matching the hole of platen and projected part of locating ring and clamp the mould fixed half with platen.

2.1.10 Mould Lifting

The three-plate mould in assembly condition is shown in Fig. 2.31. The status of mould is ready to be lifted by crane and an eye-bolt is attached to centre of mould plate. After assembly process, for safety, a tension link is attached in three-plate mould type or safety bolt or magnet in two plate type of mould which prevents mould plate from falling down when lifted up with crane.

Fig. 2.31 Mould lifting (1) is eye bolt and tension link (2) Mould is in ready position for handling.

After fabrication of the mould, it is loaded on the machine for trial production. For lifting of the mould, special arrangement is provided to maintain the safety as well as to remove the chances of damaging of the mould during shifting from machine shop to processing shop. The mould is generally lifted by chain and hook system. In mould one, strip plate is fixed in such a manner that one half of the strip is screwed with fixed mounting plate and other half with the moving mounting plate and one hook is provided at the centre. For loading on the machine, a hook is provided in both the half and through the hook by chain, the mould half is loaded on the machine.

2.1.11 Mould Clamping Method

Clamping of the mould in the machine is of two methods:

1. Direct clamping method
2. Indirect clamping method

1. Direct Clamping: In this method, clamping of the mould halves are done by direct bolting. Holes are provided in the mounting plates of both the mould halves. It is to be done with the corresponding holes in the platen of the machine shown in Fig. 2.32. The main advantage is that the maximum possible load is directly transmitted. Some of the systems are shown in Fig. 2.32. In the mounting plates, the holes or slots are made at the same distance as in the platen.

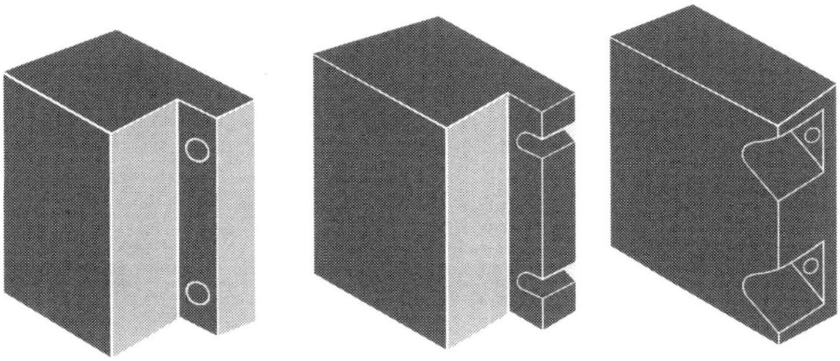

Fig. 2.32 Direct clamping.

2. Indirect Clamping: Due to the irregular size of the mould plates this method is adopted when direct bolting is not possible for use. Assembly for the indirect clamping consists of three parts – clamp plate, the bolt and the packing piece (refer to Fig. 2.33). Mould clamp plate incorporates a central slot for adjustment purpose. The required clamping force is obtained by tightening the bolt. In this method, the bolt is bolted at a distance from the plate.

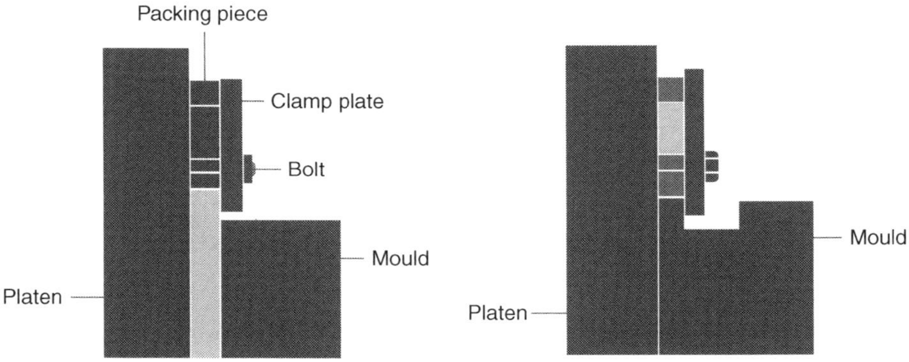

Fig. 2.33 Indirect clamping.

2.1.12 Mould Construction and Assembly

Male and female parts The injection mould has two halves, one fixed and other moving. Fixed half is fixed with the fixed platen of the machine and moving half with moving platen. The fixed half of the mould consists of female part or cavity part and moving platen has core or male part of the mould. The cavity part is made as per the external surface of the product whereas core part is internal surface of the product.

Why is the core fixed in moving half of the mould? Because as the material enters into the impression of the mould in the form of liquid, due to cooling, it changes its status from liquid to solid and the molecules move towards the centre and shrink on the core part of the mould and remains with it as the mould opens.

Fig. 2.34 Cap product.

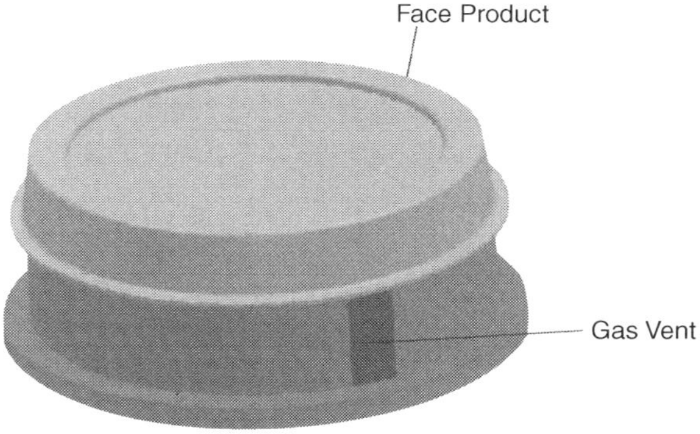

Fig. 2.35 Core side for bowl.

Product Complete core and cavity mould design is explained below.

By looking at the bowl, the internal and external surface of the bowl can easily be considered and for the above, please find below Fig. 2.35 core part of the article.

Parting line Selecting of the right parting line is an important part for the mould design. Parting line will separate core and cavity side, because the parting line is the line which is created due to the matching of cavity and core and get separated at this line. First we will take the bowl in a block in such a manner as shown in Fig. 2.36 that if the product subtract by dividing and splitting block at the edge of the bowl and the blocks for cavity and core are shown in Figs. 2.37 and 2.38.

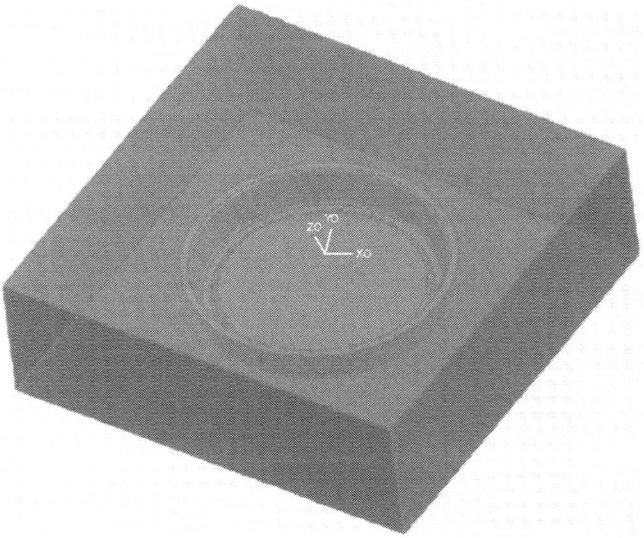

Fig. 2.36 Bowl in a block to get core and cavity side.

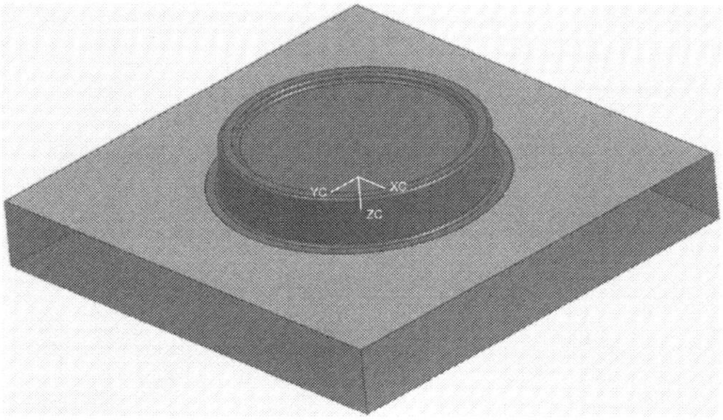

Fig. 2.37 Core side.

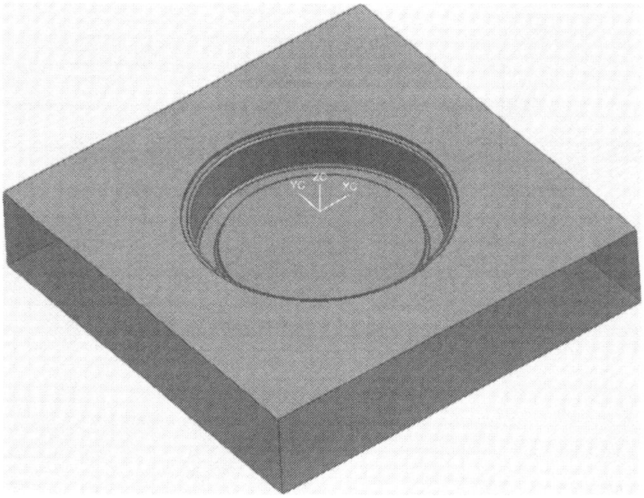

Fig. 2.38 Cavity side.

Assembly of mould Before starting assembly of the mould, the following points are to be checked:

1. Have all the mould plates, inserts, and components in one place ready for assembly.
2. Have a clean table and tools which are going to be used in good condition and the tools include Allen wrenches, aluminium hammers, etc.
3. Cleanliness is critical in mould assembly. Make sure all plates, inserts and components are clean and free from grit, debris, and chips. Cleaning should be done very carefully not to damage sharp edges, parting surfaces, shutoffs or the cavity finish.
4. Install all subinserts, gate inserts, core pins, etc., into the primary inserts. Check that all inserts and pins are marked and that they are installed in the correct location and position. Insert and grease all ejector pins, ejector sleeves, and ejector blades through the pin retainer plate, support plate, and core inserts. Install all return pins and springs.

5. Assemble the ejector housing, with support pillars, guided ejection pins, etc. Guide this assembly through the ejector and other plates and bolt it to the support plate. Insert any core pins that are mounted in the bottom clamp plate and fasten their back up plates. Lubricate the entire assembly.
6. Verify that the slide assembly moves freely; is greased; and that the slide retainer is functioning properly.
7. Move the ejector assembly forward and check if all pins, sleeves, lift cores and all other moving components spin freely.
8. Check that ejector plate can use the full length of travel.
9. Install split insert set, heel blocks, angle pins, and other components. Check that all parts are marked for split mould and that they are installed in the correct location and position.
10. Install the locating ring and sprue bushing, check that sprue radius and orifice are of the correct size and verify that the sprue bush rotation is locked and retained.
11. For three-plate moulds, verify that all latches function properly and that plate separation is sufficient for the part and the runner drop falls through clearly. Also check that all latch dowel pins are secured so as not to come loose during operation. Lubricate the whole assembly and verify that it moves freely.
12. For hot runner moulds check that all wiring is in a channel, free from damage, and free from possible 'pinches' during assembly. Check continuity of all circuits.
13. Use teflon tape or suitable thread sealant for water test.
14. Check all limit switches and close the mould.
15. Verify that the mould has a mould strap and that it is fastened correctly.

2.1.13 Standard Mould Base

The sizes and measurement of the standard components or mould bases are directly related to the platen size, the shot capacity, locking force and other parameters of the moulding machines.

A mould system is defined as an assembly of mould parts, the plates of which confirm to accepted structural shape and size. Figure 2.39 shows a standard mould base.

Generally, two-plate mould is adopted as the standard mould system because it is the most widely used design in industrial practice. The system comprises two mould plates namely core and cavity plates, an ejector system, guide pillars, guide bushes, etc.

Main advantages of standard mould base

(a) By adopting standardisation, the designer does only the essential portion of the mould design, e.g., core / cavity profile, feed system, ejection location, etc.

(b) The time spent on the manufacturing mould base can be utilised for manufacturing core and cavity details so that the efficiency of the output can be easily increased.

(c) Machinists replace highly skilled mould makers.

(d) The cost of the mould unit is known, therefore, estimation is easier.

(e) Minimum stock of steels is required, resulting reduction of investment.
(f) Interchangeability of parts will be easy.
(g) Tool maintenance is reduced.
(h) The delivery time can be reduced.

Limitations of standard mould base

(a) The number of sizes available is limited.
(b) Maximum size, depth of the mould plate may not match with the actual size.
(c) In order to avoid the deflection of the mould plate, extra support blocks may be required as the position of the support blocks may not match with the requirement.
(d) The ejector stroke may be larger than it is actually required.

Manufacturers of standard mould base Standard mould base includes all elements of a plastic mould, excepting the core and cavity inserts. Different internal national agencies manufacture standard mould system. Out of them, the following two are quite common:

1. DME Standard: In this standard, N-type ranges of mould systems are available which include bore plates plus guide pillars, guide bushes, dowels and cap screws ; assembled mould unit consisting of bored plates, push back pin, sprue bush, locating ring, etc.

2. HASCO standard: In this standard, K standard elements is used to designate their standard system. Here mould plates are supplied; bored and ground to accommodate guide pillars, guide bushes, etc.

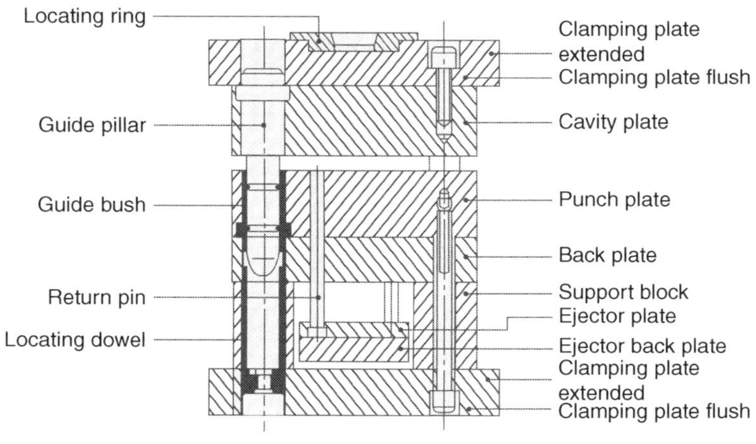

Fig. 2.39 Standard mould base.

Mould vent It is a shallow recess or hole incorporated in the design to permit entrapped air or other gases to escape freely. Most moulds do not need special design features for venting because air has sufficient possibilities to escape along ejector pins or at the parting line. Entrapment of air may cause mould defects like discolouration, sinks, incomplete filling, etc. The configuration of the moulding, its position in the mould and its gating has a considerable effect on venting. The approximate size of a vent is 0.05 mm deep by 3.0 mm wide.

Position where a vent is likely to be required are:

1. The furthest point from the gate on symmetrical mouldings.
2. At the point where flow paths are likely to meet.
3. At the bottom of projections.

2.1.14 Mould Material

Selection criteria of mould material How to select the mould steel?

Mould steel has three general types:

1. Forged steel
2. Prehardened steel
3. Tempered steel

With the consideration of below mentioned three factors, the steel material is selected.

1. Mould Life: Normally for the prehardened steel, the general mould life is 2 lakh shots. If it is required to get the mould life higher then tempered steel is selected.

2. Finishing of Plastic Parts: Plastic surface requires finishing with fine texture or mirror finished, it is suggested to use stainless steel or high nickel and chrome steel.

3. Plastic Raw Material: It is divided into two types: soft material and hard material. If plastic parts are made from a soft material its quantity is not so high, prehardened steel can be used; otherwise, use a kind of tempered steel.

Mould materials and their applications for a standard mould

1.	Back plate (fixed half and moving half)	:	Mild steel
2.	Core back plate	:	Mild steel
3.	Cavity and core	:	Tool steel
4.	Ejector back and retainer plate	:	Mild steel
5.	Ejector and push back pin	:	Carbon steel
6.	Guide pillar and guide bush	:	Case hardened steel
7.	Sprue bush	:	Case hardened steel
8.	Locating/Registered ring	:	Mild steel

Without proper mould material, proper tooling cannot be achieved. Wide ranges of mould materials are used for fabrication of moulds and dies for plastics.

Main Requirements for Selecting Steels They are:

1. Excellent machinability
2. High wear resistance
3. Good polishability
4. Excellent heat treatability
5. Sufficient corrosion resistance

The steel materials like mild steel, alloy steel, carbon steel, case hardened steel, etc., are used for the fabrication of mould. Selection of mould material should be carried out based on the requirements from product / mould designer, mould maker and mould as shown in Table 2.2.

Product design requirements Apart from functional performance of the component, the moulding is required to meet the high standards of surface finish as well as dimensional tolerance over long production run. These requirements will be met only if we have good product design, good mould design, good mould making and proper selection of tool steels for the mould.

A product designer has to consider the following for trouble free production run.

1. The surface finish (mirror/optical finish) on component.
2. Whether pattern on the moulding is required by photo etching, etc.
3. Whether moulding material is corrosive, abrasive or both.
4. Requirements of maintaining close tolerance.
5. Quantities to be produced.

Table 2.2 Steel Requirements.

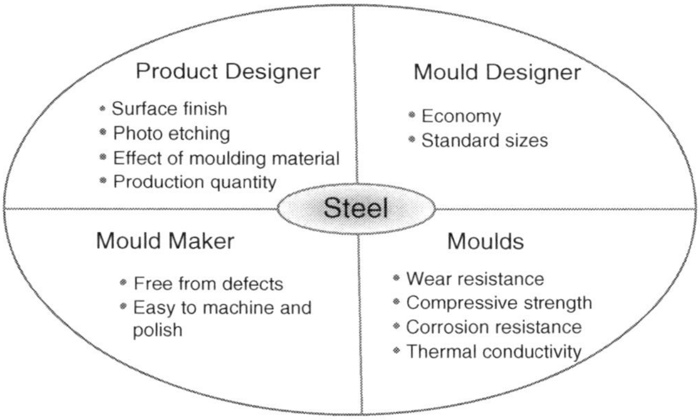

Considering the above features the designer can short-list the tool steels required for manufacturing the mould.

Mould design requirements The mould designer can economise the tooling of the mould by selecting standard steel grades, standard steel sizes and standard machined plates. The designer along with the mould maker, shares the responsibility of producing a mould, which will give reliable and economical production of the part visualised by the product designer. He has to see that the mould maker fabricates the mould as easily and economically as possible. This largely depends on specifying best mould steel and optimum hardness required for different mould parts. Selection of standard parts such as guide pillars, guide bushes, ejector pins, etc., and standard mould bases along with other machined plates considerably reduce time and cost as these parts are nowadays available at far more competitive price. This in turn helps in better planning, prompt mould delivery as well as in minimising initial machining cost and material losses.

Mould making requirements A substantial part of the total mould cost is that incurred in machining of the mould.

The mould steel should have following properties:

1. Materials are free from any manufacturing defects
2. Good machinability
3. Good polishability
4. Good hardenability
5. Good dimensional stability

Moulder requirements A processor expects certain features in the mould such as:

1. Uniform and high rate of production
2. Uniform moulding quality
3. Longer mould life
4. Low mould maintenance cost
5. Lowest possible production cost
6. Easy components and material replacement

The selection of tool steels for mould not only depends on the product/mould design but a considerable thinking has to be done from mould maker and processor point of view. If all these factors are given thorough weightage, a mould can be manufactured and processed economically and effectively.

Essential properties of mould materials

1. Excellent Machinability: The economic importance of machinability is very great indeed. Roughly 30 % of the total cost of a mould is accounted by the machining costs.

2. High Compressive Strength Combined with Sufficient Toughness: In injection mould, it is estimated that a locking force of 1 ton per 3–4 cm^2 of projected surface is required. When large items are being injection moulded the locking force calculation is very much important. There is always a risk of indentation of the parting lines, but the risk decreases with increasing compressive strength on the part of the tool steel use. High compressive strength is also required in view of the fact that the tools are liable to mechanical damage in course of transportation and installation. In some cases, the tools are nitrided to prevent the cavity from becoming scored or otherwise spoiled. The compressive strength of the steels can be improved through nitriding.

3. Capacity for Heat Treatment without Problems: A small change in hardening is unavoidable. It is nevertheless possible to limit the changes (warpage) through slow and even heating to the hardening temperature and by choosing a low hardening temperature and a suitable quenching medium.

The use of hardened and tempered steels does not require any further heat treatment.

4. Good Resistance to Heat and Wear: In order to improve the strength of the plastic product, an addition of glass fibre, asbestos, wood fibre, etc., is frequently made. These substances have an abrasive effect on parting lines, runner gates and inlet nozzles. For such products, it can therefore, be advisable to choose particularly wear-resistant tool steel.

5. High Thermal Conductivity Every plastic manufacturer is anxious to maintain as high a production pace as possible. The limitation lies in the ability of the mould to conduct heat away from the plastic item. In this context, the thermal conductivity of the material is obviously important, but so is heat transfer between tool and coolant and heat transport through the coolant.

6. Ability to Resist Corrosion Generally all moulds used for injection moulding come into contact with cooling water. Some plastics are also corrosive to the mould material. Certain plastics generate corrosive products such as hydrochloric acid from PVC plastics, acetic acids from acetate plastics and water from amino-plastics. PVC plastics are well known for their tendency to produce hydrochloric acid when heated to high temperature.

This type of corrosion attacks the surface of the mould. If the mould is then polished to high gloss, the corrosive attack necessitates expensive repolishing.

7. Good Polishability Polishing is a time-consuming and costly process. The result of polishing will depend in the first instance on the polishing technique used. The polishability of steel depends on the homogeneity of the steel and on the type, distribution and size of slag inclusions. Hard, large slag particles are particularly troublesome. Polishability is also highly dependent on the hardness level and heat treatment of the material. To receive the highest purity, the steel should be vacuum degassed or electro-slag refining.

Factors governing the choice of mould materials

1. Length of production run
2. Injection pressure required
3. Type of moulding material
4. Dimensional accuracy and intricacy
5. Method of manufacture

Cost considerations

Mould steel cost One of the major decisions that can favourably influence overall mould performance and long term maintenance costs is to specify the best possible mould steel for the job. The cost of the tool steel in a mould usually represents only 10–15 % of the total cost and offer savings made on buying cheaper mould steels turn out to be false encouragement.

Hardened steel applications

1. Moulds with long production runs
2. To resist abrasion from certain moulding materials
3. To counter high closing/injection pressures

Advantages

1. These steels are available in soft annealed and stress relieved condition.
2. They are also available in hardened and tempered to the required hardness (about 48–60 HRC).

3. These are used for core and cavity inserts.
4. Better wear resistance, resistance to deformation and indentation and better polishability.
5. Improved wear resistance is important when filled or reinforced plastic materials are to be used.
6. Resistance to deformation and indention in the cavity, gate areas and parting lines helps in maintaining proper part quality.

Corrosion resistant mould steels

Application

When a mould is likely to be exposed to a corrosive atmosphere, stainless steel is strongly recommended. The higher initial cost of the steel will be less than the cost involved in repolishing or replating of mould parts affected by corrosion.

Prehardened steels

Applications

1. Large moulds
2. Moulds with lower wear resistance requirement
3. High strength holder plates
4. Moulds with moderate production run

Advantages

1. These steels are available in the hardened and tempered condition having hardness about 25–30 HRC.
2. No heat treatment is necessary before the mould is put into use.
3. Flame hardening or nitriding can increase the surface hardness.

Effect of alloying elements in steels used for mould making

Table 2.3 Effect of alloying element on steel.

No.	Element	Increases	Decreases	Common percentage (%)
1	Carbon	Strength, Heat resistance up to 400°C, coarse-grain formation	Elongation, Ductility, Deep-drawing quality, Malleability	<1.2
2	Manganese	Strength, Ductility, Malleability, Forge weldability (with small %), Hardenability, Wear resistance, Coarse-grain formation, deoxidation	Resistance to wear and machinability	< 8

(*Contd.*)

Table 2.3 (*Contd.*)

3	Silicon	Strength, Hardenability, Resistance to scaling, Coarse-grain formation, Deoxidisation, Elasticity, Yield point	Elongation (little), Impact value (little), Cooling rate, Carbide formation, Resistance to wear, Machinability, Forgebility, Scaling, Nitrability	< 1
4	Aluminium	Resistance to scaling, Coarse-grain formation, Deoxidisation, Nitrability,	Impact value, Forgability, Scaling.	< 0.5
5	Nickel	Strength Ductility, Corrosion-resistance, Elongation, Impact value	Machinability, Forgebility, Scaling, Hardness, Yield point, Cooling rate, Coarse-grain formation, Magnetic properties	< 10
6.	Molybdenum	Strength, resistance to wear, Heat resistance, Hardenability, Retention of hardness, Creep strength, Nitrability Resistance against hydrochloric and sulphuric acid, Carbide formation, Scaling Magnetic properties	Elongation, Cooling rate, Forgebility, Machinability	< 2
7	Vanadium	Strength, High carbide formation, Impact, elasticity, Heat resistance, Resistance to wear, Forgeability, Resistance to corrosion, Nitrability, Creep strength, Deoxidisation	Temper brittleness, Cooling rate, Scaling	< 2
8	Tungsten	Strength, Hardness, Preservation of cutting edges, Temperature of heat treatment, Corrosion-resistance, Carbide formation, Resistance to wear, Magnetic-properties, Nitrability	Elongation, Coarse-grain formation, Cooling rate, Forgeability, Machinability, Scaling	< 2
9	Cobalt	Strength, Cooling rate, Resistance to wear, Preservation of cutting edges, Magnetic properties.	Temper brittleness Elongation, Forgeability, Scaling	< 2
10	Copper	Strength, Hardness, Yield point improves, Resistance to corrosion	Tendency to rust, Reduces forgeability	< 0.5
11	Sulphur	More machinability, Red-hot brittleness	Tendency to rust, Forgeability, Impact value	< 0.5

(*Contd.*)

Table 2.3 (*Contd.*)

12	Phosphorus	Strength, Heat resistance, Fluidity, Cold brittleness, Tempera-brittleness, Hardness, Yield point, Reduces elongation, Machinability improves.	Elongation, Impact value Reduces Forgeability	< 0.5

Materials and treatments for mould parts Service requirement, fabrication requirement and economical requirement of various products are different from one another. These input requirements are to be clearly understood before the selection of right material.

There are some common properties required in all parts such as good strength, less dimensional changes during heat resistance treatments, good machinability, good polishability, etc.,

Table 2.4 Materials and Heat Treatment for Mould Parts

Name of the part	Functional requirements	Materials	Treatment
1.Bolster (a) Mounting plate (b) Core and Cavity retainer and back plate (c) Ejector retainer and back plate (d) Spacerblock	To withstand bending stress High compressive strength Good dimensional stability Toughness at high temp.	Low carbon steel EN2A, Medium carbon steel, En8, C45, En 24 Prehardened steel (P 20)	Toughening, Case-hardening, and Nitriding.
2. Guide pillar and Guide bush	Good shear strength Good wear resistance	Case-hardening steels,(C 1020, C1030, etc.,) Alloy steels	Case-hardening
3. Core pin/Ejector pin/Sprue puller	Good impact strength High wear resistance Good corrosion resistance	Case-hardening steel En36, Hot work steel(H 13)	Case-hardening, Through hardening, Nitriding
4. Sprue bush	Good hardenability Good corrosion resistance Good wear resistance Good thermal conductivity	Fully hardened Steel (D2,H13, H11, etc.,)	Hardening
5. Slides	Wear resistance Shock resistance	Fully hardened Steel Prehardened steel (P20)	Hardening, Nitriding
6. Knockout rod	Good strength and rigidity Good impact strength	Medium carbon steel En8, En24	Toughening

Table 2.5 Applicable core/cavity materials according to type of plastics.

Sl. No.	Types of plastics	Properties required for core/ cavity	Type of steel needed to select
1	General purpose PE, PP, HDPE, ABS	Low strength Simple mould	Prehardened steel with 22 to 25 HRC
2	Engineering plastics nylon, PC.	Functional parts High strength Precision mould and distortion free after heat treatment	Hot die steel (H-13, H-11, WPS, etc.,)
3	Reinforced fibre glass treated	High Strength Long life mould and wear resistance	Fully hardened mould (hot die steel) above 50 H.R.C.
4.	Fire retardant additives and Vinyl chloride plastics	Corrosion resistance material	Stainless steel.
5.	Transparent plastics	Better polishability of steel	Steel with minimum impurities like 'S''Mn' and higher Cr. and Ni.
6.	Large moulds, e.g., Car dash board	High toughness	Prehardened steel up to 40 HTC (bigger size of core/cavity/difficult heat treat, where the chances of crack or break is more)

Functional requirement of materials and treatment The core and cavity inserts are the most important parts, which have to withstand various stresses continuously. There are various materials used for core and cavity inserts and the proper selection is a difficult task, since it decides the performance and life of the mould, along with other factors.

Table 2.6 Functional requirement of materials and treatment

Name of the part	Functional requirement	Materials	Treatment
Core and Cavity insert	Good strength and toughness Good compressive, impact strength Good corrosion, wear resistance Good thermal conductivity Good weldability Low cost of thermal expansion.	The most common steels are P20, H13, Nitriding Steels, Stainless steels (420) BeCu alloys	Hardening, Nitriding for prehardened steels and coatings for improved life.

Steels used for different mould parts

1. Constructional steel is used for mould bases and structural parts.
2. Components such as blocks, spacers, ejector pins and support block.
3. Pillars, Backing plates and locating rings, etc.

Material for guide pillar, bushes and ejector pins:

Table 2.7 En-18, En-19, En-24 and En-36 material composition.

Prehardened steels:

Sl. No.	Steel type	AISI	C%	Si	Mn	Cr	W	V	Co	Mo	Ni	Hardness rockwell 'C'
1	En-2A		/.12	-	/.50							
2	En-8	1040	0.35/ 0.45	0.10/ 0.35	0.60/ 0.90	S and P Each 0.06 (max)						BHN 179-229
3	En-9	1055	0.45/ 0.06	0.10/ 0.35	0.60/ 0.80	S and P Each 0.06 (max)						BHN 201-277
4	En-19	4140	0.35/ 0.45	0.10/ 0.35	0.50/ 0.80	0.90/ 1.50	S and P Each 0.05 (max)			0.20/ 0.40		BHN 201-415
5	En-24	4340	0.35/ 0.45	0.10/ 0.35	0.45/ 0.70	0.90/ 1.40	S and P Each 0.05 (max)			0.20/ 0.35	1.30/ 1.80	BHN 223-444

Core and cavity material:

Table 2.8 Mould materials – Equivalents.

Steel type	AISI	C%	Si	Mn	Cr	W	V	Mo	Ni	Hard-ness rockwell 'C'
Hot die steel	H21	0.30	0.20	0.30	2.60	8.5	0.40	----	---	42–50
	H13	0.3–0.4	0.8–1.2	0.25–0.5	4.75–5.25	1.2–1.6	1.0–1.2	1.2–1.6	---	
	H11	0.36	1.00	0.40	5.00	-----	0.40	1.10	---	38–52
Non-shrinking die steel	D3	2.00	0.30	0.30	12.00	---	---	---	---	58–65
HCHC	D2	1.70	0.30	0.30	12.00	0.50	0.10	0.600	---	58–65
OHNS	01	0.95	---	1.00	0.50	0.50	0.10	----	---	58–63
	02	0.90	0.20	1.90	0.30	----	0.10	----	---	58–64
Stainless steel	410	0.10	0.75	1.00	13.00	---	---	---	0.60	BHN 159–229
	304	0.08	1.00	2.00	18.00	---	---	---	10.0	BHN 135–185

When it comes to injection mould and blow plastic mould materials, nothing speeds production like copper alloys. Mipalloy 100's thermal conductivity is almost 10 times greater than tool steels, so it offers faster, more uniform heat dissipation. That means more than 20% shorter cycle times. But speed of production is not the only factor. Current research at Western Michigan University shows beryllium copper cores exceed most production requirements. For extreme conditions such as long runs of 30% glass-filled nylon, beryllium copper lasts longer than P-20 steel cores.

Mipalloy 100 The plastics industry has proved it as a material of choice for injection moulding and blow moulding applications.

Table 2.9 Mipalloy 100.

Mould material	Thermal conductivity (Btu/Hr/Ft2/°F)
Mipalloy 100 (C18000)	125
0.55% Beryllium copper	125
H-13	17
P-20	20
SS420	14

The injection moulding cycle is made up of a number of elements. They include the filling portion, the cooling portion and the mould open portion. The cooling portion is always the longest and is usually 65% of the overall cycle. Therefore, the longest element in the overall cycle is where the greatest benefit can be obtained in improving the injection moulding cycle and where beryllium copper works best.

In addition to its superior thermal conductivity mipalloy 100 offers the following advantages:

(a) Maintains high surface finish.
(b) Accepts etching and texturing.
(c) Requires no additional heat treatment. Supplied in heat treated condition.
(d) Can be readily machined using conventional machining practices as well as Electrical Discharge Machining (EDM).

Mipalloy manufactures and stocks beryllium copper in various shapes and sizes:

1. Round rod from 6 mm diameter to 145 mm diameter
2. Rectangular and square bars from 1"square upwards
3. Large forgings weighing more than 1000 kgs single piece

Recommended injection mould applications:
Core pins, cavity areas, sprue bushings, ejector pins and sleeves, manifold system.

Advantages

(a) Improved control of post-mould shrinkage.
(b) Better heat dissipation in areas of heavy wall sections or limited water channel access.
(c) Improved dimensional stability in multicavity tools or in large flat walled parts.
(d) Excellent wear life when mated with standard tool steels.

2.1.15 Mould Life Cycle

In the processing shop, it is clear that moulds do not last for ever. There are three basic failure modes to permanent mould tooling.

1. Thermal fatigue manifested in cracking primarily.
2. Erosion manifested in dimensional errors and/or texture problems.
3. Damage due to improper handling or maintenance.

Moulds may be retired for many reasons such as obsolescence but they fail for one or more of the three reasons above. It is also clear that moulds, even of the same design, do not last for exactly the same number of shots. So mould tooling life is not characterised by a single factor.

Mould life cycle is dependent on various factors, which are

1. Plastic material to be moulded
2. Type of moulding operation viz. injection, blow, etc.,
3. Complexity of the part
4. Type of machine used for moulding
5. Operator's care in moulding and maintenance / handling
6. Type of material used in the construction of the mould.

The table below lists the average number of mouldings to be expected, dependent on the mould material.

Table 2.10 Materials and their number of shots.

Core and cavity material	No. of shots	
Alloy steel – OHNS	2,00,000 – 3,00,000	Injection mould
Hot die steel / HCHC/WPS	2,00,000 – 5,00,000	
Prehardened steel P-20	1,00,000 – 2,00,000	

2.1.16 Heat Treatment

In heat treatment mechanical properties are altered by:

1. Changing the size of the grains of what it is composed or by
2. Changing its micro constituents

Purpose of Heat Treatment

1. To improve machinability
2. To relieve internal stresses
3. To change or refine grain size
4. To improve mechanical properties
5. To improve resistance to wear, heat and corrosion
6. To produce a hard surface on a ductile interior
7. To improve magnetic and electrical properties

Different Process of Heat Treatment

1. Annealing is generally used to soften the steel.
2. Normalising is used to eliminate coarse grain structure obtained during forging, rolling and stamping and to produce fine grains.
3. Hardening is done to develop high hardness to resist wear and enable to cut other metals.

The hardness produced by hardening depending upon the carbon content of steel. Steel containing less than 0.15% C does not respond to hardening treatment.

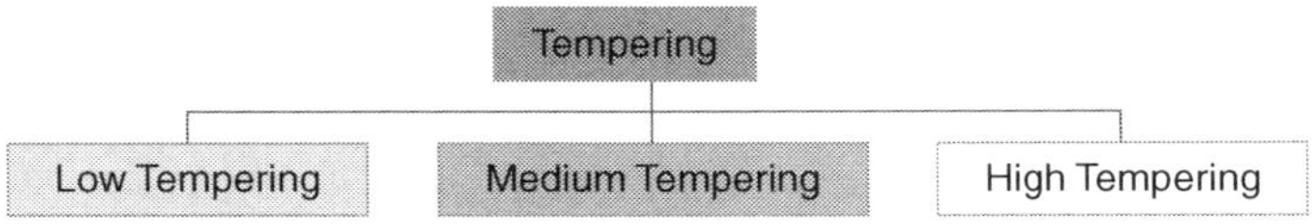

Fig. 2.41 Tempering chart.

Tempering is done to reduce internal stresses and reduce some of the hardness.

Special Hardening Techniques

1. Vacuum hardening
2. Laser hardening
3. Plasma hardening

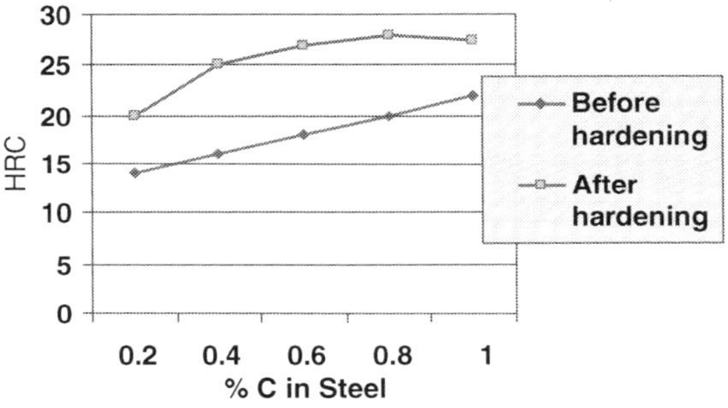

Fig. 2.42 Hardening before and after.

2.2 SELECTION OF MACHINES

2.2.1 Types of Injection Moulding Machine

The size of an injection mould to be designed depends on the moulding machine to be used. The data of an injection moulding machine are mainly:

1. Shot capacity
2. Plasticising capacity
3. Clamping force
4. Daylight of the machine
5. Injection pressure

2.2.2 Shot Capacity

It is defined as the maximum amount of plastic materials to be injected in one complete cycle of the machine.

2.2.3 Shot Weight

In a plunger type machine, shot capacity is rated as 'shot weight' (gm/oz) with polystyrene.

$$\text{Shot capacity (material } B) = \text{Shot capacity (material } A)\frac{(\text{Density of } B)}{(\text{Density of } A)} \times \frac{(\text{Bulk factor of } B)}{(\text{Bulk factor of } A)}$$

Generally, polystyrene material is used as reference material A and B is the actual material which is to be used.

The screw–type machine is normally rated in terms of 'swept volume' of the injection. Cylinder means (Cubic cm.)

Shot capacity = Swept volume $\times \rho \times c$

where ρ = Density of plastic at normal temperature (gm / cm^3)

c = Correction for percent volume expansion of the plastic at the moulding temperature.

For crystalline materials, approximately $c = 0.85$

For amorphous material, $c = 0.93$

2.2.4 Plasticising Capacity

It is expressed as the amount of material that the machine can bring per hour to moulding temperature. It means approximately the material injected by the machine in an hour. The plasticising rate of a machine of the material B can be found approximately from the machine specification chart and plasticising rate based on polystyrene material from the formula:

$$\text{Plasticising rate of the material} = \frac{\text{Specific heat of } A}{\text{Specific heat of } B} \times \frac{\text{Moulding temperature of } A}{\text{Moulding temperature of } B}$$

Plasticising rate can also be calculated from the following:

Plasticising rate (kg/ hr) = Weight of moulding (kg) × Number of mouldings / hr.

2.2.5 Clamping Force

Machines are usually characterised by the tonnage of the clamping force that they provide. The clamping force indicates the amount of force that the clamping unit can apply to the mould to keep it securely closed during the injection of the molten plastic, for example 80 tonnes means 80 tonnes of force as clamping force. Generally it is denoted on the machine as SP 80 or ST 130, etc., (80,130 are the clamping tonnage).

Formula The clamping force is proportional to the projected area of the moulding and runner and must be opposed by the clamping force. Although a proportion of the pressure produced by the injection cylinder is transmitted to the cavity, various losses will be occurring in the heating cylinder, nozzle and gate. It can be considered that force acts on the mould to make it open. It can be calculated from the following formula:

Force (kg) = Pressure (kg/cm^2) × Projected area (cm^2)

2.2.6 Projected Area

Projected area is the area that is projected from base. For complete calculation, calculate the runner projected area and the part projected area for more precision. Refer to Fig. 2.43 to get the image of projected area.

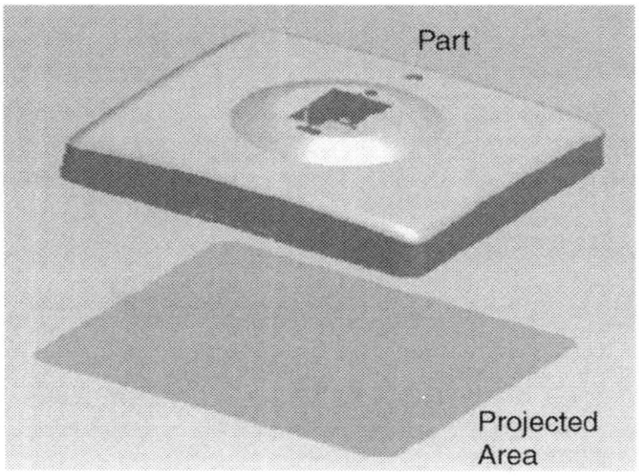

Fig 2.43 Projected area.

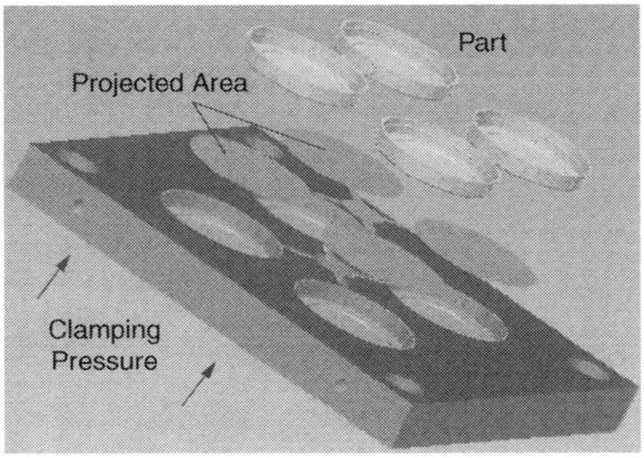

Fig. 2.44 Projected area, clamping force and injection pressure.

The table can be used to define clamping pressure.

Table 2.11 Clamping pressure for moulding.

Average part section thickness	Clamping pressure in kgf/cm^2 Flow path ratio L/t				
mm	50/1	100/1	125/1	130/1	200/1
1.0	316	491	598	703	
1/5	246	316	422	598	844
2.0	211	281	352	422	633
2.5	176	211	281	352	492
3.0	176	211	246	281	352

Clamping force (tons) = Projected area of moulding (sq. cm.) × 1/2 to 1/3 of injection pressure (tons/sq.cm.)

1. Thin section requires a high injection pressure to fill and thus needs more clamping force.
2. Relatively easy flowing material like high melt index polyethylene fills more readily and hence requires a lower clamping force.

2.2.7 Platen Details

Injection moulding machines have two platens, one is fixed platen and other is moving platen. Fixed platen is fixed on the machine with the barrel which has a nozzle for injection of the plastics material and a plasticising screw to plasticise the material inside the barrel. Fixed platen has a big hole in the centre for locating the mould to maintain the axes of the mould with the machine.

Moving platen is fixed with the moving half of the machine guided with tie bars. Back side of this platen, ejection system is provided with machine knockout rod. Both platens have the threaded holes for loading the mould half.

2.2.8 Maximum and Minimum Daylight

In injection moulding machine, two types of daylight are:

1. **Maximum Daylight**: The maximum distance between the fixed platen (injection side) and moving platen (ejection side).

   During design of the mould, maximum daylight is generally compared with the ejection stroke required for the mould to eject the components.

2. **Minimum Daylight:** The minimum distance between the fixed platen and moving platen.

   The mould height is decided on the basis of minimum daylight available in the machine. Mould height should also be more than the minimum daylight.

2.2.9 Machine Nozzle

Other parameter, which controls the flow behaviour of the plastic melt, is machine nozzle. The purpose of the nozzle is to provide a flow path for the plastic melt from the machine's cylinder to the sprue bush of mould.

There are two types of standard nozzles used in injection moulding machine and they differ only with the form of seating that is made with the sprue bush.

1. **Spherical Nose**: It has a hemispherical curvature which butts onto the sprue bush of the mould
2. **Flat Nose**: It is flat ended which butts onto the sprue bush of the mould.

The small length of the reverse taper in the bore (at the front end of the nozzle) is such that the sprue is broken just inside the nozzle. This helps to keep the nozzle face clean and assists in maintaining a leak free sealing face.

2.2.10 Dry Cycle Time

Cycle time is the period or elapsed time between a certain point in one cycle and the same point in the next.

The steps of moulding cycle in a typical injection mould are to

1. Lead the melt to the mould cavity or cavities
2. Fill the mould cavities
3. Cool the material
4. Demould and eject the moulded article

When the injection moulding machine operates without any plastic material, the cycle time is called dry cycle time. Dry cycle time has an important role in the injection mould design. It is useful for the following reasons:

1. Establishing the total shut height of the mould which are higher compared to the minimum daylight of the machine.
2. To understand the type of locking mechanism, ejection mechanism and other features available on the moulding machine.

2.3 PARTING LINE AND PARTING SURFACE

2.3.1 Introduction

Parting surface The parting surfaces are the parts of the surface where the core and cavity of a mould are matched perfectly to remove the chance of flash. The surface is those portions of both mould plates, which are adjacent to the impression and make a leak proof butting during clamping, thus, preventing material leakage from the impression.

Parting line It is the line which can be seen in the component due to the matching of core and cavity. This line during design construction of core and cavity can be easily decided. Sometimes in the article some other lines are also visible (in split mould the line due to the split matching, etc.).

2.3.2 Types of parting surface

1. Plain parting surface
2. Non-plain parting surface

1. Plain Parting Surface: In this case, the surface is totally plain. In this type of surface, the parting line lies in one plane as per the selection of the parting surface. The parting surface must be selected to make the machining easy for core and cavity.

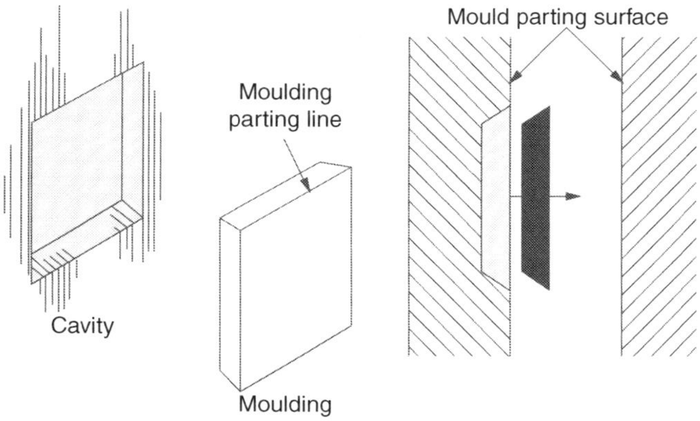

Fig. 2.45 (a) Parting line of moulding and parting surface of mould.

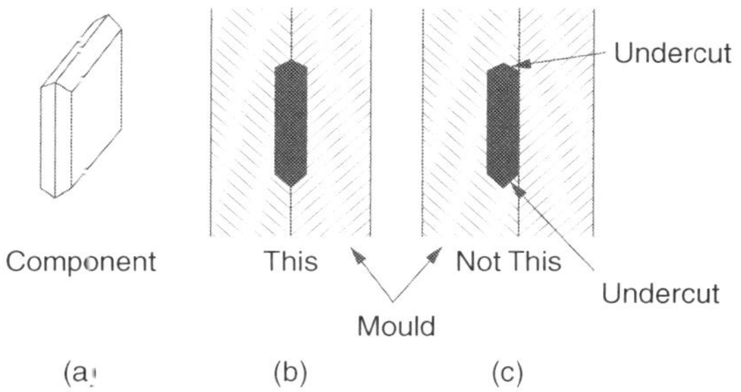

Fig. 2.45 (b) Practicable and impracticable choice of parting surface.

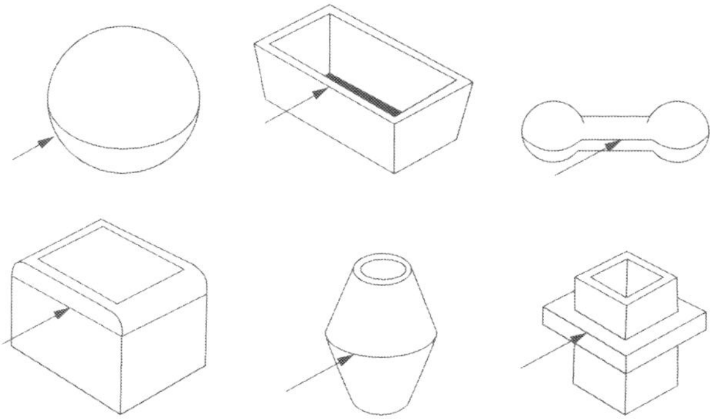

Fig. 2.45 (c) Typical moulding which permit flat parting surface to be adopted.

2. Non-plain Parting Surface: In this case, the parting line lies on curved surface and is not in a plane. Non-plain parting surface has been further classified depending on the requirement of parting surface for easy machining of the plate.

(a) Stepped Parting Surface: This type of parting surface is generally in the shape of a step. Design of core and cavity parting surface of the mould is designed in 'stepped' as per the component desired like plastic mug with handle in step.

(b) Irregular Parting Surface: As per the Fig. 2.45 (d) shown below, on the basis of component, the surface is irregular profile in shape. The parting surface is selected to do easy machining and easy opening and ejection. For multi-impression mould to reduce the machining, the parting surface is made completely across the mould plates.

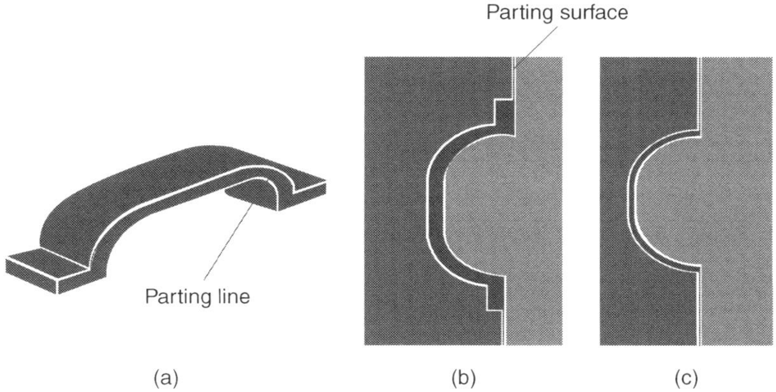

Fig. 2.45 (d) Irregular parting surface.

(c) Angled Parting Surface: The parting surface is selected for easy ejection of the component. Generally, the machining of the components is easy but with simple plastic surface (plain parting surface) designing the component has ejection problem. Directly component cannot be ejected. But by selecting angled parting surface as shown in Fig. 2.46, the ejection can be made easy. The surface is selected to help the moulding to be placed in line of draw for proper ejection.

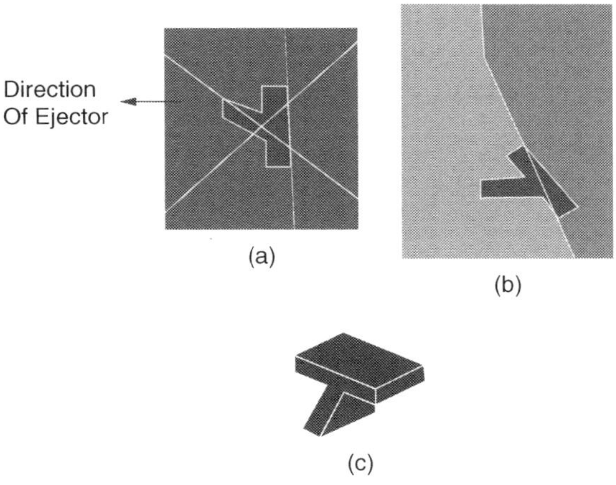

Fig. 2.46 Angled parting surface.

(d) Complex Edge Parting Surface: It is a spoon type of component in which there is no constant edge. In this type of surface, the parting line is also not in the same plane as shown in Fig. 2.47(a). For mouldings, the parting line and surface are decided on the basis of edge of the component as shown in the Fig. 2.47.

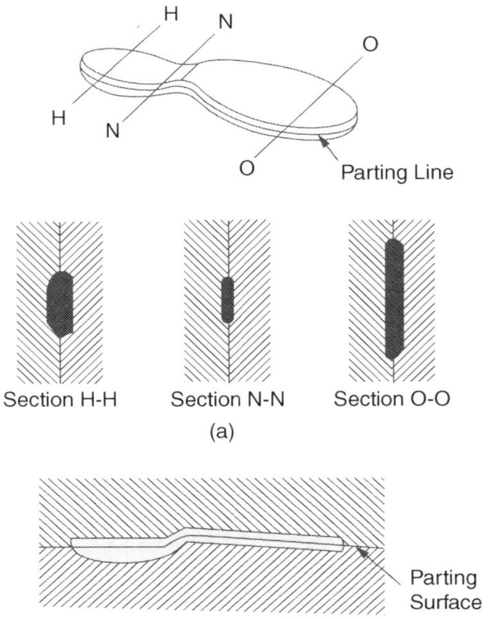

Fig. 2.47 Complex edge parting surface.

2.4 MOULD VENTING

As molten plastic enters into the mould, it quickly displaces air in the tightly sealed mould. Some of the air escapes through ejector pin, parting plane. But for complete air removal in mould design, vents are provided to remove the chances of air trap.

2.4.1 Vents on Parting Line

The vents are generally provided along the mould parting line. The pathway on the parting surface is made very easily by machining (especially by grinding, filing) to a particular size. This pathway is used as vents in the mould for air escaping from the mould.

For reducing the chance of materials to flow into the vents during filling. The approximate size of a vent is 0.05 mm deep by 3.0 mm wide. Selection of resin and processing conditions determines the vent's depth.

2.4.2 Thumb Rules of Venting

1. The size of venting depends upon the part volume and filling speed.
2. For increasing venting, it would be better to give more number of vents.
3. To avoid flash, do not increase vent depth beyond the guidelines.

For the irregular part geometries, more vents are better because the resin has flame retardants or other additives which boil at the flow front and deposit on the mould surface and vents. Due to this deposit sometimes the vents are getting blocked. Overventing and number of venting, reduce the chances of volatile or air trap in the article.

2.4.3 Vent Placement

Vents should be generally placed

1. along the runner system and
2. part perimeter

But, it is required to fill the resin up to last point of the cavity and due to the flowing of resin, the air and volatile gases is also collected at the farthest point from the gate or last point of the cavity from the gate. So the exact position of vents is at this point. If the last points of cavity to be filled are not properly vented, the air may be trapped in the mould at this point and due to high compression pressure it will cause air burnt on the part and erode the mould steel.

Trapped air can also be removed by the elector pin vents; usually it gets self clean with each ejection stroke. This type of venting requires regular cleaning of the pins after disassembling the mould. Porous metal inserts can also be used as venting for air-trap but requires periodic cleaning.

2.4.4 Importance of Venting in Injection Moulding

It is a shallow recess or hole incorporated in the design to permit entrapped air or other gases to escape freely. Some moulds do not need to go for special design for venting because air has sufficient possibilities to escape along ejector pins or at the parting line. Entrapment of air may cause mould defects like discolouration, sinks, incomplete filling, voids, etc.

2.4.5 Designing for Proper Venting

The configuration of the moulding, its position in the mould and its gating has a considerable effect on venting. The approximate size of a vent is 0.05 mm deep by 3.0 mm wide.

Positions where a vent is likely to be required are:

1. The furthest point from the gate on symmetrical mouldings.
2. At the point where flow paths are likely to meet.
3. At the bottom of projections.

2.5 FEED SYSTEM

The pathway for plastic material from nozzle to each impression is called feed system. Feed system comprises sprue, runner and gate.

2.5.1 Sprue Bush

During the injection process, the plastic material, as a melt is delivered from machine nozzle to the impression through a passage. This passage is a tapered hole with in a bush. The material in this passage is termed as sprue and the bush is called sprue bush.

During the injection process, sprue always contacts with nozzle from the injection machine and the plastics first come into contact with sprue bush. While designing the sprue consider the following:

1. No undercut in sprue
2. Diameter of nozzle
3. Aperture radius of nozzle
4. The dimensions of the sprue depend upon wall thickness. Small orifice in sprue will make sprue cool faster and make insufficient flow.

Application of sprue bush The molten plastic injected from the injector nozzle will go through a sprue (sprue bush), a runner, and a gate and fill the cavity. As the temperature of molten plastic is lowered while going through the sprue and runner, the viscosity will rise; therefore, the viscosity is lowered by shear heat generated when going through the gate to fill the cavity. So sprue bush design is important, because sprue bush is the part that contacts first time with melted plastic. Bad design of sprue bush can break the plastic product. Sprue bush is the connecting member between the machine nozzle and the mould face. In this part of the mould, sprue is formed. The application of sprue bush is to maintain the axes of nozzle and the mould axes by providing aperture radius. The radius of aperture is to maintain as per the nozzle radius.

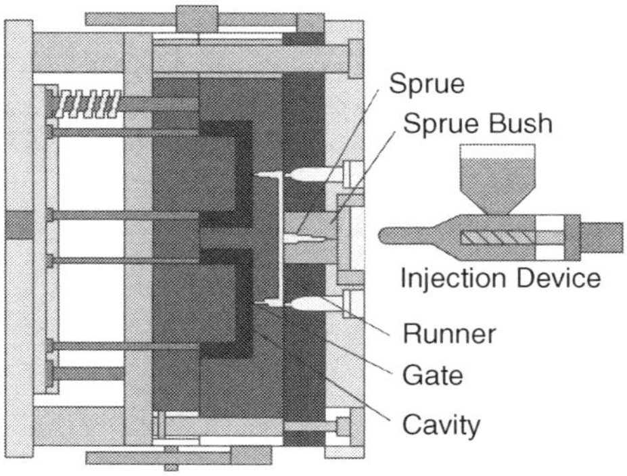

Fig. 2.48 Mould with feed system.

Actually the nozzle is set in the aperture of the sprue bush. Due to forward and backward of the nozzle during moulding process, there is a chance of wearing and tearing of the aperture of the nozzle if the aperture radius is equal to the nozzle radius. Hence, the aperture radius is generally taken slightly higher than nozzle radius.

The draft is provided in the sprue bush for easy ejection of sprue. This draft angle is generally 3 to 5. The draft angle is decided on the basis to optimise the pressure drop and temperature drop. Actually in injection moulding, the main aim is to control pressure and temperature drop of plastics materials after coming out from nozzle in cold runner mould.

1. Pressure Drop: Pressure drop means the difference of pressure at internal and external point. As per the diagram, point 1 is the inlet and point 2 is outlet of the plastics material in sprue bush.

The pressure at point 2 ($P2$) = Force / Area of point 2.

The pressure Drop = ($P1-P2$) where $P1$ is the pressure at inlet.

If the draft angle is more, the area at outlet will be more and due to more area at point 2, the pressure at point 2 will be less. So the pressure drop ($P1-P2$) in sprue bush which has more draft angle will be more as compared to less draft angle. This will not meet our requirement. Pressure drop in sprue bush should be optimised as per the requirement.

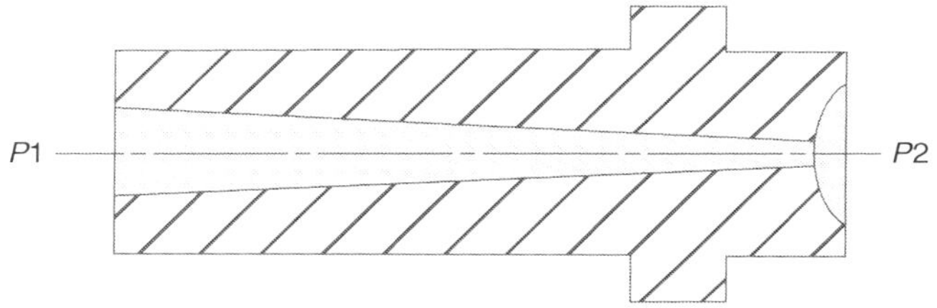

Fig. 2.49 Sprue bush.

2. Temperature Drop: The internal surface area of sprue bush will be more if the draft angle will be more. The heat transfer increases when draft angle or surface area is more. So the temperature drop will also be more in sprue bush. The draft angle is maintained at 3° to 5°.

Sprue bush and its types Sprue bush is generally of two types: spherical seating and flat seating. In spherical seating, in the centre of sprue bush, one spherical radius is created as per the nozzle radius for the perfect seating of nozzle. Whereas in flat seating, no spherical radius

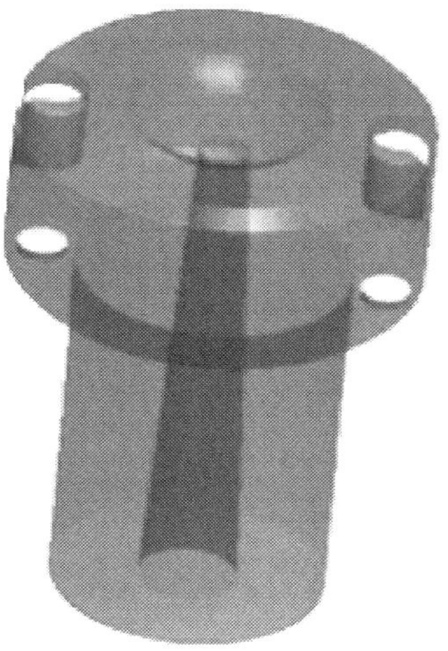

Fig. 2.50 Sprue bush – Flat seating.

Fig. 2.51 Sprue bush – Spherical seating.

is provided in the centre of the sprue bush. But in both type of seating, the matching of sprue bush and the nozzle should be perfect, otherwise, leakages may occur. This type of sprue bush is used in small size moulds only.

The other construction is 4 bolts, this sprue bush type commonly used in large mould type and mass production, more than 200 ton as shown in Figs. 2.50 and 2.51.

Sprue bush size and its design rule

1. Aperture radius (the centre spherical radius) of sprue must be larger than nozzle radius. For aperture radius, it would be better to get the nozzle radius from the machine specification chart. For example, when ball radius of nozzle of injection machine is 10 mm use ball radius sprue bush of 11 mm.
2. The internal diameter of the sprue bush mouth hole must be larger than that of the nozzle. Generally hole diameter of sprue is about 2 mm to 3 mm. The sprue mouth hole diameter is generally 0.5 to 1 mm more than the nozzle hole diameter.
3. The cavity plate thickness should be perfectly matched with the length of sprue bush. If nozzle and sprue bush are not perfectly matched, then the plastic comes out in the form of flash between nozzle and sprue. Sometimes locating ring is designed to prevent the chances of the sprue bush coming out the mould. The locating ring presses the sprue bush perfectly to keep it in position. Locating ring and sprue bush in assembled condition is shown in Fig. 2.51. As in two-plate moulds, there is no runner plate and the gate type is side gate.

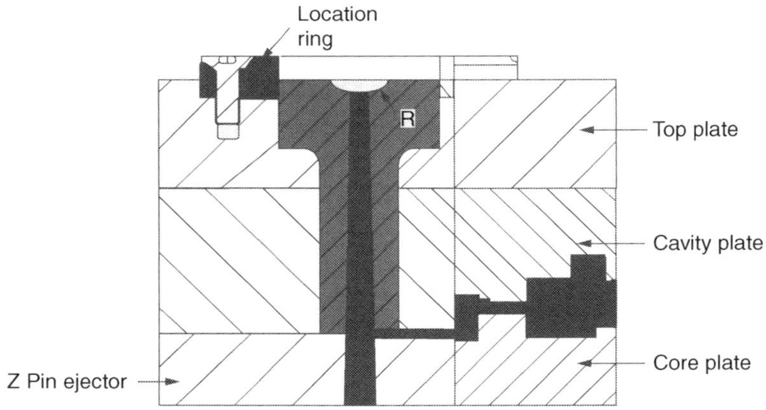

Fig. 2.51 Locating Ring and Sprue bush in Assembled Condition

Sprue puller pin ejector Sometimes to take the sprue from sprue bush, one sprue puller pin is used. This pin pulls the sprue from sprue bush during opening of the mould and after opening. It is used to push the runner and sprue together with ejectors. For extended sprue bush refer to Figs. 2.52, 2.53 and 2.54.

Extended sprue bush Sometimes for deep article for two-plate mould with side gate and multicavity three-plate mould, length of sprue bush will be longer. Due to this there is a chance of temperature drop. To solve this problem, the extended sprue bush is used. In this method, the sprue bush is made in such a manner that the nozzle is placed in the sprue bush with some depth. As per the specification of the machine, the nozzle can easily take forward maximum 30 mm, so for deep article this 30 mm depth can be controlled and reduced from overall length.

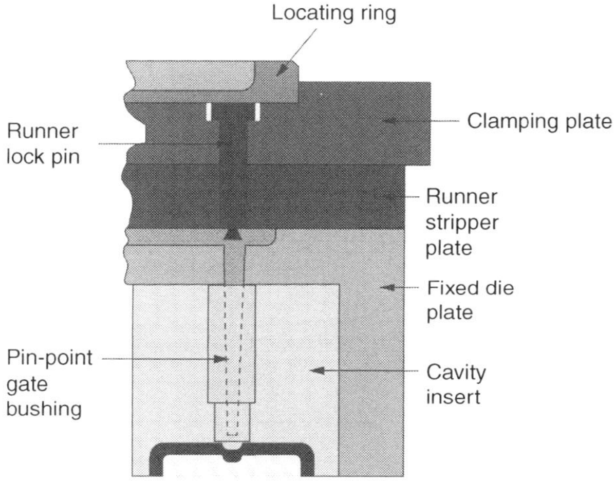

Fig. 2.52 Extended sprue bush.

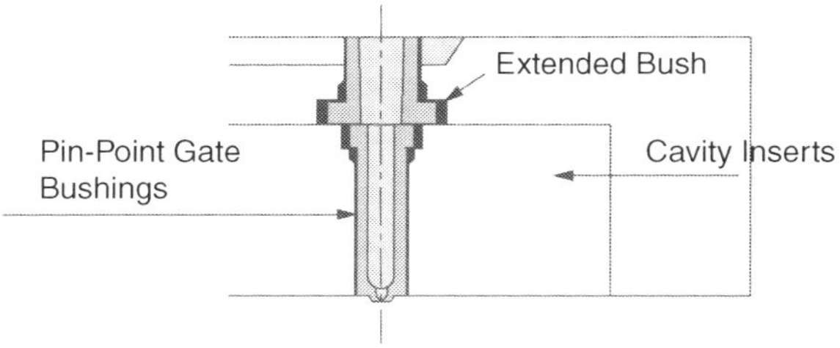

Fig. 2.53 Extended sprue bush.

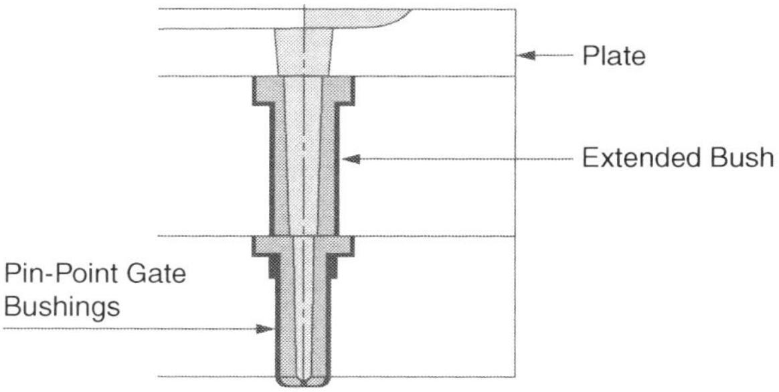

Fig. 2.54 Extended sprue bush.

Sprue bush cooling design and construction The sprue bush needs cooling to cut the cycle time of injection process, when the diameter of sprue is large, cooling in sprue bush is important. Sometimes the sprue bush is designed with cooling system containing cooling pipe as shown in Figs. 2.55 and 2.56.

Some standard moulding companies have their own methods of inner cooling of the sprue bush. In these types of sprue, sometimes the flow can be changed by changing the entrance and exit flow by using stopper baffle as shown in Fig. 2.57 or it can make the flow straight. The other cooling method in sprue is like the one shown in Fig. 2.58. In this type of cooling, the O ring can also be used.

Fig. 2.55 Sprue bush cooling.

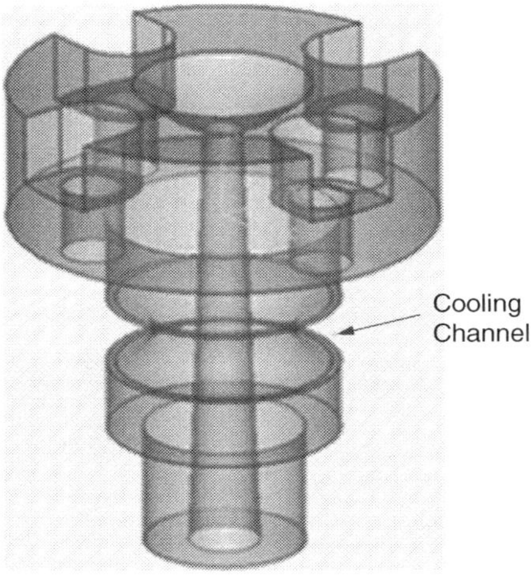

Fig. 2.56 Sprue bush cooling.

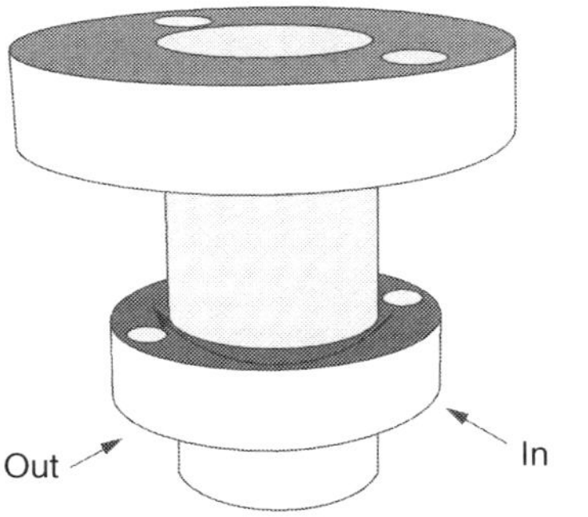

Fig. 2.57 Stopper baffle.

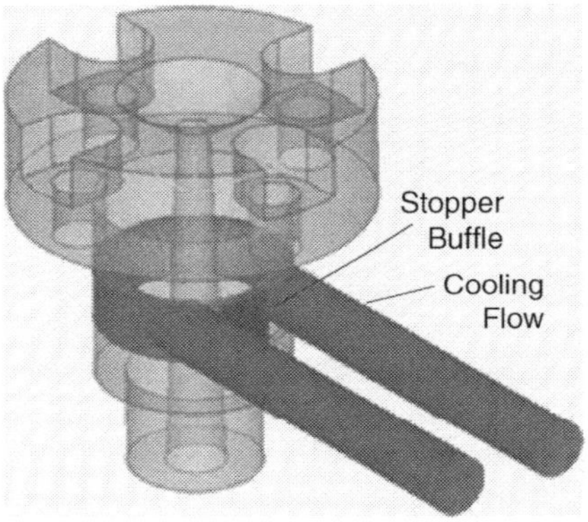

Fig. 2.58 Stopper baffle.

2.5.2 Runner

Main runner Before designing the runner system, the following points are to be kept in mind:

Determine the number of cavities, shape and gate types and arrange the cavity layout in the multicavity of mould.

Balancing of Runner Balancing ensures virtually equal flow of plastic through each gate. To achieve balancing runner layout must be suitably designed.

Subrunner In multicavity mould, the runner is subdivided into subrunner as shown in Fig. 2.59. Subrunner is the continuous flow pathway from main runner to the gate. On the basis of number of cavities, the size and number of subrunner is decided.

Cold slug When the runner layout bents, clod slug may be required to continue plastic flow of better quality to the cavities. The cold slug well helps the flow of material through the runner system by stopping colder, higher viscous material moving at the forefront of the molten mass entering into the cavity. The length of the well is usually equal to or greater about 1.5–2 time than the runner diameter.

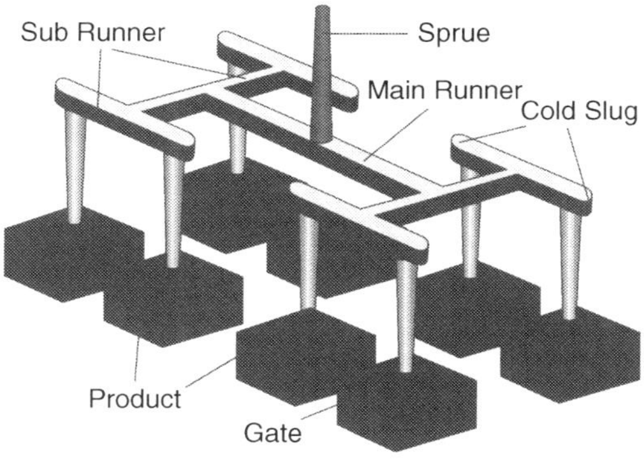

Fig. 2.59 Cold slug well.

Runner geometry There are various types of geometry that are generally used in runner system such as

1. Full round
2. Half round
3. Trapezoidal

Full round and trapezoidal shapes are recommended in various moulds. Half round runners are not recommended because of their low volume to surface ratio.

Runner dimensions The diameter of a runner depends on its length in addition to the part volume, part flow length, injection machine capacity and gate size. Generally, runner must never be smaller than the largest wall thickness of the product. It is usually made from 3 mm to 12 mm. The runner should be large enough to minimise pressure loss, yet small enough to maintain satisfactory cycle time.

Runner efficiency It is the ratio of cross sectional area of the runner to its periphery. Round and square runners have the efficiency of 0.25 of its diameter / depth, whereas semicircular runners have nearly half of the round runner efficiency. Example, for a round runner of diameter 4 mm (D), the runner efficiency is calculated as follows:

η = Cross-sectional area / its periphery

$\eta = \pi r^2 / 2\pi r$

$\eta = 3.14 \times 2^2 / 2 \times 3.14 \times 2$

$\eta = 12.56 / 12.56$

η = 1 (It is ¼ times of the diameter of the runner). It can be written as;

$\eta = 0.25\ D$

Runner layout In general, there are three types of runner layout:

1. 'H' bridge (branching) runner system
2. Standard (herringbone) runner system (conventional runner)
3. Radial (star) runner system

Table 2.12 Runner diameter and maximum runner length.

Runner diameter		Maximum runner length			
		Low viscosity		High viscosity	
(In)	(mm)	(In)	(mm)	(In)	(mm)
1/8	3	4	100	2	50
1/4	6	8	200	4	100
3/8	9	11	280	6	150
1/2	13	13	330	7	175

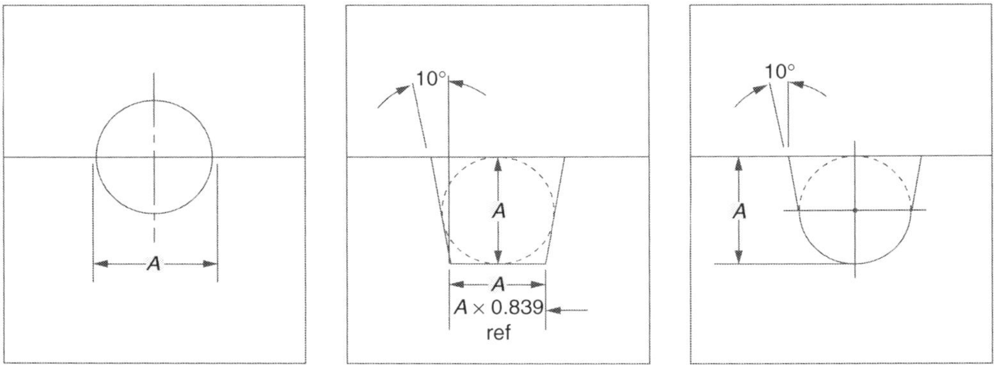

Fig. 2.60 Runner cross section.

Table 2.13 Cross-sectional area.

A	Full Round	Trapezoid	Modified trapezoid
	A	A	A
mm	mm^2	mm^2	mm^2
3.2	8.0	10.4	9.7
4.8	18.1	23.4	21.8
6.4	32.2	41.6	38.7
7.9	49.0	63.4	59.0
9.5	70.9	91.6	65.3
11.1	96.8	125.1	116.4
12.7	126.7	163.8	152.4
15.9	198.6	256.7	238.9

Runner balancing It is the balancing of runner by ensuring that all the mould cavities fill at the same rate and pressure or in other way at the same time. If the product has no similar product geometry, runner balancing is done on the basis of runner design. It is recommended to use mould analysis software like mould flow or C-mould. Refer to Fig. 2.61, shown below is a star layout runner system. Standard runner and H bridge type runner are shown in Fig. 2.62.

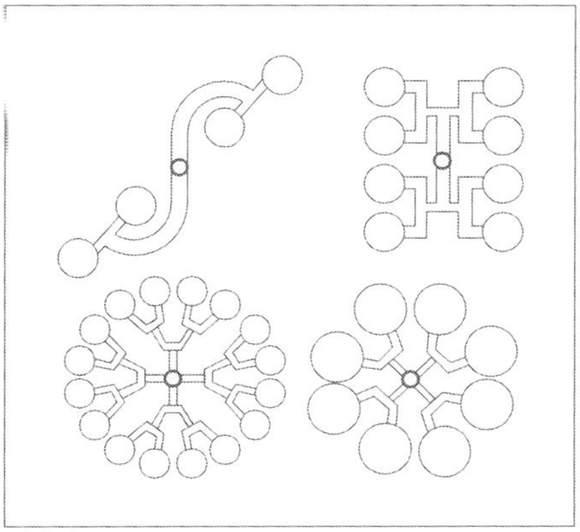

Fig. 2.61 Star layout runner system.

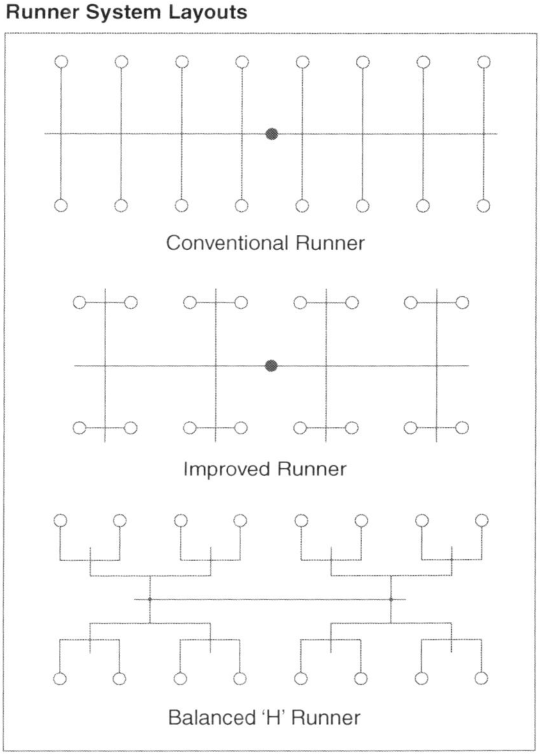

Fig. 2.62 H Bridge runner system.

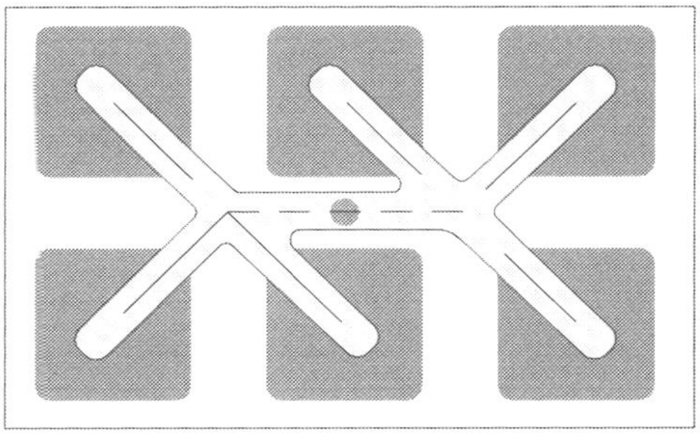

Fig. 2.63 Plan view of runner system.

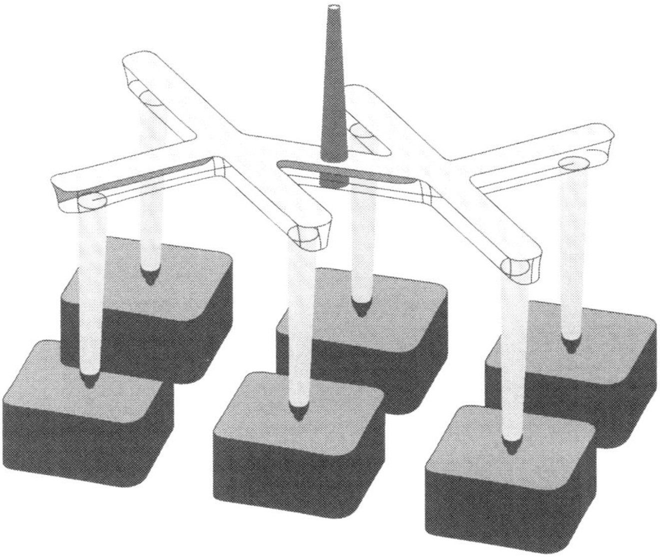

Fig. 2.64 3D view of runner system.

1. Balanced Layout Runner: Figures 2.63 and 2.64 show the layout model for 6 cavity product; star layout is used to connect three cavities, and then connect again with a straight line. Distance between one cavity and another is same. But this runner layout is balanced but it is not a final solution. By reviewing the mould flow analysis, it is found out that the flow at nearest sprue is faster than other and it fills first, although the distance from the branch to gate is same.

2. Improved balanced runner layout: Another layout which is shown in Figs. 2.65 and 2.66 is an improved runner layout for six product cavity. Although it is not the best, it is more improved than the first. But disadvantage of this runner layout is that it consumes more material. Some of the other runner layouts are shown in Figs. 2.67, 2.68 and 2.69.

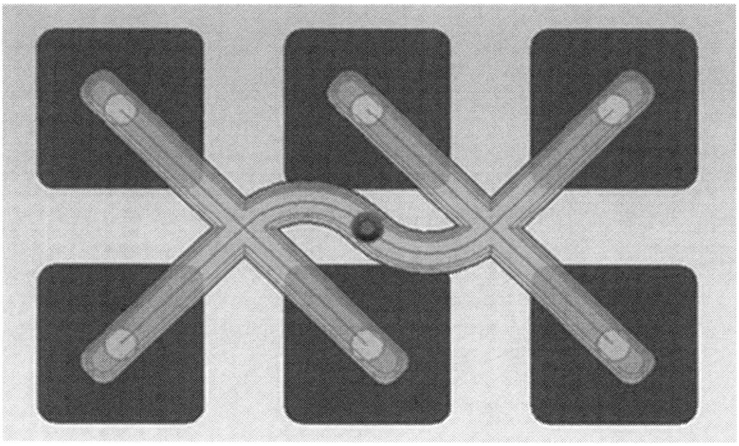

Fig. 2.65 Improved runner layout for 6 Cavity.

Fig. 2.66 3D view of six cavities runner layout.

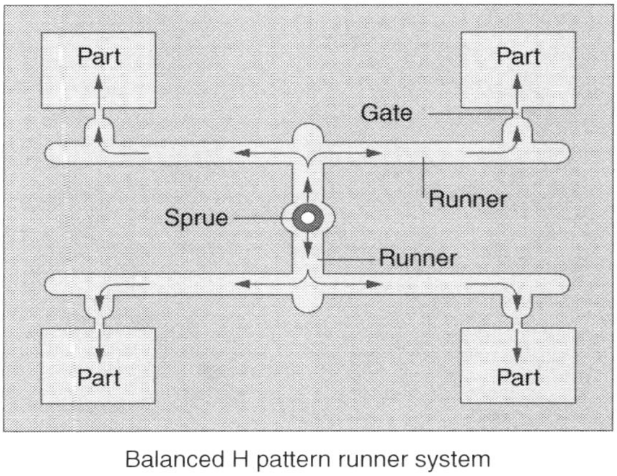

Fig. 2.67 Four cavities.

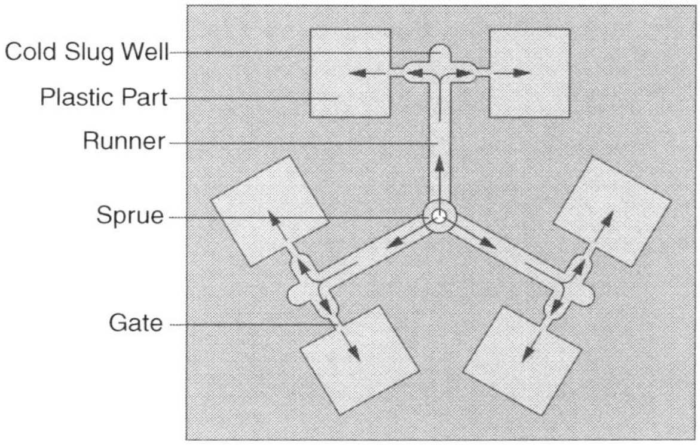

Fig. 2.68 Six cavities.

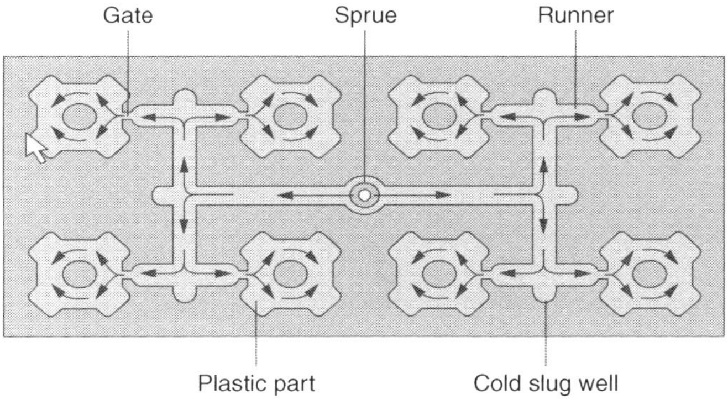

Fig. 2.69 Eight cavities top view.

Fill time runner balance on six cavity mould Runner balancing is important when the mould is designed for more than one cavity. Unbalanced runner can cause various problems like incomplete filling in mould (short shot), weld line, air traps, burning, etc.

Definition of unbalanced flow Unbalanced flow is the flow when plastic completely fills some flow paths in the mould before other flow paths have filled. As per the simulation, the filling time of cavity is shown below.

1. Unbalanced Runner System: Due to unbalanced runner system, the flow of plastic in various cavities does not fill at the same time, therefore, it can cause flashing, short shots, high cycle time, density differences throughout the part, warpage, air traps and extra weld lines as shown in Fig. 2.70. Flashing will occur in cavity that fills earlier, because cavity that fills earlier will continue to receive plastic flow until other cavities are filled. Short shots occur at cavity that fills more slowly than other cavities. Then the pressure and cycle time are not enough to fill the last cavity.

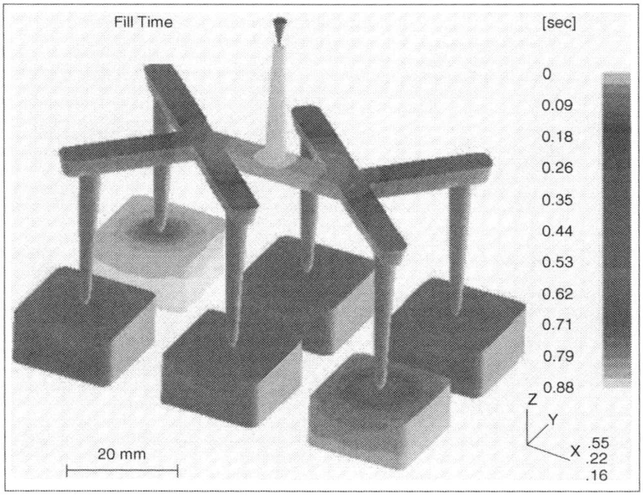

Fig. 2.70 Unbalanced runner system.

2. Balanced Runner of Six Cavities: Although this runner layout is not the best solution but it is better for six cavities layout as in this layout the fill time is same for all the six cavities as shown in Fig. 2.71 and flow is better.

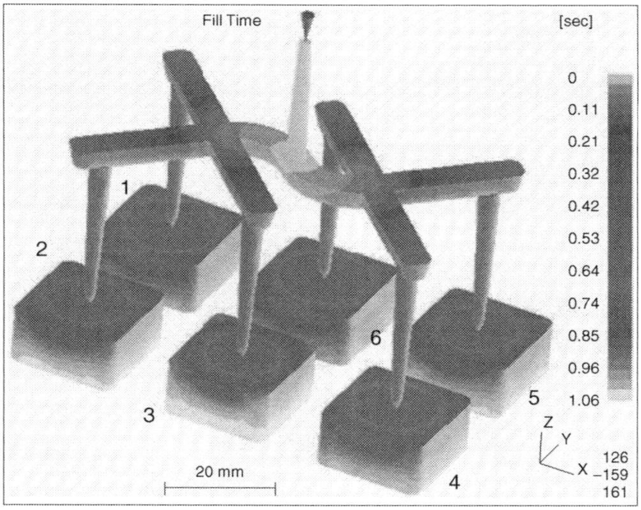

Fig. 2.71 Filling time result.

Runner system an effective part of the system Since the design of the runner system has large effect on moulding quality, moulding efficiency, cycle time, cooling time and other factor, so it is important factor of the mould.

2.5.3 Gate

Introduction

1. It is a small orifice, which connects runner to impression.
2. It is small in cross section.
3. It freezes soon after the impression is filled so that nozzle can withdraw without possibility of back flow of material, voids, etc.
4. It should degate easily.
5. Small witness mark remains on the moulding after degating.
6. Better control of filling.

Definition of gate The entrance through which molten plastic flows into the impression of the mould from the runner. It is positioned at the thickest area of a part. Other function of gate is to provide high pressure to the plastic material from runner to moulding because the runner has more volume when compared to gate and by gate only the runner is getting separated from the part. It is recommended to consider gating options during the product design.

Determination of location, shape and size of gate The factors for the determination of location, shape and size of a gate are given below:

1. **Moulding Material:** Temperature, flow characteristics, shrinkage, viscosity.
2. **Moulded Parts:** Appearance geometry, wall thickness, quality demands with respect dimensions, etc.
3. **Generalities:** Weld lines, distortion, moulding operation, ejection, costs, etc.

An eccentric positioning of gate is less expensive to accomplish. Demoulding and separation from the part is easy. The cross sectional dimensions of a gate are determined by the plastics used and the wall thickness of the moulding. It is a good practice to position the gate at the thickest part of the component because thin part will be cooled easily and faster.

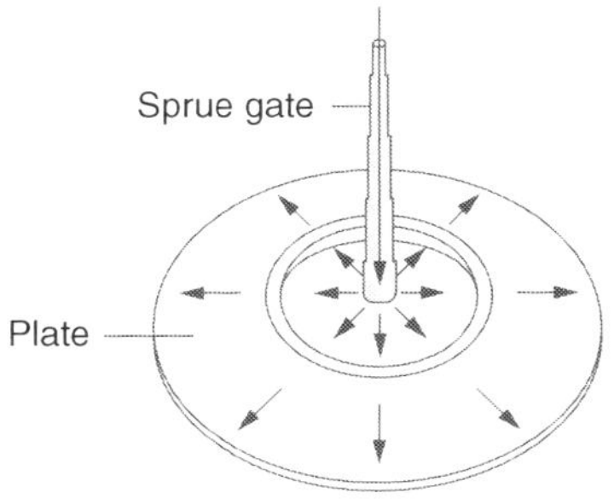

Fig. 2.72 Sprue gate.

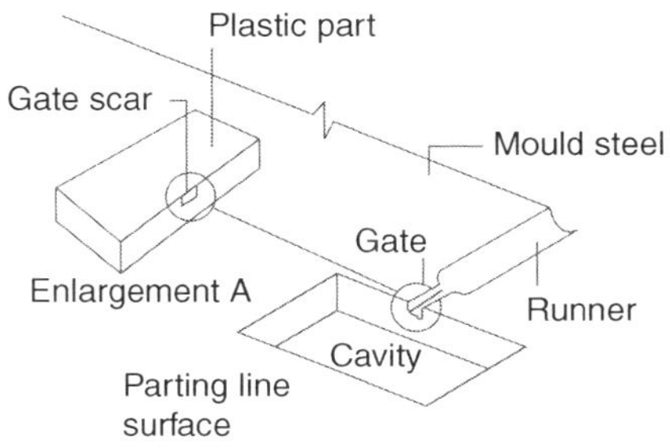

Fig. 2.73 Witness mark on moulding.

Types of gates Basic purpose of the gate is to fill the cavity easily and control the plastics. The location of gate is decided on the basis of fluidity of the material. It should be made near the centre of gravity of moulding or the line to make the flow uniform. The gate type depends on part shape, mould layout, mould system, etc.

Gate types can be divided between manually and automatically trimmed gates.

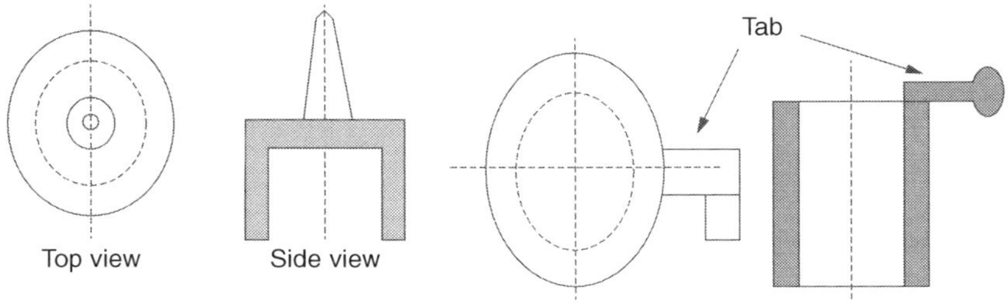

- Commonly used for single cavity moulds
- Minimises pressure drop
- Leaves gate mark on the part surface

- Commonly used for single cavity moulds
- Minimises pressure drop
- Leaves gate mark on the part surface

Fig. 2.74 Sprue (Left) and tab (Right) gates

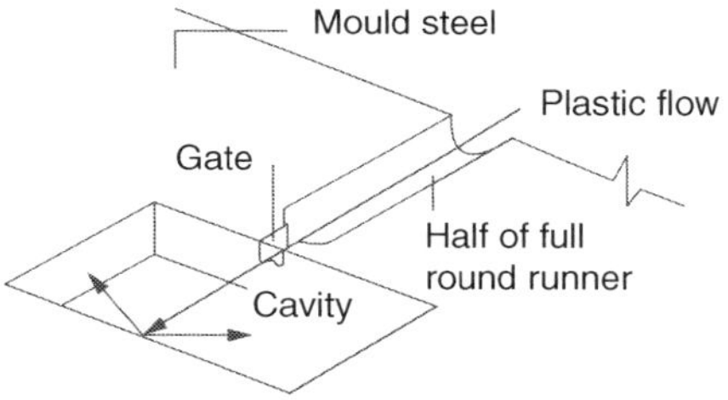

Fig. 2.75 Gate channel machined on mould plate.

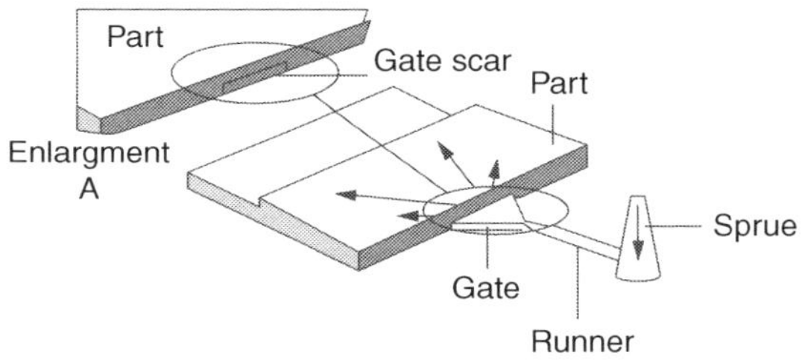

Fig. 2.76 Fan gate.

1. Manually Trimmed Gates: The manually trimmed gate means the gate which can be trimmed by operator separate parts from runners during a secondary operation.

The manually trimmed gates are given below:

1. Sprue gate
2. Edge gate
3. Tab gate
4. Overlap gate
5. Fan gate
6. Film gate
7. Diaphragm gate
8. External ring gate
9. Spoke or multipoint gate

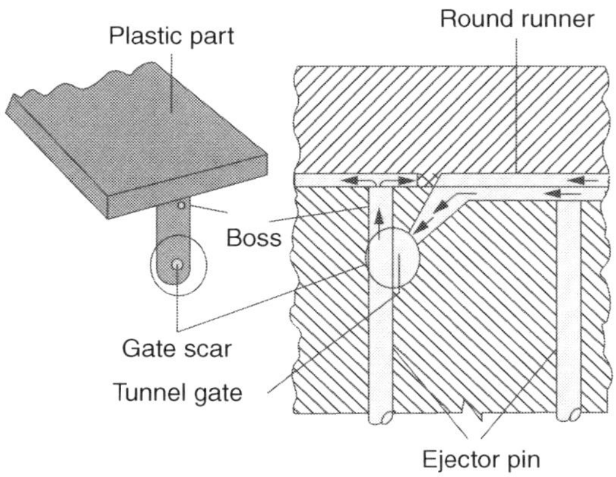

Fig. 2.77 Sub surface gate.

2. Automatically Trimmed Gates: Automatically trimmed gate means that the gates which can be trimmed automatically as the mould is opened to eject. In this type of gate, the tool some system is incorporated for automatically shearing off the gate.

For the gates the following points are generally kept in mind.

1. Avoid gate removal as a secondary operation.
2. Maintain consistent cycle times for all shots.
3. Minimise gate scars.

Gate types trimmed from the cavity automatically include:

1. Pin gate
2. Submarine (tunnel) gates
3. Hot runner gates
4. Valve gate

Details about Gates

1. Sprue Gate: This type of gate is recommended for single cavity moulds and for parts requiring symmetrical filling. It is suitable for thick sections because holding pressure is more effective. A small size sprue gate is good because rapid mould filling will be there and pressure losses will also be less.

A cold slug well should be included which is opposite of the gate. The disadvantage of using this type of gate is that the large gate mark is left on the part surface after the runner (or sprue) is trimmed off. Freeze-off is controlled by the part thickness rather than the determined gate thickness. The part shrinkage near the sprue gate will be low; shrinkage in the sprue gate will be high. This results in high tensile stresses near the gate.

This type of gate is commonly used for single cavity moulds, where sprue feeds the material directly to cavity with minimum pressure drop. Here the sprue of the mould serves as the gate. It is placed on the top side of the product. The gate has to be cut manually by a cutter. This gate type can only produce one part per shot. The moulding cost is low because the material from the runner was eliminated. The injection pressure was reduced due to direct cavity filling. The simple mould structure makes the mould cost lower.

Dimensions: Sprue diameter is controlled by the machine nozzle or orifice of the sprue bush. The sprue diameter must be 0.5 mm approximately larger than the nozzle exit diameter. Standard sprue bushes have a taper of 1.5 to 3 degrees, opening toward the part. Therefore, the diameter of the gate will be controlled by sprue length of the diameter. A larger taper wastes material and extends cooling time.

This type of gate is used for large sized products like boxes or cups like bathtub, plastic bucket and others. In Fig. 2.79, cups are produced with direct gate before cutting.

From Fig. 2.80 as per the above construction of direct gate the parting line separates core and cavity and the sprue gate is provided on the cavity. After injection process, the sprue will cut manually. Starting sprue diameter is controlled by nozzle of injection.

It should be 1–2 mm or about 1/32 inch.

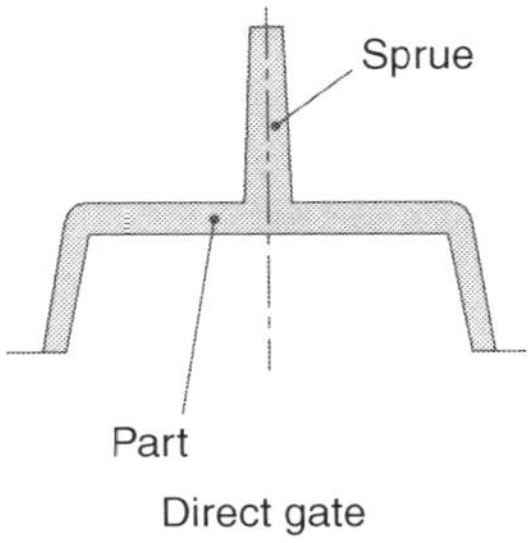

Fig. 2.78 Sprue gate line diagram.

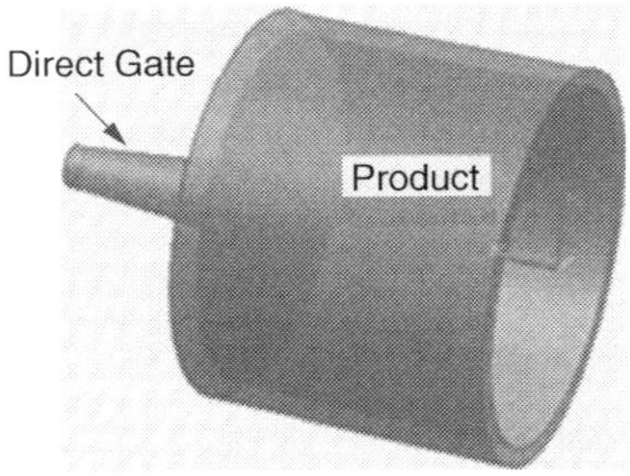

Fig. 2.79 3D view box with sprue gate.

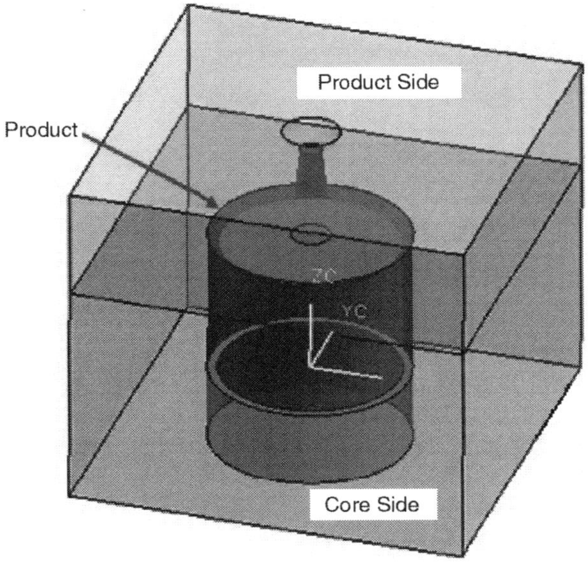

Fig. 2.80 Assembly of sprue gate.

1. Taper is provided at the sprue from orifice diameter to product about 1.5–3 degree which will make the product demoulded from the cavity side more easily.
2. In general, the diameter of the gate should be more than the thickness of the product.

Advantages

(a) Low cost of design
(b) The injection pressure was reduced due to direct cavity filling
(c) Simple way to design gate and runner system
(d) Easy design and maintenance

Disadvantages

(a) Due to excessive stress around the gate, cracking often starts around the gate
(b) Gate marks left on part surface

(c) This gate type can only produce one part per shot
(d) The gate has to be cut manually by a cutter
(e) Not applicable in small sized products

2. Side/Edge Gate: This is the most commonly used and simplest type of gates. It is normally used for two-plate mould with two or more cavities. It is placed at the side of the plastic product. It is used at the product parting line. The gate has to be cut manually by a cutter. The gate is located on the parting thick sections and can be used on multicavity. The material fills from the side, top or bottom. This type of gate has various names. The common names are standard gate, edge gate and side gate. This is the most typical gate in moulds, which is applied to almost all types of plastics. The line diagram of side gate is shown in Figs. 2.81 and 2.82. It is single side gate from sprue.

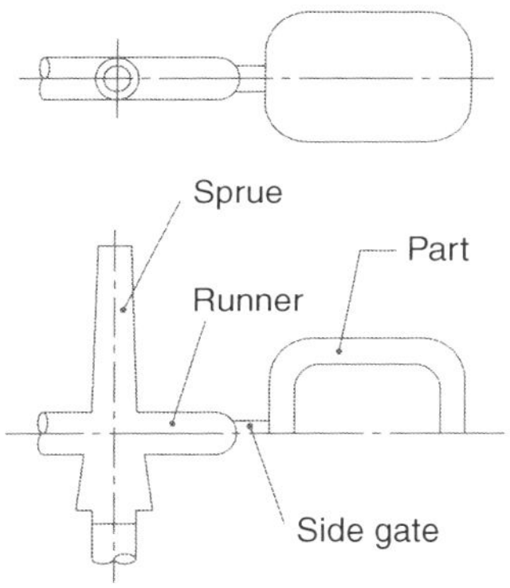

Fig. 2.81 Side gate.

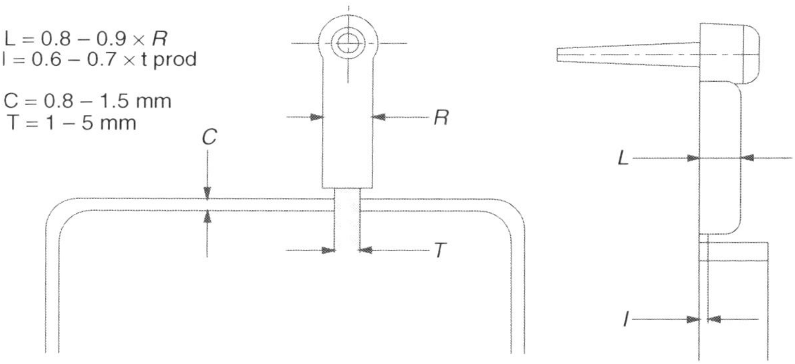

Fig. 2.82 Line diagram of side gate.

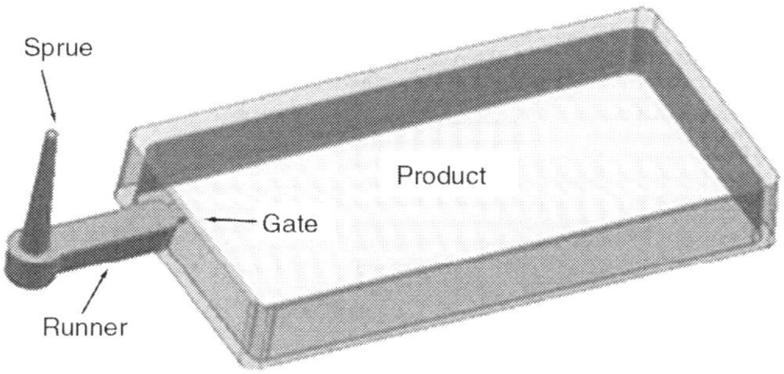

Fig. 2.83 Side gate.

Dimensions:

Runner width and height for side gate or edge gate,

Height = 0.75 of width of runner.

Width of runner depends on product size and weight,

So width = 3 to 7 mm

***T* is the gate width**, for 200 gm weight product width should be taken as 1 to 5 mm and for very large products width is up to 12 mm.
***C* is the land of gate** from runner, the land length should not be no more than 1.5 mm. Sometimes radius *R* is also provided on both sides. 3D view of side gate is given below:

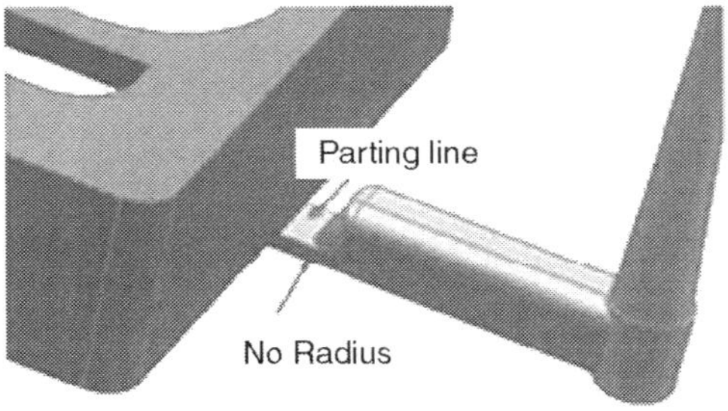

Fig. 2.84 3D view of side gate.

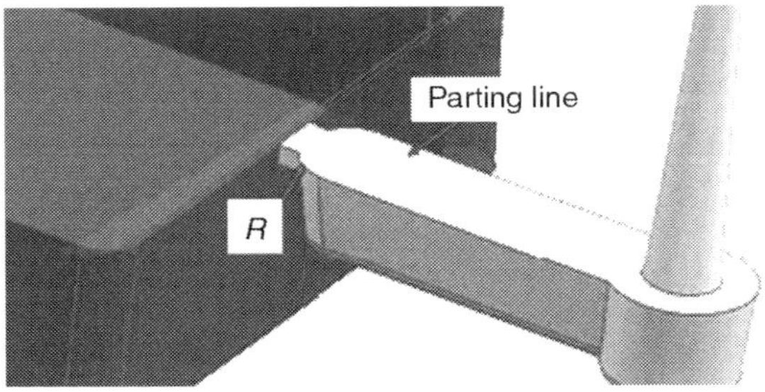

Fig. 2.85 3D view of side gate.

Both Figs. 2.84 and 2.85, show edge gate or side gate of different design on the basis of, application, parting line, taper and *R* system.

How the parting line is decided on the basis of edge gate?

These are about three methods chosen when designing parting line:

(a) Core Side Process: In this type of parting line, only core side to be machined is present. Fig. 2.86 below shows side gate design with flat parting line in core side process.

This type of parting line is useful in product without taper in side product, but when product has taper in side (as shown in Fig. 2.86), it will undercut and need slider to form the side of product.

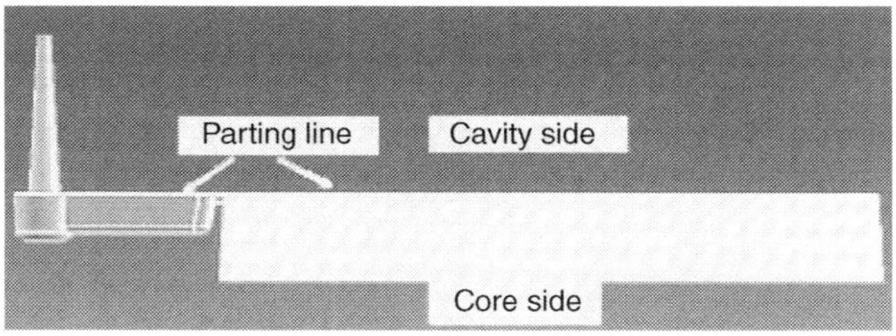

Fig. 2.86 Side gate design with parting line.

Fig. 2.87 Product with taper.

(b) Cavity Side Process: In this type of gate design, parting line is one method to hide the marks of gate after injection process. In this type runner channel is machined in cavity side, then a little gate machined at core side. This type of gate parting line is better and cheaper than number one when product has taper as shown in Fig. 2.87.

(c) Cavity and Core Side Parting Line: This type of parting line for gate and runner is not common in use. When we use this type of parting line runner and gate, cylindrical type of runner of semicylindrical type of runner must be used.

3. Tab Gate: In injection moulding machine, after nozzle, plastic material enters into the feed system and the cavity with stress affects the quality of the products. This gate reduces the stress. This gating technique is particularly used for feeding solid block type mouldings. A projection / tab is moulded of the side of the component and a conventional rectangular edge gate feeds this tab.

The melt takes a sharp right angled turn due to which the melt flow will be smooth, steady and the impression will fill uniformly. Due to the sharp right angle, the material before turning for a moment stays at the place and takes the turn and relieve some stress as shown in Fig. 2.89. Tab gate is used in various plastic materials such us ABS, PS, acrylic, PP, PVC, polycarbonate, SAN and other which have relatively low fluidity. This is also called *collision*.

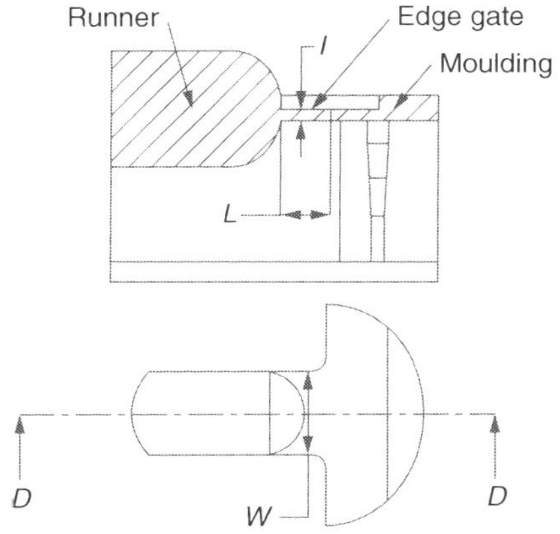

Fig. 2.88 Cylindrical type runner.

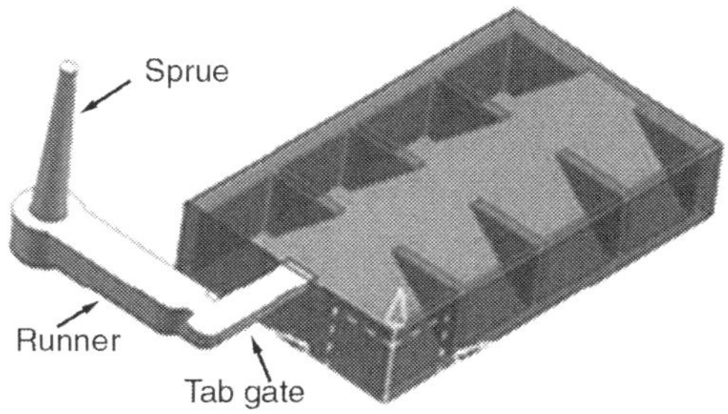

Fig. 2.89 Tab gate.

General rule of tab gate

(a) The gate is designed as centrally as possible in the product side, by considering the product size and shape.

(b) The thickness of tab gate should be the same as the part wall thickness.

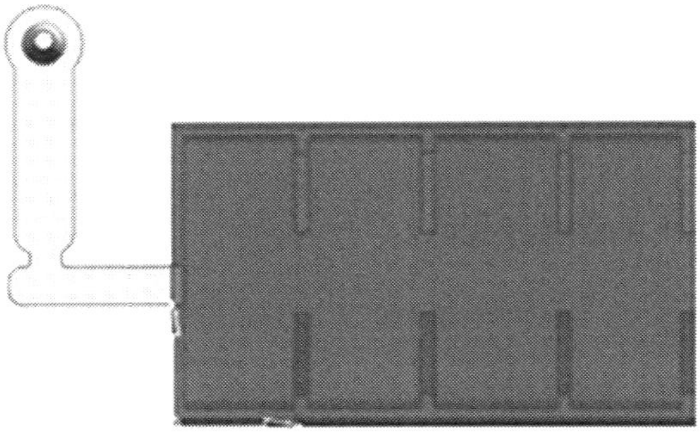

Fig. 2.90 Top view.

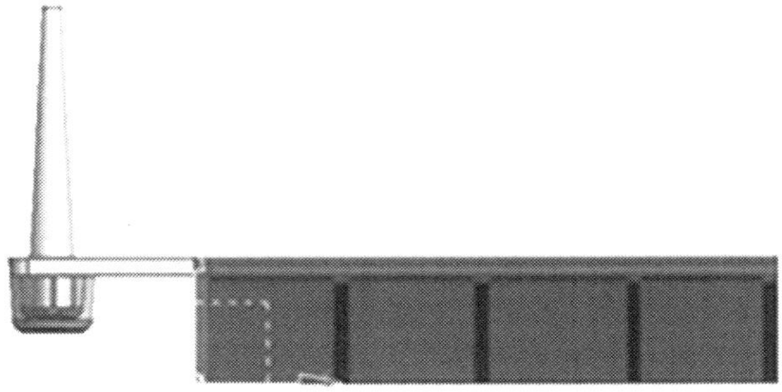

Fig. 2.91 Front view.

Dimensions: The minimum tab width is 6 mm. The minimum tab thickness is 75% of the depth of the cavity. The dimensions of tab depend upon the diameter of the runner. For tab gate, the views are shown in Figs. 2.90 and 2.91.

Use: This type of gate is useful for transparent solid block-type components.

4. Overlap Gate: For removing and eliminating any chance of jetting, the overlap gate is preferred. It is similar to an edge gate except that the gate overlaps the wall or surfaces of the products. The new type of gate has replaced the rectangular gate. In the rectangular gate, the flow of material is directly applied against the opposite surface of the product/impression. The cross section of the gate is rectangular. The disadvantage of the gate is removal of it from the product as it leaves a gate mark on the moulding.

Dimensions: The typical gate size is 1.0 mm in length × 50% of wall thickness and 1.0 to 12 mm wide.

Fig 2.92 Side view.

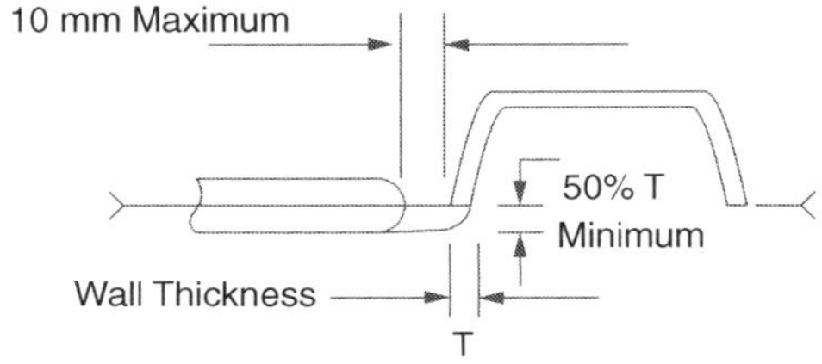

Fig. 2.93 Overlap gate.

Use: Overlap gate is used for block type of mouldings.

5. Rectangular Edge Gate: This gate feeds the material from one side into the mould cavity. The main advantages of this type of gate are:

(a) Machining cost is less.

(b) The dimensions of the gates can be easily modified.

(c) Close accuracy in the gate dimensions can be achieved.

The main disadvantage of this type of gate is visible witness mark on the moulding.

6. Fan Gate: A fan gate is a wide edge gate with variable thickness as shown in Figs. 2.94 and 2.95. This type is often used for large area thick-sectioned mouldings. Because of this a large volume of material can be injected in a short time. This is favoured for low stress mouldings or where warpage and dimensional stability are the main concerns. The gate should taper in both width and thickness, to maintain a constant cross sectional area.

Purpose of fan gate

1. The pressure is the same across the entire width.
2. The entire width is being used for the flow.
3. The melt velocity will be constant.
4. Due to the even plastic flow of the material into the cavity through a wide area, back filling is minimised and reduces imperfections and stresses in the part.
5. Used for large thin-sectioned mouldings.
6. Used where warpage and dimensional stability are the main factors.

Construction and design

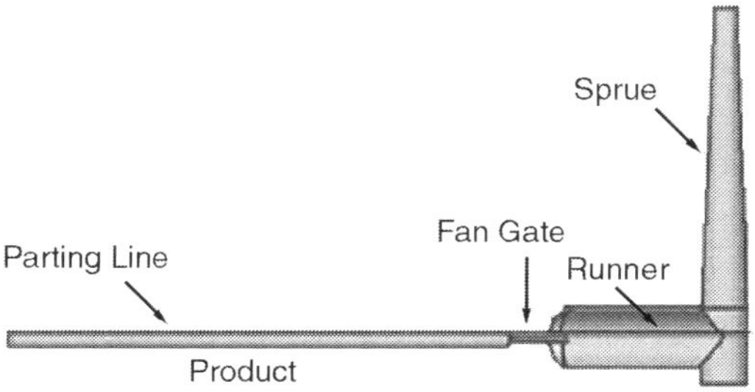

Fig. 2.94 Front view of fan gate.

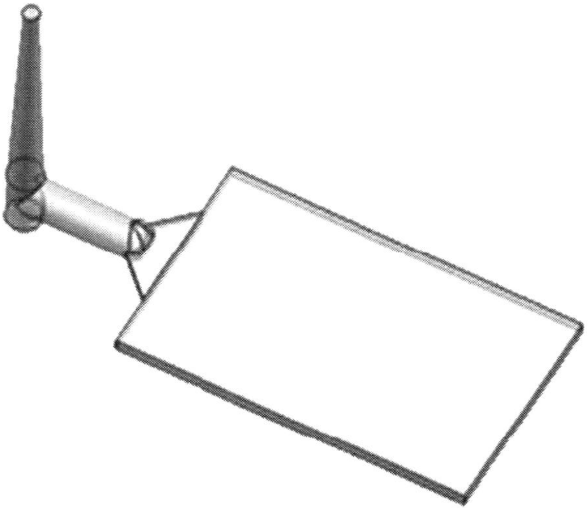

Fig. 2.95 3D view of fan gate.

Figure 2.96 shows the fan gate.

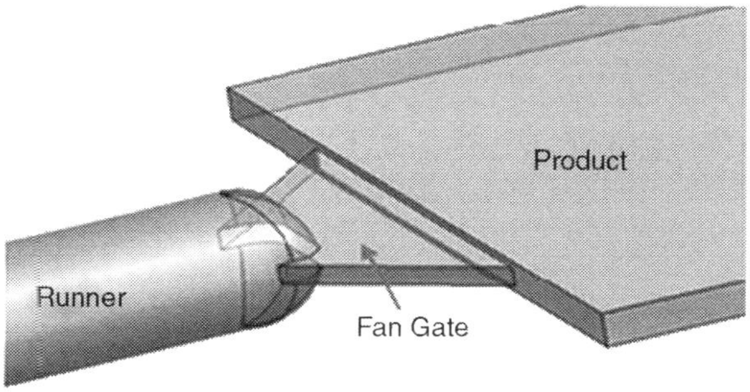

Fig. 2.96 Fan gate.

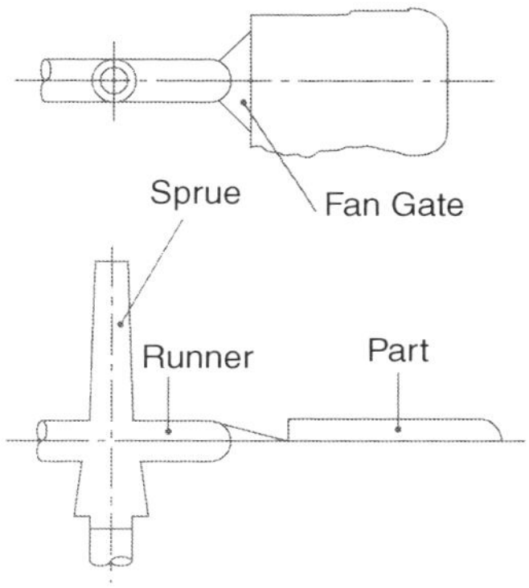

Fig. 2.97 2D view of fan gate.

Dimension: The land thickness can be very thin relative to the part thickness because the gate is wide. As with other manually trimmed gates, the maximum thickness should be not more than 80% of the part thickness. If the parts are very thin like 0.8 mm, the thickness can also be used as 0.7 mm. The gate width varies typically from 6 mm to 25% of the cavity length. In large parts, the width is often used as wide as the part itself. The recommended land length is 1.3 mm.

Use: This type of gate is used for thin walled large area moulding.

7. **Film or Flash Gate:** A film or flash gate consists of a straight runner and a gate land across either the entire length or a portion of the cavity. It is used for long flat thin walled parts and provides even filling. Shrinkage will be more uniform which is important especially for fibre reinforced thermoplastics and where warpage must be kept to a minimum. Film gate is very thin

compared to other gates. This thin gate is like a film which has parallel runner before the gate. This type of gate is used for straight edges. Figures 2.98 and 2.99 below show film gate with main parts. Film gate is used for flat mobile phone cap, Ipod cap, comb, etc. This is particularly useful for those materials which exhibit differential shrinkage for which central feeding is impractical.

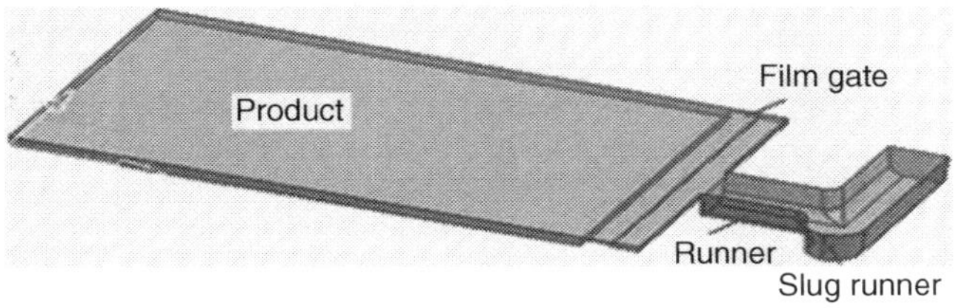

Fig.2.98 Film gate.

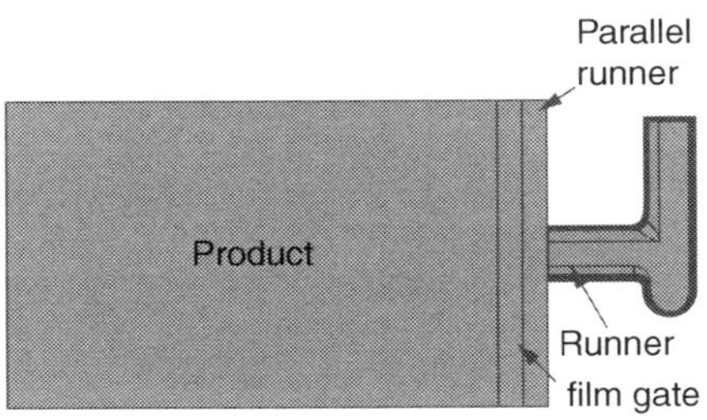

Fig. 2.99 Film gate from top view.

Dimension: Recommended dimensions are approximately 0.2 mm to 0.6 mm in thick, the land area (parallel runner) is also kept small approximately 0.6 mm–1 mm depending on the size and weight of the product.

Other Design: Figure 2.100 shows other design of film gate, which are more reliable and have an easy process.

Dimensions : The gate size is small, typically 0.25 mm to 0.5 mm thick. The land area (gate length) must also be kept small, approximately 0.5 to 1.0 mm long as shown in Fig. 2.101.

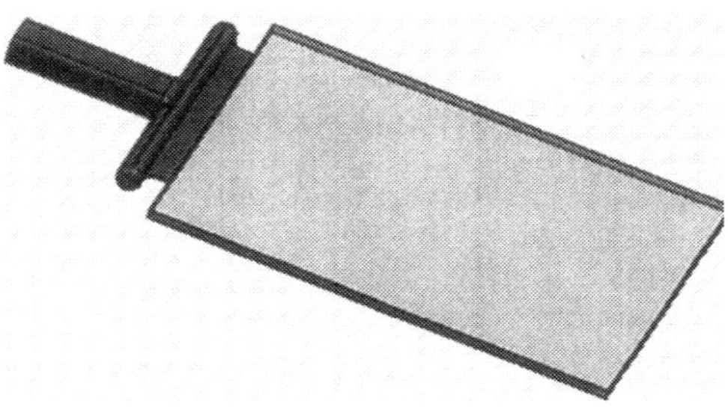

Fig. 2.100 3D view of film gate.

The gate depth $h = 0.7$ nt (n = Material constant, t = Wall thickness in mm).

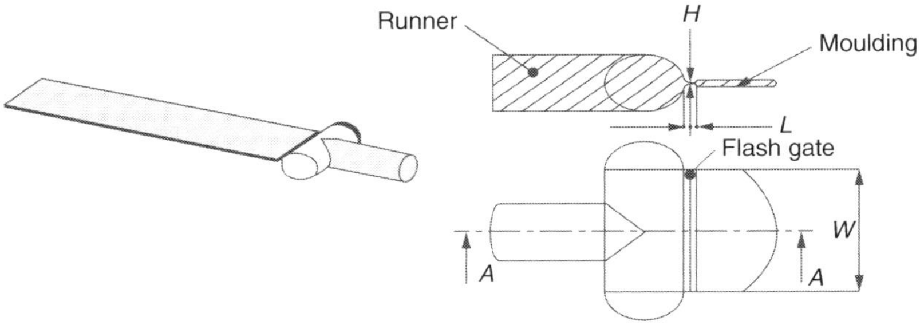

Fig. 2.101 Design of film gate.

The film gate is used for thin flat mouldings, especially those made of semicrystalline thermoplastics, where the shrinkage depends on the flow direction.

8. Diaphragm Gate: A diaphragm gate is often used for gating cylindrical or hollow round parts. It is used for single cavity moulds. It is used when concentricity is important and the presence of a weld line is not acceptable. In this type of gate, the material collects in the circular disc type recess which is slightly smaller than the inside diameter of the component. The material leads to the cavity radially from this recess. Here the recess forms a small circular runner in the centre of the component.

Sometimes it is also called *disc gate*. The basic diagram of diaphragm gate is as shown in Fig. 2.102. After moulding, this gate is trimmed off with circular disc by punching out after ejection.

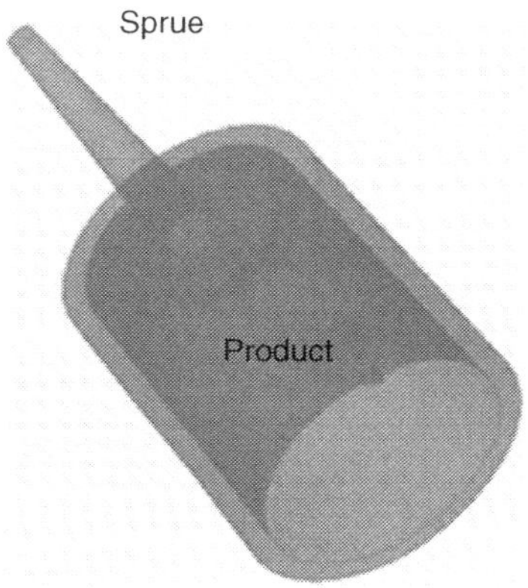

Fig. 2.102 Diaphragm gate.

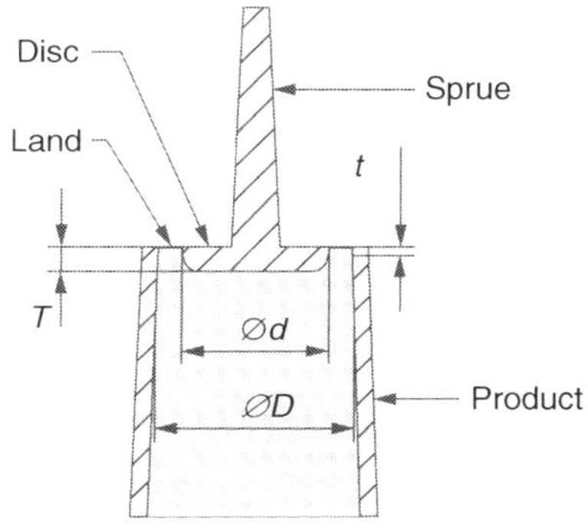

Fig. 2.103 Design of diaphragm gate.

Design and size: Figure 2.103 shows basic part and size of diaphragm gate. The **gate height (*t*)** is preferred about 0.1mm–0.15 mm. The circular **runner height (*T*)** is recommended about 5 to 10 mm depending on the size of product.

Land size **(D–d)** is the difference of internal diameter of the component and the diameter of disc which is about 1 mm–5 mm each side. Gate thickness is 0.25 to 1.5 mm.

As per the definition and position of gate, it can also be seen from the figure that this type of gate is used for the article in which external surface of the article is important for the customer because in this type of gate, the gate mark will be marked on the internal surface of the gate.

9. Ring Gate: The plastic melt flows from the sprue to a circular runner to enter the mould cavity along a film gate all round the circumference either from centre or inside as shown in Fig. 2.104 or from inside as shown in Fig. 2.105.

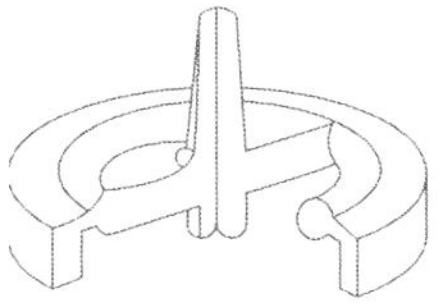

Fig. 2.104 Internal ring gate.

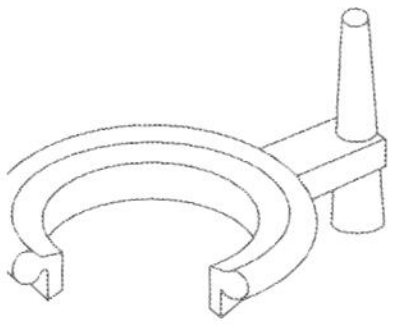

Fig. 2.105 External ring gate.

This gate is used for tubular articles in a multicavity mould or when a diaphragm gate is not practical. Material enters the external ring from one side forming a weld line on the opposite side of the runner and the weld line is not typically transferred to the part.

Dimensions: Typical gate thickness is 0.25 to 1.5 mm. This is useful when more than one impression is required in a two-plate mould.

10. Spoke Gate or Multipoint Gate: This kind of gate is used for cylindrical parts and offers easy degating and material savings. Disadvantages are the possibility of weld lines. This type of gate is round shaped and lies in the centre of the runner as shown in Fig. 2.106. The smallest diameter is about 1/3 of the diameter of the runner. The length of the gate equals the diameter (minimum 1mm) with a reverse taper towards the mould cavity.

Dimensions: The gate size ranges from 0.8 to 5 mm diameter use:

1. Three plate underfeed type of mould
2. Hot runner moulds
3. Two-plate moulds with special nozzles.

Due to the smallest dimension, automatic degating is possible which prevents any post moulding operation.

11. Pin Gates: Pin gates are only feasible with a three-plate mould because it must be ejected separately from the part in the opposite direction. The gate must be weak enough to break off without damaging the part. This type of gate is most suitable for use with thin sections. The design is particularly useful for multicavity to assure symmetric filling as shown in Fig. 2.107.

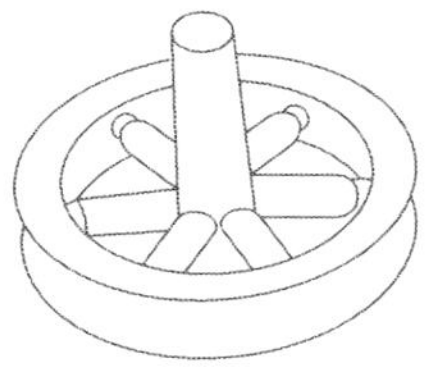

Fig.2.106 Spoke gate.

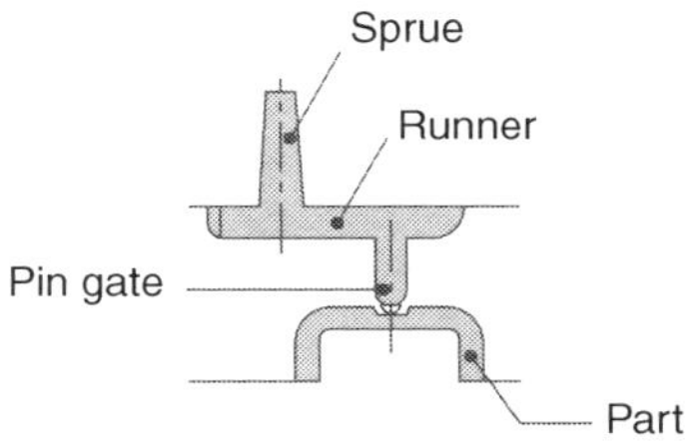

Fig. 2.107 Line diagram of pin point gate.

Dimensions: Gate diameters for unreinforced thermoplastics range from 0.75 to 7 mm. Smaller gates may induce high shear and thus thermal degradation. Reinforced thermoplastics require slightly larger gates > 1 mm. The maximum land length should be 1 mm.This type of gate is not favourable for the component which has wall thickness larger than 5 mm.

12. Submarine Gate: The positioning of this gate is flexible throughout the sides of the plastic product. It can be placed on the fixed or movable side of the mould but during design, one important point is to be thought properly so that the product will not be left inside the fixed cavity. The gate automatically detaches itself as the mould opens. So this type of gate can be used in two-plate mould with multicavity mould.

At an angle one tapered hole as a tunnel is machined from the end of the runner to the cavity, just below the parting line in such a manner that the second end of the gate should be punched as shown in Fig. 2.108 and line diagram 2.109. When the parts are ejected along with runner, the gate is sheared off at the part. The angled hole can be located either in the moving mould half or in the fixed half. To degate, the hole for gate requires a good taper and must be free to bend. This is one type of edge gate in which the material is injected below the parting line through an angled hole into the mould cavity, at a point just below the edge of the moulding. During opening of the mould, the runners and parts are ejected.

Dimensions: Typical gate size is 0.8 mm to 1.5 mm, for glass reinforced materials size would be larger. The recommended land length of this gate is 1.8 mm (minimum).

This gate is mainly used to produce smaller mouldings in multicavity moulds where the

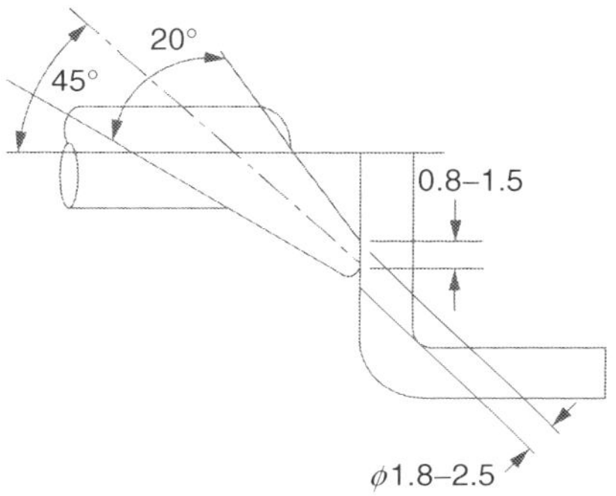

Fig. 2.108 Design of submarine gate.

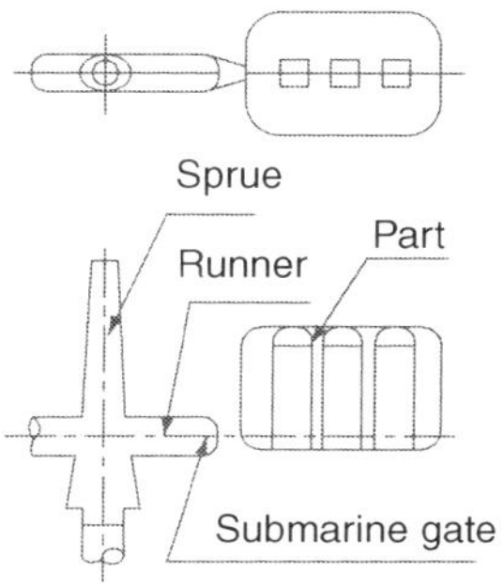

Fig. 2.109 Submarine gate.

edge can be provided. A variation of the tunnel gate design is the curved tunnel gate where the tunnel is machined in the movable mould half. This is not suitable for reinforced materials.

13. Curved Tunnel gate: The gate shown in Fig. 2.110 is one type of tunnel gates. In this type of gate the material comes out in a tunnel shape.

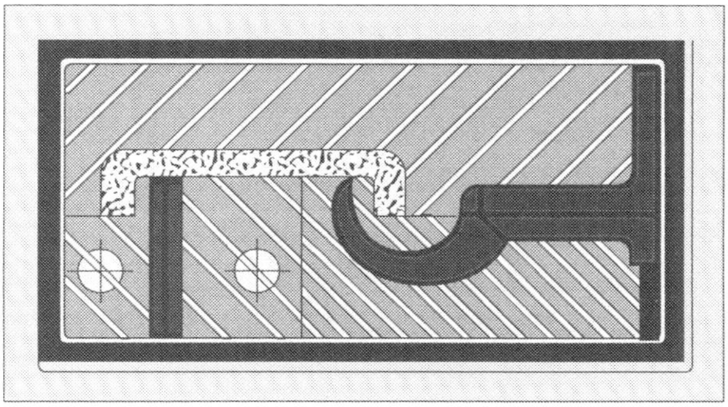

Fig. 2.110 Curved tunnel (Winkle) gate.

14. Hot Runner Gates: Hot runner gates are also known as sprueless gating. The nozzle of a runnerless mould is extended forward to the part and the material is injected through a pinpoint gate. The face of the nozzle is part of the cavity surface which will cause appearance problems (matt appearance and rippled surface). That is why the nozzle diameter should be kept as small as possible. This is most suitable for thin walled parts with short cycle times, and to avoid freezing of the nozzle.

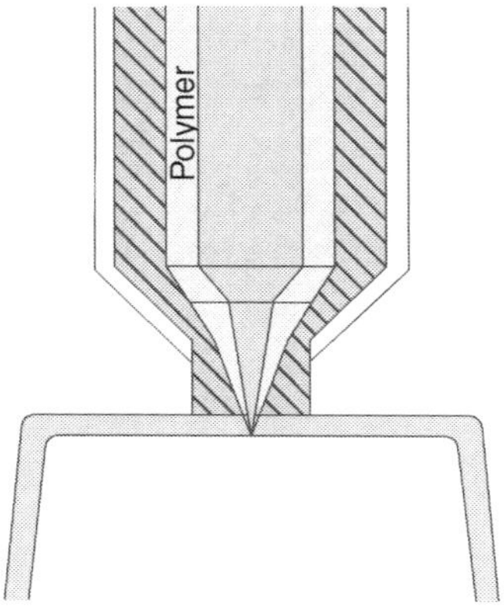

Fig. 2.111 Hot runner gates.

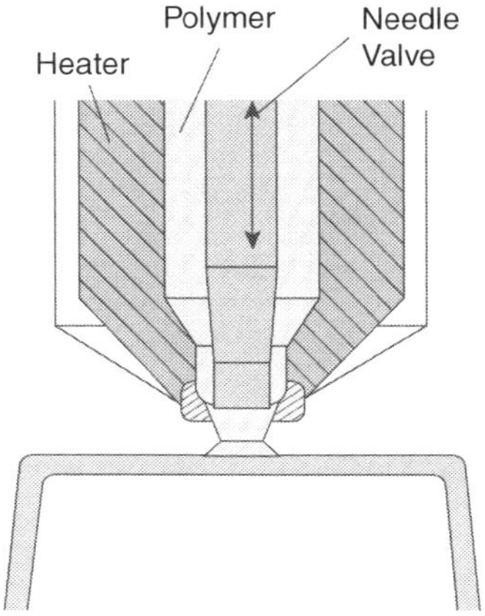

Fig. 2.112 Valve gate.

15. Valve Gates: The valve gate adds a valve rod to the hot runner gate. The valve can be activated to close the gate just before the material gets frozen. This allows a larger gate diameter and smoothens over the gate scar. Since the valve rod controls the packing cycle, better control of the cycle is maintained with more consistent quality.

Gates vary in size and shape depending upon the type of plastic material being moulded and they also depend on the size and shape of the part. Obviously, larger parts require larger gates, or even several gates.

Gate types and properties

Gate types and properties

Gate type	Runner type	Degating method	Shear rates	Resulting flow
Sprue	Cold	Manual	Moderate	Radial
Pin-point	Cold	Automatic	High	Radial
Edge	Cold	Manual	Moderate	Radial
Tab	Cold	Manual	Moderate	Radial
Flash/diaphragm	Cold	Manual	Moderate	Linear
Fan	Cold	Manual	Low	Linear
Tunnel/submarine	Cold	Automatic	High	Radial
Thermal pin-point	Hot	Automatic	High	Radial
Thermal sprue	Hot	Automatic	Moderate	Radial
Valve	Hot	Automatic	Moderate	Radial

The factors which affect the gate size The size of gate plays an important role for the quality of article. Large size gate will make unbalanced flow and defect in product. Small size gate makes the cycle time high. Short fill of product and sink mark problem will appear in product. There are various factors that affect gate size and they are given below.

1. **Mass of Part:** More mass of part requires large size of gate.
2. **Size of Part:** Large size of part requires larger size of gate.
3. **Temperature of Melt:** Sometimes when material is getting filled in the cavity with the possible maximum temperature, the cavity is still not fully filled. At that time, it would be better to go for smaller gate. Because of smaller gate more heat from the heater can be transferred to resin and as shear stress of resin increases and viscosity will decrease, then resin will flow easily with small value of viscosity.
4. **Nozzle Tip Position:** The length between nozzle tip and gate should be kept longer, because, with minimum length it will make freeze premature. If for a mould it is not possible to make longer length, it can be controlled by making the gate smaller.
5. **Viscosity of Resin:** Viscosity of resin and shear stress of resin have large effect on proper gate designing. Calculation can be done for average plastic flow by dividing shear volume by injection time. There are some other factors like cooling speed, cooling temperature and injection time which also have a valuable effect when designing the gate.

Gate balancing When the mould is designed for a large number of different components or sometimes due to not proper machining of the impression of the multicavity mould, balanced filing is done by balancing gate by varying the dimension of gate.

2.6 EJECTION SYSTEM

The thermoplastic material after cooling sticks on core or projected portion of the mould due to shrinkage. As per the injection moulding machine design, the ejection rod is provided in the moving side of the machine. It means that the ejection system in the mould is incorporated in the moving half of mould and the core as projection portion of the mould is generally fitted in moving side of the mould. Due to some mechanism like link bolt, knockout rod movement, the ejector plate moves forward and this makes the ejector pins or systems to eject the components.

1. The ejection system consists of the following:
2. Types of ejection like pins, step pins, valve pins, stripper plate, etc.
3. Ejector plate assembly consist of two plate, i.e., ejector plate and ejector back plate
4. Ejector grid place for the ejector assembly
5. Sprue puller for pulling the sprue from sprue bush button

Complete ejection system is shown in Fig. 2.113.

2.6.1 Ejector Grid

It is the most important part of the ejection system and creates the space for ejection assembly and makes them freely operate. The space is generally created by using mould bottom mounting plate and spacer block.

Types of ejector grid Ejector grid is divided into three types:

1. Circular or round shape
2. In line system
3. Frame system

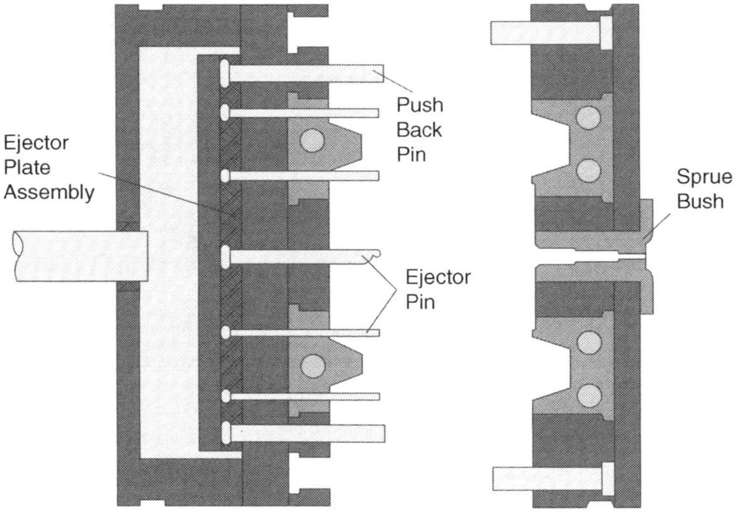

Fig. 2.113 Ejection system.

1. Circular And Round Shape Ejector Grid: For creating a space for ejection assembly in this type of ejector grid, a number of circular blocks are positioned on the mounting plate of the mould. The small round blocks are fitted between bottom mounting plate and the core back plate with the edge of plates and in the centre, the space is created for the ejector plate assembly. But the disadvantage in this type of ejector grid is deflection of the back plate due to high injection pressure on back plate.

Support Blocks: To reduce the greater deflection in the back plate due to injection pressure in large moulds, round blocks are used in the centre of the space called support blocks.

2. In Line-Type Grid: The ejector grid two rectangular blocks are mounted and clamped on bottom mounting plate. This type of ejector grid is also known as U-type ejector grid because it creates a U shape with the bottom mounting plate. In this type of ejector grid, to reduce deflection in the core back plate due to the injection pressure, the distance between two spacer block is to be kept minimum. The deflection in a simply supported beam depends upon the length of the beam. A great distance between the two blocks creates more deflection and may cause the mould plate to be distorted due to the high injection pressure. To overcome this, sometimes extra support blocks are incorporated between the two blocks.

(a) Spacer Block: It is a block which is used to create the space for the ejection system in the mould as shown in Fig. 2.114. In the mould, the spacer block in two numbers are fixed on the bottom plate by using socket-head screw.

(b) Calculation of Height and Width of Spacer Block:

(i) The Width: It is decided on the basis of size of ejector plate. The size of ejector plate is decided on the basis of placement of ejector pin.

The width of the spacer block = the width of the plate – width of ejector plate – 4mm (2 mm clearance per side).

(ii) The Height: It is decided on the basis of ejection stroke of the mould. Ejection stroke means the required forward movement of the ejector pin for ejection of article.

The height of spacer block = ejection stroke + ejector plate thickness + ejector back plate thickness + 2 mm for button.

(c) Design of Spacer Block:

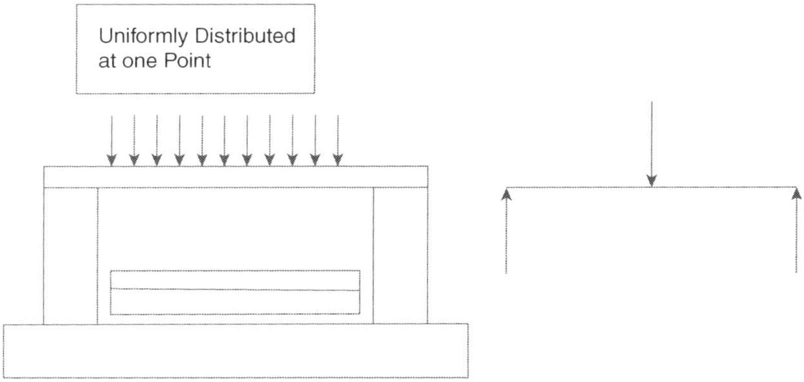

Fig. 2.114 Line diagram of spacer block.

In the line diagram shown in Fig. 2.114, spacer block is fitted to create space for the ejection assembly. When plastic material enters into the impression, the direct force is applied on core and the effect of this direct uniformly distributed force can be analysed on spacer block and back plate. As shown in the line diagram Fig. 2.114, reaction force in the spacer block will be the half of the total load of back plate.

In the mould, spacer block withstands compressive stress as the force is applied from both sides of the block. Let us assume that the size of spacer block is $H \times L \times b$ where b is to be calculated and other two can be calculated as per the ejection stroke and core back plate width and the reaction load in spacer block is k. If the two numbers M10 hole is made for fixing of spacer block, then

The area withstands the compressive load = $HLB\text{-}2(\pi/4)\ 10^2 = HLB\text{-}50\ \pi$

The compressive stress = k / (HLB - 50 π). This value should be compared with the compressive stress of spacer block material and find out the value of b.

(d) Bolt for Spacer Block

To make the position of spacer block fit between core back plate or core plate and bottom plate, bolts from bottom plate to spacer block can be used. Refer to Fig. 2.115 below. In each spacer block, add two pieces of bolts.

The other function is when disassembling the mould, spacer block and bottom plate become one part, it makes dissembling easier.

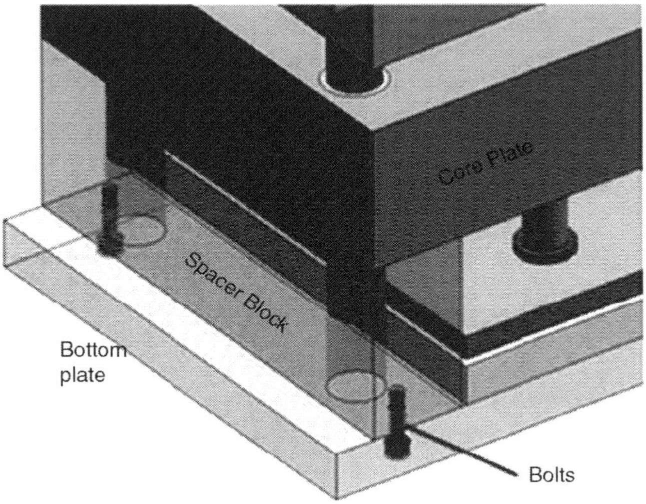

Fig. 2.115 Assembly of spacer block.

3. Frame Type Grid: It is the type of in line ejector grid except mounting of four numbers of spacer blocks. In this type of ejector grid, complete frame is machined outside and fitted with the mounting plate. Because of complete frame it creates one boundary for the ejection system. As per the design of frame, the grid can be designed as rectangular, square or round type of frame ejector grid.

The main advantages of this type of ejector grid are:

(a) Good support to the back plate from all four directions.

(b) Deflection problem can be solved up to a certain level.

(c) Manufacturing cost is less.

(d) The ejection system is completely covered by one boundary, thus it prevents outside particle to enter the system.

On the basis of shape, the frame grid is further divided as per the shape:

1. **Rectangular Type:** The blocks are placed as the shape of rectangle called spacer block.
2. **Square Type:** The frame is machined as the shape of square called square type grid.
3. **Round Type:** The blocks are machined in the round shape or well type called round type grid.

2.6.2 Ejector Plate Assembly

It is that part of the mould to which the ejector element is fitted. It normally consists of an ejector plate, retaining plate and ejector rod as shown in Fig. 2.115.

Ejector plate It is a steel plate in which the ejector system like ejector [pin, valve, step pin] is incorporated. The force is applied from the machine moving half to this plate for transmitting it to the article for ejection through the pins or other mechanism. The size depends upon the product shape or profile and the thickness depends upon the force required to striping off the product by the pin.

Retaining plate This plate is used to remove the chances of ejector pin to retain in the ejector plate and withstand the force received from the ejector pin. The plate is screwed by the socket–headed screw with the ejector plate. The plate thickness generally takes same or more than ejector plate thickness.

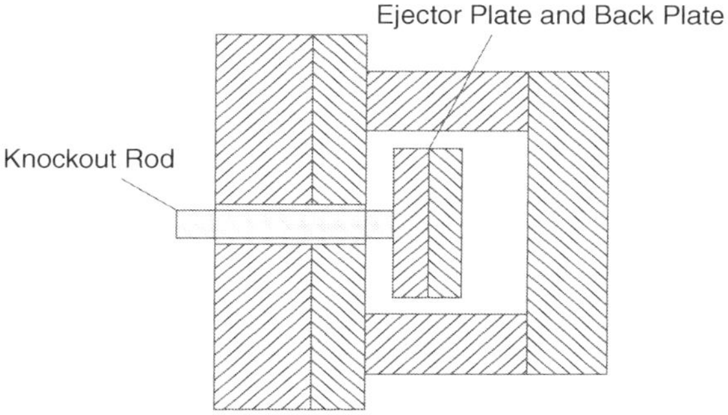

Fig. 2.115 Ejector plate assembly.

Knockout rod It is a circular steel rod which provides the actuation and guiding functions of the ejector plate. It is attached to the ejector plate by means of a thread.

Tie rod bolts Tie rod bolts are used to push the stripper plate when ejecting process begins. It requires a connector from ejector plate to stripper plate. In this case, tie rod can be used as shown in Fig. 2.116.

Button During the closing of the mould any foreign particle sticks to the bottom plate of the mould. When the ejector plate returns back, it will not seat properly on actual position. This will cause the ejector pin slightly protrude from its position. So, for getting some recess, four buttons are fixed on the bottom plate of the mould. If foreign particle sticks with the bottom plate, due to button the ejector plate will seat on its actual position.

The suitable steel material used for button is generally En-31.

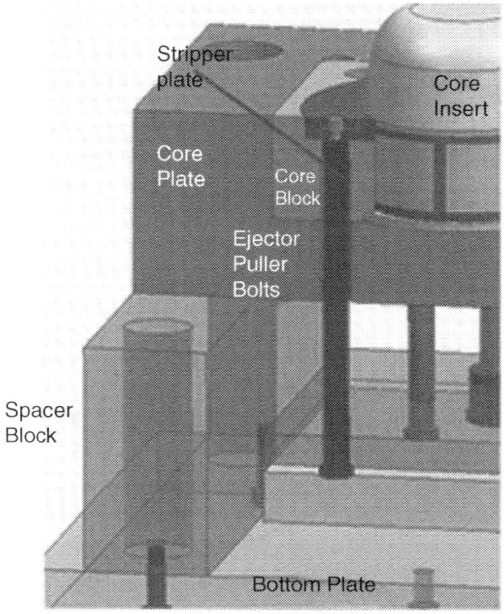

Fig 2.116 Use of tie rod for stripper ring.

Push back pin The function of push back pin is to take back the ejector assembly after ejection, when mould is being closed to get back to its original position as shown in Fig. 2.117. It is one type of ejector pin normally in size of diameter 8 mm to diameter 12 mm, which is fixed in ejector plate the same way as ejector pin. The other part of the pin is made in contact with the parting surface. Sometimes for small mould a spring is provided in the knockout rod. During actuation of the ejector assembly, the springs are getting compressed. While during closing of the mould, the spring forces the ejector assembly to its original position.

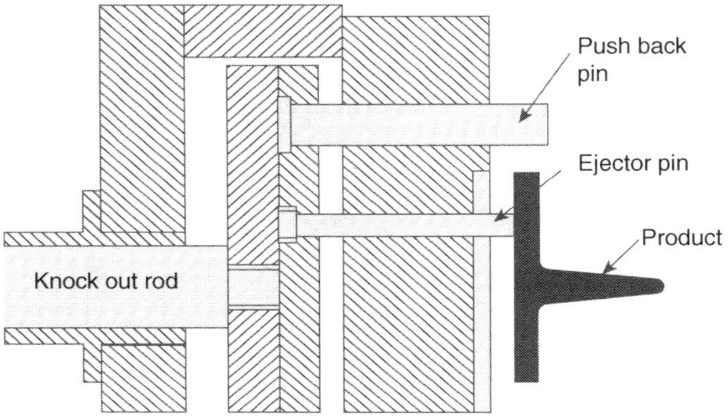

Fig. 2.117 Working of ejection system.

2.6.3 Types of Ejection

When the moulding cools after injection, it shrinks on to the core for which some positive types of ejection techniques are to be adopted. For the ejection of any component, the ejector pin is provided on wall thickness. The size of ejector pin depends upon the area of contact with the plastic material. For easy ejection, the area of contact of the pin with plastic material should be more. If it is less, there is a chance of the pin piercing the plastic material. As per the Fig. 2.118 shown below, the ejector pin shape is decided on the basis of area of contact.

There are different types of ejection systems adopted in a mould. The normally used ejection systems are as follows:

For the ejector pin, the size of the pin is decided on the basis of wall thickness as shown in Fig. 2.118.

1. Diameter of pin is nearly equal to wall thickness which is acceptable.
2. Diameter of pin is very small which will cause piercing of the material by the pin.
3. Diameter of pin is very large which is not suitable or loss of efficiency.

Therefore, example a is correct.

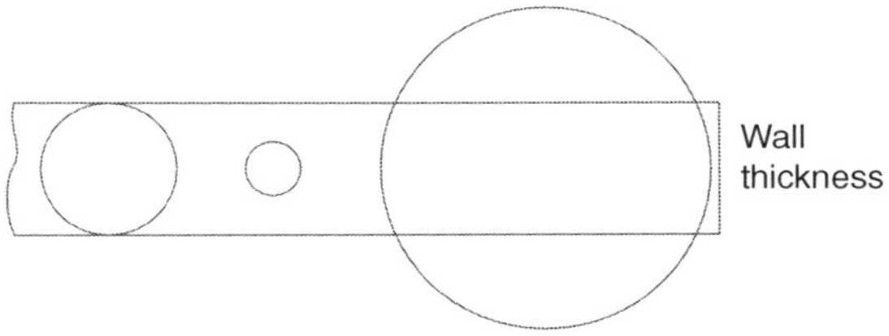

Fig. 2.118 Size of ejector pin.

1. Pin ejection:
   * Plain diameter ejector pin:
     a) Moulding face ejector pin
     b) Parting face ejector pin
   * Stepped ejector pin
   * D shaped ejector pin
2. Sleeve ejection
3. Blade ejection
4. Valve ejection
5. Air ejection
6. Stripper plate ejection
7. Stripper bar ejection

Plain diameter pin ejection This type of technique is generally used for the ejection of article. In operation, the ejector pins, which are fitted in the ejector plate assembly, are moved forward and backward due to the motion and force of knockout rod relative to the mould plate. The shapes of pins are generally circular.

The working diameter of the ejector pin is a slide fit in the core plate and clearance fit in the back plate by providing tolerance hole in the plates. The placement of pins and its number are decided as per the profile of the component and it is located in such a way that the uniform force should be applied on to the components. The plain diameter ejector pins are classified into two:

(a) Moulding Face Pin: This type of ejector pin is used where the hole of the top surface of the pin is in contact with the moulding. The moulding must be ejected from the cavity and has no internal form.

(b) Parting Face Pin: This is used mostly for standard box shaped mouldings. A part of the top surface of the pin is in contact with the moulding and the rest supports the parting surface of the closed mould. The pins push the side walls of the moulding during ejection as shown in Fig. 2.119. The steel material of the ejector pins should have good strength, wear resistance and toughness.

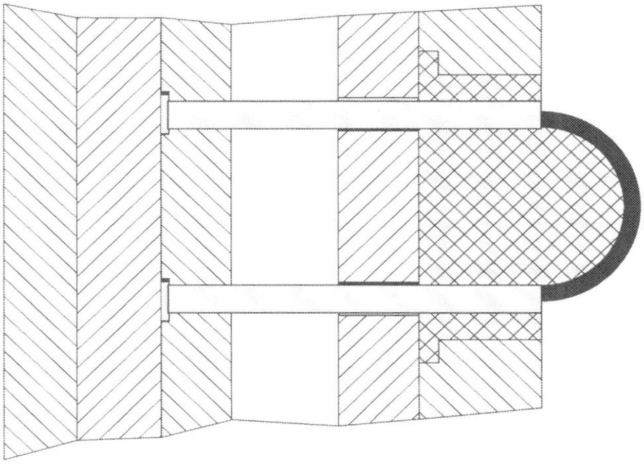

Fig. 2.119 Parting pin ejection.

Stepped ejector pin This type of pin is generally used where the wall thickness is very less and the diameter of pin is also small and which will not withstand the force applied from the machine to component. It will get buckling /bending in the hole. So for this condition, stepped pin is used and it is machined in two parts. One part in which the diameter of the pin is machined as per the wall thickness for a minimum length to withstand the force (no buckling of the pin occurs) and in other part the diameter takes more to cover entire length of ejection stroke. The total length of the pin is decided on the base of total ejection stroke.

The length of the small diameter of the pin should be kept small. This length can be calculated from the following as shown in Fig. 2.120.

$$L = L_1 + L_2 + L_3$$

where L = Total length of the small diameter of the ejector pin

L_1 = Length in contact with the mould plate which is at least 5 times the diameter of the ejector pin

This is decided on the basis of slenderness ratio (L/D) ratio.

L_2 = Ejector stroke; L_3 = A small allowance of 5 mm.

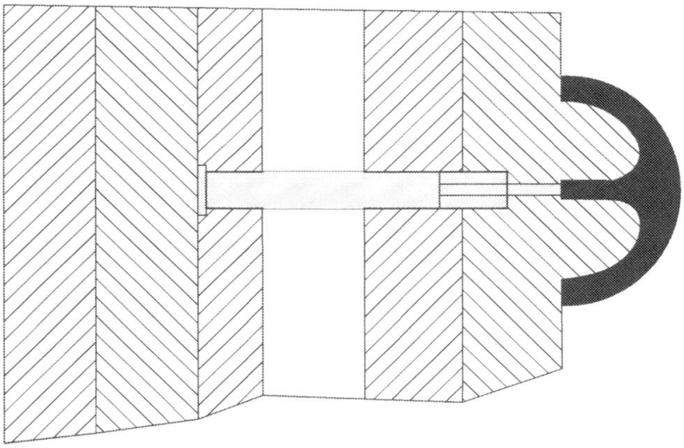

Fig. 2.120 Step pin.

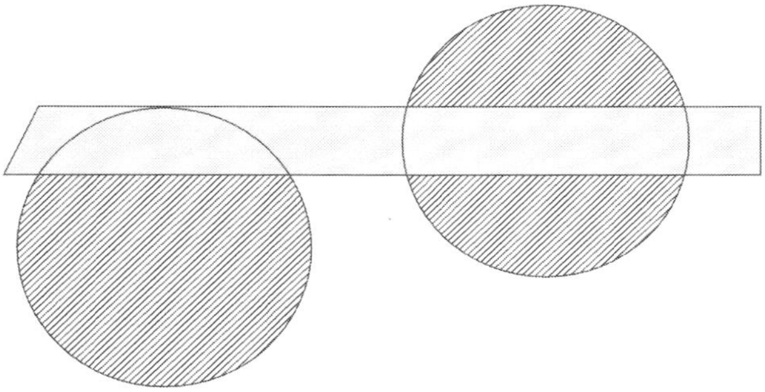

Fig. 2.121 Increasing contact area for ejection on the moulding.

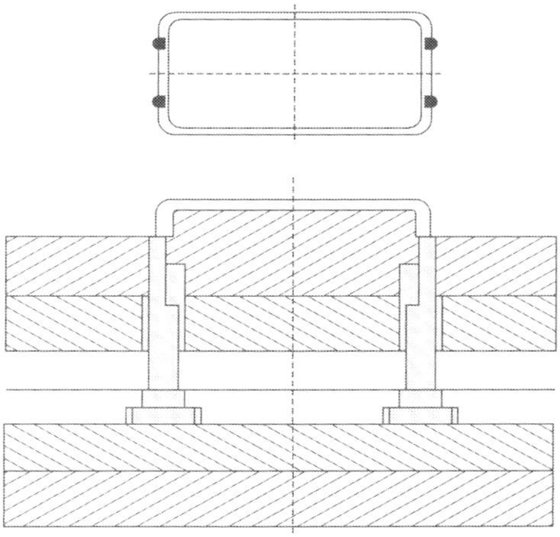

Fig. 2.122 Assembly of D-shaped pin.

D shaped ejector pin The area of contact of moulding plays an important role for the ejection of component. For increasing the area of contact and for thin wall component, the ejector pin of large diameter is made in contact with moulding as shown in Fig. 2.121. Now the area which is not in contact with the moulding is machined, removed and given the D shape. This type of pin is used advantageously due to greatly increased effective ejection area compared to standard parting surface ejector pin as shown in Fig. 2.122.

Pin blade ejection This type of ejector system is used for releasing very slender parts such as ribs, projections, which cannot be properly ejected by the standard type of ejector pins as shown in Fig. 2.121. For increasing the area of contact and for thin walled component, the ejector pin of large diameter is made in contact with moulding as shown in Fig. 2.122. Now the area which is not in contact with the moulding is machined, removed and given the rectangular shape. The blade ejector can be fitted to the ejector assembly in the same manner to that of a standard ejector pin. Very thin ejector blades (thickness less than 1 mm) are supported on one side by a slot machined into a link shaped support block.

Actually for blade ejection, the pocket of blade size cannot be easily machined in the mould back and core plate. The same size pocket can be machined only by electric discharge machine which is a costly and time taking work.

Sleeve ejection This system is used when part has a hole inside. Standard ejector pin cannot put in the product, because the ejector pin cannot be placed in the face as the area of face is very small where the ejector pin touches the product surrounding the hole. Refer to the Fig. 2.123 below.

Fig. 2.123 Sleeve ejection.

As per the Fig. 2.123 the only way to eject this part is by using sleeve ejector that will be located at circumferential face.

The mechanism of sleeve ejector is to make core of the article not to move when sleeve ejector moves to eject the hollow product. Basic construction of sleeve ejector system is shown in Fig. 2.125 and its working is shown in Fig. 2.124. In this figure, step by step the working of sleevė ejector pin is shown. The main part of the mechanism is product 1. sleeve ejector 2. core insert pin which is fitted at centre of sleeve ejector and others are mould base core, cavity insert, cavity pin insert, and cavity mould base.

1. After plastic material is injected at cavity insert then cooling water/liquid will flow to transfer heat from mould to liquid. At this position mould base closes, just seconds before ejecting process begins.

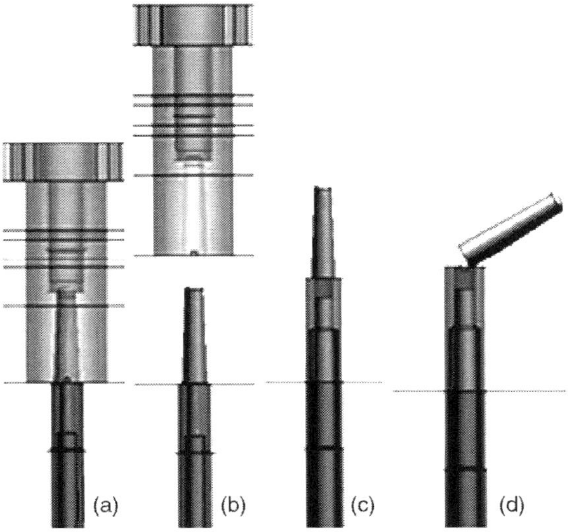

Fig. 2.124 Process of sleeve ejection.

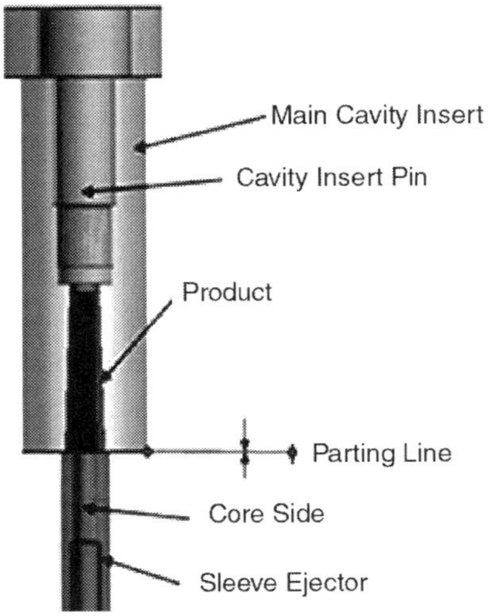

Fig. 2.125 Sleeve ejection.

2. Ejecting process begins, first mould will open, core insert move with moving half, because shrinkage of material always moves to inner side, product will stick in core insert pin.
3. Sleeve core mȯves forward when ejecting process begins, sleeve core will eject the product, and the product will leave core insert pin, because core insert pin does not move forward.

This is one type of hollow circular ejector pin which is fitted at the rear end to the ejector assembly or bottom mounting plate (refer to Fig. 2.126). The sleeve is a sliding fit in the cavity core pin. The core pin extends completely through the sleeve and is attached to the back plate. In operation, the sleeve moves relative to the core and cavity and the moulding is ejected. In this type of ejection, ejection force is applied on a large surface area.

Generally circular type of sleeves is recommended due to ease of machining and cheaper cost. This type of ejection is generally restricted to the following types of moulding:

(a) Certain types of circular mouldings
(b) Circular projections on a moulding
(c) Moulding having round hole

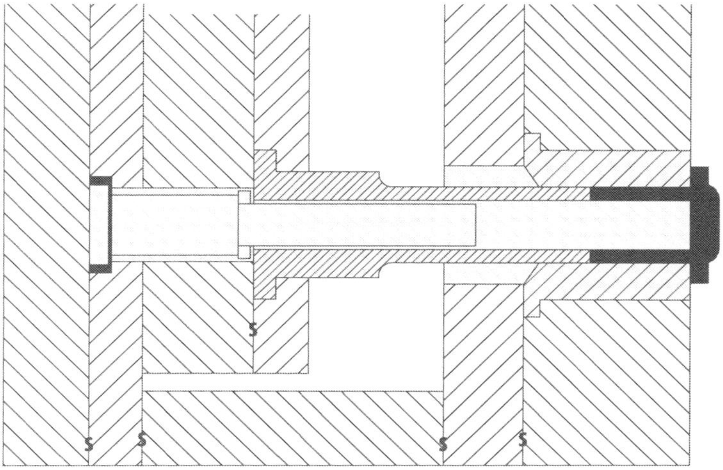

Fig. 2.126 Sleeve ejection.

When a moulding requires a sleeve with a wall thickness of less than 2 mm, then stepped sleeve is used. In this type, the diameter of the lower end is made larger in order to increase the strength of the sleeve. The length of the thin wall section is kept small. In order to reduce frictional wear, and for proper fitting the surface contact between the sleeve and the core pin is kept minimum. This type of ejection technique is recommended for limited number of impressions (generally two) of a particular component.

Stripper bar ejection In this ejection system, the ejector element, a stripper bar, pushes the bottom edge wall of the moulding. The ejection principle is same as that of standard pin type ejection. The stripper bar is fitted into the mould plate and a small angle of 10° is incorporated all around its periphery in order to minimise wear. The stripper bar is coupled to the ejector plate by a tie rod whose one end is threaded and attached to the stripper bar. A gap of about 3 mm should be provided between the ejector plate and back plate to ensure that the stripper bar seats properly in the mould plate.

Stripper plate ejection

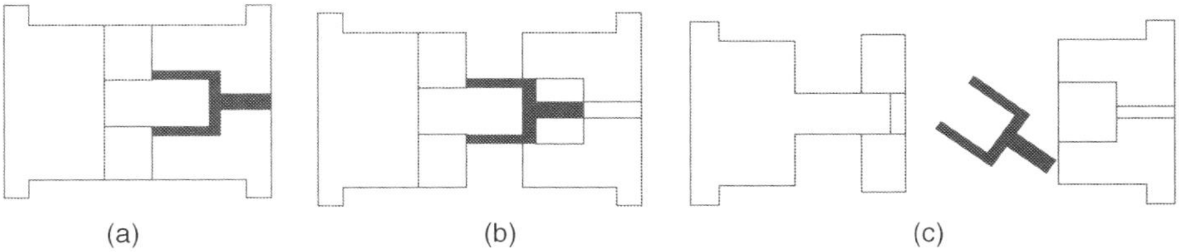

Fig. 2.127 Principle of stripper plate ejection mould.

In this type of ejection, one extra steel plate termed as stripper plate is actuated between core and cavity as shown in Fig. 2.127. In the above Fig. 2.127 (a), the plate is centrally machined as per the profile or shape of the core part. During ejection, the stripper which is floating plate and guided by guide pillar moves along with core and article (article sticks with core due to shrinkage property) when mould start opens as shown in Fig. 2.127 (b), The stripper plate has sliding fit with core. Now the article is totally free from the cavity. Now there are two types of conditions for the movement of stripper plate:

1. By providing some means to stop/arrest the movement of stripper plate and core is getting opened with moving platen of the machine as shown in the Fig. 2.127 (c).
2. As per the Fig. 2.128 in opposite direction the movement can be given to stripper plate by using tie rod which is fixed in the ejector assembly. Due to opposite direction movement, the article is released from the core and ejected. Generally the size of the stripper plate is the same as the other mould plates.

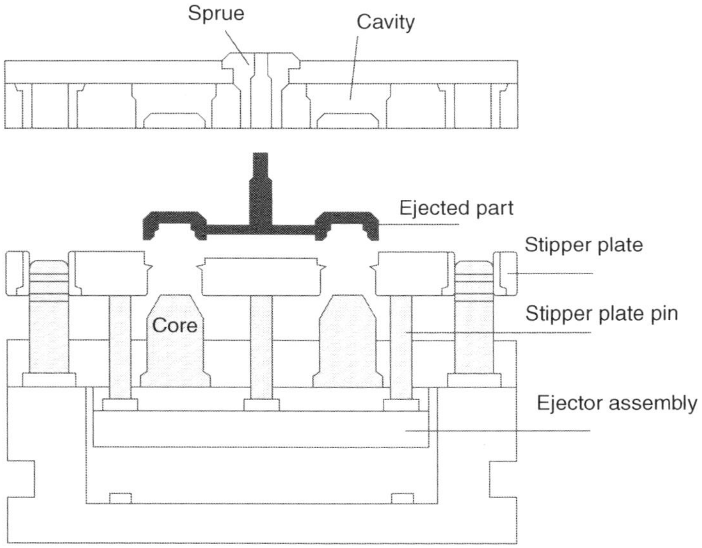

Fig. 2.128 Stripper plate ejection.

Simple stripper plate consists of three main parts:

1. Stripper plate as shown in Fig. 2.129 and its assembly as shown in Fig. 2.130.
2. Bolts for clamp
3. Ejector pin

1. Stripper Plate: The main function is to push product when ejecting, for ejection it does not require much part of contact area of the product with the plate. Only 0.8 mm product will be ejected from core insert side. Clearance and dimension tolerance below is shown in the Fig. 2.131.

(a) Between cores insert and core plate, fit tolerance about 0.010 mm with height about 15 mm from parting line.

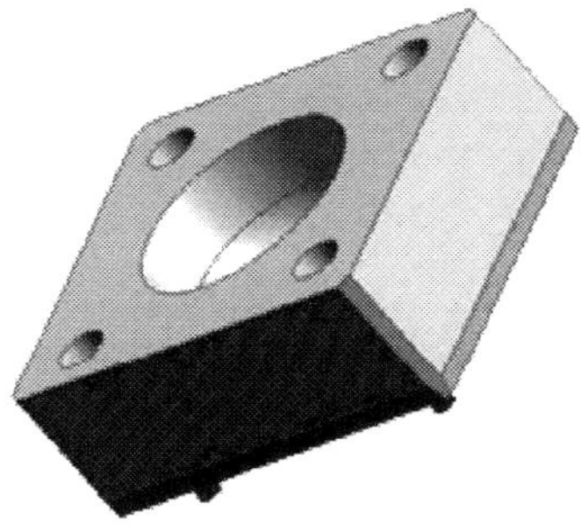

Fig. 2.129 Stripper plate.

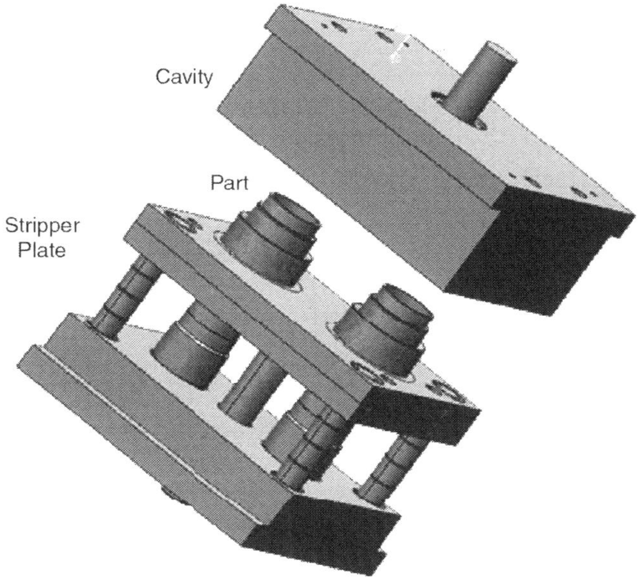

Fig. 2.130 Assembly of stripper plate.

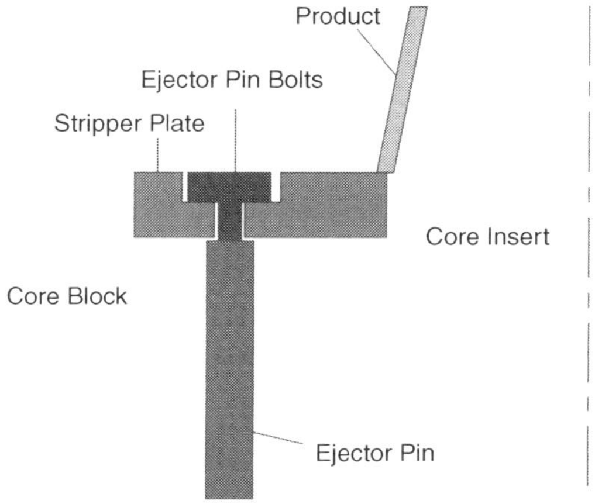

Fig. 2.131 Assembly of stripper plate with tolerance.

(b) Between stripper and core block, it is given free tolerance about 0.5 mm each side.

(c) Between stripper plate and core insert it is fit tolerance about 0.005 mm to 0.010 mm to prevent plastic flash.

(d) Between ejector pin bolts and stripper plate, it gives free tolerance from 0.5 mm to 1 mm.

(e) Ejector core pin and core block, it gives slide fit tolerance about 0.010 in top side with height 10 mm, after that gives free clearance about 0.5 mm in each side.

Different types of actuation of stripper plate are:

(i) Tie rod actuation

(ii) Length bolt actuation

(iii) Chain /External link actuation

(iv) Direct actuation

(i) Tie Rod Actuation: In this system, the stripper plate is coupled to the ejector plate by three or four tie rods. During opening of the mould, the moving half moves in backward direction and strikes the actuating rod of the machine. The movement of the ejector plate and the stripper plate is arrested and the moulding is stripped from the core. It means that the stripper plate has the movement on guide pillar by the tie rod (one end fixed with the ejection assembly and other end threaded with the stripper plate)for ejection of the component after complete opening of the mould as shown in Fig. 2.132 (a) and (b).

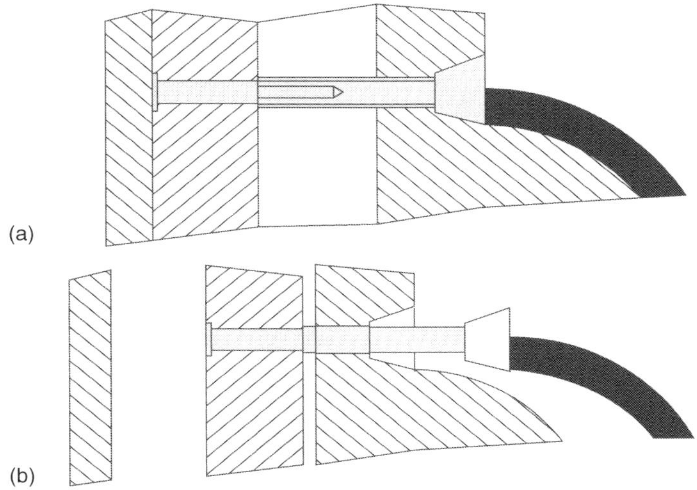

Fig. 2.132 Tie rod actuation (a) and (b).

In this method, the ejector plate does not seat on the back plate of the mould. An ejector plate and ejector grid system are adopted in this design. The working of tie rod for stripper plate is shown in Fig. 2.133.

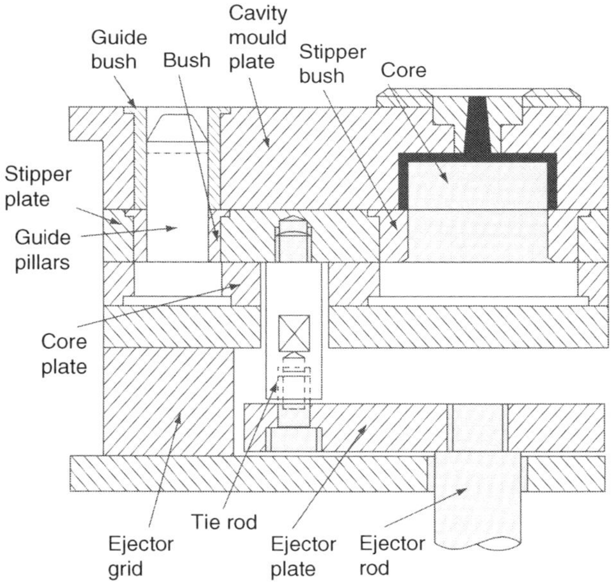

Fig. 2.133 Working of tie rod.

(ii) Length Bolt Actuation: In this method, the ejector plate is arrested by length bolts situated within the mould. The head of the length bolt is accommodated inside the fixed mould plate and the nut and locknut of the mould are accommodated in a clearance hole in the moving half. Here the ejection of component is designed in such a manner that there is no movement of stripper plate due to the restriction of length bolt,only due to the sufficient movement of core plate with the moving half of the mould to release the component fully from core. Complete calculation of length bolt actuation is shown in Fig. 2.134. This method is generally recommended for very deep moulds as the maximum stripper plate movement depends upon the mould height. To overcome this problem, an alternative design called telescopic length bolt is used even on shallow moulds.

This method is cheaper than tie rod actuation and is lighter than other processes.

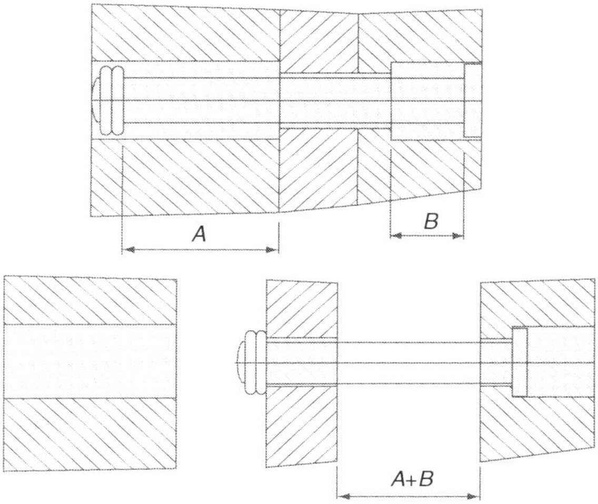

Fig. 2.134 Length bolt assembly.

Generally, three length bolts are recommended on circular type of mould whereas four on rectangular type.

(iii) Chain /External Link Actuation: In this method, chains are used to arrest the motion of the stripper plate instead of length bolts. One end of the chain is connected to the stripper plate externally and the other end to the fixed mould plate. During opening of the mould, the chains are progressively straightened until the movement of the stripper plate is arrested. One or two chains are used per side depending upon the size of the mould.

More economic and effective alternative design for chain is external link. It is a machined rectangular block having slot in it. It functions in the similar manner as that of chain.

(iv) Direct Actuation: In this method, the stripper plate moves back during opening of the mould with the moving half of the mould until it is getting arrested by the actuating rods of the moulding machine. Further movement of moving mould half causes the core to be withdrawn through the stripper bush and the moulding is ejected. In this method the length of stripper plate is taken larger and the actuating rods of the machine strike the stripper plate directly during opening of the mould and arrest the movement of the stripper plate. Core is moved with the moving half of the mould and relieved the component from the core. In this system, no ejector assembly is required.

Valve ejection This type of ejection is used for ejection of deep and large components. Actually as per the principle of ejection, 'the contact area of pin with plastics material should be more'. So for a large base area and deep component, the area of contact of the pin is less, which pierces the plastic material. To overcome this problem, a large area valve head as shown in Fig. 2.135 is machined and fitted with the ejector pin. Due to the large area of contact with the plastic material, ejection force is uniformly distributed on the surface area of the component. To ensure a material leak free joint, a small parallel portion of a length of 2.5 mm (max.) is provided at the major diameter of the valve. The included angle of the valve is between 90° and 120°.

Valve pin which is assembled with spring is actuated directly against the spring and ejects the article. When the mould gets close, injection machine's ejector operating system returns to its original position, the spring causes the valve to return to its original position. Here the valve ejector pin is machined in two ways. Once the valve head is machined separately and threaded with the pin. During assembly also in the mould the pin is first fixed with the ejection assembly and from other side valve head is threaded with the pin and other one as shown in Fig. 2.135, the ejector pin is machined with valve head and during assembly after assembly of valve head pin is fixed by socket-headed screw.

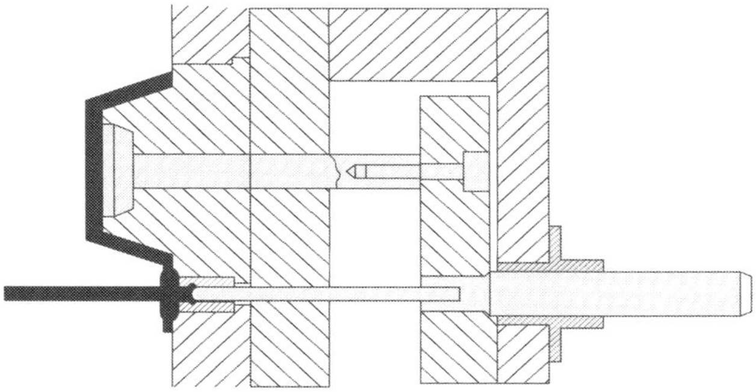

Fig. 2.135 Valve ejection.

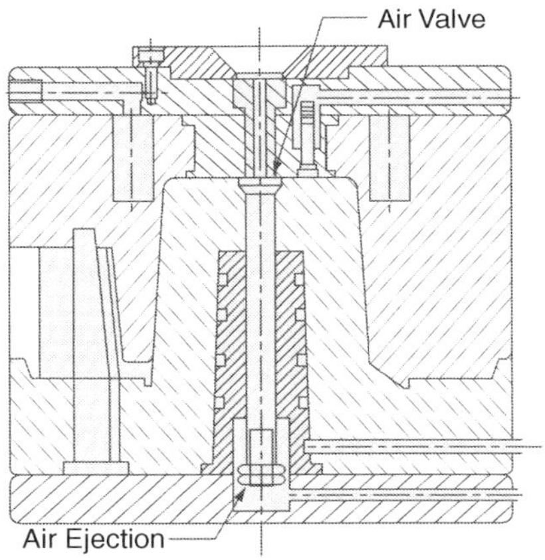

Fig. 2.136 Air ejection.

Air ejection This ejection system (Fig. 2.136) is the same as valve ejector pin. The only difference is the compressed air which is used directly on to the moulding face via a small air valve. In this type of ejection small slots are made in the periphery of the valve head. These slots have the connection with a centre hole in the pin. The compressed air passes through this line and applies a uniform pressure on the article. The amount of ejection force is dependent on the pressure of the compressed air and the area on which it acts. The valve is mounted on a spring which controls its movement. It is in contact with the moulding. When the force exerted by compressed air is larger enough than the spring force, air forces the valve through a slot machined on the periphery. The valve moves forward and ejects the moulding from the core. The control valve is used for controlling the compressed air.

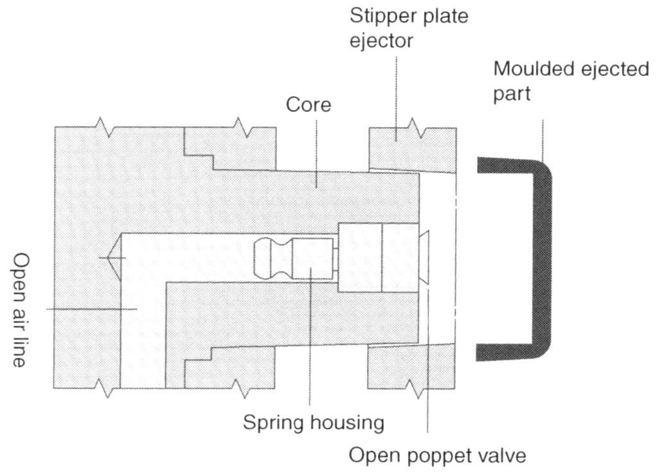

Fig. 2.137 Double ejection (Air and stripper plate ejection).

The main advantages of this ejection are:

(a) Ejector grid or ejector assembly is not required.

(b) The air ejection can be used for any type of mould and located at any half of the mould, and operated at any time during the opening of the mould.

The main limitations of this ejection technique are:

(a) Air compressor is required for this type of ejection.

(b) Useful for box type of moulding. Sometimes, the double ejection is required for the article like mineral bottle cap, etc. In this type of mould, the air ejection with valve is provided in the centre of the moulding as well as the stripper plate ejection is also provided at the periphery of the component. It is shown in Fig. 2.138 both the ejections work simultaneously.

2.6.4 Ejection from Fixed Mould Half

If the article sticks with cavity, the ejection will be required from the fixed half. In this article the sprue gate mark is not acceptable on outside surface. To design the mould, the sprue gate will be provided from fixed half and the article will remain in the fixed half after getting mould opens, for ejection of article the ejection has to be provided from fixed half. In this method, a conventional ejector assembly and ejector grid is incorporated behind the fixed mould plate which is to be actuated from the moving half of the machine.

The main drawback is that the fixed half of the mould will be very deep resulting in excessive length of the sprue. This can be minimised by special extension nozzle. Figure 2.138 shows fixed half ejection.

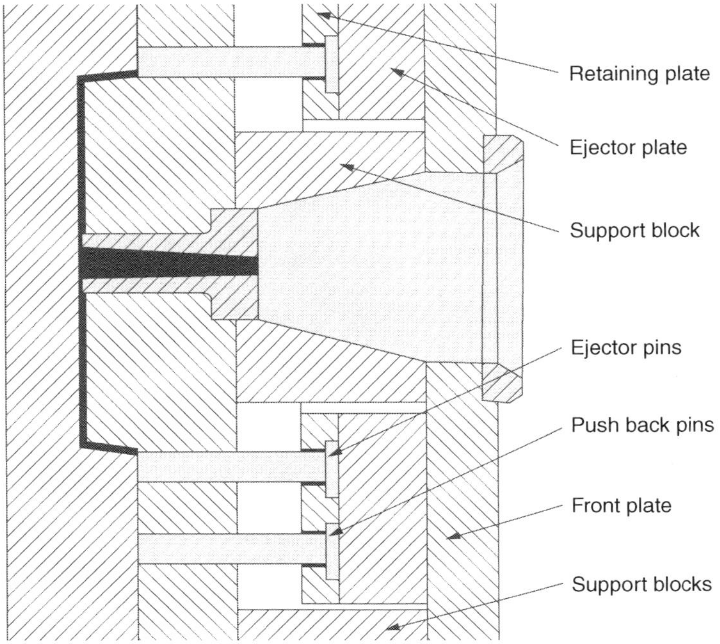

Fig. 2.138 Ejection from fixed half.

2.6.5 Calculation of Ejection Force Required

Ejection force required to strip a moulding off a male core may be calculated from the following formula:

$$F = \frac{St\, E\, A\, \sigma}{\{D(D/2t) - (Dm/4t)\}}$$

- St = Thermal expansion of the pin (cm)
- St = ⊠ $(t1 - t2)D$ here t1 and $t2$ are the temperature difference between softening and ejection temperature
- E = Modulus of elasticity
- A = Total area of contact between moulding and mould faces in line of draw(cm^2)
- σ = Coefficient of friction between plastics and steel material (for PP it is 0.33)
- D = Diameter of circle of circumference equal to length of perimeter of moulding surrounding male core (cm)
- t = Wall thickness of moulding (cm)
- m = Poisson ratio of used plastics (for PP, it is 0.4)
- F = Ejection force required (kgf)

2.6.6 Sprue Puller

It is also one type of pin which is used for removal of sprue from the sprue bush. Generally one undercut is machined in the pin placed on the opposite of the sprue entry. During filling of the material first fills the undercut of the puller pin and on solidification provides sufficient force to pull the sprue from the mould during opening. This is generally used for multi- impression moulds where the removal of sprue would be difficult during ejection. Different types of sprue pullers are used depending upon the type of ejection which are listed below:

1. Sprue puller at cold slug well (Fig. 2.139)
2. Sprue puller at grooved slug well (Fig. 2.140)
3. Z type sprue puller

All the above sprue pullers depend upon the undercut provided in the puller and this undercut of the puller is situated within the cold slug well region and below the parting surface. Generally these pullers work on the principle of ejector pin type ejection technique.

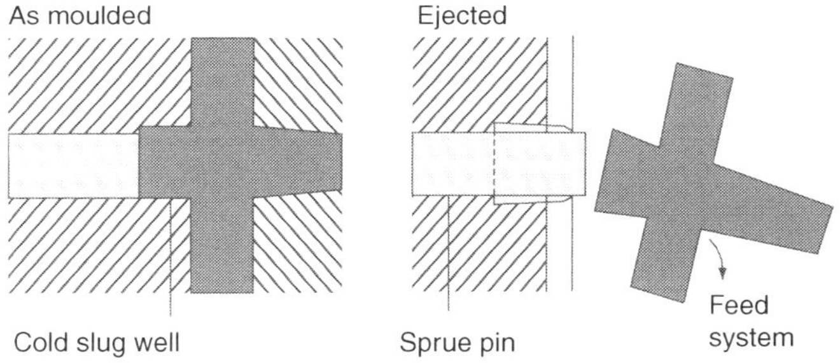

Fig. 2.139 Sprue puller at cold slug well.

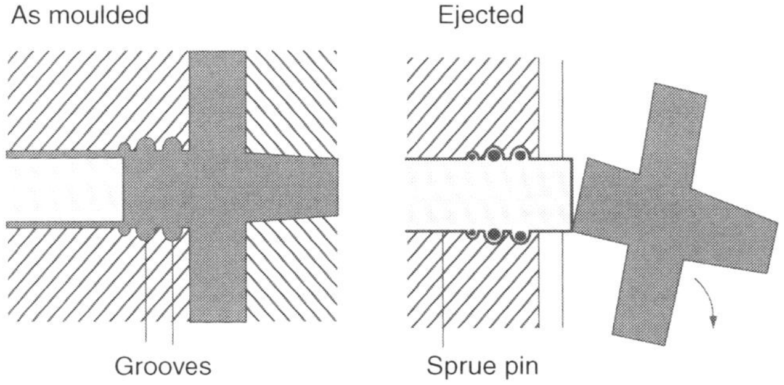

Fig. 2.140 Sprue puller at grooved slug well.

The following types of sprue pullers are used when the undercut of the sprue puller is situated above the parting surface:

(a) Mushroom headed sprue puller
(b) Reverse taper sprue puller (Fig. 2.141)

The above types of sprue puller work on the principle of withdrawing through a plate such as a stripper plate in order to eject the feed system.

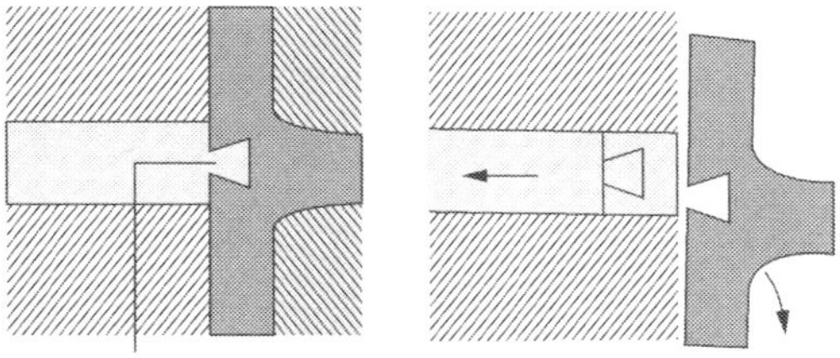

Fig. 2.141 Reverse taper sprue puller.

2.6.7 Calculation of Support Pillar Requirement

In a mould, ejector grid is U shaped. If the span between spacer blocks is long enough, forces of injection can cause a deflection in the back plates. For determination of pillars and their spacing, a formula from strength of material can be used.

Let us consider t thick plate as a beam supported at 190 mm centre distance and with a uniform load.

Stress at centre $\sigma = Pl/8z$

where P = Load on the back plate

l = Centre distance between spacer block

z = Section modulus of back plate = $wt^3/12$

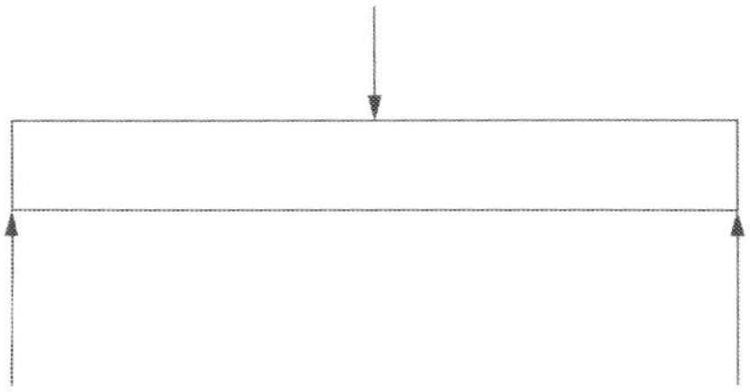

$$i.e., z = \frac{246 \times t^3}{6} = 41t^3$$

Here w = 246 mm and t is the thickness of the back plate as shown in Figure.

Safe allowable stress for the back plate material = 1000 kg/cm^2

$$\sigma = Pl/8z$$

Here P is load exerted from the cavities on the back plate. The compression force on the plate will be on the base of the injection pressure, i.e., 500 kg/cm^2.

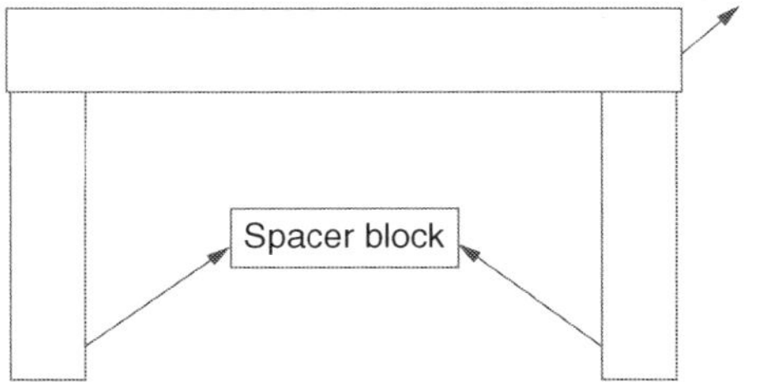

Area of the back plate = 246 × 246

So $P = 500 \times A$

$\sigma = 500 \times 246 \times 246 \times 190/8 \times 41\,t^3 = 1000$

$t^3 = 500 \times 246 \times 246 \times 190/8 \times 41 \times 1000$

t = 25.975 mm

Now if one more pillar support is added, the centre distance between the supports will also become half. Thus, thickness of the plate will be 20.65 approximately.

2.7 TEMPERATURE CONTROL SYSTEM

The cooling conditions in a mould affects the cycle time, part dimensions, surface finish, and warpage. Mould cooling time is the biggest contributor to the overall cycle time, it could be two-thirds or 70–80% of total cycle time. If an efficient cooling circuit can reduce the cooling time, besides uniform cooling improves part quality by reducing residual stress, stability of heat transfer, and accuracies of part after cooling.

2.7.1 Basic Cooling Principles and their Importance to Product Quality

Mould cooling serves to dissipate the heat of the moulding quickly and uniformly. Fast cooling is necessary to obtain economical production; and uniform cooling is required for product quality. Adequate mould temperature control is essential for consistent moulding. The layout of the cooling circuit plays an important role on the quality of the article as the cooling time is two thirds of a product cycle time.

Optimal properties of engineering plastics depend on the mould and material temperature. The mould temperature has a substantial effect on

1. Shrinkage behaviour
2. Warpage
3. Mechanical properties
4. Surface quality
5. Flow length in thin walled parts
6. Cycle time

The semicrystalline thermoplastics need to be cooled down at an optimal crystallisation rate. Parts with widely varying wall thicknesses are likely to deform because of local differences in the degree of crystallisation. Additionally, the required cooling time increases rapidly with wall thickness.

2.7.2 Heat Pipe Technology

The heat pipe technology is the passive heat transfer for achieving better efficiency.

Heat rod method This system is incorporated in cooling of slender type of core inserts where other types of cooling circuits are not practicable. A heat rod is basically a cylindrical metal rod (preferably copper) which is inserted into an accommodating hole machined in the core insert. Its purpose is to facilitate the conduction of heat from the impression.

One portion of the heat rod is always kept in contact with the water.One hole is drilled in the plate and water flows into it and the heat rod is placed in the core insert as shown in Fig 2.142. Then the heat rod absorbs the heat from the plastic material stuck with core. Heat is transferred by conduction mode from molecules to molecules. And the heat reaches the top end of the rod where water is passing through the hole and water takes heat from rod and cools plastic materials.

Heat Pipe Heat pipes offer high effective thermal conductivities (5,000 watts/meter·K to 200,000 watts/meter·K), energy-efficiency, light weight, low cost, simple and reliable operation, no moving parts, ability to transport heat over long distances and quiet vibration-free operation and it transfers heat more efficiently and evenly.

First, the heat pipe is one type of pipe filled with a small quantity of working fluid. Heat is transferred from the core part to the working fluid of heat pipes and get vapourised. This vapour transfers the heat to a cooling medium called condenser. The same condensed fluid is further returned to the evaporator by the gravity of the fluid. The fluids generally used are water, acetone, nitrogen, methanol, ammonia or sodium.

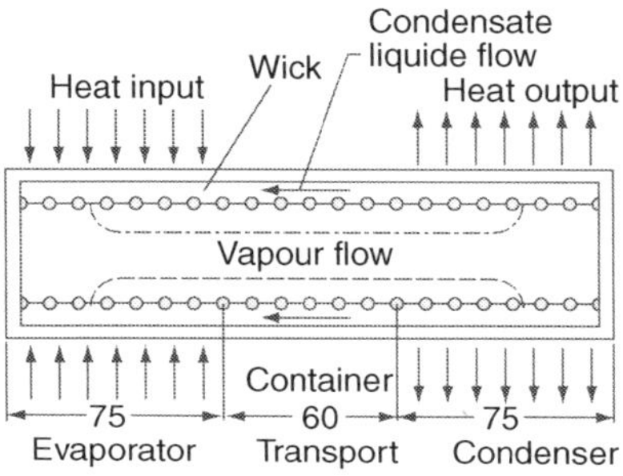

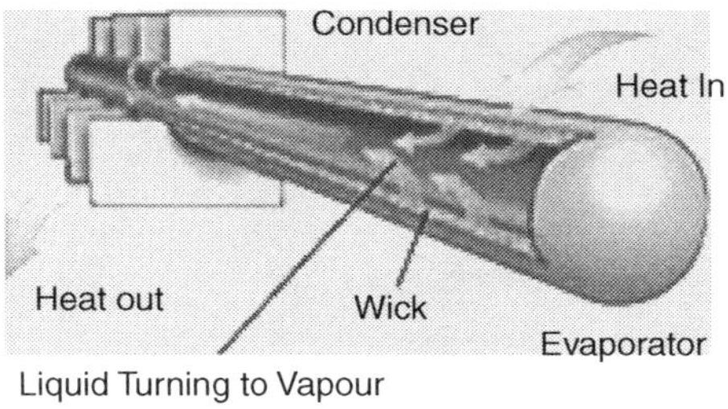

Fig. 2.142 Heat pipe.

1. **Working of Heat Pipe:** Heat pipes work on the principles on the two-phase heat transfer system; the liquid-vapour phase (boiling/evaporation and condensation) of a working fluid is shown in Fig. 2.142. While solid conductors such as aluminium, copper, graphite and diamond have thermal conductivities ranging from 250 W/m•K to 1,500 W/m•K, heat pipes have effective thermal conductivities that range from 5,000 W/m•K to 200,000 W/m•K. Heat pipes transfer heat from the heat source (evaporator) to the heat sink (condenser) over relatively long distances through the latent heat of vapourisation of a working fluid. Heat pipes typically have three sections: an evaporator section (heat input/source), adiabatic (or transport) section and a condenser section (heat output/sink).

2. **Key Components of a Heat Pipe:** The three major components of a heat pipe include:
   (a) A hollow vacuum tight-sealed shell or vessel
   (b) Working fluid
   (c) Capillary wick structure

All the above items work together for efficient heat transfer. The hollow shell inner surface is one type of wick structure lines and is saturated with the working fluid. Generally, wick structure provides the capillary action for the returning condensed fluid. Since the heat pipe contains a vacuum, the working fluid will boil and take up latent heat, at well below its boiling point at atmospheric pressure. Water, for instance, will boil at just above 273° K (0°C) and start to effectively transfer latent heat at this low temperature.

3. **Heat pipe Shell or Containment Vessel:** Heat pipes can be constructed from a variety of different materials specially which depends on the compatibility of working fluid. The

material of vessel for heat pipes are aluminium, copper, titanium, monel, stainless steel, inconel and tungsten. The most common material generally used is copper.

4. Working Fluid: The heat pipe working fluid chosen depends on the operating temperature range of the application. Working fluids range from liquid helium for extremely low temperature applications (–271°C) to silver (>2,000°C) for extremely high temperatures. The most common heat pipe working fluid is water for an operating temperature range from 1°C to 325°C. Low temperature heat pipes use fluids like ammonia and nitrogen. High temperature heat pipes utilise cesium, potassium, NaK and sodium (873–1,473°K).

Table 1.14 Heat pipe working fluids and their operating temperatures.

Heat pipe working fluid	Operating temperature range (°C)	Heat pipe shell material
Low temperature or cryogenic heat pipe working fluids		
Carbon Dioxide	–50 to 30	Aluminium, Stainless steel, Titanium
Helium	–271 to –269	Stainless steel, Titanium
Hydrogen	–260 to –230	Stainless steel
Methane	–180 to –100	Stainless steel
Neon	–240 to –230	Stainless steel
Nitrogen	–200 to –160	Stainless steel
Oxygen	–210 to –130	Aluminium, Titanium
Mid range heat pipe working fluids		
Acetone	–48 to 125	Aluminium, Stainless steel
Ammonia	–75 to 125	Aluminium, Stainless steel
Ethane	–150 to 25	Aluminium
Methanol	–75 to 120	Copper, Stainless steel
Methylamine	–90 to 125	Aluminium
Pentane	–125 to 125	Aluminium, Stainless steel
Propylene	–150 to 60	Aluminium, Stainless steel
Water	1 to 325	Copper, Monel, Nickel, Titanium
High temperature heat pipe fluids		
Cesium	350 to 925	Stainless steel, Inconel, Haynes
NaK	425 to 825	Stainless steel, Inconel, Haynes
Potassium	400 to 1,025	Stainless steel, Inconel, Haynes
Sodium	500 to 1,225	Stainless steel, Inconel, Haynes
Lithium	925 to 1,825	Tungsten, Niobium
Silver	1,625 to 2,025	Tungsten, Molybdenum

5. Wick Structures: The heat pipe wick structure is a structure that uses capillaries to move the liquid working fluid from condenser back to the evaporator section. Wick structures are generally constructed from various materials and methods. The most common heat pipe wick structures include: Axial grooves, screen/wire and 'sintered powder metal', arteries, bidispersed sintered powder and composite wick structures.

2.7.3 Types of Cooling

It is classified into two parts:

1. Cooling integer type
2. Cooling insert type

Cooling integer type is further classified into integer cavity and integer core plate.

1. Cooling Integer Cavity Plate: The temperature of an integer mould plate is controlled by circulating water through holes bored into the plate. The holes are normally interconnected to form a circuit. The number of circuits will depend upon the depth of the mould plate. Water enters into the mould plate through these holes, termed as an inlet and after circulation goes out through separate holes termed outlet. Several circuits are generally recommended. They are:

- **(a) Circuit as U type:** The holes by drill is bored as shown in Fig. 2.143 as a line diagram and Fig. 2.144 to give the shape as U letter.
- **(b) Rectangular Circuit**: Here the holes are drilled as shown in Fig. 2.145 and take the shape of rectangular, due to which the water is circulated around the cavity.

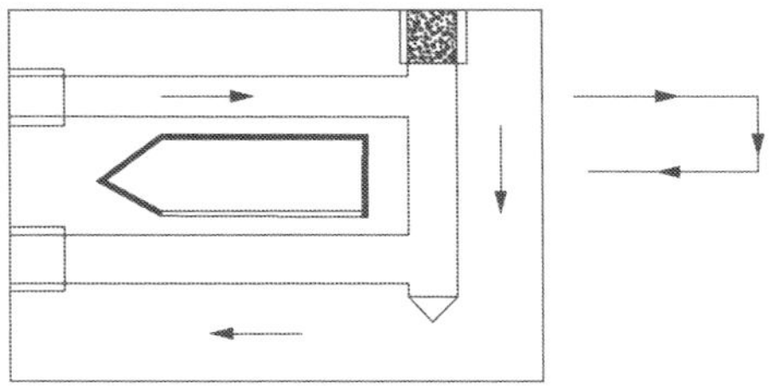

Fig. 2.143 U-Type cooling layout.

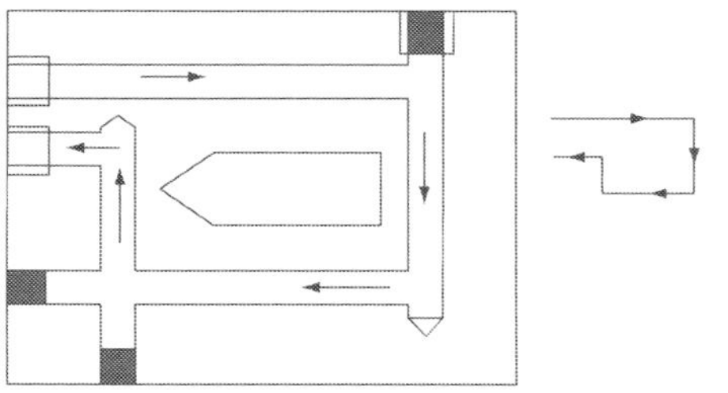

Fig. 2.145 Rectangular circuit.

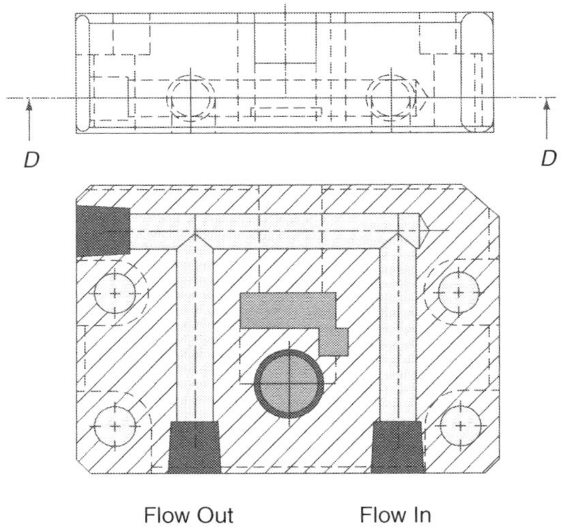

Fig. 2.144 U-Type cooling.

Serial and parallel cooling layouts in mould base As we know that cooling is the biggest part that effects cycle time of injection mould. A good design of the circuit is to provide good heat exchange media between plastics resin and the cooling fluid and reduce cycle time and give good product. For efficient cooling we can use parallel or series of cooling methods.

Parallel cooling channel: In this type of cooling basic mould base has four parts, they are:

Main cooling pipe: In this type of cooling the drilled hole is machined in the plate for the cooling channel. The drilled hole is made throughout the plate by using longer drill tools. The actual size of the hole is generally 6 to 8 mm. Figure 2.146 below shows parallel cooling channel concept. As shown in the figure, the vertical drill holes are also made which are interconnected.

Stopper plugs: It is used to prevent the water to flow to undesirable channel.

Inner stopper plugs: It is also one type of stopper plugs to prevent the water from flowing to the inner cooling.

Parallel cooling channels are drilled straight through to cooling pipe, because of the parallel flow characteristics. The flow rate along various cooling will be different, because each cooling pipe has their own flow resistance. As a result, cooling of the mould will not be uniform. So the parallel circuits of cooling in moulds are of very poor flow rate.

Series cooling: The type of cooling is machined in the same way as parallel cooling type. Only the location of stopper plug is different as shown in Fig. 2.147. In the series cooling channel types, the drilled holes are continuously connected from the coolant inlet to its outlet. The flow can be easily configured by adjusting the inner stopper plug.

When designing this type of cooling, it is recommended to use the same size of diameter cooling to maintain turbulent flow. The turbulent flow enables heat to be transferred more effectively.

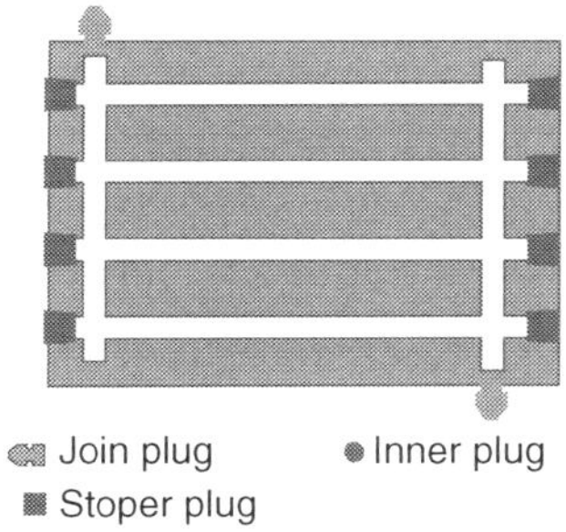

Fig. 2.146 Parallel cooling channel.

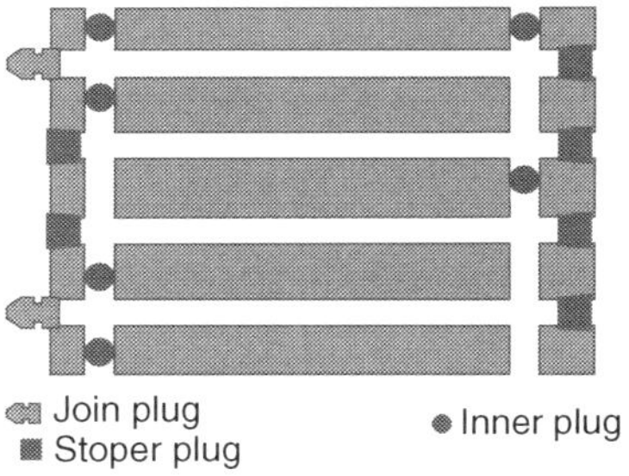

Fig. 2.147 Series cooling.

(c) Balanced Z Circuit: In this type the water is circulated as Z-type circuit. It is also the same as series cooling type circuit.

(iii) Multilevel Cooling System

Multilevel cooling system is recommended for cooling of deep integer type cavities. In this system, the circuit is machined at two or three levels to provide uniform cooling through the cavity and all levels are joined by drilled hole. On each level, it is arranged to follow the contour of the cavity as far as possible.

2. **Cooling Integer Core Plate:** Different methods of cooling integer cores are:

(a) **Angle Hole System:** In this system, the drill is made in the core block in such a manner that both side drilled holes are interconnected to each other. So water flows through the circuits, making an angle from the underside of the core plate by doing some calculation for making both the circuits interconnected.

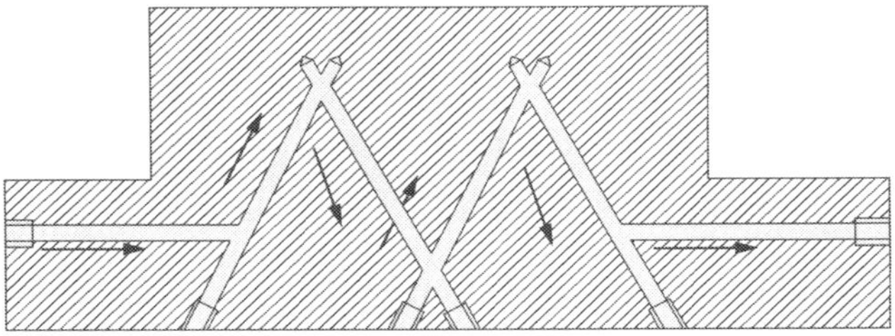

Fig. 2.148 Angle hole system.

(b) **Baffled Straight Hole System:** In this system, the holes are made on some periphery in the core block and the entire hole is interconnected. Water passes through a number of holes machined in which baffles are fitted as shown in Fig. 2.149.

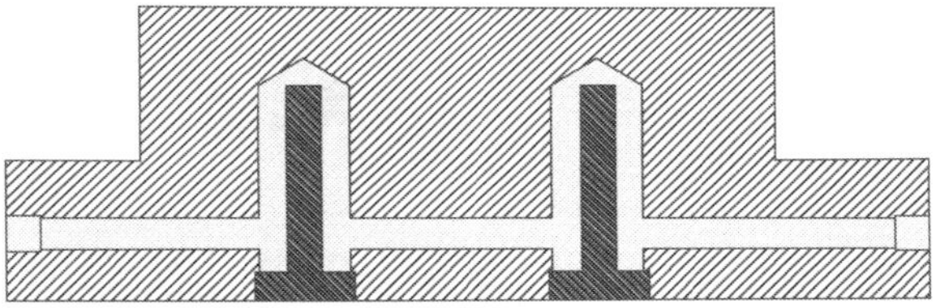

Fig. 2.149 Baffled straight hole system.

Cooling insert –bolster assembly: Why cooling is necessary in core?

As per the principle of shrinkage of plastic materials, the material in fluid form enters into the impression. When the fluid changes its status from liquid to solid, the material sticks on core. It means that the plastics material comes into contact with the cavities for a little time. Generally only $1/3^{rd}$ of total cooling time the material comes into contact with cavity. So more heat is transferred to core. Core absorbs most of the heat from the plastics. That is why cooling is necessary in core.

It is classified into the following:

(a) Cooling the bolster

(b) Cooling the insert

(a) Bolster Cooling: This is adopted for shallow type of core/cavity inserts where the cooling circuits can only be incorporated in the bolster plate (refer to Fig. 2.150). The recommended circuits are:

Rectangular circuit: The circuit is machined in the plate as shown in Fig. 2.151. The actual shape of the circuit is rectangle.

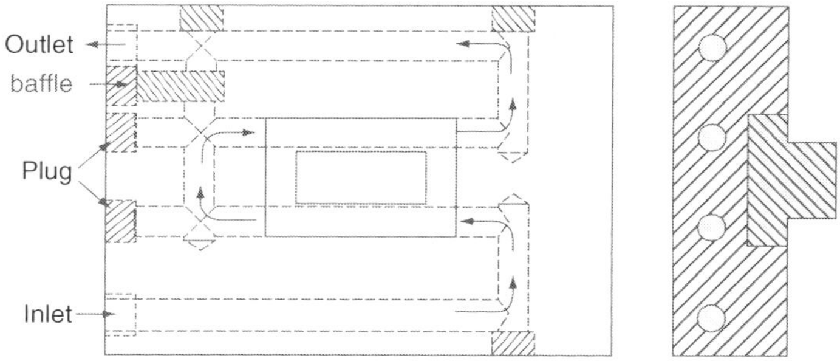

Fig. 2.150 Bolster cooling, Z circuit.

Z circuit: The circuit is machined in the bottom portion plate (Fig.2.150). The circuit shape of the cooling is Z shape.

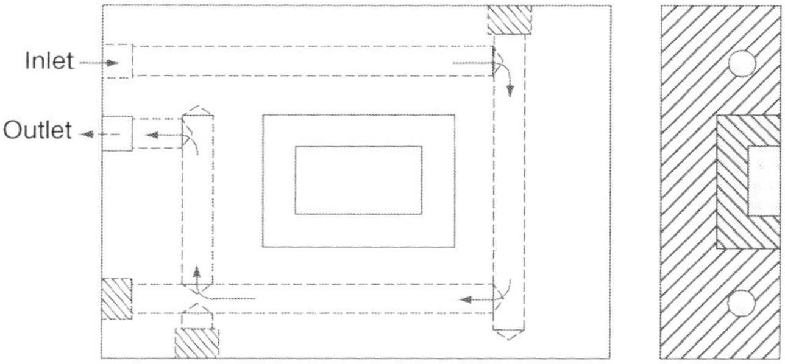

Fig. 2.151 Bolster cooling, rectangular circuit.

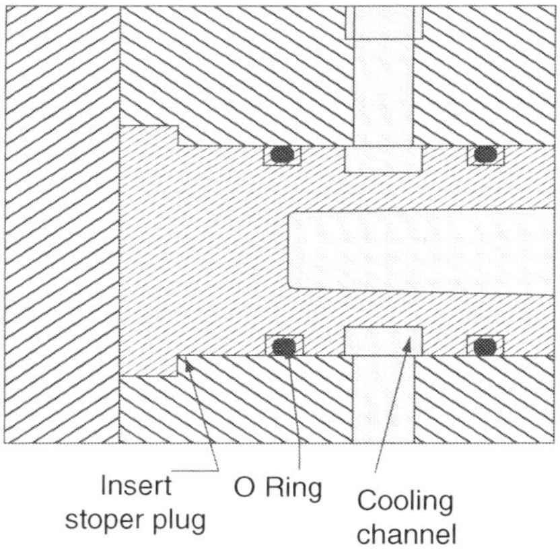

Fig. 2.152 Cooling rectangular insert.

(b) Cavity Insert Cooling:

Cooling rectangular insert: In this method, one recess is machined on the periphery of the insert as shown in Fig. 2.152. In cavity retainer plate hole is drilled in the same position of recess of the insert. Near to the recess O ring hole is machined to prevent any leakage of water.

Interconnecting groove design In this method, the coolant annulus is incorporated as groove machined into the mould plates. This design is adopted when inserts are arranged in line or on a pitch circle diameter.

(c) Core Insert Cooling: It is divided into two types:

(i) Shallow core insert cooling

(ii) Deep core insert cooling

(i) Shallow Core Insert Cooling: Most common methods are: U circuit; balanced Z circuit; spiral circuit.

(ii) Deep Core Insert Cooling: Heat transfer from the core surface can be efficiently made by circulating the coolant deeply inside the core insert. There are various methods for cooling of core insert. They are:

Deep chamber design In this method, the back face of the core insert is machined to form a deep chamber. Water from the inlet passes through the pipe fitted on the centre of the chamber. After circulation, water comes out through the outlet.

2.7.4 Alternative Cooling Devices

For areas in the mould, where it is not possible to use normal drilled cooling channels, alternative methods must be used to perfect uniform cooling. The method employed usually includes baffles, bubblers, or thermal pins.

Baffles A baffle is actually a cooling channel drilled perpendicular to a main cooling line, with a thin plate which is called baffle, that separates one cooling passage into two semicircular channels. The coolant flows in one side of the plate from the main cooling line, turns around the tip to the other side of the baffle, and then flows back to the main cooling line.

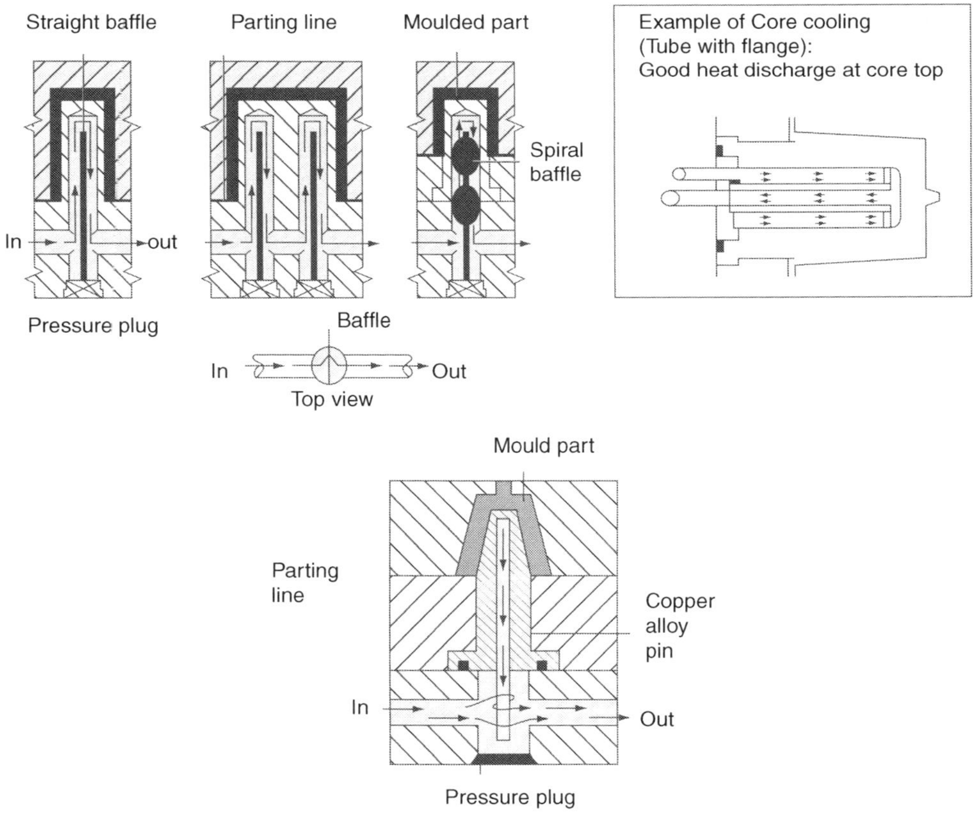

Fig. 2.153 Baffle hole system.

Baffled hole system In this design, baffled holes are used. Each insert incorporates a chamber which connects with the inlet and outlet passage as shown in Fig. 2.153. Baffles are incorporated in each insert chamber at right angle to the inlet outlet passage. This method is limited to a small number of inserts as the temperature of water entering the last insert is higher than that of the first insert which results in non-uniform cooling which is the disadvantage of baffled cooling.

Baffles and bubblers are good sections of cooling lines that divert the coolant flow into areas.

Example of core cooling by means of baffles

Less effective heat discharge at core top

Baffles cooling are an easy method to provide cooling in the core, but it is difficult to keep the baffles by dividing exactly in the centre. Because of this cooling effect will be different in both sides of the core. The use of a helix baffle as shown in Fig. 2.154 will solve the problem by conveying the coolant to the tip and back in the form of a helix. It is useful for diameters of 12 to 50 mm, and makes for a very homogeneous temperature distribution. Another logical development of baffles is single or double-flight spiral cores as shown in Fig. 2.154.

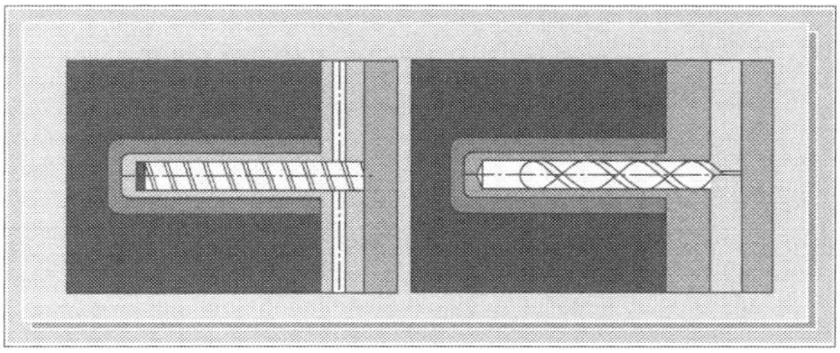

Fig. 2.154 Single and double flight spiral cores.

Bubbler As in the baffle cooling system, same water is circulated in all cavities of multicavity mould. So uniform cooling cannot be maintained. To overcome this problem, the bubbler cooling is used. A bubbler cooling is similar to a baffle except that the blade is replaced with a small tube. The coolant flows into the bottom of the tube and 'bubbles' out of the top, like a fountain. The coolant then flows down around the outside of the tube to continue its flow through the cooling channels.

For slender cores this is the most effective form of cooling. The inner and outer diameters must be adjusted to maintain same flow resistance in both cross sections. The condition for this is:

Inner Diameter / Outer Diameter = 0.7

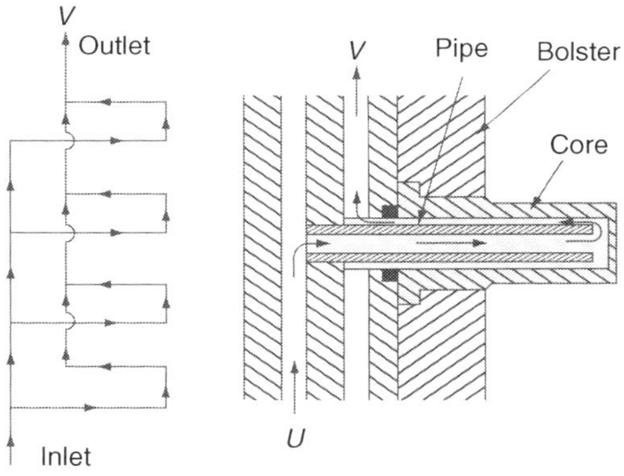

Fig. 2.155 Bubbler cooling.

In this system, the inlet and outlet passages are different (Fig. 2.155 and 2.156). Due to this, the temperature of water entering the first insert is same as that of the last insert. This gives the uniform cooling of the core inserts.

Example of core cooling (tube with flange): good heat discharge at core top

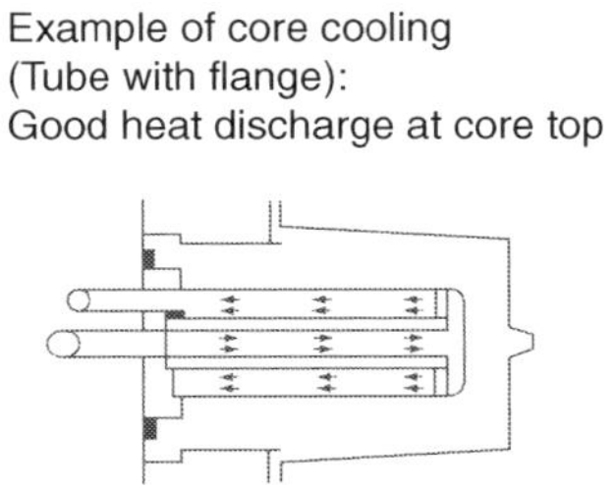

Fig. 2.156 Core cooling (tube with flange).

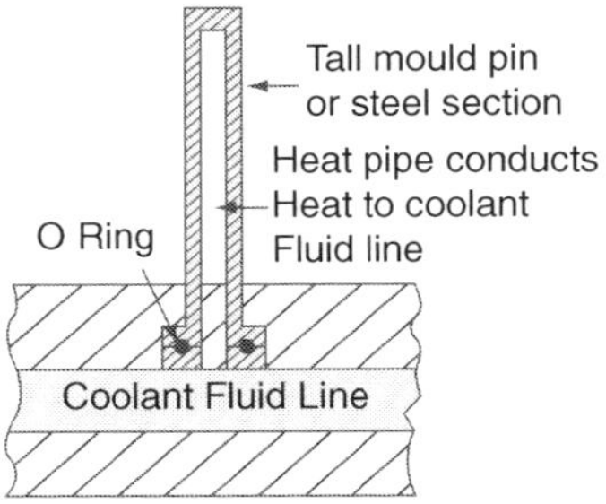

Fig. 2.157 Thermal pin.

Thermal pin A thermal pin is an alternative to baffles and bubblers. It is a sealed cylinder filled with a fluid. The fluid vapourises as it draws heat from the tool steel and condenses as it releases the heat to the coolant. The heat transfer efficiency of a thermal pin is almost ten times as greater than a copper tube. For good heat conduction, avoid an air gap between the thermal pin and the wall or fill it with a highly conductive sealant as shown in Fig. 2.157.

Cooling of large cores For large core diameters (40 mm and larger), a positive transport of coolant must be ensured. This can be done with inserts in which the coolant reaches the tip of the core through a central bore and is led through a spiral to its circumference, and between core and insert helically to the outlet. This design weakens the core significantly.

Cooling of cylinder core Cooling of cylinder cores and other round parts should be done with a double helix, as shown in Fig. 2.158 below. The coolant flows to the core tip in one helix and returns in another helix. The wall thickness of the core should be at least 3 mm in this case.

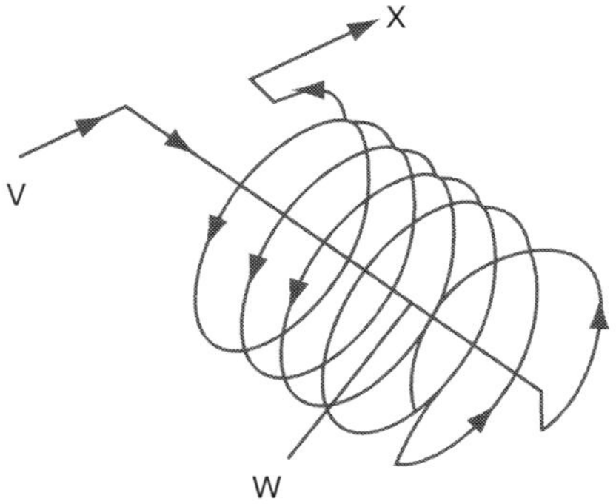

Fig. 2.158 *X* - Outlet port; *Y* – Inlet port; *W* – Water flow through a pipe.

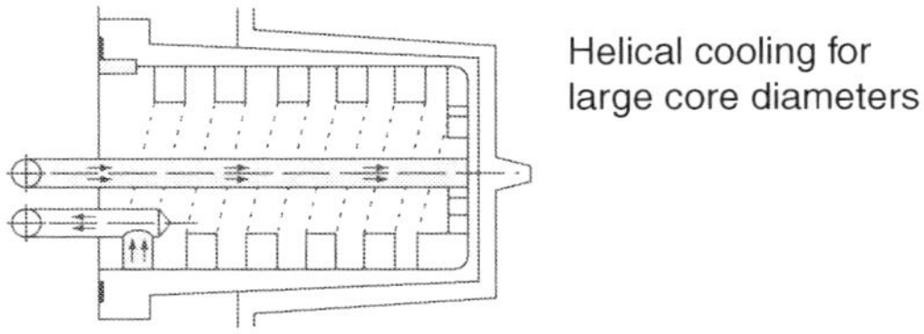

Fig. 2.159 Helical core insert cooling.

Helical channel design This design is adopted to ensure that the coolant follows a precise path and no dead water is possible which results in rapid transfer of heat from the moulding. Here the water follows a helical path machined into a steel or brass block which is fitted inside the chamber as shown in Fig. 2.158 and Fig. 2.159.

If the diameter or width is very small (less than 3 mm), only air-cooling is feasible. Air is blown at the cores from the outside during opening or flows through a central hole from inside as shown in Fig. 2.160. In this procedure the maintaining of exact temperature is very difficult.

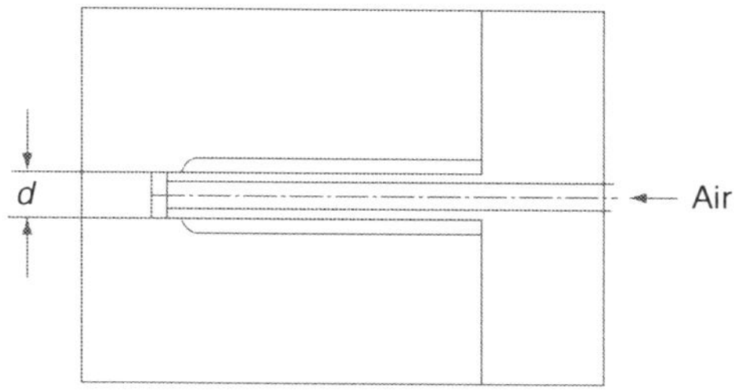

Fig. 2.160 Cooling of slender core insert.

Better cooling of slender cores (those measuring less than 5 mm) is accomplished by using inserts made of materials with high thermal conductivity, such as copper or beryllium-copper materials. Such inserts are press-fitted into the core and extended with their base, which has a cross section as large as is feasible, into a cooling channel as shown in Fig. 2.160.

2.7.5 Calculation of Rate of Heat to be Removed by Cooling Circuit

$Q = M \times [Cp\,\{T_1 - T_2\} + L]$

where Q = Heat to be transferred from mould per hour (cal/ hr)

C_p = Specific heat of material (cal/gm/ °C)

M = Mass of plastic material injected into the mould per hour (gm/hr)

L = Latent heat of fusion of plastic materials (cal/gm)

T_1 = Injection temperature of material (°C) ; T_2 = Temperature of mould (°C)

2.7.6 Ejection Temperature

The operating temperature for a particular mould will depend on the following factors:

1. Type and grade of plastic material to be moulded
2. Length of flow within the impression
3. Wall section of the moulding
4. Length of the feed system

Maintaining the optimum temperature in a mould is very important as:

1. It improves the cycle time
2. It improves the quality of the product
3. It avoids distortion of the product

The holes or channels through which water/coolant passes are termed as flow ways and the complete system of flow ways is called circuit.

2.7.7 Weight of Water to be Circulated per Hour

The following formula gives weight of water to be circulated per hour to dissipate the heat;

$Q = k\, m_1\, (T_{out} - T_{in})$

where Q_w = Rate of heat extracted (kcal/hr)

k = Constant to allow for heat transfer efficiency

m_1 = Weight of water passed (gm/hr)

$T_{out} - T_{in}$ = 5°C

For the component shown in the Fig. 2.2, the value of Q can be calculated as follows:

$Q = 2316 \times [0.5 \times (250–60) + 20.1]$

$= 266571.6$ cal/hr. = 266.57 kcal/hr (Assuming single impression and 1 min. cycle time)

where C_p = 0.5 cal/° C/gm for PP; M = Shot wt. of the product × No. of shot/hr. = 38.6 gms. (including 10% feed system) ; L = 20.1 cal/gm. For PP; T_1 = Injection temp. of PP; T_2 = Mould temp. for PP

Weight of water to be circulated per hour is $266571.6 = 0.64 \times m_1 \times 5$; $m_1 = 83303.625$ gm/hr = 83.3 kg/hr.

2.7.8 Cooling Time

Theoretically, cooling time is proportional to the square of the heaviest part wall thickness or the power of 1.6 for the largest runner diameter. That is given in Table 2.15.

Table 2.15

$$\text{Cooling time} = \frac{(\text{Heaviest wall thickness})^2}{\text{Thermal diffusivity of polymer melt}}$$

$$\text{Cooling time} = \frac{(\text{Largest runner diameter})^{1.6}}{\text{Thermal diffusivity of polymer melt}}$$

where the thermal diffusivity of polymer melt is defined as

$$\text{Thermal diffusivity} = \frac{(\text{Thermal conductivity})^{1.6}}{(\text{Density})(\text{Specific heat})}$$

In other words, doubling the wall thickness quadruples the cooling time

2.7.9 Cooling Channel Configuration

In general, the cooling system will be roughly drilled or milled. Rough inner surfaces enhance turbulent flow of coolant, thus, providing better heat exchange. Turbulent flow achieves 3 to 5 times as much heat transfer as does non-turbulent flow. Cooling channels should be placed close to the mould cavity surface with equal centre distances in between. The mechanical strength of the mould steel should be considered when designing the cooling system.

Basic principles of cooling channels Proper care should be taken in the correct placing of seals. It may be damaged by the sharp edges of the pocket when the mould insert is mounted. Seals or O rings should be resistant to elevated temperatures and oils.

Sealing and cooling channel layout The complete layout structure is shown in Fig. 2.161.

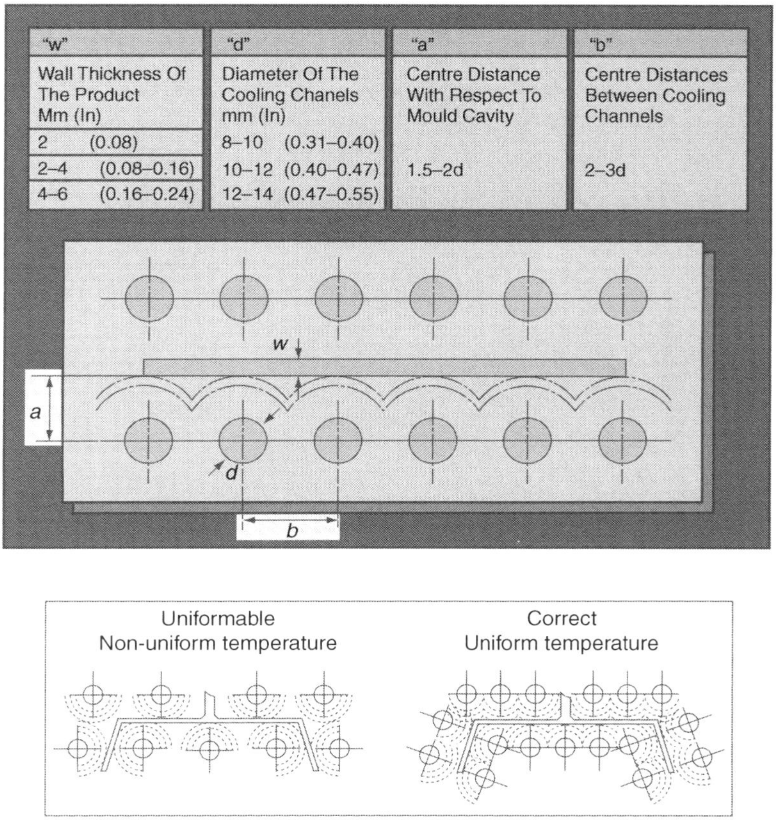

Fig. 2.161 Cooling channel layout.

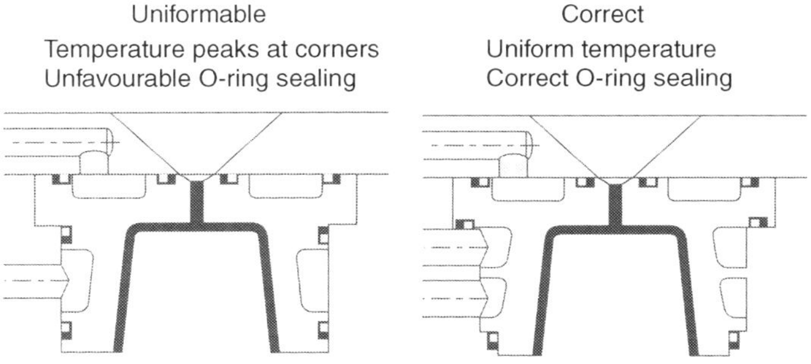

Fig. 2.162 Sealing of cooling channel.

2.7.10 'O' Ring

Function of 'O' Ring O ring is elastomer with a round (o-shaped) cross section used as a mechanical seal or gasket as shown in Fig. 2.163. They are designed to be seated in a groove and compressed during assembly between two or more parts, creating a seal at the interface. **O-rings** are doughnut-shaped objects made from elastomeric compounds such as natural or synthetic rubber, and are used to seal mechanical parts against fluid movement (air or liquid). O ring drawing is shown in Fig. 2.163.

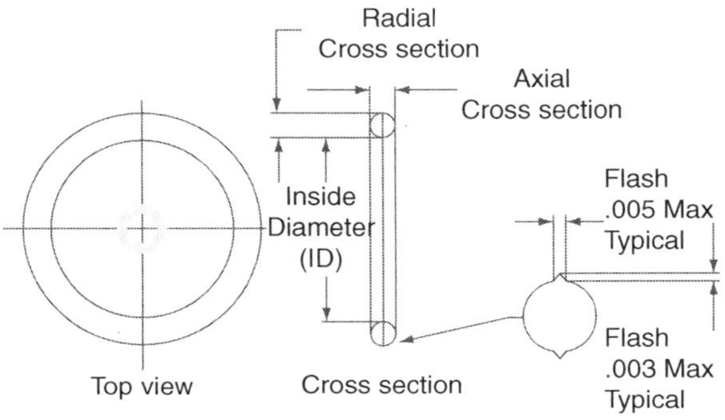

Figure. 2.163 O – Ring.

Static Seals exist where there is no relative motion between the mating surfaces being sealed. Figure 2.164 below shows the fitting of O-ring and its placing.
Basic rules from the picture are:

1. T is height of O-ring
2. Some part of O-ring always comes out form its groove that will receive pressure from cavity plate. 15–30% from its height is best. This little pressure prevents leak when water flows in cooling system.

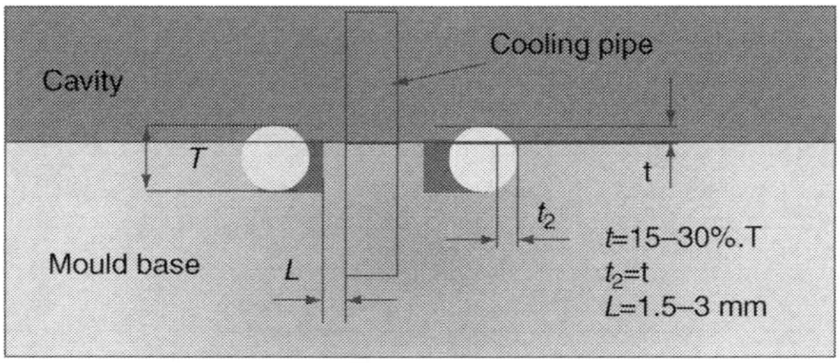

Fig. 2.164 Fitting of O-Ring.

3. t_2 value is same with t, but t_2 is horizontal, because O ring is always made from elastomeric plastic; by little pressure we can add O ring to O ring hole.
4. L is gap between cooling channel and O ring hole.

O Ring design considerations

1. Proper squeeze
   (a) Compression expressed as a percentage
   (b) $cs > \dfrac{B_d - G_d}{2}$
   (c) Face seal: 20–30%
   (d) Static male/female: 18–25%
   (e) Reciprocating: 10–20%
   (f) Rotary: 0–10%
2. In static seals, where the O-Ring is not in axial motion in the bore, the recommended maximum compression is approximately 40%.
3. The O-Ring must be compressed by a predetermined amount, and this compression determines the O-Ring cross section diameter.
4. The O-Ring inner diameter is typically chosen to be close to the groove's inner diameter; by selecting it to be slightly less than the groove's inner diameter, the O-Ring will stretch and hug the groove.
5. The groove width must be larger than the diameter of O-Ring cross section, to accommodate the radial expansion of the O-Ring when it is axially compressed in the gland stretch.
6. Excessive stretch can overstress material,
7. Sharp corners make R in corner to prevent damage during seal installation.
8. Pressure and clearance gap. Most elastomeric seals are designed to operate within ambient pressure to about 1,500 psi. At very high pressure, the seal must have sufficient strength to resist extrusion into the clearance gap.
9. DO NOT use lubricant of the same material as the O-ring. For example, a silicone lubricant should NOT be used with a silicone O-ring.

2.7.11 Water Connectors

Adaptors: It is a standard mould pipe fitting which is fitted to the inlet and outlet holes of the water cooling system.

Long and short nipple joint plug assembly methods Considering mould base material strength, closer distance between two nipple plug are not advisable, as shown in Fig. 2.165. Assembly between long type and shorts type is

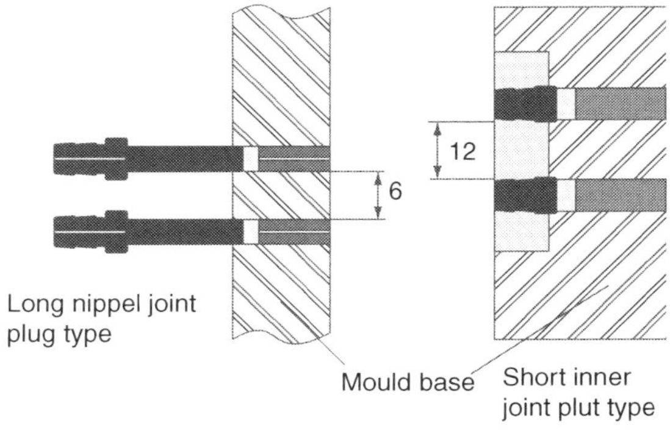

Fig. 2.165 Long and short nipple joint plug assembly

different, by using long type minimum length from both side is 6 mm, except for little diameter hole, less then 5 mm, smaller length spacing may be acceptable.

Tips for the design of cooling pipe for cavity

1. The side of cooling pipe should have enough length from insert. It should be a minimum of 3 cm.
2. Suitable diameter must be chosen for the hole like 6 mm, 8 mm or 10 mm.
3. When injection process begins make sure using calculation to get turbulent flow in cooling pipe.
4. The pipe should be longer to get turbulent flow.

2.8. METHODICAL APPROACH TO MOULD DESIGN

2.8.1 Basic Principles of Mould Design

1. The part design, quantity, appointed time, moulding material, surface finish, tolerances, etc., to be considered.
2. Moulding machine specifications like number of cavities and arrangements of cavities to be calculated.
3. Type of mould to be designed, selection of parting surface, development of core and cavity should be taken care.
4. Incorporation of feed system like sprue, runner, gate must be proper.
5. Incorporation of ejection system must be proper.
6. Provision for temperature control system.
7. Alignment of mould.
8. Provision for mould location.
9. Arrangement for mould lifting and clamping.
10. Preparation of bill of materials.
11. Comparison with mould design check list.

Classification of injection moulds In general two types of moulds are common. They are:

1. Two-plate mould
2. Three-plate mould

In a standard two-plate mould, it opens at the parting surface. This is called single daylight mould.

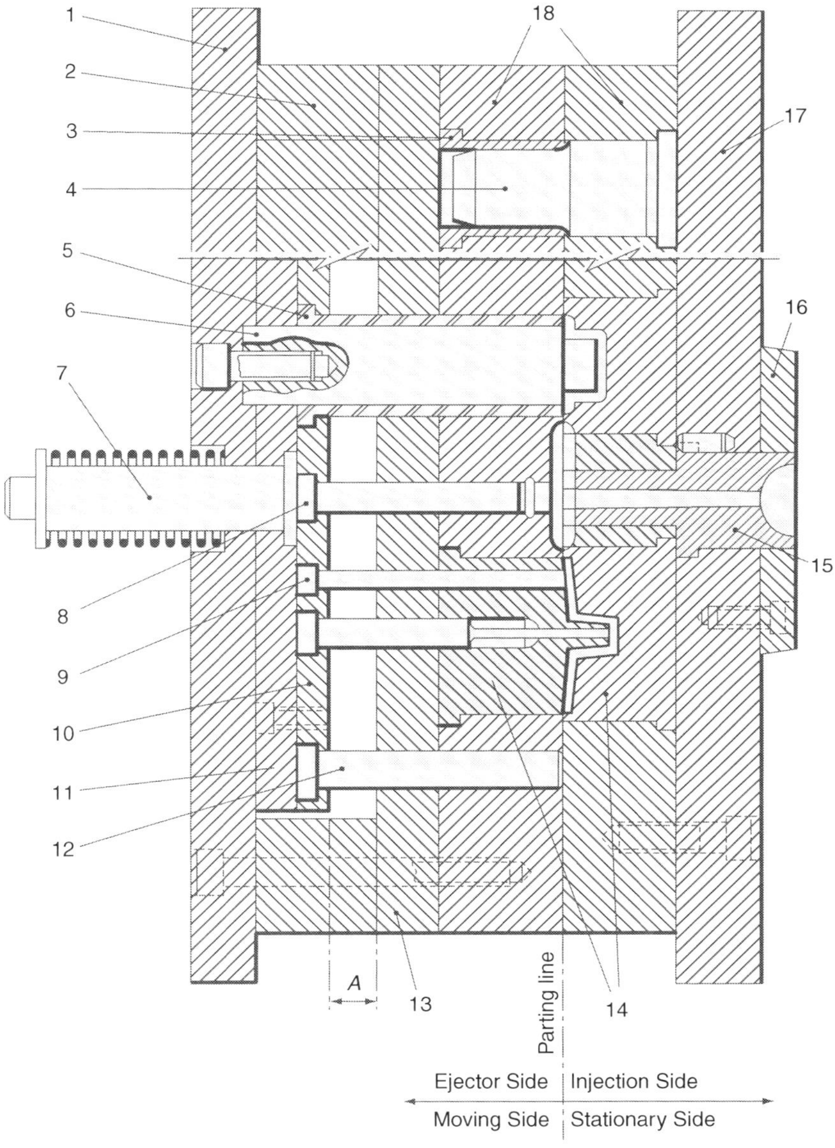

Fig. 2.166 A standard two-plate mould.

1 – Back Plate (Ejection side) ; 2 – Support Pillar ; 3 – Guide Bush ; 4 – Guide Pillar ; 5 – Sleeve Ejector ; 6 – Core Pin ; 7 – Ejector Return Mechanism ; 8 – Sprue Puller; 9 – Ejector pin ; 10 – Retaining plate; 11 – Ejector plate ; 12 – Push Back Pin ; 13 – Core Retainer Plate ; 14 – Core and Cavity Inserts ; 15 – Sprue Bush ; 16 – Locating Ring ; 17 – Back Plate (Injection Side) ; 18 – Core and Cavity Plates.

A standard three-plate mould is a mould plate assembly consisting of three mould parts. In this type, mould opens at two parts. An example of an underfeed mould—one opening will be at the parting surface and the other opening at the floating cavity plate situated in between the cavity plate and core plate.

The main applications of this type of mould are:

1. Centre feeding of components in multi-impression tools
2. Area feeding of components with multiple restricted gates, etc.

The recommended gates are inherently self degating like pin point gate since the gate breaks off as soon as the mould opens and the product is pulled out of the cavity. Cross section of the recommended runners is a trapezoid cut into the cavity plate. The runner ejector plate is always flat with only the sprue puller projecting at the drops.

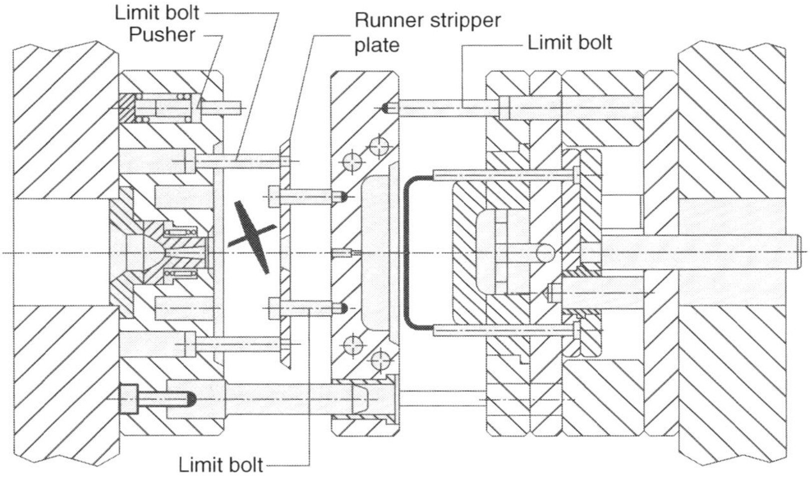

Fig. 2.167 Three-plate mould.

Elastic Modulus of Steel 'Elastic modulus' is a material property that indicates the strength or elasticity of the steel materials used for making mould parts. The elastic modulus is also called the 'Young's modulus'. The elastic modulus is the coefficient of proportionality between the 'strain' and the 'tensile stress' when the steel material is pulled. This relationship can be expressed by the following equation:

$$\sigma = E \times \varepsilon$$

Table 2.16 Unit of important factors.

Description	Unit
Elastic Modulus: E	kgf/cm^2 or Pa
Strain: ε	
Tensile stress: σ	kgf/cm^2 or Pa

ε : Epsilon σ: Sigma

In other words, 'stress is proportional to strain'. The physical value of the elastic modulus is determined by the type of the metallic material. In general, a material with a larger value for the elastic modulus has a higher tensile stress or rigidity. The data of the elastic modulus is shown in Table 2.17 for some typical metallic materials.

Table 2.17 Elastic modulus.

Description	(kgf/cm^2)
Low carbon steel	210 × 104
S50C	210 × 104
Prehardened steel (SCM440 series)	203 × 104
SDK11	210 × 104
Brass	63 × 104
Copper	105 × 104
Aluminium	68 × 104
Super duralumin	73 × 104

2.8.2 Rigidity and the Rigidity of Mould

A plastic injection mould is subjected to high internal pressure at the time of filling the molten plastic, and also, it is subjected to high compression stress at the time of clamping of the mould. In addition, if the mould becomes large, it can also be subjected to a bending stress due to its own weight. In order to make sure that the mould does not get deformed or broken due to the external stresses or stresses due to its own weight, it is necessary to strengthen the rigidity of the mould. Rigidity is the resistance to deformation when subjected to a load. The modulus of longitudinal elasticity E and the modulus of transverse elasticity G of the material affect the rigidity. A material for which the value of E or G is large can be said to have a high rigidity. In other words, it exhibits strong resistance to bending or twisting. For a rigid material, the material is difficult to bend, and also has a very small deflection.

For example, while the value of E for prehardened steel is 2.03×10^6 (kgf/cm^2) but the value of E for cold rolled die steel is 2.10×10^6 (kgf/cm^2). It can be said that cold rolled die steel is more rigid.

'Bending rigidity' is particularly more important in the case of the moulds for plastic injection moulding. Bending rigidity (flexural rigidity) indicates the resistance to bending due to I. (I is moment of inertia of area.) In order to select I large, which causes increase of the bending rigidity, so it is necessary to select a material with a large value of the modulus of longitudinal elasticity E and also adopting a cross sectional shape that makes the moment of inertia I large results in making the bending rigidity high. If the structure has a high bending rigidity, even the deflection becomes small and it is also possible to resist breakage due to bending deformations.

2.8.3 Section Modulus of Mould Components

The section modulus of mould components which is very important for predicting the bending stress is explained below:

At the time of calculating the strength against bending of a mould for plastic injection moulding, the term 'section modulus' plays an important role. The 'section modulus' depends on the cross-sectional shape of the part. Therefore, there is no direct effect of material of the part. For example, if the cross sectional area of the sample is same, the value of the 'section modulus' will also be same. This principle is correct for any type of material whether the material is non-heat treated steel, tempered steel, or even wood.

Definition: A section modulus is the value of the moment of inertia 'related to the neutral axis of the crosssection of a beam multiplied by the distance from the neutral axis to the outer surface'. Therefore, the relationship between the section modulus Z and the moment of inertia of area I is as expressed by the following equation. Section Modulus $Z = I / Y$ (unit is mm^3 or m^3).
where

I = The moment of inertia of area

Y = The distance from the neutral axis to the outer surface

The symbol Z is used for the section modulus. In general, as the section modulus becomes larger, the strength against bending also becomes larger. Regarding bending, the maximum bending stress σ acting on the outer surface of the part can be calculated using the following equation.

$$\sigma = M / Z$$

where

σ = Maximum bending stress (kgf/cm^2) and

M = The maximum bending moment (kgf * m)

If the crosssectional shape is rectangular or circular, the basic equation for calculation becomes clear as is shown in Table 2. 18.

2.8.4 Moment of Inertia of Mould Components

The moment of inertia of mould components is given below:

At the time of carrying out the strength calculations of a mould for plastic injection moulding, the moment of inertia is used for several calculations. Moment of inertia is a value that is identified by the cross-sectional shape of the part. This is used frequently for estimating the amount of deflection due to the bending moment or injection pressure. The moment of inertia depends only on the cross-sectional shape of a part. Therefore, it has no relationship with the material. For example, if the cross-sectional shape is same, the value of the moment of inertia is also the same whether the material is a non-heat treated steel, tempered steel, or even wood. The definition of moment of inertia according to mechanics is as follows:

'When a cross section is divided into an infinite number of differential areas dA and the distance from one axis X is a taken as Y, the moment of inertia of area is the sum over the entire area the product of the differential area and the square of the distance'.

This can be expressed in the form of an equation as follows:

Cross section second order moment is $I = \Sigma Y^2 \times dA$ unit is mm^4or m^4

The moment of inertia is usually represented by the symbol I. In general, the strength against bending becomes larger as the moment of inertia of area becomes larger. If the cross-sectional shape is rectangular or circular, the basic equation for calculation becomes clear.

Table 2.18 Moment of inertia and modulus of Section - I.

Cross section shape	Moment of inertia area	modulus of section Z
Rectangle 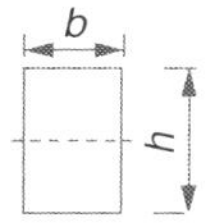	$I = \frac{bh^3}{12}$	$Z = \frac{bh^2}{6}$
Square 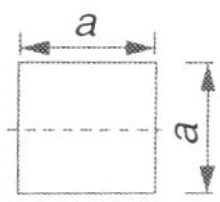	$I = \frac{a^4}{12}$	$Z = \frac{a^3}{6}$

Table 2.19 Moment of inertia and modulus of Section - II.

Cross section shape	Moment of inertia area	modulus of section Z
Circle 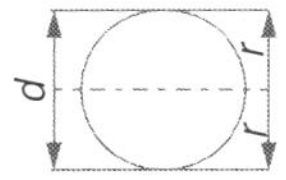	$I = \frac{\pi d^4}{64} = \frac{\pi r^4}{4}$	$Z = \frac{\pi d^3}{32} = \frac{\pi r^3}{4}$
Hollow circle 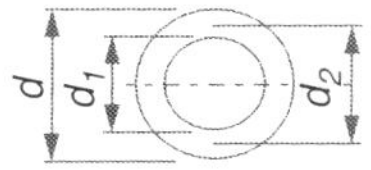	$t = \frac{\pi(d^4 - d_1{}^4)}{64}$ Thin wall $I = \frac{\pi}{8} t \cdot dm^3$	$Z = \frac{\pi(d^4 - dr^4)}{32d}$ Thin wall $Z = \frac{\pi}{4} t \cdot dm^2$

2.8.5 Thermal Expansion of Mould Components

The basic knowledge about the thermal expansion of components in the moulds for plastic injection moulding is discussed below:

In the case of a mould for plastic injection moulding, the appropriate cavity surface temperature, should be maintained at 30 to 150°C. On the other hand, molten plastic flows into the sprue, runner, and cavity, which receive heat from the plastic at temperatures in the range of 180 to 300°C. Metals generally undergo thermal expansion when the temperature rises. Therefore, even the constituent parts of a mould for plastic injection moulding undergo thermal expansion. Thermal expansion can disrupt the mating between the guide post and guide bush, or can cause bad movement of the slide core, or can enlarge the dimensions of the core pins. The basic changes in the dimensions due to thermal expansion can be calculated using the following equation:

$$\Delta = \alpha l_o (T - T_a)$$

where Δ is the expansion (in mm) in the dimensions expected to thermally expand, α is the linear thermal expansion coefficient (mm/mm) of the metal, l_o is the initial length (mm), t is the initial temperature (°C), and T_a is the temperature after heating.

The linear thermal expansion coefficients for typical metallic material used in moulds are given below:

Table 2.20 Linear thermal expansion coefficient.

Material	α Linear thermal expansion coeff. (mm/mm)
S50C	11.7×10^{-6}
SKD11	11.7×10^{-6}
Prehardened steel (SCM440 series)	11.5×10^{-6}
18-8 Stainless steel	$(17\sim18) \times 10^{-6}$
36% Nickel steel	0.9×10^{-6}
Super duralumin	23.4×10^{-6}
Brass	$(18\sim23) \times 10^{-6}$
Copper	16.5×10^{-6}

Thermal expansion examples Let us see a case study about the thermal expansion of mould components described in the last course. The overall length of a core prepared in a 27°C was 26.43 mm. When this core is heated to 162°C, how long is the amount of thermal expansion? The material of the core is prehardened steel of the SCM440 series.

Sample answer The equation for calculating the thermal expansion of metals is the following:

Let us substitute the values in this equation.

$$\Delta = \alpha l_o (T - T_a)$$

where Δ is the elongation (mm) of the core pin due to thermal expansion and α is the linear thermal expansion coefficient of the metal (mm/mm). In the case of a pre-hardened steel,

$\alpha = 11.5 \times 10^{-6}$ mm/mm

l_o is the initial length of the core pin (mm)

$l_o = 26.43$ mm

T is the initial temperature (°C), $T = 27$°C

T_a is the temperature after heating (°C), $T_a = 162$°C.

Therefore, $\Delta = 11.5 \times 10^{-6} \times 26.43 \times (162 - 27) = 0.04103$ (mm).

2.8.6 Deflection Calculation

To calculate the deflection in guide pillar when it receives the weight of a mould base plate on its one end. The pillar may deflect at the end due to a large amount of bending moment. To choose right diameter on the base of its length, this deflection is due to the weight of mould base plate. Refer to Fig. 2.168 given below:

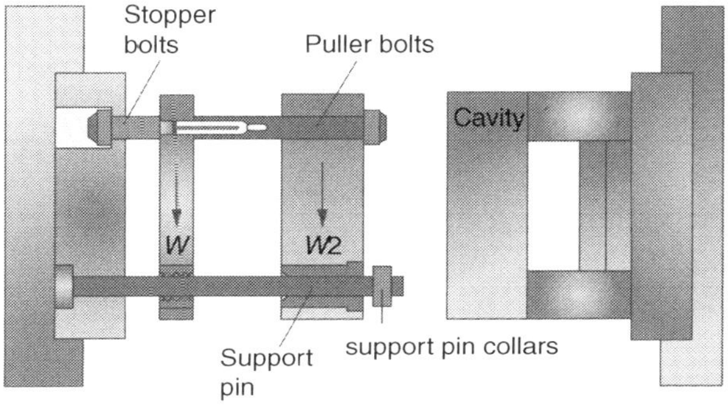

Fig. 2.168 Assembly of guide pillar.

In three-plate-type mould, minimum load is from weight of stripper plate and cavity plate, with simple calculation of deflection from strength of material, the following formula can be used for getting approximate value of deflection.

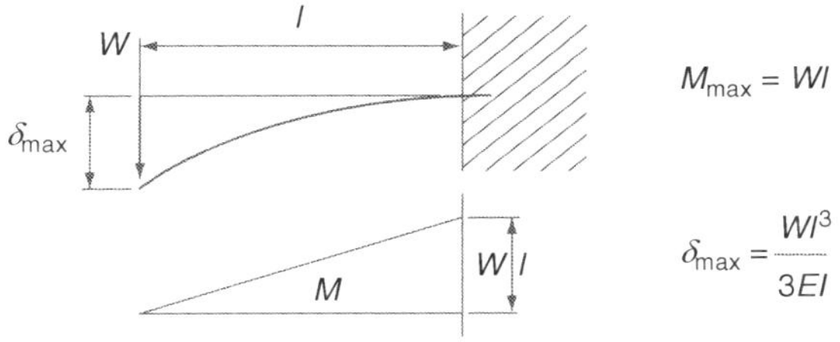

Fig. 2.169 (a) Bending of core pins due to injection pressure.

The basic equations for calculating the bending deformation of core pins due to the injection pressure are explained. In injection moulding, due to high filling pressure inside the cavity, the core pins, insert pins and long thin parts may get deflected which may cause accidents. The force acting on the pins depends upon the flow pattern of the molten plastic in the mould.

1. When a concentrated load acts upon the tip of the core pin

$$\delta_{max} = \frac{W l^3}{3EI}$$

Here, δ max is the maximum amount of bending (cm), W is the concentrated load (kgf), E is the longitudinal elastic modulus (kgf/cm^2), and I is moment of inertia of area (cm^4).

2. When a uniformly distributed load is assumed to act on the side surface of the core pin

$$\delta_{\max} = \frac{W l^4}{8EI}$$

where W is the uniformly distributed load (kgf/cm^2)

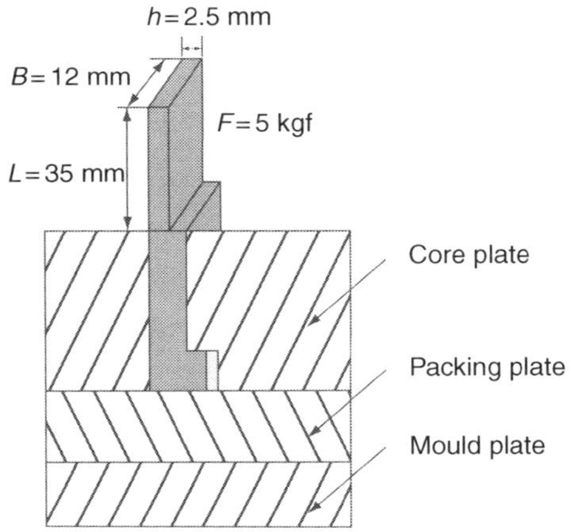

Fig 2.169 (b) Placing of core insert.

2.8.7 Procedure for Determining the External Dimensions of a Cavity

Step 1: Calculating the minimum wall thickness

A cavity is the external shape of the product which is machined inside a block of steel material, with the thickness 'h' of the wall. The steel material between the shape of the moulded product and the external shape of the block has a certain thickness. The thickness will be selected to withstand deformation due to the filling pressure of the plastic. It is possible to obtain the recommended value of this thickness by theoretical calculations by applying the equations of the field of strength of materials.

The appropriate equation should be selected on the basis of cavity machining.

1. The external shape of the cavity (cubical or cylindrical).
2. The structure of the cavity (unified or separated).

$$t = 3\sqrt{\frac{5 \times P \times l^4 \times A}{32 \times E \times b \times \delta \max}}$$

Here, p = Injection pressure
A = Area of the wall
E = Modulus of elasticity of steel material
b = Width of the wall
l = Length of the wall

The data is substituted in the equation and the thickness of steel material is determined. A margin of safety is also considered for the value of 't' obtained by the calculation.

Step 2: Cavity

After the calculation of *t* at the time of fixing the cavity to the mould plate, some flange is provided or to make some extra space for drilling hole for the screw, etc. In such cases, dimensions should be determined so that any one of these can be placed, and the final cavity dimensions are determined using even integer numbers that are round numbers (for example, 50 mm, 80 mm, etc.).

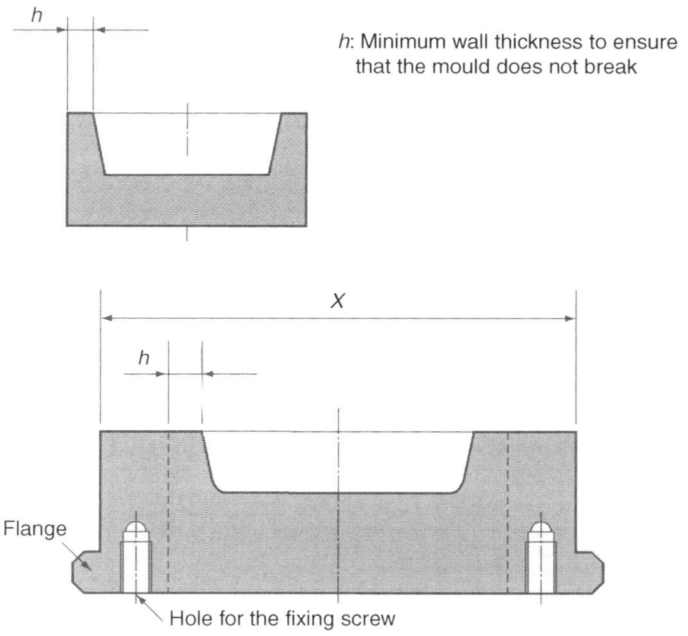

Fig. 2.170 Cavity wall thickness.

Cavity wall thickness, examples

Let us assume
Two no. of cavities and the cavities sizes are 50 mm long, 60 mm wide and 40 mm deep,
Projected area = $5 \times 6 = 30$ cm^2
Projected area of runners and sprue = 3 cm^2(10% of total projected area)
Tons capacity = $2(30 + 3) \times 700 / 1000 = 46.20$ ton,
Here, 700 kg/cm^2 is assumed injection pressure at the tip of nozzle. (1/2 of injection pressure in hydraulic cylinder)
For this tonnage next tonnage which is available = 80 tons
For the 80 tons
Contact area of the cavities = 80×1000/allowable stress = $80000/700 = 114.3$ cm^2

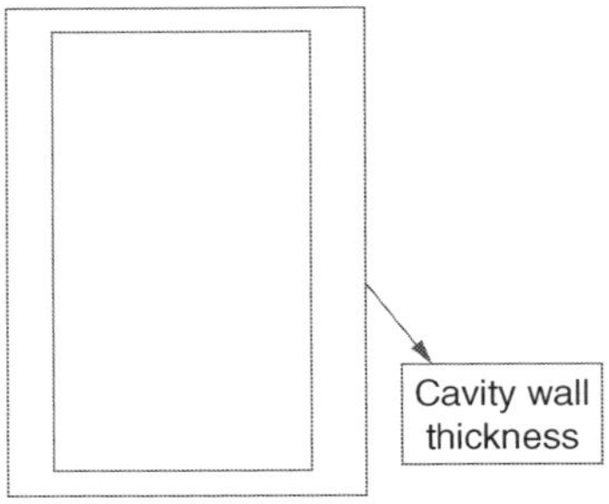

Area for one cavity = 57.15 cm^2
and as per the figure

$$2 \times 5 \times T + 2 \times 4 \times T + 4T^2 = 57.15$$

$$4T^2 + 18T - 57.15 = 0$$

So, $T = 21.48$ mm

Second method

The fluid pressure that would be exerted on the 50 mm long, 40 mm deep cavity blocks
$= 5 \times 4 \times 1400 = 28{,}000$ kg (1400 kg/cm^2 is the injection pressure)
Now with the strength of material
The deflection = $W\,l^3/192\,E\,I$
Here $w = 28{,}000$ kg, $l = 50$ mm, $E = 2\text{x}\ 10^6$ kg/cm^2, $I = bt^3/12$
For the cavity, more 0.02 mm deflection is not acceptable to receive flash free sample.

$$0.002 = 28{,}000 \times 5 \times 5 \times 5/192 \times 2000000 \times I$$

So $I = 28{,}000 \times 125\ /192 \times 2000000 \times 0.002$

$$I = 4.55\ \text{cm}^3 = bt^3/12$$

$$t^3 = 12 \times 4.55/4 = 13.65$$

$$t = 23.9\ \text{mm}$$

By the above two methods cavity wall thickness can be calculated.

2.8.8 Guide Pillar Diameter Calculation

The guide pillar can be assumed as a beam which is fixed at one end and free at other end. The deflection as per strength of material is given below.

$$\Delta = Pl^3/3EI$$

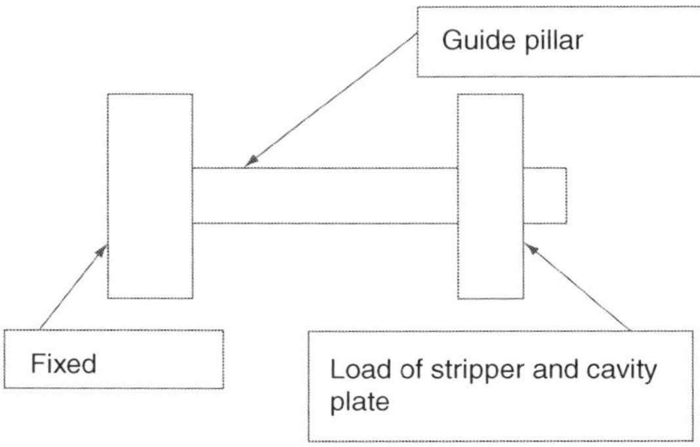

For this equation,

P = Weight of floating plate,
L = Length of guide pillar = 150 mm (assume)
E = Modulus of elasticity = 2.1×10^6 kg/cm^2
I = Moment of inertia = $\pi d^4/64$

Let us assume that the size of floating plate is 246 × 246 × 18. Then
Weight of plate = 8.6 kg (approx.)
There are four pillars. So each pillar withstands equal load.
Load on one pillar = 2.15, d = diameter of guide pillar
The deflection should not be more than 0.02 mm

$$\Delta = Pl^3/3EI = 2.15 \times 150 \times 150 \times 150 \times 64/3 \times 2.1 \times 1000000 \times \pi \times d^4$$

$$d^4 = 2.15 \times 150 \times 150 \times 150 \times 64/3 \times 2.1 \times 1000000 \times \pi \times 0.002$$

d = 18.51 mm which can be compared with the actual guide pillar.

2.8.9 Checking of Floating Stripper Plate Weight on Guide Pillar

As the guide pillar is fixed at one end and free at other end so the maximum deflection at the free end

$$\Delta = Wl^3/3EI$$

where

W = Weight of the stripper plate = l(length) × w(width) × h(height) × 7.85 units

7.85 g/cc is taken as density of steel material
For the stripper plate of size 246 × 246 × 26 mm^{3,} the weight would be 12.35 kg

l = Length of guide pillar = 106 mm (assume)
E = Modulus of elasticity = 2.1×10^6 kg/cm^2
I = Moment of inertia for guide pillar = $\pi D^4/64 = 1.149$
where D = Diameter of guide pillar = 22 mm

$$\Delta = Wl^3/3EI = 12.35 \times 10.6 \times 10.6 \times 10.6/3 \times 2.1 \times 1000000 \times 1.149$$

$$\Delta = 0.0020 = 0.02 \text{ mm which is up to certain level acceptable.}$$

2.8.10 Thickness of Steel Material Between Plastic Material and Cooling Channel

Thickness of steel material between plastic material and cooling channel (L) can be calculated by referring to the figure. Here it is considered as a fixed beam at both end and load at the centre is considered as the strength of material.

The deflection at this point where the cooling channel is made

$$\partial = Wl^3/192EI$$

The hydraulic pressure exerted in the cylinder is approximately 1400 kg/cm^2

and the pressure drop inside the cavity is about 350 kg/cm2
So the pressure inside the cavity will be 1050 kg/cm^2

W = LOAD on $1cm^2$ of hole opening = 1050 kg

L = Minimum distance between cooling channel and the cavity wall

$$b \times l = 1cm^2$$

l = length of the beam = diameter of the channel (D)-0.8 cm

$b = 1/D = 1/D$

$E = 2.1 \times 10^6 kg/cm^2$

$I = bh^3/12 = bL^3/12$

$\partial = 0.0002$ mm = 0.00002 cm

$0.00002 = 1050 \times D^3/192 \times 2.1 \times 10^6 \times (L^3/12D)$

$0.00002 = 1050D^4/33.6 \times 10^6 L^3$

$672 L^3 = 1050 D^4$

$L^3 = 1050 D^4/672 = 1.5625 D^4 = 0.254D$

$L = 1.1604 D^{4/3}$

Here if $D = 0.8$ cm, $L = 0.86$

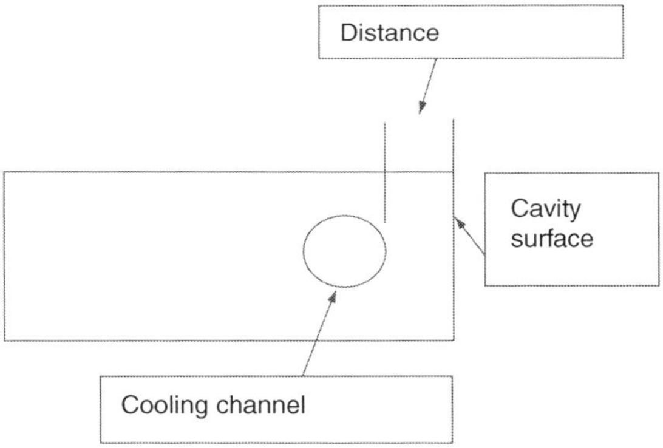

If D = 1.2cm, L = 12 so the steel material thickness should be equal to the diameter of water channel.

2.8.11 Spacer Block Design

Most attention is the contact area of the spacer block. The stress on the area of spacer block should be such as to prevent the embedding of the blocks into the plates. Actually the weak area in the spacer block is at the clamping slot area. Sometimes, in the spacer block, no slot is provided. The area of the block is $a \times b$ and width of the slot is w given in the block is w.

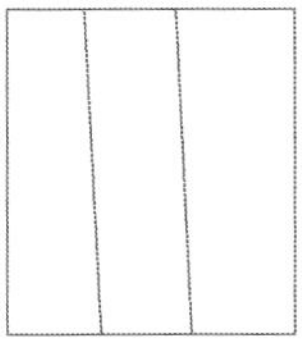

The effecting area = $(a-w)$ b with slot and $a \times b$ without slot, there are two blocks

Area × Allowable stress = Compressive force

The allowable stress for low carbon steel is 1400 kg/cm^2

So the compressive force = $1400b\,(a-w)$ with slot
= $1400ab$ without slot

This will be compared with the force comes to the spacer block due to injection pressure. Let us assume that w is the injection force applied on the cavity.

The load/force comes on the spacer block which is explained by the line diagram = $w/2$

So $w/2 = 1400b\,(a-w)$ or $1400ab$, from here a can be calculated.

2.8.12 Flow Length/Wall Thickness Ratio

Another criterion pertaining to the machine is the ratio between flow length and wall thickness. According to Hagen–Poiseuille's law, the ratio between flow length L and the square of the wall thickness of the moulding H^2 is determined by the injection pressure P_{inj}, quantity of.the machine, and the viscosity of the melt, if the velocity of the melt flow is given.

For thermoplastics there are certain optimum values for the velocity which are determined by the orientation, to which the molecules are subjected. They are around V_{inj} 30 cm/s.

2.8.13 Sprue Bush Pressure Drop

$$\text{Pressure drop } \alpha \ \frac{1}{(\text{Area at external pathway})^2} - \frac{1}{(\text{Area at internal pathway})^2}$$

As per the above sprue bush, the molten plastic material enters into the sprue bush

So as per the Bernoulis theorem

$$P_1/Y + V_1^2/2g + Z_1 = P_2/Y + V_2^2/2g + Z_2 + \text{losses}$$

As per above equation

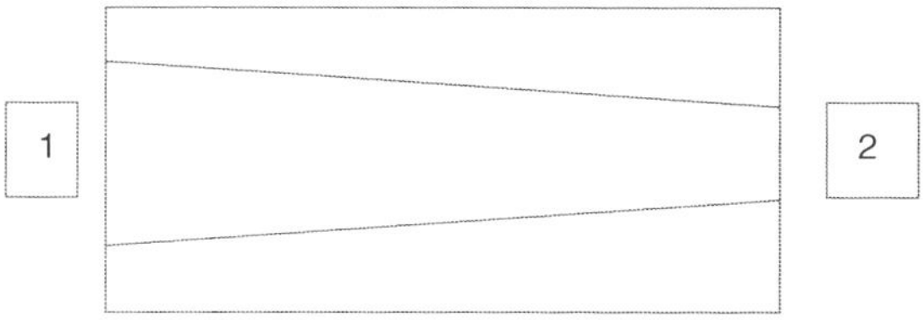

Let us consider losses are negligible and during processing Z_1 and Z_2 are equal

$$P_1/Y + V_1^2/2g = P_2/Y + V_2^2/2g$$

$$(P_1 - P_2)/Y = (V_2^2 - V_1^2)/2g$$

As per continuity theorem

$$Q = A_1V_1 = A_2V_2;\ V_1 = Q/A_1 \text{ and } V_2 = Q/A_2$$

$$A_1 = \pi/4d_1^2 \text{ and } A_2 = \pi/4d_2^2$$

By placing the values of A_1 and A_2, we may prove that

$$\text{Pressure drop } \alpha \; \frac{1}{(\text{Area at external pathway})^2} - \frac{1}{(\text{Area at internal pathway})^2}$$

2.8.14 Design of Step Ejector Pin

The critical force that will cause buckling of the ejector pin can be calculated as follows:

$$P = 2\pi^2 EM/l^2$$

where P = Critical load (Kg), E = Modulus of elasticity (Kg/cm$^{2)}$
M = Moment of inertia (cm^4), and l = Critical length (cm)

And stripping force which is required for the stripping of the moulding from the core

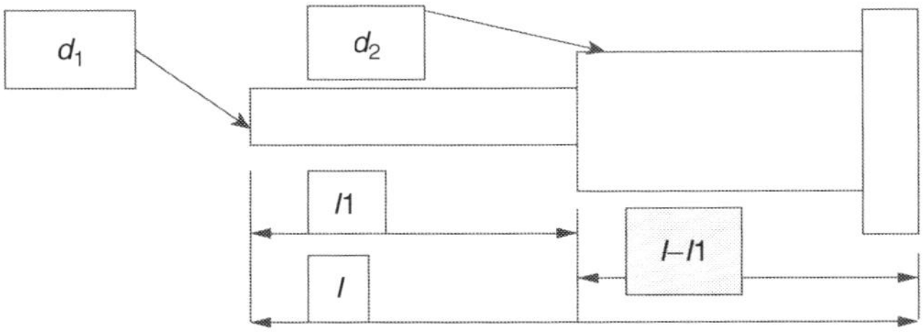

For the ejection;

Stripping force $= P = 2\,\pi^2 EM/l^2$

$= 2\pi^2 E\,\pi d^4/64l^2 = \pi^3 Ed^4/32l^2$

Stripping force $= P = (S_t \times E \times A \times f_p) \,/\, [d \times (d/2t - d \times P_p/4t)]$

St = Coefficient of thermal expansion × temperature difference between softening point and ejection temperature × d (cm)

E = Elastic modulus (kgf/cm^2)

A = Total area of contact between moulding and mould faces in line of draw (cm^2)

f_p = Coefficient of friction between plastics and steel

d = Diameter of circle of circumference equal to length of perimeter of moulding surrounding male core (cm)

t = Thickness of moulding

P_p = Poisson`s ratio of plastics

In the step pin, the diameter (d_1)of the portion which is in contact with the plastic material is a known quantity as per the wall thickness of moulding. We have to calculate the length (l_1) of this portion that can be done by placing d_1 and find the length $l_1$from the above equation.

After finding l_1 from the above equation, the diameter of other portion d_2 of the step pin can be calculated by placing length of this portion l_2 (total length – l_1) in the above equation.

2.8.15 Heaters

The calculation is done to calculate number of heaters required for manifolds

Manifolds dimension is $a \times b \times c$

Now it is required to come up the manifold up to temperature to 30 min. The processing temperature is t°F and the starting point is T°F (preheat temperature)of plastics material.

1. Required capacity for bringing the job up to operating temperature in the desired time

$$Q = m\, c_v\, \delta t$$

where Q = Heat content in kcal per hr, m = Weight of manifold = $abc\,\partial$ units
c_v = Specific heat of manifolds material = 0.12 (generally for steel), δt = Temperature rise $(T - t_1)$, ∂ = Density of manifolds material.

So,

$Q_1 = a \times b \times c \times \mathrm{s} \times c_v \times (T - t_1)$ units /hr.

2. Heat loses at processing temperature

   a) Heat loss by radiation and convection

The exposed area = $2(bc + ac)$ units
Heat loss by radiation $Q_2 = \sigma A€\{(t + 273)^4 - (t_1{}^4 - 273)\}$units per hr.
Here σ = Stefan Boltzman constant = 5.76×10^{-12}, A = Exposed area, € = Emissivity,
t = Processing temperature and t_1 = Room temperature in Kelvin
Heat loss by convection $Q_3 = h\,A(t - t_1)$ units per hr
Here h = Heat transfer coefficient, A = Exposed area,

3. Losses by conduction through insulation $Q_4 = kA\,(t - t_1)$ units per hr.

where A = Conduction area (area of contact with the plates) = $2ab$
K = Thermal conductivity of insulation material

4. Heat required for plastic material

$$Q_5 = m_1\, c_{v1}\, \delta t$$

where m_1 = Weight of plastic material, c_{v1} = Specific heat of plastics material,

$$\delta t = (t - T)$$

Total heat required = $(Q_1 + Q_2 + Q_3 + Q_4 + Q_5)$kcal per hr
Kilowatt = $(Q_1 + Q_2 + Q_3 + Q_4 + Q_5)/867$
As per the above we may decide the number of heaters required for manifolds.

2.8.16 Determining the Economical Number of Cavities

Determination of the Number of Cavities in Moulds with Economical Way.

Let us assume that

$1,00,000pcs = n$ = The total number of parts are required by the party
X = Number of mould cavities to be fabricated
r = 15% of wages/hr = Overhead cost
w = Rs. 25 per hour = Wages of the workers
A = Production and maintenance cost of one cavity = 10,000/-
B = Production cost of mould casing
T = Cycle time = 60 second

S = Total manufacturing cost of the parts

By the above we may conclude that

No. of cycle required = n/x

Total time required = $n\,T/x$ seconds = $nT/60\,x$ minutes

Total worker cost with overhead = $n\,T\,w(1+r)/x$

So $S = \{nT$ The temperature difference between mould plates and other plates from platen side or platen $w(1+r)/60x\} + Ax + B$

This is the equation for the total cost of the parts. By differential calculation we try to get the very economical number of cavities by assuming minimum manufacturing cost. For getting this differentiates the equation with respect to x.

$$d/dx(S) = d/dx[\{nTw(1+r)/60x\} + Ax+B]$$
$$= d/dx[\{nTw(1+r)/60x\} + d/dx\,(Ax) + d/dx\,(B)]$$
$$= d/dx\{nTw(1+r)/60x\} + A$$
$$= d/dx\{nTw(1+r)x^{-1}/60\} + A$$
$$= -nTw(1+r)x^{-2}\,/60 + A = 0$$
$$A = nTw(1+r)x^{-2}/60$$
$$x^2 = nTw(1+r)/60A$$
$$x = \sqrt{}nTw(1+r)/60A \text{ by putting the values}$$
$$x = \sqrt{\{1,00,000 \times 60 \times 25(1+0.15)/60 \times 10,000\}}$$
$$x = 16.9 = 16 \text{ number of cavities}$$

For the above data, 16 is the economical number of cavity. Now we have compared the number with the technological number of cavity. Whichever is less, we have to design the mould for that number.

2.8.17 Cost Reduction through Design Arrangement

1. Cost of the mould: The cost of a particular mould is expensive. To reduce the cost of the mould, it would be better to go for multicavities injection mould by which the cost of one product can be reduced. Take one example.

2. By the production of large quantities: The approximate cost of a mould that may make a lunch box moulding may be Rs. 50,000 to Rs. 1,00,000. The capital cost is of course expensive but consider that a single mould may produce ten lakhs of mouldings. Therefore, the unit cost of each moulding can be very low because they are manufactured in such large quantities.

3. By the fabrication of multi-impression mould: DME can be further reduced by fabricating the mould in multicavities. In multicavities mould the cost of mould is increased by 1.4 to 1.5 times but the production cost is reduced due to number of cavities per cycle time.

4. By the fabrication of family type of mould: The mould cost is also reduced by fabricating family type of mould.

5. By quick change insert system: By fabrication of one mould with different type of its insert, cost of the mould can be reduced by quick- change tooling systems. These kits offer inserts to modify existing moulds rather than making new ones every time. This has also reduced the setting up times of the injection moulding machine. This is called the quick change injection moulding technique.

2.8.18 Effect of Draft on Design

For easy removal or movement of one part from another, it is required to provide the draft in the mould parts like sprue bush, core and cavities, etc. These drafts are used to give proper and easy movement to the parts.

Like in sprue bush, the draft is provided for easy ejection of sprue from sprue bush. In sprue bush 1.5° to 3° angle draft is generally given. Same way in deep core or cavities, the draft is provided as per the depth of the components.

2.9 SPLIT MOULD

2.9.1 Introduction

This mould is used for the components which have projection or undercut on the surface of component. Due to this projection undercut, the component cannot be ejected from the mould. Actually, recess or projection of component prevents its removal from the cavity. That can be removed only by the opening of half cavity on either direction as shown in the Figs. 2.171 and 2.172. (In closed and opened conditions of split) split mould is used for this type of components. In this type of mould two blocks are guided together and are kept it with perfect matching by using locking heels during injection and it is called split mould. It is required to split the cavity insert into parts and open at right angles to the line of draw, to relieve the undercut before the moulding is removed. A parting line can be visible on the components due to the matching of split blocks. For perfect matching and opening and closing of the split, the splits are fixed in guide rails on a flat mould plate and they are actuated by some mechanism. The splits are possible to mount on either the moving or fixed mould plate. A screw-threaded component or undercut/recess component is shown in Fig. 2.171.

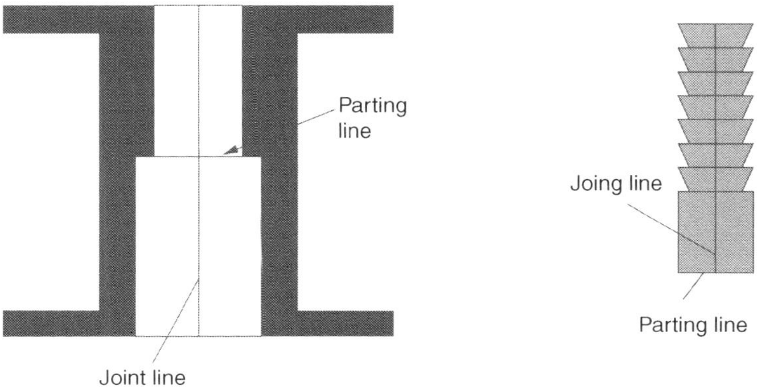

Fig. 2.171 Split moulding.

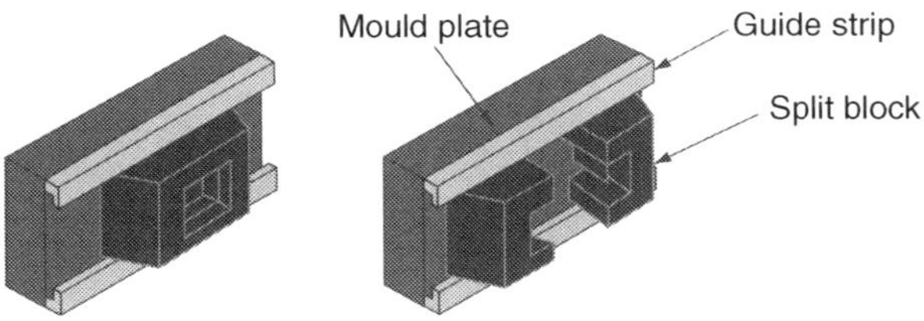

Fig. 2.172 Split in closed and open position.

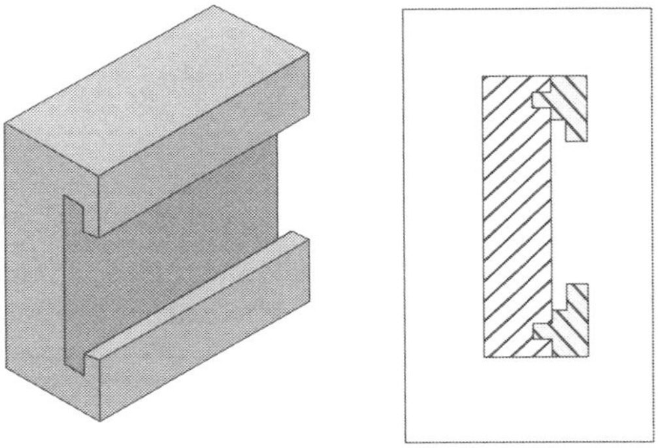

Fig. 2.173 Guide rail.

2.9.2 Guiding and Retention of Splits

The guide rail shown in Fig. 2.173 and assembly of splits are shown below in the Fig. 2.174.

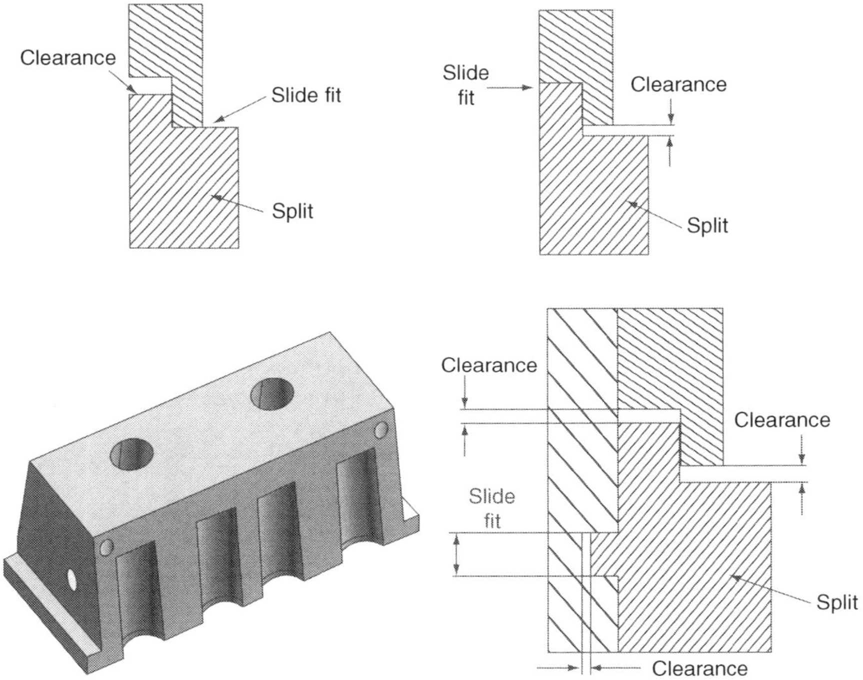

Fig. 2.174 Split block with assembly of splits.

2.9.3 Split Mould Design Requirements

1. Amount of splits movement required.
2. Amount of delay period and length of delay period required.

3. Whether moulding inserts are to be incorporated or not.
4. Whether the available machines are programmed for ancillary cylinder control or not.
5. Whether a short or long production run is required.

2.9.4 Types of Split Mould Actuation Methods

1. Finger cam actuation method
2. Dog-leg cam actuation method
3. Cam track actuation method
4. Hydraulic actuation method
5. Spring actuation system
6. Angled-lift splits.

1. Finger Cam Actuation Method: Mould has a linear motion due to machine platen. In this method, for the operation of split on guide rail, the linear motion of mould half is used to change the direction of motion for split opening. This is done by using a round bar fixed at an angle with fixed mounting plate or fixed half of the mould. The round bar is called finger cam which is made out of hardened and tempered tool steel.

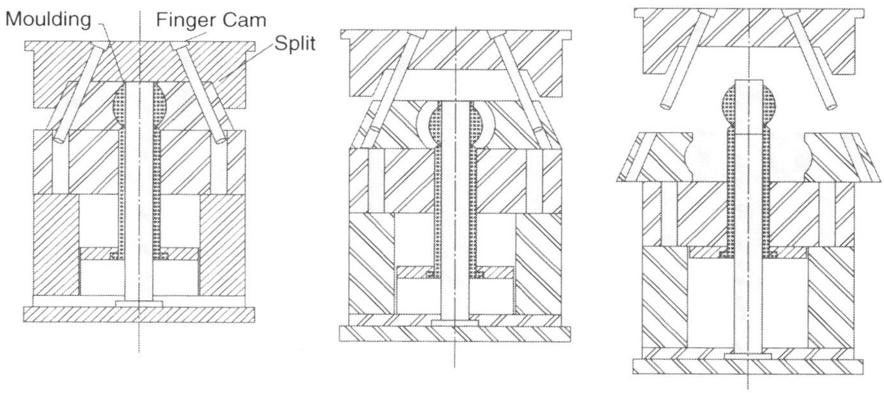

Fig. 2.175 Split actuation.

As the mould opens, the finger cam first comes in contact with the split hole wall and forces the split to move outwards and sliding on the mould plate as per the guide rail provided on the cavity plate. Once the contact with the finger cam is lost, the split's movement stops immediately. After required opening of the moving half, the ejector system operates and ejects the moulding as shown in the Fig. 2.175. On closing of the mould, the reverse action occurs. The finger cam re-enters the hole in the split and forces the split to move inwards. The final closing of the splits is achieved by the locking heels and not by the finger cams as the finger cam is assembled in the free hole of split. The finger cam movement can be computed by the formula: $M = (L \sin\theta) - (C / \cos\theta)$.

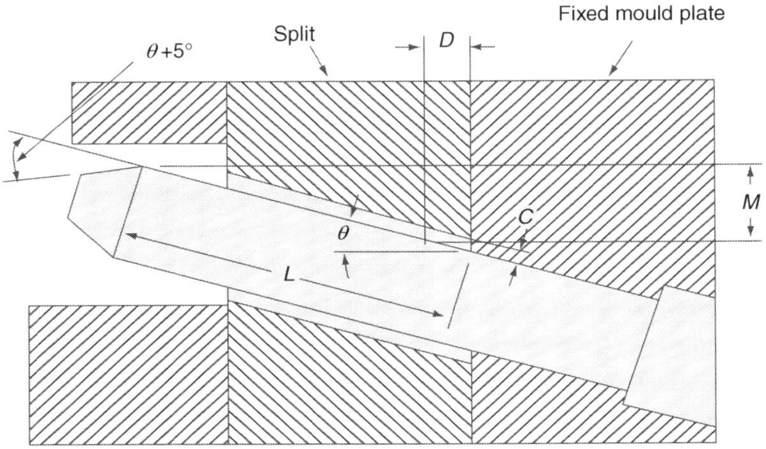

Fig. 2.176 Finger cam pin actuation.

If the required movement is known from the amount of component undercut, the following formula is used to determine the finger cam length.

$L = (M / \sin\theta) + (2C / \sin 2\theta)$

where M = Split movement
θ = Angle of finger cam (10–25°)
L = Working length of finger cam
C = Clearance

The clearance 'C' provides for

1. No direct force on finger cam pin.
2. Permits to open the mould to open a predetermined amount before the splits are actuated.
3. The amount of delay (movement 'D') before the splits are actuated is determined by the formula: $D = C/\sin\theta$

The finger cam angle is generally selected from 10° to 25°. If the mould height is higher, long finger cam is required and hence, the angle can be increased up to 25°. For actuating small splits, a finger cam diameter of 13 mm is suitable, but for large splits or where greater than 10° angle is used for split, the diameter should be increased accordingly. The lead-in angle at the front end of the finger cam is taken normally (θ + 5°) which allows the re-entry of the finger cam into the splits.

Calculation of finger cam pin diameter: For a finger cam operation, the diameter of finger cam can be calculated as follows:

Finger cam is a member which is fixed at an angle (20° with vertical) at one place and free at other end. The finger cam passes through the split hole for giving the movement to split.

Let us assume that the weight of split = w

For the movement of split, the force on finger cam (at 70° with finger cam) = μw where the coefficient of friction between split and base plate and this load is uniformly distributed on the finger cam

The actual load which will cause to bend the finger cam = $\mu w \sin 70$

As per the strength of material the maximum deflection in the finger cam

$\Delta = Wl^3/8EI$, where w = Load on finger cam = $\mu w \sin 70$, l = Length of finger cam, E = Modulus of elasticity and I = Moment of inertia for finger cam = $\pi D^4/64$ (D = Diameter of finger cam)

Here the deflection should not be more than 0.02 mm after substituting the values,

$\Delta = 0.3 \times 12.5 \times \sin 70 \times 15 \times 15 \times 15 \times 64/8 \times 2.1 \times 1000000 \times 3.14 \times D^4$ here we have assumed Weight of split = 12.5 kg,

$\mu = 0.3$(as per the web site)

$E = 2.1 \times 1000000$, length of finger cam- 150 mm

$D^4 = 0.3 \times 12.5 \times 0.969 \times 15 \times 15 \times 15 \times 64/8 \times 2.1 \times 1000000 \times 3.14 \times 0.002$

$D = 16.15$ mm

2. Dog-Leg Cam Actuation: It is also one type of actuation which is similar to finger cam. In the place of round bar of finger cam, the cam of dog-leg cam is in the shape of rectangular section and fixed with the mounting plate. In this type of cam actuation, the delay period is decided on the basis of eccentric length of dog-leg cam and it is more when compared to finger cam. Each split having a rectangular hole allows entering the cam for actuating inward movement of the splits.

The dog-leg cam actuation system is shown in Fig. 2.178. The mould is in closed condition and the splits are locked together by the locking heels of the fixed mould plate. As the mould opens, no movement will be in the split on either direction which is known as delay action of the split. After sometime, the splits cause a delay period to start open when the mould halves are parted because of the straight portion of the dog-leg cam. The moulding, which is made within the splits, will thus be pulled from the stationary core.

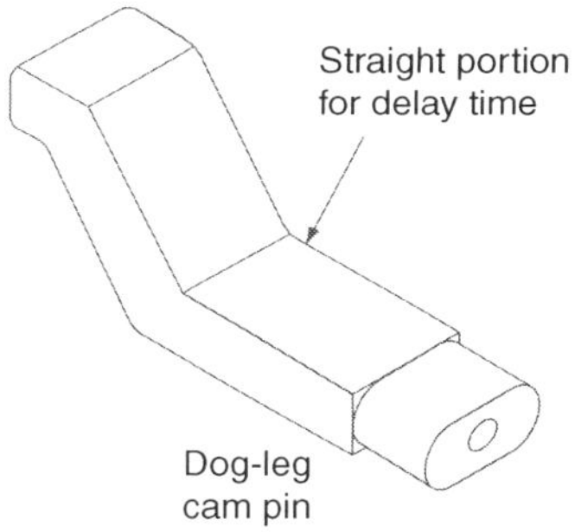

Fig. 2.177 Dog-leg cam.

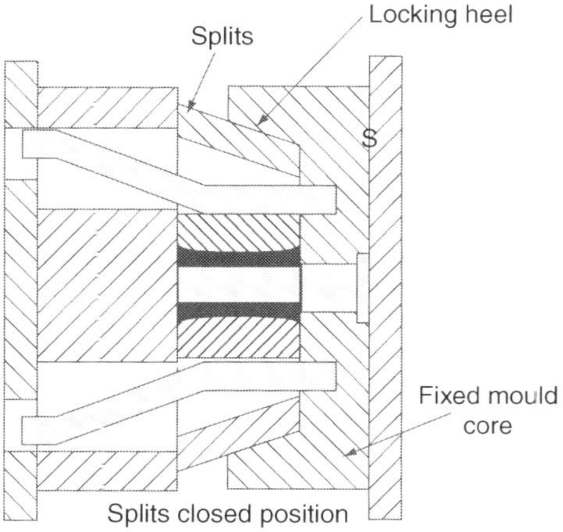

Fig. 2.178 Dog-leg cam actuated closed split mould.

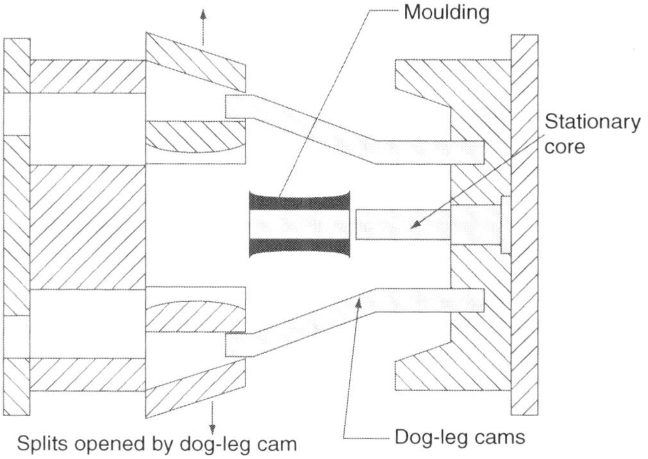

Fig. 2.179 Dog-leg cam actuated opened split mould.

Further movement of the mould half causes actuation of the splits by the dog-leg cams, thereby releases the moulding shown in Fig. 2.179. The reverse action occurs when the mould is closed. The cross section dimensions of a dog-leg cam are 13 mm by 18 mm and the angle θ is 10° to 25°.

The relevant formula for calculating the opening movement, the length of cam, and the delay period are given by

$$M = L_a \tan\theta - C,\ L_a = (M + C)/\tan\theta,\ D = (L_s - e) + (C/\tan\theta)$$

where M = Movement of each split, L_a = Angled length of cam, L_s = Straight length of cam, θ = Cam angle, C = Clearance, D = Delay, e = Length of straight portion of the hole.

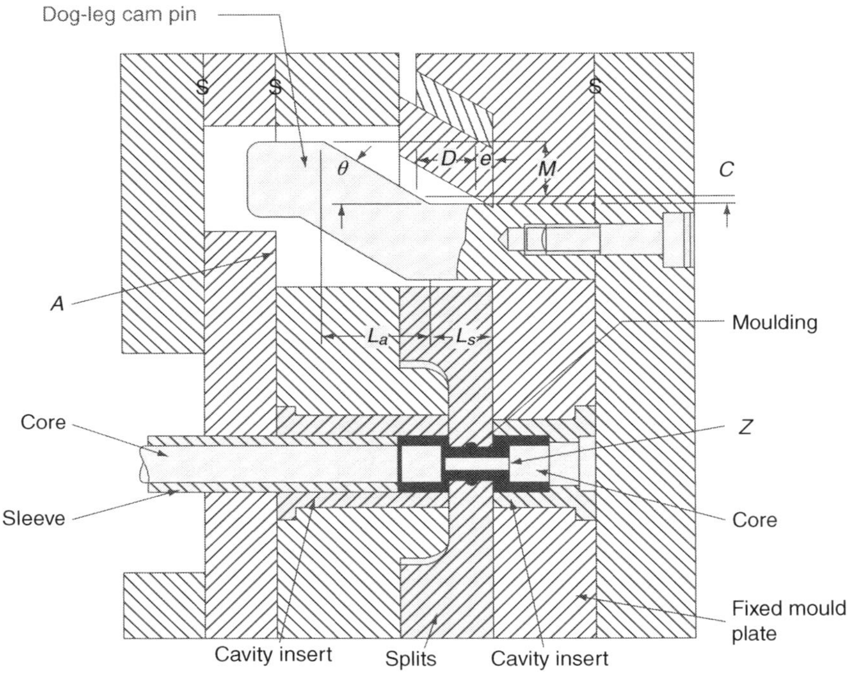

Fig. 2.180 Dog-leg cam actuated mould.

3. Cam Track Actuation: For getting more delay, cam track actuation method is used. In this method, the cam is fixed outside of the mould. The cam track (Fig. 2.181) is machined into a steel plate, attached to the fixed mould half. A boss fitted to both sides of the split runs in this cam track. The movement of the splits is accurately controlled by specific cam track design shown in Fig. 2.182. A radius or taper should be provided at the entrance for the boss as it re-enters the track.

The splits are mounted on a mould plate and the bosses are fixed into the side faces of the split, projected into the cam track plates. A clearance of 1.5 mm is provided between the cam track plate and the moving mould half. The permissible angle of the cam track plate is between 10° and 40°. The formula for calculating the distance required for each split, the length of cam track, and the delay period are as follows:

$$M = L_a \tan\theta - c$$

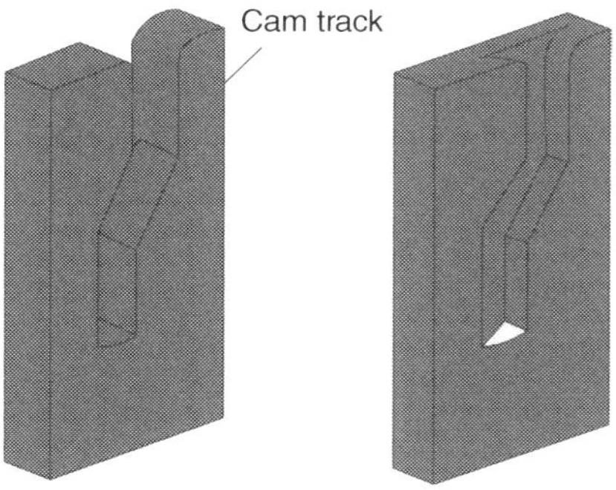

Fig. 2.181 Cam track plate designs.

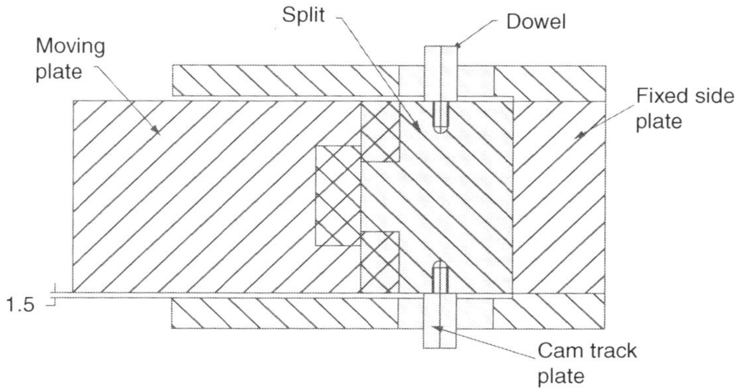

Fig. 2.182 Assembly of cam track actuation closed point.

$$L_a = \frac{M+c}{\tan\theta}$$

$$D = L_a + \frac{c}{\tan\theta} + r\left(\frac{1}{\tan\theta} - \frac{1}{\sin\theta}\right)$$

where M = Movement of each split
L_a = Angled length of cam track
θ = Cam track angle, 10 – 40°
D = Delay
L_s = Straight length of cam track
c = Clearance, 1.5 mm
r = Radius of boss

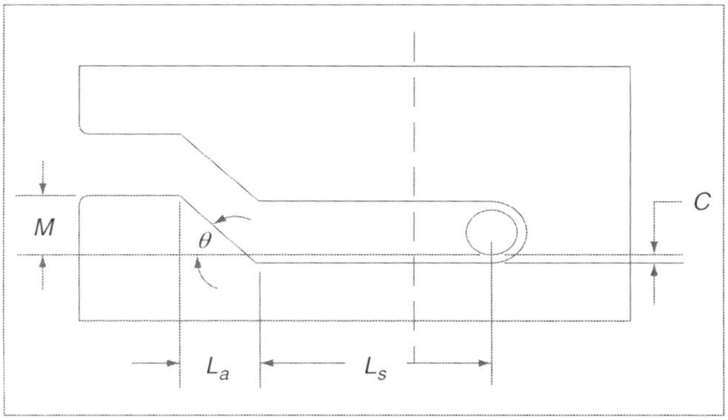

Fig. 2.183 Cam track actuation.

4. **Hydraulic Actuation:** In this method,
   (a) The splits are actuated by hydraulic mechanism.
   (b) The hydraulic actuation system is used for design of large shape split mould components as more locking force is required to keep the splits closed during the injection phase.
   (c) In this case, more delay movement and large split movements can be achieved.

(d) The splits having a projection on the underside to which the ram of the hydraulic actuator is attached.

(e) The hydraulic actuator is fitted in a mounting plate, which is fixed to the side wall of the mould plate.

(f) The splits are closed by the locking heels of the chase bolster.

(g) To reduce the cycle time, the splits should be opened while the mould is opening.

(h) On return strokes, the splits should be closed before they re-enter inside the chase bolster of the mould shown in Fig. 2.184.

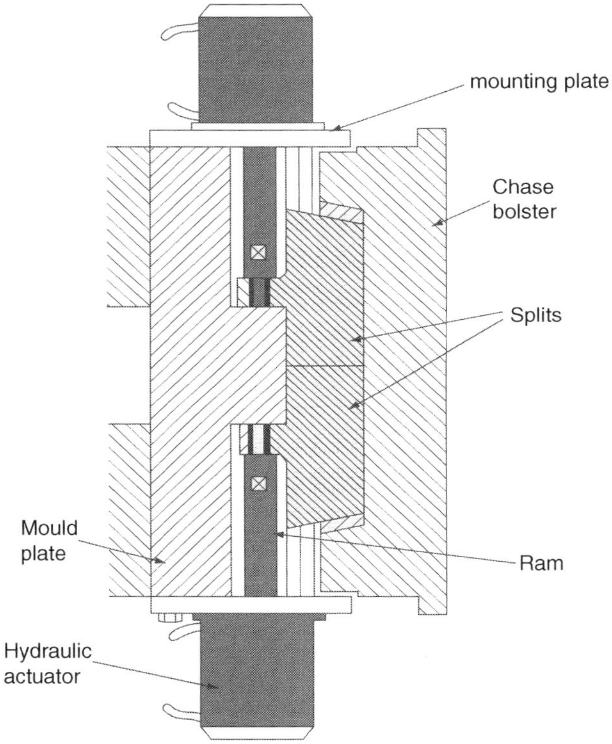

Fig. 2.184 Hydraulic actuation.

Disadvantages

(a) The mould is more bulky when compared with the other designs and makes the mould setting more difficult.

(b) The hydraulic system has to be connected each time when the mould is set up.

(c) Mould cost is high.

5. **Spring Actuation:** Sequence of operation:

(a) First, the chase bolster holds the splits during the injection phase.

(b) The compression springs exert a force to split halves immediately when the mould starts to open.

(c) The stud reaching the end of the slot in the mould plate stops the split movement.

During the closing stroke, the splits re-enter the chase bolster and are progressively closed. The formula for calculating the splits opening movement is

$$M = \tfrac{1}{2} H \tan \theta$$

where M = Movement of each split (approximately $M = 0.2$ H)

H = Height of locking heel

θ = Angle of locking heel

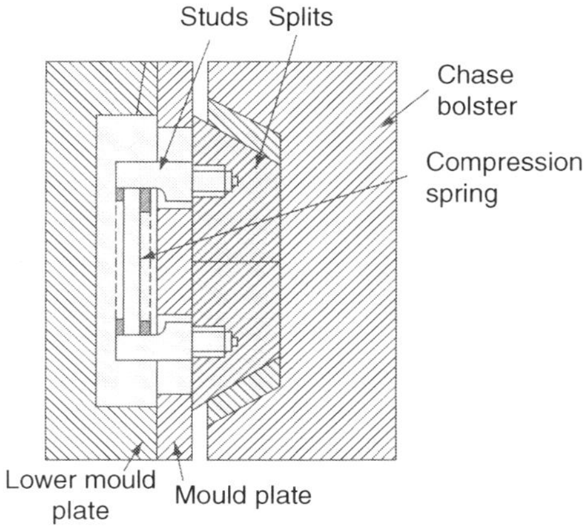

Fig. 2.185 Spring actuation.

A suitable angle for the locking heel is 20° to 25°.

The splits open immediately the mould parts without any delay period and the moulding remains in the moving half so that it can be positively ejected. It is advised to provide one stud per split which is under 76 mm width.

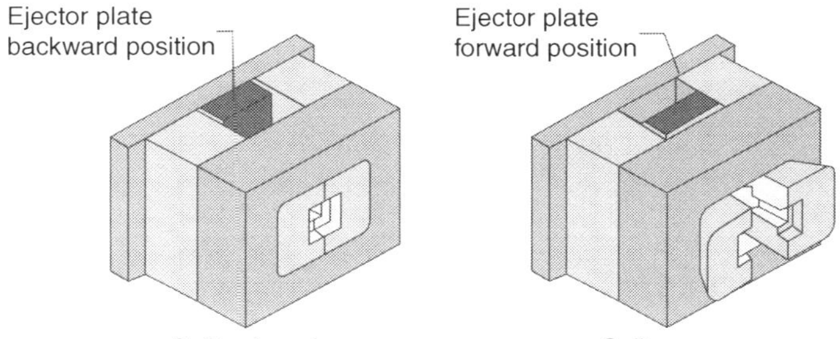

Fig. 2.186 Angled lift splits.

6. Angled-Lift Splits: In this system, the splits are mounted in a chase-bolster, and it is opened by moving out an angular pin of which one part is attached with ejector plate and second part is fixed with split of the mould, which relieves the undercut portion of the moulding as shown in the Fig. 2.186. The alignment of the split is accomplished by their being seated in the chase-bolster.

Angled guide dowel actuating system In this design, the guide dowels are fitted at an angle to the underside of each split. These guide dowels are passed through holes machined at an angle in the chase-bolster. When the ejector system is actuated, the relative movements between the ejector plate and the enclosed chase-bolster cause the guide dowels to

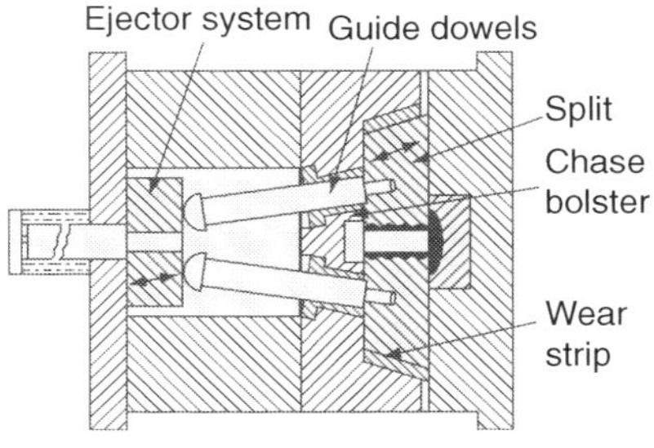

Fig. 2.187 Actuation of angle lift.

move forward at an angle corresponding to the splits. They open as shown in Fig. 2.188.

A convenient angle for the guide dowel is 10° but this may be increased if large opening movement is required. The opening movement of each split is computed from:

$$M = E \tan \theta$$

where E = Effective ejector plate movement

θ = Guide dowel angle

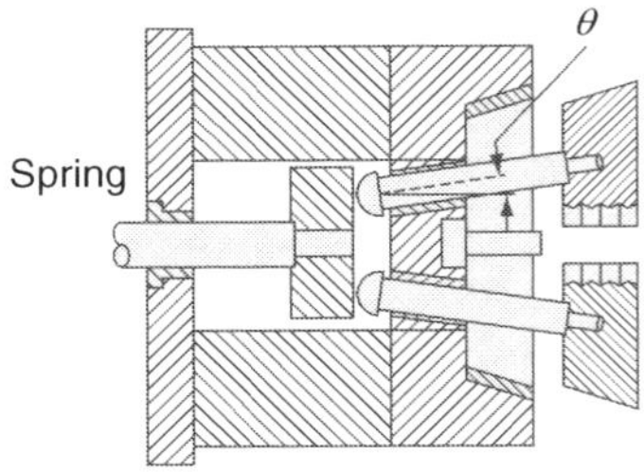

Fig. 2.188 Actuation of angle lift.

2.9.5 Side Cores

It is a local core, which is generally mounted normal to the mould axis for forming a hole or recess in the side of a moulding. This side core prevents the in-line removal of the moulding and some means must be provided for withdrawing the side core prior to ejection.

Internal side core assembly In this design, the side core is fixed in a T-shaped slide block (carriage) which is mounted on guides, now as per same principle during opening of the mould, due to cam actuation, the slide block is moved with side core pin, relieved the components. The slide block is locked in the forward position by the locking heel.

Types of actuation The internal side core assembly can be actuated by means of

1. Finger Cam for short delay, the length of finger cam depends on the relieving movement of side core pin
2. Dog-leg cam actuation
3. Spring actuation

1. Finger Cam Method: The finger cam method is already discussed in the previous topics in detail.

2. Dog-Leg Cam Method: The dog-leg cam method is already discussed in the previous topics in detail.

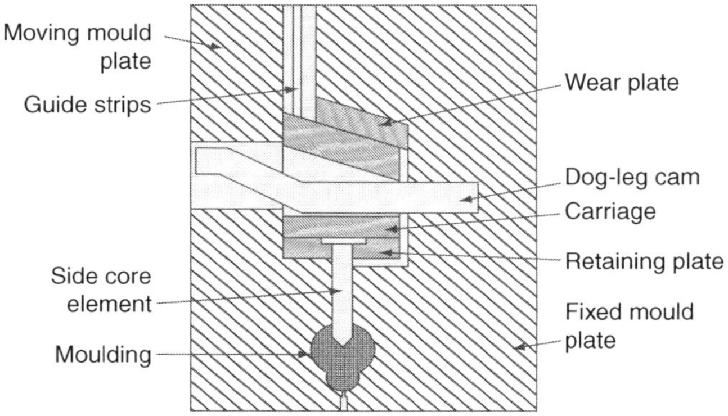

Fig. 2.189 Dog-leg cam method of actuating side core assembly.

3. The Spring-Loaded System: It is an operating method confined to moulding has very shallow undercuts or projections. As shown in Fig. 2.190, a stud is fixed to the underside of the split block and it is accommodated in a slot machined in the mould plate. A spring or springs are fitted to the slot and cause the side core assembly to withdraw the mould opens immediately. The locking heel is used to progressively return the assembly when the mould is being closed.

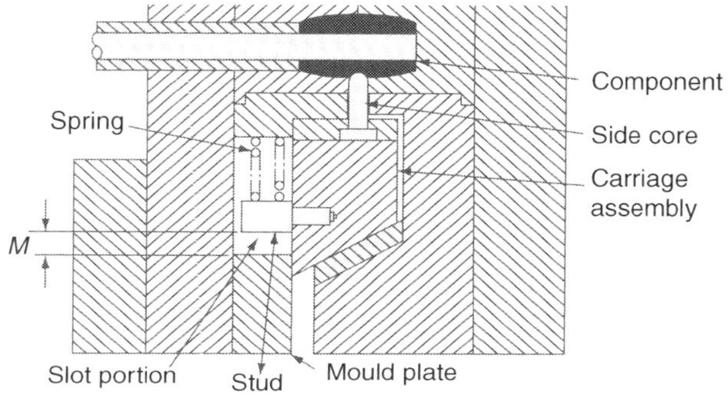

Fig. 2.190 Spring loaded system for actuating side core assembly.

2.9.6 External Undercuts

Threaded Splits This method is adopted when automatic production is required for an externally threaded component, where the thread form is such that it cannot be stripped. Extreme accuracy is to be maintained while manufacturing and fitting of the splits; otherwise flashing may occur along the joint line which impairs the efficiency and quality of the produced component. In the case of a component which has an interrupted thread form as shown in Fig. 2.191, the joint line can be positioned on the plain section, thereby avoiding the necessity of requiring such extreme accuracy in the fitting of the split.

A split mould with finger or dog-leg cam operation can be used for an externally threaded component.

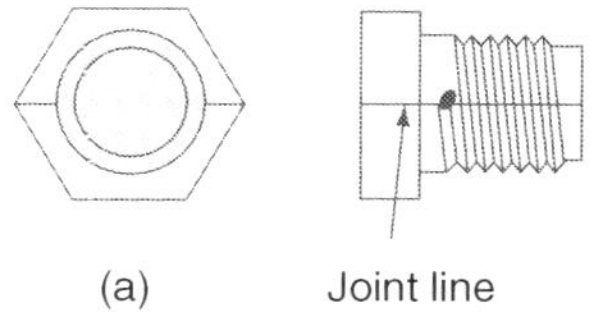

Fig. 2.191 Joint line on externally threaded components in splits.

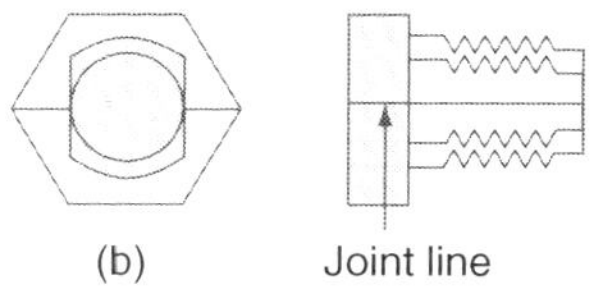

Fig. 2.192 Split mould for externally threaded component.

Questions

1. Write down the advantages of gas injection mould.
2. Write down the general principle of gas injection mould.
3. What are steps involved in quality moulding?
4. How mould design depends on machine specification?
5. What is short capacity?
6. Write short notes on the following:
   a) Guide pillar **b)** Guide bush
7. What are the design guide lines for proper alignment of guide pillar and guide bush?
8. How can the strength of guide pillar be calculated?
9. Write brief note on spigotted guide pillar and guide bush.
10. Draw a support pin and explain its parts.
11. Write a short note on mould lifting.
12. Why is the core fixed in the moving half of the mould?
13. How do parting lines influence moulding?
14. Name the two manufactures of standard mould base.
15. What is the significance of mould venting on mould?
16. What are the selection criteria of mould material?
17. List out the mould materials and their applications for a standard mould.
18. What are the main requirements for sleeping steels?
19. Explain the essential property of mould material.
20. Write the advantages and application of hardened steel of mould design.
21. Write the advantages and application of prehardened steel of mould design.
22. Explain the application of corrosion registrant mould steel.
23. Write short notes on the following:
    a) Parting line **b)** Parting surface
24. The vents are generally provided along the mould parting line. Justify.
25. Write thumb rule of venting.
26. Describe vent placement.
27. What should be the design criteria for proper venting?
28. Define feed system.
29. Define sprue bush.
30. Explain extended sprue bush with a neat sketch.
31. Why does balancing of runner play a vital role in injection mould?
32. Define cold slug.
33. Classify runner according to their cross section. Explain with a neat sketch.
34. What do you mean by runner efficiency? How can it be calculated?

35. List out the three general types of runner layout.
36. Explain various types of runner methodology used in injection mould.
37. What is unbalanced flow?
38. Describe gate.
39. Describe the determination of location shape and size of gate.
40. What is sprue gate? Write its advantages and disadvantages with the help of a neat sketch.
41. How is the parting line decided on the basis of edge gate?
42. Explain tap gate with a neat sketch.
43. Explain fan gate and its purpose.
44. What is rectangular edge gate? Write its three advantages.
45. Explain the following:
    a) Film or flash gate **b)** Diaphragm **c)** Spoke gate
46. Explain the following:
    a) Pin gates **b)** Submarine gate **c)** Curved tunnel gate.
47. Explain the following:
    a) Hot runner gate **b)** Valve gate.
48. List out the factors which affect the gate size.
49. What is ejection system? Describe the parts which ejection system consists of.
50. How do you calculate the height and width of a pressure block?
51. Explain the following:
    a)Ejector plate **b)** Ejector retainer plate **c)** Knock out road **d)** Tie rod bolts.
52. Describe the function of bottom plate in ejector plate assembly.
53. Describe the size of ejector pin decided on basis of wall thickness.
54. Explain the following with a neat sketch.
    a) Pin ejection **b)** Sleeve ejection **c)** Blade ejection.
55. Explain the following with neat sketch.
    a) Valve ejection **b)** Air ejection
56. Explain the following with a neat sketch.
    a) Stripper plate ejection **b)** Stripper bar ejection.
57. How can the ejection force requirement be calculated?
58. How can the support pillar requirement be calculated?
59. Why does mould need cooling?
60. Describe basic cooling principles and their importance to product quality.
61. Write short notes on following:
    a) Parallel cooling channel **b)** Series cooling
62. How is cooling is done in integer cavity plate?
63. How is cooling done in integer core plate?

64. Why is cooling necessary in core?

65. Explain the following with a neat sketch:
a) Bolster cooling **b)** Cavity insert cooling **c)** Core insert cooling

66. What is the use of baffles in cooling system of a mould?

67. What is bubbler cooling? Explain with the help of a neat sketch.

68. What is the use of thermal pin in cooling system of a mould?

69. How can large cores be cooled?

70. How can slender core be cooled?

71. Explain helical channel design.

72. What are the factors on which, the operating temperature of a mould will depend?

73. Write the calculation for weight of water to be circulated per hour?

74. Draw a simple sealing and cooling channel layout?

75. What is 'O' ring? Explain its function with a neat sketch.

76. How can 'O' ring size be determined?

77. What is the design consideration for 'O' ring?

78. What is the role of water connector in mould?

79. What should a designer consider before designing of cooling pipe for cavity?

80. How can diameter of guide pillar be calculated?

81. Derive the calculation for thickness of steel materials between plastic material and cooling channel.

82. Derive the equation for pressure drop in sprue bush.

83. Derive the equation for designing step ejector pin.

84. What is the effect of draft in mould design?

85. What design arrangement can be made in mould for cost reduction?

86. What is split in mould?

87. What are the requirements of split design?

88. List out the types of split mould actuation methods?

89. Explain the finger cam actuation method with a neat sketch.

90. Explain the cam track actuation method following with a neat sketch.

91. Explain the spring actuation system with a neat sketch.

92. Explain the angled – lift split system with a neat sketch

93. What is threaded split? Where is it used?

94. How can flank thread angle be calculated?

95. Explain hydraulic actuation method with a neat sketch.

96. Explain dog-leg cam actuation method with a neat sketch.

97. Explain the stages of injection moulding process with a neat sketch.

98. Classify the injection mould based on cavity layout, based on injection moulding process and based on materials to be injected.

99. Draw a two-plate mould sketch and show its various parts and explain.
100. Explain the types of opening in three-plate moulds.
101. Write short notes on following:
 a) Stepper injection mould b) Split mould
102. Write short notes on following:
 a) Collapsible core and Cavity mould **b)** Hot runner mould
103. Draw a schematic diagram of stack mould and explain.
104. How are number of impressions calculated in injection mould?
105. Describe the types of locating with a neat sketch.
106. What rules should a designer follow before designing a locating ring?
107. Explain the mould clamping methods.
108. List out the checking points before starting assembly of a mould.
109. Explain standard mould base and list out its advantages and limitations.
110. Describe the materials and heat treatments of various mould parts.
111. Give a brief idea on mipalloy100 and write its advantages.
112. What is mould life cycle?
113. Write down the various factors which depend on mould life cycle.
114. What is heat treatment process and write its purpose.
115. What are different types of heat treatment process?
116. Explain types of surface with a neat sketch.
117. Write the application of sprue bush. Explain pressure and temperature draw in sprue bush.
118. Give a brief idea on design rule for size of sprue bush.
119. Classify types of gate. Explain any five with neat sketches.
120. What is ejector grid? Explain types of ejector grid with neat sketches.
121. Explain the types of ejection with neat sketches.
122. Explain different types of actuation of stripper plate ejection.
123. What is sprue puller? Explain types of sprue puller with neat sketches.
124. Explain the heat rod method with a neat sketch.
125. Explain the types of cooling with schematic diagram.
126. What are the procedures for determining the external dimension s of a cavity? Explain with example and required calculation.
127. Derive and explain the calculation for Finger CAM actuation with a neat sketch.
128. Derive and explain the calculation for Dog-leg CAM actuation with a neat sketch.
129. Derive and explain the calculation for CAM track actuation with a neat sketch.
130. Derive the calculation for number of heaters required for manifolds.
131. Explain the heat pipe method with a neat sketch.

References

1. CIPET, *Technical Manual*, Chennai.
2. Dominick V. Rosato. P. E. and Donald V. Rosato, *Injection Molding Handbook*, CBS Publishers and Distributors, New Delhi.
3. DuBois, J. H. and Pribble, W. I., *Plastics Mold Engineering*, Reinhold Publishing Corporation, New York.
4. Gastrow, *Injection Molds*, Hanser Publishers, New York.
5. Irvin I. Rubin, *Injection Molding : Theory and Practice*, Wiley Interscience Publication. New York.
6. Laszlo Sors, *Plastic Mould Engineering*, Pergamon Press, London.
7. Pye, R. G. W., *Injection Mould Design*, Affiliated East-West Press. Pvt. Ltd., New Delhi.
8. Rees, Herbert, *Mold Engineering*, Hanser Publishers, New York.
9. Tim A. Osswald, Lih-Sheng Turng and Paul Gramann, *Injection Molding Handbook*, Hanser Publishers, Munich.
10. Wordingham, J. A. and Reeboul, P., *Dictionary of Plastics*, Littlefield Adams and Co., New Jersey.

CHAPTER 3

Compression Mould Design

3.1 INTRODUCTION

Compression moulding is the process for moulding thermo-set polymers like phenolics, melamines, ureas, diallyl phthalates (DAPS), unsaturated polyesters, silicones, epoxies, alkyds, etc. Unlike thermoplastics, thermo-set plastics experience an irreversible and exothermic chemical reaction during moulding called polymerisation, brought about by heat (approximately 160°C) and pressure (between 1800 to 4500 lb/in^2). Following polymerisation, the rigid thermo-set moulded parts retain their physical, chemical, and electrical properties through a temperature range from 70°C to 160°C. The thermo-set moulding is of low cost, temperature, solvent and chemical resistant having good electrical insulation properties.

3.2 COMPRESSION MOULDING PROCESSES

Compression moulding is the most commonly used method by which thermo-set plastics are moulded. Thermo-set plastic materials are one type of polymeric resin that are capable of forming chemical cross linking of the molecules which is called *curing of the materials.* By this cross linking of the molecules, the structure gets the shape of wire frame model which cannot be further melted.

In this method, the plastic, in the form of powder, pellet, or disc, is dried by preheating up to a temperature of 70°C–80°C to remove the moisture from the material and the preheated thermo-set material is loaded directly into the mould cavity. Further, the temperature of the mould cavity is held at 150°C–200°C, depending on the material. The mould is then partially closed, and the plastic, which is liquefied by the heat and the exerted pressure, flows into the recess of the mould. At this stage, the mould is fully closed, and the flow and cure of the plastic are complete. Finally, the mould is opened, and the completely cured moulded part is ejected as shown in Fig. 3.1.

Compression-moulding equipment consists of a matched mould, a means of heating the plastic and the mould, and ejector mechanism for exerting force on the mould halves. For severe moulding conditions, moulds are usually made of various grades of tool steel, cavity and core are polished to improve material flow and overall part quality.

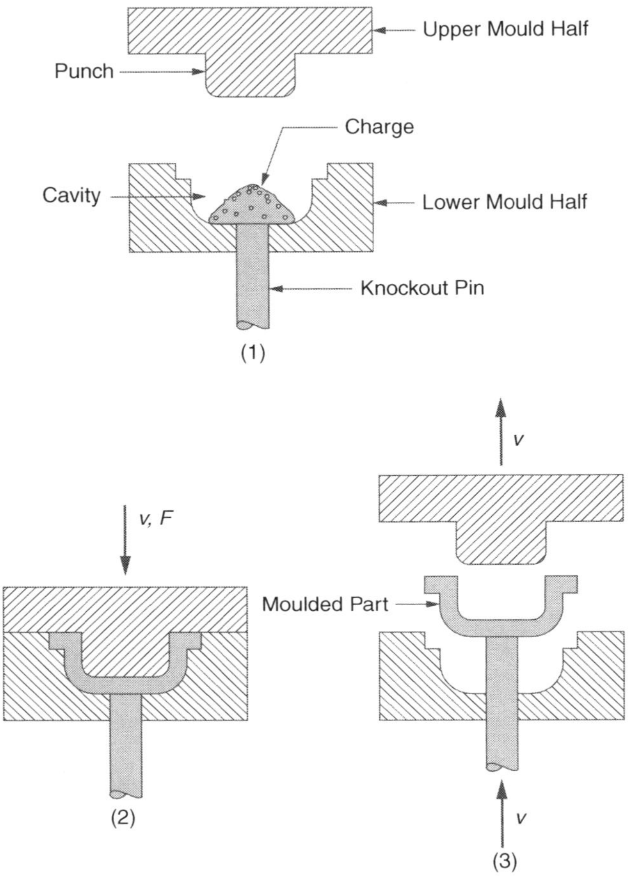

Fig. 3.1 Compression moulding process

In compression moulding, a pressure of 158 kg/cm^2 to 211 kg/cm^2 is suitable for phenolic materials. The lower pressure is adequate only for easy-flow materials and a simple uncomplicated shallow moulded shape. For a medium-flow material and where there are average-sized recesses, cores, shapes, and pins in the moulding cavity, a pressure of 3,000 psi (211 kg/cm^2) or above is required. For moulding urea and melamine materials, pressures of approximately one and one-half times needed for phenolic material are necessary.

3.3 THERMO-SET PLASTICS MATERIALS AND PRODUCT APPLICATION

Thermo-set materials are chemical compounds made by processing a mixture of heat reactive resin with fillers, pigments, dyestuffs, lubricants, etc. These materials or moulding compounds are in powder, granulated or nodular form, having bulk factors ranging from 1.2 to 10. Some are used in the form of rope, putty or slabs.

Phenol formaldehyde (PF) is a common resin and catalyst combination, generally mineral like mica; the moulded part will have good electrical properties. If the filler is glass fibers of one quarter inch long, the moulded part will have good impact strength. Small hollow glass micro balloons have been used as fillers to make low density parts. A very wide range of properties can be obtained from PF because of their compatibility with a variety of reinforcements and fillers, so the properties are dependent on the fillers used. Wood flour filled PF is regarded as general purpose thermo-set material.

Applications of thermo-set materials: Melamine formaldehyde (often used in plastic dinnerware), urea formaldehyde (common in white or pastel heat resistant handles for kitchen ware, or outlet sockets for household use), alkyds and polyesters (often used in high voltage insulators in TV sets, or for arc resistance and insulation in circuit breakers and switch gear), dialkyl epoxy (housings for electronic components), and silicone (high temperature requirements to 600°F). The common fillers used in thermo-set moulding are silica, glass, wood flour, natural or synthetic fibers and combination of these.

Fig. 3.2 Compression-moulded products: electric plugs, sockets, and housings

Thermo-set mouldings are glossy, opaque, dark coloured and inherently flame retardant. They are stiff, hard having low elongation properties and possess good creep resistance.

3.4 TYPES OF COMPRESSION MOULDS

Compression moulds are made out of high carbon and high chromium steel, hot-die steel; nickel alloy steels are case-hardened in order to withstand the high temperature and pressures of compression moulding. The core and cavity should be heat treated to the required hardness for maintaining the dimensional accuracy of the product and the life of the mould.

Types of Compression Moulds

- Hand Compression Mould
- Semi–Automatic Moulds
  - Semi-automatic open flash mould
  - Semi-automatic fully positive and landed positive mould
  - Semi-automatic semipositive moulds
  - Semipositive, Vertical flash type mould
  - Semipositive, Horizontal flash type

3.4.1 Hand Compression Mould

Hand compression moulds are used for smaller production runs or prototypes, experimental jobs that require lower mould costs and parts having open tolerances and less intricacy.

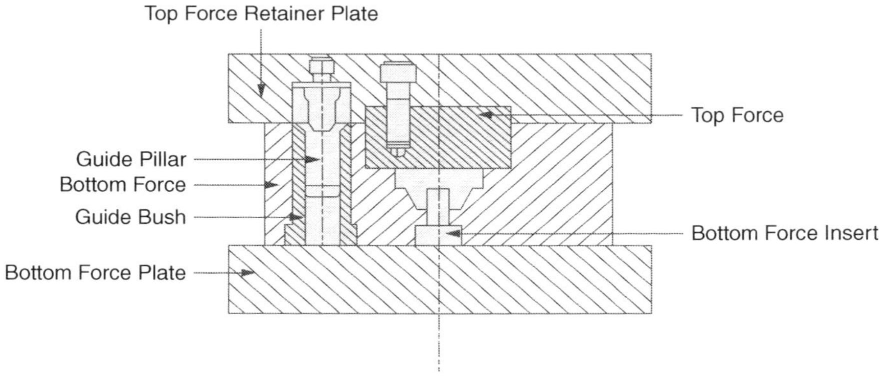

Fig. 3.3 Hand compression mould

These moulds are used advantageously for complex parts incorporating number of loose pull pins and wedges. Hand moulds weigh less than 15 kg for easy manual handling and the operations are fully manual. Hence, the hand moulds are slow in operation which requires longer cycle time and is labour intensive, adding to production cost as compared to other type of moulds. Moreover, the moulds are more easily damaged by misalignment, mishandling, etc., which may result from improper mould operation and closing of the mould. Flash must be removed in the land areas and additional pressure pads should be provided outside the cavity to have better mould life. The hand moulds are less costly and recommended for simple components only as shown in Fig. 3.3.

The overall height of the moulding and density is controlled by land areas on mating surfaces and the clearance between the top force and cavity. Generally, high bulk materials are used for processing large deep drawn parts of maximum density.

3.4.2 Semi–Automatic Moulds

Semi-automatic moulds are fastened in the compression moulding press for the duration of the run. In the process of loading the thermo-set material into a compression mould, the charges of the material to be moulded are some what smaller in length and width but are thicker than the final part and are loaded near the geometric centre of the mould cavity. The closing of the mould spreads the preform / powder to fill the cavity and the component is produced. The semi-automatic moulds are used for mass production of jobs and complex components can be processed. They are classified into the following types:

1. Semi-automatic open flash mould
2. Semi-automatic fully positive and landed positive mould
3. Semi-automatic semipositive mould

Semi-automatic open flash mould The flash-type compression mould is used to produce shallow shaped components and in this mould a slight excess of moulding powder is loaded into the mould cavity as shown in Fig. 3.4. On closing the top and bottom platens, the excess material is forced out and flash is formed. The flash blocks the plastic remaining in the cavity causes the mould plunger to exert pressure on it. Gas or air can be trapped by closing the mould too quickly, and finely powdered material can be splashed out of the mould.

Since the only pressure on the material remaining in the flash mould when it is closed results from the high viscosity of the melt which did not allow it to escape, only resins having high melt viscosities can be moulded by this process. Because of lower pressure exerted on the plastic in the flash moulds, the moulded products are usually less dense than when made using other moulds. Moreover, because of the excess material loading needed, the process is somewhat wasteful as far as raw materials are concerned. However, the process has the advantage that the moulds are cheap, and very slight labour costs are necessary in weighing out the powder. However, the moulds are suitable for building up into tools containing multi-impressions.

Semi-automatic fully positive and landed positive mould The positive type of compression mould in which the plunger telescopes within the cavity, compressing the thermo-set material and the moulding is formed as shown in Fig. 3.5(a) and 3.5(b). There is very little clearance between the plunger and the cavity wall. In the positive mould, almost all the pressure is exerted on the material and a very little material is allowed to escape as flash, the clearance between the plunger and cavity varies between 0.035 mm to 0.135 mm per side, depending on the size of the mould and the material to be moulded. The flash is formed vertically on the type of mould shown. The disadvantage of the positive type of mould is that after frequent operation the cavity walls become scored and ejection of piece parts is difficult. Flash is formed on every piece part moulded by the compression method. The thickness and position of this flash depends on the design of the mould, type of material being moulded, and accuracy of the mould. Flash is removed by filling, sanding, and tumbling. The positive mould is used primarily with material containing coarse fillers and the amount of material placed into the mould cavity must be measured accurately as there is very limited means for the excess material to escape.

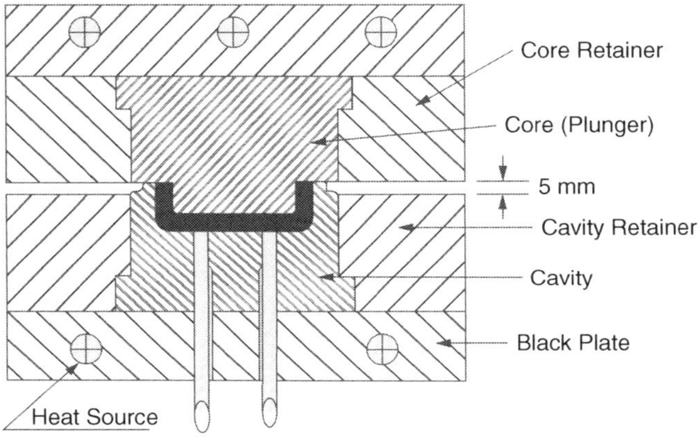

Fig. 3.4 Semi-automatic open flash mould

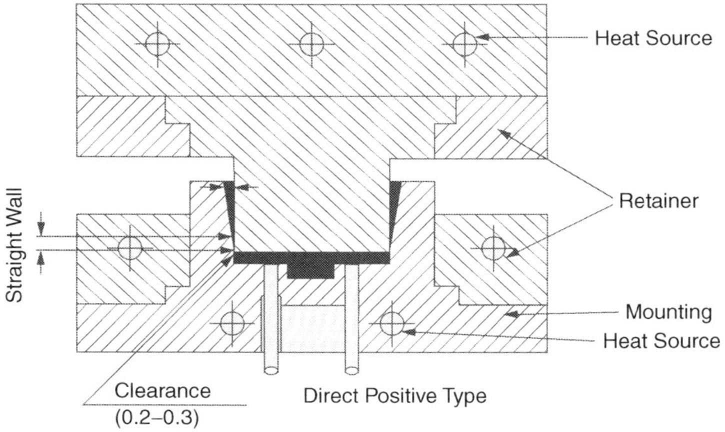

Fig. 3.5 (a) Semi-automatic fully positive

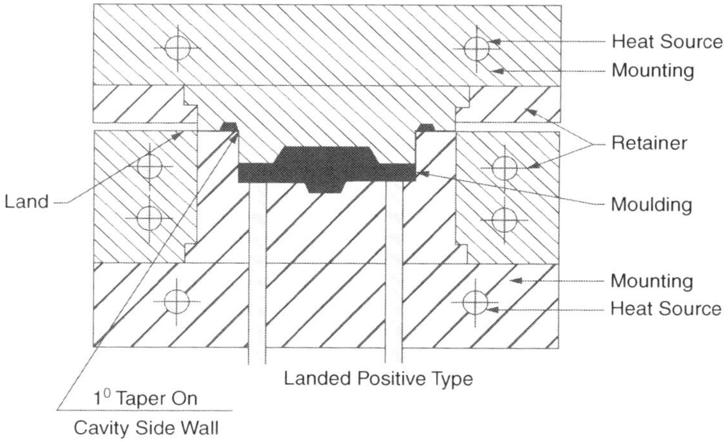

Fig. 3.5 (b) Semi-automatic landed positive mould

Semi-automatic semi-positive moulds The semipositive mould is a combination of the features of open flash and fully positive moulds and allows for excess powder and flash. It is also possible to get both horizontal and vertical flash. Semipositive moulds are more expensive to manufacture and maintain than the other type compression moulds. Semipositive moulds are suitable for producing components to close tolerance and better surface finish. A clearance of 0.025 mm per side, for a diameter of 25 mm between the plunger (top force) and the cavity is maintained for satisfactory operation of mould. Moreover, the mould is given a 2° taper on each side of the cavity for allowing the flash to flow on and the entrapped gases to escape along with it, thereby producing a clean, blemish-free mould component.

Semipositive, vertical flash-type mould These types of moulds are suitable for moulding components of high-density and critical dimensions as related to cavity and top force. Easy removal of flash on large parts and leaves no flash line scars on the side of the parts.

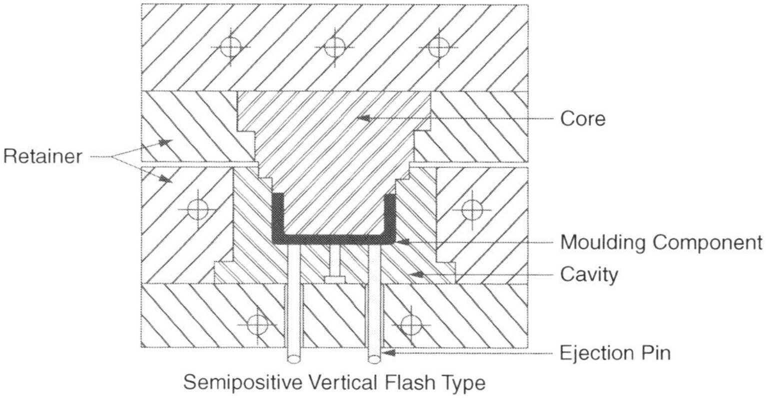

Fig. 3.6 (a) Semipositive vertical flash type

Semipositive, horizontal flash type These types of moulds are recommended for processing of components having close dimensional tolerance and higher accuracy. The component produced out of the mould assures minimum flash finish.

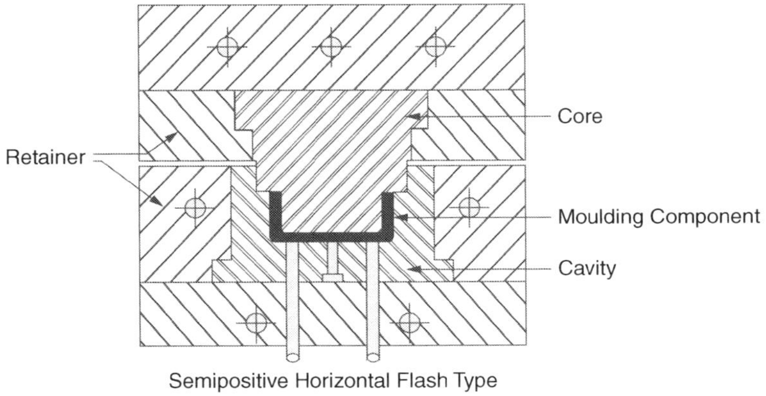

Fig. 3.6(b) Semipositive horizontal flash type

3.4.3 Subcavity Mould

Compression moulds can be further classified by the number of piece parts produced in each cycle. Single cavity moulds producing one piece are generally used in moulding large cavities. Multiple cavity moulds come in several types. Identical cavity moulds contain a separate cavity, top force and loading well.

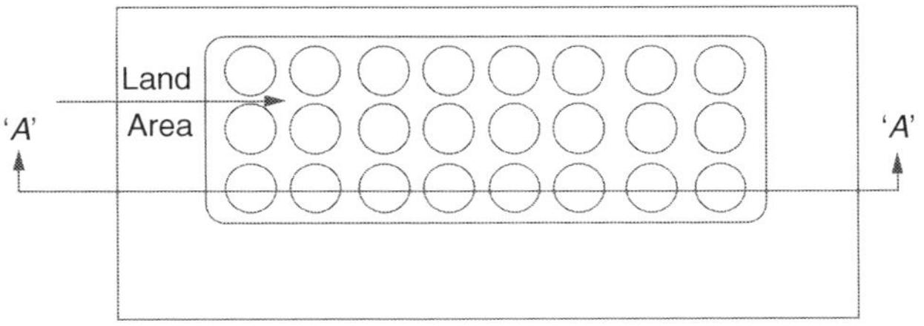

Fig. 3.7 Illustrates the plan view of cavities

For each piece part the number of cavities varies depending on the size of the piece part, locking force and pressure of the press, ease of loading and ejection of the piece parts. Often two or more cavities moulding differently shaped piece parts are mounted in the same mould base, of the plunger of a subcavity mould.

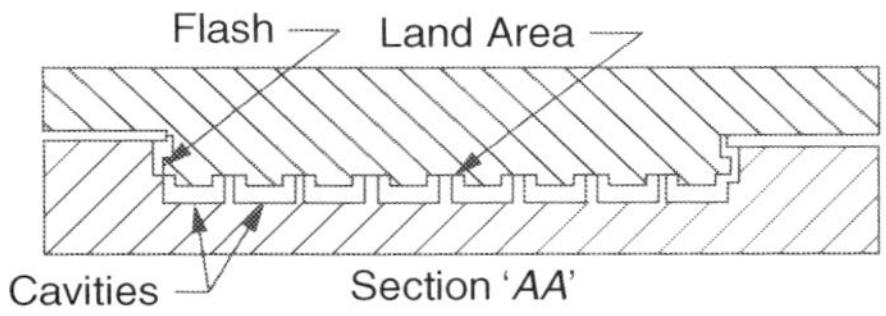

Fig. 3.8 Shows a cross section of a subcavity mould

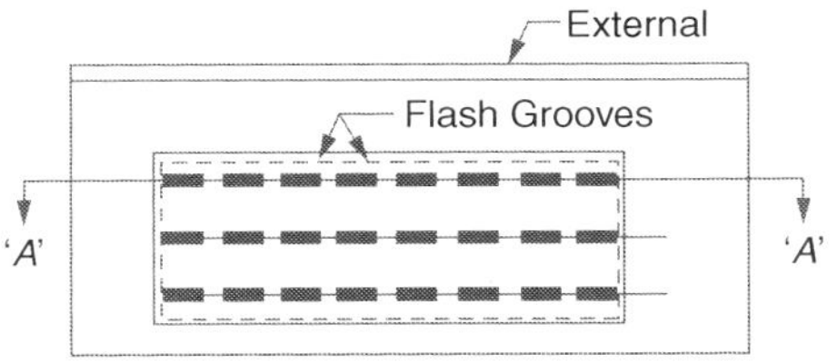

Fig. 3.9 Shows the plan view of the plunger

Another type of multiple cavity mould is called the subcavity mould, or gang mould. Contains common well or loading chamber and common top force and a number of identical parts are moulded. The subcavity mould is used for moulding small parts where loading of individual cavities would be impractical. Measured quantity of powder or material is placed in the common well to produce all the pieces. The top force should not touch the bottom on the land surface of the loading well and an external land provided to keep the top force 0.05 mm to 0.13 mm from the inner land surface. This clearance between the land surface and the top force forms a thin flash between the cavities, which breaks off easily. The land area between the cavities should be kept at a minimum. If the land area between cavities is too large, much of the moulding pressure is wasted on moulding and not enough pressure is exerted on the piece parts, resulting in shorts moulding and parts of low density.

Flash grooves of 0.2 mm deep, 2.5 mm wide are spaced above 12.5 mm apart on the outside edge of the top force to vent the mould and to facilitate the removal of excess material.

It is recommended that the moulding area should be approximately three times greater than the land area. This type of mould is recommended when moulding small and shallow parts using the more free flowing thermosetting materials.

3.5 BULK FACTOR

It is the ratio of the volume of the loose plastic powder to the volume of the moulding. The bulk factor varies for different thermo-set plastic materials.

Let us assume, the weight of the part is w_1

The volume of the same part = w_1/ρ

where

ρ - Density of the plastic material

If, K - Bulk factor of the material, then;

The loose powder volume required for the moulding = $w_1\ K/\rho$

The flash allowance will also be taken into account that will be 10 to 20% extra.

With extra material the loose powder volume = (1.1 to 1.2) $w_1\ K/\rho$

The inclusion of different fillers, affect the bulk factor of the material. The bulk factor for general purpose material and most mineral fillers varies between 2 and 3. In the case of light fabric filler the bulk factor will increase between 8 and 15 and use of a heavy fabric or string filler can increase the bulk factor to between 12 and 24.

The size of the charge, in unmoulded powder form, regulates the size of the 'powder well' or space provided above the cavity impression, in the case of compression moulding.

In general, the dimension of the powder well in a compression mould should be kept to a minimum. The 'land' dimension, i.e., the distance between the actual cavity dimension and the vertical wall of the powder well should be between 4 mm to 9 mm on normal sized mouldings. On large compression mouldings, such as large table model radio cabinets, containers, etc., a land dimension of 9 mm to 16 mm can be used.

Two advantages are obtained by keeping the powder well area as small as possible:

1. Moulding pressure is increased.
2. Less obstruction is put in the way of the escaping excess moulding material (flash)

Keeping in view the mould height dimension and press 'daylight' consideration, the mould powder well should be made deeper rather than have its area increased to take the charge of moulding material. On deep box type or container mouldings it is not necessary to provide a powder well, as the volume of the moulding cavity itself is large enough to contain the charge. But it is a good design practice to include a shallow powder well even in case of bigger mouldings, of say 9 mm deep, as this provides a positive location between the two mould halves.

For the materials having high bulk factor, the pressing of such material often requires an extra high loading chamber and therefore, when pressing bearing bushes or similar parts the punch being not only long but also thin, hence a removal loading chamber extension is fixed to the mould as shown in Fig. 3.10.

The loading chamber has to be designed with a flare of 1/3–1°. Generally the start of the flare should lie only 4–6 mm above the compacted moulding. The upper edges should be rounded off with a radius R=2–5 mm as shown in Fig. 3.11. The different design of loading chamber used in compression moulds are illustrated in Fig. 3.12 and 3.13.

3.6 TYPE OF LOADING CHAMBER DESIGN

Plastic powder fed to the mould need more space in a loose state than in the pressed moulding. The mould cavity design must allow for sufficient space for the loose powder.

Calculation of loading chamber depth

$$D = (V_T - V_C) / A$$

where

D – Depth of loading space from top of cavity to pinch-off land

V_C –Volume of actual cavity space (cm^3)

V_T –Total volume of loose powder (cm^3)

A -Projected area of the loading chamber (cm^2)

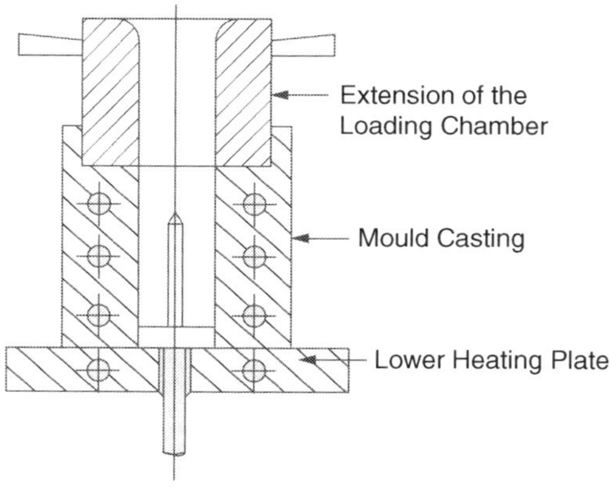

Fig. 3.10 Extension of the loading chamber

where V-Total volume of part including flash factor 10 to 20%. In this way, the depth of loading chamber can be determined.

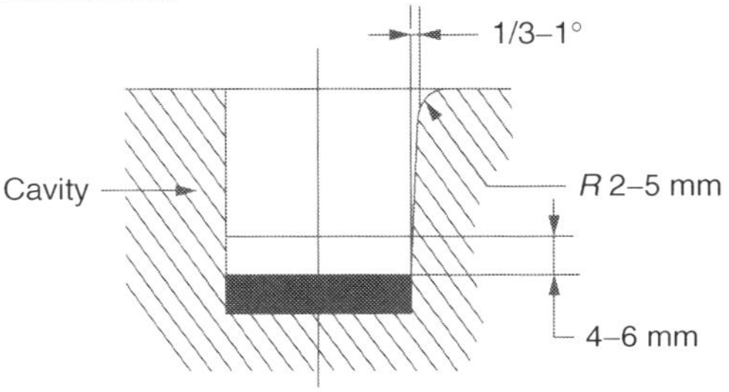

Fig. 3.11 Loading chamber design

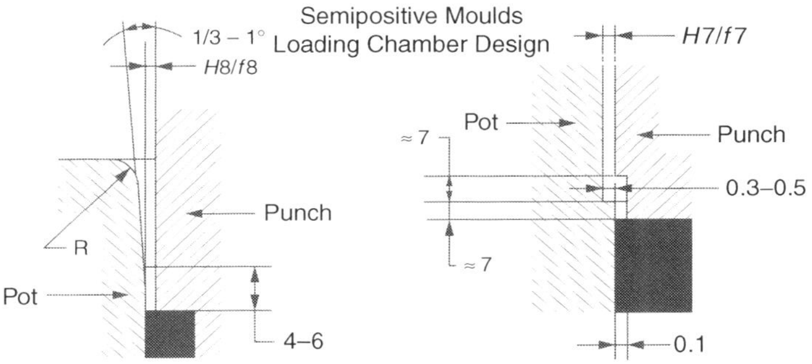

Fig. 3.12 Vertical flash type

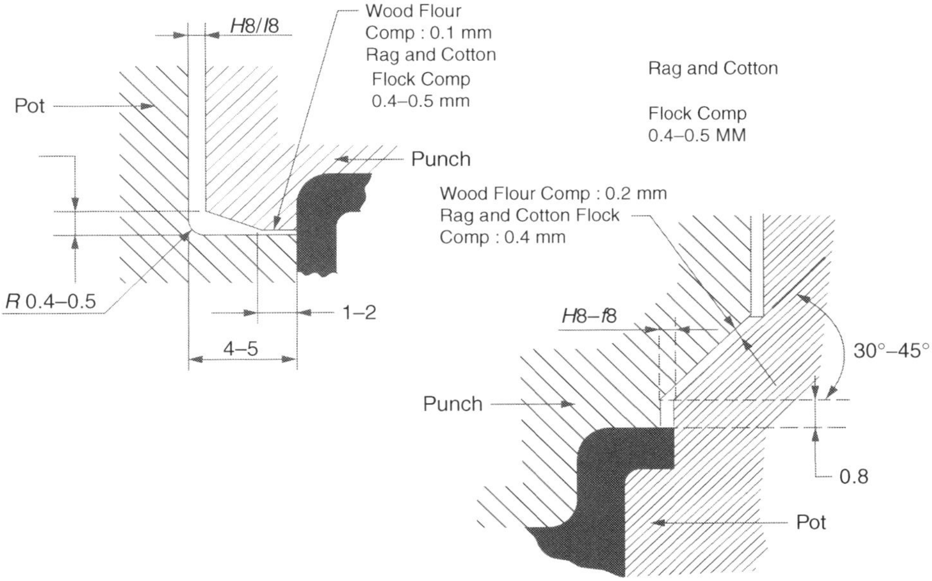

Fig. 3.13 Inclined flash type (flash may be easily removed by grinding)

3.7 FLASH THICKNESS

The excess material flow out from the cavity and form a thin layer of plastic film called flash over land portion of the mould before solidifying the material as shown in Fig. 3.14.

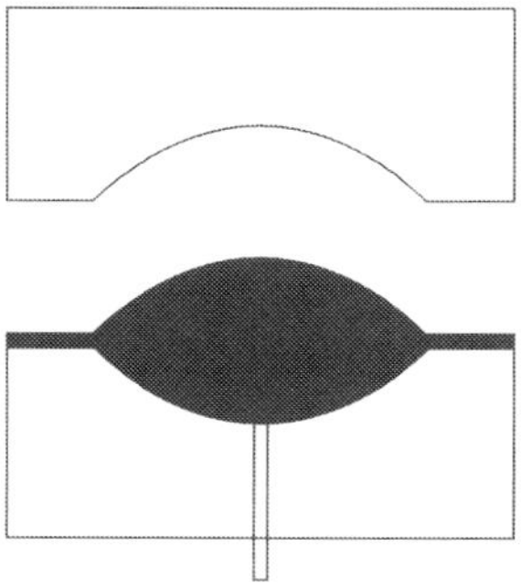

Fig. 3.14 Line diagram of flash on mould

The fitting arrangement of punch and loading chamber of semipositive mould and the space between punch and loading chamber wall through which the plastic material comes out in the form of thin layer as flash as shown in Fig. 3.15.

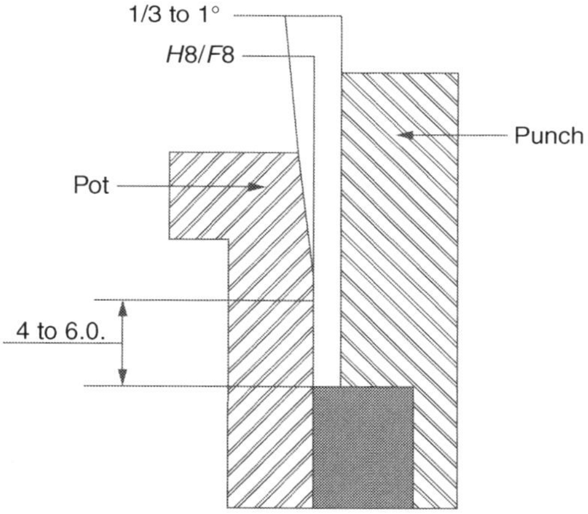

Fig. 3.15 Location of punch and cavity for semipositive mould

3.7.1 Flash Thickness Allowance

Allowances for flash thickness in compression moulds, using thermosetting compounds are:

Rag-filled high impact compound	-	0.25 mm
Cotton – flock compounds in large moulds	-	0.2 mm
Wood –flour compounds in small moulds	-	0.1 mm

All other moulds and for all other compounds allow 0.13 mm (as previously noted).

Because of the flash thickness that we are considering in the mould design the depth of cavity become:

Depth of cavity = Minimum dimension of moulding + Shrinkage of compound

The flash thickness adds to the total thickness of the part and this thickness must be subtracted from the basic cavity depth in order that the finished piece may have the desired thickness.

3.8 CLAMPING FORCE

Clamping force is the force which is required to hold both the mould half in closed position press against the compression pressure.

$$\text{Clamping force (kgf)} = \text{Projected area of the moulding (cm}^2) \times \text{Compression pressure of the plastic material (kgf/cm}^2)$$

For vertical flash or positive type mould where there is no horizontal land, the pressure will be acted fully on the plastic materials but for horizontal mould, the horizontal land is provided for horizontal flash. So during calculation for number of cavities this horizontal flash should be added in the projected area.

3.8.1 Projected Area

The projected area is the total area of the moulding, when viewed in the direction of the moulding in the plane normal to the press opening. View (A) in Fig. 3.16 illustrates a side evaluation of a component and the view (B) a plan view of the same component. The moulding has a central cut-out opening of dimensions ($w \times h$). Thus the area of the cavity which sets up pressure opposing the clamp is the projected area ($W \times H$) less the area of the opening ($w \times h$). The total projected area is shown shaded in the view (C), and this is equal to $(W \times H) - (w \times h)$.

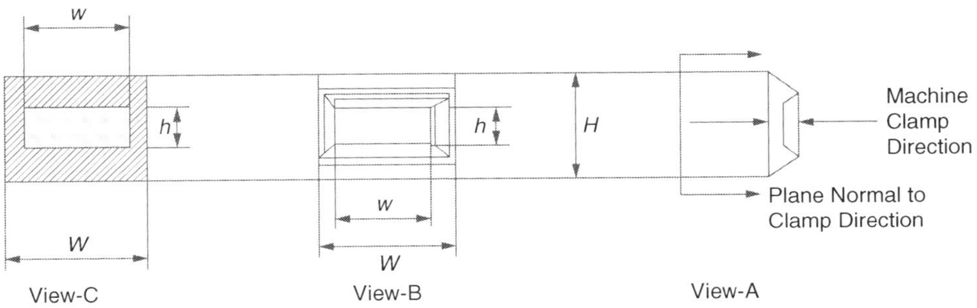

Fig. 3.16 Projected area

3.9 DETERMINATION OF THE NUMBER OF CAVITIES

Number of cavities can be determined in two ways. They are:

1. Technological determination
2. Economical determination

3.9.1 Technological Determination

During calculation of the number of cavities or impressions by technological method for multicavity moulds, the following parameters should be considered:

1. Compression moulding press (machine) – available machine clamping force, and size of the platen to be considered.
2. Moulding material – The compression pressure of the thermo-set material and the component projected area should be taken into consideration during calculation.
   The calculation is as follows:

   Clamping force (kgf) = Projected area of the moulding (cm^2) x
   Compression pressure of the plastic material.

3. The projected area can vary depending on the size of the component as well as on the design of cavity or loading chamber and the compression pressure also can vary based on the type of plastic materials.

   For example, in a vertical flash or positive type of mould, there is no need of horizontal land. But in the case of horizontal flash type of mould the flash width should be taken in to account for the determination of the projected area.

   So the projected area in the case of vertical flash-type mould is same as the projected area of the component. But, in the case of horizontal flash type, 20% of the projected area of the component should be taken into account for the flash.

   Therefore, the actual projected area

   = 1.2 × Projected Area of the component

4. The compression pressure must be regulated in order to produce satisfactory parts economically. Pressure needed to mould a particular article depends on the flow characteristics of the material, the cavity depth and the projected area of the piece part. Generally, it is recommended that minimum moulding pressure of 240 kg/cm^2 of projected area be used. However, in practise, about 300 kg/cm^2 of projected area is used to compensate for any variables that may be encountered.
5. After finding the clamping force required for one impression, the number of impressions can be determined from the actual clamping force available for a particular machine:

   No. of impression = Clamping force available on the machine / Clamping force required for an impression

3.9.2 Economical Determination

Let us assume that

1, 00,000 pcs = n = The total number of parts are required by the party.

x = Number of mould cavities to be fabricated

r = 15% of wages/hr = Overhead cost

w = ₹ 25 per hour = Wages of the workers

A = Production and maintenance cost of one cavity = ₹ 10,000/-

B = Production cost of mould casing

T = Cycle time = 60 seconds

S = Total manufacturing cost of the parts

By the above we may conclude that

No. of cycle required = n/x

Total time required= $n\,T/x$ seconds = $nT/60x$ minutes

Total worker cost with overhead = $n\,T\,w(1+r)/x$

So,

$S = \{nT$ The temperature difference between mould plates and other plates from platen side or platen

$$w(1+r)/60x\} + Ax + B$$

This equation derives the total cost of the part. By differential calculation, we try to get the very economical number of cavities by assuming minimum manufacturing cost. For getting this, differentiate the equation with respect to x.

$$d/dx(S) = d/dx[\{nTw(1+r)/60x\} + Ax + B]$$
$$= d/dx[\{nTw(1+r)/60x\} + d/dx\,(Ax) + d/dx\,(B)]$$
$$= d/dx\{nTw(1+r)/60x\} + A$$
$$= d/dx\{nTw(1+r)x^{-1}/60\} + A$$
$$= -nTw(1+r)x^{-2}/60 + A = 0$$

$$A = nTw(1+r)x^{-2}/60$$
$$x^2 = nTw(1+r)/60A$$
$$x = \sqrt{nTw(1+r)/60A} \text{ by putting the values}$$
$$x = \sqrt{\{1{,}00{,}000 \times 60 \times 25(1+0.15)/60 \times 10{,}000\}}$$
$$x = 16.9 = 16 \text{ number of cavities}$$

For the above data, 16 is the economical number of cavity. Now we have compared the number with the technological number of cavity. Whichever is less, we have to design the mould for that number.

3.10 FACTORS THAT INFLUENCE THERMOSET MOULDING

The following three important factors considered in thermoset moulding are:

1. Temperature
2. Pressure
3. Cure time

3.10.1 Temperature

The thermo-set compounds are generally heated approximately up to a temperature of 190°C for optimum cure and the temperature for moulding of the various materials can be determined by experimentation or the data received from the manufacturer of the particular material.

Higher temperatures may degrade some of the physical properties or electrical characteristics of the materials, may cause the materials to precure before the cavity is completely filled. High temperatures also cause blisters and burn spots on the finished articles.

Low temperature does not allow the material to flow properly and result in incompletely cured piece parts of poor consistency, thus reducing the productivity of the cycle. There is generally an optimum temperature which produces the best flow characteristics for the particular material and cavity.

The mould temperature not only varies with the material used and the type of the mould but also varies with the geometry of the moulded components and type of plastic material whether it is loose powder or preheated forms.

Preheated material generally flows more rapidly during the actual moulding process and the time of complete the cure in the mould cavity is shortened, generally yielding a more economical overall cycle.

3.10.2 Pressure

In compression moulding process, the pressure plays a vital role for ideal filling of the cavity so as to get a defect free component. The thermo-set plastics require greater pressure due to the following reasons:

1. To ensure that the plastic fills all the cavities and has relatively uniform density throughout the moulding. The pressure causes the cavity to fill and resists the tendency of internal gases to form voids or gas pockets and to overcome resistance of the plastic to flow.
2. To ensure better heat transfer to the material, i.e., higher pressure produces a higher density, which generally means faster thermal conductivity.

In compression moulding a pressure ranging from 158 to 211 kg/cm^2 is suitable for processing of shallow moulded shape components, whereas pins in the moulding cavity where the material has to flow in to small intricacies and orifices to produce a good quality hard packed and dense moulding, a pressure of 211 kg/cm^2 or above, is necessary.

For deeper moulding cavity, more pressure is required and the thumb rule is to add approximately 19.3 kg/cm^2 per cm of depth in excess of 2.54 cm of cavity depth (maximum up to 30 cm depth) for the material without preheat and for the material with preheat, the pressure approximately 70 kg/cm^2 or above is required.

For moulding urea and melamine material, pressure of 2 times that needed for phenolic material are necessary, (i.e.), approximately 315 kg/cm^2, again adding 19.3 kg/cm^2 per cm of depth in excess of 2.54 cm is added for material without preheat.

3.10.3 Cure Time

The period required to harden thermosetting material to partial or complete polymerisation is called the *curetime.* Many compounds produce parts that are hard enough, blister free and apparently cured, yet the polymerisation of the resin system is not complete and a post back cycle may be required to optimise properties.

To achieve the minimum cure time, the material must be at the maximum temperature when it is loaded in to the moulds.

Material may be preheated by using infrared lamps, radio frequency preheaters and extrudates formed from screw feed material in a heated barrel.

3.10.4 Material Type

General-purpose materials having wood flour, cotton flock, cellulose, paper and so on offer the greatest moulding latitudes. Mineral and glass-filled materials are more heat sensitive and are more difficult to mould. Adjustments of preheat temperature, plasticity and mould temperature must be studied to obtain the optimum mould cycle.

3.10.5 Cross sectional Area

The cross sectional area or wall thickness of the moulding will determine the cure-time required to produce the part. A cross section up to 2 mm to 4 mm thick will cure in a matter of second, where as increased wall sections may require minutes. Parts having thicknesses or cross sectional areas in excess of 9.5 mm to 13 mm may be difficult to mould by compression moulding, hence, transfer moulding process will be ideal.

3.11 HEATING OF COMPRESSION MOULDS

Thermosetting materials which are used in compression moulding are cured by heat and pressure. Heating of compression moulds is an important phase in the moulding operation.

Heat must soften the material sufficiently to allow it to flow under the influence of the compression pressure into the cavity opening of the mould.

Calculated amount of heat must be applied to bring about the chemical change or polymerize the material into its hard, infusible finished state.

The moulds are generally heated electrically, but steam or hot water is also used. The electrical heating is more effective than steam or hot water heating. The mould temperature can be controlled more easily and over a wide range with electrical heating than with steam or hot water heating.

The maintenance of the switching and control devices in electrical system is easy compared with maintenance of complicated equipments in steam heating.

In electrical system, the heating coil should be placed at equal distance from the moulding surface, as local overheating or under heating can occur more easily than with steam or hot water system.

With all heating methods the proper dimensioning of the heating elements is an essential condition for economic production. Electric resistance heating is besides electric induction heating the most frequently used method. It is easily controllable and clean. Frames, tapes and cartridges are used as heaters.

Heater tapes have diameter from 50 mm to 500 mm and are available with different widths. Their heating power amounts up to 3 watt/cm^2. The cartridge heaters are fitted tightly inside the cylindrical holes provided in the compression mould and the usual cartridge diameters are 17 and 19 mm, but sometimes 12.5, 16, 20, 24, 32, 40 and up to 50 mm diameters are also used depending on the size of the mould.

The cartridge length is generally 100 mm, but in special cases 300 mm, or at most 1200 mm may be used based on the mould specification. The required number of cartridges may be calculated from the current rate.

Actually, by empirical way, the power requirement is calculated on the basis of weight of the mould, i.e., 25 watt/kg steel is taken as standard. Medium sized and bigger moulds are sometimes induction heated. Insulated copper conductors of large cross section are embedded in the mould. Alternating current of 50 Hertz at 20 volt and several hundred ampere cause eddy currents that heat up the mould.

When moulding thermo-set compounds by compression or transfer methods, the mould is maintained at a constant temperature set for optimum polymerization of the material at each cycle. Such temperatures range from 300–400 F (149–204°C). For optimum moulding operations, temperature must be uniform across the surface of the mould and in the cavity areas, ideally to ±2 °F (±1.1°C). Electric heating is presently the most common technique, utilising multiple electric heating cartridges inserted in the top and bottom halves of the mould, positioned to supply heat to all cavity areas. On larger moulds, temperature controllers and sensing elements are often used in several zones. Steam has the advantage of rapid temperature recovery because of its tendency to condense in the steam channels when any lower temperature occurs, rapidly releasing the heat of condensation. It requires excessively high pressures when higher mould temperatures are required.

By contrast, hydraulic oil is also used for mould heating which can easily bring the mould temperatures up to 400 F (204 °C) or higher. Self contained oil heating and cooling systems are used for mould heating, although the cost of fluid-heated moulds is generally higher than the cost of electric cartridge-heated moulds.

3.11.1 Heat Losses in Compression Mould

1. Heat loss by conduction
2. Heat loss by radiation
3. Heat loss by convection

3.11.2 Heat Loss by Conduction

During compression or transfer moulding process heat from the mould will be transferred by conduction from the mould plates to the mould platens thereby causing loss of heat due to conduction. This heat loss can be calculated by using the formula,

$$Q_c1 = (k\ A\ \Delta T)/t$$

where
- K: Thermal conductivity of asbestos or other used insulation
- A: Total area of the mould plates through which heat is transferred towards the platen
- ΔT: The temperature difference between mould plates and other plates from platen side or platen
- t: Total thickness of the asbestos material

3.11.3 Heat Loss by Radiation

Initially compression or transfer moulds are heated in closed position so the only radiation losses are from the vertical faces of the mould plates. However, when the mould opens horizontal faces are also exposed for a certain period of time so this should also be taken into consideration. This heat loss can be calculated by using the formula,

$$Q_r2 = 1.38 \times 10^{-9}\,(T_2 + 460)^4 \times A_2$$

where
- Q_r2: Heat loss by radiation
- **1.38×10^{-9}**: Steffan Boltzman constant for rough finished tool surface.
- A_2: Surface area of exposed tool faces.
- T_2: Temperature of mould in °C or °F

3.11.4 Heat Loss by Convection

Heat in compression will be transferred by convection from the vertical faces of the mould. During heating up of the mould, normally the mould is in closed condition and convention heat transferred from vertical faces only.

$$Q_c3 = (0.7 + \Delta T_3/375)\ \Delta T_3\ A_2$$

where
- Q_c3: Heat lost by convection
- A_2: Area of vertical faces of the mould
- ΔT_3: The temperature difference between mould plates and room temperature

The heat required for heating of the mould

$$Q_r = Q_c1 + Q_r2 + Q_c3 + Q4$$

The heat required for maintaining the mould at operating temperature for curing thermo-set plastic material

$$Q_r = Q_c1 + Q_r2 + Q_c3 + Q5$$

where

Q_r : The heat required for heating of the mould
Q_c1 : Heat loss by conduction
Q_r2 : Heat loss by radiation
Q_c3 : Heat loss by convection
$Q4$: Heat required to raise the temperature of the material to operating temperature
$Q5$: Heat required for heating plastic material

The initial heat required to raise the tool from room temperature to operating temperature, considering the heat losses, can be calculated

$$Q_4 = m_1 \times C_{p1} \times \Delta T_4$$

where

m_1 = Weight of mould (lb or kg)
C_{p1} = Specific heat capacity of mould steel
ΔT_4 = Temperature rise from room temperature to operating temperature (°F or °C).

Heat required to cure the moulding material is given by

$$Q_5 = m_2 \times C_{p2} \times \Delta T_5$$

where

m_2 = Weight of mould (lb/h or kg/h)
C_{p2} = Specific heat capacity of moulding material
ΔT_5 = Temperature rise required from room (or preheat) temperature to moulding temperature (°F or °C)

3.12 MOULD VENTS

During the process of compression moulding, gases are formed as the chemical reaction takes place in the material. The provisions should be made to get rid of the air; volatile gasses from the mould cavity, otherwise poor piece parts will be produced. The air acts as gas pockets in the mould which causes incomplete shots or blistered in the piece part.

One method of getting rid of the gases is to allow the mould to breathe; that is, the mould is closed and then opened again for about 3 mm to get rid of the gases and then closed again.

The other method is to incorporate `openings` located at mould parting surface. The size of the vent depends upon the viscosity of the thermo-set material. Small grooves of size 0.05

mm to 0.25 mm deep and 3 mm to 6 mm wide ground on the periphery of the plunger that telescope into the cavity as shown in Fig.3.17.

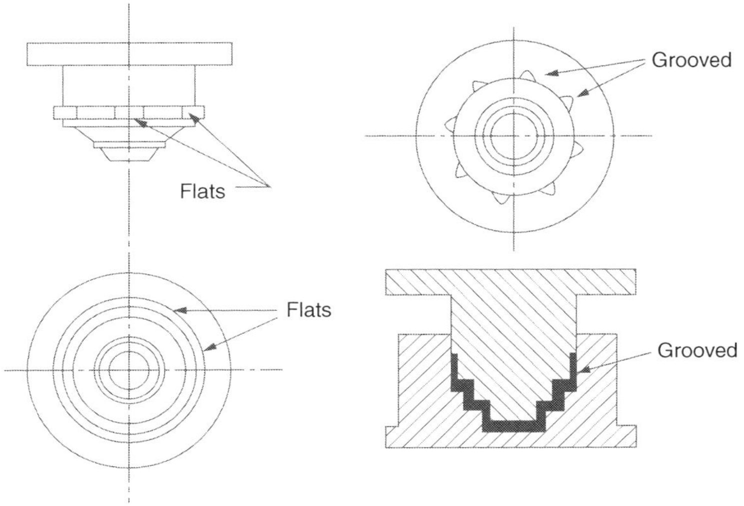

Fig. 3.17 Types of vents

3.12.1 Types of Vents

The recommended vent size of phenolic parts should be 6 mm wide and 0.08 mm–0.09 mm deep and for polyester parts 6 mm wide and 0.05 mm–0.06 mm deep.

The vent should be approximately 25 mm long to allow pressure to build in the cavity after the material in the vent cures.

All the moulding surfaces should be polished and chrome plated including the vents for free flow of material and release of gases from the mould.

Knockout pins often provide a means for venting and requires recessed section, such as "flats" ground on the diameter that will allow venting in a compression mould.

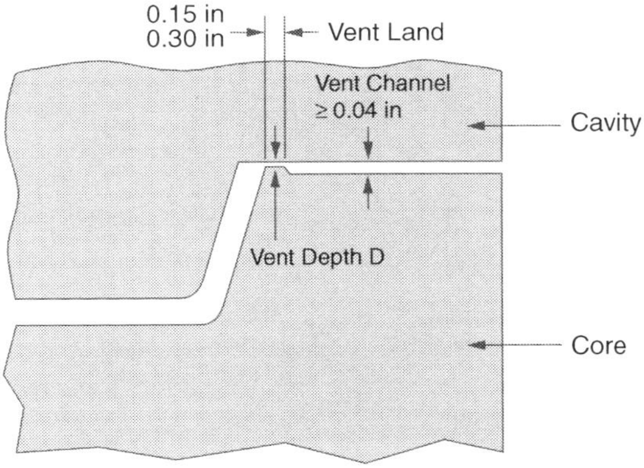

Fig. 3.18 Vent grooves

Vents should be located in the movable half of the mould, wherever a gas void or knit line is seen on a part.

3.13 MOULD CONSTRUCTION

The design and manufacturing of compression mould plays a significant role in dimensional integrity of the final product. The proper choice of material for construction of core and cavity is paramount to quality, performance and longevity. Desirable properties are good machinability of metal parts, material that will accept the desired finish, ability with the mould to transfer the heat rapidly and evenly, capability to sustained production without frequent maintenance.

Mould materials can be classified according to the size and the number of cavities required in a compression mould. Tool steels, high quality nickel alloy steel are recommended for manufacturing of core and cavity for long production run. Generally, the mould materials like P-20 (C-0.3%, Cr-1.65%) prehardened alloys steel used for making mould base plates, large cavity and core, slides and interlocks.

H-13, hot die steel with 5% chromium, higher hardness than P-20 steel, it has good toughness and polishability, used for manufacturing of top punch and cavities.

A-2 cold die steel (C-1%, Cr-5%) used for manufacturing of core, cavity with high hardness and abrasion resistance, long wearing compression characteristics.

D-2 cold die steel (C-1.55%, Cr-11.5%) has highest abrasion resistance and corrosion resistance used for manufacturing of small compression moulds.

High carbon high chromium (HCHC) steel are preferred due to it's high compression strength, good polishability, corrosion resistance, toughness to withstand pressure, uniform thermal conductivity and good machinability characteristics.

414 stainless steel (C-0.3%, Cr-12%, Ni-2% and Cd-1%) used for manufacturing of mould base and large mould back plates.

In a compression mould, the inserted cavities and cores are encouraged due to the reason that in the event of an individual cavity or core being damaged, that particular cavity can be removed from the mould and repaired while the rest of the mould is put back into service. Having individual cavity also allows for insert changes that make it possible to run the multiple versions of the same basic part simultaneously.

The top force and cavity generally case hardened to have tough core as well as a wear resistance surface after heat treatment. The high surface hardness and di-oxidation methods offer the best condition for polishing.

3.13.1 Polishing and Plating

The moulds manufactured in conventional machining process shows cutter marks on the core and cavity surfaces. The non-polished areas will generate frictional heat in the plastic material during its flow and causes degradation of polymer melt. Hence, the mould surfaces need to be mirror polished like core, cavity, vents, ejector pins, guide pillar, guide bush, etc. Core, cavity, vents need to be polished in the direction of material flow and they should have the same degree of polish in all the areas uniformly.

After the mould is completely polished, the core and cavity surfaces should be hard chrome plated to a thickness of 30 to 40 μ which provides the best part release and protects the moulding surfaces. After the mould is plated, it is necessary to repolish the chrome plated areas to avoid sticking of the components.

3.14 LIMITATIONS OF COMPRESSION MOULDING

In case of intricate design articles containing undercuts, side draws and small hole, the compression moulding method is not practicable.

Articles of polyesters require very careful adherence to all rules for draft angle. Ejector pins should be ideally located in a mould to avoid fracture on the component during ejection. Hence, compression moulding of thermo-set materials is not suitable for production of polyester articles having extremely close dimensional tolerance.

Multicavity moulds, particularly in relation to non-uniformity of thickness at the parting line of the moulded articles cannot be processed in a compression mould.

Thermo-set materials having high bulk factor is not suitable for compression moulding. Curing rate of heavy section (or wall thickness) is longer hence not advisable in compression moulding process.

Part design incorporating undercuts or side draws are troublesome because of high maintenance cost.

Questions

1. Write a brief note on compression moulding.
2. List out the different types of compression moulding.
3. Explain hand compression mould with a neat sketch.
4. Explain semi-automatic open flash mould with a neat sketch.
5. Describe semi-automatic fully positive mould with a neat sketch.
6. Explain semi-automatic, semipositive mould with a neat sketch.
7. What is subcavity mould?
8. Explain bulk factor in compression mould.
9. Explain loading chamber in compression mould with a neat sketch.
10. Why is flash thickness allowance required?
11. How does mould vent act in compression mould?
12. How is polishing done in compression mould?
13. Explain the process of plating in compression mould.
14. What are the limitations of compression moulding?
15. Describe landed positive mould with a neat sketch.
16. What is compression mould? What are the different types of compression mould? Explain any one with a neat sketch.

17. What is a semi-automatic mould? What are the different types of semi-automatic mould? Explain any one with a neat sketch.
18. What is the technological determination of the number of cavities in compression mould?
19. What is the economical determination of the number of cavities in compression mould?
20. How do temperature, pressure and cure time play a vital role while moulding thermo-set material in compression mould?

References

1. Bruce A. Davis, Paul J. Gramann, Tim A Oswald and Antoine C. Rios. *Compression Moulding*, Hanser Publishers. Munich.
2. Charles A. Harper, *Handbook of Plastic Processes*, Wiley Interscience Publication, New Jersey.
3. CIPET, *Technical Manual*, Chennai.
4. Dominick V. Rosato, P. E., *Plastics Processing Data Handbook*, Second Edition.
5. Dominik V. Rosato, *Plastics Processing Data Handbook*, Chapman and Hall, London.
6. Harry DuBois and Wayne I. Pribble, *Plastics Mold Engineering Handbook*, Reinhold Publishing Corporation, New York.
7. Manas Chanda, Salil K. Roy, *Plastics Technology Handbook*, CRC Press, London.

CHAPTER 4

Transfer Mould Design

4.1 INTRODUCTION

Compression moulds are not suitable for moulding intricate sections, thick mould walls, long through holes, fragile inserts and components having close tolerance. Due to high compression pressure, there are chances of bending of pins which causes poor quality of product. In case of transfer moulds, the liquid plastics enters and fills the impressions into the two half of the mould through runner and gate at the parting line of the mould, like injection mould. Transfer moulds are used frequently when the moulding sections are delicate and component thickness more than 3.2 mm.

Transfer moulding process Transfer moulding is a process where the predetermined quantity of thermo-set plastic material is inserted inside the cavity before the moulding takes place. The moulding material is preheated and loaded into a chamber known as the 'pot'. A plunger is then used to force the material from the pot through channels (known as a sprue, runner, and gate) system into the mould cavities. The mould remains closed as the material is inserted and curing takes place and finally the mould is opened to release the part. The mould walls are heated to a temperature above the melting point of the moulding material; this allows a faster flow of material through the cavities.

It is a process in which measured quantity of preform thermo-set polymer material is heated in a pot and it is transferred into the heated mould. Due to increase of pressure the polymer melt is forced through runner and gate, leading to homogenisation and causes cross-linking of the polymer and solidification of the part. Generally transfer moulding is used for the production of precise shaped electrical and electronics items.

4.2 TYPES OF TRANSFER MOULDS

Transfer moulds are of semi-automatic type and are made in single and multiple cavities. The types of transfer moulds are detailed below:

Types of Transfer Mould

- Pot-type transfer mould
- Plunger-type transfer mould
  - Top plunger transfer mould
  - Bottom plunger transfer mould

4.3 POT TRANSFER MOULDS

The pot type transfer mould and its function used in conventional transfer moulding press are shown in Figs. 4.1 to 4.4.

4.3.1 Designs of Pot and Plunger

The pot and plunger are made out of good quality wear resistant tool steel which is heat treated and ground and are of square, rectangular or round in shape. The shape is determined based on the shape of the piece part, number of cavities, and available space in the mould base. Round shape pots and plungers are preferred because less machining difficulties are encountered. A clearance of 0.02 to 0.07 mm per side is provided between the pot and plunger to avoid abrasion.

The area of the pot should be 20 to 30% higher than the projected area of all the cavities and runners. The dimensions of the pot either round or square, can be calculated once the cavity area is known.

To determine the volume of the pot, the total volume of all the piece parts, the runner and the sprue, plus a small amount for a 0.35 to 0.70 mm thick cull, is calculated. At least twice this volume is to be used for the pot volume. Knowing the area and the volume, the depth of the pot is derived by dividing the volume by the area. The additional volume in the pot is provided to compensate for the bulk factor of the preforms used and to allow the plunger to enter the pot a short distance before exerting pressure on the material. High bulk factor materials are generally not preferred in transfer moulding. The bulk factor of the preforms used in transfer moulding is approximately 1 to 2. Figure 4.5 illustrates the construction details of the pot and plunger.

For sufficient strength, horizontal distance '*Y*' should be equal to the depth of the pot '*Y*', and 1.5 to 3 mm radius is provided at the top edge of the pot, 1.5 mm radius is machined at the bottom of the pot to facilitate the flow of the material and to simplify the machining of the corner. A 2.5 to 3 mm radius is machined at the bottom of the plunger. The difference in the radii on the plunger and the bottom of the pot results in a clearance so that the plunger will not wedge in the pot but will land on the flat surface of the pot. In assembly, a small clearance between the plunger and the transfer pot is maintained.

Fitting the plunger to the cold pot size-the plunger staying relatively cold during operation but the pot having to be heated-the clearance widened between them allowing material outflow.

Adjusting the plunger to the chamber when expanded by heat prevented its entrance in the cold state. So the plunger is forced into the pot strongly on clamping the mould and there is cracking and other damage occurs. This problem is solved by adjusting the plunger to the cold pot with running fit, but permitting material flow around the plunger so as to form a collar when solidified according to the hot pot size, thus preventing material flow out, during moulding. A sealing groove approximately 2.5 mm wide and 0.8 mm deep is cut in the perimeter or periphery of the plunger.

During the operation of the mould, groove fills with the moulding material and acts as a natural seal, allowing very little material to escape past the bearing surface of the plunger. Flats or grooves are ground on the bearing surface of the plunger for venting purposes. A clearance of 0.75 mm per side is machined above the bearing surface of the plunger. This clearance keeps the bearing surface narrow to prevent galling, and allows flash and excess material to escape. The sprue and the interior of the pot are polished so the material can flow easily. The sprue has a taper of 2° to 3° per side. The large diameter of the sprue varies in size from 9 to 12 mm with a 1.5 to 3 mm radius at the entrance of the sprue. The small diameter at the runner or piece part varies from 3 mm to 6 mm depending on the size of the piece part. Wedge-shaped slots called cull pick up are machined in the plunger. The thick or heavy section of the cull pick up is located directly above the sprue as shown in Fig. 4.5. The length of the cull pick up varies from 3 to 6 times the diameter of the sprue, depending on the location of the sprue and the diameter of the pot. The width is generally 2 to 3 times the diameter of the sprue.

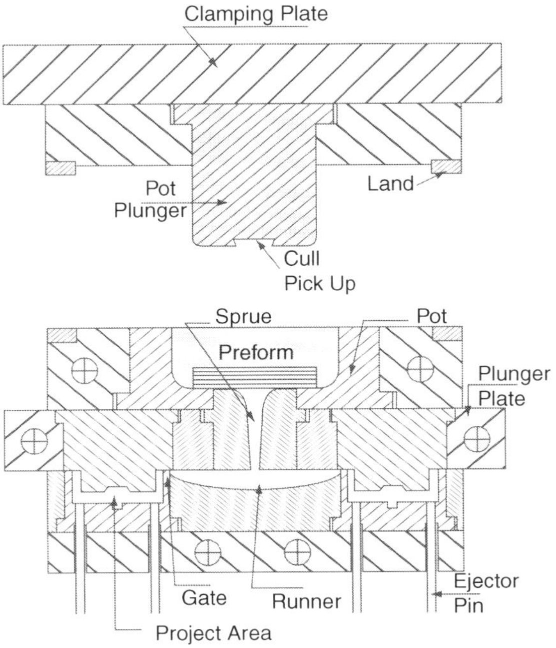

Fig. 4.1 Pot transfer mould (Preform loaded position).

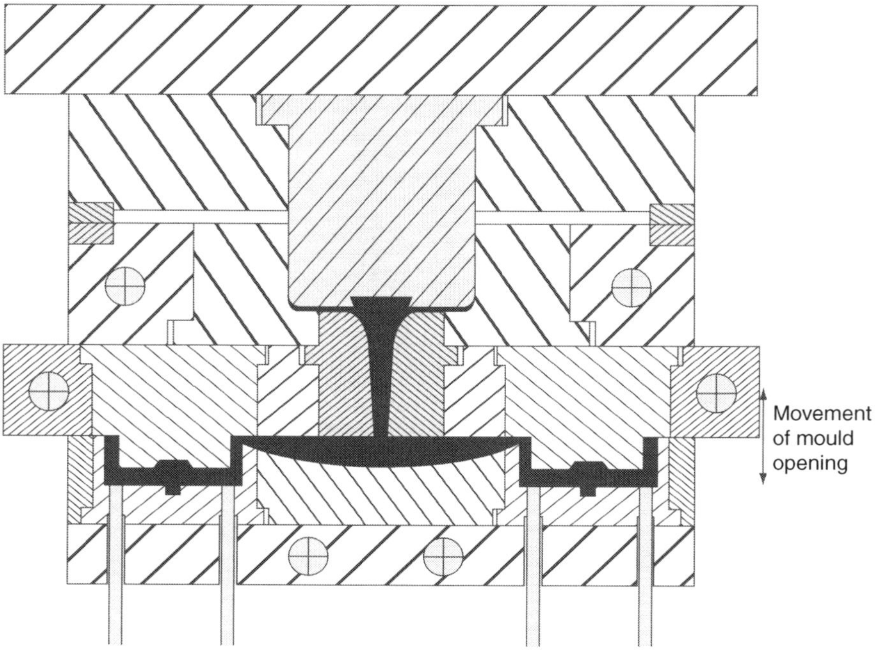

Fig. 4.2 Pot transfer mould fully closed position.

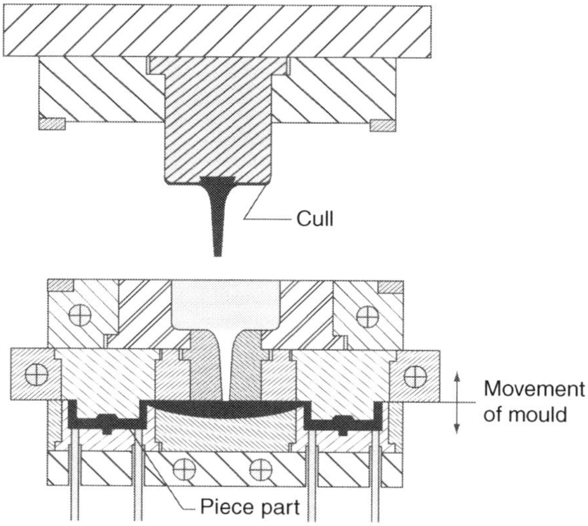

Fig. 4.3 Pot transfer mould with plunger open position.

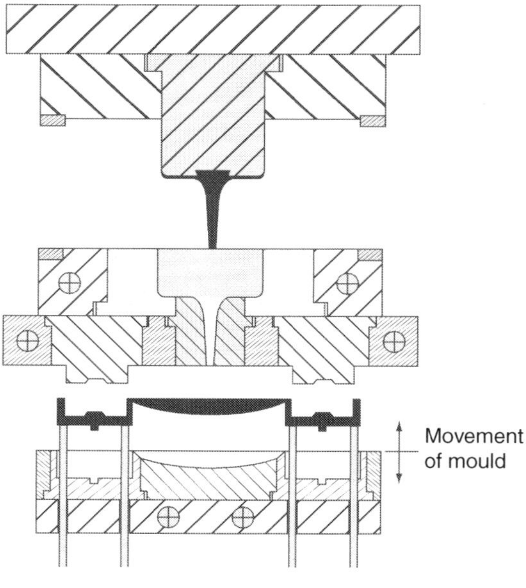

Fig. 4.4 Pot transfer mould fully open position.

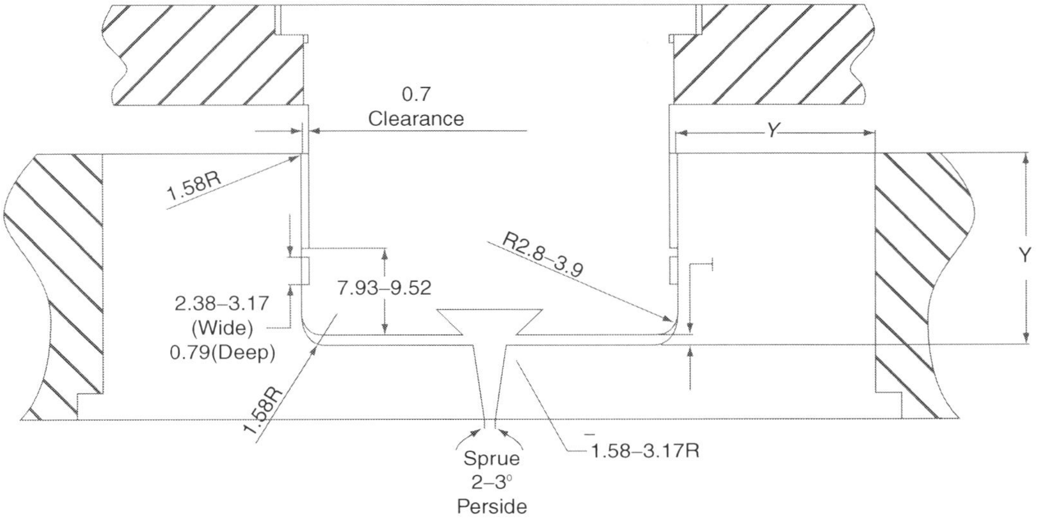

Fig. 4.5 Construction details of pot and plunger.

4.4 PLUNGER TRANSFER MOULD

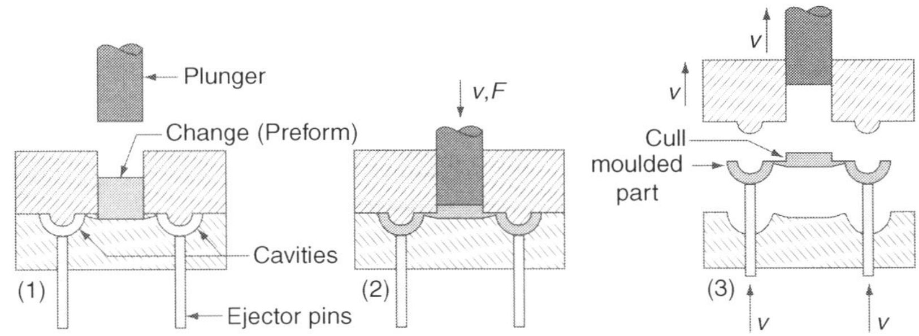

Fig. 4.6 Plunger transfer moulding:
(1) Charge is loaded into pot (2) Softened polymer is pressed into mould cavity and cured (3) Part is ejected.

Plunger type of transfer mould shown in closed position with the preforms in the target area and the plunger on its downward stroke. The combination of the heat of the mould and the pressure of the plunger on the preforms causes the material to become fluid and to flow through the runners and gates into the cavities. The plunger transfer differs from the pot transfer in that the plunger is part of the moulding press and not a part of the mould itself. By the use of this plunger, the sprue is eliminated and very thin cull of small area is formed above the target area, thus reducing the loss of material. The top clamp plate is fastened to the stationary platen of the press. The arrow at the right is shown in Fig. 4.6 (3) indicates the movement of the mould base at the parting line.

The preheated preforms are placed or stacked in the target area as shown in Fig. 4.6. As the mould is closed, the preforms are lifted into the transfer sleeve. Once the mould is completely closed, the plunger is activated for its downward stroke.

The mould is closed, and the preforms are loaded into the closed mould through the opening at the top of the transfer sleeve. The plunger is then activated, forcing the material throughout the mould.

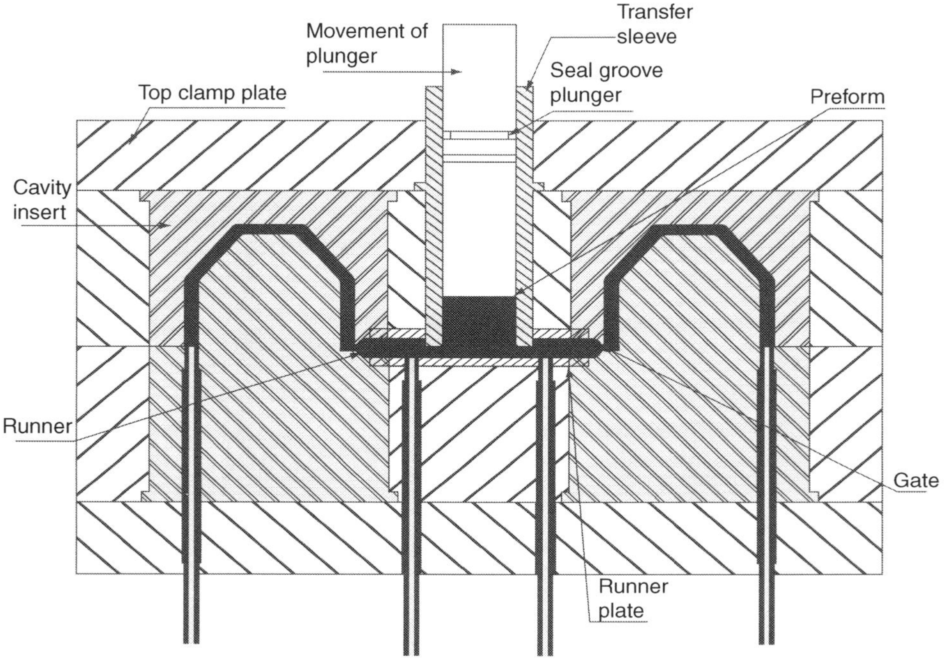

Fig. 4.7 Plunger-type transfer mould.

A two cavity plunger type transfer mould illustrated in Fig. 4.7 feeding the material into the runners, round type of runner is used. In order to maintain a constant volume flowing into the runner system, the runners are machined at an angle in the runner plate. The type of loading depends on the height and size of the press used and clamping pressure of 700 to 800 kg/cm^2 is used to keep the mould together at the parting line. All transfer moulds are vented to allow air to escape from the cavities.

4.4.1 Top Plunger Transfer Moulding

Top plunger mould is a two plate mould in which the preforms, tablets are placed into the space provided in the mould. The plunger is the part of press and not to the part of mould. The press plunger moves downward and applied the pressure on the charge. Due to heat and pressure in the cylinder, the thermo-set material changes its state from solid to liquid. This liquid is forced into the cavity by runner and gate and the sprue is not required as pot type transfer mould. The material gets cured in the cavity and the press is opened and the part is got ejected by knockout pins as shown in Figs. 4.8 and 4.9.

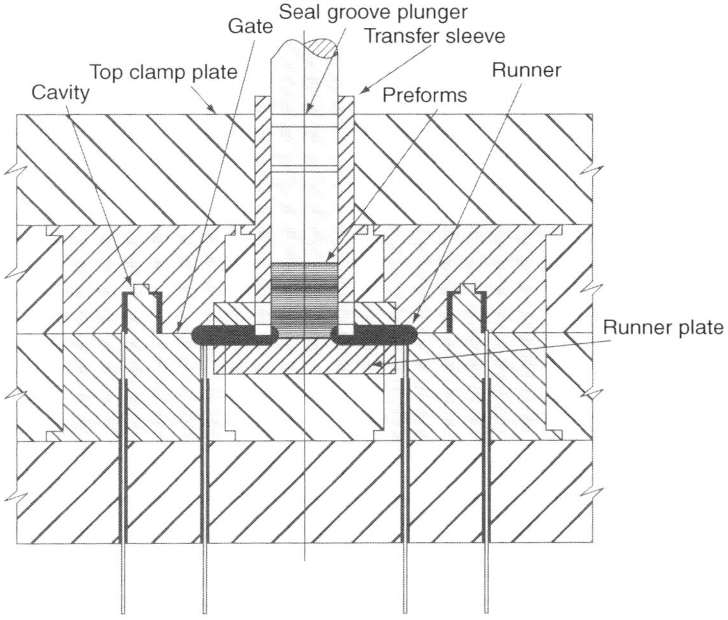

Fig. 4.8 Top plunger transfer mould.

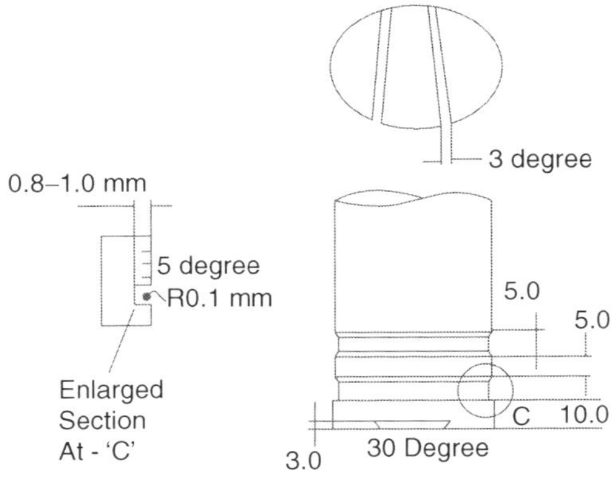

Fig. 4.9 Standard top plunger design.

4.4.2 Bottom Plunger Transfer Moulding

Bottom plunger mould is a two plate mould. The preforms, tablets are placed into the sleeve; cylinder or space provided in the mould and the plunger is the part of press and not to the part of mould. The bottom plunger moves upward and applies the pressure on the charge. Due to the heat and pressure in the cylinder, the charge changes its state from solid to liquid. This liquid forced into the cavity by runner and gate and the material gets cured in the cavity. Then the press is opened and the part is ejected by knockout pins.

The plunger in the bottom half of the mould, the process consists of placing preheated preforms in the transfer sleeve or cylinder, closing the two halves of the mould, and activating the plunger, which forces material out through the passage known as runners, and the gate area into the mould halves. When the cavities are completely filled, the excess material remains as a cull at the face of the plunger. After the material is cured, the press is opened at the parting line, parts are removed and the gate, runner and cull. This moulding process is commonly called the bottom plunger transfer moulding. In this process, all the activity of the process is automatic, and auxiliary devices are used to load the preheated preforms, and unloading trays are utilised to receive and separate the parts, runner gate and plunger culls.

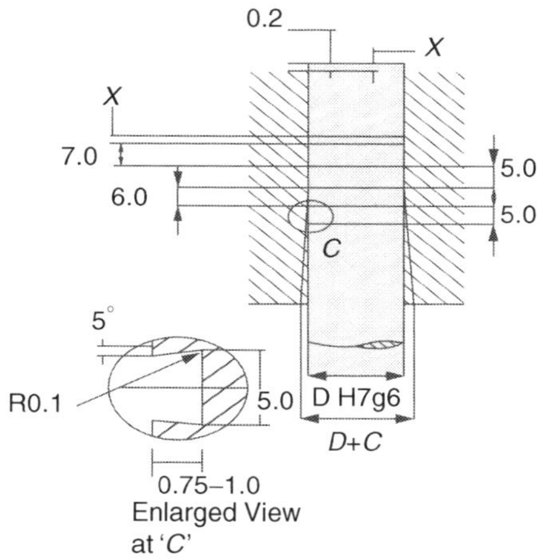

Fig. 4.10 Design of lower plunger.

4.5 TRANSFER - POT CALCULATIONS

The clamping pressure provided by the loading chamber is an important factor in transfer mould. If the total cavity area is greater than the total pot area, the hydraulic pressure exerted by the plastic compound would tend to open the mould at the parting line. To ensure perfect mould locking, the area (A_p) should be 20%–25% greater than the combined area of the moulding surface and the area of all runners and sprue.

The dimensions of the pot, if it is round or square can be calculated once the area is known.

Total area of pot A_p = Total projected area of cavities, runners and sprue + 25–30% of total projected area.

Volume of pot V_p = Total volume of all the piece parts, the runners and the sprue plus approximate volume of a small amount for a 0.5 to 1 mm thick cull multiplied by bulk factor of the compound.

$$\text{Depth of pot} = V_p / A_p$$

Calculation aspects:

Ram load

1. Transfer ram load required (tonf) = Area of transfer plunger (cm^2) × Pressure required to transfer moulding material ($tonf/cm^2$)

The pressure required for transfer moulding for the particular material may be obtained from the material manufacturer's data, e.g., for Phenolic 3–6 $tonf/in^2$ (472–945 kgf/cm^2).

2. Main ram load required to clamp mould (tonf) = Pressure developed by transfer ram (tf/cm^2)

$$\times \left(\begin{array}{l} \text{Projected area of} \\ \text{moulding at flash} \\ \text{face, including} \\ \text{runners } (cm^2) \end{array} + \begin{array}{l} \text{(bottom transfer} \\ \text{Area of transfer pot } (cm^2) \end{array} \right)$$

A safety margin of clamp over transfer pressure of 10–20% is added.

3. To maintain balance where transfer moulds employ an integral pot:

Area of pot (cm^2) = Projected area of mould cavities and runners (in^2 or cm^2)

With adequate gates and runners the projected area of the mould may sometimes exceed that of the pot, although conservative practice adopts the reverse-the area of the pot exceeding that of the mould by amounts varying from 10 to 25%.

Press tonnage required for integral pot moulds is given by

Main ram load (tonf) = Area of transfer pot plunger (cm^2) × (Pressure available to transfer moulding material (tf/cm^2)

4.6 MOULD HEATING

A uniform mould temperature is to be maintained in each half of the mould well within 3°C for all locations, heated by oil or steam. Moulds that are heated with electric cartridge heaters can vary by as much as 6°C. A mould with a uniform temperature will fill easier and produce parts with less warpage, improved dimensional stability and a uniform surface appearance.

The mould heating by steam or oil will have a uniform mould temperature because the heat source maintains a constant temperature. But steam heating is more effective than oil heating. Therefore, when using oil to heat a mould, it is necessary to set the oil temperature higher than the desired mould temperature.

Electrically heated moulds are more difficult to maintain at a uniform temperature because the cartridge heaters are constantly cycling on and off. When they are on, they generate a great deal of heat at the source but this heat must be distributed throughout the mould in a way that produces a uniform mould temperature.

The amount of wattage needed to heat a mould is generally 1¼ kilowatts for every 45 kg of mould steel.

Locating a heater on the centreline of the mould is not recommended, because the centre of the mould is normally hot enough without adding any additional heat. The cartridge heaters are located in the support plates, with a distance of 60 mm between the heaters and the deep draw moulds requires heaters in the retainer plate. The thermocouples are provided to control the temperature of each half of the mould. The thermocouples should be located in the 'A' and 'B' plates, between two heaters and at a distance of 32 mm to 38 mm from the respective cartridge heater. This distance is to be measured from the edge of thermocouple hole to the edge of the cartridge heater hole. The distance from the thermocouple to the heater is important because a heater that is too close will cause the thermocouple to turn off the heat before the mould is at temperature. A heater that is too far away from the thermocouple will result in a mould that overheats and then gets too cool. The thermocouple should be located 35 mm to 50 mm inside the mould, since the temperature taken there, is less susceptible to outside influences and therefore, more stable.

Electrical energy requirements to heat the mould The empirical formula to heat the mould is 20–40 Watt/kg of mould.

The heat required to raise the mould to operating temperature is given by

$$Q_5 = Q_1 + Q_2 + Q_3 + Q_4$$

where

Q_1 = Conduction losses through asbestos insulation from mould to platens (cal/h)

Q_2 = Radiation losses from mould faces (cal/h)

Q_3 = Convection losses from mould faces (cal/h)

Q_4 = Heat required to raise temperature of metal to operating temperature (cal/h)

Q_5 = Heat required for heating plastic material (cal/h)

1.

$$Q1 = KA_1\Delta T_1 / L$$

where k = Thermal conductivity of asbestos insulation (cgs units)

A_1 = Total area of mould top and bottom faces in contact insulation (cm^2)

L = Total thickness of top and bottom asbestos insulation (cm)

ΔT_1 = Temperature difference between mould and press (°C)

2.

$$Q_2 = 1.38 \times 10^{-9} (T_2+460)^4 \times A_2$$

where

$1.38 \times 10^{-9} (T_2+460)^4$ is the modified Stefan's constant for a rough finished mould surface

T_2 = Temperature of mould (°C)

A_2 = Area of exposed mould faces (cm^2).

The radiation losses are from the vertical faces of the mould, however, the horizontal faces area exposed for a proportion of the time and hence an additional allowance must be made when determining the heat requirements during moulding.

3.

The heat lost by convection from the vertical faces is

$$Q_3 = \left(0.7 + \frac{\Delta T_3}{375}\right)\Delta T_3 A_2$$

where ΔT_3 = Temperature difference between mould and surrounding air (°C).

During initial heating, the mould is normally closed and the only convection losses are from the vertical faces considered above. However, the horizontal faces are exposed when the mould is open so that the losses from these must be taken into account when determining the heat requirements during moulding.

The heat lost by convection from horizontal faces lying upwards is

$$Q_3' = \left(0.7 + \frac{\Delta T_3}{375}\right)\Delta T_3 \times 1.1 \times A_3$$

The heat lost by convection from horizontal faces lying downwards is

$$Q_3'' = \left(0.7 + \frac{\Delta T_3}{375}\right)\frac{\Delta T_3}{2} \times A_4$$

where A_3 = Area of mould face lying upwards (cm^2)

and A_4 = Area of mould face lying downwards (cm^2)

4. The initial heat required to raise the mould from room temperature to operating temperature, additional to the heat losses listed, is $Q_4 = m_1 \times C_{p1} \times \Delta T_4$

where m_1 = Weight of mould (kg)

C_{p1} = Specific heat capacity of moulding material

ΔT_4 = Temperature rise from room temperature to operating temperature (°C).

5. Heat required to cure the moulding material is given by $Q_4 = m_2 \times C_{p2} \times \Delta T_4$

where m_2 = Weight of mould (kg/h)

C_{p2} = Specific heat capacity of moulding material

ΔT_5 = Temperature rise from room (or preheat temperature) to moulding temperature (°C).

4.7 SELECTION OF MOULDING METHOD-COMPRESSION OR TRANSFER

Table 4.1 Comparison of compression and transfer moulds.

Factor to consider; advantages-limitations	Compression	Transfer
Close tolerance, projected area		•
Close tolerances, over flash line, minimum flash		•
Lowest mould shrinkage	•	
Uniform shrinkage, all directions	•	
Maximum uniform density	•	
Reduced cure, thick sections		•
No weld lines, less moulded in strains	•	
Small holes, longer length, through holes		•
Extremely thin mould sections, telescoping		•
No venting problems	•	
Impact strength	•	
Moulds with movable sections or cores		•
Moulded in inserts		•
Large projected area parts	•	
Lowest mould-flash scrap	•	
Generally less mould maintenance		•
Gate or sprue removal necessary		•
Maximum number of cavities per clamps force	•	
Mould erosion, sprues, runners, gates		•
Generally higher mould cost		•

4.8 ECONOMIC DETERMINATION OF THE NUMBER OF CAVITIES

n = The total number of parts to be processed

x = Number of mould cavities

t (minute) = Period of pressing + Change of piece using a mould with x cavities

G (₹) = Production cost of the mould casing

A_1(₹) = Production and maintenance cost of a cavity

R = Workshop overheads calculated with the wages in decimal fractions

A (₹/hr) = Wage per hour of a worker

K = a (1+r) (₹/hr) - The hourly wage plus operating costs

P (₹) = Total manufacturing cost of the n pieces to be produced (including mould production costs)

$$P \approx \frac{n}{x}.t.\frac{K}{60} + G + A_1 \qquad \text{............(4.1)}$$

It is evident that

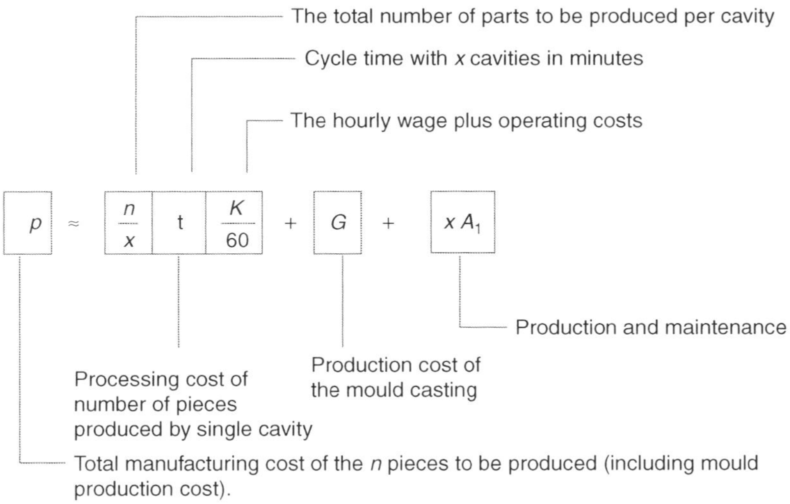

This cost, as a function of the number of cavities has a minimum value when,

$$\frac{dp}{d(x)} = 0$$

$$\frac{dp}{d(x)} = \frac{dp}{d}\left(\frac{nt\,K}{60}x^{-1} + G + X\,A_1\right) = 0$$

where n, t, k, G, and A_1 are constants for a particular case.

$$\therefore \quad \frac{dp}{dx} = \frac{ntK}{60}(-1)x^{-2} + 0 + A_1 = 0$$

$$\frac{dp}{dx} = \frac{-ntK}{60x^2} + A_1 = 0$$

$$\therefore \quad \frac{-ntK}{60x^2} + A_1 = 0$$

$$A_1 = \frac{-ntK}{60x^2} \quad ; \quad x^2 = \frac{-ntK}{60A_1}$$

The most economical number of cavities is

$$x = \sqrt{\frac{-ntK}{60A_1}} \qquad \text{............(4.2)}$$

The results of the formula (4.2) can be considered as an approximate value and the economical number of cavities should be checked technologically. For example, there may not be enough clamping power available in the machine, therefore, a mould with a limited number of cavities is used.

This method of calculating the number of impression is practically impossible during design stage. So the number of impression can be determined only by the technological method.

4.8.1 Moulding Pressure

In transfer moulding, both clamp pressure and transfer pressure are to be considered. In case of pot-type transfer, the transfer force ranges from 35 to 69 N/mm^2 of pot. The clamp pressure must be sufficient to keep the mould from flashing or opening. The pot area must therefore, be equal to or slightly greater than the projected area of the moulded part.

Plunger-transfer operating or transfer forces are in the 41–52 N/mm^2 area of plunger range. The clamping loads must be greater than the load developed by the transfer pressure acting on the area of the plunger, runners, gates and projected area of the parts. Low-pressure materials will require much less transfer pressure.

4.8.2 Transfer Moulding Relates to Pressure, Cure and Quality

Transfer moulds are charged with preheated material in a preform state and the preheat temperatures in the range of 93 to 127°C. These result in lowering transfer pressures and reduces cure, especially where the part design has heavy or thick cross sections. The proper selection of the type of gate, gate size and location helps the part to fill with a minimum knit line. Large flat parts are difficult to mould free of flow lines that appear on the moulded surfaces. Venting of moulding is very important to avoid burning of material caused by trapped air and volatiles. Vents also prevent weld or knit lines around pins or openings near the outside edges of the part.

4.9 DESIGN OF SPRUE, RUNNER AND GATE

The ideal design of sprue, runner and the gate is very important in the success of transfer moulding.

The sprue leads directly to the mould cavity in a single cavity mould. In multicavity moulds material flows through the runner from the sprue to the mould cavities. There is a 'gate' located at the end of the runner before the cavity. The kinetic energy of the material is transformed into heat and the heating of the material is most rapid at the gate.

The shortest possible runner is always an ideal design. A bigger diameter runner is desirable for promoting material flow and a low transfer pressure, but on the other hand, material is deposited in the runners after each injection, reducing it to the minimum possible diameter is advantageous.

Gates, a degate of the moulds serve a double purpose: first the uniform heating of the material, secondly to ensure a degate of the moulded part without any trace. For this, a gate with small diameter is desirable.

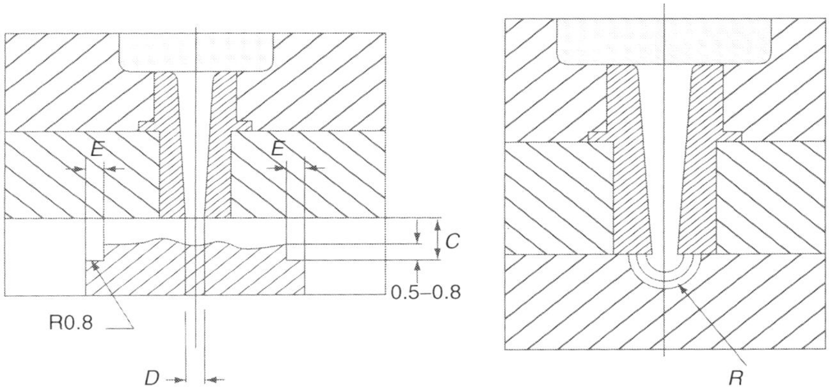

Fig. 4.11 Sprue, runner and gate design.

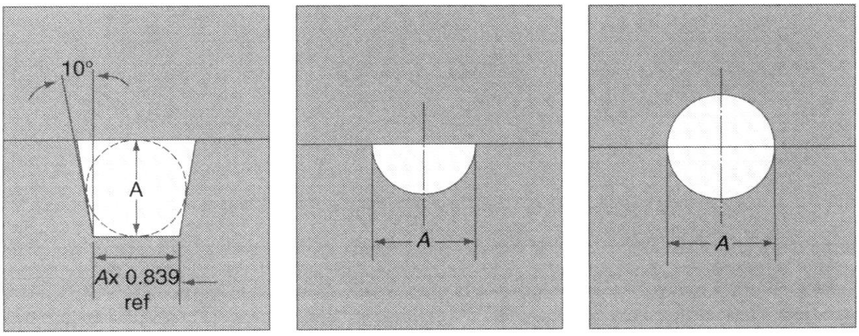

Fig. 4.12 Sprue, runner and gate design – Cross section.

The amount of pressure needed for forcing the thermo-set material to flow are considerably increased by decreasing the runner diameter, thus limiting diameter reduction. Viscosity of plastic material plays an important role to determine the gate and runner flow conditions from the laws and equations of fluid mechanics. The viscosity of the fluidised plastic powder is not only a function of the resin quality and its filler content, but of the instantaneous temperature which fluctuates owing to friction and the state of curing.

The runner and gate dimensions as shown in Figs. 4.11 and 4.12 as recommended are listed in Table 4.2.

Table 4.2 Runner and gate dimensions.

Type of plastic powder	Minimum diameter dmin.	R	E	Cmin	K
Wood flour phenolic	4	1/2(D+5)	2.5	0.8	0.28–0.30
Cotton flock phenolic	5–6	1/2(D+5)	3	1.2	0.32–0.36
Fabric-filled phenolic	6–7	1/2(D+7)	4	1.6	0.37–0.42

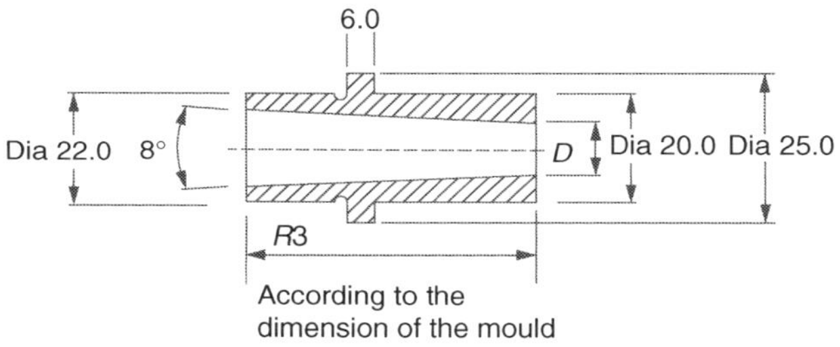

Fig. 4.13 Sprue bush.

The minimum sprue area or with multicavity moulds, the sum of gate areas may be determined from the formula.

$$A = GK$$

where A (mm^2) = The minimum runner area and the sum of gate areas.

G(gm/stroke) = Weight of pieces produced

K = Value to be found in Table 4.2

It is also possible to determine optimum cross-sectional areas of runner and gate, utilising the weights of moulded part and the weight of material to be transferred through runners and gates.

Various gate designs used in transfer moulds are rectangular gate, half round and full round gate. The runner and gate design depends on the following factors:

1. Geometry of the moulded part
2. Location of the gate
3. Moulding pressure required
4. Number of impressions
5. Plasticity and heat sensitivity of the material selected for the application
6. Physical properties required in the moulded part

The half round runner and rectangular gate system require 35% higher injection pressure to fill the cavity in the same time. The half round runner and gate system required 15% higher injection pressures to fill the cavity in the same amount of time. In general, greater the injection pressure, greater the wear on erosion of runners and gates. Therefore, the full round runner and gate should be used whenever possible.

4.9.1 Runner Design

When designing runners for moulds, there are number of possible approaches. These include the standard full round with a centreline. This is the most efficient runner, but in some cases it is necessary for the runner to be in only one half of the mould. Round runners are the most efficient types but are the most expensive because they must be machined into both plates of the mould.

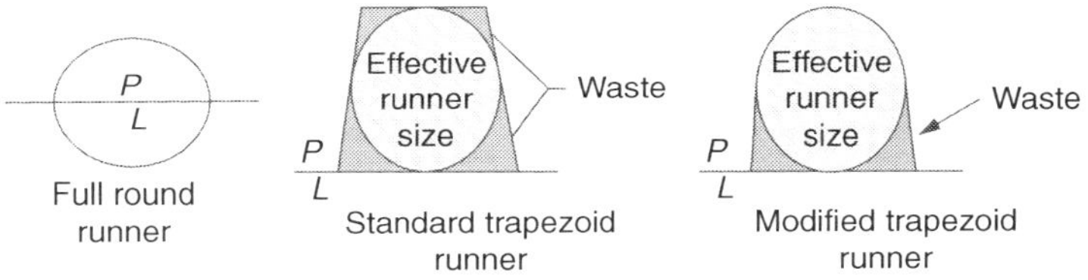

Fig. 4.14 Runner design.

A standard trapezoid runner is used in transfer moulds and it is machined in one half of the mould only. Due to sharp corners the material movement is restricted hence the application of such type of runner is very limited. To reduce the amount of scrap in the runner, a modified trapezoid runner design is suggested. This design is more efficient and avoids the flow restriction due to the radius provided in corner areas in the runner. *Effectiveness of this runner:* These runners are machined into only one mould plate but reduce the pressure and temperature losses observed with half-round runners.

Gates The gates for thermo-set moulds are the high wear areas of the mould. The gate should be made using a replaceable insert so that when the gate becomes badly worn, it can easily be replaced. A gate should be made of good quality tool steel, that do not wear easily. The materials commonly used for gate inserts are OHNS, hot die steel, *D*-2 steel, *EN*-24 steel, etc.

A gate should be large enough to allow the part to fill without using excessive transfer pressures or requiring long transfer times.

4.9.2 Core and Cavity

It is preferred to use the inserted cavities and cores in a transfer mould due to its better interchangeability. The primary reason for this is in the event of an individual cavity or core being damaged, that particular cavity can be removed from the mould and repaired while the rest of the mould is put back into service. Having individual cavities also allows for insert changes that make it possible to run multiple versions of the same basic part simultaneously.

When the parts are very small and there is a large number of cavities, individual cavity inserts might not be feasible. In those cases, the cavity inserts of 3 or 4 cavities can be machined in a single insert. The materials most commonly used for cavity inserts are *H*-13 and *S*-7, *P*-20; prehardened steel, OHNS, *EN*-19 and *EN*-24 steel, hot die steel, etc. The core and cavity steel is hardened to Rockwell 52 to 54 HRC and can be polished to produce an excellent surface finish on the parts.

4.9.3 Guide Pins and Ejector Pins

Ejector pins are used to eject the component from the mould once it is cured. The ejection technique is used both in compression moulding and transfer moulding. The guide pins are sufficiently long to enter the guide bushings before the transfer plunger enters the chamber and when anchored in the cavity plate, it must be long enough to enter the chamber. The guide pins when anchored in the cavity plate must be long enough to enter the guide bushing before the horizontal core pins enter their tapered seat.

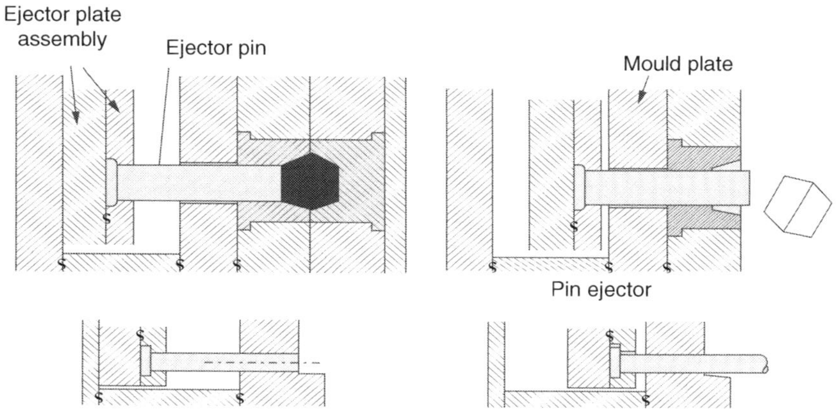

Fig. 4.15 (a) Transverse movement of misaligned ejector pin.

The guide pins are provided in the transfer plunger retainer plate. It provides the alignment between the cavity sections and between the transfer chamber and plunger. Pins of this type must have sufficient length to project far enough through the loading plate to enter the guide bushings in the lower cavity plate before these two plates come together. Guide pins always enter their guide bushings before any of the other mould parts are engaged.

Safety pins are used on transfer moulds in the same manner as on compression moulds. When used in conjunction with side springs, they give positive assurance that the ejector bar and pins will seat properly. When the press is not equipped with side springs below the platen, the safety pin can be mounted directly on the mould.

Ejector pins are relatively short, since the flat type of parting line is used. The pins must have enough length of permit use of a stripping fork when required. A good minimum amount of ejector pin travel will lift the moulded part about 25 mm above the flash line. The ejector pins also serve as insert holding pins, as this combination will usually necessitate the use of a stripping fork for removal of the part in case of hollow components. The ejector pin, guide pins are made out of good quality silver steel.

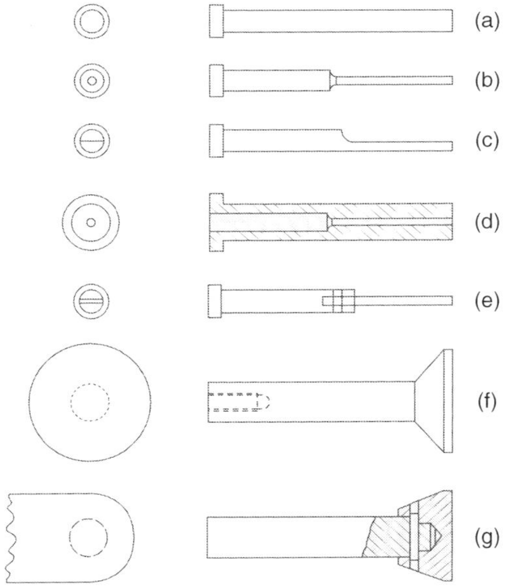

Fig. 4.15 (b) Types of ejector elements: (a) Pin (b) Shouldered Pin (c) D-pin (d) Sleeve (e) blade (f) Valve headed type (g) Bar.

4.9.4 Guide Bushes and Guide Pillars

Guide pillar and guide bushes are used for concentric alignment of core and cavity of a mould so as to get a uniform wall thickness component in a mould.

Guide bushes A guide bush is incorporated in a mould to provide a suitable wear-resisting surface for the guide pillar and to provide replacement in the event of wear and damage.

The guide bush is shown in Fig. 4.16. The internal bore is designed as a slide fit on the adjacent guide pillar, while the external diameter is a press fit into the mould plate. A radius is made at the front end of the bore to provide a lead-in for the guide pillar. The rear end of the bush is often counter bored to a greater diameter than the working diameter.

On each stroke the guide pillar should ideally pass through the working diameter of the bush, its end passing well into the counter bore. If the counter bore is not incorporated, the guide pillar will operate over a limited part of the internal bore and causes uneven wear on long production runs, may cause a ridge to occur inside the bush.

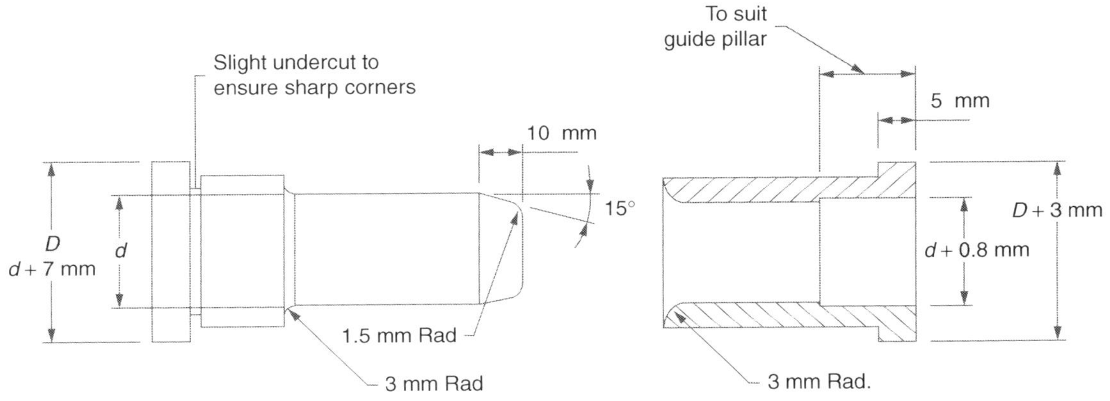

Fig. 4.16 Guide pillar and guide bush design.

Guide pillar There are different types of guide pillars viz. leader pins, standard, spigotted, surface fitting, pull-back type.

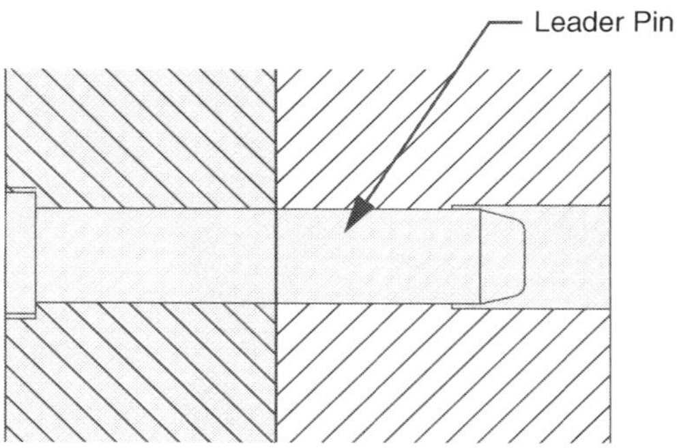

Fig. 4.17 A non-stepped guide pillar - 'Leader Pin'.

Leader pins The transfer mould consists of two plates, a cavity plate and a core plate. The alignment between the two plates was achieved by incorporating shouldered pins in one half and by machining accommodating holes in the other half. These pins are called *'leader pins'*

The guide pillar, guide bush, leader pins are made out of good quality steel viz: EN- 24, EN-31, EN-18 steels. These are hardened and tempered to 42 to 45 HRC to have better wear resistance, improved polishability.

Polishing and plating The transfer mould parts like core and cavity are being machined that show cutter marks on the surface of the parts. The non-polished areas will generate frictional heat in the material as it passes over these areas. This added heat causes the material to cure prior to filling the part. The unpolished areas change the filling pattern of the material, which results in gas being trapped in locations that cannot be vented. For these reasons, all the moulding surfaces are to be polished to a minimum of SPI #2 rating. The mould surfaces to be polished include the cavities and cores, vents, gates, runners, transfer pot and the entire parting surface. The mould polishing should be done in the direction of draw around the moulding areas and the vents to be polished in the direction of material flow and they should have the same degree of polish as the cavity and core. Flat surfaces that have no influence on the part removal. They can be polished in any direction.

After the mould is completely polished, it is to be plated. The defects in the steel surface will not be covered by the plating, but will be accentuated by it. While there are a number of different types of plating available, whereas the hard chrome plating provides the best part release, surface finish and able to sustain high temperature. Because some materials have fillers that are incompatible with nickel, the use of nickel or electro-less nickel to plate moulding surfaces is discouraged. In addition, nickel plating lack the wear resistance of chrome plating.

The surfaces to be plated in a mould are the cores, cavities, core pins, ends of the ejector pins, runner blocks, vents, and the entire parting line. To protect the moulding surfaces and to insure good part release, it is necessary to plate all the surfaces that were polished. After the mould is plated, it will be necessary to repolish the chrome because unpolished chrome plating causes sticking.

Centre supports The moulds built to run thermo set materials having no support in the middle, results heavy flash around the sprue and parts that vary in thickness from the sprue side to the side opposite. To avoid this problem substantial support pillars down the centre of the mould between the parallels are provided called the centre support.

High centreing of the mould Sometimes the centre of a mould will have heavy flash even with good centre support. In these cases, it is necessary to provide steel shim of size 0.05 to 0.07 mm on the support pillars in the centre of the mould, which will cause the moving side of the mould to be slightly domed called high centreing of the mould.

Side locks If the alignment of the mould halves is critical to meet the quality-requirements then the non-tapered side locks are provided on all four sides of the mould. The overall design of progressive components side locks is very good, since they have a longer engagement and are thicker.

4.10 VENTING

In transfer moulding when the thermo-set material polymerisation process takes place, it produces volatile gases along with the trapped air already within the cavity chamber. Then it is heated to a temperature of 375°C to 425° C. If the gases are not allowed to escape through the vents, it oxidises the surface and the lubricant creates burn marks on the part. Improper venting results voids in the moulding, which causes dimensional inaccuracy and less physical and electrical strength.

Vents should be located in the movable half of the mould and must lead to the atmosphere. Vents for phenolic parts should be 6 mm wide and 0.08 to 0.09 mm deep and vents for polyester parts should be 6 mm wide and 0.05 to 0.06 mm deep. The vent should initially be cut to a depth of 0.08 mm and extend to atmosphere. The vent should be approximately 25 mm long to allow pressure to build in the cavity after the material in the vent cures. After this point, the vent can be relieved to a depth of 0.25 to 0.50 mm and the corner of the vent at the part edge can be made radius or chamfered.

The 'dead' areas of the mould are vented with vented ejector pins. Before adding the vents, the ejector pins should fit the hole in which it will operate within 0.025 mm. A flat is then ground on the diameter not deeper than 0.13 mm for a distance that will take the vent 3 mm below the fitting diameter of the ejector pin and the fitting diameter should be 13 to 16 mm long. In addition, the stroke of the ejectors should be long enough for the entire vent plus 3 mm to come up above the bottom of the cavity.

All vents are to be mirror polished in the direction of flow, as well as the cavities and the cores. They should be polished for their entire length including the relieved distance. If a mould is to be chrome plated, all the moulding surfaces should be polished and plated including vents.

Air or volatile gases entrapped within a mould are the cause of many moulding problems. In the case of transfer moulding, the location of vented knockout pins, proper venting at activated mould sections, and placement of vented activated pins at dead end sections of the mould will provide adequate venting. A breathe cycle, wherein the mould is opened slightly for a second or two shortly after closing and then is closed again for the cure, allows easy escape of volatiles that otherwise would be trapped in the moulded part, creating voids.

If entrapped air and volatiles are not permitted to escape, they will prevent the filling of the cavity, will produce porosity in the part, which may cause warpage and dimensional problem, and may contain weld lines causing weak parts. The trapped air, being heated by compression, causes a burning of the material. Vents a minimum of 3/16 inch in width and 0.003 to 0.005 inch in depth should be ground in the face of the sealing or land area of the cavity.

Location of Vents

1. At the far corners
2. Near inserts or thin-walled sections where a knit line will be formed
3. At side-pull pins which form holes in bosses
4. At the point where the cavity fills last
5. In insert holding pins
6. Around ejector pins

4.11 NUMBER OF CAVITIES

Pot calculation For pot calculation, cavity area is a one important factor of consideration. The cavity area should be less than the pot area, otherwise the mould gets opened. So the pot area should always be greater than 1.3 times of the total surface area plus runner area.
Let us assume that the pot is in the shape of round with the diameter *D* and height *H*.

Number of cavity = n

Component weight = W (gm)

Sprue, runner and gate weight = 15% to 25% of total cavity area = (0.15 to 0.25) nW gm

Volume with sprue, runner and gate = (1.15 to 1.25) nW/ρ mm^3

Cull volume = 10 to 20 % of the total volume

Total volume with cull = (1.1to 1.2) (1.15 to 1.25) nW/ρ mm^3

Bulk factor of the material = K

Total loose powder volume = (1.1to 1.2) (1.15 to 1.25) nWK/ ρ mm^3

This volume is required to keep in the pot. So pot volume should be more than these volumes.

Required round pot volume = $\pi r^2 h$ = (1.1to 1.2) (1.15 to 1.25) nWK/ρ

Pot area × Pot depth = (1.1to 1.2) (1.15 to 1.25) nWk/ρ

'The pot area should always be greater than 1.3 times of the total cavity surface area + runner area' so,

1.3 × (Cavity surface area + Runner and gate area) (Pot depth)

= (1.1 to 1.2) (1.15 to 1.25) nWK/ρ

$$\text{Pot depth} = \frac{(1.1 \text{ to } 1.2)\ (1.15 \text{ to } 1.25)\ n\,WK/\rho}{1.3 \times (\text{Cavity surface area} + \text{Runner and gate area})}$$

4.12 ADVANTAGES AND DISADVANTAGES OF TRANSFER MOULDING

4.12.1 Advantages

(a) Loading a preform into the pot takes less time than loading preforms into each mould cavity.
(b) Higher tensile and flexural strength achieved in a component with transfer moulding process.
(c) Mould maintenance cost is low, although gates and runners are susceptible to normal wear.
(d) Due to closing of the mould before the process starts, delicate inserts and sections can be easily moulded.
(e) Longer core pins can be used and can be supported on both ends, allowing smaller diameters.
(f) Automatic degating of the components in transfer mould by provision of tunnel gates.
(g) Tolerance in part can be maintained up to 0.05 mm
(h) Larger structures can be processed.
(i) Increased flexibility of design features.
(j) Lower cost of production.
(k) Reduces cure time.
(l) Smaller diameter holes with longer lengths can be easily moulded.
(m) The actuation of side core for any side projection or undercut can easily be actuated.
(n) The process for insert moulding article can be easily done.
(o) Minimum flash lines on moulding.

4.12.2 Disadvantages

(a) Gate or sprue removal necessary unless pinpoint or tunnel gate is used.
(b) Moulded parts contain knit lines in back of pins and inserts.
(c) Reduces the moulded part's impact strength near to gate or runner.
(d) Improper gate locations cause warpage due to uneven fill or lack of density.
(e) Depending on gate location, circular parts may be considerably out of round.
(f) Large flat parts have flow lines.
(g) Mould vents are required.
(h) As the material gets waste in sprue or cull, runner, gates and vents in each cycle; material loss is higher.
(i) More stress which causes uneven shrinkage, warpage and weakness in the part.

(j) Number of cavity may be limited to runner length.

(k) Part weight and size limited on the base of press tonnage capacity.

(l) Mould costs are more than compression moulds.

4.13 COMPARISON OF COMPRESSION AND TRANSFER MOULDING

Table 4.3 Comparison of compression and transfer moulding processes.

S.No.	Description	Compression moulding	Transfer moulding
1	Cycle Time	Longer	Shorter
2	Shapes	Problem arises to develop complicated shapes	Capable to produce complicated shapes
3	Moulding	Not suitable or problem arises for moulding with inserts	Suitable for moulding with ceramic or metallic inserts
4	Tolerance	Close tolerance cannot be maintained	Possible to maintain tolerance up to 0.05 mm
5	Parallel smaller holes of longer depths	Due to lengthy insert pins, chances of bending of the pins	Easily moulded
6	Perpendicular holes	Easily not possible	Side core can be easily assembled
7	Maintenance of the mould	More care required for maintenance	Less maintenance
8	Strength	Strength cannot be achieved as injection moulding	Higher strength can be archived in the direction of flow
9	Flash	Higher flash, not controlled	Minimum flash
10	Gate mark	Gate is not provided	Gate mark available
11	Gate location	Gate is not provided	Cause warpage due to improper filling
12	Weld lines	No weld line, melt lines	Weld line will be noticed on the article
13	Mould vents	Breathing and vent both provided	Breathing and vent both provided
14	Pressure	Pressure not uniform, used for flow of plastic materials	Uniform pressure applied
15	Scrap	Scrap is less in the form of flash only	More scrap in the form of sprue, gate cull, runner, vents
16	Moulded in stress	Very less	More chance, causes uneven shrink age, warpage, etc.
17	Number of cavities	No such type of limitation	Limited to runner length
18	Mould cost	Cheaper cost of the mould	Costlier than injection mould

4.13.1 Mould Design Data for Thermo-set Moulding

Table 4.4 Pressure table: Pressure, psi of projected land area.

	Conventional phenolic		Low pressure phenolic	
Depth of moulding (in.)	Preheated by high frequency	Not preheated	Preheated by high frequency	Not preheated
0–3/4	1000–2000	3000	350	1000
3/4–3/2	1250–2500	3700	450	1250
2	1500–3000	4400	550	1500
3	1750–3500	5100	650	1750
4	2000–4000	5800	750	2000
5	2250–4500	*	850	**
6	2500–5000	*	950	**
7	2750–5500	*	1050	**
8	3000–6000	*	1150	**
9	3250–6500	*	1250	**
10	3500–7000	*	1350	**
12	4000–8000	*	1450	**
14	4500–9000	*	1550	**
16	5000–10000	*	1650	**

* Add 700 psi for each additional inch of cavity depth.
** Add 250 psi for each additional inch of depth.

4.13.2 Recommended Moulding Pressures for Compression Mould

Table 4.5 Types of fillers and recommended moulding pressures.

Type of filler	Moulding pressure (psi)
Phenolic : Wood flour	1500–3500
Cotton flock	1500–4000
Macerated cotton	2000–5000
Tire cord	2000–5000
Sisal	2000–5000
Rubber	1500–3000
Mica	1500–3000
Asbestos	2000–4000

(Contd.)

Table 4.5 *(Contd.)*

Type of filler	Moulding pressure (psi)
Mineral	2000–3000
Glass : Diced squares	2000–10000
Chopped roving	2000–6000
Bulk	1000–6000
Urea : Alpha cellulose	4000–8000
Alpha cellulose plus plasticiser	2000–4000
Melamine :Alpha cellulose	2000–8000
Asbestos	2000–8000
Mineral	2000–6000
Glass : Chopped roving	2000–8000
Bulk	2000–6000
Alkyds :Mineral (granular)	1000–1500
Glass : Extruded rope	600–1000
Putty	400–800
Bulk	1000–2500
Diallylphthalate (DAP) : Synthetic (nylon, orlon, dacron)	500–2000
Glass fibre	500–2000
Asbestos	500–2000
Polyester : Glass fibre (premix)	50–500
Sisal	50–500
Mineral (clay)	50–300
Epoxy	50
Mineral	100–1000
Glass fibre	100–2000
Silicone : Asbestos	1000–8000
Glass fibre	1000–5000
Mineral	1000–5000

Mould pressure (tons) = Moving press (Psi) × Projected area (in^2)

4.13.3 Specific Gravity and Bulk Factor of Plastics Material

Table 4.6 Specific gravity and bulk factor of thermo-set materials.

Material	Filler or reinforcement	Specific gravity	Specific weight (lb/cu.in)	Specific volume (cu.in/lb)	Bulk factor
Phenol formaldehyde	Cellulose	1.32–1.45	0.047–0.052	21.0–19.1	2.1–4.4
	Mica	1.65–1.92	0.059–0.069	16.98–14.4	2.1–2.8
	Glass	1.60–2.20	0.058–0.079	17.3–12.6	2.0–10
	Asbestos	1.45–1.90	0.052–0.068	19.1–14.6	2.0–14
	Macerated fabric	1.36–1.43	0.048–0.051	20.4–19.4	3.5–18
Urea formaldehyde	Cellulose	1.45–1.55	0.052–0.056	19.1–17.9	2.2–30
Melamine formaldehyde	Cellulose	1.45–1.55	0.052–0.056	19.1–17.9	2.2–2.5
	Asbestos	1.70–2.00	0.061–0.072	15.4–13.9	5.0–12
	Glass	1.80–2.00	0.065–0.072	15.4–13.9	5.8–12
	Macerated fabric	1.50–1.55	0.054–0.056	18.5–17.9	5.0–10
Epoxy (cast, unfilled) moulded	Glass	1.11–1.40 1.60–2.00	0.040–0.050 0.057–0.072	25.0–19.8 17.3–13.9	-- 2.0–7
Polyester and DAP (cast, unfilled) moulded		1.12–1.18	0.040–0.042	24.7–23.5	--
	Clay	1.40–1.60	0.050–0.058	19.8–17.3	2.0–4
	Glass	1.35–2.30	0.049–0.083	20.5–12.0	2.0–10
Silicone	Asbestos	1.60–1.90	0.057–0.068	17.3–14.6	6.0–8
	Glass	1.68–2.00	0.060–0.072	16.5–13.9	6.0–9
Alkyd	Powder	1.60–2.30	0.057–0.083	17.3–12.0	1.8–2.5
	Putty	1.60–2.30	0.057–0.083	17.3–12.0	1.0–1.2

4.13.4 Processing Parameters

Table 4.7 Processing parameters.

S. No.	Material	Compression moulding temp. °C	Compression moulding pressure kg/cm^3	Mould linear shrinkage%
		DAL moulding compounds		
1.	Mineral filled	-	-	0.5–0.7
2.	Synthetic fibre filled	-	-	0.9–1.1
		Epoxy resins		
1.	Glass fibre filled	150–170	-	0.1–0.5
2.	Mineral filled	120–170	-	0.2–0.8
3.	Low density	125–150	7–105	0.6–1.0
		Polyester		
1.	SMC	130–650	21–85	0.05–04
2.	BMC and TMC	150–190	35–141	0.05–04
	BMC	150–190	-	0.05–04
	Silicon epoxy	180	28–70	0.5–06
		Cellulosic		
1.	Cellulose acetate	130–220	565–2260	0.3–1.0
2.	Cellulose acetate butyrate	130–200	565–2260	0.3–0.9
3.	Cellulse nitrate	90–130	180	-
4.	Polyimide unifilled	330–370	210–352	-
5.	40% Graphite filled	370	210–352	-
6.	50% Glass fibre reinforced	240	210–706	0.2
		Melamine-formaldehyde moulding compounds		
1.	No filler	130–165	140–350	1.1–1.2
2.	Alpha cellulose filled	140–185	105–560	0.5–1.5
3.	Cellulose filled	145–180	105–420	0.6–0.8
4.	Flock filled	140–165	280–560	0.6–0.7
5.	Asbestos filled	140–170	70–490	0.5–07
6.	Macerated fabric filled	140–165	280–560	0.3-0.4
7.	Macerated fabric filled (phenolic modified)	150–175	140–560	0.3–06
8.	Glass fibre filled (including bodular)	140–170	140–350	0.1–0.4
9.	Melamine phenol moulding compounds	150–175	140–280	0.4–1.0

(Contd.)

Table 4.7 *(Contd.)*

S. No.	Material	Compression moulding temp. °C	Compression moulding pressure kg/cm3	Mould linear shrinkage%
	Phenol formaldehyde and phenol-furfural moulding compounds			
1.	No filler	115–160	140–350	1.0–1.2
2.	Wood flour and cotton flock filled	145–190	140–350	0.4–0.9
3.	Asbestos filled	145–190	140–350	0.2–0.9
4.	Mica filled	135–175	70–420	0.2–0.6
5.	Glass fibre filled	135–175	140–280	0.0–0.4
6.	Macerated fabric and cord filled	140–190	140–280	0.2–0.9
7.	Plup performed and moulding board	145–175	-	0.18–0.8
	Compounded with butadiene acrylonitrile copolymers			
1.	Wood flour and cotton flock filled	150–180	140–280	0.4–0.9
2.	Asbestos filled	155–180	140–280	0.4–0.7
3.	Rag filled	155–180	140–280	0.2–0.4
4.	Metal filled (iron or lead)	140–160	140–350	0.3–0.4
	Urea formaldehyde moulding compound			
1.	Ally to resins and monomers	130–160	140–560	0.6–1.4
2.	DAP moulding compounds glass filled	150–180	140–420	0.20–0.6
3.	DAP moulding compounds mineral filled	140–180	170–300	0.19–0.6

4.14 GENERAL MOULD DESIGN CHECK LIST

All mould components contacted by the moulding compound—runners, gates, cavities, land areas—should be made of through-hardened tool steels, hardened to 65 to 68 on the Rockwell C scale, highly polished, and hard chrome plated.

Because most thermo-set compounds are slightly soft at the time of ejection from cavities, ejector pins should have an adequate correctional area to minimise the possibility of distorting or puncturing the moulded plastic at this point in the cycle.

In automatic moulds, it is to ensure, with part design, undercuts, or hold-down pins, that the moulded parts, during mould opening, consistently remain in the desired half of the mould, so that when the parts and the runner system are ejected, the extractor will always 'find' them and effect positive removal.

Flash removal from the mould, each cycle is critical for successful automatic moulding. The flash should be ejected with the moulded part. An air blast, directed appropriately to cavities and land areas each moulding cycle when the mould or press to further ensure the absence of flash each cycle.

Moulds should be of uniform temperature in the cavities, and should have adequate heating capacity to ensure maintenance of the desired temperature despite continual heat extraction by the relatively cooler moulding compound each cycle.

Temperature sensors and heating cartridges must be judiciously placed to provide this uniformity of temperature. Insulating blankets may prove helpful to minimise mould heat losses and variations due to local air currents around moulds and presses.

To minimise local temperature variations in large moulds, heating cartridges often are grouped in zones, with each zone having its own temperature controller and sensor.

The 'mould over temperature' sensor, which will cut off power to the heating cartridges are provided whenever it senses a mould temperature more than a few degree over the desired mould temperature.

Excessive mould temperature not only will result in rejecting parts, but also anneals the mould steels and warp critical mould components.

An adequate moulding press should be used considering the required tonnage capacity of mould. Over tonnage application damages the mould.

For a transfer mould, the pot dimension must be adequate to the required volume of loose power for plastic material feeding.

4.15 INTRODUCTION TO THERMO-SET PLASTIC MATERIALS

Thermosetting materials are chemical compounds made by processing a mixture of heat reactive resin with fillers, pigments, dyestuffs, lubricants, etc., in preparation for the final moulding operation. These materials or moulding compounds are in most cases, in powder, granulated or nodular form, having bulk factors ranging from 1.2 to 10. Some are used in the form of rope, putty or slabs.

The material of power bulk factor for usually those having wood flour or mineral compounds as fillers, while those of higher bulk factor have as fillers cotton or nylon flock, rag fibres, pieces of macerated rag, tire cord, sisal and for very high impact strengths, glass roving. Phenol formaldehyde is the single most common resin and catalyst combination, generally a mineral such as mica, the moulded part will have good electrical properties. If the filler is glass fibres, say one quarter inch long, the moulded part will have good impact strength. Small hollow glass micro balloons have been used as fillers to make low density parts.

Other resin systems include melamine formaldehyde (often used in plastic dinnerware), urea formaldehyde (common in white or pastel heat resistant handles for kitchen ware, or outlet sockets for household use), alkyds and polyesters (often used in high voltage insulators in TV sets, or for arc resistance and insulation in circuit breakers and switch gear), dialkyl epoxy (housings for electronic components), and silicone (high temperature requirements to 600 F or more). Common fillers include silica, glass, wood flour, natural or synthetic fibres and combination of these.

Although most thermosetting formulations are dry and granular at room temperature, some are putty like, some in the form of dry or moist matted fibres and some a fine powder.

When subjected to heat, thermosetting formulations first become liquid then undergo an irreversible chemical reaction called cure or polymerisation. If polymerisation occurs under mechanical pressure, as in a closed mould, the resulting material is a dense solid. Polymerisation is generally a time temperature relationship with shorter cure times when higher temperatures are used. Typical pressure, temperature and time values, for a phenolic wall socket, in semi-automatic compression moulding, might be 3000 psi, 300 F, and 1-1/2 min.

4.15.1 Phenol Formaldehyde

Common name(s): Phenolics: phenolic moulding materials
Abbreviation: PF
Alternative Names: Bakelite; phenoplast; phenolic resin holding compounds; novolak resin moulding compounds.

Material properties Phenolic resins are based on the resinous material formed when a phenol type material is reacted with an aldehyde. Phenol itself is widely used as the resins produced have good mechanical strength and cure quickly. Cresols may be used for more acid resistant products and the use of phenol/cresol mixtures lowers the cost and controls the flow in processing, for example, in compression moulding, the use of 20% cresol may be employed. Where improved alkali resistance is required, xylenols may be used. Formaldehyde is far and away the most widely used aldehyde.

The unreinforced resin is very brittle and requires extensive modification with fillers to produce useful products. A wide range of properties can be obtained from PFs because of their compatibility with a variety of reinforcements and fillers, so the properties are very dependent on the fillers used. Wood flour gives reasonable properties at an acceptable cost, and so wood flour filled PF is regarded as a general-purpose (GP) material. The use of more fibrous, organic filler (cellulose fibres) gives improved toughness and impact strength.

Modification with material fillers yields increased rigidity, improved dimensional and thermal stability, results in a higher UL index of use, lower water absorption and thermal expansion. Glass fibre (GF) addition can improve the dimensional stability and rigidity even more; the UL index of use can reach 180°C/356°F. Because of density differences, organic fillers are used in a weight ratio of 1:1 with the resin and inorganic fillers (mineral and glass) used in the ratio of 1.5:1.

Mouldings are glossy, opaque and dark coloured but have also tremendous advantage of being inherently flame retardant, low smoke systems which do not require halogen or phosphorous-based additives. They are stiff, hard, have low elongations and possess good creep resistance.

The materials are commonly supplied in powder form and have a useful combination of low cost, ease and versatility of moulding, temperature resistance, solvent and chemical resistance and, good electrical insulation properties. They have better water resistance than melamine formaldehyde and possess a more stable melt rheology than amino-plastics, i.e., they are not so temperature dependent. Tracking resistance under conditions of high humidity is poor and their impact resistance is not very good. Electrical properties, especially tracking resistance is inferior to amino-plastics. Melamine phenolics (MPF) have superior electrical properties to phenolics and have a wider colour range; they are used in decorative and electrical application that is in areas which are beyond PFs. The reduced post-moulding shrinkage of this class of materials is a significant advantage over MFs.

Powder grades are available in differing granulations to suit the process. For example, fine powders are used in compression moulding where limited flow is envisaged but a high gloss is required. Fine free grades are preferred in automatic compression moulding, so as to avoid premature melting of the fines in the hopper. When injection moulding, choose a granulation that is large enough to flow freely from the hopper and gives consistent pick up on the screw.

Mould consideration High quality nickel-chromium steels used, should be suitably hardened and chrome plated. Maintain the plating and polish as wear occurs so as to ensure lack of mould staining and ease of ejection. During the transfer moulding, mould temperatures of 160 to 180°C are usual. Vents are tapered with a narrow end at the mould face; the vent then widens as the cavity is left behind. The sprue should be short and well tapered (up to 5 degrees), the gates should be machined on the inserts so that it can be easily replaced as wear occurs, and ion implantation and titanium nitriding help to minimise the wear.

Bulk factor of general purpose compression grades is approximate 2.3. High impact grades can reach 3.6. Powders are generally easy to palletise.

Ease of flow PF is available in range of flows soft (easy), medium and hard (stiff). It is hard, stiff materials which require high pressure to get smooth mould filling and having more stable melt rheology than amino-plastics. The flow, and cure rate, of a moulding material is determined by the amount of resin, the degree of condensation of the resin, the residual volatile content and the moisture content. If the resin is only lightly condensed during manufacture then, the resulting moulding material will flow easily but will require comparatively long cure times; the moulding will exhibit high shrinkage and will not exhibit the best mechanical properties. Increasing the degree of condensation will worsen the flow but improve the properties of the mouldings.

Shrinkage For GP compression moulding grades, it is about 0.8% that is 0.008 mm/mm; that of glass filled grades may be reduced to 0.3%. The shrinkage of acid resistant and minimum odour grades will be higher than GP. Injection grades (GP) will range from 1 to 1.7%, but impact and heat resistant grades could be as low as 0.8%.

Applications Phenolic resins are used in place of polyester resins to allow the production of large, glass fibre reinforced mouldings by low pressure processes such as hand or spray lay up. Complex mouldings like train cab components may be produced without the need for heat, pressure and expensive forming tools. The big advantage is that phenolics are inherently fire resistant and in a fire do not produce copious quantities of smoke and fumes.

4.15.2 Urea Formaldehyde

Common name: Urea formaldehyde moulding material
Abbreviation: UF
Alternative name: Amino-plastic; urea; urea resin moulding compounds or materials.

Material properties It is an amino-plastic plastic material formed by the reaction between materials containing amine (-NH_2) or amide groups and aldehydes. The two, commercially important members of this group are UF and MR. To make a UF material, a low molecular weight resin is produced by reacting urea and formaldehyde and this syrup is combined with fillers, lubricants, hardeners, etc., to give a moulding powder. UFs are thermosetting materials which are available in fine powder or granular form. A wide colour range is possible (because of the lack of resin colour). When wood flour is used as the filler, brown mouldings will result; this is masked by intense colouring so as to give black or brown shades. Such moulding powders have a well balanced range of properties and are cheap. They are used in electrical and closure applications. If bleached wood pulp (paper) is used as the filler, then a wide colour range is possible as the resultant compound can be translucent; bright, intense, colours are possible including pastel shades and white. Such colours are light fast. The properties are similar to wood flour filled grades and, the mouldings are widely used in electrical fittings for their low cost, wide colour range, rigidity and good electrical properties.

The heat resistance of UF's is lower than that of PF's but their resistance is higher and curing time is faster. The water absorption is also significantly higher than for either PF or MF. In order to obtain improved water resistance, the resin may be fortified with melamine; acid, alkali; heat resistances are thus improved.

Colour The materials are supplied in a wide range of colours. White tints (from semitranslucent to fully opaque), pastel shades and intense colours (from white to black) are available. Light fast in both white and pastel shades.

Moulding Condition Moulding temperatures are slightly lower than those recommended for MF. Too low an injection barrel temperature should however, be avoided as this will tend to cause procure due to the development of large amounts of frictional heat. Such heat is unlikely to be uniformly distributed and so may result in highly stressed mouldings. As with other thermo-sets, a reasonably fast injection speed is suggested as to get rapid cure and a good surface finish; speeds should be slightly lower than those used for PF.

If mould temperatures are too high, blisters will occur on moulding surfaces. If porosity is observed the injection speed is too high, the mould temperature is too low (under 135°C/275°F). Temperatures must be held accurately—more accurately than for PF. UF can also discolour through over curing of delicate shades.

When compression moulding, a cure time of about 20 seconds for 0.078 in thick moulding at a temperature of 150°C/302°F; a 4 mm/0.157 inch thick moulding will require about 50 seconds. If the material is being preheated by infrared, keep it well stirred or keep in very shallow layers; when preheated to high temperatures (60 to 90°C/140 to 194°F) by this method, care must be taken so as to prevent powder 'lumping' occurring. With UF it is very easy to confuse under cure with over cure as both may result in the appearance of blisters.

Mould consideration When injection moulding amino-plastics, the cavities should be electroplated and due allowance is made for after shrinkage. During the hardening or setting process, a large amount of volatile material, i.e., water is produced. Adequate mould venting must be provided - through pins, or through wide shallows slots around the mould periphery. Venting to atmosphere is essential; chrome plating and draw polishing of vents is also essential as these allow an easier passage for the escape of gases and for the removal of excess flash.

The technique known as 'breathing' helps just before the mould fills, injection is momentarily stopped and the mould is slightly parted or 'bumped'. After final mould closing, injection is completed under second stage pressure.

Ease of flow UF comes in a range of flows or grinds, such as easy, medium and stiff for compression and transfer moulding. 'Easy' is used for transfer moulding, medium for general purpose compression moulding and 'stiff' for the compression moulding of large and deep draw articles. Special injection moulding grades are available.

Shrinkage This may reach 1% but is usually 0.5 to 0.38%. After shrinkage, up to 1% will occur on heating, Test mouldings, by heating them at 80°C/176°F for 48 hours- about 0.5% after shrinkage should result. (Dimensional stability of amino-plastics is not very good at elevated temperatures for example over 80°C/176°F). Shrinkage may be reduced by using a highly condensed resin to make the moulding powder to maintain ease of flow, a plasticiser, such as glyceryl-tolyl-ether, may then also be added.

Applications Moulding powder use is only a small outlet for UF resins. They are more widely used as adhesives in, particle board and plywood.

UFs are used extensively in the electrical industry for insulating parts in plugs, sockets, switches, connectors and lamps. In domestic applications, UFs find use as handles and knobs (furniture and kitchen utensils), buttons, buckles and closures for jars and bottles; this is because of its good resistance to solvents, etc., and excellent surface finish. They are widely used for toilet seats, because of the bright, attractive appearance possible.

Table 4.8 Suggested processing conditions for UF.

Process	Setting	Unit	Value MF
Compression	Preheat temperature	°C	< 95
	Tool temperature	°C	135 to 165
	Pressure	MNm^{-2}	> 15 eg 30 to 60
	Curing time	S	30 to 180
Transfer	Tool temperature	°C	150 to 170
	Transfer pressure	MNm^{-2}	> 50 eg 60 to 120
	Curing time	S	30 to 120

(Contd.)

Table 4.8 *(Contd.)*

Process	Setting	Unit	Value MF
Injection	Cylinder temperature	°C	65 to 95
	Nozzle temperature	°C	85 to 110
	Melt temperature	°C	120 to 140
	Screw temperature	°C	50 to 80
	Tool temperature	°C	145 to 165
	Thin section < 3 mm		145 to 155
	Thick section > 3 mm		135 to 145
	Injection pressure	MNm^{-2}	100 to 248
	Back pressure	MNm^{-2}	< 1 (eg 0.5)
	Curing time	S	15 to 80

4.15.3 Melamine Formaldehyde

Common Name: Melamine formaldehyde: melamine formaldehyde moulding material
Abbreviation: MF
Alternate Name: Melamine resin moulding material or compound. MF like UF is an amino-plastic that is the moulding powders based on an amino formaldehyde resin.

Material properties Melamine may be made from urea and like urea it contains amine ($-NH_2$) groups. These react with formaldehyde to form a resin which is combined with fillers, hardeners, pigments, etc., to make a moulding powder. As with colourless UF resins, it is possible to produce moulding powders of virtually any colour when white alpha cellulose, or paper fillers, is used. Mouldings are bright and attractive but unfortunately they are more expensive than UF or PF. They have lower water absorption than UF, are less flammable, maintain their electrical properties better in conditions of high humidity (particularly when mineral filled), are harder and resist staining and heat to a greater degree.

MF is supplied as granular powders which have a bulk density of about 2:1. Particle size and size distribution, is adjusted during manufacture so that the material may be loaded either by weight or by volume or it may be pelleted and used with or without preheating. Grades are chosen to suit the process and application, for example, compression, transfer moulding. Special grades are available for the compression moulding of plates; the use of such grades gives an even, hard, scratch resistance surface. These grades are usually in the form of fine powder rather than a granular material.

Colour possibilities There is an unlimited choice of colour, with a wide range of colours offered by manufacturers. MF gives very even pigmentation and may be colour coded according to RAL system. MPF materials are also available in white and pastel shades colour stability is adequate for many uses.

Moulding conditions When injection moulding the highest mould temperature are for the thinnest sections below 3 mm thickness. Thicker section requires lower temperatures, around 155°C, but longer cure times. The screw speed low about 50 rpm and fill the mould as quickly as possible consistent with avoiding gas trapping. Avoid over packing use up to 30% of the injection pressure as packing pressure bring this in when the screw is about 10 mm, away from bottom. At this point, it is possible to arrange for mould breathing to occur although MF is not as gassy as UF, good venting and/or breathing is still necessary. When processing melamine phenolics, screw surface speeds of up to 0.3 m/s, approximately 1ft/s, are permissible for plasticising.

In compression moulding, MF is usually run hotter than UF at around 160°C/320°F. At this temperature a 2 mm or 0.079 in, thick component will need a cure of 30 seconds, a 4 mm/0.157 in, section will need 60 seconds. To prevent cracking around large inserts, warm them to the mould temperature before use. If the material is preheated (HF) then the equipment must be powerful enough to raise the temperature from 100°C to 212°C, in less than 60 seconds to avoid moulding problem. Preheating improves appearance, electrical properties, dimensional stability and output. Compared to UF materials, MF is normally pelleted and preheated prior to introduction into the mould. Materials handling can be improved by producing a heated slug of material on an in line screw unit, this saves pelleting and preheating.

Ease of flow MF materials can be of comparatively low viscosity. Such flow properties permit the moulding of large complex handles and the bowls which give design freedom.

Shrinkage About 0.7% or 0.007 in/in, or 0.007 mm/mm. Like UF after shrinkage will occur if the moulding is held at high temperatures. Such shrinkage is usually identified by testing at 110°C for 48 hours, after this time, the MF moulding will shrink up to 1.2%.

MF mouldings are resistant to fuels, oils, greases, common organic liquids and organic solvents such as acetone and alcohol. Also resistant to cold dilute acids and alkalis; MF is more chemically resistant and more stain resistant than UF that is, with better resistance to weak acids, alkalis and water..UF mouldings are attacked by boiling water; fully cured MFs are more resistant, being attacked only by concentrated acids and alkalis or by hot dilute acids.

Applications MF resins are widely used in laminates as by their use it is possible to produce a very wide range of attractive, and durable, patterned laminates - the pattern printed on the paper base shows through the colourless resin. The laminate core is made of cheaper, PF coated paper. Such laminates have excellent light stability. MF resins are also used for textile treatments and for paint manufacture.

Because moulded components can be free from odour and are bright, attractive, scratch and stain resistant, they are widely used in dinnerware and kitchen utensils. Such items can be made more attractive by moulding in foils which carry a legend or picture.

MF is also used for handles, knobs, household appliances and electricity insulating parts - particularly where these must withstand high temperatures. It can withstand repeated cycling from 150°C (302 F) into cold water. Compared to other thermo-sets, MFs are expensive and are only selected where appearance is of prime importance. Where improved impact strength is required, glass fibre is used as filler - it gives high mechanical strength and heat resistance. The moulding process goods are track resistance and burn with difficulty.

Table 4.9 Suggested processing conditions for MF.

Process	Setting	Unit	Value MF	Value MPF
Compression	Preheat temperature	°C	< 100	< 100
	Tool temperature	°C	140 to 180	150 to 175
	Pressure	MNm^{-2}	15 to 60	20 to 60
	Curing time	S	40 to 210	20 to 40
Transfer	Tool temperature	°C	150 to 170	140 to 170
	Transfer pressure	MNm^{-2}	60 to 120	80 to 150
	Curing time	S	30 to 120	60 to 120
Injection	Cylinder temperature	°C	85 to 105	70 to 80
	Nozzle temperature	°C	95 to 120	110 to 120
	Melt temperature	°C	120 to 140	120 to 140
	Screw temperature	°C	65 to 80	65 to 80
	Tool temperature	°C	145 to 180	165 to 180
	Injection pressure	MNm^{-2}	100 to 248	100 to 175
	Back pressure	MNm^{-2}	< 1 (eg 0.5)	8 to 12
	Curing time	S	15 to 80	25 to 50

4.16 TROUBLE SHOOTING AND QUALITY ASSURANCE

Table 4.10 Moulding problems and its remedies.

S.No.	Trouble/Problems	Remedies
1.	Excess flash on mouldings	Pressure pad dimensions to be rectified
2.	Poor surface finish	Processing parameters and mould temperature is to be checked
3.	More cycle time	Processing parameters and mould temperature is to be checked
4.	Burn marks on moulding	Processing parameters and venting is to be verified
5.	Mismatching on parting line and variable wall thickness	Mould alignment to be rectified
6.	Variable part dimension	Powder/material, feed system is to improve
7.	Unsafe mould temperature	Thermocouple and temperature controller is to be checked
8.	Component brakes during ejection	Types of ejection system and ejector pin dimension is to be rectified.
9.	Heater burnt out frequently	Poor heating element or improper heater capacity used in the mould
10.	Uneven shrinkage	Uniform mould temperature required with good quality of plastic material used

4.17 MOULD MATERIAL AND ITS SELECTION

Table 4.11 General characteristics and applications of mould steels.

AISI type	General characteristics	Applications
P-20	Medium carbon (0.30%) and chrome (1.65%) alloy available prehardened or annealed form.	High grade mould base plates, large cavity and cores, gibs, slides and interlocks
H-13	Hot die steel with 5% chrome	Higher hardness than *P*-20. Good toughness and polishability. Used for punch and cavities
A-2	Cold die steel with hot carbon (1%) and chrome (5%)	Punch and cavity with high hardness for abrasion resistant, long wearing compression, transfer moulds, limited to small size
D-2	Cold die steel, high carbon (1.55%) and chrome (11.5%)	Highest abrasion resistance, difficult to machine susceptible to stress cracking used for small moulds
414	Stainless steel with 12% chrome, 2% Ni, 1% Br Cd, low carbon 0.03%	Stainless steel version of *P*-20 with similar properties and use
420	Stainless steel, 13% chrome, 0.8% Ni, medium carbon 0.30%	Stainless steel version of *H*-13. Similar property and use
4145	Medium carbon 0.50%, chrome 0.65%	Low cost steel used for mould bases and large mould back plates not suitable for high surface finish

Table 4.12 Typical heat treatment of mould steel and uses.

AISI SAE Steel designation	General characteristics and uses	Typical heat treatment	
AISI 1030 carbon steel	Mould bases and structural components such as blocks, spacers, ejector housings, clamping plates	Normalising Annealing Hardening Quenching Tempering	870–915°C 675–745°C 855–900°C water or oil optional
AISI 1040 carbon steel	Support pillars	Normalising Annealing Hardening Quenching Tempering	855–900°C 845–885°C 815–855°C water or oil 315–750°C, RC 20–26
AISI 1095 carbon steel	Backing plates		

(Contd.)

Table 4.12 (*Contd.*)

AISI type	General characteristics	Typical heat treatment
AISI 4130 alloy steel, generally supplied preheat treated	Mould bases and structural components such as cavity retainer plates, support plates, clamping plates	Normalising 870–925°C Annealing 800–845°C Hardening 870–900°C Quenching oil Tempering 315–750°C, RC 18–44
AISI 4140 alloy steel	Structural components such as retainer plates, support plates, clamping plates	Heat treated to approx. RC 32
AISI 6150 alloy steel	Sprue bushings	Normalising 870–900°C Annealing 855–900°C Hardening 870–900°C Quenching oil Tempering 455–565°C, RC 17–46
AISI 8620 alloy steel	Cast steel	Normalising 925–955°C Annealing 540–650°C Hardening 800–830°C Quenching oil Tempering 455–565°C, RC 24–49
AISI S1 tool steel	Master hobs	Annealing 800°C Hardening 900–925°C Quenching oil Tempering 455–565°C, RC 58–40
AISI P2 tool steel	Hobbed cavity inserts	
AISI P20 tool steel usually supplied prehardened	Machined and hobbed cavity inserts, stripper plates	Annealing 915°C Carburising 925°C Tempering 175–230°C, RC 64–58 Forging 1090–925°C Normalising 850°C Soft annealing 800°C, RC 20 Stress relieving 550°C Carburising 850–940°C Quenching oil Tempering 180–650°C,RC 28–52 Nitriding 525°C
AISI 420 tool steel generally supplied fully annealed	Cavity inserts	Soft annealing 780°C RC 18 Stress relieving 650°C Preheating 600–850°C Hardening 980–1050°C Quenching air Tempering 180–450°C, RC 58–62 Nitriding : Not recommended

4.17.1 Typical Applications for Steels in Moulds

Table 4.13 Typical applications of steels in moulds.

AISI SAE Steel designation	General characteristics and uses	Typical heat treatment	
AISI S7 shock resistant tool steel	Interlocks, latches, sprue, bushings	Forging : Normalising : Annealing : BHN : Stress relieving : Carburising : Preheating : Hardening : Quenching : Tempering :	1120–925°C Do not normalise 815–845°C 197 (max) 675°C Delivered with RC 55–58 590–700°C 925–955°C Air for smaller than 60 mm oil for sections larger than 60 down to 540°C then air 180–540°C, RC 50–58
AISI 01 tool steel	Master hobs, plate for gate cutting, small inserts	Annealing : Hardening : Quenching : Tempering :	790°C, RC 62–57 Oil
AISI A2 medium alloy tool steel	Master hobs, slides, plates for gate cutting, cold work	Annealing : Hardening : Quenching : Tempering :	900°C 925–980°C 175–540°C, RC 62–57
AISI A6 tool steel	Master hobs, plates for gate cutting, for cavity requiring optical finish	Annealing : Hardening : Quenching : Tempering :	760°C 830–870°C air 175–540°C, RC 60–54
AISI D2 high carbon, high chromium tool steel	Master hobs, slides, lifters, cold work	Annealing : Hardening : Quenching : Tempering :	900°C 980–1025°C Air 150–425°C, RC 61–54
AISI H13 hot work tool steel, chromium base, generally supplied fully annealed	Cavity plates and inserts, hot hobs, ejector pins, core pins, leader pins, return pins, sprue pullers	Soft annealing : Stress relieving : Carburising : Preheating : Hardening : Quenching : Tempering : Nitriding :	850°C RC9 650°C 900–980°C 600–850°C 980–1080°C Air 180–500°C, RC 57–60 525°C

(Contd.)

Table 4.13 (*Contd.*)

AISI SAE Steel designation	General characteristics and uses	Typical heat treatment	
AISI H23 tool steel	Hot hobs for beryllium copper hobbing	Annealing : Hardening : Tempering :	880°C 1090–1260°C 650–815°C, RC 47–30
AISI 1020 carbon steel	Ejector plates, ejector retainer plates	Normalising : Annealing : Hardening : Quenching : Tempering :	900–955°C 540–730°C 870–910°C water or oil 120–205°C

4.17.2 Mould Materials - Equivalent Standards

Table 4.14 Mould materials and its equivalent standards.

S.No.	ISI	BS	DIN	AISI/SAE	W.nr. (German)	JIS
1	T35Cr5Mo1V30	BH 11	X38CrMoV51	H 11	--	--
2	T55Cr5MoW1V30	BH 12	X37CrMo W51	H 12	--	--
3	T35Cr5MoW1	BH 13	--	H 13	--	--
4	--	EN 110	34CrNiMo6	--	--	--
5	40Ni2Cr1Mo28	EN 24	30CrNiMo4	* 9840	--	--
***6	13Ni3Cr28	EN 368	14NiCr14	* 3318	--	--
**7	--	BF 1	120WA4	F 1	--	--
(a)8	T105Cr1	EN31/B13	100Cr	13	--	--
(b)9	T103	BW 18	C10 5W1	W1	--	--
(c)10	C10 and C14	EN 2A	--	*1006	--	--
11	C35Mn75	EN 8	--	*1040/ C1040	--	--
12	40Cr1	EN 18	--	*5140	--	--
(d) 13	50Cr1V23	EN 47	50Cr84	6152	--	--
(e) 14	50Cr1V23	EN 48	--	5152	--	--
15	T55Ni2Cr65Mo30	--	--	--	--	--
16	--	--	--	420 improved	1.2083	SUS 420

*SAE specification; ** silver steel; *** case hardening; (a) bearing steel; (b) carbon tool steel; (c) mild steel; (d) 1% chromium spring steel; (e) corrosion resisting, direct quenching steel for moulds used for moulding corrosive plastics used for thermo-set moulds, for gate, runners and cavities.

4.17.3 Applications for the Typical Mould Steel

Table 4.15 Applications for the typical mould steels.

Type of steel	Typical uses in injection moulds
4130/4140	General mould base plates
P-20	High grade mould base plates, hot runner manifolds, large cavities and cores, gibs, slides, interlocks
4414 SS, 420 SS (preheated)	Best grade mould base plate (no plating required), large cores, cavities and inserts
P5, P6	Hobbed cavities
01	Gibs, slides, water plates
06	Gibs, slides, water plates, stripper rings
H -13	Cavities, cores, inserts, ejector pins, sleeves
S7	Cavities, cores, inserts, striper rings
A2	Small inserts in high water areas
A6	Cavities, cores, inserts for high wear areas
A10	Excellent for high wear areas, gibs, interlocks, wedges
D2	Cavities, cores, runner and gates inserts for abrasive plastics
420 SS	Best all round cavity, core and insert steel, best polishability
440C SS	Small to medium size cavities, cores, inserts, stripper rings
250, 350	Highest toughness for cavities, cores, small unsupported inserts
455M SS	High toughness for cavities, cores, inserts
M2	Small core pins, ejector pins, ejector blades (up to 5/8 in.diameter)
ASP 30	Best high strength steel for tall, unsupported cores and core pins

4.17.4 Mould Material and its Applications

Table 4.16 Mould material and its applications.

ISI	Applications
T35Cr5Mo1V30	Cavities, cores, ejector pins, guides, wear pads of moulds
T35CrMo1 W1V30	Die casting dies, plastics moulds
T35Cr5Mo1 V1	Cavities, cores, ejector pins, guides, wear pads of moulds
40Ni2Cr1Mo28	High tensile load applications max. strength when hardened to 58–60 Rc used for cavity housing core and cavity back plate, shoulder screws, clamp
13Ni3Cr80	Cavity and core
T105Cr1	Used for guide pillar, bush and bearing purposes
T103	Used as a wear plate, backing pad
C10 and C14	Bolster, support block, plates, backing plate, holder plates

(Contd.)

Table 4.16 (*Contd.*)

ISI	Applications
C35Mn75	Moulds with a shorter run and on non-accurate cavities
40Cr1	Pillar, bush, sprue bush, locating ring, bigger diameter ejector pins
50Cr1 V23	Used for coil and plate springs
T55Ni2Cr65Mo30	Core and cavity

Table 4.17 Spectrum of materials used in building moulds - Arranged in order of surface hardness.

Material class	Surface hardness	Core hardness
Carbides	68–75 RC	68–75 RC
Steel, nitriding	> 68 RC	> 38 RC
Steel, carburising	60-65 RC	20–42 RC
Steel, water hardening	67 RC	40–55 RC
Steel, oil hardening	62 RC	40–60 RC
Steel, air hardening	60 RC	60 RC
Nickel cobalt alloy	45–52 RC	45–52 RC
Steel, prehardened	44 RC	44 RC
Beryllium-copper	28–42 RC	44 RC
Steel, prehardened	28–32 RC	28–32 RC
Kisksite (zinc alloy)	80–105 BN (1)	80–105 BN (1)
Aluminium alloy	60–95 BN (1)	60–95 BN (1)
Brass	50 BN (1)	50 BN (1)
Sprayed metal	< 50 BN (1)	< 50 BN (1)
Epoxy, metal filled	85 RM	85 RM
Silicone rubber	15–65 Shore 'A'	15–65 Shore 'A'

Legend: RC = Rockwell hardness = 'C' scale; RM = 'M' scale; BHN = Brinell number 3000 kg load; BN (1) = Brinell number 500 kg load.

Questions

1. Write about transfer moulding process.
2. Explain transfer –pot calculation.
3. How does mould heating occur in transfer mould?
4. How does moulding pressure act in transfer mould?
5. How does transfer moulding relate to pressure, cure and quality?
6. What is transfer moulding? Classify transfer moulding and explain any one.
7. Explain pot-type transfer moulding with a neat sketch.
8. Explain various types of plunger transfer moulding with required sketch.
9. Describe the electrical energy requirements to heat the mould.
10. How sprue, runner and gate influence a successful moulding in transfer mould?
11. Write down the advantage and disadvantage of transfer moulding.
12. What is the difference between compression moulding and transfer moulding?
13. List out the mould design check list to be considered in transfer moulding.
14. What are the factors that influence while designing runner and gate?

References

1. Charles A. Harper, *Handbook of Plastic Processes*, Wiley Interscience Publication, New Jersey.
2. CIPET, *Technical Manual*, Chennai.
3. Dominick V. Rosato, P. E., *Plastics Processing Data Handbook*, Second Edition, Chapman and Hall, London.
4. Harry DuBois and Wayne I. Pribble, *Plastics Mould Engineering Handbook*, Reinhold Publishing Corporation, New York.
5. Manas Chanda, Salil K. Roy, *Plastics Technology Handbook*, CRC Press, London.

CHAPTER

5

Advanced Injection Mould Design

5.1 MOULDS FOR THREADED COMPONENTS

5.1.1 Introduction

The threaded components are used where two mating parts – assembly and disassembly are required. The threads must have sufficient strength to withstand the pressure without wear.There are various types of threads used in injection moulding components like: (a) blind and open female threads, (b) male threads of V-shape, roll thread and modified square threads, etc. The threads are formed by various processes like the drilled holes made in plastic component with either thread cutting or thread forming using self-tapping type screws. Parts can be moulded with local inserts, moulded threads are formed by using unscrewing threaded inserts method.

Due to the increased complexity of the mould, extra mechanism is required to unscrew the threaded core pins in each cycle. The moulded component, which incorporates a thread, is classified as external thread and internal thread and according to thread shape, it is subdivided into continuous and discontinuous threads.

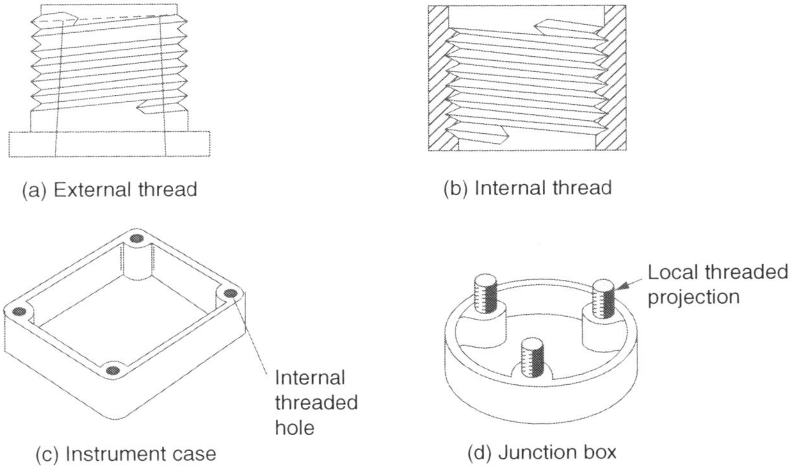

Fig. 5.1 Components incorporating threads.

The small-diameter internal thread below 8 mm diameter is normally incorporated in a moulded part by means of metal inserts for attachment with other parts. The larger internal threads are moulded without the metal insert and can be used in conjunction with moulded male threaded components. Different types of threaded components are shown in Fig. 5.1.

The external and internal threads can be directly moulded into the part, eliminating the need for mechanical thread forming operations.

5.1.2 Thread Profile

The Unified Thread Standard is best for moulded plastic threaded parts as it eliminates the feathered edge at both the tip and root of the thread. Other thread profiles such as acme or buttress threads are used with good results.
The Unified Thread Standard is divided into three categories:

1. **Class 1A, 1B**: Adequate for most threaded nuts and bolts.
2. **Class 2A, 2B**: Provides a tighter fit than Class 1 with no looseness in the thread.
3. **Class 3A, 3B**: Used in precision work and requires extreme care during the moulding operation.

'A' refers to external thread, 'B' to internal. Threads finer than 32 pitches are difficult to mould successfully, hence they are avoided. Sometimes a small interference is placed between two threaded parts, which will prevent loosening under mechanical vibration.
Parts should be designed so that threads terminate a minimum of 0.8 mm from the end as shown in Figs. 5.2 and 5.3.

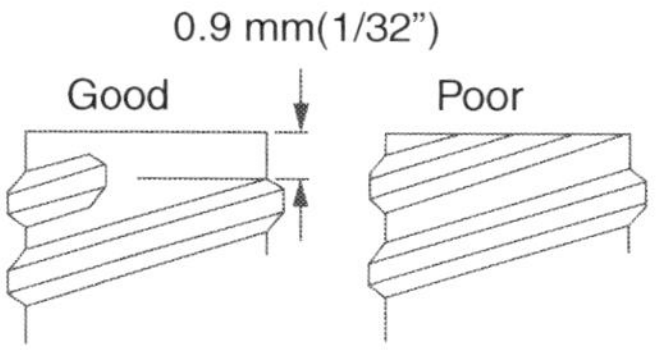

Fig. 5.2 Correct termination of threads.

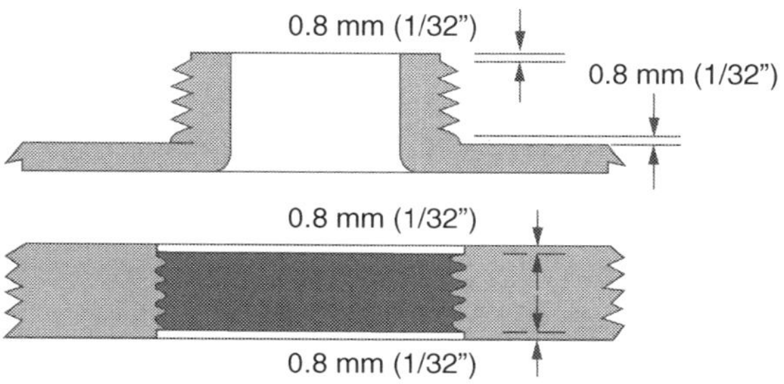

Fig. 5.3 Suggested end clearance on threads.

This practise helps reduce fretting from repeated assembly and disassembly, and eliminates compound sharp corners at the end of the thread. It also prevents cross-threading of finer threads when assembled to a mating metal thread.

Threads-effect of creep When designing threaded assemblies of metal to plastic, it is preferable to have the metal part external to the plastic. In other words, the male thread should be on the plastic part. However, in a metal/plastic assembly, the large difference in the coefficient of linear thermal expansion between the metal and plastic must be carefully considered. Thermal stresses are created because of this difference, which will result in creep or stress relaxation of the plastic part after an extended period of time if the assembly is subject to temperature fluctuations or if the end use temperature is elevated. If the plastic part must be external to the metal, a metal back up sleeve may be needed as shown in Fig. 5.4.

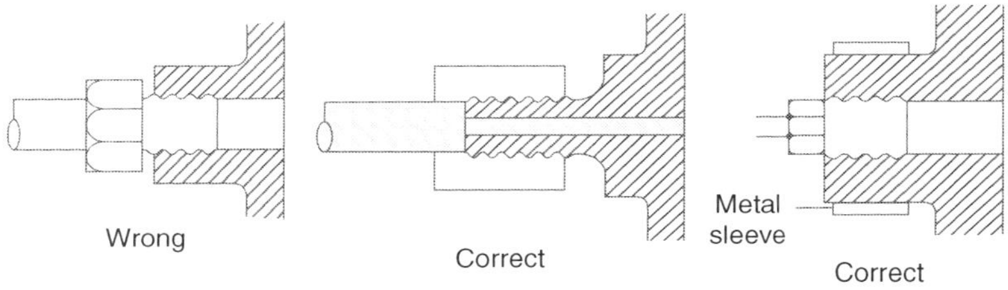

Fig. 5.4 Different thread assemblies.

5.1.3 Application of Shrinkage Allowance on Thread Forms

In all injection moulds shrinkage allowance is added to the dimensions of core and cavity to achieve the actual size of the component after cooling. Shrinkage value varies with the processing conditions and grade of plastic material used for a given process.

Generally, higher the material shrinkage, the larger the cavity and core sizes have to be compensated. Often there is cavity or core forms that will restrict the full amount of shrinkage taking place and moulding a screw thread undergoes such restrictions. If the thread form is a continuous helical undercut then it will prevent free, unrestricted shrinkage in the component. Hence, it is suggested that a lower shrinkage factor should be applied to the pitch than to the thread diameters, where such restrictions do not exist. This becomes increasingly important as the thread accuracy requirements increase, for such applications it is suggested that a reduced shrinkage allowance of 0.7 mm of the normal shrinkage allowance to be applied.

5.1.4 Moulds for Internally Threaded Components

Generally, the threaded components form a restriction during straight draw removal of the moulding from the core. Different types of mould designs adopted for withdrawing the threaded components from the core are detailed below:

1. Stripping (jumping-off) thread design
2. Fixed threaded core design

3. Loose threaded core design
4. Collapsible core design
5. Unscrewing mould designs

Stripping internal threads The internally threaded components are stripped from the core using the stripper plate design for roll threads and the plastic material has sufficient elasticity during the ejection phase. The same principle is applied for stripping any internal undercut components, i.e., the moulding must be free to expand during ejection to permit the moulded undercut to ride over the restriction on the core as shown in Fig. 5.5. This means that the outside form of the component must be such that it can be formed in a cavity, which is fully, contained in one half of the mould. This method is recommended for polyethylene (PE), polypropylene (PP).

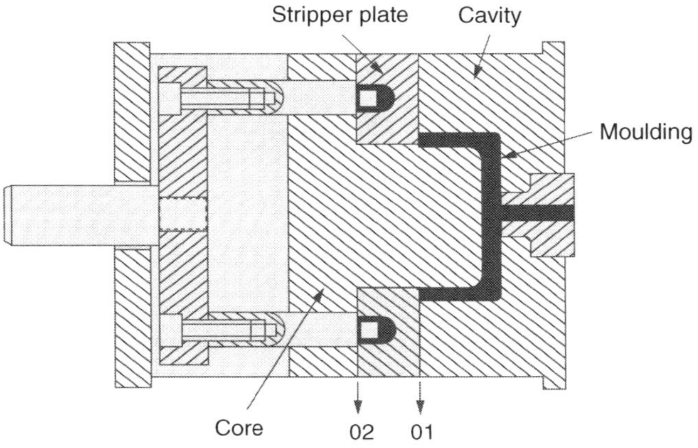

Fig. 5.5 Stripping internally threaded components.

This figure shows a single impression mould for an internally threaded cap. The moulding is formed by the cavity and core and the ejection is by means of a stripper plate actuation system.

In the stripper plate mould after the injection and cooling stage are complete, the mould opens at the parting surface and the component sticks to the core, then the stripper plate moves forward by the action of ejector rod and the core withdraws the moulding through the stripper plate.

The through hole in the stripper plate must be slightly larger in diameter than the major diameter of the thread in order to prevent scoring of the thread. This principle is applicable for single and multi-impression moulds.

Fixed threaded core design In this method, the thread form is incorporated on a non-rotating core fixed to the moving mould plate and an integer type cavity forms the external shape of the moulding as shown in the Fig. 5.6.

In operation, when the mould is opened, the moulding remains on the core and is afterward unscrewed by the operator or using release devices.

The advantages of this design, compared with the unscrewing type mould design are as follows:

1. Mould cost is cheaper than the unscrewing mould design because of the non-requirement of ejection mechanism.
2. Maintenance costs are minimum as there is no moving part within the mould.

The major disadvantage of this method particularly for multi-impression moulds, as the individual mouldings must be unscrewed manually, which increases the moulding cycle time.

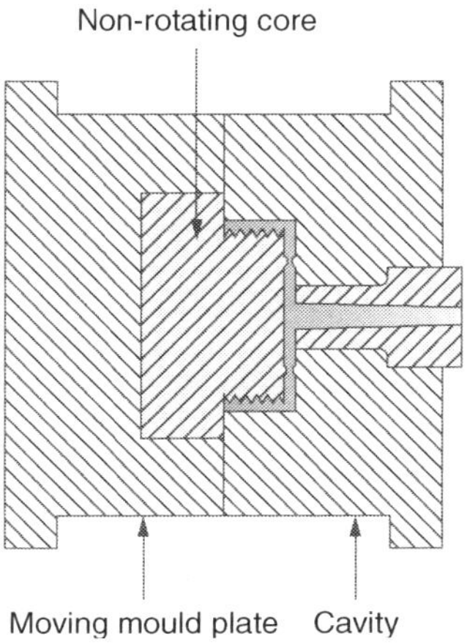

Fig. 5.6 Fixed threaded core.

Loose threaded cores This method is suitable for large component which incorporates a local internally threaded hole or has several internally threaded holes in close proximity to each other. The loose threaded core technique should be considered. This technique prevents automatic unscrewing thereby considerably reducing the cost of the mould. This design is not suitable where number of holes are closely spaced as it requires space to adopt like gear arrangement in automatic unscrewing operation.

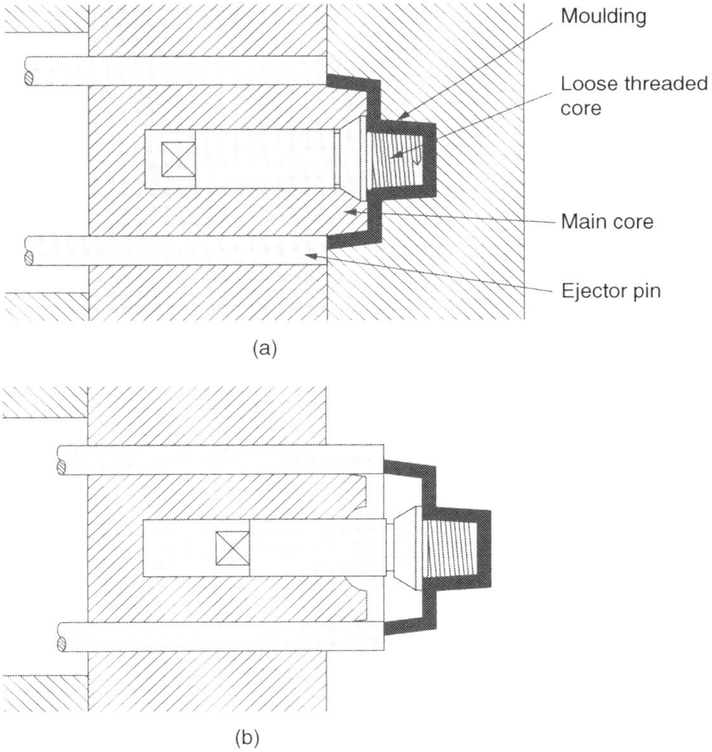

Fig. 5.7 Loose threaded cores: (a) In moulding position, (b) In partly ejected position.

The basic principle of the loose threaded core design is shown in Fig. 5.7 (a), the mould is shown in closed position. The loose threaded core is forming the threaded hole in the moulding. This loose core, which has a valve head type seating, is accommodated in the pocket machined into the main core. When the mould is opened shown in Fig. 5.7 (b), the loose threaded core is ejected with the moulding, and afterwards it is unscrewed manually by using secondary devices like spanner. Two sets of loose cores are used during production. At the end of the first moulding cycle, the second set of cores can be inserted into the mould and the next cycle started. During this cycle the first set of loose cores can be removed from the first moulding and made ready for insertion into the mould immediately the next component is ejected.

Collapsible core This is another widely used method for producing screw threads and a very convenient one for the internal undercuts and threads. In this the core block is machined in such a manner that during mould closing and opening, it works as expansion of core to take the shape of the internal article and reducing the size to remove the formed article. In collapsible core design, two parts are there one as mandrel and second as collapsible sleeve. The collapsible sleeve is slotted through a greater part of the length, forming a number of individual segments. These segments are moving inwards and outwards for reducing and expanding the size of core for ejection.

Types of collapsible core

There are two basic types of collapsible coring:

1. Two-segment core
2. Multisegment core

1. Two-Segment Core: In this method, discontinuous thread in a component is produced like bottle closures use the partial threads.

Two plate moulds are not suitable for collapsible core as longer thread length components are required. In this mould two-segment cores are used and collapsible cores are like mini side core units placed at an angle to the axis of the core.

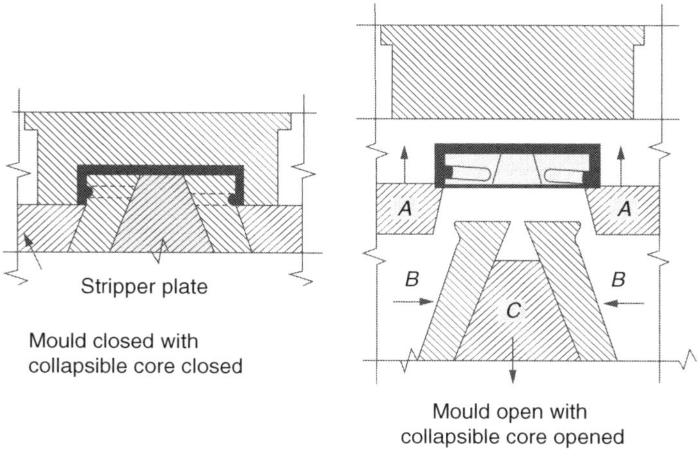

Fig. 5.8 Two-segment collapsible core details.

In Fig. 5.8, the central core is withdrawn downward (action C) away from the moulding. As it moves it forces the two small side cores to move inwards (action B). This clears the cores from the moulded thread. Finally, the part is stripped off from the mould face (action A). An undercut form cannot be placed on the non-side core areas, as it would not be released when the central core withdraws.

2. Multisegment Collapsible Cores: The multisegment collapsible cores consist of a number of segments, which are constrained to move inwards or outwards by the action of a tapered sleeve expanding and collapsing all the individual segments. This system is very similar to the two-segment system but in this the undercut thread form must be machined in a number of segments.

The advantage of collapsible core is to eliminate the need for complex unscrewing mechanisms. This design is applicable for components having internally threads and internal undercuts. The important advantage of this method over the rotating threaded core designs is that it eliminates the need of unscrewing mechanisms using the gear mechanism. The principle of the collapsible core is shown in Fig. 5.9.

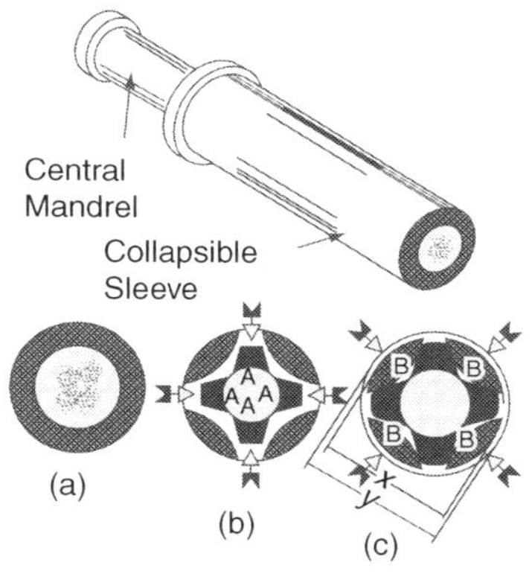

Fig. 5.9 Collapsible core: (a) Hollow core shown in the moulding position, (b, c) showing progressive 'collapse' of core.

The assembly consists of two primary parts, a centre mandrel and a collapsible sleeve. The collapsible sleeve is slotted through a greater part of its length, forming a number of individual segments. The mechanism, upon which this 'collapsible core' operates, is shown in Fig. 5.9 (a), (b) and (c). The segment in their extended moulding position is shown in Fig. 5.9 (b). For explanation purposes, let us assume that the 'black' segments '*A*' can be moved inwards towards each other as shown at Fig. 5.9 (b). This creates a space, which allows the larger shaded segments 'B' to collapse inwards, as shown at Fig. 5.9 (c). As the central mandrel is removed the segments collapse to their final positions. The comparison between the diameters of the collapsed state and uncollapsed state is shown as dimension '*X*' and '*Y*' in Fig. 5.9 (c).

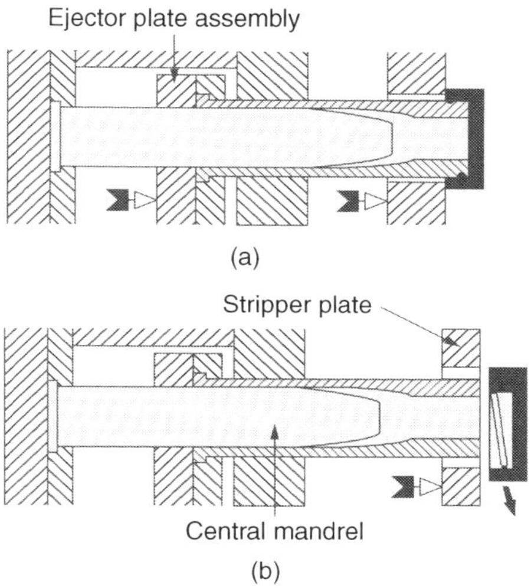

Fig. 5.10 Section view: (a) Ejector plate assembly operated causing collapse of core, (b) Stripper plate operated to eject the moulding.

The collapsible core of the mould is installed in the ejector plate assembly as shown in Fig. 5.10 (a) comes through the support plates and the stripper plate, into the cavity inserts area to mould the inside of a cap. The centre core pin (mandrel) is attached to the back clamp plate of the mould. In moulding sequence, after the mould opens, the ejector plate and the stripper plate come forward together, during which the segments collape away from the internal threads (since the centre core pin retracts). At the end of the ejector plate travel, the stripper plate is actuated to lift the moulded part of the collapsed core as shown in Fig. 5.10 (b).

These units can be used for single or multicavity moulds and are available in six sizes, from 25 mm diameter approximately to 90 mm diameter, with corresponding collapses per side of approximately 1.1 mm for the smallest and 3.5 mm for the largest, at the top of the core. These collapses can be reduced by 0.5 mm along segment length within usable moulding length.

Advantages and disadvantages of collapsible core

Advantages: This type of mould is available as standard units with reasonable range of sizes and an undercut form in the component is up to 5 mm. It is cost effective and suitable for small numbers of impression.

Disadvantages: This type of mould is suitable for limited thread length components. Mould maintenance cost is more and cooling of core is difficult.

Rotary unscrewing Rotary unscrewing is used where split line witness cannot be accepted on the component when larger numbers of impressions are required in a mould. The basic method for rotary unscrewing is either to unscrew the part from the core or to unscrew the core from the part as shown in Fig. 5.11.

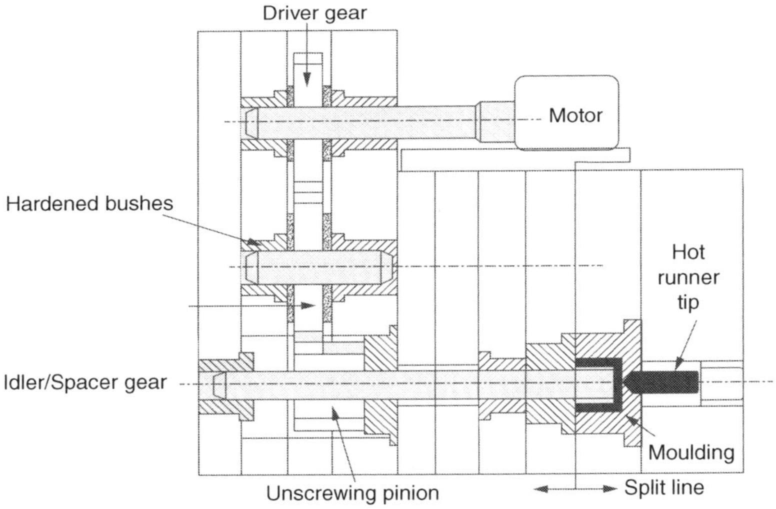

Fig. 5.11 Rotating core unscrewing tool.

Rotating cores or cavities can be actuated by a rack-and-pinion system, with a rotating core or with a rotating cavity.

Core unscrewing Core unscrewing is used to free internal screw thread forms and falls into two categories:

1. Core remains fixed with respect to the mould
2. Core moves axially with respect to mould

Fixed core systems: Fixed core systems are the most widely used method for automatic unscrewing. With this system, the core rotates in a static plane, simply revolving with no axial displacement of the core relative to the mould tool.

Types of unscrewing moulds: In an unscrewing type mould, either the cores or the cavities are rotated to automatically unscrew the mouldings from the mould. The unscrewing mechanism is fitted behind the moving mould plate in place of the conventional ejector unit which provides required rotational motion. From the impression construction stand point; various designs are followed like

1. The axially fixed core design in which the threaded core is rotated to remove the moulding.
2. The extractor plate design in which an extractor plate is actuated at the same time as the threaded core is rotated.
3. The withdrawing rotating core design in which the threaded core, being rotated and simultaneously.
4. The rotating cavity design.

Types of impression layouts in an unscrewing mould is 1. pitch circle diameter (PCD) layout 2. in-line layout.

1. Manually powered transmission system: In this method, the moulding is unscrewed manually by the operator using the rotating handle positioned either at the top or the front of the mould. The handle assembly is usually mounted on an extension of the back plate and the

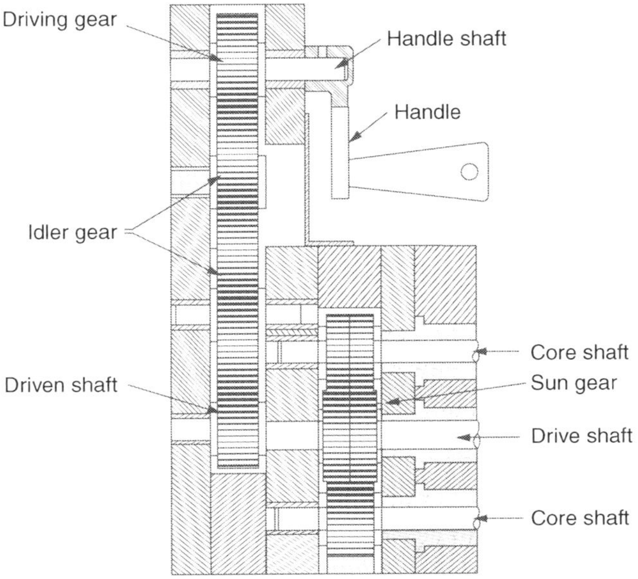

Fig. 5.12 Manually powered fixed rotating core design with gear transmission.

assembly consists of a U-shaped handle block locked to the handle shaft by a grub screw. The manually powered systems differ with respect to the transmission system used to connect the handle assembly to the sun gear drive shaft.

(a) Manually powered fixed rotating core design with gear transmission In axially fixed rotating core design, the driving gear is mounted on the handle shaft and a driven gear on the drive shaft. The idler gears are provided between the driving and the driven gears to complete the gear train. The manual rotation of the handle causes the sun gear, pinions and core shaft to rotate which in turn causes the moulding to ride up their respective threads and thereby the component is ejected. All the shafts which are rotated are mounted in bearings, as shown in Fig. 5.12.

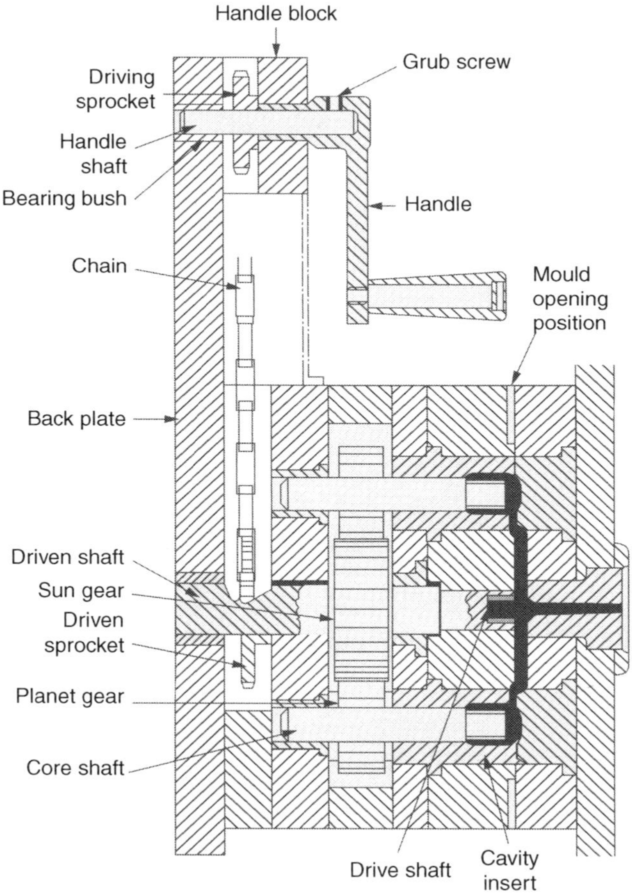

Fig. 5.13 Manually powered design with fixed rotating core and chain and sprocket transmission.

In manually powered design with fixed rotating core and chain and sprocket transmission system, the mould opens at the parting surface and the moulding remains at the cavity. The feed system is pulled by the puller incorporated in the drive shaft. As the mould opens the operator rotates the driving sprocket by the handle and the handle shaft. A chain connects the driving and driven sprockets, the latter mounted on the drive shaft. The sun gear also mounted on the drive shaft meshes with the planet gears or pinions secure to the individual core shafts. The rotation of handle causes the individual core shafts to rotate, which in turn causes the mouldings to be ejected.

(b) Manually powered withdrawing core design, with rack-and-pinion transmission In this method, the unscrewing mechanism is operated by rack-and-pinion transmission system.

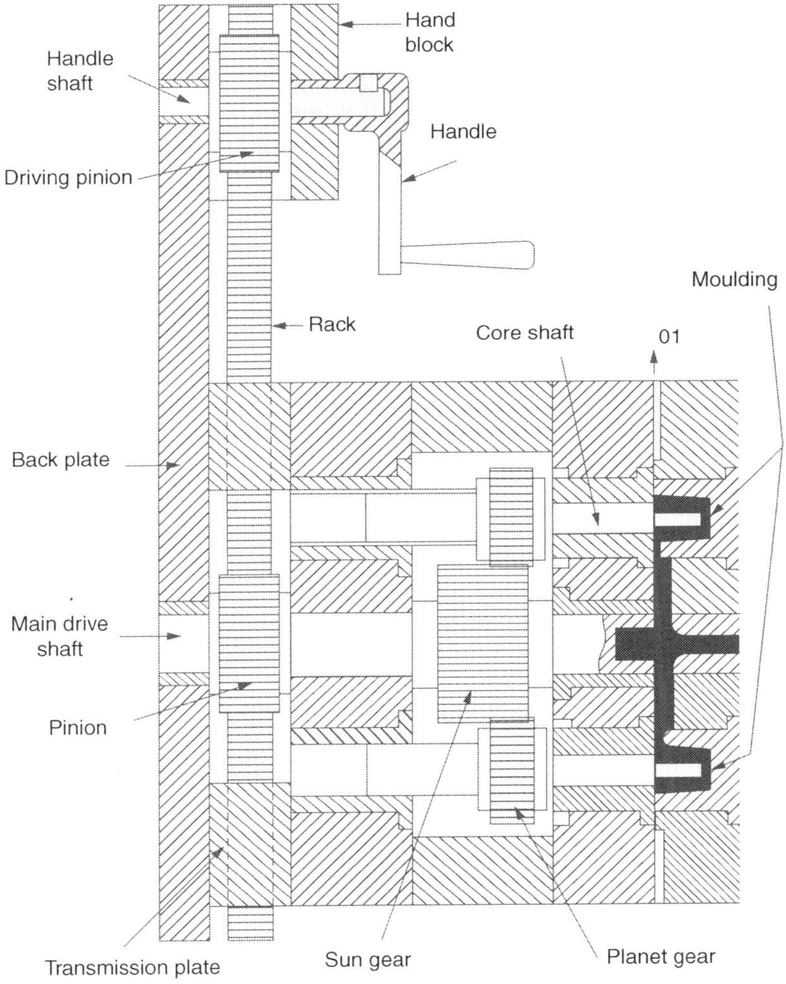

Fig. 5.14 Manually powered withdrawing core design, with rack-and-pinion transmission.

A pinion is secured to the main driving shaft on which the sun gear is mounted. A rack suitably guided within the transmission plate, meshes with the pinion, and also with the driving pinion located within the handle block assembly. Manual rotation of the driving pinion by the handle causes the rack to move upwards, which in turn causes the driven pinion and hence the sun gear rotates. The planet gears are thereby rotated, which causes the core shafts to be progressively withdrawn through their respective bushes and the mouldings ejected.

2. Machine-powered system: The machine powered system is suitable for an in-line layout of gear arrangements. In this mechanism two racks are mounted at right angle to each other as shown in Fig. 5.15. The horizontal rack and pinion transform a linear motion of the platen to rotary motion and the vertical rack transmits the rotary motion to the core shaft pinions.

When the mould is initially open, the mouldings are withdrawn from the cavities as a slight delay is provided on the front part of the rack. Immediately, this delay movement has been taken up the horizontal rack is stopped by the stop bar mounted on the fixed mould plate assembly, the horizontal rack passes through a

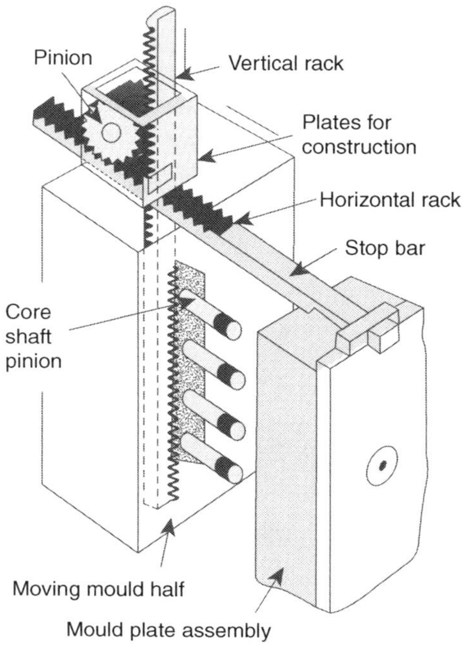

Fig. 5.15 Machine powered in-line unscrewing system.

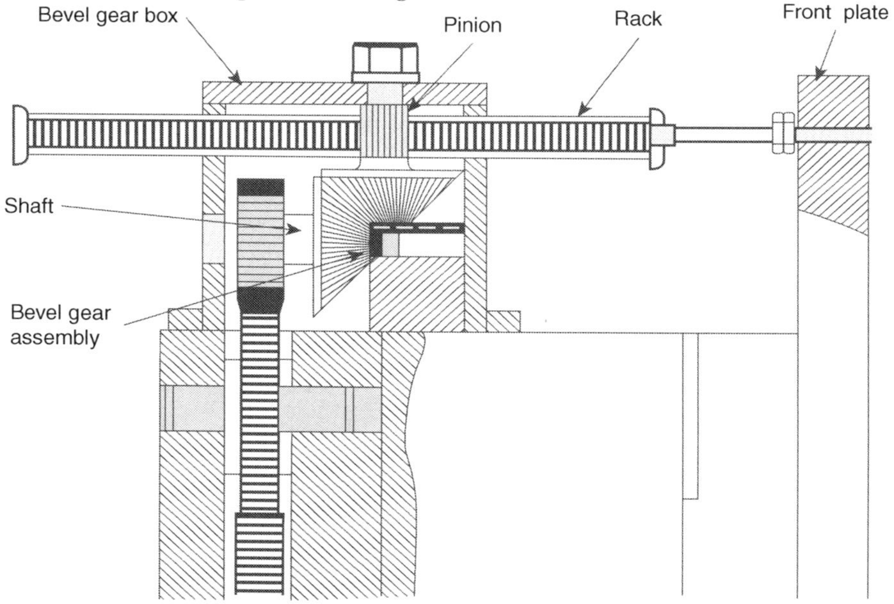

Fig. 5.16 Machine powered, horizontal rack design.

slot machined in the back plate. The continued movement of the moving mould half causes the operating pinion to rotate. The rotation of the pinion causes the vertical rack to move upward, thus the vertical movement of the rack causes the core shaft pinion to be rotated and the mouldings are unscrewed.

3. Rotating helix spindle method: In rotating helix spindle mechanism, a multiple start helix spindle is mounted in bearings located in the moving half of the mould, while a complementary multistart helix nut is attached to a rear machine platen or auxiliary platen shown in Fig. 5.17. The helix spindle remains in contact with the fixed helix nut and during the mould opening stroke, the moving mould half moves to the left and relative longitudinal motion occurs between the helix spindle and the helix nut. This causes the helix spindle to rotate. The sun gear is attached to the helix spindle and rotation of the gear wheel operates a number of planetary gears. Closing the mould causes the helix spindle to rotate in the opposite direction, thereby returning the planet gears to their original positions.

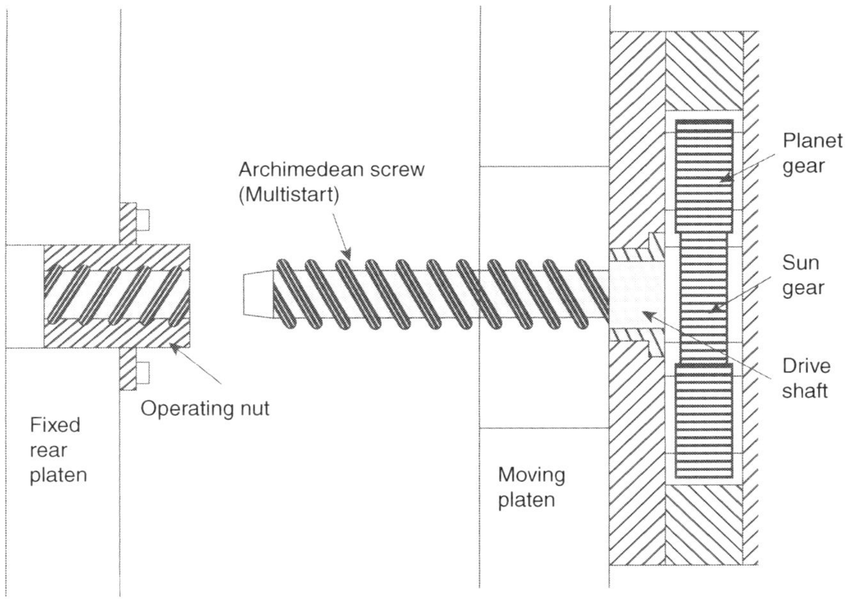

Fig. 5.17 Machine powered, archimedean screw design.

Fixed helix spindle In this method, the helix spindle is attached to the fixed mould half, and the helix nut is rotated within the moving mould half during the opening stroke. During the opening stroke, the moving mould half moves to the left, relative movement occurs between the helix spindle and helix nut causing the helix nut and sun gear to rotate. This in turn rotates the planetary gears thereby unscrewing the mouldings. The mould closing stroke causes the helix spindle to rotate in the opposite direction and returning the individual gears and core shafts to their previous position.

4. Hydraulically or Pneumatically Powered Systems: In this method, the actuation of rack and pinion is done hydraulically. It is advancement of the manual rack-and-pinion design which allows the unscrewing operation to be made fully automatic. The operating handle assembly is replaced by an actuator, the ram of which is directly coupled to the rack. The actuator is mounted on a platform which is supported on columns suitably secured to the moving half of the mould. The rack extends through the mould is coupled to the ram of the actuator and independent control of the actuator causes the unscrewing operation.

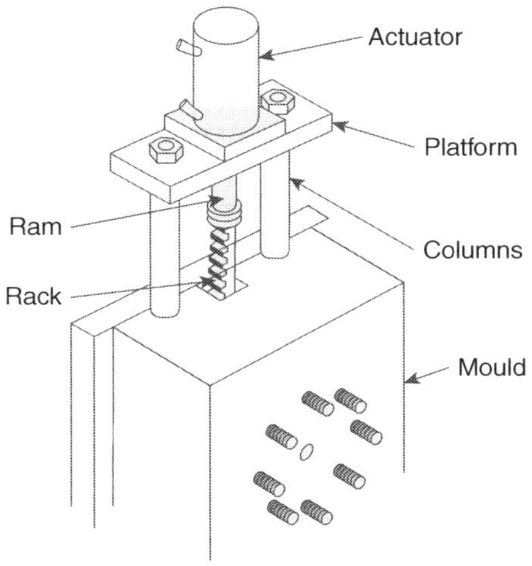

Fig. 5.18 Hydraulically powered systems.

Hydraulically or pneumatically powered systems This design is used for power transmission by using rack-and-pinion method which allows the unscrewing operation to be made fully automatic. The handle assembly is operated by hydraulically or pneumatically powered systems with the help of an actuator, which is directly coupled to the rack. The rack extends through the mould is coupled to the ram of the actuator and the actuator causes the unscrewing operation shown in Fig. 5.19.

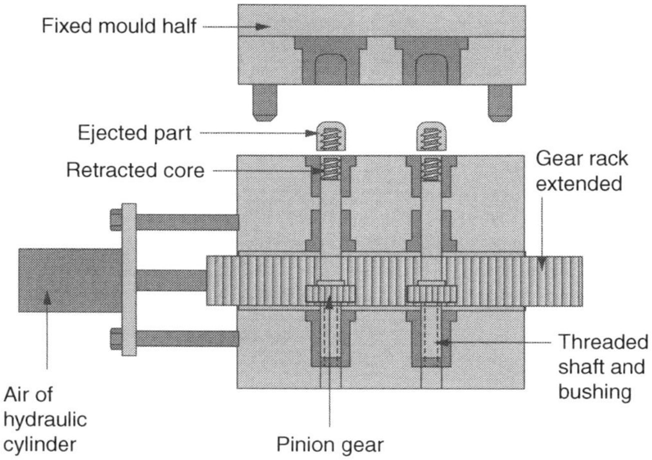

Fig. 5.19 Unscrewing with hydraulic system.

5.1.5 External Threads

The components with external threads can be moulded in two ways. One method is to design the split cavity mould in which the parting line is located on the centreline of the thread and two cavity halves move oppositely to release the part as shown in Fig. 5.20. The second method of moulding and unscrewing the external thread in the direction of mould opening is done by manual and automatic thread unscrewing devices.

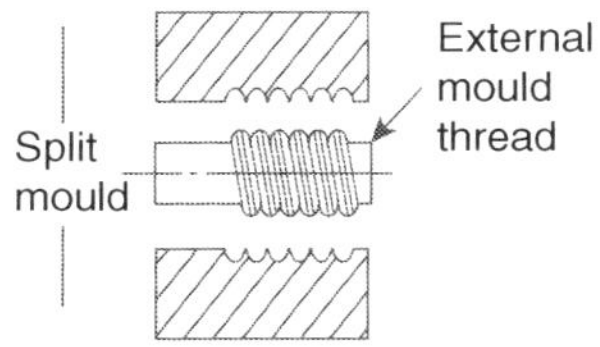

Fig. 5.20 Split mould.

5.1.6 Moulds for Externally Threaded Components

The externally threaded moulding is considered as an undercut while designing the mould and it prevents straight draw removal of the moulding from the cavity during ejection. The threaded type of moulding can be released from the cavity by unscrewing mechanism, stripper ejection method or by incorporating split mould technique.

The unscrewing mechanism requires rotary motion within the mould to perform the unscrewing operation of the component automatically. The stripping method makes the production cycle faster, but it is limited to those components which incorporate roll threads only. In case of split mould design, the thread shape is machined in the splits and it is opened by using finger cam or dog-leg pin method for releasing the moulding from the cavity.

Fixed threaded cavity design This type of mould is used for producing externally threaded component in which the threaded portion is machined directly into the cavity insert shown in Fig. 5.21 (a). When the mould is opened the moulding is unscrewed from the mould core plate shown in Fig. 5.21 (b). The unscrewing method can be manual or electrically power assisted.

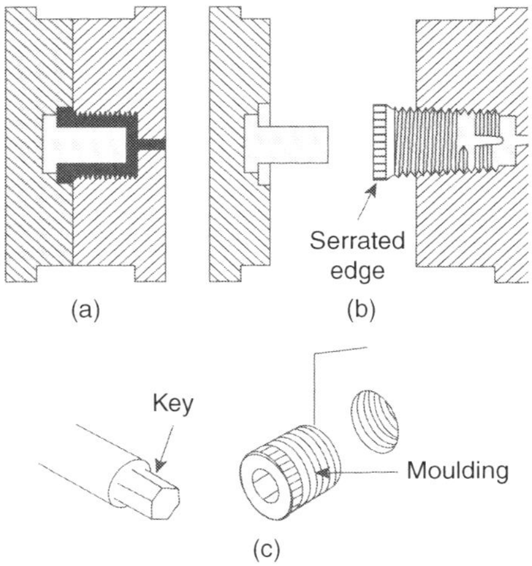

Fig. 5.21 Fixed threaded cavity design: (a) Mould closed, (b) Mould open, (c) Moulding and release key.

The components should have sufficient grip for unscrewing. The parting line has been chosen to allow the head of the component to project above the mould's parting surface and the edge of the head is serrated to provide a grip. An alternative method is to design the component with either a square or a hexagonal aperture. This permits the use of a key as shown in Fig. 5.21(c) for unscrewing the moulding.

Advantages of fixed thread cavity design over the automatic splits or rotating cavity designs:

The mould cost is cheaper compared to the automatic design and the joint line is not visible in case of automatic unscrewing method, as the disadvantage associated with the split design in fixed thread cavity mould.

Since there are no moving parts within the mould the maintenance cost is low compared with automatic moulds.

The major disadvantage of the fixed thread design is that the moulding cycle is relatively long because it involves a manual unscrewing operation.

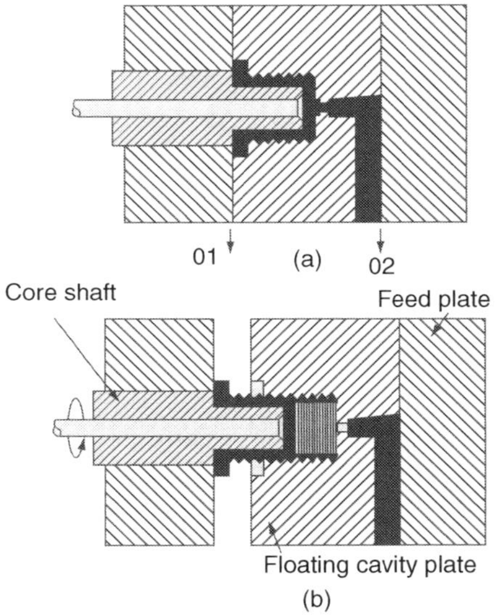

Fig. 5.22 Automatic unscrewing for externally threaded components.

Automatic unscrewing of externally threaded components This design is the same as the automatic unscrewing method for internally threaded components. However, for externally threaded components the cavity is fixed and the core is rotated for unscrewing action.

The operating sequence of the automatic unscrewing mould as follows:

After the cavity is filled the moulding is cooled sufficiently, further the mould is opened at the parting surface as (01) shown in Fig. 5.22 (b) and the core shaft is rotated. Thus the moulding is progressively unscrewed from the cavity and the gate is broken in the process. While moulding is being unscrewed the second opening occurs between the floating cavity plate and the feed plate. This allows for the removal of the feed system and the moulding is ejected from the core.

The rate at which the moulding is unscrewed must be synchronised with the opening movement of the mould and this can be achieved by the use of the screw jack. It is essential that the shape of the core is such that it acts as a key when the core shaft is rotated, to unscrew the moulding. For example, small projecting ribs on the internal bore of a component provide a suitable grip for unscrewing purposes. The general build-up of the moving half of the mould is basically the same as for the fixed rotating core arrangement and similar layout, power unit and transmission systems are adopted.

Stripping (jumping) external threads This method is applicable for the components having external threads in rolled form and the moulding material should be sufficiently elastic to re-turn to its original shape after being deformed.

The stripping design of mould avoids the unscrewing method of moulding, or using splits, and fully automatic operation. The stripping mechanism is shown in Fig. 5.23 in which the external shape of the moulding is machined in the cavity insert and it is mounted in the moving mould plate, the core is mounted in the fixed mould plate. When the mould is opened the component is retained in the cavity by the threads and is, therefore, pulled from the cavity. Relatively large ejector pin is positioned below the lower face of the moulding and when the ejector assembly is actuated the moulding is ejected from the cavity. The pocket in the component, formed by the core, permits slight contraction, which allows the moulded threads to ride over the complementary cavity threads.

Figure 5.24 shows the roll thread form of the moulding. The recommended plastic material is PP, HDPE, LDPE and LLDPE. This method is not suitable for moulding conventional 'V' or square threads as the sharp edges will be damaged during ejection phase.

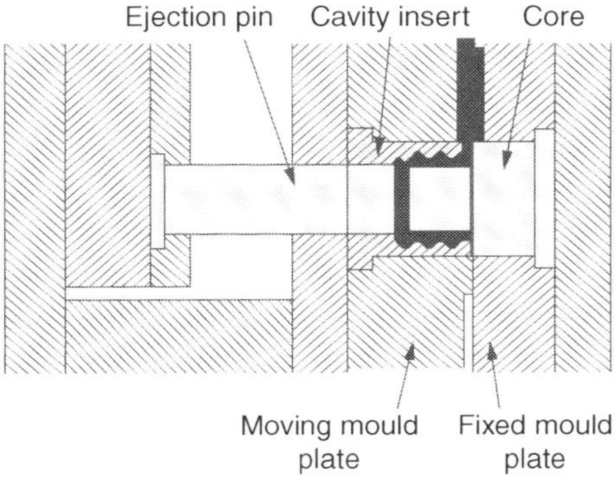

Fig. 5.23 Stripping (Jumping).

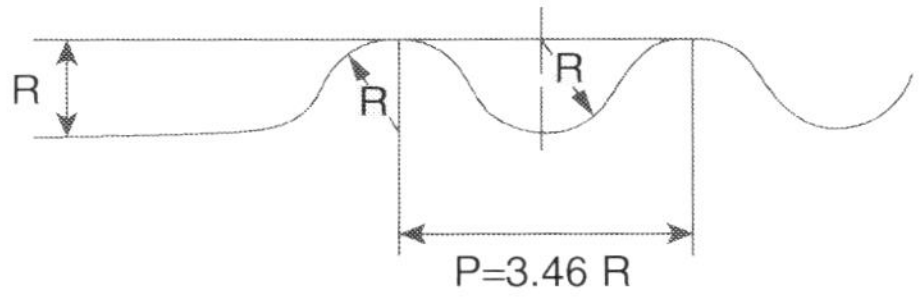

Fig. 5.24 Typical roll thread.

The joint line impairs the efficiency and quality of the produced component. In the case of a component which has an interrupted thread, from the joint line can be positioned on the plain section, thereby avoiding the necessity of requiring such extreme accuracy as shown in Fig. 5.25.

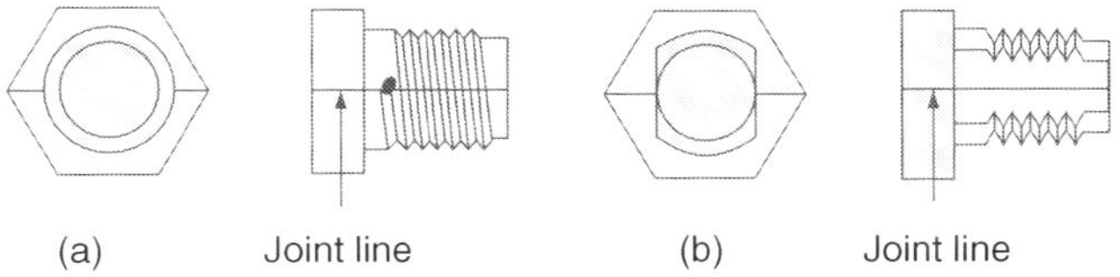

Fig. 5.25 Joint line on externally threaded components in splits.

Threaded splits This method is adopted when automatic production is required for an externally threaded component, where the thread form is such that it cannot be stripped. Extreme accuracy is to be maintained while manufacturing and fitting of the splits; otherwise, flashing may occur along the fitting of the spilt. A spilt mould for an externally threaded component is shown in Fig. 5.26.

In this design the dog-leg cam operating method was chosen to permit the use of simple pin ejector system. The dog-leg cam provides sufficient delay, before the splits are opened, to allow the moulding to be pulled clear of the core. The square thread profile will always create an undercut of varying severity according to the pitch and depth of thread shown in Fig. 5.27.

The point '*X*' is situated on the minor thread diameter at the joint line. The complementary point on the splits must travel distance '*Y*' before it is clear of the thread shown in Fig. 5.27. The point '*Z*' at which the split loses contact with the moulding is higher than point '*X*', therefore, an undercut must result. To avoid this, the thread profile should be changed to a trapezoidal form having a flank thread angle θ. From the formula a value for θ can be calculated.

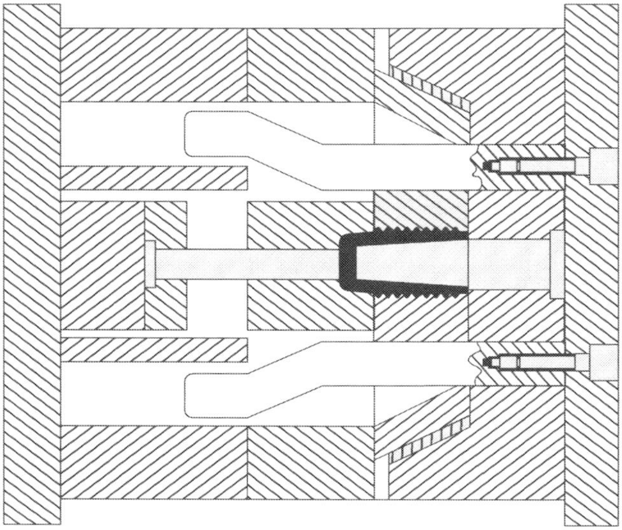

Fig. 5.26 Split mould for externally threaded components.

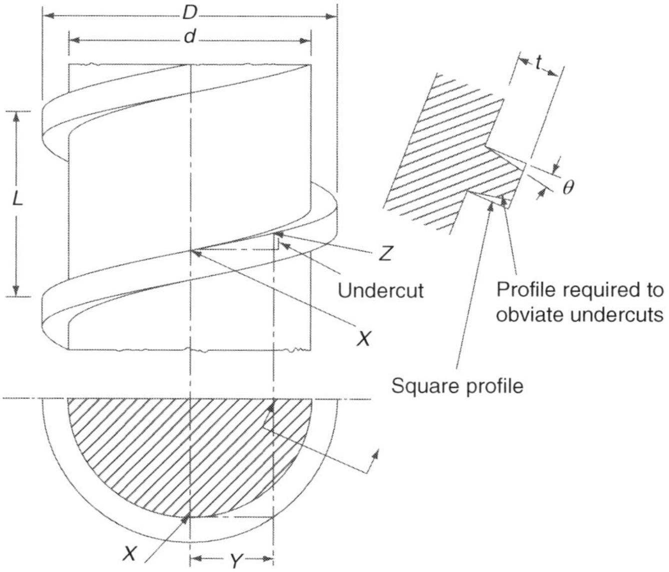

Fig. 5.27 Design of thread.

$$\tan \theta = \frac{L}{\pi D t} \sqrt{\frac{D^2 - d^2}{2}}$$

where L = Lead of thread
D = Major thread diameter
d = Minor thread diameter
t = 0.5 (D–d)

5.2 HOT RUNNER MOULD

5.2.1 Introduction

The hot runner mould contains a heated runner manifold block within the mould and the manifold block is insulated from the rest of the mould, is maintained at a closely controlled elevated temperature to keep the runner permanently as a melt. The polymer material is directed to the mould cavities without loss of heat and pressure. The hot runner unit is mounted adjacent to the cavity plate. The polymer material enters via a centrally positioned sprue bush via secondary nozzle to the impression. The hot runner system represents the most extensive technological development from a conventional runner system to a system ready to be installed. Large mouldings such as automotive dashboards, bumpers, computer housing, etc., are produced using hot runner moulds. Almost all the plastic materials can be moulded in hot runner system, even reinforced plastic and structural foam can also be processed.
The injection moulds are classified in to the following types:

1. Cold runner system
2. Hot runner system

1. Cold runner mould: A cold runner mould is a conventional mould in which the feed system, i.e., the sprue, runner and gate is cooled and ejected with the part. Every production cycle, the component and runner are produced. There are two major types of cold runner moulds, i.e., two plate and three-plate mould.

The advantage of cold runner mould is the mould requires less maintenance and the design is simple, much cheaper than a hot runner system. Less skill requires setting up the mould and operation. Colour changes are also very easy, since all of the plastic in the mould is ejected with each cycle.

The disadvantage of cold runner mould is the plastic waste that is generated and the runners are either disposed of, or reprocessed with the original material. The regrind plastic material will increase variation in the injection moulding process, and decreases the mechanical properties of polymer.

5.2.2 Runnerless Moulding

Runnerless moulding can be described as a system of moulding in which the conventional melt from the barrel is brought directly to the cavities. Thus, the molten materials are delivered through auxiliary heated passages and into the parts that are being moulded.

The advantage of runnerless moulding is that there are no runners to regrind and consequently no need for use of a mixture with regrinds in moulded parts. The use of regrind in a mix with virgin material increases the rejection rate, which is higher compared with the use of virgin material alone.

The melt is delivered to the cavity at optimum flowability. Each plastic has its limited flow length, and, when runners are long, the flow in the cavity may be working in the tail end of polymer flow length and thus requires high pressure for filling the cavity.

Polymer melt is used at lower range of the temperature, thus providing the potential of lower cycles. Furthermore, only solidification of the part rather than that of the runner determine the duration of the cycle. Press plasticising capacity is improved because no volume is required for sprue and runner and available heated volume contributed by the auxiliary manifold system is increased and moulding problems associated with sprue and runners are eliminated.

Injection-pressure requirements are lowered because the melt fluidity is maintained right up to the gate. Good fluidity of melt at the gate reduces injection-pressure values, which in turn is reflected in lower clamp-pressure requirement. Greater freedom of gate location is afforded as a result of the melt being fully fluid at the entrance to the part.

Mouldings requiring long flow paths can be made with a single gate, whereas, in conventional runner moulding multiple gating is required because of loss of heat in the runner. Since there is no wastage of feed system, requirements for plastic material grinders are reduced and handling of regrind is eliminated.

The disadvantage of runnerless moulding is the mould cost is higher compared to conventional injection moulds. In this process, the controls for manifold and nozzle temperature are required and the initial debugging time is longer than in a conventional mould.

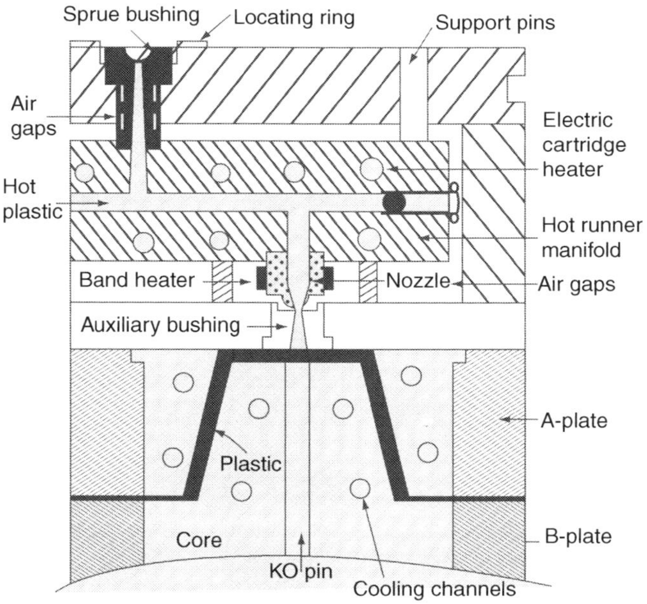

Fig. 5.28 Hot runner mould.

The runnerless moulding is accomplished by the use of

1. Hot runner manifold
2. Insulated manifold

5.2.3 Principle of Hot Runner Mould

In a three-plate cold runner mould, the runner system is reground and the material is re-used. In case of hot runner mould, it eliminates the solidification of runner system entirely by keeping it fluid. The material is kept plasticised by the hot runner manifold, which is heated with heating element like electric cartridge heaters. The manifold block and the band heaters mounted round the nozzle are thermostatically controlled. The plastic is kept fluid and the injection pressure is transmitted through the hot runner manifold.

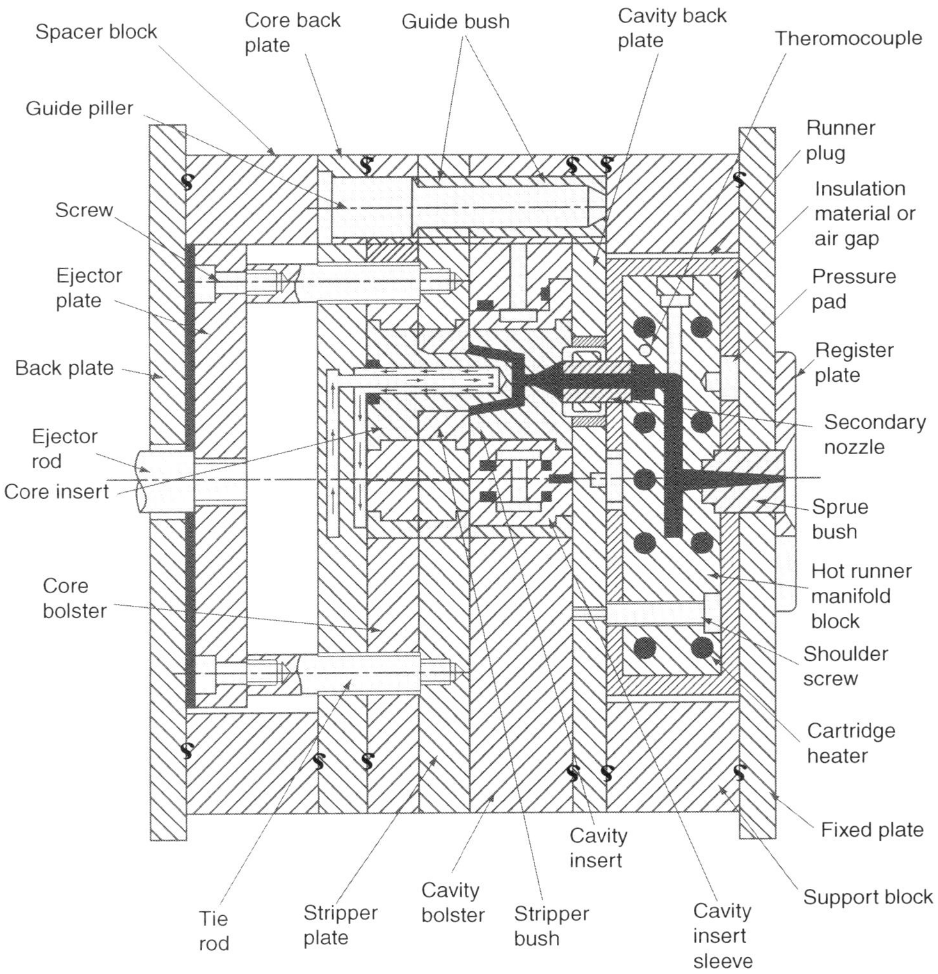

Fig. 5.29 Assembly parts of hot runner mould.

The hot runner mould takes more starts up time and in case of multicavity moulds, balancing of gate, the melt flow and preventing drooling are difficult. A keen balance between no drooling and freezing up of the nozzle must be achieved. The hot runner mould is highly affected to tramp metal, wood, paper, and other contaminants, which quickly block the nozzle. Cleaning out plugged hot runner mould is a long process. These moulds are more expensive compared to cold runner moulds.

Hot runner mould with externally heated manifold and drops A two plate hot runner mould with a heated runner system inside one half of the mould is shown in the Fig. 5.30 in which the runner system is divided into two parts are manifold and drops. The manifold has channels that convey the plastic melt on a single plane, parallel to the parting line, to a point above the cavity and the drops are situated perpendicular to the manifold convey the plastic from the manifold to the part. The heating system in hot runner mould is categorised as internal and external heating of manifolds and drops.

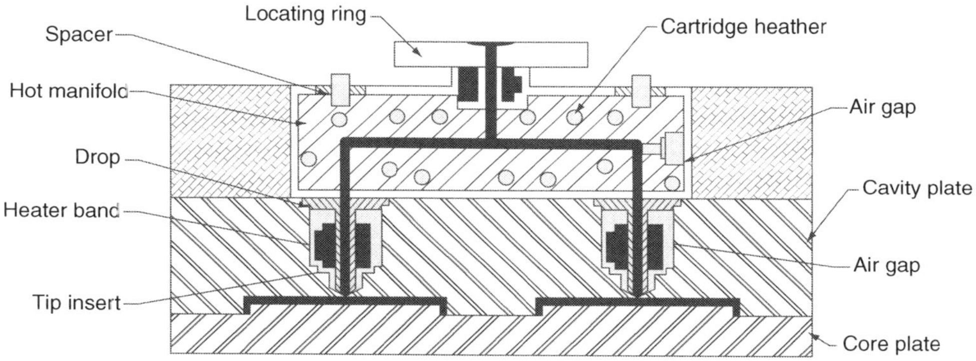

Fig. 5.30 Hot runner mould with externally heated manifold and drop.

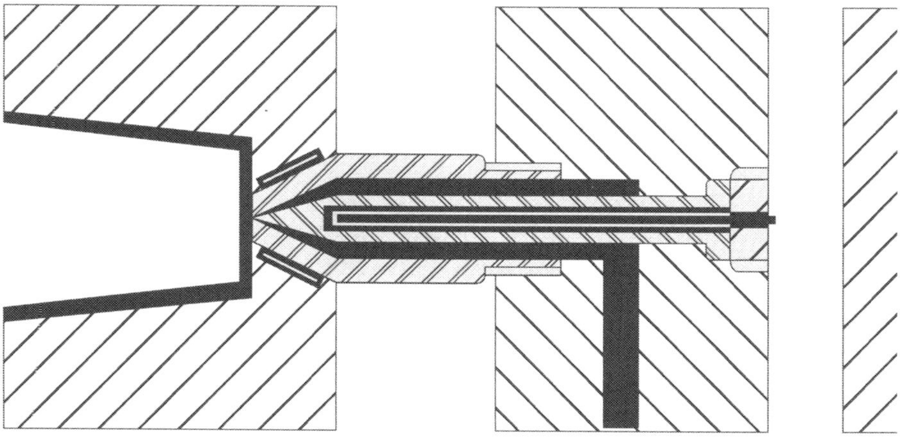

Fig. 5.31 Internally heated-core rod-secondary nozzle.

Externally heated hot runner channels have the lowest pressure drop of any other runner system, because there is no heater obstructing the melt flow path and they are better for colour changes. There are no places for material to hang up and degrade, so externally heated systems are good for thermally sensitive materials.

Internally heated runner systems require higher moulding pressures, and colour changes are very difficult in this process. There are many places for material to hang up and degrade, so thermally sensitive materials should not be used. Internally heated drops offer better gate tip control and internally heated manifold separates runner heat from the mould because an insulating frozen layer is formed against the steel mould wall on the inside surface of the flow channel.

5.2.4 Design Guidelines of the Hot-Runner System

The following are the design considerations of a hot runner mould:

1. The insert plugs and close off manifold bores should be smooth for free flow of material.
2. The pressure drop from sprue bushing to tip should be not more than 25 per cent of the maximum plastic fill pressure, with the resulting temperature increase and the runner volume should be 25 per cent of the part volume.
3. Sufficient heaters should be incorporated in the mould so that the hot-runner unit heats quickly to the required moulding temperature from cold.
4. Sufficient heat energy must be supplied to heat the hot-runner unit to compensate the heat losses by conduction, convection and radiation.
5. Suitable location of the heating elements is essential to ensure that the temperature of the melt in the flow-way is maintained to keep it molten. Considerable production time may be lost if a heater fails, hence proper choice of heating element; its location and the facilities for removing it must be ideally done. The layout of the wiring system of heaters should be neat and easily traceable and heating element wires, which are subject to, heat or abrasion attack should be protected. While fitting the heating elements, cartridge type heaters are recommended and necessary clearance must be provided. The cartridge-heating element does not release heat over its entire length as there are end 'cold spots'. Flat-type heating elements and induction heating elements should be completely enclosed within the unit for maximum efficiency.

Allowance is provided as the hot-runner unit increases in size in three dimensions when heated and it should be insulated from the rest of the mould structure. For many materials, close control of the temperature of the melt is vital hence, careful consideration to the location of thermocouples is essential.

To minimise degradation, colour-changing and material-changing problems the melt-flow-way must be as streamlined as possible, without sharp corners, edges or other stagnation points which tend to hold back the polymer melt for extended period.

Applications of hot-runner mould In hot runner moulds the pin point gating of mouldings on multi-impression types of mould is adopted and it also allows for multipoint gating on single-impression and multi-impression moulds. It is feasible for side or film gating of large mouldings and it permits the semirunnerless design to be adopted, where small groups of impressions are fed from secondary sprue.

5.2.5 The Manifold Block

The manifold block and secondary nozzle is the main part of hot runner mould. The design of manifold parts varies depending upon the moulding size and material properties. The function of manifold is to convey the melt to the cavities without affecting polymer properties. It is made from hardened hot worked tool steel (P20), OHNS, etc., which can withstand highest moulding pressure and temperature.

The manifold block is designed to withstand highest melt pressure of 1800–2500 bar and it should be correctly sized, smooth flowing and balanced melt channel layout to be made. It must be durable and easily replaceable heating elements with quick heat-up system to be fitted.

The manifold block should have low radiation losses and temperature variation as low as possible. Embedded or surface mounted thermocouples should be provided and influence of thermal expansion on the position of gate in the cavity occurs.

The manifold block can be classified according to its cross-sectional shape; rectangular or circular. While the rectangular type is usually manufactured as a one-piece structure and the circular cross-sectional manifold block is of composite structure.

Two types of manifold blocks are shown in Figs. 5.32 and 5.33. The difference between these two is the method adopted for heating the manifold blocks. In the rectangular design, cartridge, coil or flat-type heating elements are fitted into suitably shaped holes or recesses within the manifold block, whereas in the cylindrical design, band or coil-type heating elements are fitted on the outside of the manifold block.

The efficiency of the manifold is determined by estimating the amount of heat loss to surrounding mould components. A manifold with a relatively low surface contact area through spacers and runner bushings and with a relatively large gap on all sides of the manifold will be more efficient than that with a high number of spacers, etc.

Rectangular manifold block The rectangular cross sectional manifold block is one among the number of alternative shapes as shown in Fig. 5.32(a). The right prism manifold block is manufactured from a rectangular block of steel in which holes are bored to form the flow way system. The outlet apertures should be in one line to permit the individual holes to interconnect by straight drilling.

However, for widely spaced outlet points or for applications where the outlet points are not symmetrical about the mould's centreline, the use of an alternative manifold block shape should be designated by *X*-shaped unit and an *H*-shaped unit shown in Fig. 5.33.

Circular manifold block The circular cross section manifold block consists of an assembly of cylindrical sections, an inlet block and outlet blocks. The simplest version of the circular cross-sectional manifold block is illustrated in Fig. 5.32 (b). It is manufactured from a cylindrical bar of steel and incorporates a central flow way drilling and the outlet flow-ways is bored to interconnect with the central drilling.

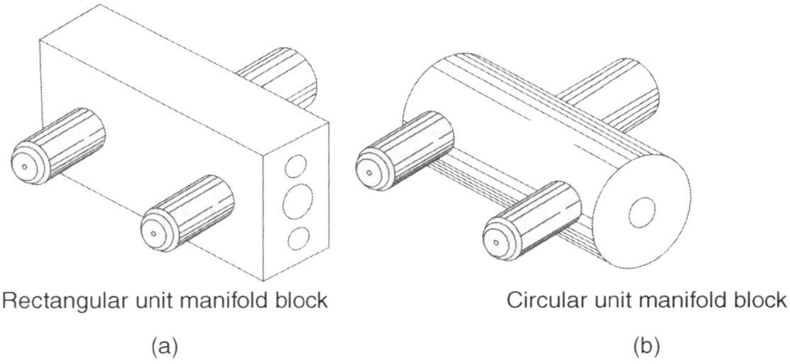

Fig. 5.32 (a) Rectangular unit (b) Circular unit manifold blocks.

5.2.6 Heating of Manifolds

Incorrect selection of heaters or their mounting create problems connected with the hot runner system, for which the calculation is required to get the heater size for heating up the manifold.

Let us assume that the length, width and height of the manifold is $a \times b \times c$, respectively. The time required to heat up the manifolds up to the required temperature = 30 min.

The processing temperature will be T_2 and the starting temperature is T_1.

The amount of heat for the job is determined by the need for bringing the manifold to operating temperature in the required time and by the amount necessary to maintain the temperature under operating conditions whichever be the higher will be the one selected for the job.

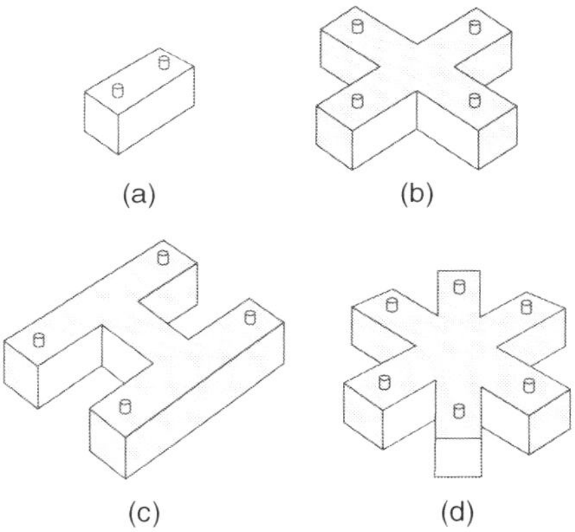

Fig. 5.33 Different types of rectangular manifold design.

1. **Required heaters size for bringing the manifolds up to the processing temperature T_2 in 1 hour from T_1**

   Heat required = Weight of manifolds × Specific heat × Temperature difference

   $$= abc\rho \times k \times (T_2-T_1) = abck\rho\,(T_2-T_1)\ \text{kcal/hr} = Q\text{(assume)}$$

2. **Heat losses at T_2**

   (a) Losses due to convection and radiation

   Exposed surface area = $2(ac + bc)$

   Losses due to convection $Q_c = h\,A\,(T_2-T_1)$ where h = Heat transfer coefficient and A=Exposed area of manifolds

   Losses due to radiation $Q_r = \epsilon\sigma A\,(T_2^4-T_1^4)$

   here ϵ = Emissivity

   σ = Stefan Boltzmann Constant

Heat losses due to convection and radiation ($Q_c + Q_r$), by putting the values, Q_c and Q_r can be calculated.

(b) Losses by conduction

Q con = $K A (T_2 - T_1)/L$

Where K= Thermal conductivity of insulation, A= Area of conduction and L= Thickness of insulation generally air may be selected.

Area of conduction= $2ab$

By selecting the values, calculate the Q con

Total heat required per hour = (Q+Q_c+ Q_r+ Q con) kcal/hr

Watt required = Total heat required in (Kcal/ hr)/ (864).

Criteria for heaters into the manifold The holes for the heater should be reamed to a smooth finish and size so that the clearance per side should be 0.125 mm.

The heater should be spaced so that the lead end is flush with the opening.The heater should be retained in place by a clip attached to the manifold so that it will not move out of opposition from either direction.

The leads should be protected by armoured covering and held in position to keep away from vibration.

The wires extending to the leads should be the heat resisting type attached firmly with crimped tubing so that there is no chance of poor contact.

Capacities of wires, terminal block, connector plugs, etc., should be carefully checked out for current carrying ability and correct voltage.

The melt-flow-way The melt-flow-way is the series of drilled holes formed in the manifold block through which the melted material passes from the primary nozzle ,i.e., injection machine nozzle to the various secondary nozzles in the multicavity mould. The path of the melt -flow-way should be smooth otherwise the polymer material will be held up in the way, difficulties arise when changing colour or type of material, and also degradation of the stagnant material may lead to discolouration of the moulding. There are various possible designs for the flow-way, shown in Fig. 5.34.

Runner plug The runner plug is made from stainless steel and it is fitted in the flow path of hot runner system to direct the plasticised materials. It has a larger thermal expansion than the manifold material. As the manifold is heated to operating temperature, the stainless plugs will expand at a greater rate making the possibility of leakage impossible.

5.2.7 Secondary Nozzles

The secondary nozzle provides a connecting flow path from the manifold block to the cavity plate impression in a hot runner mould. The function of secondary hot nozzle is to convey the melt through a gate into the space between core and cavity. If the secondary nozzle can not control the gate temperature accurately, the gate may string from being too hot or freeze off. Hence, the nozzle must provide just enough heat to the gate to keep it molten without drooling. Therefore, the heat supplied to the gate must be very carefully controlled by either

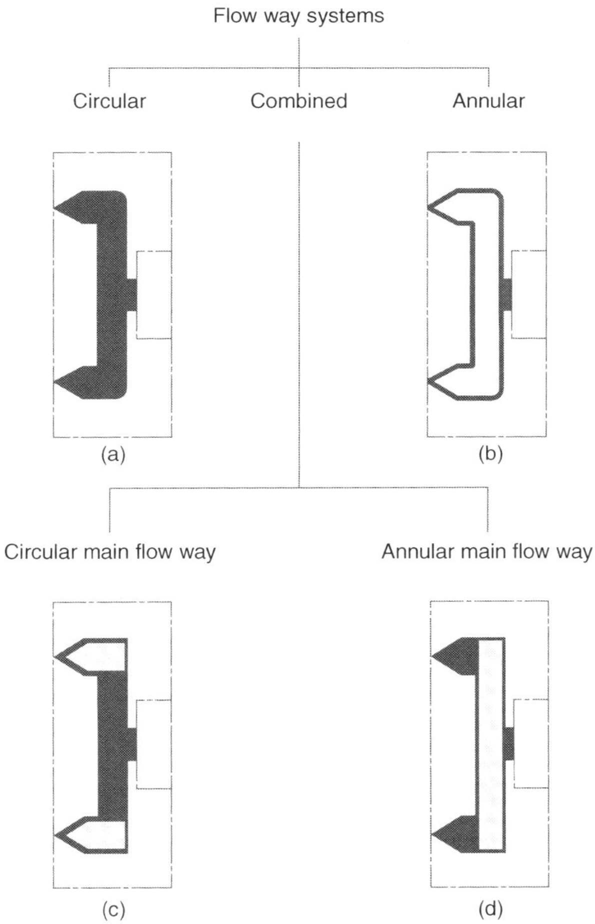

Fig. 5.34 Melt-flow-way system.

fluctuating power to the gate or precisely balancing the cooling and heating in the gate area. In order to separate the hot nozzle tip from the cooled cavity plate, a space called insulating bubble is machined into the gate detail. The insulating bubble fills with the molten melt and acts as an excellent insulator.

It should also have low heat losses to reduce unnecessary heating of nozzle and the tip of nozzle should be as close as possible to the moulding surface. Nozzle should be made of wear resistance material with high thermal conductivity and a wear resistance coating helps in increasing the life of nozzle. Positioning of the heater fitting either internally or externally depends primarily on the design of secondary nozzle.

Standard secondary nozzle It is a type of secondary nozzle shown in Fig. 5.35. Its front face is in sliding contact with the cavity plate. When expansion of the manifold block occurs in the heating-up phase, the secondary nozzle slides across the surface of the cavity plate.

When calculating the position of the secondary nozzles allowance must be provided for expansion in the manifold block so that, at its working temperature, the corresponding holes in the secondary nozzles and mould plate are in line.

In this type, the conventional sprue gate is adopted to connect the secondary nozzle outlet to the impression. Alternatively, when the hot-runner unit is being used as a semirunnerless system, a conventional sprue is used to connect the secondary nozzle outlet with the runner system.

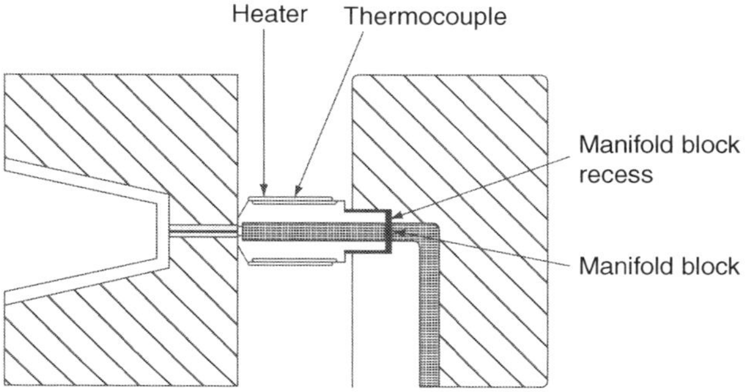

Fig. 5.35 Standard secondary nozzle.

Barb nozzle Generally, in case of conventional sprue gate system the gate mark is visible at the injection point on moulding. In order to reduce the gate mark to minimum barb nozzle is used. It is a special nozzle in conjunction with a reverse-tapered sprue shown in Fig. 5.36. The barb nozzle is similar to standard nozzle except that there is a projection at the front, which incorporates barbs. In this nozzle, the projected portion is accommodated in the reverse-tapered sprue ensuring that the flat face of the nozzle seats on to the sprue bush and prevents leakage of material.

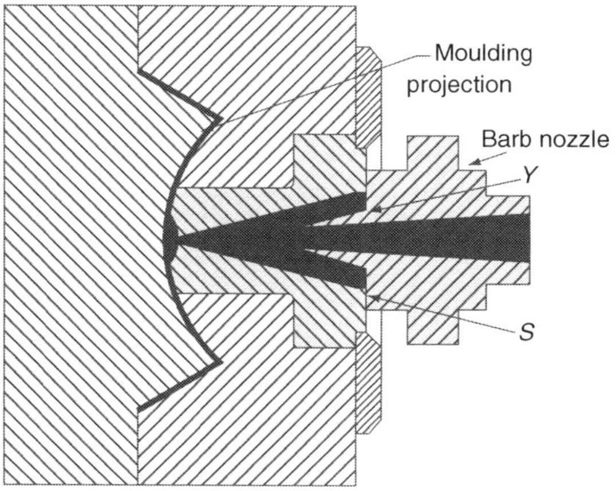

Fig. 5.36 Barb nozzle.

In operation, the plastic material flows through the nozzle, sprue, and gate, and so into the impression. After the material has solidified, the sprue is pulled from the sprue bush by the barbs as soon as the nozzle is withdrawn. The mould is opened and the moulding ejected in the normal way. To permit the removal of the sprue from the nozzle a larger than normal sprue break is required; hence it is normally used on injection machines incorporate a sliding carriage, which can be withdrawn automatically as part of the machine's normal cycle of operation.

The sprue is normally removed manually from the nozzle, which lengthens the moulding cycle. An alternate design incorporates a local stripper plate, to strip the sprue from the barbs.

Extended nozzle This type of extended nozzle protrudes into a pocket machined in the mould plate to prevent the undue cooling of the plastic material shown in Fig. 5.37. A resistance type band heater is provided which is controlled by means of thermocouple and the thermocouple is fitted into a slot machined into the barrel section of the nozzle directly beneath the heater band. To minimise the transfer of heat from the heated nozzle to the mould, a circumferential clearance of at least 5–7mm to be provided between the two parts.

The main advantage of this type of nozzle is that it minimises the length of the sprue gate as short as possible. The pressure drop across the gate is minimised and it avoids the post process of sprue removal.

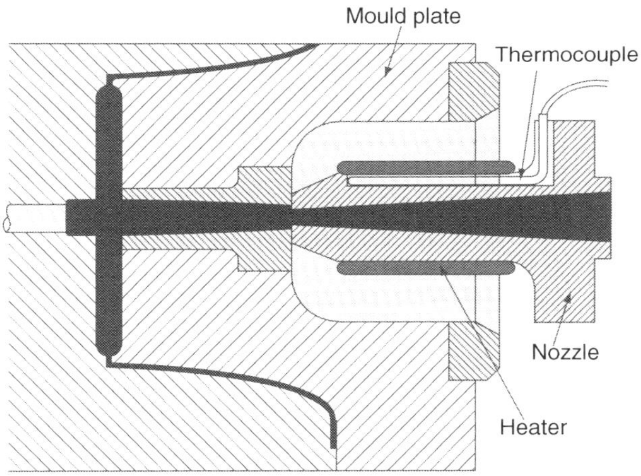

Fig. 5.37 Extended nozzle.

Antechamber secondary nozzle In this design, the nose of the secondary nozzle is bell-shaped and it extends into the cavity plate directly behind the impression. The body of the secondary nozzle, is large in diameter, and touches against the face of the cavity plate. A ring, manufactured from an insulating material, is usually incorporated at the junction between nozzle and local cavity insert for the transfer of heat from the secondary nozzle to the cavity plate is minimised. As the insulating material is slightly compressed due to the expansion of the manifold block, a good leak-free material seal is achieved.

During processing, the antechamber is filled with a small volume of polymer melt. The polymer melt passes from the manifold block, through the central hole of the secondary nozzle and into the antechamber via two or three small holes machined at an acute angle in the bell-shaped nose. Further the melt flows into the impression via the gate. The principle of the antechamber design is that, because polymers are poor conductors of heat, the material adjacent to the wall of the secondary nozzle nose remains sufficient fluid to allow the melt to pass intermittently into the impression via the gate. The layer of material adjacent to the cavity plate acts as an insulator, preventing undue loss of heat from the secondary nozzle to the cavity plate.

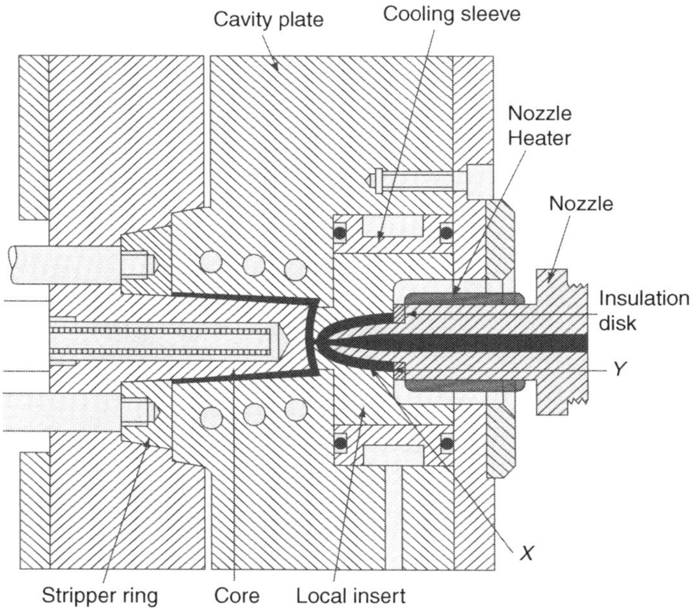

Fig. 5.38 Antechamber design.

Internally heated sprue bush In this design, the heating element is incorporated on the centre line of the sprue bush in the flow-way between the injection machine nozzle and the gate into the impression. By this means, the polymer material is kept at an elevated, controlled temperature up to a position relatively close to the gate. A standard injection machine nozzle, with either a flat or a radius end, is used in this method.

Internally heated sprue bush consists of basic parts torpedo, cartridge heater. The cartridge heater is housed within the torpedo assembly shown in Fig. 5.39. The heater wires pass through one of the torpedo legs via an internal insulator and adapter. The torpedo assembly is mounted inside part of the sprue. The front end of the sprue body-outer is conical in shape, with a small parallel nose which fits into a complementary hole machined in the cavity plate. The insulation gap between the body-outer and the cavity plate should be provided to minimise heat transfer.

When an internally heated sprue bush is incorporated in design, careful consideration must be given to the disposition of the coolant flow-way system in the cavity plate to minimise local heating effects.

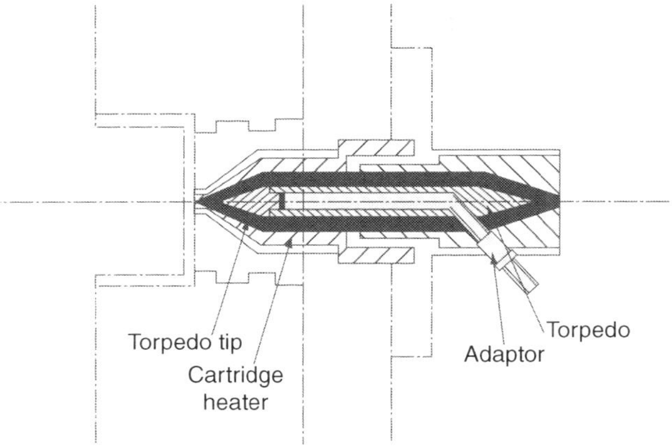

Fig. 5.39 Internal flow way heater-cartridge heater-type sprue bush.

5.2.8 Heating System

The ideal temperature required for heating the hot runner unit plays a vital role for getting stress free component. Generally the heating element used in hot runner mould is cast in pure copper into the manifold to provide uniform and constant temperature profile. The heater should be durable, easily replaceable and efficient. Heaters in the manifold should be located at a minimum and consistent distance from the melt channels and it should not cross the melt channels. Heater size should be about 2–3 Watts / cm^2 of manifold plus calculated heat loss per contact point. The temperature variation of melt in the channels should be as low as possible. It should be less than 5°or at the most less than 10°.Temperature uniformity is primarily achieved by reducing the heat losses and also by heating the manifold uniformly. Heat losses are reduced by designing suitable back up insulators like, pressure pad or spacers. An air gap of 5 mm to 10 mm is provided around the manifold, gives good insulation. Electrical devices are normally preferred for heating the manifold block although other systems such as circulating hot oil, high-pressure hot water and stream have been used. Nickel-chromium alloy, heavy gauge copper wires are used in electrical heaters.

Heating methods Different electrical devices are used for heating the manifold block although other methods like circulating hot oil, high pressure hot water and steam are circulated. There are various types of electrical systems such as (1) standard voltage resistance heating (2) low voltage resistance heating (3) induction heating.

1. Standard Voltage Resistance Heating: In this method, the manifold block is heated by means of voltage resistance heating elements. The heating element is embedded in a refractory material and enclosed in a metal casing. Generally, cartridge heating element, the flat heating element and the band heating element are commonly used.

2. Low Voltage Resistance Heating: In this method, the heating element is a length of relatively thick wire made with high resistivity value, 80/20 Ni-Cr alloy. The wire is formed into a shape such as a coil or a zigzag to facilitate it into the manifold block when the current is passed through the wire; it heats as a result of the resistance to the flow.

3. Induction Heating: In this case, when a conductor is placed with in a magnetic field carrying an alternating current, eddy current is induced into the conductor, causing it to heat. The surface is heated immediately and further the current maintains the heat right through the body. The coils used in induction heating are made out of copper wire.

The external flow-way heating The external flow way heating of manifold blocks is shown in Fig. 5.40, positioned in the centre of the rectangular and cylindrical type. The flow-way is not necessarily central in the manifold block.

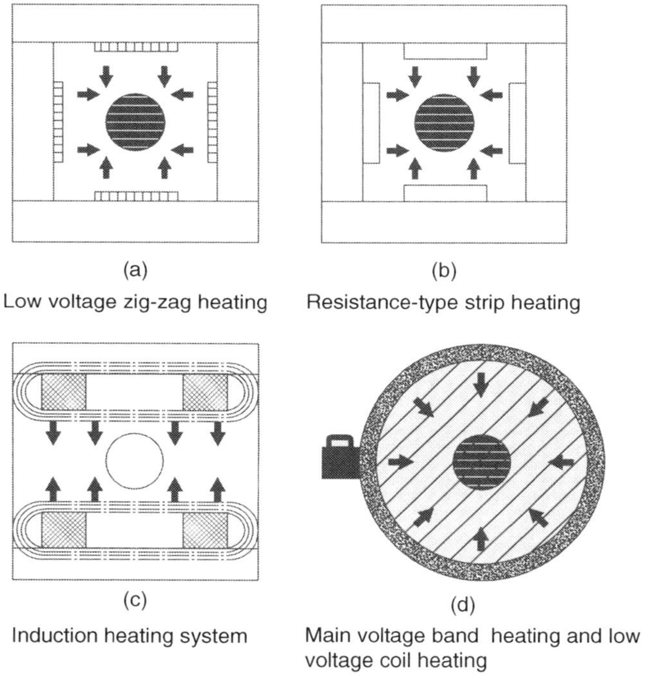

Fig. 5.40 External flow-way heating systems: (a) Induction heating, (b) Band heating and coil heating, (c) Resistance-type strip heating, (d) Low voltage zig-zag heating.

The arrows in the manifold block indicate the conduction heat flow path from the heating element to the flow. The relative distance between heating element and flow-way can be decided based on the type of heater provided in the mould.

There are different methods used for external flow-way heating, viz. cartridge heating element, low voltage coil (internal) heater, flat heating element, low voltage zigzag heating, induction heating coil, band heaters, low voltage coil external heater.

1. Cartridge Heating Element: Cartridge heaters are available as a standard part in various diameters which ranges from 6.5–19 mm. The cylindrical cartridge heating element is fitted into a hole bored and reamed through the manifold block.

2. Low Voltage Coil Heater: In this method, a relatively thick wire is wound onto a cartridge heater form a spring like coil. Terminals are attached to the coil for low voltage supply. The wire used in this process should have high resistivity value to obtain the desired heating effect. The coil heaters are fitted into holes bored through the manifold block near to the flow way.

3. Flat Heating System: It is a rectangular section having shallow heating capacity. It is available as a standard part in a range of sizes and wattages. The flat heating element is accommodated in a slot machined into the side of the manifold block and clamped with a cover plate. The cover plate should have good surface to surface contact between heating element and manifold block to have better heat conduction. The flat heating elements are commonly fitted in all four sides of the manifold block.

4. Low Voltage Zig-zag Elements: In this method high resistivity wire is fitted in a zigzag configuration and terminals are fitted in either end of the wire to facilitate its connection to the low voltage supply. The total length of the wire required depends upon the resistivity value of the wire, the voltage used and power input required. The zigzag element is fitted into a recess in the sidewall of the manifold block and secured by a cover plate.

5. Induction Heating Coil: In this method, heavy gauge copper wire is wound onto a heater and then bound with tape to form a coil. The induction heating coil is fitted into a recess machined into the sides of the manifold block. When the current is passed through the coil, eddy currents are induced into the surface of the manifold block and heat is transferred from the coil to the melt-flow-way by conduction.

The above external flow-way heating methods are used for rectangular cross section manifold blocks. For heating cylindrical manifold blocks the band heaters and the low voltage coil heaters are used.

6. Band Heating Elements: In this type, the main voltage resistance type of element is enclosed within a casing which is in the form of split hollow cylinder. It is mounted in the external surface of the cylindrical manifold block. The band heaters are available as a standard part in a variety of diameters, widths and wattages. The band heaters are also used for heating of secondary nozzles and manifold bush.

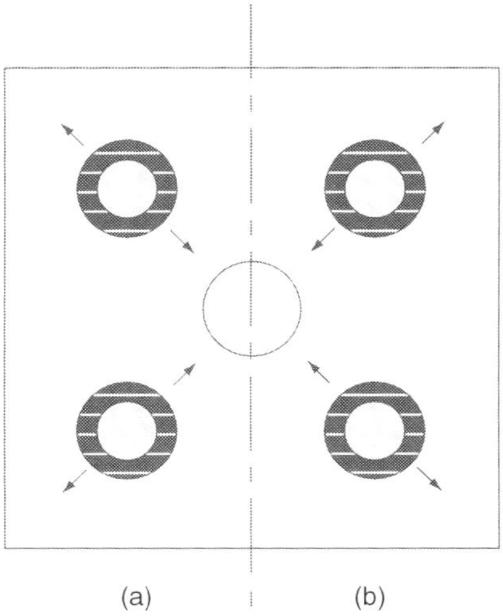

Fig. 5.41 (a) Cartridge (b) Low voltage coil heating element.

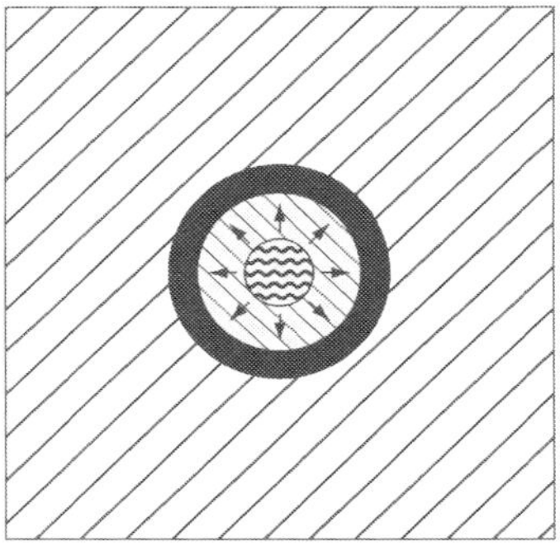

Fig. 5.42 Main voltage cartridge heating.

The internal flow-way heating system In internal flow-way heating method, the annular flow-way system is adopted. The annulus is formed by the inner surface of a relatively large central bore hole and the outer surface of a heater tube. A heating element is fitted down the centre of the tube shown in Fig. 5.42; the heating element is in close proximity to the relatively thin shell of polymer melt.

This concept makes use of the excellent insulating properties of polymer materials. The low thermal conductivity values reduce the flow of heat from the polymer to the cooler manifold block. Two types of heating elements are used in conjunction with the annulus system, the cartridge heat element and the low voltage coil heaters.

A hot-runner unit is an insulated heated assembly, which is mounted within the structure of a mould to provide a flow-path for the polymer, melts from the injection machine's nozzle to the gate entry into the impression.

5.2.9 Expansion Problems in Hot Runner Mould

In a hot-runner mould, when the manifold block is heated by the heating elements, the metal expands in all directions. The distance between the centres of the secondary nozzles will increase with respect to the 'fixed' distance between the centres of the impressions machined in the cavity plate. While designing the hot runner mould, allowance must be provided for the expansion to ensure that the centrelines are in line during production.

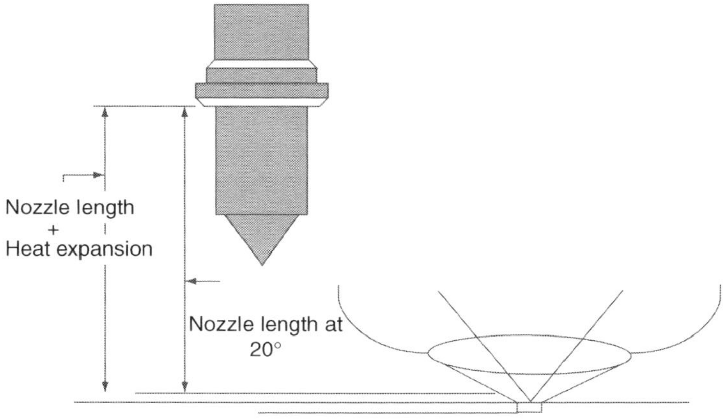

Fig. 5.43 Expansion of the manifold block.

NOZZLE LENGTH	200°	250°	300°
48 mm 68 mm	0.13 0.20	0.17 0.20	0.21 0.30

The general equation for calculating expansion is as follows:

$$e = L \times \alpha \times \Delta T$$

where e = Expansion (mm or in) ; L = Length dimension (mm or in)
α = Coefficient of thermal expansion (mm/mm°C or in / in°C)
ΔT = Increase in temperature (°C)

5.2.10 Advantages and Disadvantages of Hot Runner System

Advantages The hot runner system completely eliminates the runner scrap, so there are no feed systems to sort from the parts and no runners to throw away or regrind into the virgin material.

Hot runner moulds are more effective for mass production parts, especially with a multicavity injection mould.

Disadvantages Hot runner mould is much more expensive than a cold runner mould, mould maintenance cost is more and skilled workforce is required to operate.

Colour changes of material with hot runner moulds are difficult, since it is practically impossible to remove all of the plastic from an internal runner system.

5.2.11 Limitations of Hot Runner System

Limitations

(a) The mould setting time is generally higher than a cold runner mould.
(b) An extended period waiting and the initial debugging of a new hot-runner unit mould is usually more extensive than with a standard mould.
(c) The cost is higher than that of a standard two plate mould and an underfeed mould.
(d) The area of the moulding adjacent to the gate may be blemished with surface heat marks.
(e) Polymer melt leaking from the hot-runner unit creates problems.
(f) Polymer material at the gate if solidified then creates interruption in production.
(g) Certain materials have a tendency to drool from the gate into the impression when the mould is open. This causes blemishes on the subsequent moulding.
(h) Some materials have the tendency to string when the moulding is extracted from the cavity. This has adverse effects similar to the above.
(i) Replacement of the heating element is sometimes difficult and can be time consuming.
(j) There are degradation problems if the runner flow path and temperature of the melt are not maintained properly
(k) Changing the colour and the grade of plastic material is difficult.

5.2.12 Insulated Runner Mould

The insulated runner mould in which the melt flows through a large diameter runner machined in the butting surfaces of the cavity plate and feed plate of a three–plate mould. The polymer melt flows from a standard or extended nozzle into a large diameter runner and finally into the impression via the reverse taper sprue and gate shown in Fig. 5.44. The insulated runner is not heated; the runner channels are extremely thick and stay molten during constant cycling. This system is very inexpensive, and offers the flexible gating advantages of other hot runners and the elimination of gates without the added cost of the manifold and drops of a heated hot runner system.

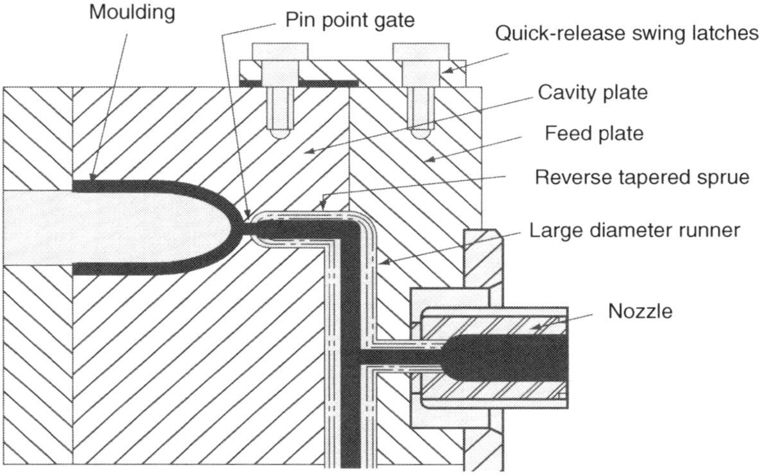

Fig. 5.44 Insulated runner mould.

The insulated runner mould consists of a large diameter runner, generally used in a three plate mould, the outside periphery freezes and the inside area remains molten and act as a runner. After the first shot, when the system is filled with molten material, the centre portion of the materials fluid due to the passage of freshly plasticised material with each cycle. The outer portion of the material in the manifold forms a solidified shell or tube, through which the molten material delivered with the shot, will flow into the cavities. The heat insulating characteristics of plastic material are used to good advantages by having the outer portion of the runner, which is frozen, insulated the hot molten materials.

The two halves of the runner are machined in the respective mould plates, when the component is cooled the runner system freezes. During mould opening stage, the two plates are separated and the runner system is removed. As soon as the runner reaches equilibrium, the latch is closed and the mould is operated. Usually a heated torpedo individually controlled is inserted into each gate area and kept on continually and the wattage is controlled and set pragmatically. The gate size will vary with the material, but it should be somewhat smaller than the standard recommendations because of the high fluidity of the resin. The land should be about half of the gate diameter. For some materials, it is advantageous to provide a chamber on the tip of the nozzle should not only have a very fine finish, but should also offer smooth transition from one area to another. Insulated runner moulds are more difficult to start and operate than three-plate moulds, but are considerably easier than a hot runner mould. In insulated runner mould there is no runner system to regrind.

5.3 OUTSERT AND INSERT MOULDING

In outsert moulding process, one or a number of plastic functional components are injection moulded in one operation onto both sides of a base plate made out of ferrous or non- ferrous metals as shown in Fig. 5.45.

The metal insert as shown in Fig. 5.45, is placed in the injection mould and the plastic is then injected into the mould cavities and hence into the appropriate cut-outs in the plate as shown in Fig. 5.46.

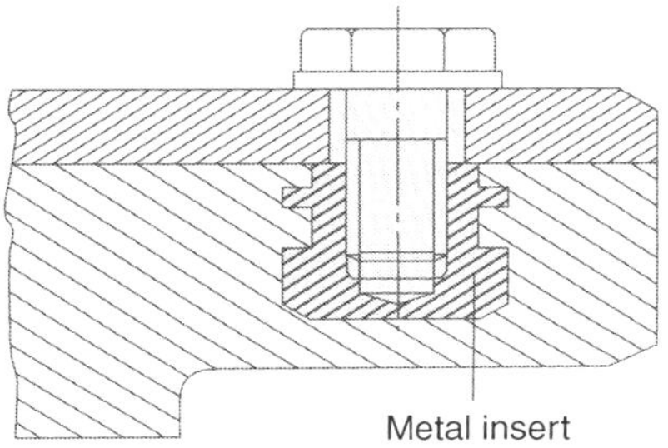

Fig. 5.45 Plastic moulding with metal insert.

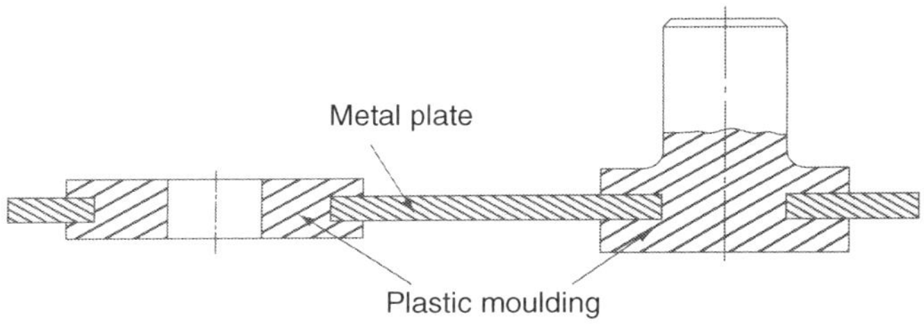

Fig. 5.46 Plastics mouldings anchored in a metal plate by injection moulding (Outsert moulding).

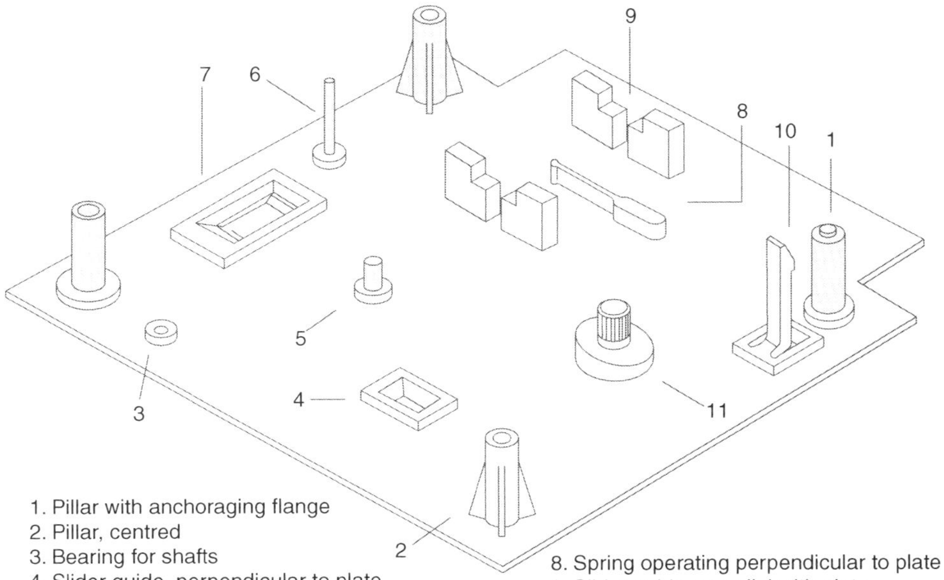

1. Pillar with anchoraging flange
2. Pillar, centred
3. Bearing for shafts
4. Slider guide, perpendicular to plate
5. Pin (All plastic)
6. Metal pin in plastic anchorage
7. Spring
8. Spring operating perpendicular to plate
9. Slider guide, parrallel with plate
10. Snapfit hook
11. Rotating component

Fig. 5.47 Basic components of outsert moulding.

5.3.1 Objectives of Outsert Moulding

As manufacturing of metal components individually by turning, drilling, milling, etc., and then assemble them on a base plate is a time-consuming, labour-intensive operation, hence the outsert moulding is preferred. The aim of outsert moulding is to produce at a lower cost engineering assemblies which match or exceed the quality of those manufactured by traditional methods. Reduction in manufacturing costs would be possible with assemblies that are still produced entirely from metal and comprises a number of functional components like plastics or metal individually mounted on base plate.

5.3.2 Application of Outsert Moulding

Component requirements in outsert moulding depend primarily on the function to be fulfilled. The following individual precision engineering components are used either alone or in combination as shown in Fig. 5.47, viz. bearings, pins, pillars, rotating parts, guides, slide ways, springs, hooks, etc.

The materials for outsert moulding components should have special properties like high mechanical load-bearing properties, low creep, good sliding properties, low coefficient of friction, good spring properties, high dimensional stability, no environmental stress cracking and high chemical resistance.

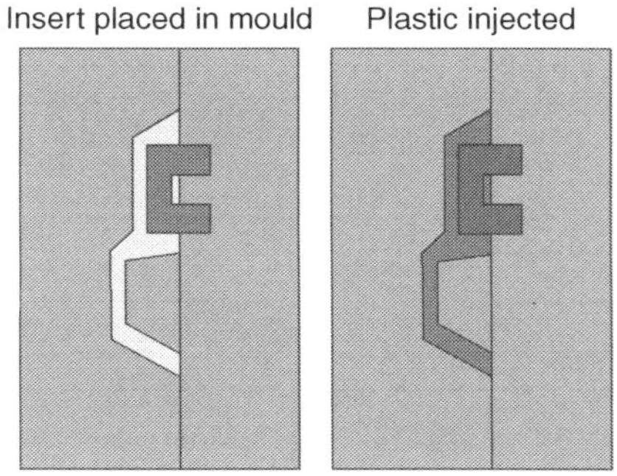

Fig. 5.48 Insert mould.

5.3.3 Insert Moulding

It is an injection moulding process in which the plastic material is injected into a cavity and around an insert piece placed into the same cavity just prior to moulding.

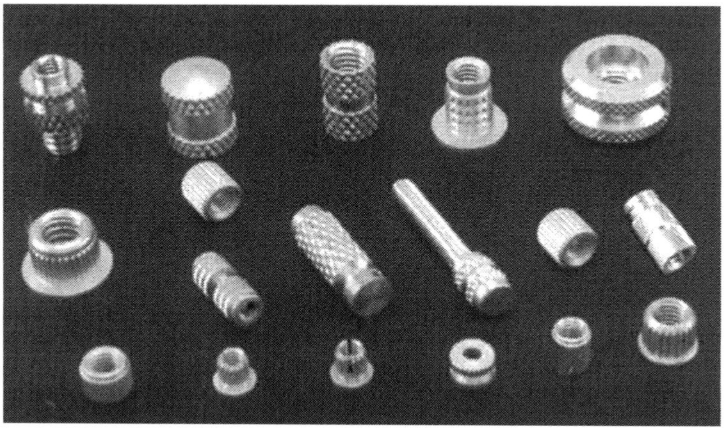

Fig. 5.49 Brass inserts

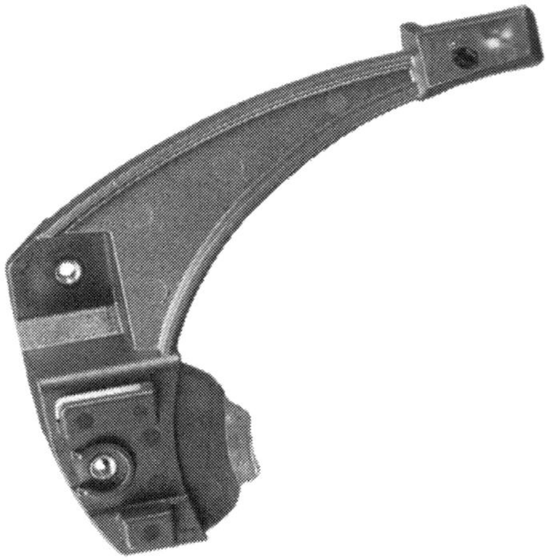

Fig. 5.50 Component with moulded inserts.

In this type of design, the plastic parts containing metal inserts are produced using injection moulding. Generally inserts made out of ferrous or non-ferrous metal are placed inside the mould prior to the injection of plastic. The polymer melt flows around the inserts and fixes their position.

The advantage of insert moulding is that, a single piece with the insert is encapsulated by the plastic. Insert moulding was initially developed to place threaded inserts in moulded parts and to encapsulate the wire-plug connection on electrical cords. Insert moulding is also used quite extensively in the manufacture of medical devices. The typical applications of insert moulding include insert-moulded needle hubs and bifurcations, as well as encapsulated electrical components and threaded fasteners, etc.

5.4 MULTICOLOUR INJECTION MOULDING

In multicolour injection moulding, the parts are produced that are made from different coloured plastics. The individual components are not merged, but are processed in separate colours. The visual effects can be created by combining areas of different colours or even transparent and optically conductive areas. As a result, the moulded parts are resistant to external influences such as chemical, thermal and mechanical loads.

Fig. 5.51 Components of multicolour injection moulding.

5.4.1 Mould Concepts

The multicolour moulding is processed by following two different types of mould concepts. The first method is transferring the premoulded part to another cavity by means of a rotary movement, by hand or by using a robotics system, so that it may then be encapsulated by the second component. This method is switching the component from one mould cavity to another.

The second method is the use of internal movement of the component inside the mould to free up space for the next set of components, i.e., only one cavity is used.

5.4.2 Rotary Moulds

In this method, the first step to be performed when using rotary moulds is the initial production of a premoulded part in one of the cavities. Subsequently, during the next step, the part is manually shifted into the required position by transferring it into a second cavity by means of a rotary movement of 180°, so that it can be encapsulated by the second component.

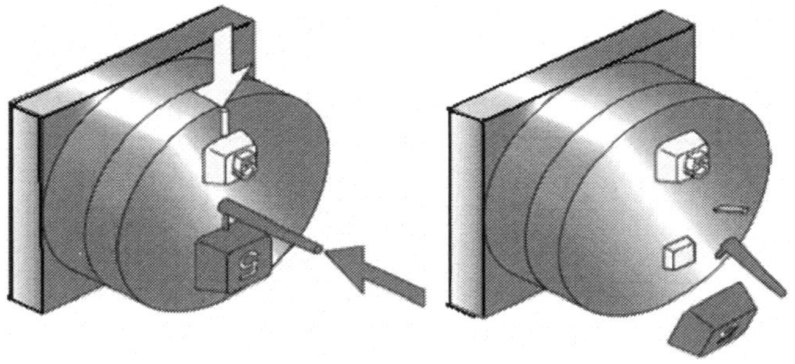

Fig. 5.52 Rotary moulds.

The advantage of using rotary moulds is that premoulded parts can be produced at the same time that parts are encapsulated by the second component.

Based on the geometry of the part, different design systems are available for performing the rotary movement, whereby either the entire mould half or individual parts of the mould are rotated like rotation of a stripper plate, rotation of a platen by means of an ejector movement, rotation of an insert, rotation of the moveable mould half, rotation of a central platen.

5.4.3 Altering the Mould Using an Internal Movement Inside the Mould

The other method is by shifting the cores or sliders inside the mould as part of the composite injection moulding process. The hollow spaces can first of all be closed off and subsequently reopened and the polymer melt is injected inside the cavity accordingly as shown in Fig. 5.53. The moulds are configured by lifting and lowering movements and by sliding movements.

The main advantage of this process is that parts can be produced without the need for intermediate opening of the mould and without further transport of the premoulded part. The production is performed as per sequence, whereas with rotary moulds simultaneous operations are possible.

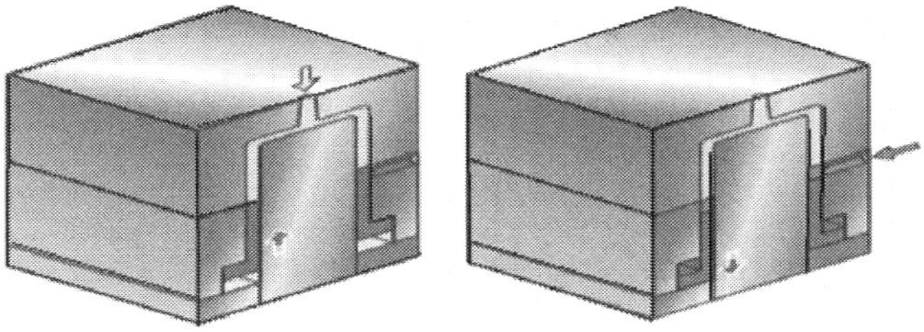

Fig. 5.53 Cores or sliders.

5.4.4 Transfer by Hand or Robotic System

The alternate method of rotary movement is that the premoulded part can also be transferred from the first cavity to a second one ready for the final injection stage, either by using conventional means, i.e., by hand or a robotic system.

The transfer is advantageous when processing the cross-linking materials such as liquid silicones (LSR), because the mould can be divided into two distinct halves, each of which is completely thermally insulated from the other.

5.4.5 Simultaneous Direct Injection

In this method, the polymer melt is directly injected inside the cavity simultaneously to form two components in a single cavity, without having to alter the mould. However, in this process an uneven dividing line is formed between the two materials with the result that it is only used for parts whose ultimate visual appearance is not of prime concern.

5.5 GAS-ASSISTED INJECTION MOULDING

The gas-assisted injection moulding is a process in which hollow rigid parts can be produced which are free of sink marks and weigh less thereby saving resin material. The process begins with a partial or full injection of polymer melt into the mould cavity. Compressed gas is then injected into the core of the polymer melt to help fill and pack the mould as shown in Figs. 5.54 and 5.55.

Other advantages include shorter cooling cycles, reduced clamp force tonnage and part consolidation. The process allows high precision moulding with greater dimensional stability by eliminating uneven mould shrinkage and makes it possible to mould complicated shapes in single form, thus reducing product assembly work and simplifying mould design.

The formation of thick walled sections of a moulding can be easily achieved by introducing gas in the desired locations. The gas channels which are formed also effectively support the flow of resin. It allows reducing the moulding pressure. The effect of which, is that internal stresses are reduced. Moulding gets uniform mould shrinkage, and reduces sink marks and warpage.

5.5.1 Working Principle of Gas-Assisted Injection Moulding

Standard injection-moulding machine is used for gas-injection moulding process. Only additional equipments are fitted to the moulding machine to inject the nitrogen gas parallel or in series with the injection of the melt as shown in Fig. 5.54.

The gas is injected through the same nozzle in which the melt flows in the machine nozzle, or via one or more special gas injection needles located at the runner or where there are more material concentrations, generally in thicker walls. Special machine nozzles are designed and used to ensure reliability.

The gas-injection moulding process starts with injection of plastic into the cavity as shown in Fig. 5.54 and Fig. 5.55. When the cavity is 50 to 95% full, depending on the shape of the part the barrel is closed by a 'shut-off' nozzle and gas injection starts. It can be controlled by pressure or by volume. The gas expands in the cavity, pushing the plastic in front of it until the cavity is filled. Then the gas-pressure is reduced by withdrawing the injection nozzle from the sprue, so that the gas can escape. In some designs the gas is allowed to escape from the cavity via the injection needle, so that the machine can recover the gas for re-use. If the gas is injected through the same nozzle as the melt, a second injection of plastic is made to seal the hole in the part.

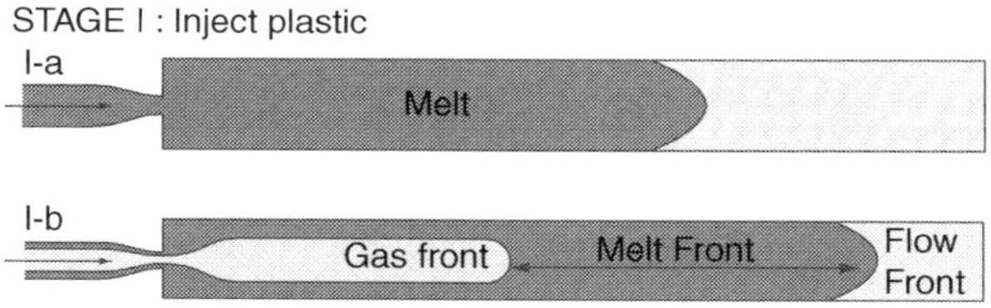

Fig. 5.54 Injection of plastic into the cavity.

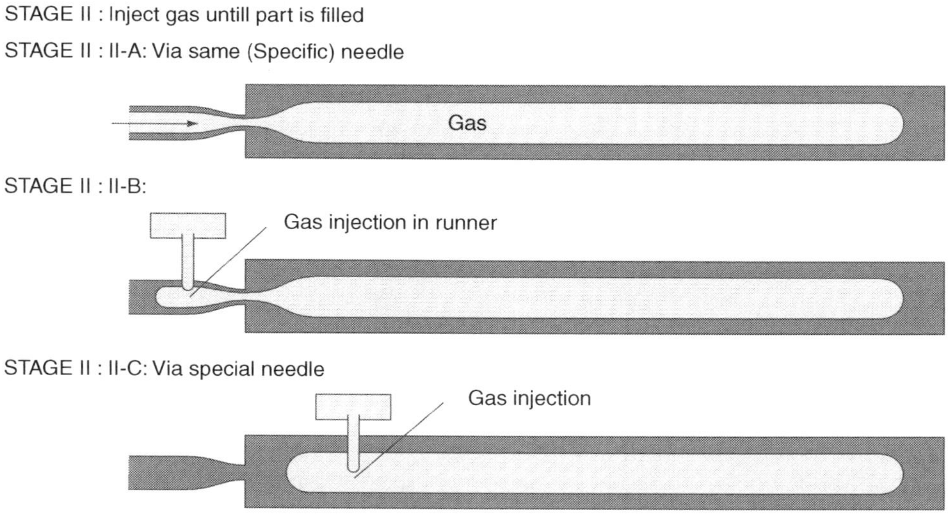

Fig. 5.55 Injection of gas into the cavity.

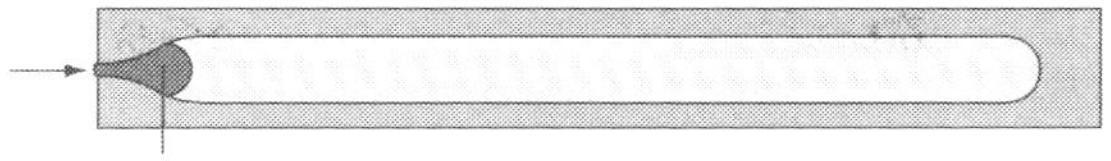

Fig. 5.56 Close gate with plastics.

Different gas-assist injection moulding techniques are followed as detailed below:

1. The short-shot moulding, i.e., prefill of cavity with melt, followed by gas injection.
2. The full-shot moulding, i.e., complete fill of cavity with melt, followed by gas injection.
3. The full-shot moulding with overspill, i.e., complete fill of cavity with melt, followed by gas injection and opening of overspill cavities.
4. The full-shot moulding pushes resin back into barrel, i.e., similar to the overspill technique using the machine barrel as the overspill cavity.
5. The external gas moulding, i.e., adding a layer of nitrogen gas to the part surface adjacent to the cosmetic surface after complete fill of the cavity.

Other than the above processes, there are some methods that use low-temperature nitrogen gas to create a gas flow through the part for an additional cooling effect. For some resins, the short-shot method cannot be used because of cosmetic surface requirements. Full-shot processes that have additional cavities, into which the resin in the core is evacuated, generally provide more control over the process.

5.5.2 Design Considerations.

The most ideal designs for gas-injection moulding are parts in which the gas flow is in one direction only, i.e., not around corners or bends.

There are special rules applied to parts designed for the gas-injection moulding process. The basic rule is that the gas always pushes the plastic away at the locations with the best flow conditions that is, where resistance to the melt front is the lowest. For this reason, gas channels will tend to appear in sections with large cross sectional area and higher melt temperatures.

In Fig. 5.57, the Fig. 5.57 (a) shows a design with a sharp corner and melt accumulation; the Fig. 5.57 (b) shows an improved design with rounded corners.

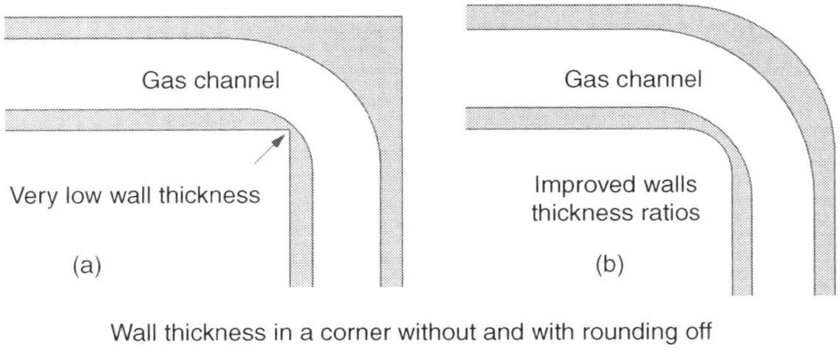

Fig. 5.57 (a) Design with sharp corner, (b) Improved design with rounded corners.

The gas in the cavity of gas injection moulding fulfils two main functions:

1. It produces a hollow cavity by the gas which causes the weight reduction of the component.
2. It provides constant pressure throughout the part in order to compensate volume shrinkage after the cavity has been filled.

The main objective of this process is to avoid sink marks and to obtain a smooth surface in the moulding. Gas-assisted injection moulding is used for producing thin parts with a relatively large cross section, such as handles, weight saving items.

Gas channel incorporated in the mould is the main feature of this design. For shallow parts with stiffening ribs, the main reason for using gas-injection moulding is the processing the components with a smooth surface without sink marks. The ideal location of gas channels is very much required to avoid sink marks in moulding. As shell structures in housings normally do not include sections which can be used as gas channels, channels are to be specially designed. This is done at corners for the shell or at ribs or shell junctions as shown in Fig. 5.58.

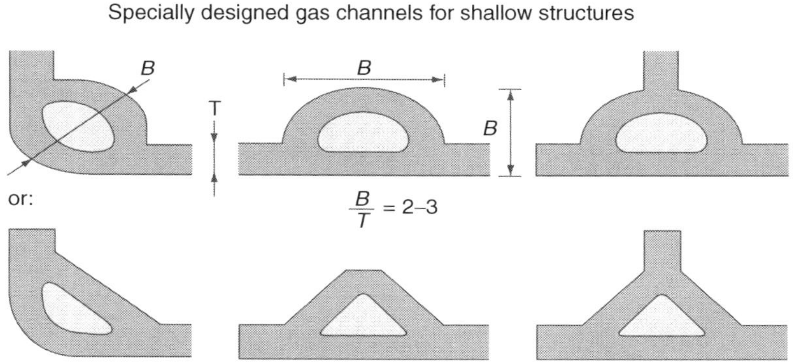

Fig. 5.58 Specially designed gas channels.

The ribbed components with gas channels avoid sink marks due to the link of the ribs in a network by an incorporated gas channel which goes around the geometry of the part. In this method it reduces warpage in the part. The size of the gas channel depends on the volume shrinkage of the material and on the size of the shell.

The following considerations are of importance:

The gas pocket size is smaller compared to the total flow length and the sooner the gas pressure has to be activated in order to avoid the gas channels freezing off. This means that the gas channel will try to compensate the volume shrinkage of the flat part, by pushing melt into flat, shrinking sections. A cross section over a part with a gas channel that is too small results such as those shown in Fig. 5.59. Because of lack of melt, gas is pushed into the shell, producing 'notches' which are defect in the part and unable to resist high dynamic loads.

If the part has an application of integrated function, such as carrying liquids, or if it has connectors attached to it, the gas channel has to extend over the full length of the part, leaving only a thin skin of plastic as shown in Fig. 5.59.

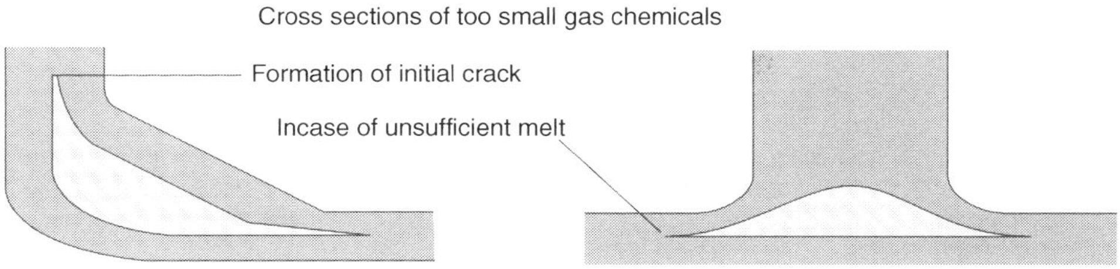

Fig. 5.59 Cross sections of too small gas channels.

In this case, the size of the gas channel will depend on part design and determined by process simulation studies, which reduces the need for practical trials. Inserts are a challenging design problem for gas-injected parts. Inserts have to be properly surrounded by plastic material; the gas channel has to be placed at a certain distance from it. Usually, this can be achieved by using a separate gas-injection needle placed downstream from the insert as shown in Fig. 5.60, if the machine nozzle is used for gas injection. More design effort will be needed to ensure that the insert remains properly surrounded by plastic.

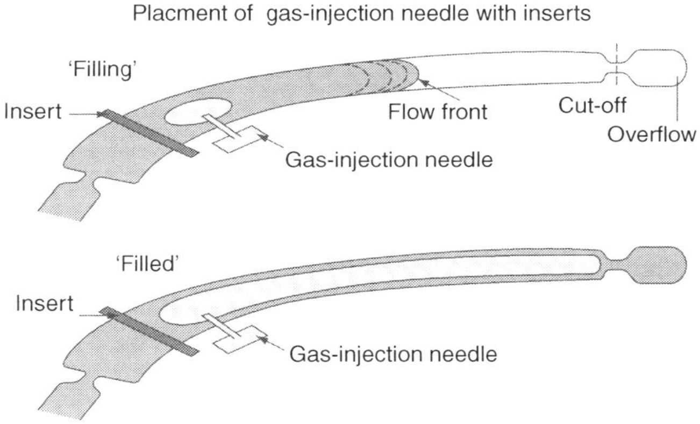

Fig. 5.60 Placement of gas-injection needles.

5.5.3 Gate

The design of gate in gas-injection moulding is different from gates for conventional moulding. The gas-injection is done via the machine nozzle, the gate and runner dimensions have to be about twice as big as for conventional moulding. The gate should be positioned so that the melt is injected in a wide, even flow along the cavity wall, as in the extrusion process; split flows and confluences, which can lead to unwanted turbulence and should be avoided.

5.5.4 Advantages of Gas-Injection Moulding

The gas-injection moulding process requires lower clamping force; the pressure drop during the moulding is less. Large-size mouldings can be produced with greater flow length. The mould is simple and cheaper compared to hot runner moulds.

In terms of design of mould: Lower part weight (reduction up to 40%), less sink marks, less shrinkage across the direction of flow, higher torque resistance, more design freedom with non-uniform wall thicknesses.

5.5.5 Limitations

Processing

1. Extra equipments are attached in injection moulding machine.
2. Special nozzle design/gas-injection needles are provided.

Design

1. Achieving exact wall thickness of component is difficult.
2. Cross section of gas channels less than 15 to 20 mm is preferred.
3. Increased shrinkage in the direction of gas channel flow.

Material

1. Material properties usually lower than in equivalent part processed by conventional injection moulding process.
2. Surface quality of part depends on plastic material used.

5.5.6 General Applications

1. **Automotive:** Sunshine roof trim, external mirror housings, handles, wiper blade, carriers, covers and panels, seat frames, head rests, pedals, steering wheels, knobs and gears, filter housings
2. **Furniture:** Armrests, chair bases, back shells, seat-pans, housings, and bathroom products
3. **Packaging:** Boxes, pallets
4. **Appliances:** Handles, lawn mower housings
5. **Sports:** Racquets, hockey sticks, ski bindings, ski boots

5.6 STACK MOULD

In stack mould, two injection moulds are operated in the same moulding machine and these moulds are used for moulding shallow, small parts in large quantities like tape cassettes ,etc. The cavities are located in two planes corresponding to two parting lines and are filled. Clamping force required for filling the mould is 15% higher than a standard moulds.

It is the simplest mould design, from a runner point of view. It is a single-cavity, single-face mould arranged in back to back configuration of same impression or different impressions.

Fig. 5.61 Stack mould.

The machine nozzle injects plastic directly into the mould cavity. The single-face mould can also be extended to a multicavity layout.

The machine nozzle injects the melt into a runner system that feeds each individual cavity.

This type of mould is suitable for processing thin-wall injection moulding with hot runner designs.

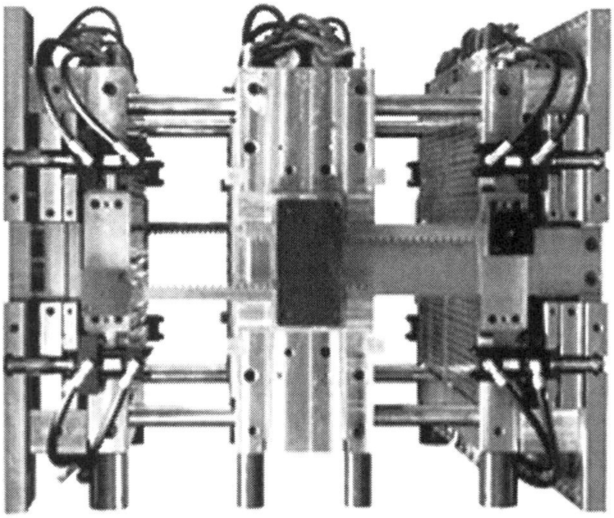

Fig. 5.62 Stack mould.

5.6.1 Types of Stack Mould

1. Two-face or two-level stack moulds
2. Three-face or three-level stack moulds

Two-Face (Two-Level) stack moulds This type of stack moulds quadruples the output over single-face moulds, and are suitable for very high production volumes of shallow parts.

Quick change moulds switch from one product to another in less than an hour in both single-face and two-face stack mould applications.

Two-cavity slack moulds produce moulding of large parts in a back-to-back configuration, thus doubling the machine capacity. The two-face stack mould is two-stack moulds placed back-to-back and it increases the mould shut height.

It operates in the same moulding machine with same cycle time as a conventional stack mould.

Shot-to-shot change over time of less than one hour is required for both mechanically and air ejected parts in the stack mould. Machine shut height remains the same from one product to the other.

Three-Face (Three-Level) stack moulds This type of mould permits moulding of deep-draw or tall pans to maximise the productivity of machine shut height. It is the development of a proprietary melt transfer system to pass the plastic across the mould parting line. The system avoids drooling on mould opening due to the self-decompression of the central hot runner system.

The three-level stack mould uses triple Valueless Melt Transfer System (VMTS) cross over nozzles to provide equal pressure and flow characteristics to the plastic melt to each cavity. This stack mould is used when a two-level mould cannot produce enough pans and a four-level mould is too large for the machine to process.

Three-level stack mould is used to produce both shallow draw parts like packaging lids and deep draw pans such as tall containers. It can be combined with quick-change systems to give added flexibility to high-production tooling.

5.7 REACTION INJECTION MOULDING (RIM)

5.7.1 Introduction

The reaction injection moulding process is based on a chemical reaction. A reactive liquid mixture usually polyol and isocyanate is injected or poured into a mould where a chemical reaction takes place. The finished part is produced after an exothermic reaction and heat is generated during the process. Depending on the chemical formulation, the end product can take on a wide range of physical characteristics like foam or solid, highly rigid or very flexible. This method saves material and energy costs compared to thermoplastics, where high heat and pressure is required to melt the resins. The parts are formed from two liquid components that chemically react inside a mould and it requires significantly less heat, clamping pressure and tooling cost.

This process is less complicated with a reduced initial cost of investment and it requires lower-tonnage presses than thermoplastic moulding. RIM process needs less equipment and floor space than injection moulding and the mould manufacturing cost is less compared to injection moulds.

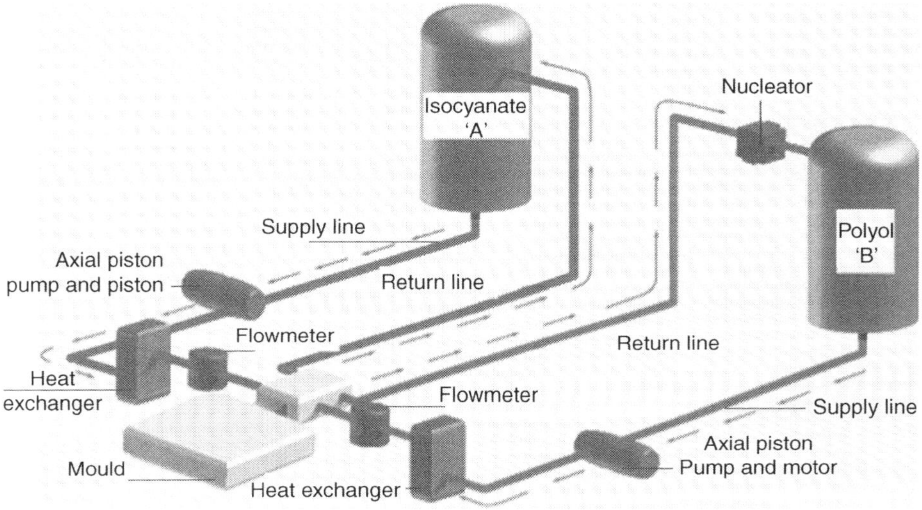

Fig. 5.63 Reaction injection moulding (RIM) process

The RIM process involves three steps viz; material storage, dispensing and moulding.

5.7.2 Material Storage

In the RIM operation, the raw materials are stored in bulk storage tanks before processing. The materials are temperature-controlled to the optimum processing temperature as per the material specification. This results in a consistent manufacturing environment and provides the desired physical properties or cell structure. If the materials have other ingredients like fillers or pigments that need to be evenly dispersed throughout the chemical system, stirring devices or tank agitators are often incorporated into the tanks to prevent settling or chemical separation. Recirculation of the materials is continuously circulated at low pressure by the pumping system and through the mixing head. When the materials reach the mixing head, they are recirculated back to the day tanks and then through the same path again back out to the mixing head. This low-pressure recirculation can be used to maintain temperature, nucleation, and will help keep added ingredients such as fillers or pigments evenly dispersed.

5.7.3 Dispensing

Dispensing is a process in which the two reactive materials, polyol and isocyanate, are kept separate until they reach the mixing head. When it is time to dispense a shot or make a pour, the machine automatically switches from recirculation to dispense mode. At this point, the metreing pumps precisely deliver the materials to the mixing head at the required volume, ratio, flow rate and temperature. The chemicals are then mixed by either high-pressure impingement or in a high shear dynamic mix chamber. The mixture is then injected into a closed mould or poured into an open mould or cavity.

5.7.4 Moulding

In moulding process, the chemical reaction takes place inside the mixing head, with a continued exothermic reaction inside the mould cavity as the curing process progresses. When processing foams, significant forces created inside the mould must be resisted to ensure the integrity of the part. The clamping pressure required can be up to many tons depending on the size, expansion rates, and the desired density of the part, along with other material factors. Mounting the mould in a pneumatic or hydraulic press provides the force required to keep the mould tightly closed during the curing process. The elastomeric materials require very little clamping pressure as they do not expand or develop internal mould forces.

5.7.5 Benefits of RIM

Very large, lightweight parts are produced by RIM process and the 'flowability' of polyurethane components allows for even distribution of the material within the mould. Bigger size mouldings can be moulded due to low pressure requirement which is not possible with injection moulding machine.

The mould manufacturing cost is less and generally mould materials like steel, aluminium, Kirksite alloys, nickel, epoxy, silicone and fibreglass are used.

Complex shapes or highly detailed parts with intricate design features at relatively low tooling and capital equipment costs can be designed. Monolithic parts or components with varying wall thicknesses are designed into the same moulded part.

The prototype moulds can be developed with less lead time at a cost much less than traditional injection moulding. This also helps in functionality testing prior to machining actual high-pressure injection moulds. RIM process is ideal for shorter production run. The surface finish of RIM parts produces high-gloss that matches with painted metal parts.

Table 5.1 Comparison of thermoplastic moulding with RIM.

Conventional thermoplastics moulding vs reaction injection moulding		
	Thermoplastic moulding	**RIM**
Material	Thermoplastics in pellet form	Low viscosity liquids
Processing Temperature	350° to 450°F(176° to 232°C)	Low processing temperatures 90° to 105°F (32° to 40°C)
Mould Temperature	350° to 450°F (176° to 232°C)	Low mould temperatures 90° to 105°F (32° to 40°C)
Mould Pleasure	Multiple tons of pressure	Low internal mould pressure 50 psi (3.4 bar) and up

(Contd.)

Table 5.1 (*Contd.*)

Conventional thermoplastics moulding vs reaction injection moulding		
	Thermoplastic moulding	**RIM**
Floor Space	Equipment and moulds require more floor space	Equipment requires less floor space
Energy	More energy to make a product	Less energy to make a product
Investment	High initial investment	Low initial investment

5.8 CASTING

5.8.1 Introduction

Casting is the process in which the polymer material in liquid form is poured into a box as per the required shape of the component called a mould and is kept for becoming solid mass. This type of process is found in civil construction, casting of steel, etc.

The modified monomers, powders, solvent solutions in liquid form are poured into a mould or die and allowed it to polymerise to become a solid plastic shape. The liquid plastic changes its status by evaporation, chemical action, cooling or external heat. In general no pressure is applied in casting.

The most widely used casting processes are simple casting, slush casting, rotational casting and solvent casting.

5.8.2 Simple Casting

This process is used for moulding rods, tubes, cylinders, sheets, etc. Liquid resins are mixed with catalyst or molten plastics and are poured into moulds and this is allowed to polymerise and become a solid mass. The plastics part is then removed from the mould for further finishing if required. In simple casting, wood, metal, plaster of paris, glass, elastomers are used for making the mould. The acrylic sheets are made by pouring the polymerised resins between two parallel glass surfaces. After full polymerisation in the oven, the sheet is separated from the glass surface and the internal stresses are relieved by reheating.

5.8.3 Slush Casting

In this process, the polymer melt is poured into a heated hollow mould. During pouring, the materials strike the wall of mould and get solidified with wall.By regular pouring of the material, the wall thickness of the moulded part is increased as the temperature increases. The pouring continues until the desired thickness is reached. The mould is then placed in an oven until the plastic fuses together or evaporation of solvent is completed. After water cooling, the mould is opened and the product is removed. Aluminium, ceramic, steel or plastic material are used for making the mould. Sometime vibrating, spinning or the use of vacuum chambers may also be used to remove the air bubbles in the product. Rain boots, shoes, hollow toys, and dolls and automotive products ,etc., are made by this process.

5.8.4 Rotational Casting

The rotational casting is sometimes named as centrifugal casting. A predetermined quantity of plastics powder or monomers is placed in one half of the mould. The mould is closed by keeping the other half and placed in a hot oven and rotated into axes. Because of the bi-axial rotation and the centrifugal force, the plastic material sticks with the wall of the mould and gets cooled. The solidified plastic product is removed. This process is used for making balls, toys, containers, fuel tank, large pipes and tubes ,etc. On the base of rotation speed, the wall thickness in different areas is controlled.

5.8.5 Solvent Process

Solvent process is used to make sheets and films of uniform sizes. Plastics are dissolved in a solvent along with additives plasticisers and colourants. Further the mixtures are poured on to a stainless steel belt. The film or sheets, made by process is free from stresses and will be uniform in thickness. This method is an easy way to make films and sheets. The materials used for making sheet are cellulose acetate, cellulose butyrate, PVC, PMMA, PC, PVA and other copolymers.

5.9 RESIN TRANSFER MOULDING (RTM)

5.9.1 General Process

In this process, the dry unimpregnated reinforcement is preshaped and oriented into skeleton of the actual part known as the preform, which is inserted into matched die mould as shown in Fig. 5.64. The mould is then closed, and a low-viscosity reactive fluid is injected into the tool. The air is displaced and escapes from vent ports placed at the high points. During this time, known as the injection or infiltration stage, the resin 'wets out' the fibres. Heat applied to the mould activates polymerisation mechanisms that solidify the resin in the mould known as cure, as shown in the Fig. 5.65.

The resin cure begins during filling and continues after the filling process. Once the part develops sufficient green strength, it is moved or demoulded. Green strength refers to the strength of apart before it has completely cured. The green strength is an indication of how well it holds its shape until it is completely cross linked as shown in the Fig. 5.66.

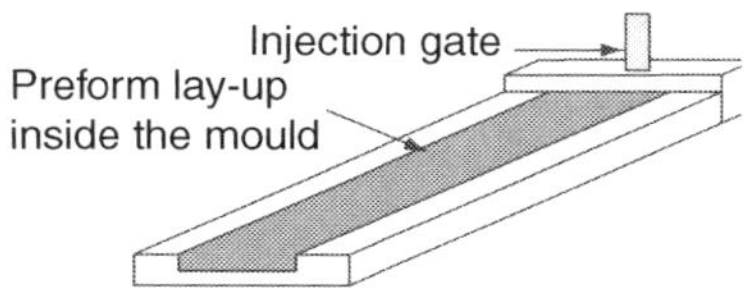

Fig. 5.64 Loading of preform.

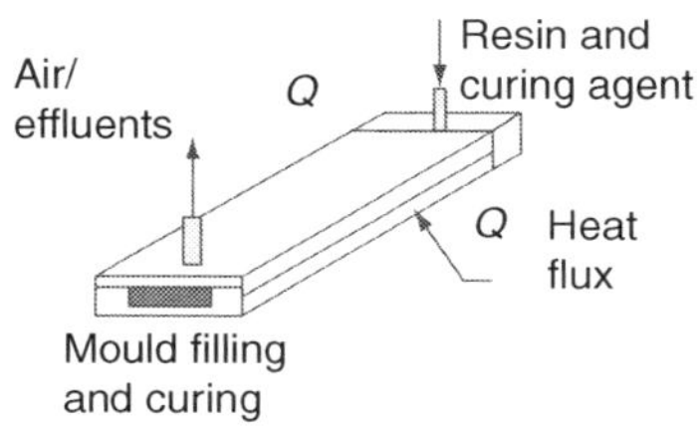

Fig. 5.65 Solidification by using heat.

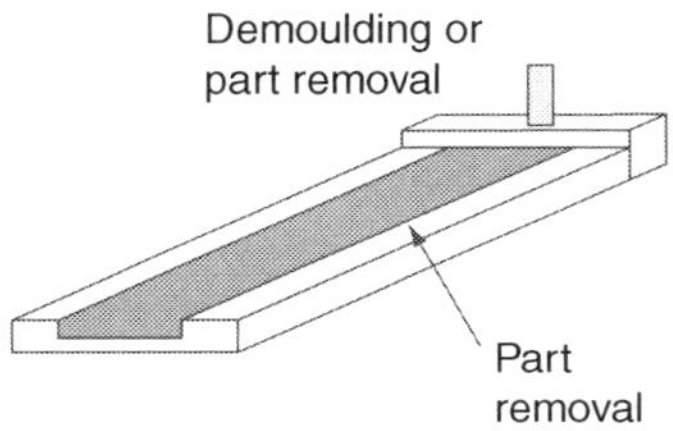

Fig. 5.66 Demoulding.

5.9.2 Advantage of Resin Transfer Moulding (RTM)

The advantage of RTM relative to other polymer composite manufacturing techniques is the separation of the moulding process from the design of the fibre architecture.

This process is of low capital investment, good surface quality is achieved and flexibility in tooling, large and complex shapes can be moulded. The components can be processed with ribs and parts integration is easier. Range of resins can be processed and re-enforcement is easier.

5.10 STRUCTURAL FOAM MOULDING

5.10.1 Introduction

This is a low pressure injection moulding process where the plastic melt is foamed with nitrogen gas or with a chemical blowing agent. Once the plastic fills the mould, the wall of the part solidifies against the cold mould wall and a thin layer of plastic solidifies along the mould wall. This thin layer forms a skin structure over the foamed inner core. The thin outer solid wall is supported by the interior cellular foam structure. This process can be used on many different sized parts, small to large, to produce a stronger and potentially longer lasting part. Many large parts require thicker walls than standard injection moulding can produce.

Structural foam moulding allows for the injection moulding of large parts through the addition of a foaming agent. Any injection mouldable plastic can be used in the structural foam moulding process. Structural foam moulding allows for the injection moulding of such parts through the addition of a foaming agent.

The lower viscosity of this mixture and the thicker cross section of the moulded part, results in less injection and clamping pressures required and less stresses created during injecting the plastic into the mould. As low pressures and forces are involved in processing, it allows more economical moulding equipment and tooling to be utilised. Structural foam moulding is suitable for mass production of multiple parts in the same machine cycle.

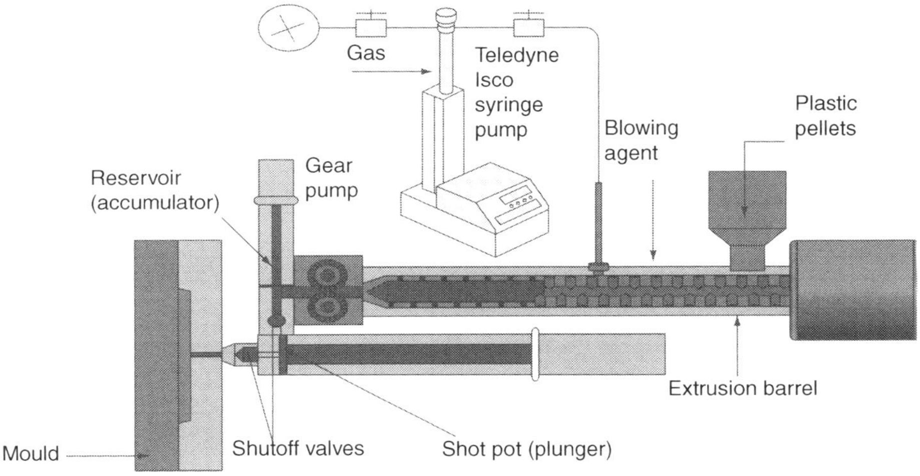

Fig. 5.67 Structural foam moulding.

5.10.2 Advantages

The advantage of this process is that, the mould cavity pressures are 10–20 times less than the conventional injection moulding and the part weight can be reduced to 15–30%.Large mould parts with high rigidity with reduced stress and warpage is produced, multiple parts can be moulded in a single cycle. Low cavity pressure permits use of lower cost aluminium moulds. Smooth and uniform surface finish is achieved; reduced cycle time is required for producing very large parts and high stiffness-to-weight ratio. Thicker parts and tubular shapes are processed without sink marks.

5.11 LAMINATION

The laminate is a material that can be constructed by uniting two or more layers of material together. The process of creating a laminate is called lamination.

There are different lamination processes, depending on the type of materials to be laminated. The materials used in laminates can be the same or different, depending on the processes and the object to be laminated.

5.11.1 Hot Roll and Belt Lamination

The hot roll and belt lamination uses heat and pressure by means of bonding shown in figure below, the adhesive film and substrate are drawn onto heated rollers where the materials are heated and pressed together. The heat activates the adhesive film, creating a bond when pressed against the substrate material. Hot roll and belt lamination of adhesive films allow for continuous in-line lamination and an even distribution of adhesive.

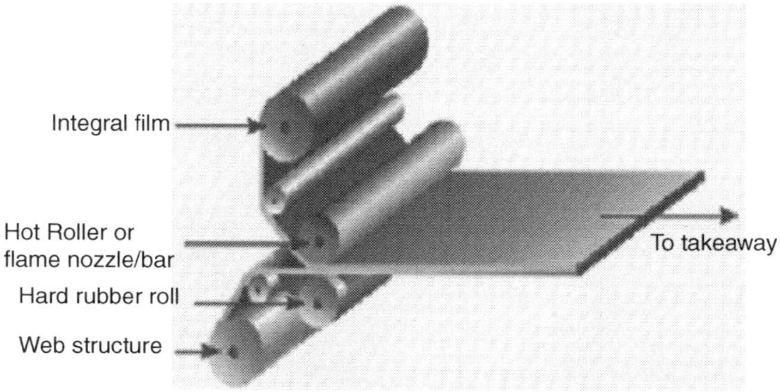

Fig. 5.68 Hot Roll / Belt Lamination

5.11.2 Flame Lamination

The flame lamination is generally used to bond film and fabric to soft polyurethane foams. The process is shown in figure below, involves the passing of the soft foam over an open flame, which creates a thin layer of molten polymer. The film and/or fabric are quickly pressed against the foam while it is still in the molten state. The strength of the bond depends upon the film, fabric and foam selected and the processing conditions ,i.e., gas type, flame height and spread, foam burn-off and nip pressure.

Flame lamination is a continuous process that, depending on the equipment, adheres fabric or film to one or both sides of the foam in a single pass.

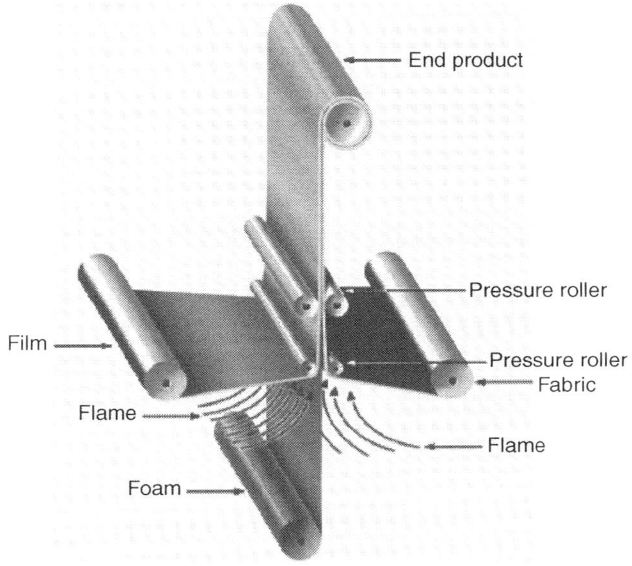

Fig. 5.69 Flame lamination.

5.11.3 Calendar Lamination

The calendar lamination of adhesive films allows for continuous in-line lamination and provides an opportunity for an even distribution of adhesive. Calendar lamination, similar to hot roll

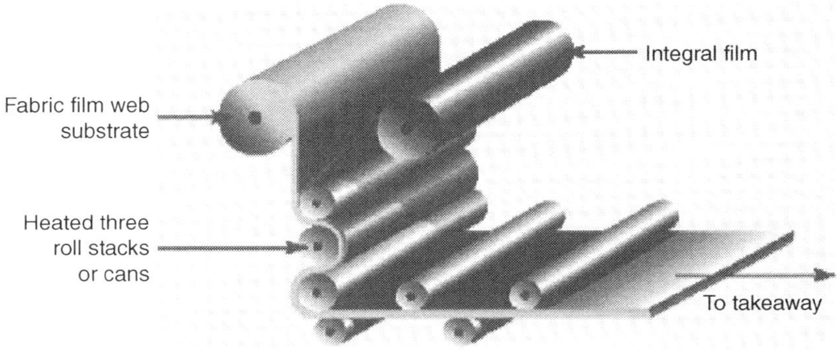

Fig. 5.70 Calendar lamination.

lamination, uses a heated three-roll stack to heat and activate adhesive films. The adhesive film and substrate as shown in Fig. 5.70 are drawn into a stack of heated rolls where the film is heated, activated and applied to the substrate. Material selection is critical in providing a strong, long-lasting bond between the film and the substrate.

Questions

1. Where are threaded components used?
2. What are the categories of unified thread? Explain briefly.
3. How does shrinkage allowance play an important role in thread forms?
4. Write the various types of mould design techniques for withdrawing the threaded components?
5. How does 'stripping internal thread design' work?
6. What is the difference between fixed threaded core and loose threaded core?
7. List out the advantages of fixed threaded core over unscrewing mould design.
8. What is collapsible core? Write down the advantages and disadvantages of collapsible core.
9. Explain the rotary unscrewing.
10. How does 'manually powered fixed rotating core design with gear transmission' mechanism work?
11. What is the role of chain and sprocket transmission in manually power design with fixed rotating core mechanism?
12. Explain briefly rotating helix spindle method with a neat sketch?
13. List out the types of mould design techniques for withdrawing the external threaded components.
14. How does automatic unscrewing mechanism work to release external threaded component?
15. Explain stripping (jumping) method to release external threaded component.
16. How does a split work to release external threaded component?
17. What is hot runner mould?
18. Explain cold runner mould.
19. Explain runnerless moulding.
20. Classify runnerless moulding and explain any one.
21. Explain the application of hot runner mould.
22. How does manifold block play an important role in hot runner mould?
23. What is the difference between rectangular manifold block and circular manifold block?
24. Explain the selection criteria of heaters for the manifold.
25. What do you mean by melt-flow-way in manifold?
26. Why is runner plug required in manifold?
27. What role does secondary nozzle play in manifold?
28. Explain standard secondary nozzle.

29. Explain the working principle of internally heated sprue bush with a neat sketch.
30. How does expansion problem occur in manifold? Explain its remedy with formula.
31. Explain outsert and insert moulding with a neat sketch.
32. Write the objective and application of outsert moulding.
33. What is insert moulding? Explain with a neat sketch.
34. Write notes on dispensing.
35. Describe the design consideration in gas-assisted injection moulding process.
36. What is reaction injection moulding (RIM)?
37. Write notes on material storage.
38. Describe the benefits of RIM?
39. Describe structural foam moulding?
40. Write down the advantages of foam moulding.
41. What are the different types of mould design techniques for withdrawing the threaded components? Explain any one.
42. How does collapsible cores mechanism work to release undercuts from mould? Explain the different types of collapsible coring techniques.
43. What are the types of unscrewing mould? Explain briefly.
44. Explain the principle of hot runner mould with a neat sketch.
45. What are the design guidelines to be considered while making hot runner moulds?
46. How does heating of manifold take place in hot runner mould? Explain with formula.
47. Explain barb nozzle with a neat sketch.
48. Draw an neat sketch of an antechamber secondary nozzle and explain its working principle.
49. Describe the heating system in hot runner unit and explain its method.
50. Explain the external flow way heating system with a neat sketch.
51. Explain the internal flow way heating system with a neat sketch.
52. Write the advantages, disadvantages and limitation of hot runner system.
53. Explain the principle of insulated runner mould with a neat sketch.
54. Describe the working principle of gas-assisted injection moulding process with a suitable sketch.
55. Write the advantages, limitation and application of gas-assisted injection moulding process.
56. What is stack mould? Explain the types of stack mould.
57. What is casting? Explain various types of casting.
58. Describe the process and advantages of resin transfer moulding (RTM) with a suitable sketch.
59. What is lamination? Explain various types of lamination.
60. Write notes on the following:
    (a) Multicolour injection mould and **(b)** Rotary mould.

References

1. CIPET, *Technical Manual,* Chennai.
2. Jones, Peter, *The Mould Design Guide,* Smithers Rapra Technology, Ltd. UK.
3. Pye R. G. W., *Injection Mould Design,* Affiliated East West Press Pvt. Ltd. New Delhi.

CHAPTER 6

Blow Mould Design

6.1 INTRODUCTION – BLOW MOULDING PROCESS

Blow moulding is a plastic moulding process where a thermoplastic material is heated to its forming temperature, which is below the plasticising temperature of the plastic materials being used to form a hollow tube called a parison. This heated homogeneous plastic material is then placed between two female mould halves. The two female mould halves are then closed and compressed air enters via an open end of the parison via a blow pin or needle, the air blown into the closed female mould halves forces the parison to take the internal shape of the closed mould thereby forming the blown component. The mould halves are cooled through suitable medium through the entire process so that the blown parison when comes in contact solidifies immediately. The two female moulds are then separated and shaped hollow component is then ejected or allowed to drop out and the cycle is repeated again.

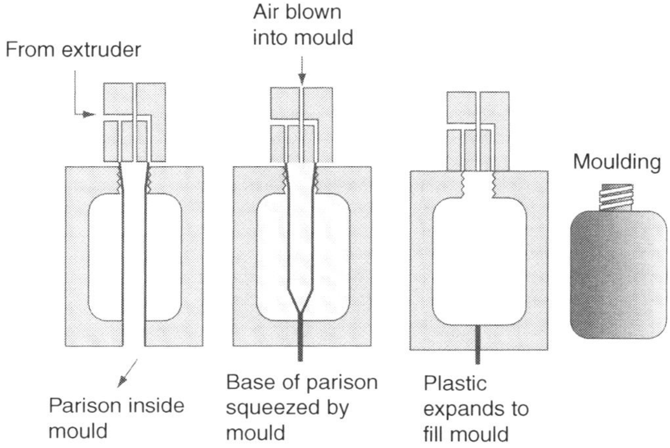

Fig. 6.1 Blow moulding process.

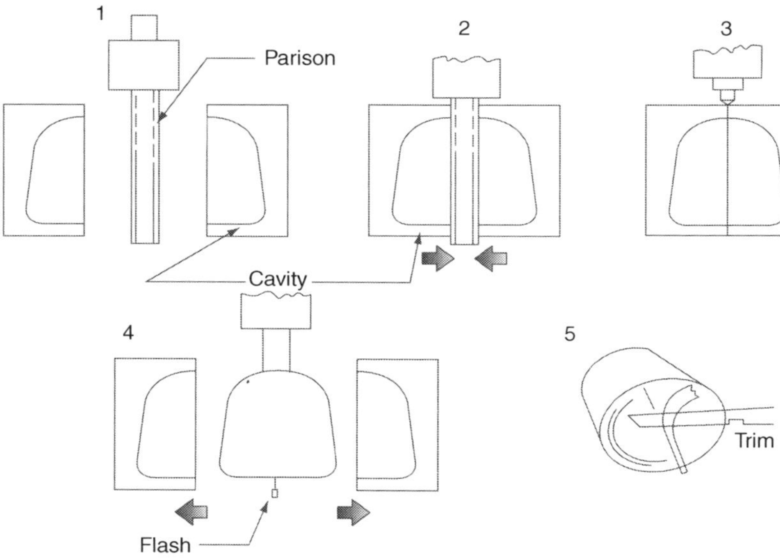

Fig. 6.2 Five stages in blow moulding process.

Stages in blow moulding process:

1. Plasticising or melting the resin
2. Parison or preform production
3. Inflation of the parison or preform in a mould to produce the end part
4. Ejection of the part
5. Trimming and finishing of the part

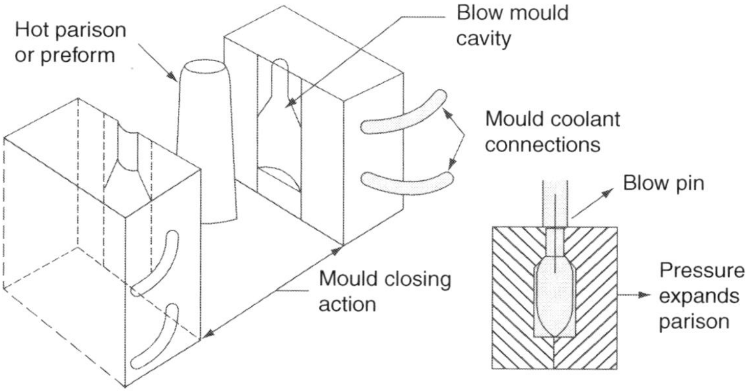

Fig. 6.3 Basic blow moulding process.

6.2 BLOW MOULD TERMINOLOGY

Blow mould terminology

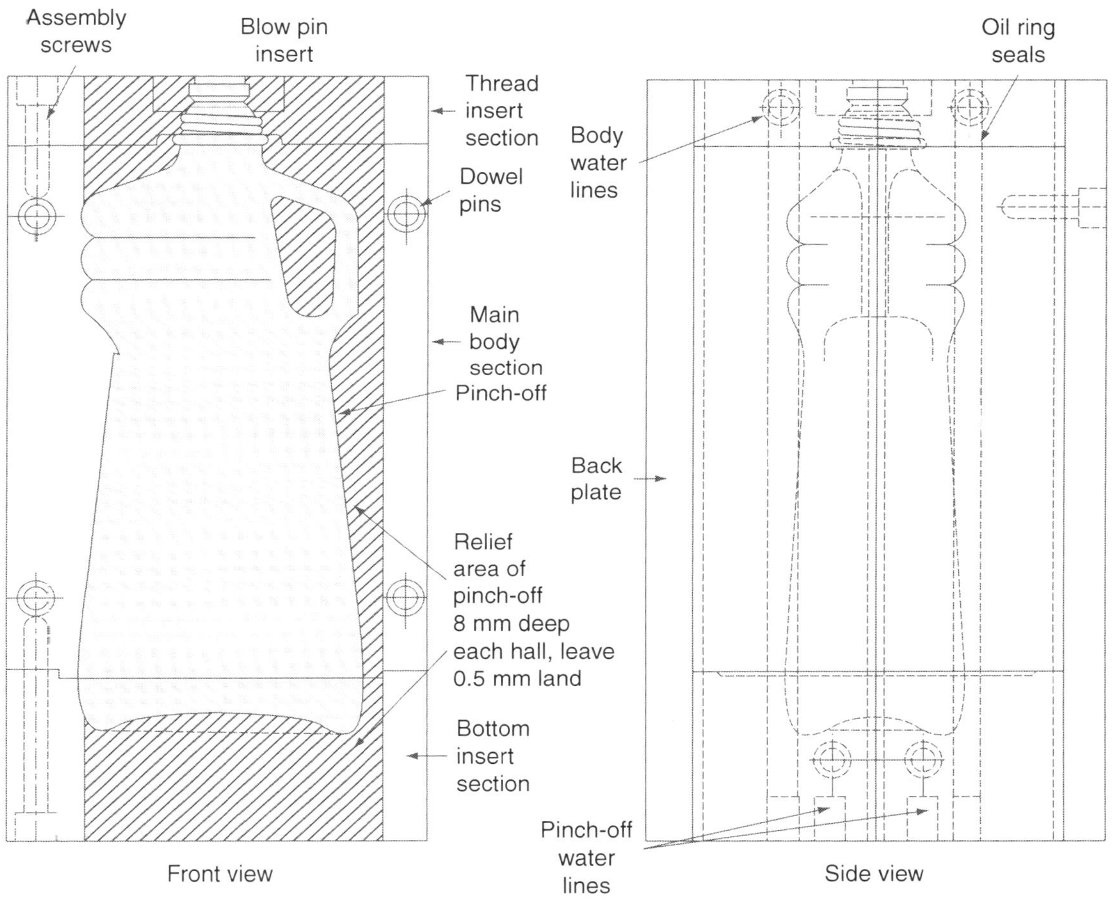

Fig. 6.4 Extrusion blow mould showing different parts.

6.3 APPLICATIONS OF BLOW MOULDED PARTS

Application of blow moulding components is given below:

1. Packagings for milk, fluids, medicines, cosmetics, etc.
2. Automotive fuel tanks, oil bottles, air-ducts, and seat-backs, etc.
3. Consumer products like toys, house-wares, sports goods, etc.
4. Drums for chemical industries.
5. Bellow shaped shields and double-walled carrying cases.

Fig. 6.5 Applications of blow moulded parts.

6.4 BLOW MOULD PART DESIGN PARAMETERS

Design of a blow-moulded part involves the selection of proper plastics, appearance of the product, end use properties and cost economics.

The property requirements vary with the specific application of the product like crack-resistance, impact-resistance, chemical-resistance, etc.

The appearance of the part covers basic design along with factors such as gloss, smoothness, wall-thickness, colour, etc.

Design of a blow moulded bottle and other shapes requires consideration of the factors like:

1. Material to be blown
2. Size and weight of the product and mould
3. Contours on the part
4. Surface texture and engraving
5. Sharp corners and straight edges
6. Blow opening available and locations
7. Parting lines

Blow moulded articles perform better with rounded, slanted and tapered surfaces. Squares and flat surfaces with sharp corners should be avoided.

The blow-ratio of a blow moulded part gives the amount of stretch for a given combination of parison size and part size.

For cylindrical containers, blow-ratio = (Mould or part diameter / Parison diameter)

This ratio is usually between 1.5 to 3.0, it can be maximum up to 7.

6.5 BLOW MOULDED PART DESIGN CONSIDERATIONS

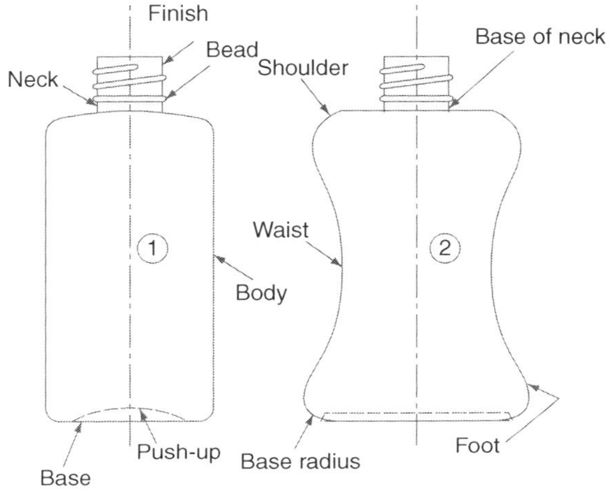

Fig. 6.6 Terminology of blow moulded parts.

6.5.1 Corner and Edge Rounding

Wall thinning in corner areas should be considered, as it creates weaker areas in the moulding.

Practical guideline for designing corner and edge-rounding is given below:

1. Rectangular containers : at least $1/3^{rd}$ depth of mould half
2. Cylindrical containers : at least $1/10^{th}$ depth of mould half

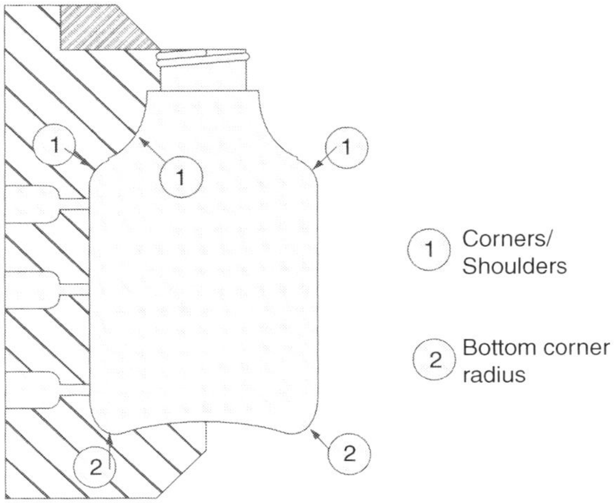

Fig. 6.7 Corner and edge rounding.

6.5.2 Volume

For designing the mould cavity, overall cavity volume is calculated as

$$V_{oc} = V_c + V_b - V_s - V_r$$

where

V_{oc} = Overall cavity volume
V_c = Usable volume of container
V_b = Volume increase due to sidewall bulging
V_s = Volume loss due to shrinkage of container
V_r = Volume of resin in the part

Bulging is usually more in flat-sided containers. This can be reduced by using ridges or grooves.

Volume adjustment can also be done by using changeable inserts in the mould, for side walls. The depth of these inserts can be changed for adjusting the volume.

6.5.3 Neck, Spouts and Other Openings

Each part must be designed with an opening, which may be utilised for blow and this opening is utilised as neck or spout.

The important dimensions of a threaded neck finish are shown in Fig. 6.8.

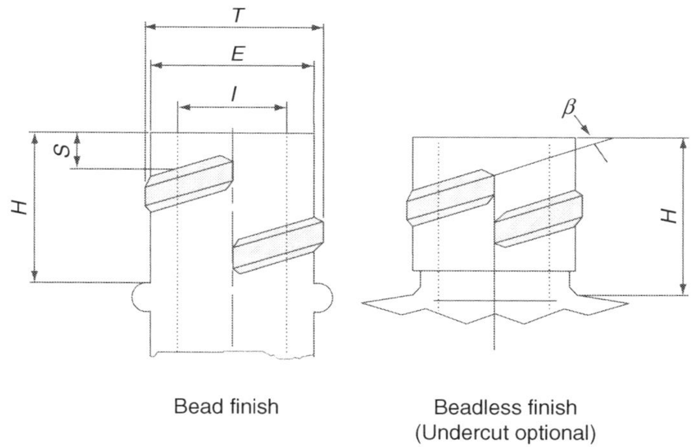

Fig. 6.8 Neck finish of blow moulded parts.

Wide mouthed containers are blown through a chamber type cavity near the neck area. This chamber is later trimmed-off with a rotating knife.

6.5.4 Base Design

On stretch blown PET bottles, the base should be spherical due to internal pressure. Petaloid type base provides a self-standing container with several egg-shaped feet on which it balances.

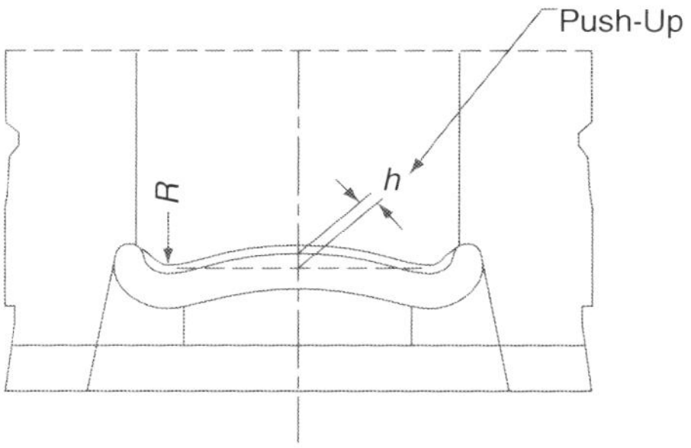

Fig. 6.9 Push-up for stability of part.

6.6 CONTAINER DESIGN FEATURES

The basic shapes of containers are round, square, oblong, or oval. The following are the factors to be considered before proceeding for design.

1. The type of plastic material to be used
2. The physical size and shape of the container
3. The neck finish size and design
4. The shape of the container
5. Type of thread
6. The undercuts
7. The parting line location, the surface finish
8. The fill point and weight of the container

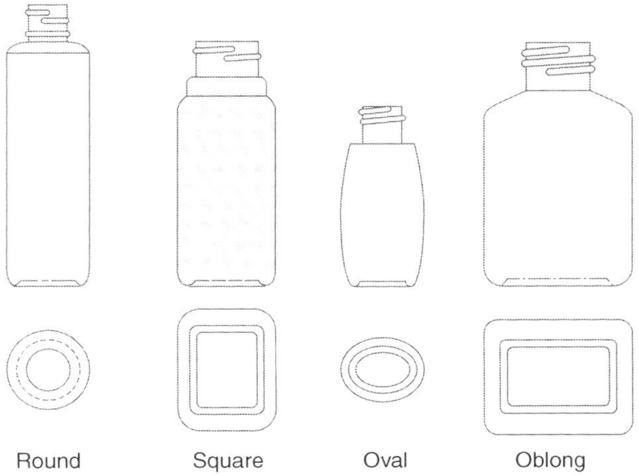

Fig. 6.10 Types of containers.

6.6.1 Rib design

The round containers are provided with horizontal ribs to improve hoop stiffness, and vertical ribs for compression stiffness, as shown in Fig. 6.11.

The square containers reduce stiffness, thus losing both top load strength and bulge resistance as well. Square containers are shown in Fig. 6.12 and truss groove as shown in Fig. 6.13.

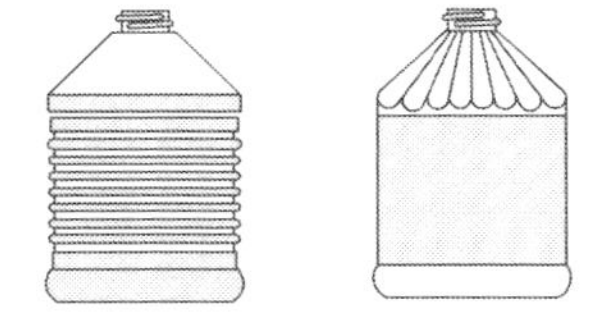

Fig. 6.11 Horizontal and vertical ribs.

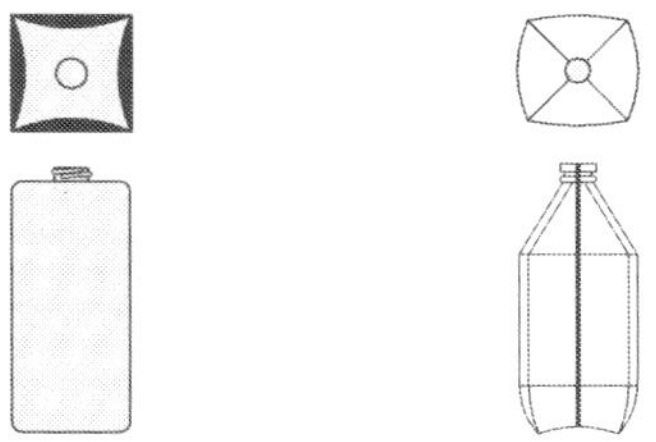

Fig. 6.12 Square containers.

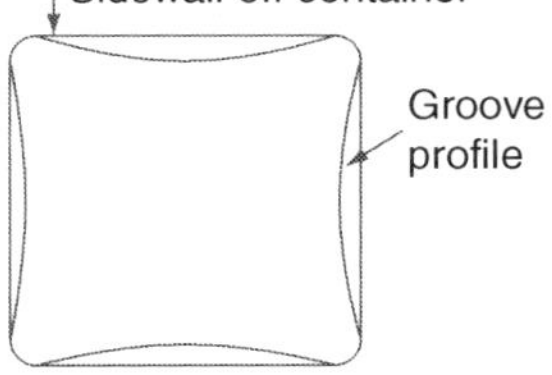

Fig. 6.13 Truss groove.

6.7 DESIGN DETAILS

One of the major advantages with plastic processes is the ability to add strength to the container by incorporating bosses, snap-fits, inserts and gussets in the mould/components.

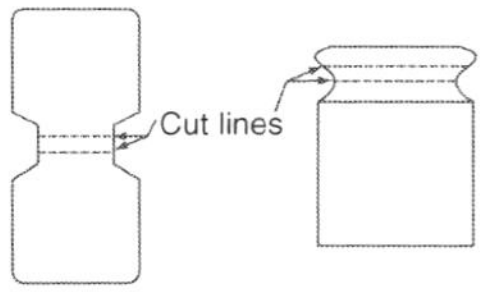

Fig. 6.14 Lid with snap fit.

1. **Mould-in Inserts or Components:** Blow moulded components have certain limitations based on its size, shape, type of material, etc., during the moulding process. Low pressure is required during blowing and also to eject the moulding from the mould cavity.
2. **Snap-Fits:** Open top container with lids such as waste or trash bins are moulded with the lid and container in one piece and cut apart in a post moulding operation as shown in Fig. 6.14. When the component is cut apart the lid will snap fit to the container.
3. **Multiple/Combination Cavities:** To reduce the cost of producing an open top product, making two parts from the same parison is to be considered. In two siamese moulding, the mould is built with open ends together with the moulded part cut in two halves. A short transition between the two halves is desired so that the parison to be blown through a hollow needle, which pierces and blows through this area leaving no hole in the part. Two cuts are made, separating the moulding into two containers leaving the transition which is later reground.
4. **Tack-off's:** Stiffening ribs are added to a blow moulded part by allowing the inflated parison to compress, in local areas as shown in Fig. 6.15.

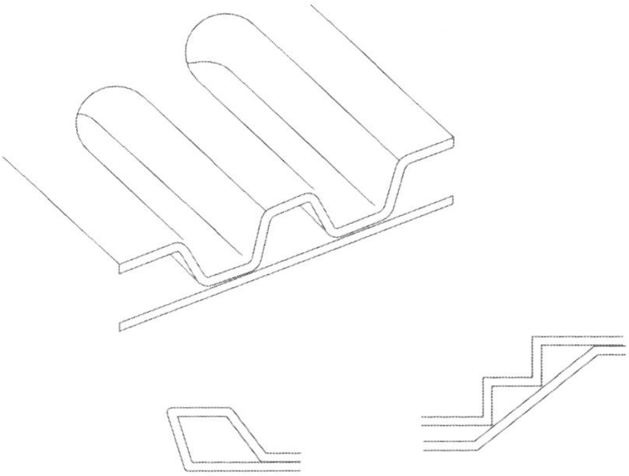

Fig. 6.15 Tack-off ribs.

5. **Container Configuration:** The effect of warpage in square and rectangular shape containers is high, the flat sides tend to warp, and the warpage is exaggerated after the part is cut due to stress set up by shrinkage and wall thickness variations. Hence a shallow curvature is recommended in the containers.

Warpage is experienced when the container is cut to create the opening. This is because the lip tends to fold inward. To avoid this phenomenon a corrugated or cross section design must be considered as shown in Fig. 6.16.

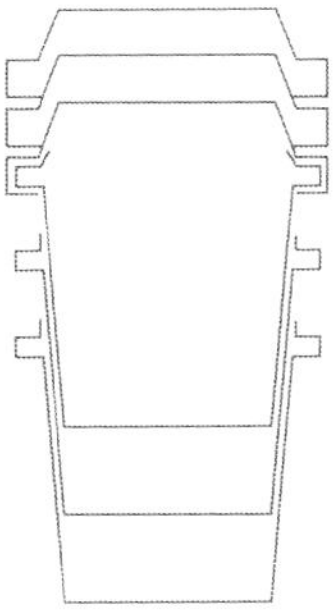

Fig. 6.16 Corrugated cross section.

6.8 SPECIAL CONSIDERATIONS FOR BOTTLE DESIGN

The most important structural and mechanical considerations in a bottle design include:

1. Vertical strength
2. Wall thickness uniformity
3. Highlight deflection
4. Push-up strength
5. Label considerations
6. Rigidity
7. Shape
8. Hot-fill capacity

If the bottle is subjected to vertical loadings, horizontal corrugations or bellows on the part should be avoided.

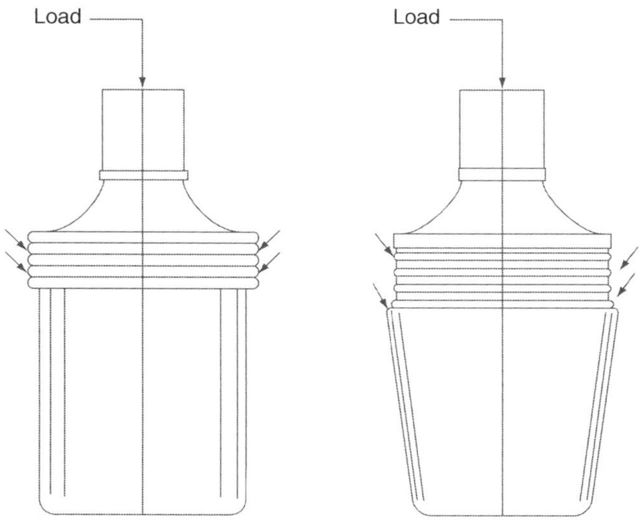

Fig. 6.17 Loading consideration for a blow moulded part.

6.9 PLASTICS MATERIALS FOR BLOW MOULDING

The following properties are suitable for blow moulding applications:

1. Good thermal stability
2. Good flowability of homogenous melt
3. Sufficient stretchability of parison
4. Excellent repeatability of parison weight and length
5. Smooth parison surface
6. Compatibility with additives
7. Sufficient wide processing range
8. Excellent lot to lot consistency

Blow mouldable polyolefin's materials

LDPE	:	Low Density Polyethylene
LLDPE	:	Linear Low Density Polyethylene
HDPE	:	High Density Polyethylene
EVA	:	Ethylene Vinyl Acetate and Ethylene copolymers
PP	:	Polypropylene and Polypropylene copolymers

Table 6.1 Blow moulding resins grade.

S.No.	Resin	Melt Index Range (gm / 10 min)	
1.	LLDPE	< 1 to 2	
2.	LDPE	< 1 to 2	
3.	HDPE	< 1 to 2	
4.	EVA	< 1 to 3	
5.	PP	< 1 to 4	** MFR

Melt index describes the flow behaviour of a resin at a specified test temp (190°C), and a specified test weight (2,160 gm). Higher value indicates easy flow of the melt.

** Melt Flow Rate (MFR) describes the flow behaviour of polypropylene resins at a specified test temp (230°C) and a specified test weight (2,160 gm).

HDPE: Blow moulding grade

High density polyethylene grades are suitable for general purpose extrusion blow moulding applications.

Articles blown from these grades exhibit good stiffness. The resin offers good melt strength, ESCR and impact resistance and typically used for packaging of oil, vanaspati, general purpose containers, jerry can, etc.

Table 6.2 Physical characteristics.

Physical Characteristics				
S.No.	**Property**	**Unit**	**Test Method**	**Value**
1.	Density	g/cc	ASTM D 1505	0.956
2.	MFI (2.16 kg)	g/10 min	ASTM D 1238	0.30
Typical Properties				
S.No.	**Property**	**Unit**	**Test Method**	**Value**
1.	Tensile strength at yield	MPa	ASTM D 638	26
2.	Elongation at break	%	ASTM D 638	550
3.	Flexural yield strength	MPa	ASTM D 790	28.5
4.	Flexural modulus	MPa	ASTM D 790	900
5.	Hardness	Shore D	ASTM D 2240	69
6.	Vicat softening point	°C	ASTM D 1525	128

Processing parameters

1. Melt temperature in range of 175–205°C are recommended.
2. Temperature of 190–205°C will result in optimum ESCR properties.

6.10 TYPES OF BLOW MOULDING

Blow moulding components are produced either by extrusion or injection process. The breakdown of the subsidiary methods are given below:

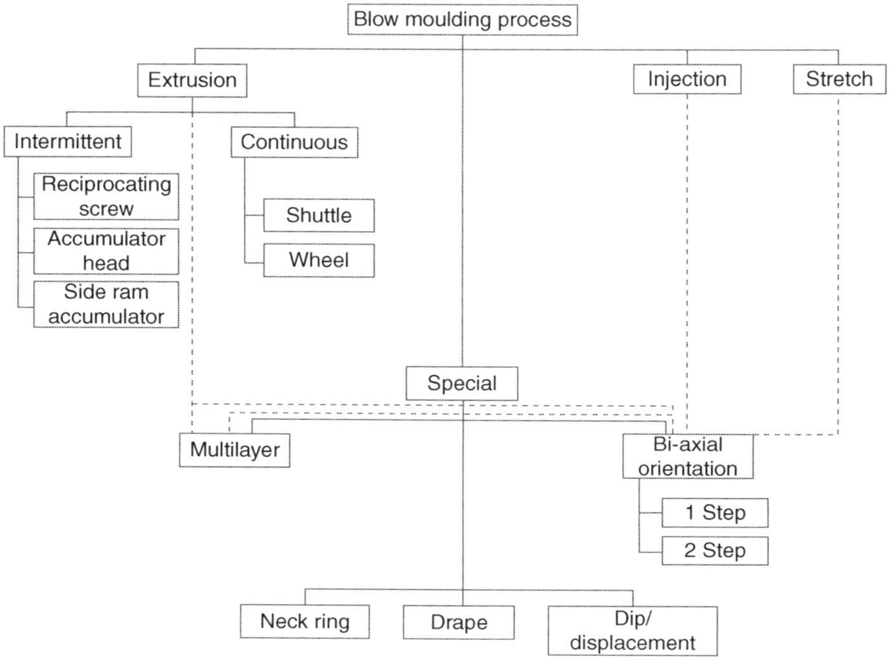

6.11 EXTRUSION BLOW MOULDING PROCESS

In this process heat and pressure are applied to melt the resin and force it through an accurately dimensioned die to produce the desired shape. There are several main parts to an extrusion blow moulding machine, viz. hopper, screw, barrel, feed section, compression section, metreing section, screen pack, breaker plates, adapter, die head, core, mandrel, and die tip. The resin is melted in various sections and plasticised, and then the melt is delivered to the die at proper temperature in a uniform rate as shown in Fig. 6.18.

Advantages of extrusion blow moulding It is a process for producing containers and hollow parts preferred for moulding high volume containers.

Disadvantages of extrusion blow moulding Uneven wall thickness components are produced and difficult to achieve close dimensional tolerances; relatively low accuracy of surface finishing is achieved.

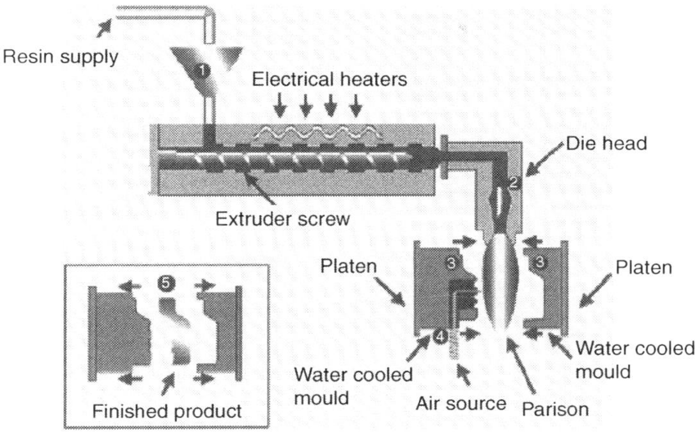

Fig. 6.18 Cross section - extrusion blow mould machine.

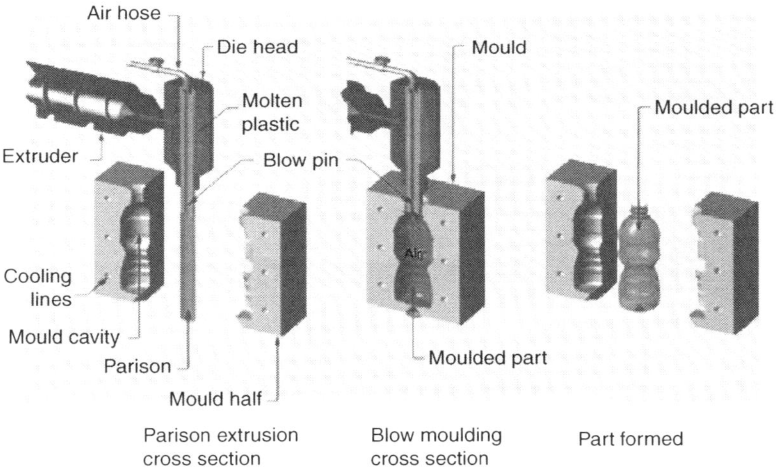

Fig. 6.19 Various stages of extrusion blow moulding process.

6.11.1 Blow-ratio

Blow-ratio is one of the most important factors for designing a blow moulded part.

In blow moulding, the blow-up ratio is the ratio of the mould cavity diameter to the parison diameter.

The materials of different blow grades exhibit differing stretch behaviour. The blow-ratio of a moulding is a way of representing the amount of stretch involved for a given combination of parison size and part size.

Blow-ratio is a measure of the amount of stretch the parison will experience when it is blown into the part's shape.

For cylindrical containers,

Blow-ratio = Mould diameter / Parison diameter

In general, this value is between 1.5 to 3, and maximum up to 7.

From the hollow parison, the product is formed by expansion of air. The blow-ratio is the maximum part finished outer diameter divided by die outer diameter of the parison.

It depends on the parison temperature and thickness.

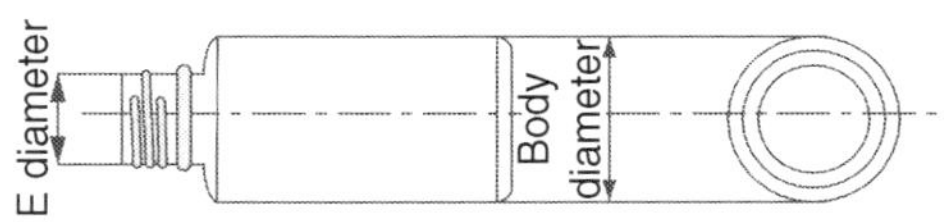

Fig. 6.20 Blow up ratio.

6.11.2 Dies for Parison

After the molten polymer melt leaves the extruder manifold, it enters the die-head shown in Fig. 6.21.

The polymer melt forced inside the die assembly and split into two streams around a mandrel- sleeve or flow-divider and meet on the opposite side where it welds together. This is the first step in forming a hollow tube or parison.

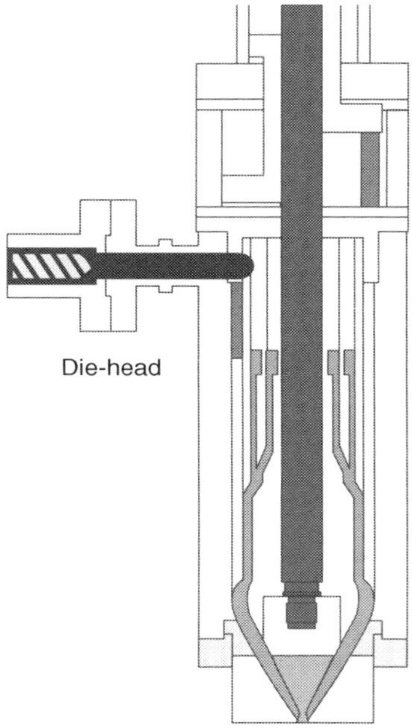

Fig. 6.21 Die for producing parison.

6.11.3 Die-Head Design

In case of bottle weight where no significant parison draw-down occurs, the relationship between bottle weight and die/mandrel dimension may be determined approximately from the following formula:

$$W = C\,\pi\, d_m t l s^2$$

where
- W = Weight of moulding including neck flash (g)
- I = Length of moulding a parison including pinch-off flash (cm)
- d_m = Mean diameter of die annulus (cm)
- t = Annulus gap (cm)
- S = Die swell ratio of polymer concerned
- C = Constant for polyolefins = 0.78 (metric units)

This formula applies directly to landed (fixed gap) dies only. On machine with parison control (variable die gap) appropriate modifications have to be made to calculation in accordance with the parison thickness profile adopted.

6.11.4 Parison Programming

Parison programming is the control of the wall-thickness, from top to bottom, of the parison as it emerges from the die-head during extrusion.

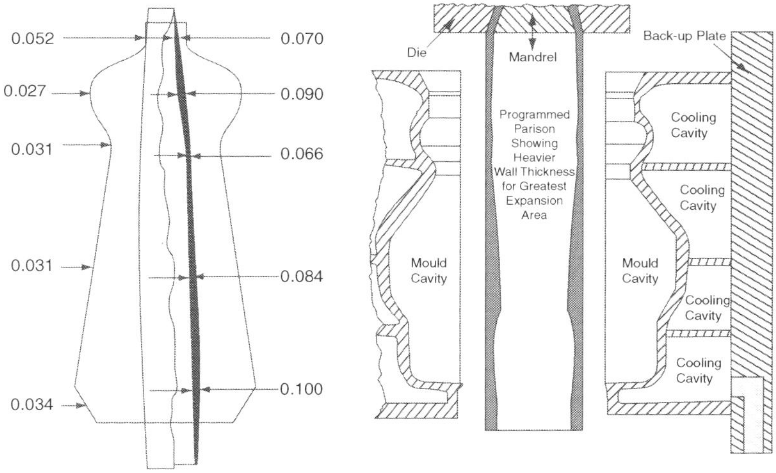

Fig. 6.22 Wall-thickness variation in parison.

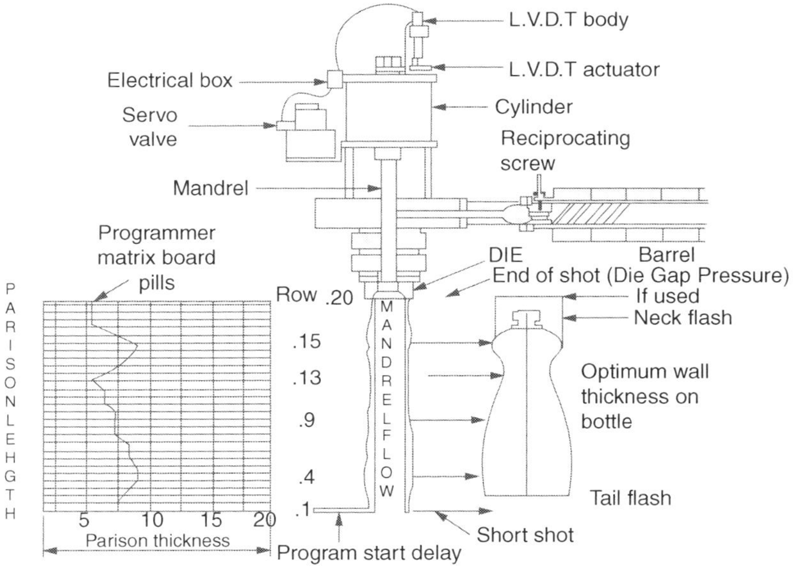

Fig. 6.23 A parison programming device.

6.11.5 Parting Lines

The parting line is the line at which the two halves of a mould meet when the blow mould closes. It should be decided in a mould suitably to create two equal or mirror image mould halves. For unsymmetrical or awkwardly shaped angled articles the location of parting line is determined by the ideal blowing position and avoiding the ejection difficulties.

Cylindrical articles are parted along the axis of rotation and elliptical shapes along the large diameter. Cube shaped articles can be parted along the diagonals or parallel to the side faces.

Various types of parting lines like pinch, dammed and flat parting surfaces are used in extrusion blow moulds.

Pinch parting line Pinch parting line is used in the areas where the parison is to be pinched together, creating flash. A pinch design is shown in the Fig. 6.24 in which pinch land length is 0.50–0.75 mm with a pinch angle of 45 degrees. Pinch relief depth is 1.5–2.0 times the wall thickness.

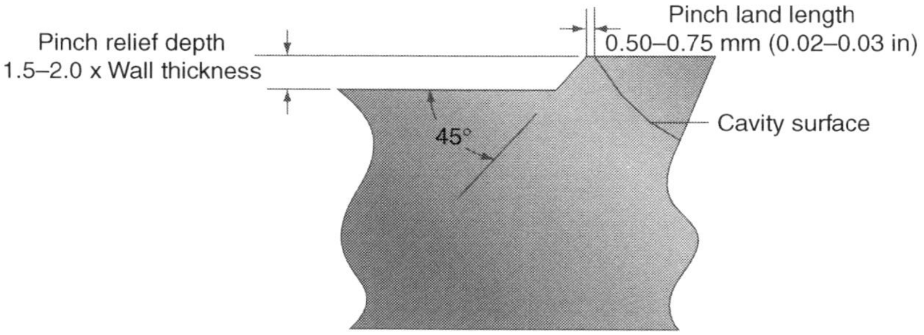

Fig. 6.24 Pinch parting line.

Dammed parting line This is used in areas where the parting line on the inside requires additional material. It is often used when an inside diameter must be machined smooth to contain no voids.

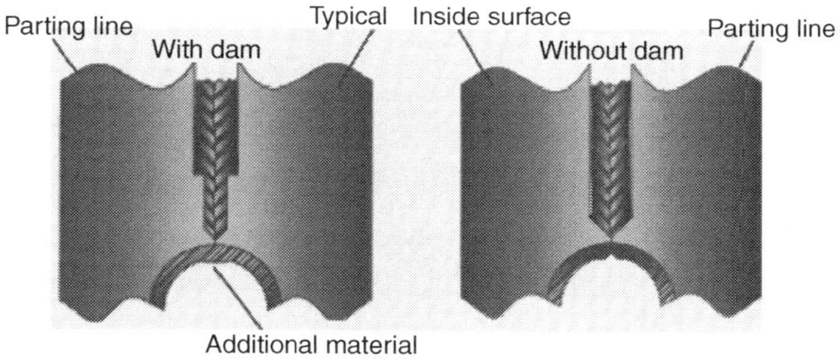

Fig. 6.25 Dammed parting line.

Flat parting line A flat parting line is used where the parison is captured inside the cavity and will not contact the parting line until blowing.

6.11.6 Clamping Force

The clamping force in a blow mould plays an important role to keep the two mould halves closed so as to produce a defect free component.

The clamp and platen mechanism in a blow moulding process serves the following functions:

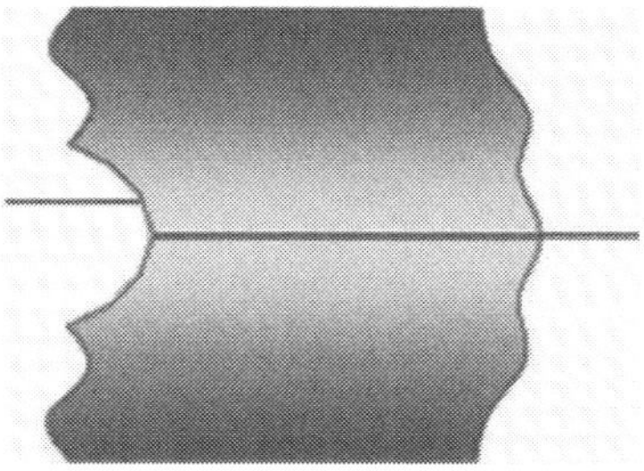

Fig. 6.26 Flat parting line.

1. Holding and aligning the mould halves properly.
2. Cycle the mould from open to closed position.
3. Holding the mould closed against the pressure of blowing air.
4. Large platen areas are desirable to provide uniform pressures to reduce the tendency of the mould to bend and to take different mould sizes. The clamping force required to keep the mould closed during blow moulding must exceed the cavity pressure and the total projected area of the impressions.

$$\text{Clamp forces(kg)} = 1.25 \times \text{Projected area of component (cm}^2) \times \text{Blow pressure (kg/cm}^2)$$

The required clamping force should be 25% higher than the projected area of the component and the blow pressure to keep the mould closed so as to avoid flash between the mould halves, as well as surface contact between the mould halves during the pinch-off.

The blowing pressure is usually in the region of 0.21 to 2.1 MPa, preferably 0.5 to 1.0 MPa. This requires a clamping force of at least approximately 10 tons per unit pressure and square metre of the projected area on the mould mounting platens. These data are of particular importance for the design of blow moulds when the weight of the mould is of prime consideration (steel, aluminium, etc., has different compression loading limits).

6.11.7 Pinch-off Design

The pinch-off areas pinches the ends of the plastic parison and seals the edges together when the mould closes. These surfaces are subjected to more wear than other parts of the mould, the metal with high thermal conductivity are preferred for blow moulds, such as aluminium and copper alloys are less wear resistant than steel. Steel inserts made out of hard and tough steel are often used for the pinch-off areas of blow mould. Due to the comparatively high pressure and mechanical stress exerted on the mould bottom half's during closing of mould, it pinches one end of the parison together. The pinch-off section does not cut off the excess parison 'tail'. Its protruding edges are cut nearly through, creating an airtight closure by pinching the parison along a straight line which makes it easy later to break off or otherwise remove the excess 'tail' piece. A high quality pinch-off of a thick-walled parison is more difficult to obtain than that of a thin-walled parison. It depends on the construction of the pinch-off insert.

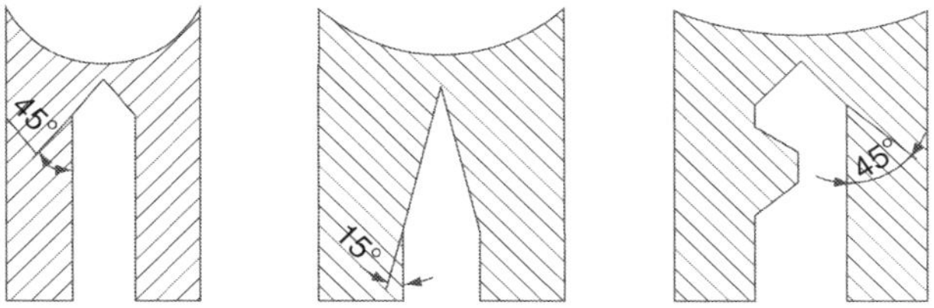

Fig. 6.27 Various classification of pinch- off design.

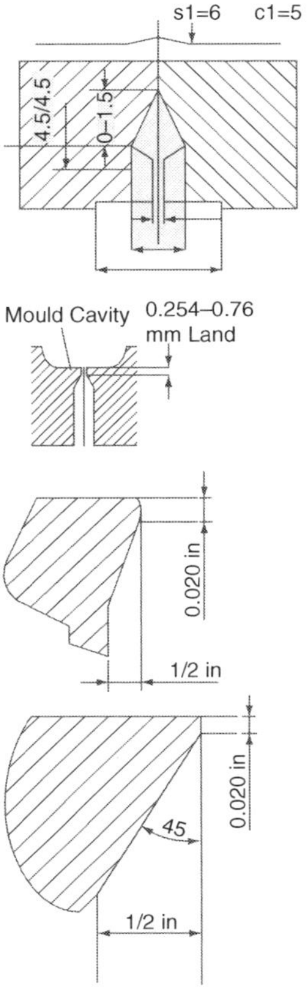

Fig. 6.28 Dimensions pinch-off design.

The pinch-off sections are very critical parts of blow mouldings. It should have good thermal conductivity for rapid cooling and good toughness for long production run. The pinch-off is shown in Fig. 6.27. 'A' is one of the most widely used designs. In case of the part is large relative to the parison diameter in the area of the pinch-off, the plastic will thin down and sometimes leave a hole on the parting line. To prevent this, design 'B' is used.

This design which has shallow angle of 15° has a tendency to force the plastic to the inside of the blown part, thereby increasing the thickness at the parting line. Another method which has proven to be successful is to use a design similar to that shown in design 'A' and install a dam in the relieved area of pinch-off has shown in design 'C'.

Pinch-off is the bottom part of the mould where parison is squeezed and welded/sealed together.

Pinch-off design for thin bottles and parts The pinch-off should not be knife-edged, to be formed by lands about 0.1 to 0.5 mm long. The total angle outward from the pinch-off should be acute, up to 15°.·Because of the land and acute angle a welding line in the bottom part of

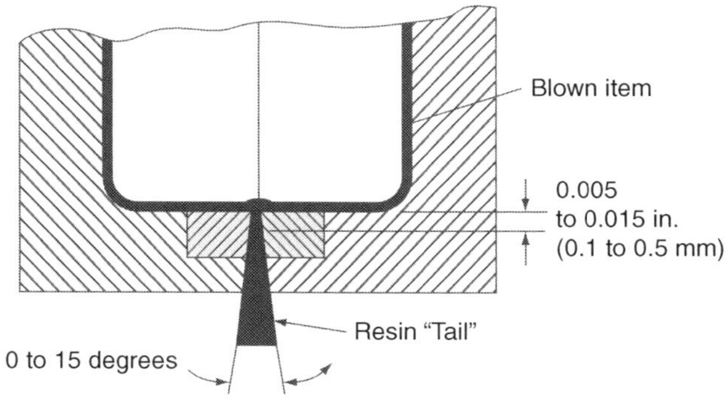

Fig. 6.29 Pinch-off design for thin bottles and parts.

the component is formed. It should be smooth on the outside and forms a float elevated line or a low bead inside the component. There should not be formation of groove which weakens the bottom along the seam.

Pinch-offs and inserts design Pinch-off is the part of the mould that welds the ends and the interior portions of the parison and also cut it to facilitate its removal. As the plastic material is thicker at the pinch-off areas, the heat load is higher in this zone. Beryllium-Copper is often used as mould material at pinch-off area, which is highly conductive and heat sensitive in nature. Interchangeable inserts are preferred for pinch-off area which facilitates rapid changing of brand name, designation, and date or worn out inserts.

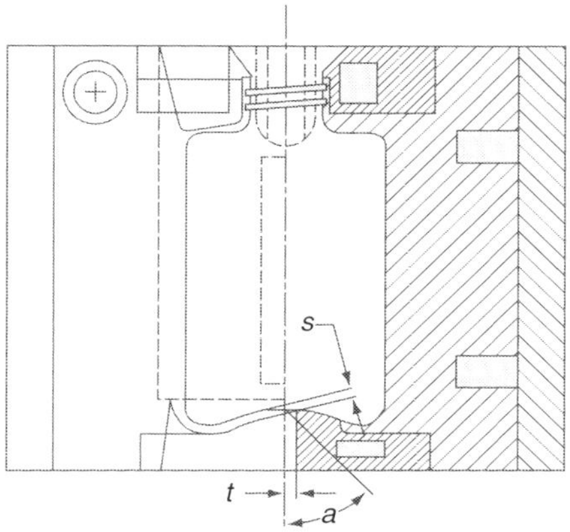

Fig. 6.30 Pinch-off angle showing styles of pinch-off pockets and welding edges
s = Edge width; a = Opening angle of pinch-off pocket; t = Width of pinch-off pocket.

The width of the pinch-off edge depends on the plastic material, parison-thickness and part-size. The thumb rule is, the width of edge in mm = $(\text{volume})^{1/3}$.

Flash-pockets are the areas in the parting surface of the mould that allow the mould to close, without parison holding it apart. The depth of these pockets should be large enough to slightly compress the parison to cool it adequately.

Inserts are fitted for pinch-off areas, to utilise a tougher and thermally conductive material than the rest of the mould. Changeable inserts are fitted to facilitate rapid changing of brand name, designation, date or other information.

Functions of pinch-off The main function of pinch-off is to withstand the pressure of plastic material, sustain the repeated pressure during closing of the mould and the thermal conductivity should be more for instant cooling. The design should be in such a manner that some plastics material moves inside the part to the thick bottom area and makes it full proof sealing joint. A break point should be provided to cut it from the parison.

Generally two types of pinch-offs are used in extrusion blow moulds as described below:

1. Double angle pinch-off
2. Compression pinch-off

Double angle pinch-off The double angle Pinch-off withstands more pressure and a land of 0.25 mm is provided in the pinch off area. A clear break point is formed with 30° angle is machined in the pinch-off insert for sending the melt to increase the thickness at the weld area.

Compression pinch-off The compression pinch-off is used where the parison is stretched too much by the mould before it is squeezed and welded and the weld line is usually thin and weak.

The land length is 3.17 mm to 6.35 mm which can push more amount of material in to interior of the part.

6.11.8 Flash Pocket

Factors determining depth of flash pocket depend on approximate density of plastic material, the weight of the part, approximate weight of the parison and estimate of parison diameter.

If the flash pocket depth is too shallow the flash will be squeezed with too much pressure which create undue certain on the moulds, mould pinch off area, and clamping system. If the pocket depth is too deep the flash will not be in contact with mould surface for proper cooling.

6.11.9 Flash Trimming

Automatic trimming is essential in large-volume container production. The trimming can be performed in the mould or in separate stations.

The mould wear and tear can be minimised by trimming in separate stations.

Blow pin is designed to cut the neck and retain the shoulder flash until the bottle has been separated.

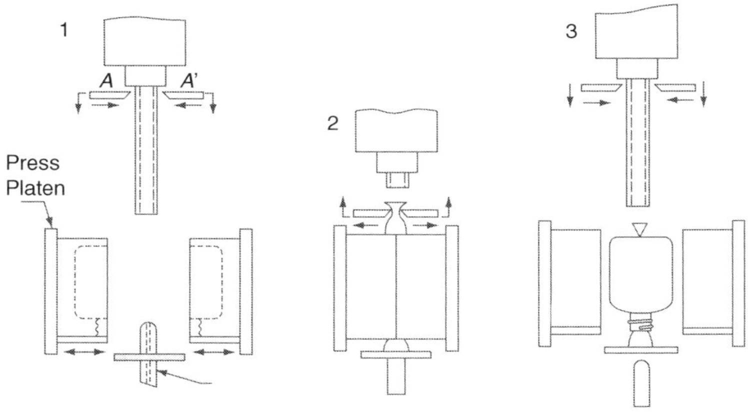

Fig. 6.31 Flash or parison trimming.

6.11.10 Handles

Moulded handles must lie along the parting line of the mould. The central cutout or eye of the handle must be pinched-out, and should therefore, have a proper pinch-off or flash-pocket design. This section may be designed as an insert.

The cross section of the handle should be a rounded square or rectangle for optimal wall thickness.

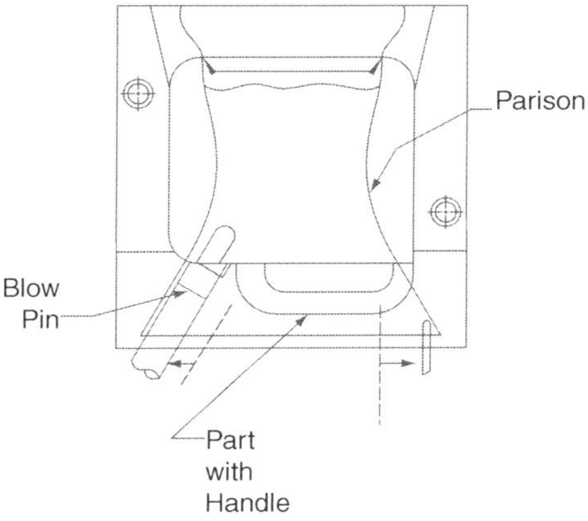

Fig. 6.32 Bottom blowing after spreading the parison.

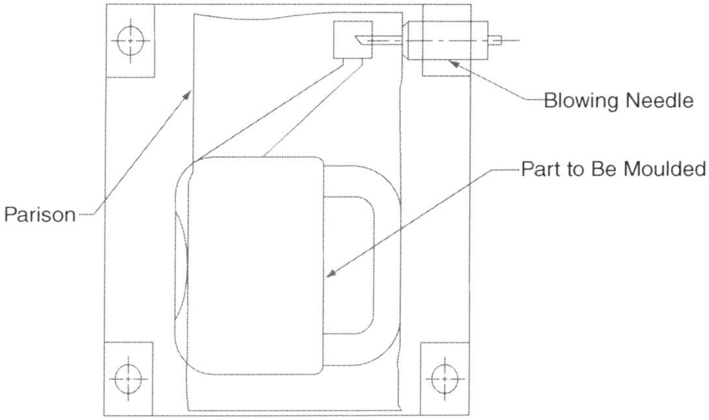

Fig. 6.33 Needle blowing the parison.

6.11.11 Shrinkage

Shrinkage allowance: In all mould manufacturing, allowances must be made for shrinkage of the material being moulded. Shrinkage value varies with the processing conditions and particular grades of material used for a given process. The shrinkage values for various plastic materials are shown in Table 6.3. As in injection moulding, in blow moulding process also shrinkage is to be added in the cavity dimensions while designing the mould. But due to shrinkage the bottle dimension as well as its volume is also changed. As blow moulded parts cools in the mould under air pressure, it will cool to the mould dimension. When the part is decompressed and ejected from the mould, it will shrink more than the nominal values and may not be within the desired final part dimensions. HDPE bottle shrink, with 80–90 percent of the shrinkage taking place in the first 24 hours.

Table. 6.3 Recommended mould shrinkage value for different plastic materials.

Materials	Longitudinal shrinkage
Acetral	0.025–0.035
Cellulose acetate	0.003–0.007
Nylon	0.01–0.02
Phenoxy	0.003–0.004
Polycarbonate	0.005–0.01
Polyethylene	
Low density	0.015–0.03
High density	0.08–0.04
Polypropylene	0.02–0.035
Polystyrene	0.002–0.008
Toughened	0.003–0.01
Polyvinyl chloride	0.003–0.01
Styrene acrylonitrile	0.003–0.004

Shrinkage numbers will vary with different grades of same material manufactured by a different process. Operating conditions such as cooling time and temperature will greatly influence the amount of shrinkage that occurs after the part is moulded.

Table 6.4

Shrinkage (%)	Shrinkage (%)	
	Polyethylene	PVC
Longitudinal shrinkage	2.1 ± 0.1	0.36 ± 0.04
Transverse shrinkage	1.9 ± 0.2	0.13 ± 0.05
Neck length shrinkage	3.7 ± 0.1	0.98 ± 0.16

The concave bottom is best suited to compensate for shrinkage in blow moulded articles. It is therefore, preferred to be round basis, roof shaped or inclined base designs are recommended for other shapes. The height of the curvature depends on the rigidity of the plastic material as well as on the geometry of the hollow object.

6.11.12 Venting

The air that first occupied the cavity area must escape more rapidly than the rate at which the hot plastic is blown to fill the product area within the closed blow mould is known as venting.

Moulds are vented through the parting line, with face vents and with small holes. The venting is incorporated in one mould half and this type of venting can be used on all sizes of moulds. When certain areas of the mould cavity are prone to air trap, core vents can be used. Venting in the mould cavity should be anticipated in the mould design and layout of the cooling channels so that provisions can be made for their locations.

The air between parison and mould must be expelled as completely as possible during blowing so that the moulded part receives the correct shape and cool down when contact with the mould walls. The surface quality of the moulded article is very much improved by the venting of the mould, in addition to the condition of the mould surface and the blowing pressure.

The mould surface should be mirror polished and scratch free. Chrome plating is usually not recommended as it could flake-off around pressurised edges.

Types of Venting

Parting Line Venting Parting line venting is used in the areas of the mould containing flat parting lines. It is added to only one side of the mould. Vent depths range from 0.05–0.08 mm with a land of 6.0 mm. Beyond the land, vent depth is increased to 0.25–0.40 mm through channels that lead to the atmosphere.

Cavity Venting Cavity venting is added to areas inside the cavity containing deep draws and flat surfaces. Slotted vents are used for non-cosmetic parts.

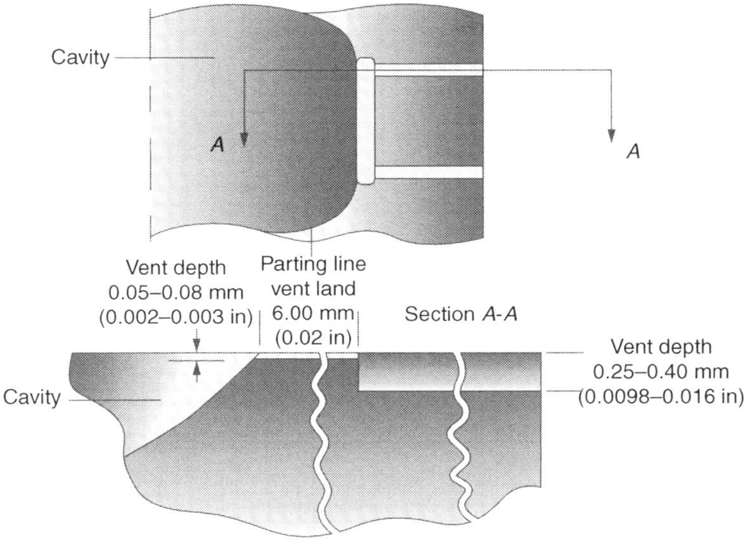

Fig. 6.34 Parting line venting.

Slotted Venting Aluminium or brass slotted vents are available in a variety of sizes. They are installed from the cavity side after the cavity is cut, and they are benched to match the cavity contour. Slot widths should be in the range of 0.40–1.25 mm.

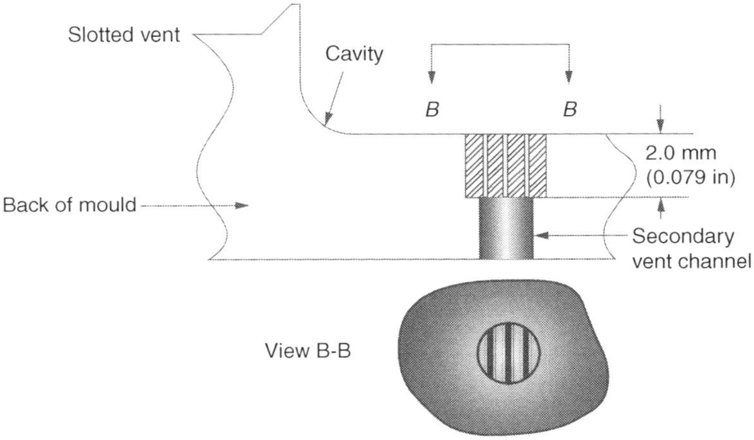

Fig. 6.35 Slotted venting.

Pinhole Venting Pinhole vents are typically used for cosmetic parts. A pinhole vent can consist of one or a group of holes anywhere from 0.40–1.25 mm diameter. A secondary vent channel is drilled from the back of the mould block to within 2.0 mm of the cavity surface. The pinhole vents are then drilled into the secondary vent channel from the cavity side.

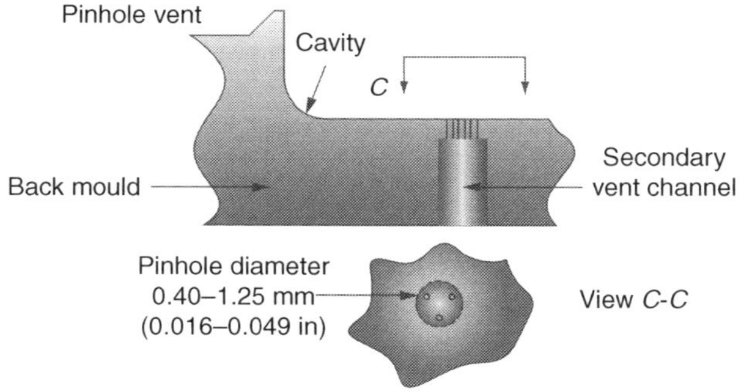

Fig. 6.36 Pinhole venting.

Surface Venting In order to effectively vent the air from a mould, the surface of the mould should be either sandblasted or lightly textured. This is necessary to allow the air contained between the parison and the mould surface to migrate through the valleys of the mould surface finish and exit through the vents.

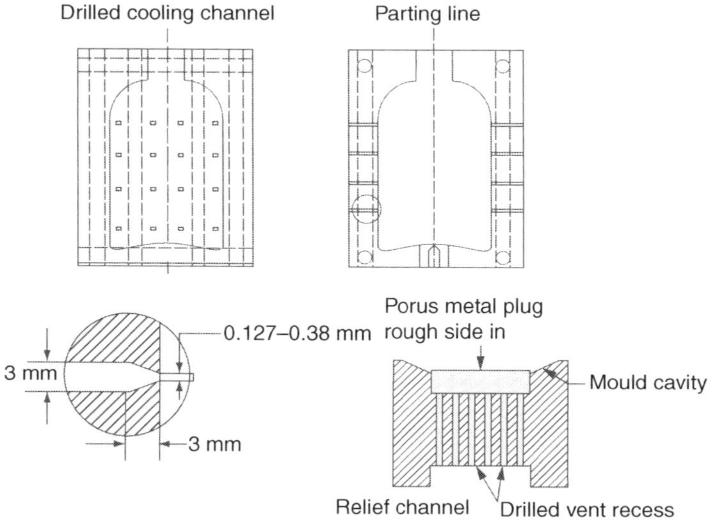

Fig. 6.37 Positions for venting in a blow mould and venting positions on a blow mould.

6.11.13 Blow Mould Design Check List

Part Description :- --	
Part Number :- -------------------------	Material :- -----------------------------
Material Shrinkage :- -------------------	Wall thickness :- ---------------------
Number of Cavities :- ------------------	Centre Line Distance :- -------------
Press Size :- ------------------------------	Platen Size :- -------------------------
Mounting Holes (Size) :- --------------	Location :- ---------------------------

(*Contd.*)

(*Contd.*)

Shut Height :- Max : --------------------	Min : ----------------------------------
Type of Blow :- -------------------------	Blow Pin :- ---------------------------
Parting Line Location :- ---------------	
Relief Requirements :- ----------------	Orientation of Part :- ----------------
Pinch-Off areas :- -----------------------	Depth of Relief :- --------------------
Cavity Construction :- ------------------	Material :- ----------------------------
Machined :- ------------------------------	Cast :- ---------------------------------
Model Required :- -----------------------	CAD :- --------------------------------
Type of Cooling :- -----------------------	Size, In/Out Connectors :- -----------
Venting :- Parting Line------------------	Within Cavity ------------------------
Inserts :-----------------------------------	Secondary Action :- -----------------
Cavity Finish :- -------------------------	Texture :- ----------------------------
Engraving :- -----------------------------	General Notes :- --------------------

6.11.14 Mould Mounting, Set up and Guiding

Mounting the mould onto the machine is facilitated by laterally protruding base plates or strips. The mould is fastened to the platens by bolts or clamps. Heavier blow moulds are provided with tapped holes for eyebolts.

The blow mould should be centrally aligned with the parison.

The mould is guided by the respective pins and their corresponding bushings of the machine platen. Flush clamping of the mould halves is ensured by sinking the guide pins and bushings below the surface by 1 to 2 mm. Mould opening on the work-bench is facilitated by milling 3 to 4 mm wide deep slots into the parting line.

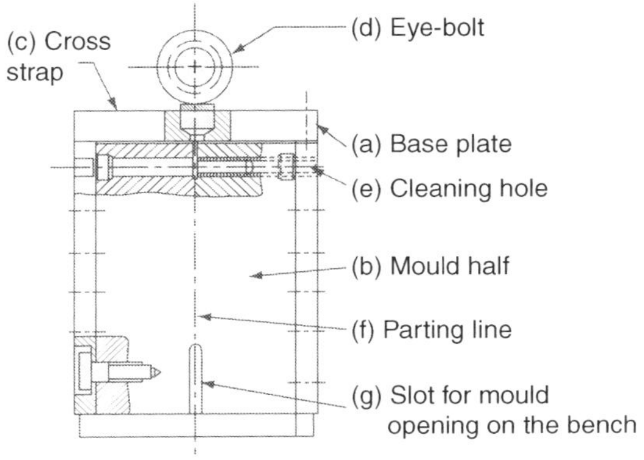

Fig. 6.38 View of a closed blow mould, ready to be loaded on the machine.

6.11.15 Moving Section Blow Moulds

The technique used to make a threaded neck off the parting line of a mould is done with a reciprocating plug. The sequence is to blow the parison against the extended plug and retracting the plug during the blowing operation. Moving sections moulds are used to make water cooler lids having integral handles.

6.11.16 Blow Mould Construction

Materials for Blow Mould Construction

1. **Aluminium Alloy:** Aircraft grade aluminium which contains zinc, magnesium, copper and chromium is most popular choice for blow moulds. It has high thermal conductivity, good machineability, light weight and resistance to corrosion. Another grade of aluminium may be cast directly into the desired shape, for large moulds. Cast moulds may have low cooling efficiency due to porosity.

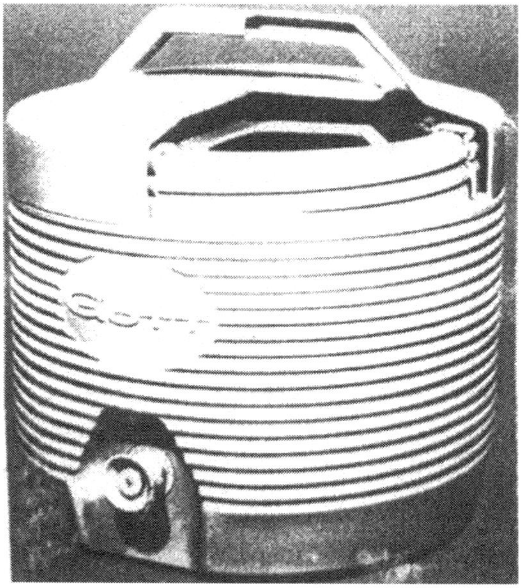

Fig. 6.39 A handled cap made by using moving-section blow mould.

2. **Beryllium-Copper (Be-Cu) Alloy:** These alloys are used extensively due to excellent thermal conductivity, corrosion-resistance and mechanical toughness. Their drawback is high cost and poor machineability as compared to aluminium.

   In some cases Be-Cu pinch-off inserts are used, to its hardness and to reduce mould cost. With corrosive plastics, the entire mould may be produced from Be-Cu.
3. **Steel:** It is mostly used for blow moulds for PVC or engineering resins, due to its corrosion resistance and extreme toughness. Excellent surface finish can be obtained by etching process.

   AISI-P20 prehardened steel is widely used. For corrosive resins, AISI-420 stainless steel is used.
4. **Miscellaneous Materials**: Zinc alloy (Kirksite) can be used for casting large moulds or large quantities of small moulds. It has good thermal conductivity.

Synthetic plastics such as acrylates, polyesters and epoxies can be cast to produce low cost moulds and prototypes.

6.11.17 Mould Cooling

Effective part cooling can be achieved by observing heat transfer related to

1. The type of plastic being moulded
2. Wall thickness of the moulded part
3. Type of mould material used to construct the mould
4. Mould wall thickness between cooling channels and the cavity wall
5. Temperature of the cooling water
6. Volume of mould cooling water
7. Design of mould cooling water zone
8. Blow air pressure
9. Surface roughness of mould cavity

Cooling of a blow moulded part consists of the following heat transfer mechanisms:

1. Conduction of heat in the wall of the part.
2. Conduction of heat in the mould wall.
3. Convective transfer of heat in cooling fluid.
4. Cooling of flash is also important to effectively trim the part.

External mould cooling In a blow mould the cooling fluid channels are drilled, interconnected in tube-like passages; drilled passages are plugged and cross-drilled with other passages. These passages drilled in the mould halves should be easy to clean and reroute during mould cleaning and reconditioning.

The cooling lines centre-to-centre distance and distance from mould surface should be at least twice the diameter of drilled holes.

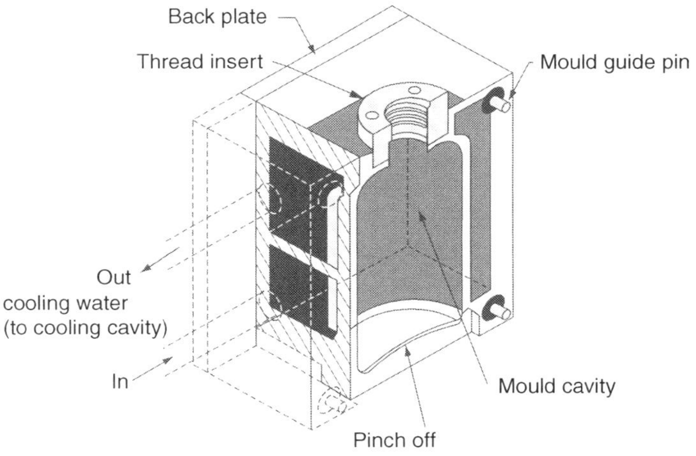

Fig. 6.40 Blow mould half with cooling water channels.

Cooling channels produced during machining must be impregnated with a sealant to prevent leakage of water. The back plates are fitted with gasket on to the rear end of the cavity block where the cooling channels are machined. This backing plate may be removed for easy cleaning of the cooling channels.

The location of cooling channel fluid entrance should be at the bottom of the mould and the exit at a higher level to eliminate air trapping.

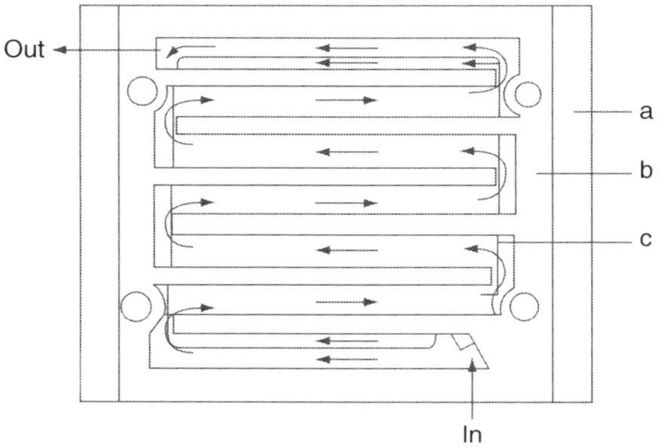

Fig. 6.41 Channel System with Labyrinth-type Water Flow Produced by Baffles (a) Mounting Plate (b) Blow Mould (c) Visible Contours of Mould Cavity for a Canister with Fillers Connections at One Corner.

The water circulates through the hollow mould halves and to create the most useful flow. Water channels are machined in the rear side of the cavity halves. Sometimes, the copper tubing system is cast into the mould for better cooling efficiency.

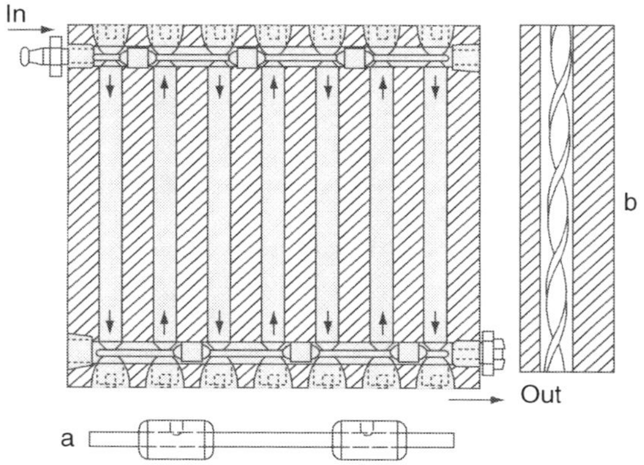

Fig. 6.42 Cooling channels with labyrinth-type water flow (a) Continuous rod with stoppers, (b) Inserted copper spiral.

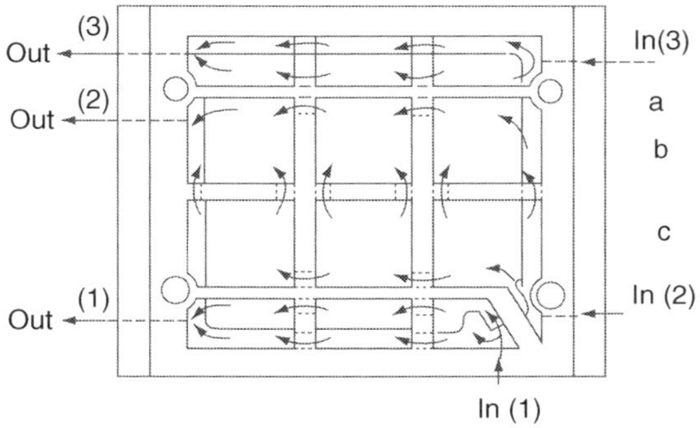

Fig. 6.43 Water circulation.

Cooling a blow mould by means of three cooling circuits and cooling

1. Mounting plate
2. Blow mould
3. Visible contours of mould cavity for a canister with a filler connection

Well-placed cooling channels will ensure that the cooling water comes as close to the mould shown in figures below cavity as is feasible.

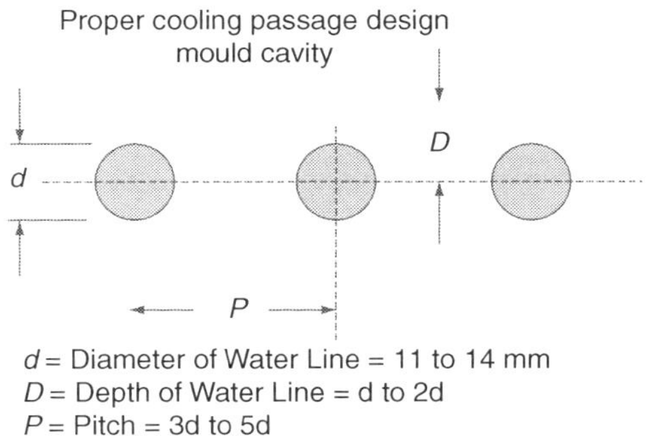

Fig. 6.44 Cooling water lines spacing.

Internal cooling Sometimes an additional means of cooling is required to cool the inside of the part. The following methods may be employed:

1. Venting of blow air to create turbulence inside the part
2. Blowing with a cryogenic liquefied gas to quickly cool the inside of the part
3. Blowing with a fine mist of water or ice

These internal cooling methods results in faster cooling time, less stresses and warpage in blown parts.

6.11.18 Blow Mould Ancillary Elements

Base plates Blow moulds are usually mounted on the base plates that are larger than the mould plates in size. The mould is attached to the base plates or backing plates by means of bolts or pins. Sometimes base plate can be used to mount more than one blow mould.

Base plates can also be used as a manifold for coolant lines. O rings are used to seal the mating surfaces between mould and base plates.

Alignment Pins These are provided to ensure accurate positioning of the mould halves as they are closed. These steel parts may be hardened to 55–60 HRC.

The surface of the guide pins and bushings are usually recessed into the mould to allow flush closing. To allow removal of plastic that is trapped into the alignment bushings, the bush-fitting hole is drilled completely through the mould.

Striker plates Mould inserts are generally used to produce the neck of a bottle. A striker-plate is fitted across the top of the insert to cut the flash from the neck.

Ejectors When the blow-pin inserted into the top of the mould, ejection mechanisms can be arranged to grasp the tail flash, the bottle can be moved onto a cooling bed or conveyor belt for subsequent handling.

For a bottom-blow pin type of mould, the part is retained on the pin as the mould indexes over to grasp the next parison. The part is dropped on the collection bed when the blow pin is retracted.

Miscellaneous Features A hole should be drilled and tapped into the top of heavy moulds to allow fitting an eye-bolt for ease of lifting and fitting onto machine. Mechanical arrangement should be provided to lock the mould halves together to prevent accidental opening of the mould.

6.12 INJECTION BLOW MOULDING

6.12.1 Introduction

The injection blow moulding is a process used for mass production of plastic containers ranging from 1 ml to 2 litres capacity. These containers meet the exact dimensional tolerance and consistent to weight and volume. It has got wide applications from pharmaceuticals to toiletries, automotive to household use.

Injection blow moulding machine cost is almost similarly with extrusion blow moulding machine, but with injection blow moulding it is possible to produce more bottles per cycle and more bottles per hour without deflashing, trimming, regranulation and remixing of scrap, at a constant weight and with injection moulding tolerances. Injection blow moulding does not produce a significant amount of scrap polymer like extrusion blow moulding.

The injection blow moulding is a process in which a triangular rotary table indexes in 120° steps. Core rods mounted on the face of the table form the inside of the hot parison (or preform), later blown into the finished container.

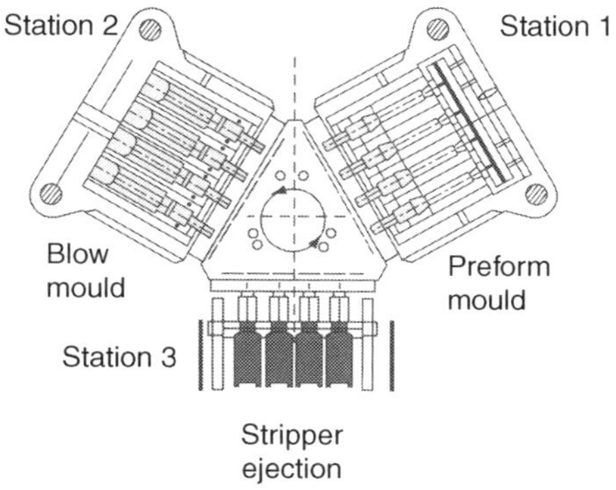

Fig. 6.45 Injection blow moulding process.

Station 1 In preform mould, the molten material is injected under low pressure into the mould cavity, where it forms a parison around the core rod. In this stage, the neck section is injection moulded to close tolerances and after suitable conditioning, the moulds open and the parison is transferred on the core rod to station 2.

Station 2 The parison blow takes place in station 2 and the container is formed as per the shape and finish of the cavity. In this process, the parison is blown with air fed internally through the core rod. As the blown plastic contacts with the cold blow mould, the final moulding is produced. The mould opens and the finished bottle is transferred on the core rod to station 3.
Station 3 Finally, the bottle is ejected from the core rod.

6.12.2 Plastic Materials for Injection Blow Moulding

Generally the plastic materials like LDPE, HDPE, PP, PS, SAN, EVA, PVC, PC, PET, etc., are used for manufacturing of injection blow moulding components.

6.12.3 Parison Layout

The parison layout requires the knowledge of injection blow process, material behaviour, and swell and shrinkage factors. The outside configuration of the parison is formed by neck-ring and the parison mould and the inside shape is formed by core-rod.

Parison volume is defined as the weight of container divided by the material density.

Parison volume calculations involve the optimal relationship between a minimum wall thickness of approximately 2 mm and a maximum blow-up ratio of approximately 3:1.

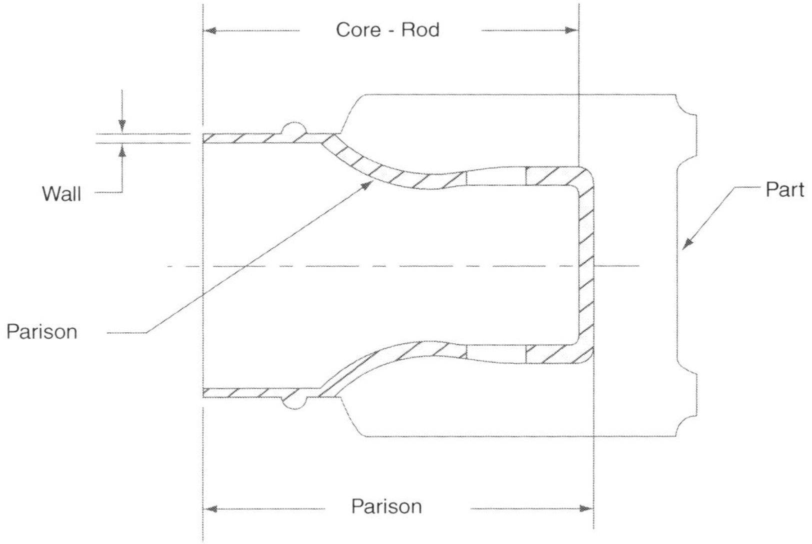

Fig. 6.46 Parison layout.

6.12.4 Parison Mould

The parison mould comprises of the body and the neck-ring. Moulds are made with cooling lines in a V-shape perpendicular to the parison. The neck-ring forms the thread is hardened to HRC 40–45 and it locates the core-rod in its pocket. All surfaces that come in contact with plastics are highly polished and chrome plated for ease of fill and release of component.

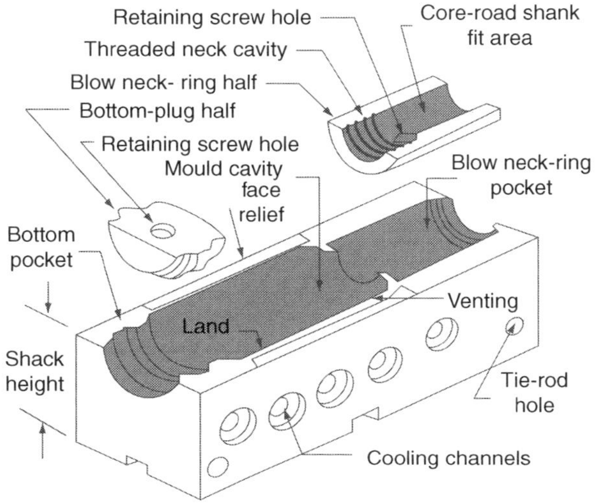

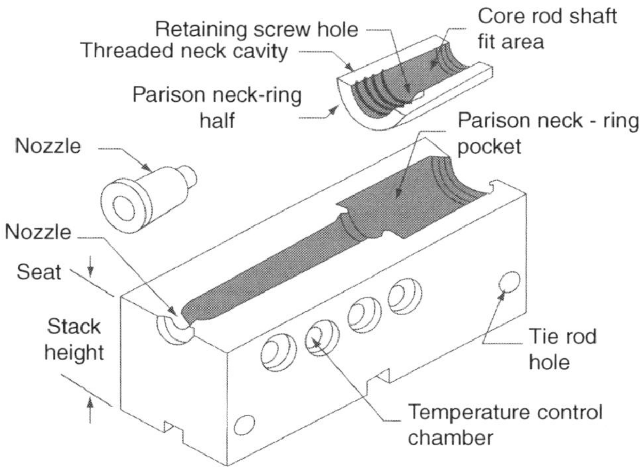

Fig. 6.47 Parison mould.

6.12.5 Injection Blow Mould Core Rods

The blow mould forms the final shape of the container. For polyolefin materials, the mould body and bottom plug are made from aircraft aluminium alloy.

For hard resins, such as PS or PC, air hardened tool-steel is used, and for PVC beryllium-copper alloy or stainless steel is used. The cooling lines are drilled in a V-shape to allow maximum circumferential cooling.

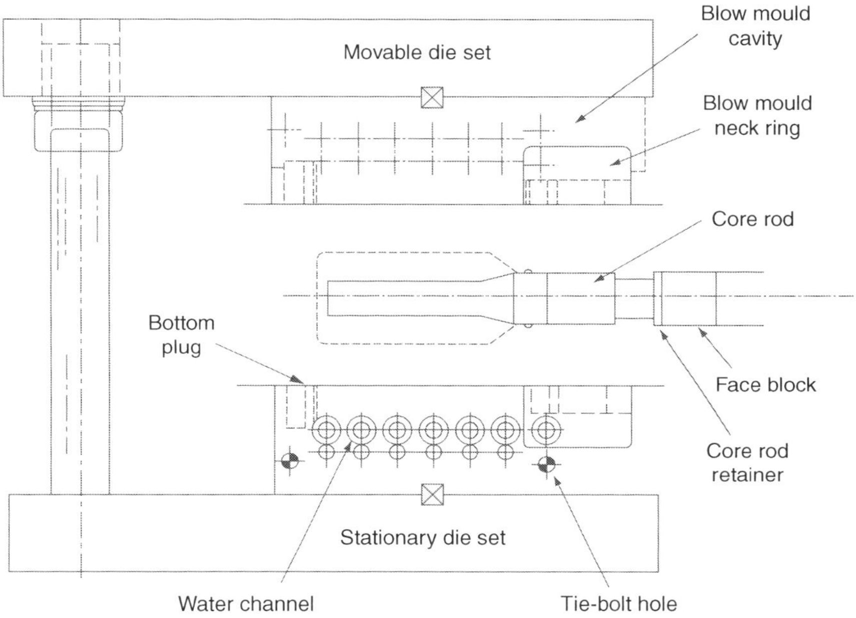

Fig. 6.48 Injection blow mould.

The core-rod forms the internal diameter of the neck and parison, when sitting in the parison mould. After conditioning the core-rod transfers the parison into blow mould.

There is a valve mechanism in the core-rod that allows blowing the parison into its final shape and cooling it.

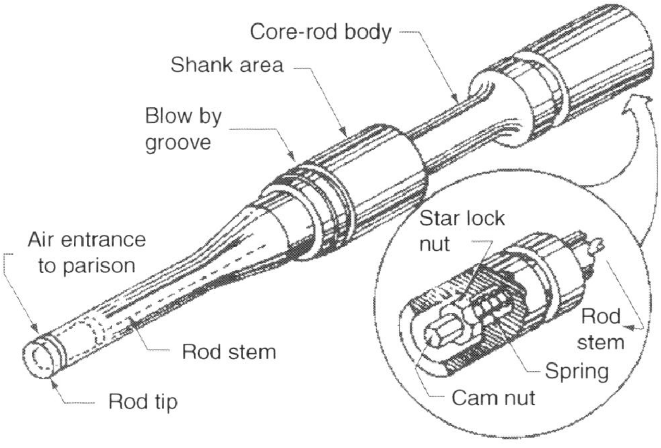

Fig. 6.49 Core rod assembly.

6.12.6 Core-Rod Holder

The core-rod holders are mounted on the machine's indexing head.

6.12.7 Manifold Assembly

The manifold assembly is similar to a hot runner mould used in injection mould. The assembly is mounted on a die set and is made up of a base, clamps and nozzle clamps.

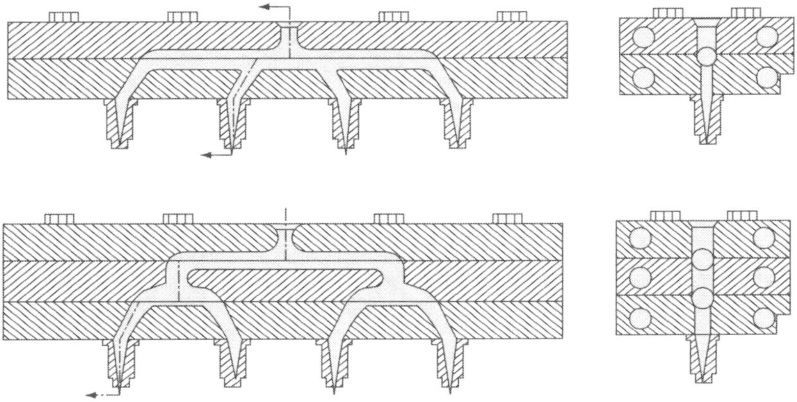

Fig. 6.50 Manifold assembly section view.

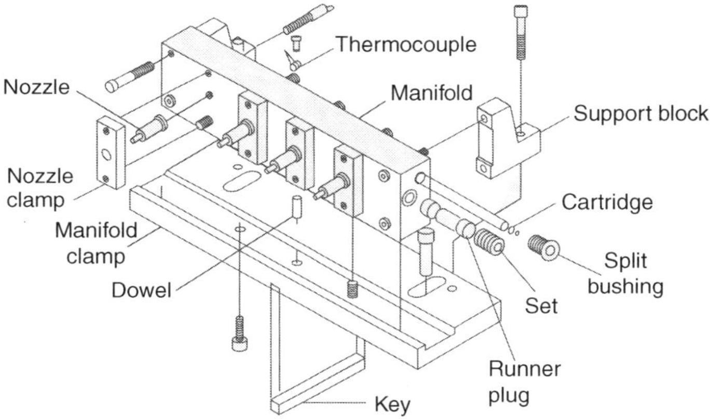

Fig. 6.51 Manifold assembly.

The manifold assembly is mounted on the parison mould die-set.

6.12.8 Die-Set

The parison and blow mould cavities are assembled on individual die-sets. Die-sets facilitate mould alignment in the machine and mould change-over.

6.12.9 Mould Mounting, Set up and Guiding

With injection blow moulding the precise set up of the mould is of prime importance to ensure that transfer of the preform takes place accurately and reliably to prevent damaging the preform and the mould.

6.12.10 Injection Blow Mould Construction

Materials for mould construction Core rods are typically hardened steel, polished and hard-chromed, parison moulds are made from P-20 prehardened steel. Blow moulds are made from steel, Be-Cu alloy and aluminium.

Mould cavity production techniques The advanced machining methods using CNC and conventional machining operations are involved for manufacturing of injection blow moulds. Proper heat treatment processes like hardening and tempering of steel inserts are required to sustain the pressure and resistance to wear, which is highly essential for durability. The final phase of mould involves polishing, and surface treatment processes like sandblasting or photo-etching the surface.

Injection nozzles Injection nozzles are a critical interface between the manifold and the parison mould and it is made from P-21 prehardened tool-steel. Because the front face of the nozzle comes in contact with the parison, it is important for it to be polished and chrome-plated for easy release.

The gate dimensions of a nozzle vary according to preform dimensions. The follow-up pressure used in the injection stage, the amount of pressure as well as the time hold is dependent on the size of gate and type of plastic material used. The various types of nozzles used in injection blow moulds are shown in the Fig. 6.52 below:

The different gating designs are:

1. Cylindrical
2. Identical cone
3. Double cone
4. Opposed cone
5. Tunnel gate
6. Needle valve nozzle

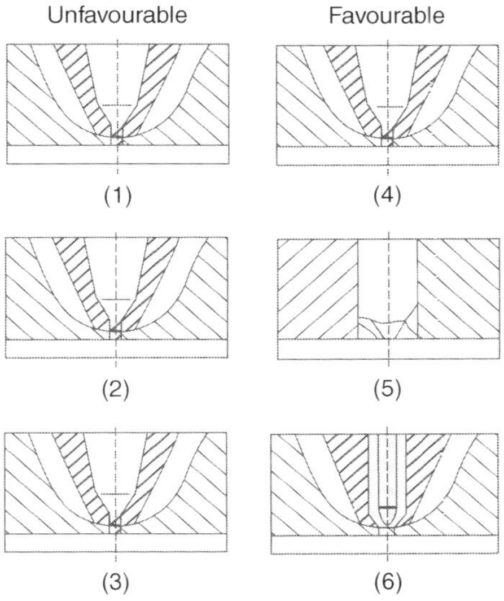

Fig. 6.52 Injection nozzles.

Stripper plates The stripper plate is shaped to form the shoulder of the container; it pulls the finished bottle off the core-rods after blowing. After stripping the bottle the stripper plate most often remains in the extended position and compressed air is blown against the tips of the core-rods.

6.13 INJECTION STRETCH BLOW MOULDING PROCESS

Injection stretch blow moulding is used for the production of high quality containers. The process involves the following steps.

6.13.1 Injection

Molten polymer flows into the injection cavity via the hot runner block, to produce the desired shape of the preform with a mandrel (the core pin) producing the inner diameter and the injection cavity forms the outer shape.

The injection moulds and core pins part after the set time and the preform held in a neck carrier is rotated 90°.

6.13.2 Stretching and Blowing

The preform is conditioned to the desired temperature for stretching and blowing to the finished shape. Once the preform is within the blow mould area the moulds close and a stretch rod is introduced to stretch the preform longitudinally using two levels of air pressure, the preform is blown circumferentially.

6.13.3 Ejection

After the cooling cycle, the moulds open and the preform is removed via drop chutes or robotics. These stages are carried out concurrently using a revolving carousel of moulds.

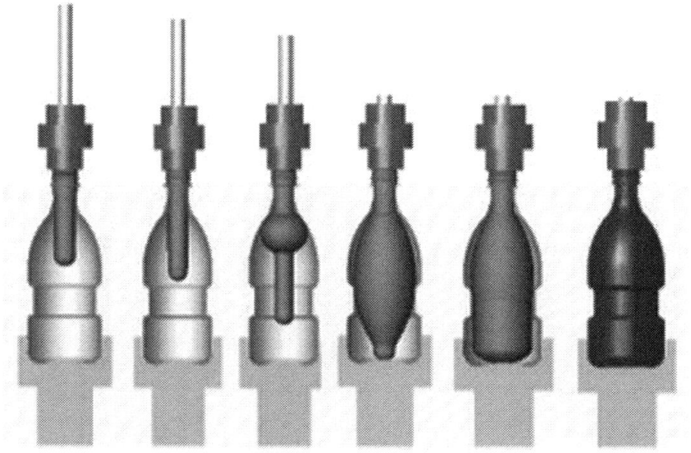

Fig. 6.53 The stretch blow moulding process.

6.13.4 Products and their Applications

Injection blow moulded products are produced for various applications like carbonated and soft drink bottles, cooking oil containers, agrochemical containers, health, oral hygiene products, bathroom and toiletry products, etc.

The injection blow moulding process produces bottles of superior visual and dimensional quality compared to extrusion blow moulding. The process is ideal for both narrow and wide-mouthed finished containers without flash. The identification of injection blow moulding is the seam where the two halves of the mould meet.

Fig. 6.54 Examples of stretch blow moulded parts.

This process involves the production of hollow objects, such as bottles, having biaxial molecular orientation. Bi-axial orientation provides enhanced physical properties, clarity, and gas barrier properties, which are all important in products such as bottles for carbonated beverages.

Typical plastic materials used

1. Polyethylene-terephthalate (PET)
2. Polyolefins (PE)
3. PVC

Polyethylene terephthalate (PET) is a thermoplastic resin of the polyester family that is used to make beverage, food and other liquid containers, synthetic fibres, as well as for some other thermoforming applications.

6.13.5 PET Preform and Bottle

6.13.5.1 The One-Step Method In this method, all processes starting from PET granule to finished bottle are completed on one integrated machine. The injection moulded preform is withdrawn from the injection cavity in hot condition and stretch blown to form the bottle. No extra heating of preform is required for the stretch blow process. Since preforms are not stock-piled to be blown at a later date, there is no risk of surface damage from preforms knocking together during storage or transportation.

One-step method is highly suited to small and medium scale production lines.

Two-step method The two-step method requires two separate machines. The preform is injection moulded on the first stage, then reheated and blown on the second stage. The two-step system uses an injection moulding machine for making the preforms, and a reheat blow moulding machine to reheat the preforms from cold and blow the bottles. The requirement for a preform heating system means the two-step process has a lower thermal efficiency.

This method is most suited to medium to large-scale production.

Usually two-stage method is adopted to produce PET bottle. It involves the following steps:

PET absorbs moisture from the atmosphere. This must be removed by a dehumidifying drying before processing named as PET drying.

Dried PET pellets are compressed and melted by a rotating screw for plasticising purpose.

Molten PET is injected into the injection cavity and cooled rapidly to form the test-tube-like form from which bottles are blown is known as a preform.

The heating of PET preform is done by adjusting the temperature to achieve the correct profile by blowing.

The hot preform is simultaneously stretched and blown thereby orienting the crystals and strengthening the PET into a blow mould to form a tough, lightweight container. PET is heated to a temperature where its chain-like molecules are sufficiently mobile to uncoil instead of breaking when extended, can be oriented by stretching. Stretching applied from two directions at right angles, as in stretch blow moulding, gives bi-axial orientation. Oriented PET contains closely packed chains aligned in the directions of stretch. The material is stronger because the molecules act together instead of individually stretching. The tensile strength of oriented PET is several times higher than that of the unstretched material and the impact strength, barrier and chemical resistance are also significantly improved, so bottles can be lighter without sacrificing the performance. The finished container is ejected after the blowing process completes.

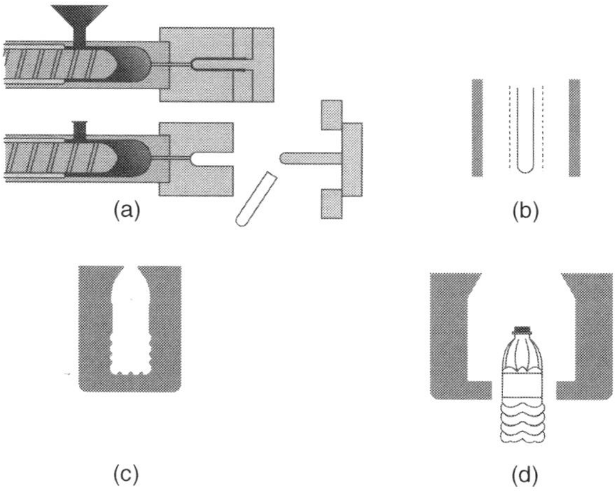

Fig. 6.55 A typical two-stage method of producing parts.

6.13.6 PET Preforms

A wide range of PET preforms for PET bottles and containers can be produced, which includes preforms for mineral water, carbonated soft drinks, and hot-fill drinks such as orange juice, milk, tea and isotonic beverages, and preforms for wide mouth jars and containers.

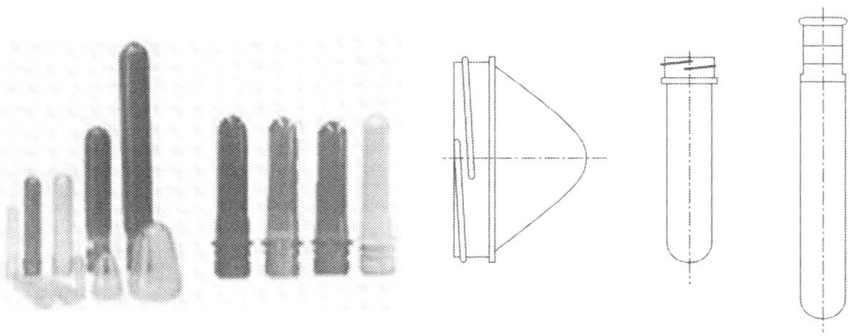

Fig. 6.56 Preforms used for stretch blow moulding.

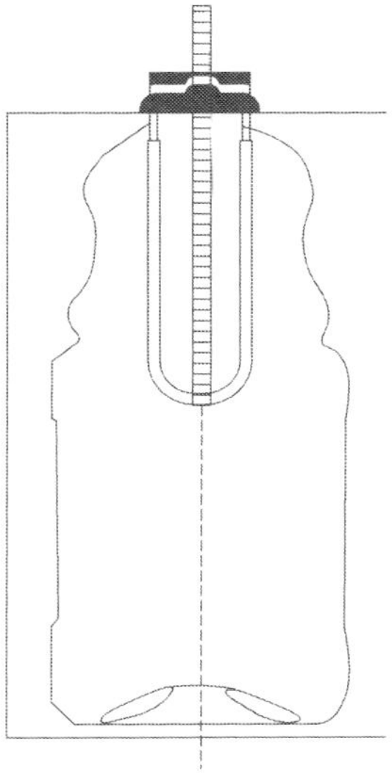

Fig. 6.57 Stretching a preform.

Preform quality requirements

1. The preform should be gate-free.
2. Low preform eccentricity, no more than 0.10 mm.
3. Low weight variation between cavities, +/–0.2 g.
4. Low acetaldehyde level.
5. Low preform temperature at exit to avoid preform scratches.

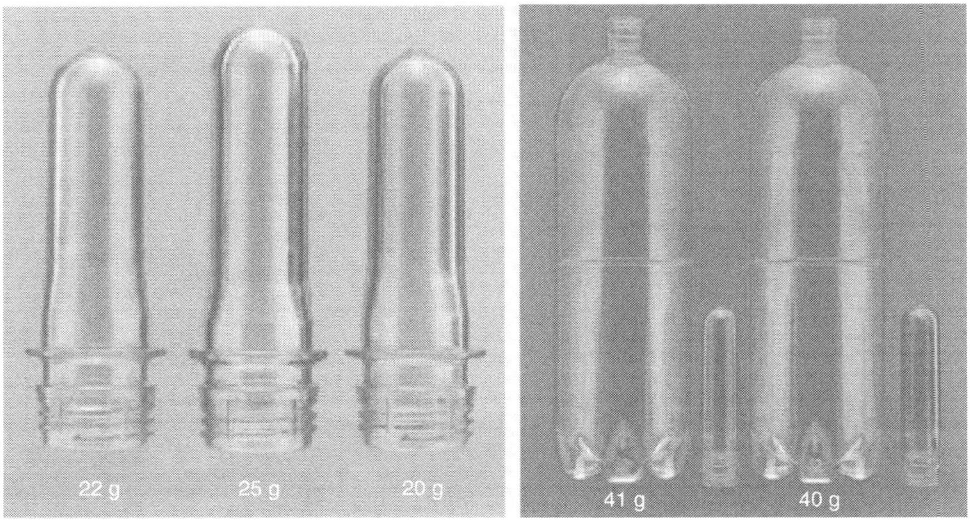

Fig. 6.58 Preforms.

Preform design The container shape is determined based on the shape and size of the preform. To maintain the bottle weight and minimum wall thickness the preform will be smaller in diameter which in turn increases the blow-up ratio. When considering preforms for an oval container, the parison is usually ovalised in the direction of the container depth. To prevent weld lines, the maximum wall thickness should be kept to the minimum, the preform across section should be less than 1.5. Because it is difficult to adequately condition in the preform cavity during the blowing process, a wall thickness of less than 0.35 mm is considered unsuitable.

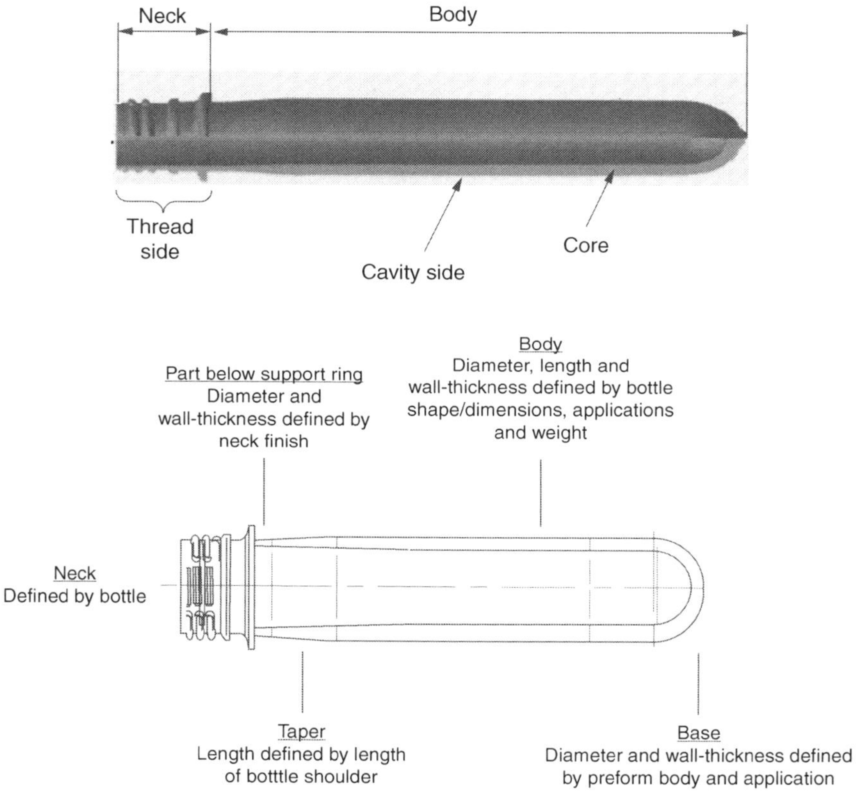

Fig. 6.59 Preform details.

6.13.6.2 Preform Moulds

The preform mould is made out of:

1. Stainless steel mould plates for corrosion resistance and easier mould maintenance.
2. Part ejection sensors that monitor preform transfer to a take-off plate for cycle optimisation.
3. Auxiliary tapers that reduce mould wear by pre-aligning the mould halves prior to the engagement of the neck-ring tapers.
4. Standard interchangeable mould components'.

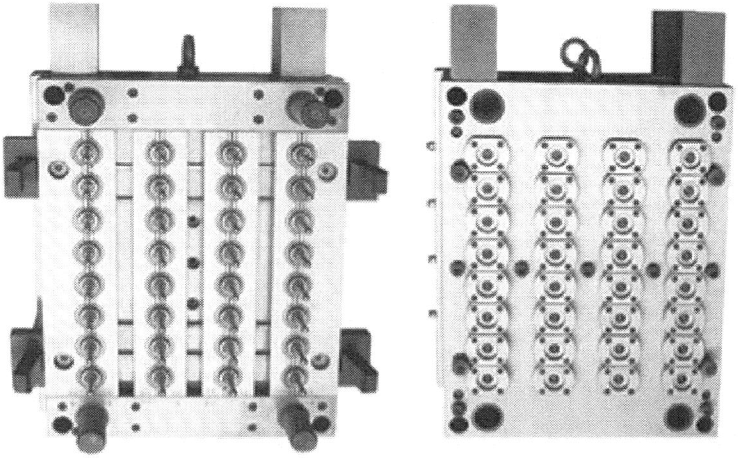

Fig. 6.60 PET preform hot runner moulds.

5. Water hoses are manifolded at bottom of mould, eliminating leaks and water marks on preforms.
6. Roller cams actuate both the opening and closing actions on the thread splits, a design that requires fewer components and simplifies mould maintenance.

6.13.6.3 PET Moulds with Hot Runner Systems

1. It reduces thermal variation between cavities through better flow path design, improves preform consistency and provides easier start ups.
2. Longer intervals between preventive maintenance through reducing the operating temperature in the piston area.
3. Three-level design for precise geometric balancing and reduced cavity-to-cavity variation.

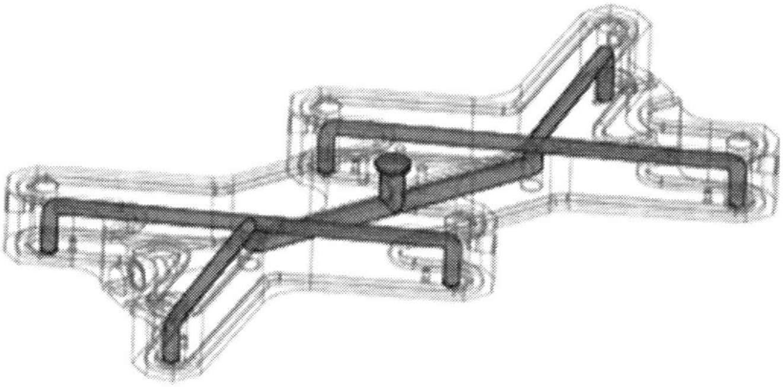

Fig. 6.61 Examples of some PET preform hot runner moulds.

Fig. 6.62 Balanced six-cavity hot runner design for preform moulds.

Common neck finish for pet preforms

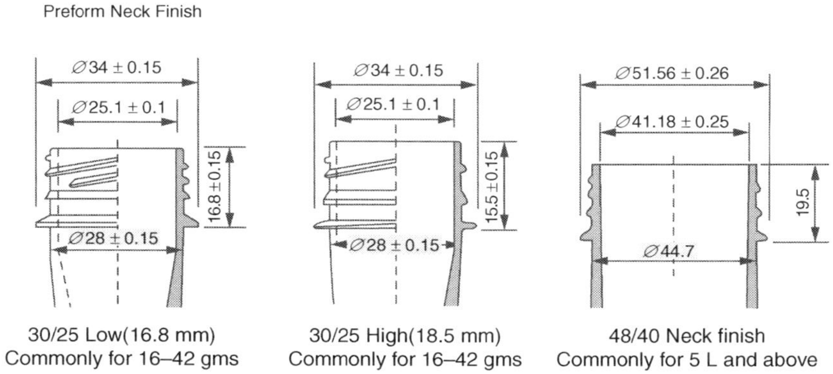

Fig. 6.63 Preforms for water bottles.

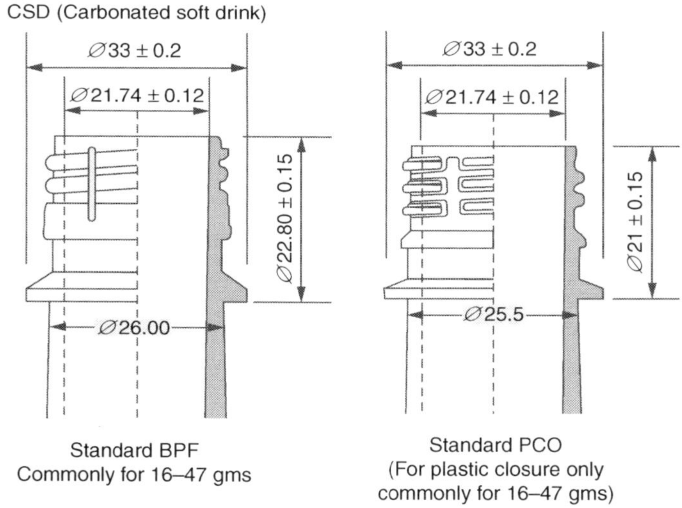

Fig. 6.64 Preforms for soft drink bottles.

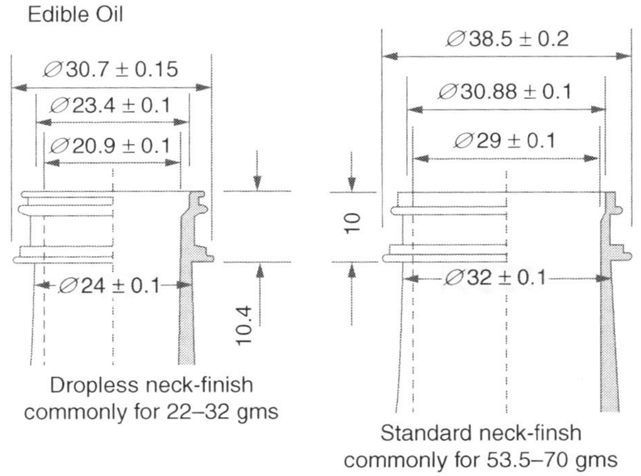

Fig. 6.65 Preforms for edible oil bottles.

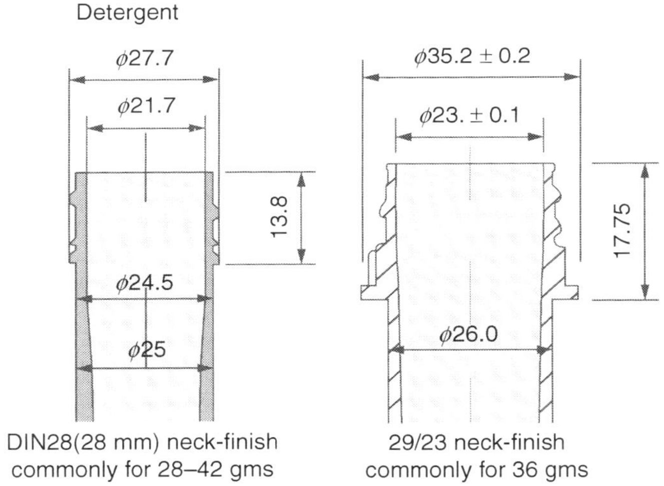

Fig. 6.66 Preforms for liquid detergent bottles.

Questions

1. Write the stages of blow moulding process.
2. Briefly explain the application of blow moulding.
3. What are the design considerations before designing a blow mould?
4. What are the practical guidelines for designing corner and edge rounding in blow mould?
5. Why are truss grooves required in blow moulding?
6. What are the important structural and mechanical considerations for blow moulding?
7. What should be the properties of plastic materials used in blow moulding?

8. What are the blow moulding materials?
9. What is blow up ratio? Explain with a neat sketch.
10. What is parison programming?
11. Explain various types of parting line in blow moulding.
12. What are the functions served by the clamp and platen mechanism in blow moulding?
13. What is the function of pinch-off?
14. Write sort notes on flash pocket.
15. How does shrinkage play a major role in blow moulding?
16. Write sort notes on mould mounting, set up and guiding.
17. Explain the materials used for blow mould construction.
18. Describe the ancillary elements in blow mould.
19. Explain the injection blow moulding process.
20. What are the plastic materials used for injection blow moulding?
21. Explain the parison layout with a neat sketch.
22. What is the function of core rods in injection blow mould?
23. Explain various types of injection nozzles used in injection blow mould with a neat sketch.
24. Describe injection stretch blow moulding process.
25. Write down various injection stretch blow moulding products and its application.
26. What are the preform quality requirements?
27. Write the advantages of PET moulds with hot runner systems.
28. Explain the parison mould with a neat sketch.
29. Write short notes on flash trimming.
30. Explain various types of pinch-off in extrusion blow moulds.
31. Explain extrusion blow moulding process. Write its advantages and disadvantages.
32. Explain pinch-off design with a neat sketch.
33. What is the significance of venting in blow mould? Explain types of venting used in blow mould with a neat sketch.
34. Describe the types of cooling systems used in blow moulding with line diagram.
35. Explain the steps involved to manufacture PET preform and bottle.
36. Describe the various features of preform with a neat sketch and explain the following:
    a) Preform design   b) Preform moulds
37. Define pinch-off. Explain various types of pinch-off.
38. What is the purpose of venting? Explain different methods of it.
39. Explain the nomenclature of blow mould.
40. Explain different types of materials used for manufacturing blow mould.

References

1. CIPET, *Technical Manual*, Chennai.
2. *Extrusion Blow Molding Guide for Thermoplastic Rubbers and Thermoplastic Elastomers.*
3. Norman C. Lee, *Practical Guide to Blow Moulding,* Van Nostrand Reinhold, New York.
4. Rosato D. V., Rosato A. V. and David P. Dimitia, *Blow Moulding Handbook,* Hanser Publishers, Munich.
5. Samuel L. Belcher, *Practical Guide to Injection Blow Molding,* CRC Press, Taylor and Francis Group, Boca Ratan, London.

CHAPTER

7

Extrusion Dies

7.1 INTRODUCTION

Polymer products are primarily categorised into two types. The injection, compression, transfer moulded products, which are in form of fixed dimension with changing cross-sectional shapes as per the mould dimensions called *intermittent process* and second, which has fixed cross section and has long length called *continuous process*. First one is produced by moulding and second one by extrusion process.

In extrusion process, the extruder pushes the polymer melt in semifluid form through the die, mandrel and the product is cooled to get required shape. This conversion or processing operations primarily involve four stages, viz; the first stage is converting plastics granules into homogeneous melt: the next stage is forwarding the melt to achieve a high output rate in a desired shape from the extruder through the die; the third stage is forming the product outside the die; and the final stage makes the product form stable.

In order to produce a quality product in an optimum condition with high output rate, the following factors are to be considered with relevance to processing.

1. The characteristics and property of plastic material
2. Extruder characteristics and types
3. Die geometry with respect to the shape of the product
4. The melt flow behaviour along die geometry and die swell
5. Extruder - die system characteristics
6. Heating system and temperature control
7. Die strength and material of construction

7.2 CHARACTERISTICS AND PROPERTY OF PLASTIC MATERIALS

The relevant property of polymers under different physical states varies during the conversion process. Initially the plastic material is in the solid form and is further converted to a melt. From the melt stage, the product shape is obtained in the subsequent processes by proper forming and cooling. The properties such as flow melt rheological, thermal properties, etc., influence the processing operations and these properties under operating conditions must be known for the polymer to be processed.

The rheological and thermal properties of polymer melt play a vital role for design of extrusion die. Thermoplastic resins are extruded, based on the properties of the material and specific grade of thermoplastic resins is used in the extrusion process. According to the basic resin, formulation of the compound and the operating condition properties vary considerably.

The viscosity of the polymer melts causes a pressure drop, as the melt flow in the die. Temperature and shear rate influence the range of viscosity. Pressure drop, therefore will vary according to the operating conditions. Similarly, elastic properties of the polymer melt causes additional pressure drop certain die shapes as also swelling of the extrudate. Elasticity is also influenced by similar operating conditions. Certain thermoplastics are prone to degradation and discolouration due to over heating stagnation. Smooth melt flow and temperature controls are essential. Some thermoplastics are likely to corrode the material of construction of the dies. This must be taken into consideration in making a choice of the material.

7.3 FLOW PROPERTIES OF SOLID POLYMERS

Usually solid polymers in the form of powder, granules or pellets are used during the initial stage of conversion of polymers. A particulate flow is essential and in this type of flow, the controlling properties are bulk density of bulk factor, interparticle cohesion and particle-to-metal adhesion.

A high bulk density is required in order to facilitate easy flow of polymer melt. The bulk density is largely influenced by the shape, size and size distribution of the particles. The cohesion or agglomeration of particles is the result of surface forces. Low cohesion forces and smooth surfaces favour easy flow. A small average particle size increases surface-to-volume ratio and favours cohesion. Adhesion between powders and metal surfaces of equipment is undesirable. The major factor causing adhesion is static electricity. This may be generated by friction between the particles as they flow.

In particulate flow problems, any pressure exerted on a volume of powder will tend to pack the particles and increase cohesive forces and therefore, lower mass flow. Similarly, flow of powder through a long tube is greater than through a short orifice of the same diameter. These characteristics of particulate flow differ from a normal fluid flow.

7.4 FLOW PROPERTIES OF POLYMER MELTS

The most important characteristic of polymer melts is that they are highly viscous. The viscosity decreases when the velocity of melt flow is increased or if temperature increases and the viscosity increases when pressure on melt flow is increased. Liquids such as water and common organic solvents do not exhibit such reduction or increase in viscosity. Another interesting property of polymer melts is elasticity. Whenever a polymer melt is extruded through a capillary, the extrudate diameter is greater than the capillary diameter, indicating recovery of elastic strain. This is also exhibited by polymer melts in their ability to generate normal stresses. These melts are therefore, known as *visco-elastic fluids.*

The parameters that define the rheological behaviour of visco-elastic polymer melts are viscosity, elasticity under simple shear and simple tension. Viscous deformation is permanent while elastic deformation is recoverable.

7.4.1 Viscosity under Simple Shear

The stress deformation behaviour under simple shear can be considered by imagining two parallel plates of large area A, separated by a distance by the fluid. A shear force F, is applied to the top plate and it is made to move with a uniform velocity (V). The fluid in contact with the plate will move with a velocity V, while the fluid in contact with the lower stationary plate will not move at all. This will give rise to shearing of the fluid. An ordinary fluid will obey the following relationship:

$$T = n\,y$$

T is shear stress, y is strain rate and n is the coefficient of viscosity. In normal fluids, viscosity n, remains constant with shear rate y. These are known as *Newtonian fluids* and when polymer melts, viscosity changes with shear rate. This behaviour is quite often represented by a relationship known as *power law*.

$$T = k(y)^n$$

where k is the consistency index and n is the flow behaviour index. If $n < 1$, viscosity decreases with shear rate (shear thinning). Such fluids are non-Newtonian and are known as *pseudo plastics*. If $n > 1$, viscosity increases with shear rate, these are known as *dilatant fluids*.

7.4.2 Viscosity under Simple Tension

Polymer melts can be made to flow by uni-axial stretching as in the case of fibre spinning. Other examples include parison sag and blow moulding. This is known as *extensional flow*. Such flows are possible only for liquids having a sufficiently high viscosity. During such flows polymer melts exhibit viscosity which is known as *traction viscosity.*

The tensile melt property of thermoplastic at given temperature and pressure can be expressed as tensile viscosity (λ), defined as ratio of tensile stress to tensile strain rate.

$$\lambda = \sigma / \varepsilon$$

At low stress, the viscosity under tension is independent of stress and equal to three-time viscosity under simple shear

$$\lambda = 3\,\eta$$

At high stress, this viscosity increases, decreases or be independent of stress.

7.4.3 Elasticity under Simple Shear

In addition to viscosity polymer melts also exhibit elasticity. The extruded swelling manifests the effect of elasticity when the melt is extruded through the capillary. Similarly, when a rod is rotated in the melt, the liquid melt climbs up. This phenomenon is due to an elastic stress generated in the normal direction.

Elasticity in simple sheer can be assessed by the phenomenon of die swell, in which the cross section of melts increases as it flows out of a capillary. In the capillary, the melt is under stress, and is therefore, deformed, but the elastic component of the deformation can recover on the removal of stress.

7.4.4 Elasticity under Simple Tension

When the polymer melt is subjected to uniaxial stretching, a tensile stress is created along with an elastic strain.

Usually the modulus under tension E is equal to three times the modulus under shear G. Elasticity plays a larger part in extensional flows than in simple shear flows of the same magnitude.

Various plastic materials can bear different level of stress and strain. This value is measured based on the required processing temperature.

7.5 EFFECT OF TEMPERATURE AND PRESSURE ON VISCOSITY

The increase in temperature of a polymer melt reduces its viscosity. Among various polymer melts, acrylic has a viscosity most sensitive to temperature while polyethylene and polypropylene are the most insensitive. Similarly, viscosity at high shear rates is not as sensitive to temperature as at low shear rates. The problem of temperature dependency of viscosity, at lower shear rates therefore, becomes an important consideration in the extrusion process and increase in pressure increases viscosity of the melt.

7.6 THERMAL PROPERTIES

7.6.1 Specific Heat

The specific heat is an important parameter in processing of plastics. From the theoretical point of view a curve of specific heat against temperature is extremely valuable in the calculation of thermodynamic properties like enthalpy, entropy, free energy, etc., and changes in such properties influence such physical process as crystallisation, melting, quenching and cooling.

The melting point of polymers is not sharply defined, since crystallinity is either absent in polystyrene, PVC or only partially present in polyethylene, PTFE, etc., ethyl-temperature relationship of amorphous and crystalline polymers, from which it may be seen that the specific heat of crystalline polymer tends to increase much more rapidly below transition temperature than that of amorphous polymer, and above the transition point the specific heat remains fairly constant with temperature.

Polymer melts show an increase in specific heat with temperature. With certain polymer this increase is irregular, showing a sudden rise in specific heat and subsequent drop. Relationship between specific heat and temperature must be known over the processing temperature range.

7.6.2 Thermal Conductivity

The thermal conductivity of polymers plays a vital role, since most of the polymers are heated and cooled during processing. Amorphous polymers such as polystyrene, polyvinyl chloride and polycarbonate show relatively small changes in thermal conductivity of partly crystalline polymer may either decrease with temperature or increase slightly.

7.6.3 Thermal Diffusivity

As the polymers are heated or cooled by the use of external means, both specific heat and conductivity are important properties influencing many processing operations. A useful property which incorporated both the above properties is the thermal diffusivity which is defined by the equation.

$$\text{Diffusivity} = \text{Conductivity} / \text{Density} \times \text{Specific heat}$$

7.7 PRINCIPLES OF EXTRUSION

7.7.1 Screw Extruder

In extrusion process of polymer products, the function of an extruder is to convert a solid polymer, to a homogeneous melt and transfer it thoroughly and deliver it at a constant rate and at uniform temperature and pressure. There are two basic types of extruders, viz. single and twin screw extruder. A properly designed screw of appropriate length rotating in a heating barrel will fulfil the following functions like transporting the solid plastic material, which may be in the form of powder, granules, pellets, chips, film scrap, etc., and heating and plasticising the material.

The homogenising, metreing and building up pressure for pumping of the melt through the die is done in this process.

Extrusion is a continuous process of manufacturing long products of constant cross section like rods, sheets, pipes, films, wire insulation coating, etc., forcing softened polymer through a die with an opening.

Plastic material in form of pellets or granules is fed into an extruder through a hopper. The material is then conveyed forward by a feeding screw and forced through a die, converting to continuous polymer product. The heating elements, placed over the barrel, soften and melt the polymer and the temperature of the material is controlled by thermocouples. The product going out of the die is cooled by blown air or in water bath. It is a continuous process used to produce both solid and hollow products that have a constant cross section like window frame, pipe, hose pipe, profiles, etc.

The two main components of an extruder are barrel and screw. Die is not an extruder component.

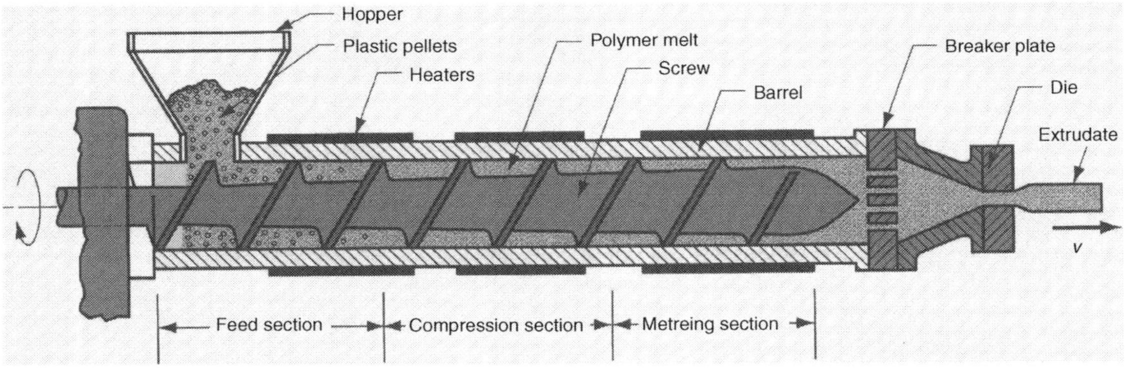

Fig. 7.1 Extrusion process.

7.7.2 Single Screw Extrusion Process

In order to study the single screw extrusion process, it is convenient to consider the screw as divided into three different sections, viz. feed section, transition section and metreing section. The feed section has the deepest channel and it picks up the solid material from a hopper and conveys it forward to the transition section. The channel depth decreases over the length of the transition section and the loosely packed material in first compacted and then melted as it passes through it. The heat that causes melting is initially conducted from the hot barrel surface, but as melting proceeds considerable heat is dissipated by shear within the newly molten viscous material. On leaving the transition section the molten material enters the shallow channel metreing section of the screw. After passing through the metreing section of the melt is forced through the die head assembly which includes the screen pack and breaker plate.

Feed section The solid plastic material in the form of powder, granules, pellets, etc., fed to the extruder is compacted to a solid mass as soon as it enters the screw channel. The heaters provided in barrel wall heat the material trough conduction. The heat conducted may be just sufficient to increase the temperature of the solid mass without any melting or it may heat the material to a temperature approaching the melting point. The material will have no relative movement of the layers and in some case a film of the melt will be formed at the barrel surface and the material conveying will be controlled by viscous drag mechanism.

Transition section In the transition section the material is compacted due to a granule reduction in channel depth. The compacted solid polymer is called solid plug or solid bed. The major portion of melting occurs on the side in contact with the heated barrel surface. A melt film is formed near the surface of the barrel, which is scraped off by the advancing flight and accumulates in the channel space adjoining the solid bed.

The heat is generated due to viscous shear of the melt and as the melt volume increases, it exerts a pressure on the solid bed and forces it towards the barrel surfaces. Since the melting process requires sufficient time it is essential to compress the material gradually over a sufficient length, otherwise the channel may get plugged and may give rise to pressure surges.

At some stage during the down channel movement, the solid bed becomes physically unstable and breaks up. The unmelted material floats in the melt pool and moves with it. The length of the transition section is the controlling factor in screw operation. The length should be enough to complete the melting process by the time the material reaches the end of the section.

Metreing section In the metreing section the melt along with some unmelted material is sheared and generates heat, which helps to produce a homogenous melt. Special mixing device are sometimes corporate to facilitate melting and homogenisation, so that no insufficiently molten solid particles reach the end of screw.

A normal metreing section has constant depth shallow channel, in which the melt flows by the drag of the barrel, and moves towards the discharge end, the resistance to the melt flow to the die result into a back flow or pressure flow through the channel.

7.7.3 Single Screw Extruder Characteristics

In single screw extruder, the output is the head pressure developed due to die resistance characteristics for screws with different metreing section depths operating under identical condition at same melt viscosity and temperature. The shallow channel metreing screw is less sensitive to change in back pressure and higher speed gives higher output.

The output of the section is also pressure dependent and the output of the solid transport section should match with output of the metreing section. If the output of the feed section is greater than the metreing section, a pressure will be built up within the extruder, resulting in an increase in pressure flow.

7.7.4 Modified Single Screw Extruders

Modifications to single screw extruders have been adopted to improve the performance of the conventional single screw extruder. Some of these modifications are as follows:

Devolatising type extruder In extrusion operations, it is necessary to process polymers which contain volatiles in the form of moisture, solvents, or adsorbed gases. Unless these are removed, the product may contain bubbles. There are two approaches used to remove volatile from the polymer. In the conventional method Avnet port is provided in the barrel wall at the appropriate position. In the other method the volatile are extracted through a hole drilled into the bottom of the screw channel to the hollow core of the screw. In both methods, the extraction of the volatile takes place in the zone of the screw in which the channel is only partially filled.

Feed section modification In feed section, it compacts the granules, pellets, chips or powder feed into it and convey the solid plug so formed against high feed pressure. The performance of the feed section can be improved by the use of the special trapped bushes with axial groves fitted inside the feed section barrel. These groves prevent the polymer plug from rotating with screw, so that the plug is moved forward like a nut held at its circumference, on a screw rotating inside.

Mixing device In a conventional single extruder, the improved mixing and homogenisation mixing devices are incorporated. These devices help to create a uniform melt of constant temperature and discharge pressure. They are usually adopted especially for large diameter screws and for high-speed operation according to the type of mixing action and other characteristics.

7.7.5 Twin Screw Extruder

In twin screw extruder, two parallel screws are placed in a barrel with figure of eight cross sections. Generally the arrangement of the screws can be divided into two major categories. In non-intermeshing extruder type, the separation between the screw axes is at least equal to the screw outer diameter. This configuration can be regarded more or less as two single screw extruder and should be more appropriately called *double screw extruder*.

When the screws are intermeshing, the separation between the screw axes is somewhat less than the outer screw diameter, in the limit the screw surface can be in mutual contact. According to the screw rotation intermeshing screws are known as contra-rotating and corotating screws. The shape of the screws and the pattern of the material conveyance in each case are as follows:

Contra-rotating screws In case of contra-rotating screws, the screw rotations and the helix angle of the flights are in opposite directions. The plastic material is carried by the screw flights in such a way that all the material is forced on the centre where the two screws meet forming a build-up or accumulation and the consequent creation of high pressure zone. Due to the gap between the screws certain amount of material will pass through it and rotate with the parting screw flights. It is subjected to an extremely high degree of shear. According to the extent of the gap, it is possible to adjust the amount of shear and mixing. The clearance between screws prevents the flight of one screw from completely wiping the channel of the other screw.

Corotating screws The screw rotation and the helix angle of the flights are in the same reaction for these screws. When the flights and channels of the screw intermesh without leaving any gap, the screw profiles are said to be conjugating. Except for mechanical tolerance, no clearances are left. When the material contained within a channel of one screw reaches an intermeshing point finds no passage to the other screw, since the intermeshing flight of the other screw penetrates the channel. The flank of that flight acts as a wedge and forces the material to leave the channel and move into the adjacent channel of the other screw. The transfer of material from one screw to other creates a movement around both screws in a figure of eight pattern. Each time the material reaches the intermeshing point, it moves from screw to the other, but it is also displaced half a pitch in axial direction. This creates a distributive mixing action. Due to very close tolerance between screw and barrel as well as between screws, wiping of screws is complete.

7.7.6 Twin Screw Extrusion Process

Similar to single screw extruder, twin screw extruders have three sections, a solid conveying section near the hopper known as feed section, a pump section near the end of the extruder and a transition section in between, where melting takes place. The extruder with closely intermeshing screws can be conveniently represented by two series of C-shaped chambers-one for each screw. These chambers convey the plastic material positively from hopper to the die by the rotation of the screws.

Feed section It is usual to feed the plastic material by a mechanical feeder regulated independently of screw speed, which partly feeds the screw channel. The material in the feed section is moved axially and positively down the barrel by the intermeshing screw flights. The screw feed zone temperature must be well below the adhesion temperature of the plastic material, so as to avoid sticking of the material to the screws and passing through the gaps.

Transition section The melting mechanism in a twin screw extruder differs considerably from that in single screw extruders. The melting takes place within five or six chambers. This means that the screw length between the chamber in which no melt has yet been formed and the chamber in which the entire polymer molten is of the order of one screw diameter. Moreover, the process has a repetitive character. Complete melting sequence takes place within one chamber. The head pressure of the die affects both the length and the position of the melting zone, i.e., the transition section.

Pumping section In the pumping section of the extruder, the chambers are generally filled with the melt. In a plasticating extruder screws are fully filled with the melt. The idealised theoretical volumetric output is given by

$$Q = 2NV$$

where N = Revolutions per unit time; V = Chamber volume.

This output is reduced due to leakage through the following gaps between screw flights:

Over the flights

At the flight root

Through the tetrahedron shaped spaces between the flight walls

Through the equatorial gap between the sides of the intermeshing flights

All the above leakage varies linearly with screw speed and is controlled by the axial pressure gradient.

7.7.7 Twin Screw Extruder Characteristics

The output vs head pressure characteristic is completely independent of operating conditions. It is greatly dependent on details of the screw geometry. The size of the gaps and the pressure gradient are responsible for the reduction in output at increased head pressures.

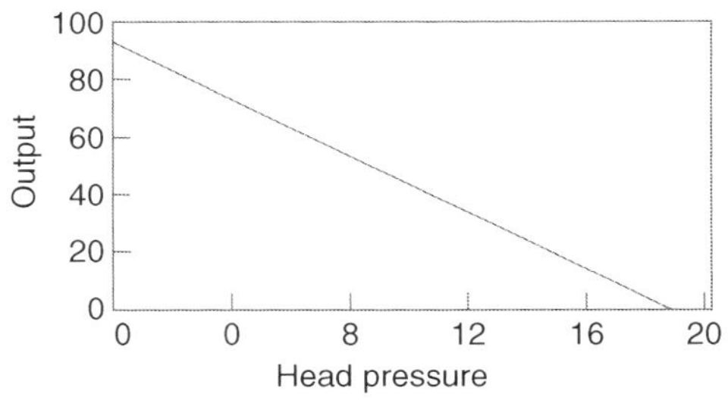

Fig. 7.2 Twin screw extruder characteristic.

7.8 CLASSIFICATION OF DIES AND DIE GEOMETRY

7.8.1 Introduction

Extrusion dies are used to streamline the flow path to give shape to melt pumped by the extruder. As per the die design guidelines, the abrupt change in cross section should be avoided. Stagnation areas should be avoided in the flow path and flow should be smooth. The melt should be maintained compact by ensuring sufficient pressure on melt possibly by path restriction. The die construction should be able to withstand internal pressure and the die temperature should be maintained.

7.8.2 Classification of Dies

According to the extrudate

1. **Solid Extrudate Dies** are used for manufacturing rod, sheet, and profiles, etc.
2. **Hollow Extrudate Dies** are used for manufacturing pipe, wire coating components, etc.

According to direction of flow of melt

1. **Straight through or inline die:** In this type of die the flow in die is in line with extruder.
2. **Cross-Head Dies:** In this case the flow is diverted perpendicular to extruder either upward (blown film), downward (cast film, parison die) or sidewise (wires and cable) by the die.
3. **Offset Dies:** Flow is diverted by right angle twice and is maintained offset to extruder flow axis. (in certain pipe).

According to the construction

1. **Integer Dies:** Integer die consists of a single piece die head, which has a streamlined tapered approach section followed by a land section orifice, shaped to produce appropriate product cross section. It is mainly used for production of sections, which have regular and simple cross sections such as rods.
2. **Split Dies:** This type of die is used for extrusion of sections such as tapes, belts, gaskets and for profiles of more complex shapes, dies have to be made in to pieces by splitting them either transversely or longitudinally. The approach section and land section may be made as two separate pieces. An alternate arrangement is to make the approach section as a truncated conical piece, which merely reduces the diameter and a second piece, which has an abrupt transition from the reduced diameter to the requisite profile followed, by the land orifice section. The two pieces are finally clamped together by bolts or for small circular dies a threaded retainer ring may be used. In addition, dies can also be split longitudinally to form two separate halves, which are then clamped together to form a single die. Dowel pins can be used to maintain perfect alignment between split sections during and after assembly. Split construction facilitates easy machining and alterations.

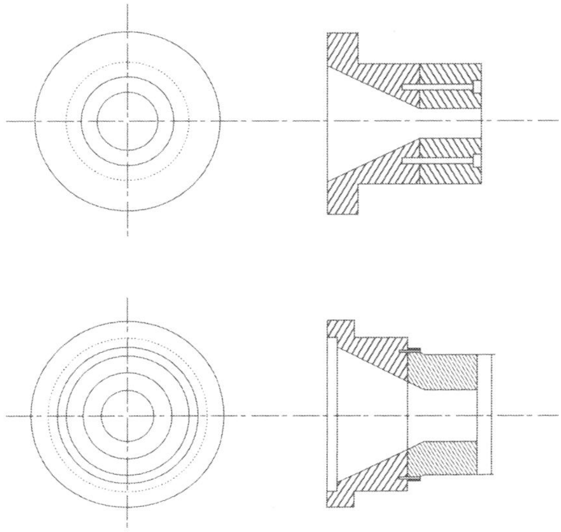

Fig. 7.3 Split dies.

3. **Plate dies:** These dies are made up of two or more plates which incorporate the approach section and the land section. Each plate has a limited thickness, which facilitates the machining of the flow channels. In this type of die, abrupt changes in the flow channels are common, but streamlining is effected in the final plate. The plate dies offer many advantages, such as simple construction, low cost, easy adjustment and mounting. The plates are clamped together by axial screws. These dies are only used for heat-stable materials. The land orifice plate can be changed so as to obtain interchangeability of shapes.

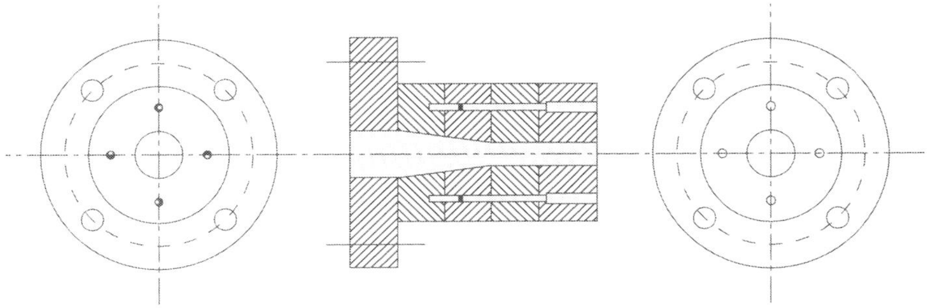

Fig. 7.4 Plate dies.

7.8.3 Die Design Procedure

Die design procedure involves the following steps:

1. While designing a die the extruder performance regarding flow output, output pressure, and linear velocity of melt and the plastic material melt property and its behaviour should be considered.

2. Based on the product shape or geometry and the flow, profile of melt in die is designed.
3. The minimum wall thickness of the die is decided based on the strength of material and actual dimension thereof depending upon assembly requirements.

7.8.4 Operation of Die

In case of extrusion process, both the extruder and the die performance affect the output performance. The graph below shows performance of an extruder and a die at specified temperature.

If extruder runs at pressure P_c extruder is extruding melt at '*c*' rate and die supply also at '*c*' rate. Thus, the overall process output is '*c*', which is optimum output rate for this particular set up of extruder and die. If extruder runs at pressure P_i, although extruder can extrude melt at '*a*' rate higher than '*c*' but die can supply only at '*b*' rate, which is lower than '*c*'. Thus, the overall process output is '*b*' which is lower than '*c*'. If extruder runs at pressure P_o, although die can supply only at '*d*' rate higher than '*c*' but extruder can extrude melt at '*e*' rate which is lower than '*c*'. Thus, the overall process output is '*e*' which is lower than '*c*'.

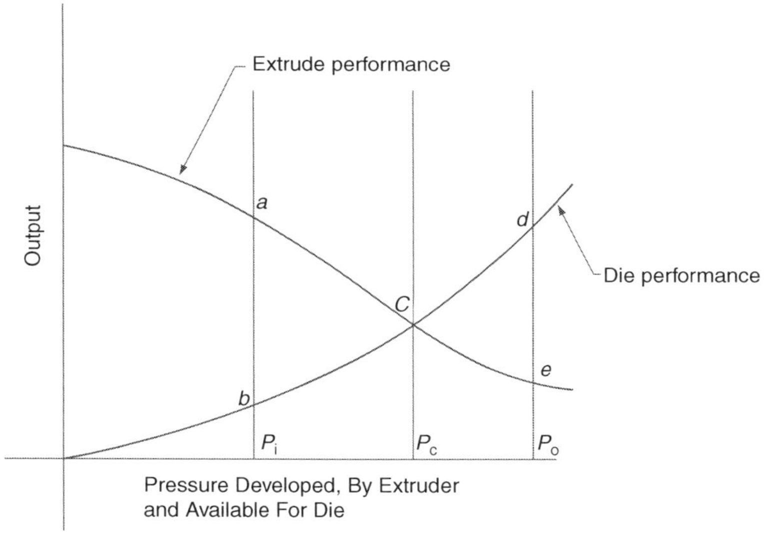

Fig. 7.5 Die and extruder characteristics curve.

7.8.5 Die Geometry

The function of the die is to change the cross section of the flow channel, i.e., the circular shape of the extruder barrel to a required cross section which is very close to the product section or profile. This is usually done by a streamlined die channel, which is formed to obtain a gradual change of cross section. The flow in a die channel is not uniform. The material at the centre of the channel moves at a higher velocity than at the wall. The layer of the melt, which is in direct contact with the die wall, is likely to adhere to the walls and degrade. It is therefore, advisable to provide short and narrow channels to minimise the residential time of the plastic melt in the die.

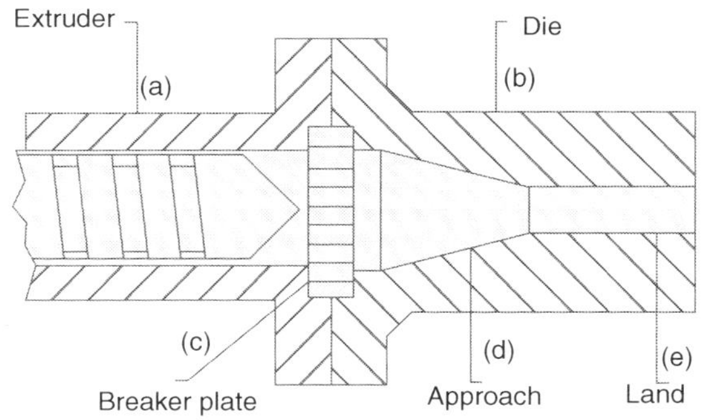

Fig. 7.6 Die extruder (a) Extruder (b) Die (c) Breaker plate (d) Approach (e) Land.

The die channel is made up of two sections, namely 1. Approach section 2.Forming section or land section.

Approach section The approach section is a convergent channel, which helps to change the cross section of the extruder to the desired shape and create a compression of the melt, as it flows. Compression of the melt, due to narrowing of the convergent channel is essential to create a uniform pressure and eliminate stagnation. The angle of the convergent section is usually between 20 to 30 degrees depending on the viscosity of the melt. The convergent approach section has 30° to 90° included angles. For viscous melts such as acrylic or polycarbonate, the angle may be reduced. Similarly, for the production of small size rods the angle should be considerably reduced. For heat sensitive materials the approach should be fully streamlined to avoid stagnation and consequent degradation. For higher viscosity a smaller angle is preferred. Judicious use of tapered channels permits considerable increase in output rates of the product free from defects.

Die land design The land section has the cross section of the desired product, and is therefore, the most vital section of the die. It should be long enough to produce a sufficient back pressure, so that the exit of the extrudate will be smooth and even. The major pressure drop takes place along the carefully dimensioned land, thus achieving the desired metreing effect. Properties of the product usually improve with increased land length and compression ratios. Back pressure must be within stipulated range and frictional heat should be controlled. Usually, the purpose is to allow the material to relax in the land, before leaving the die; the length of the land section depends on the shape and size of the product and also on the material to be extruded. It is necessary to make the land orifice oversized because a certain amount of tension is needed to pull the extrudate away from the die, resulting in a draw down of the cross section, i.e., the residence time of the melt should be at least equal to relaxation time. The land should help to bring about a perfect appearance and correct dimensions of the product. The length of the land depends on the extruder output, type of resin and the size of the product, it is assumed as 15 to 50 times the product wall thickness.

The choice of the die channel dimensions depends on several factors, such as the plastic material, additives, product shape, variation in wall thickness, die swell, extrusion rate, temperature, cooling system, shrinkage, etc.

Die swell As soon as melt emerges out of the die, it swells due to its elastic property. This is called die swell. It is influenced by temperature along with following factors:

1. Shear rate, as shear rate increases die swell increases.
2. Length to diameter ratio of die land, as length to diameter ratio increases die swell decreases
3. Ratio of extruder diameter to land diameter, as ratio of extruder diameter to land diameter increases die swell increases.

Table 7.1 Die swell ratio with respect to shear rate.

Die swell ratio	Approximate shear rate (s^{-1})			
Polymer	**10**	**100**	**400**	**700**
Acrylic, high impact	1.17	1.27	1.35	1.42
Polyethylene low density	1.45	1.58	-	-
Polyethylene high density	1.49	1.92	2.15	2.35
Polypropylene copolymer	1.52	1.84	2.1	-
Polypropylene homopolymer	1.61	1.9	2.05	-
Polystyrene (G.P)	1.37	1.7	1.88	-
Polystyrene (toughened)	1.22	1.4	-	-
Polyvinyl chloride (rigid)	1.35	1.5	1.52	1.53

Die swell, or more properly called extrudate swell, is the actual material swelling as it exits the die polymer molecules or chains oriented in the flow direction in the die land area relax and re-entangle once the material exits the die due to the polymer elastic component. This relaxation and re-entanglement of polymer chains causes distortion of the extrudate cross sectional area compared to the die cross section.

When polymer exits the die of the extruder it will swell, this action is called *die swell*. Die swell occurs because the sudden release of pressure causes the polymer chains to relax and remember its previous shape when in the larger cross section of the extruder, tries to return to it after leaving the die orifice since polymers are visco-elastic and display time dependent stress relaxation.

Die swell or extrudate swell depends on the following:

1. Shear rates in the die

2. Melt temperature
3. Die land length
4. Reservoir length

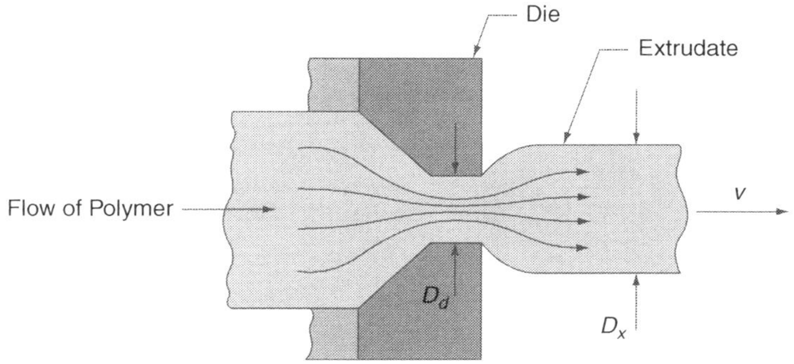

Fig. 7.7 Die swell.

High die shear rates and low melt temperature create more die swell, while longer die land lengths and lower reservoir-to-land length lead to less extrudate swell or die swell. Combinations of these factors can create different conditions, giving the same cross sectional profile. The Fig. 7.8 shows distortion in a square cross section and the die shape required to produce a square rod. Table 7.2 indicates the information on cross sectional guidelines for different materials.

Table 7.2 Cross-sectional guidelines.

Polymer	Increase in orifice size relative to cross-sectional area of part, %
Polyethylene	15–20
Rigid PVC	12–15
Flexible PVC	Small Profile: 8–10
Flexible PVC	5–10
Flexible PVC	Large Profile: 3–5
Polystyrene	8–10
Polyamide	15–20

On the left, a profile obtained from the die is shown above. The hot liquid profile as it leaves the die, has dimensions which exceed those of the die, there is a change of 12%.

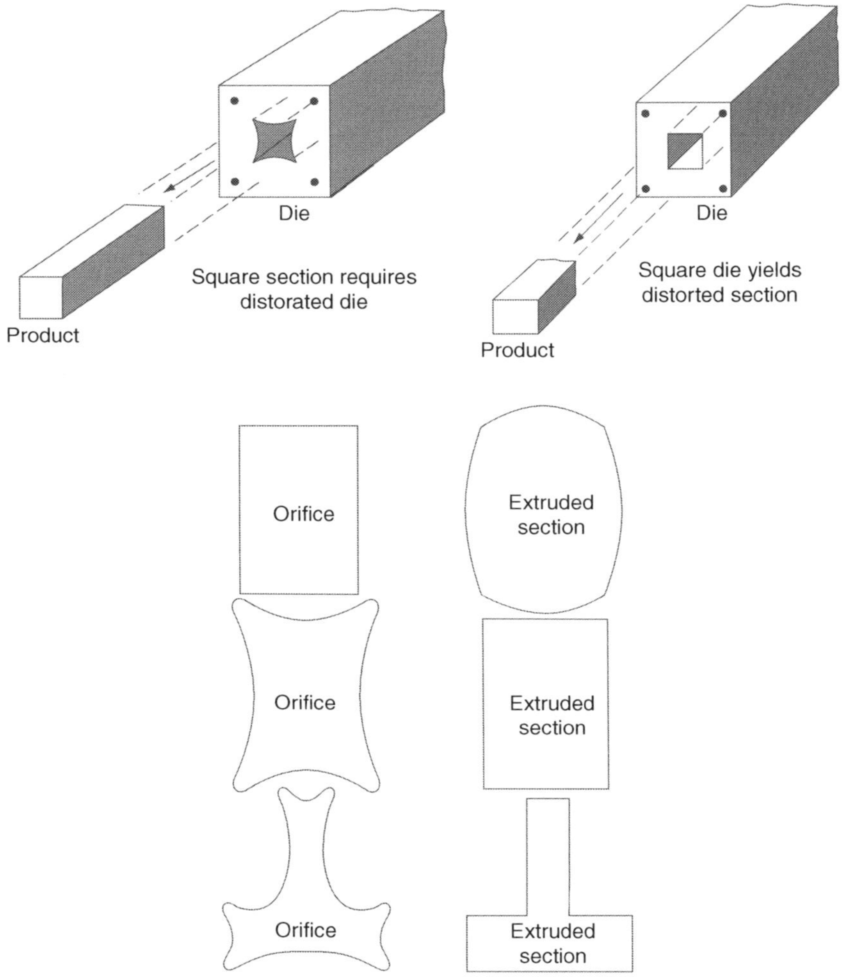

Fig. 7.8 Extrudate distortion due to die swell.

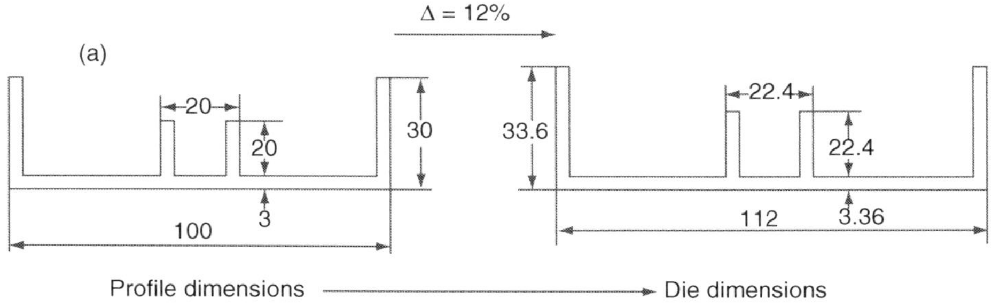

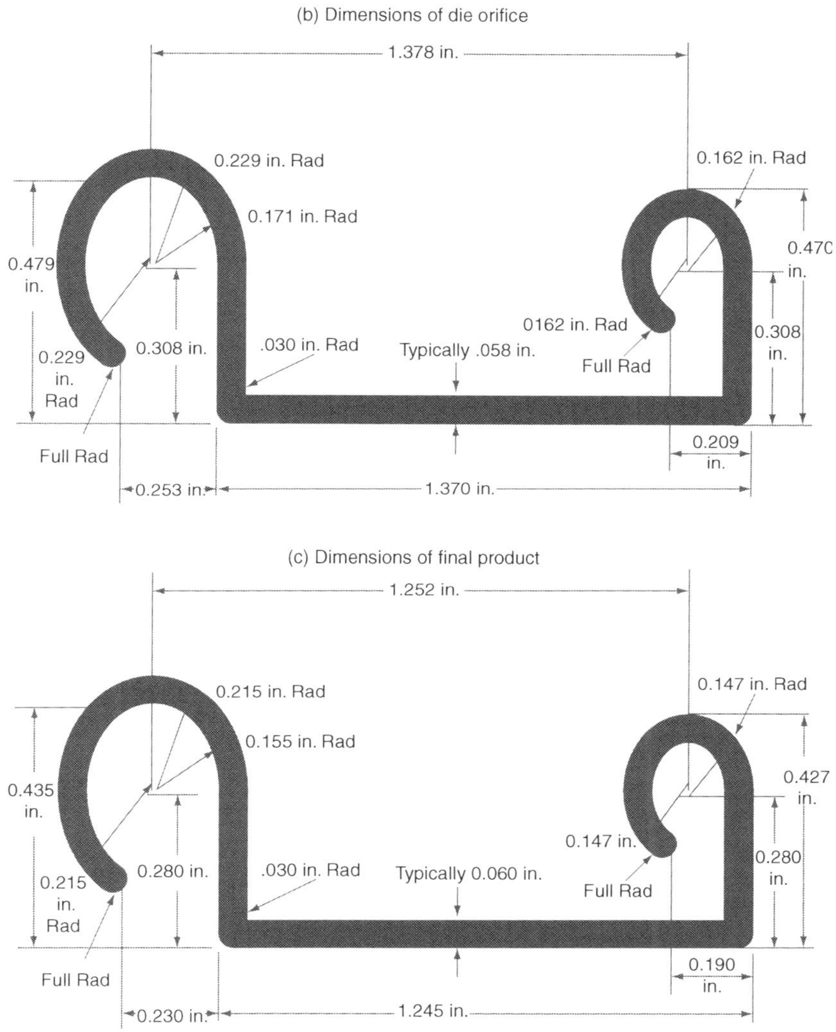

Fig. 7. 9 (a), (b) and (c) Example dimensional changes for a PVC profile shape from the die orifice to the product.

7.9 TYPES OF DIES

7.9.1 Introduction

Various types of solid sections are extruded from different thermoplastic materials such as rod, tape, strip, etc., that have a simple and uniform cross section. Others known as profiles are irregular and complex in cross section. The design and fabrication of dies for complex sections requires skill, along with experience. Further, the plastic to be processed influences die dimensions and these dies are designed especially for extrusion of specific plastics and their compounds.

7.9.2 Rod Die

The rod die has a simpler geometry and it consists of an approach section with a taper of 60° to 90° included angle. The length of the land depends on the material extruded, rod diameter, extrusion rate, etc., and special care to be taken while designing dies for crystalline polymers such as nylon and polyacetal. In these materials, rapid volume change takes place during post extrusion cooling and solidifying, which will create voids in the central portion of rods. With amorphous polymers such as polycarbonate and PVC, the change in specific volume is more gradual and limited.

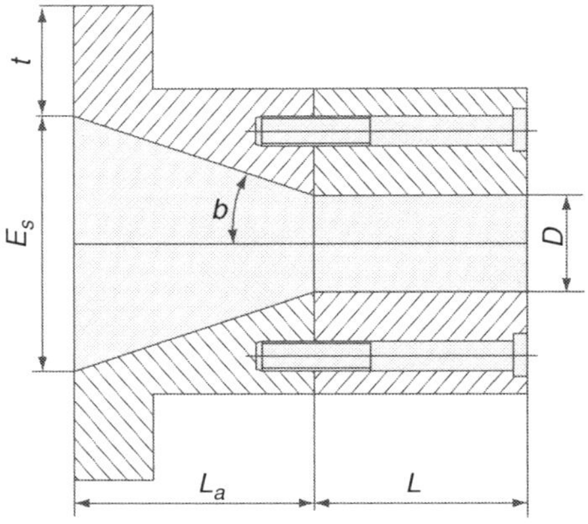

Fig. 7.10 Rod die.

The method of extruding rods in nylon is free extrusion into a water quench bath which requires a circular die land that has a diameter 1.7 to 2.0 times the diameter of the desired rod. A closely balanced cooling system is necessary to obtain a round, void-free rod. Shrinkage voids are usually eliminated by shortening the length of quench bath, by air quenching or by a hot oil bath. This method is suitable for making rods up to 12.5 mm diameter. For larger diameter rods or where precise control of diameter is desired, a forming box method is used. The molten mass is extruded under pressure through a short die land, then through a converging section and finally through a cylindrical water-cooled tube. The rod is thus shaped by the pressure from the extruder, which supplies the melt continuously to the central cone and eliminates voids due to shrinkage.

Example for design of die for a rod

Data given or assumed

Extruder size [E_s]	-	45 mm
Extruder output (Q_m)	-	30 kg/hr
Rod diameter[P_d]	-	15 mm

Material- L. D. polyethylene

Properties

Solid density (ρs)	-	0.923 gm/cm^3
Melt density (ρm)	-	0.76 gm/cm^3 at 170°C
Power law constants (at 170°C)		
(n)	-	1/3
(k)	-	3250 kg (f) sec^{-1}/ m^2
Extensional viscosity (λ)	-	0.822 * 10^4 kg (f) sec/ m^2
Critical extensional strain rate (ε_e)	-	18 sec^{-1}

Permissible tensile strength of die material (f) - 1400 kg/cm^2

Design Calculation

1. Volumetric output of extruder

$$Q_v = Q_m \times 1000 / \rho m$$
$$= (30*1000)/\ 0.76 = 39500\ \text{cm}^3/\text{hr}$$

2. Velocity of melt

$$V = Q_v/A = (Q_v/60) * (4 / \pi Es^2)$$
$$= (39500/60) * (4/\pi 0.45^2) = 372\ \text{cm/min} = 3.72\ \text{m/min.}$$

Land section

Assuming land length to be twice the product wall and land wall to be 10% larger than product to accommodate draw down.

Land section (L) = 30 mm

Land diameter (D) = 16.5 mm

Designing for length of approach section

$$PL = [((3n + 1)/n)\ (Q_v/\ \pi R^3)]\ n\ (2kL / R)$$

where Q_v = Volumetric flow rate

R = Radius of land

L = Length of land

PL = Pressure drop in land

K and n are power law constants

$$PL = [(((3 \times 1/3) + 1)/\ (1/3)) \times (39500/\ (60 \times 60)) \times (1 / (\pi \times (1.65/2)^3))]^{1/3} \times [((2 \times 3250)/ 10^4) \times (3 /(1.65/2))]$$
$$= 6.25\ \text{kg / cm}^2$$

Shear stress

$$\tau_w = (RP)/2L = (1.65 \times 6.25) / (2 \times 2 \times 3.0)$$
$$= 0.859\ \text{kg / cm}^2$$

Shear rate

$$\gamma_w = ((3n /+ 1) / n)\ (\ Q_v / \pi R^3\ )$$
$$= 6 \times (39500/(60 \times 60))\ (\ 1/(\ \pi (1.65/2)^3\ )$$
$$= 37.74\ \text{sec}^{-1}$$

Approach section

To calculate approach angle α

$$\tan\ \alpha = 2\ \varepsilon_c / \gamma$$

where ε_c = Critical extensional strain rate

γ = Shear rate

$$\tan\ \alpha = 2 \times 18/37.74 = 0.954$$
$$\alpha = 43^\circ 40'$$

Extruder size = 45 mm

$$\text{Tan}\ \alpha (45 - 16.5) / (2 * L_a)$$

where L_a = Length of approach section

L_a = 15.1 mm.

Calculation for Minimum Thickness of Die Wall

Pressure drop in the approach section

$$\begin{aligned} Ps &= (2\tau/3n \tan\alpha)\,[\,1 - (R_0/R_i)^{3n}\,] \\ &= (2 \times 0.859)/(3 \times 1/3 \times 0.954)\,[\,1 - (16.5/45)^{3 \times 1/3}\,] \\ &= 1.14\ \text{kg/cm}^2 \\ Pe &= ((2\lambda\varepsilon_c)/3)\,[\,1 - (R_0/R_i)^3\,] \\ &= ((2 \times 0.822 \times 10^{4 \times} 18)/3)\,[\,1 - (16.5/45)^3\,] \\ &= 9.38\ \text{kg/cm}^2 \end{aligned}$$

So, pressure drop in approach section = 1.14 + 9.38 = 10.52 kg / cm^2

Assuming no pressure drop in the breaker plate, screen pack and the adapter.

Total pressure drop in die (Pt) = Pressure drop in land + Pressure drop in approach

$$Pt = 6.25 + 10.52 = 16.77\ \text{kg / cm}^2$$

Minimum die wall thickness (t)

From Eq. $t = (D/2)\,[\,((f + Pt)/(f - Pt))^{1/2} - 1\,]$

where, D = Land diameter

f = Permissible tensile strength of die material

Pt = Total pressure drop in die

So, $t = (16.5/2)\,[((1400 + 16.77)/(1400 - 16.78))^{1/2} - 1\,]$

$= 8.25 * 0.01 = 0.0825$ mm.

Actual wall thickness is to be decided depending upon assembly situation.

7.9.3 Pipe Die

Pipe is basically a hollow and symmetric round cross sections and a pipe extrusion line is very similar to a profile line with a vacuum sizing cooling unit, puller, and saw or on cutter. Products can be rigid or flexible and vary from something very small, such as a catheter tube used in medical applications, to large-diameter pipe used to transport water or other fluid. Pipe or tubing can be wound up as a continuous product or cut to length. Pipe and tubing can be extruded for added value and to meet specific end-use requirements.

There are different types of die used for manufacturing of pipes. Various types of dies are in line or straight through or offset type.

Function of die

The extrusion head or die is determined by the diameter and wall thickness required in the final product. The extrudate enters directly from the extruder through the breaker plate into the die. Large dies require a die stand to support the die weight and prevent distorting the extruder barrel. An adapter is present between the extruder and the pipe or tubing die, versus the direct connection the entrance cone distributes the melt uniformly around the mandrel. The mandrel or centre section is held in place by spokes radiating out from the mandrel called a *spider ring* with the individual spokes referred to as *spider legs*. The spider legs support the mandrel weight and the shearing forces from the high upstream pressure.

The number of spider legs is determined by the die size and the individual spider leg size. The load on each spider leg is calculated similar to the extruder thrust bearing load, where the head pressure is multiplied by the cross sectional area, divided by the number of spider legs. Equation below can be used to calculate the force on each spider leg:

$$L_{sl} = \frac{P_d \times \pi \times R^2}{N}$$

where

L_{sl} = Spider leg load
P_d = Die pressure
R = Die radius
N = Number of spider legs

As the number of spider legs is increased, their size can be decreased. Each spider leg divides the polymer flow, causing a knit or weld line that is a potential weakness in the final pipe or tube.

Spider legs are shaped to minimise the melt stream disturbance. Sufficient die length has to be present after the spider ring for the polymer chains to re-entangle and provide proper strength at the weld line.

Molecular chain re-entanglement is encouraged by

1. Higher melt temperature (more molecular motion)
2. Higher die pressure (forces more molecules to pack into a given area)
3. Longer die and land length (provides more time)

A larger melt channel area after the spider ring followed by a flow channel restriction (increased mandrel diameter or decreased bushing inside diameter) builds pressure to encourage molecular re-entanglement.

The mandrel is centred in the die and the bushing is centred around the mandrel. To make adjustments, the bushing clamping ring is loosened and the die adjustment bolts are used. If the bushing is improperly centred, the pipe or tube concentricity will be incorrect, with one side having a thicker wall than the other side. Through one spider leg, there is an air inlet tube that supplies air to the centre of the pipe. Air is critical in obtaining the correct pipe or tube diameter in the calibration tank. The air supply is normally at atmospheric pressure. The die area where the melt channel in a constant diameter is called the *die land area* and is used to shape the molten polymer into a pipe or tube before it exits the die. Die land lengths need to be proper for the polymer being processed. Longer land lengths may generate too much shear, while shorter land lengths may not provide sufficient molecular orientation and lead to extrudate swell.

Die development over the years has focused on minimising weld lines caused by the spider ring. Two approaches have been successful in eliminating weld lines. First, the spiral dies are used in large-diameter pipe applications. Second, a melt pool is generated directly after the spider ring by expanding the flow channel followed by a restriction. Expanding and contracting the flow channel one or two times forces the polymer chains to re-entangle.

Regardless of the die used, the polymer melt channel has to be properly streamlined to prevent stagnation, which can generate degraded or cross-linked polymer. Degraded polymer in the melt channel can cause a weld line with no mechanical mechanism to encourage the polymer molecules to recombine. This causes a weak area in the pipe. Larger extruders can be used to produce multiple pipes by splitting the melt stream as it exits the extruder and feeding two dies. Instead of using end-fed dies, the dies are side-fed. As the polymer melt converges on the back side of the mandrel, it has to mix to allow the polymer molecules to properly entangle. Melt impinging on the mandrel cannot deform or move the mandrel in the die. Multiple dies are used when small-diameter pipe or tubing is produced on large extruders with excessive capacity for one small-diameter product.

By varying the rotation speed with the puller speed or extruder screw speed, the inside diameter and outside diameter of the tube can be changed independently, using the same tubing die without changing any of its components:

Advantages of this die are:

1. A single die can be used to manufacture different outside diameter and inside diameter.
2. Tubing is weld line free.
3. Tubing with different outside diameter and inside diameter can be produced by changing the rotation speed during tube production.
4. Tubing is more uniform.

In-line or straight-through dies These dies are most widely used for the production of hollow articles. The components of die are designed to suit specific material and product size and accordingly the shape of the components is modified. The adapter is attached to the end of the extruder. With the cone shaped spider, it forms the flow channel of increasing volume compression. The degree of taper is determined by the size of the channels formed in the spider. The axial length of the flow channel, formed prior to the spider, is made large compared to

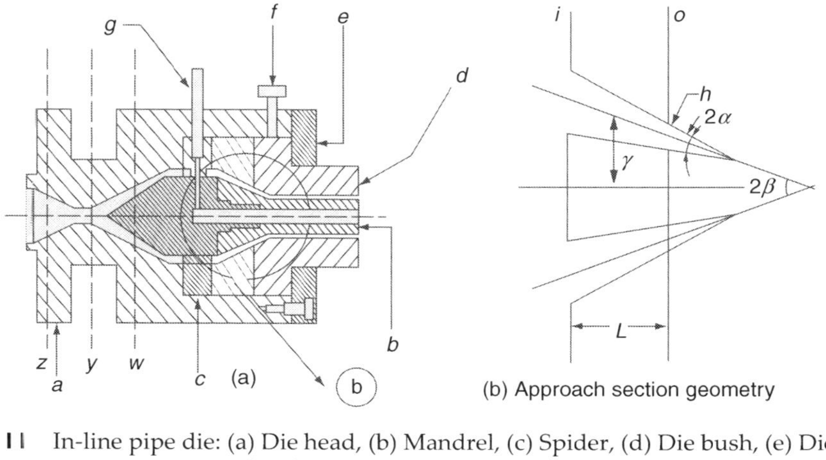

Fig. 7.11 In-line pipe die: (a) Die head, (b) Mandrel, (c) Spider, (d) Die bush, (e) Die ring, (f) Adjusting screw, (g) Air inlet.

the dimensions of the flow channel in the spider. This helps to reduce the obstruction to melt flow in the spider channels. High melt temperature and frictional drag in the spider passage should be avoided. The ratio between the land length and the annular gap which forms the wall thickness varies between 10 to 30. The annulus is usually designed to oversize to allow for some draw down and an excess diameter of 10 to 30 per cent is usual. The mean diameter of the annular gap should be 5 to 25 per cent larger in the case of pipes. The lower values refer to rigid PVC and larger values to polyolefin's.

Cross section of flow should gradually decrease from section '*z*' to '*y*' to '*w*' (see Fig. 7.11).

Example for design of in-line pipe die

Data given or assumed -

Extruder size	$[Es]$	–	65 mm
Extruder output	(Q_m)	–	60 kg/hr
Outside diameter of pipe		–	90 mm
Thickness of pipe		–	4.5 mm
Material and its property		–	H.D. Polyethylene
Density of solid	(ρs)	–	0.955 g /cm^3
Density of melt	(ρm)	–	0.81 g /cm^3 at 170°C
Power–law constants			
K		–	631 kg(f) sec$^{-0.56}$/m^2
n		–	0.56

Design calculation

1. Volumetric output of extruder

$$Q_v = Q_m \times 1000 / \rho m$$

$$= (60 \times 1000) / 0.81 = 74074 \text{ cm}^3/\text{hr}$$

2. Velocity of melt $V = Q_v/A = (Q_v/60) \times (4/\pi Es^2)$

$$= (74074/60) \times (4/\pi 0.65^2)$$

$$= 372 \text{ cm/min} = 3.72 \text{ m/min.}$$

Land section

Assuming land length to be 15 times the product wall thickness and land wall to be 2% larger than product to accommodate draw down.

Outside diameter of annulus	–	90 * 1.02 = 91.8 mm
Inside diameter of annulus (mandrel diameter)	–	91.8 – (4.5 * 2) = 82.8 mm
Land length	–	15 * 4.5 = 67.5 mm

Shear rate in land

$$\text{Shear rate } (\gamma) = (6\,Q_v) / [\,\pi(R_0 + R_i)(R_0 - R_i)^2\,]$$

$$= (6 * 74074) / [\,\pi(4.59 + 4.14)(4.59 - 4.14)^2 \times 3600\,]$$

$$= 22.24 \text{ sec}^{-1}$$

where Q_v = Volumetric output of extruder
R_i = Inlet radius of taper
R_0 = Outlet radius of taper

Pressure drop in land can be obtained from (assuming thickness very less compared to radius)

$$Q_v = (n / (2n + 1)) * (\pi R_0{}^3) [(R_0 P) / (2KL)]^{1/n} ((\beta - 1) / \beta)^{((2n + 1) / n)} ((1 + \beta) / 2\beta)$$

where P = Pressure Drop in Land
β = $R_0 / R_I = 4.59 / 4.14 = 1.11$
k, n = Flow behaviour index
R_i = Inlet radius
R_0 = Outlet radius
L = Land length
Q = $\{(0.56/2.12) \times \pi \times (4.59)^3) [(4.59\Delta P)/(2(631/10^4))]^{1/0.56} ((0.11-1.11)^{(2.12/0.56)} (2.11/2.22)\}$
= $0.012 \times (5.388\ P)^{1.785}$
$P^{1.785}$ = $22.24 / (0.012 \times 20.21) = 91.71$
P = $12.57\ kg / cm^2$

Approach section

Assuming angle of approach (convergence) $\beta = 20°$

$2\beta = 40°$

and assuming $2\alpha = 10°$

Assuming the length of approach section = 100 mm

Pressure drop in approach section = Due to simple shear flow (P_s) + Due to Extensional Flow (P_e)

P_s = $(K\gamma^n / H_o) (1 / (U - V)) \log_e [(1 + UL) / (1 + VL)]$
P_e = $[1 - (R_o H_o{}^2 / R_i H_i^2)] \lambda / 2$

where K, n = Power law constant; γ = Shear rate;
λ = Viscosity under simple tension
R_o = Mean radius in outlet of approach section
R_i = Mean radius in inlet of approach section
H_i = Half Gap in inlet of approach section
H_o = Half Gap in outlet of approach section
L = Length of approach section
U = $((\tan \alpha \sec \beta) / H_o)$
V = $n [(\tan\beta / R_o) + 2U]$
U = $((\tan \alpha \sec \beta) / h_0) = (0.0875 \times 1) / ((0.45/2) \times 0.94) = 0.41$ cm
V = $n [(\tan\beta / R_0) + 2U]$
= $0.56 [(0.364 / 4.59) + (2 \times 0.41)]$
= 0.56×0.899
= 0.5 cm

$$P_s = (631/10^4) \times ((22.24)^{0.56}/0.45) \times (1/(0.41 - 0.5)) \log_e [(1 + (0.41 \times 10)) / (1 + (0.5 \times 10))]$$

$$= (631 / 10^4) \times ((22.24)^{0.56}/0.45) \times (-(1/0.09)) \times (-0.1625) = 1.438 \text{ kg/cm}^2$$

P_e = Assumed negligible

The total pressure drop

$$P_t = P_{\text{approach}} + P_{\text{land}} = 1.438 + 12.57 = 14.008 \text{ kg/cm}^2.$$

Minimum die wall thickness (*t*)

$$t = (D/2) [((f + Pt) / (f - Pt))^{1/2} - 1]$$

where D = Land diameter; f = Permissible tensile strength of die material

P_t = Total pressure drop in die

$$t = (91.8/2) [((1400 + 14) / (1400 - 14))^{1/2} - 1] = 0.23 \text{ mm.}$$

Actual wall thickness is to be decided depending upon assembly situation.

7.9.4 Offset Dies

Offset dies are used in certain special cases like production of pipes. The flow is diverted twice by right angle to offset flow parallel to axis of extruder.

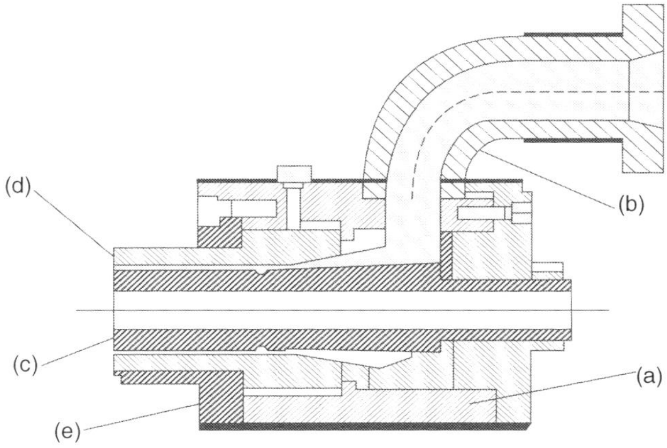

Fig. 7.12 Offset pipe die: (a) Die head, (b) Adapter, (c) Mandrel, (d) Die bush, (e) Die ring.

7.9.5 Co-extrusion In-Line Pipe Die

Co-extrusion dies extrude pipes with two layers of different materials, by use of special dies, which are fed from two separate extruders. The melts from the extruders are supplied to different manifolds in the die. These manifolds are usually placed at an angle of 90°. One of the manifolds is in line with one of the screw extruder, while the other feeds the material similar to a cross-head die. The annular flow paths are therefore, provided around two concentric mandrels. One of the flow paths is between the inner mandrel and the surrounding outer mandrel and the other flow path is between the outer mandrel and the die body. The flow of each melt therefore, takes place independently till it reaches close to the die land. The laminating of melt layers then occurs just upstream of the die land. In other respects, the die is similar to a standard in line pipe die.

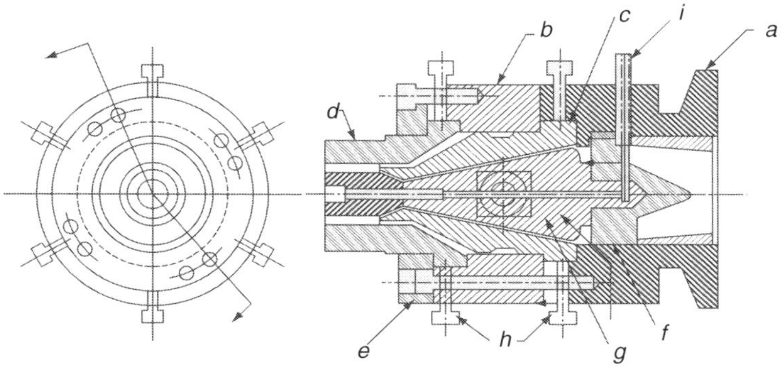

Fig. 7.13 Co-extrusion die: (a) Adapter, (b) Die head, (c) Hollow mandrel, (d) Die bush, (e) Ring, (f) Spider, (g) Mandrel, (h) Adjusting screws.

7.9.6 Profile Die

Profile sections sometimes have varying thickness along the cross section. The cross section is divided into number of subsections depending on its area and shape. Different orifice section needs different land. Longer land gives better surface finish but more pressure drop occurs. Table 7.3 and 7.4 shows general requirement of land length in respect to profile thickness. Non-uniform swelling of extrudate from rectangular and T-section causes problem in the final product. It can be overcome by concave curvature in die as shown by dotted line in Fig. 7.14.

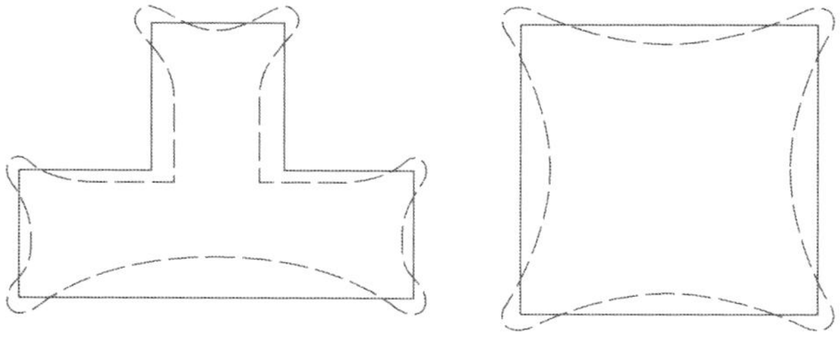

Fig. 7.14 Curvature in die for T-section and rectangular cross section.

Table 7.3 Typical size of land section and slit section.

Type	Die land / profile thickness	Thickness of slit upon product
UPVC	20 – 30	0.9 – 1.0
Soft PVC	6 – 9	1.1 – 1.2
PE	15 – 20	1.1 – 1.2
Nylon	10 – 20	1.5 – 2.0

Table 7.4 Typical ratio of die length and various thicknesses.

S.No.	Types of plastic	Land length / Profile thickness	Channel profile wall thickness / Product profile wall thickness
1.	Plasticised PVC	6–9	1.1–1.2
2.	Rigid PVC	20–30	0.9–1.00
3.	PVC dry blend	15	0.9–1.1
4.	Polyethylene	15–20	1.1–1.2
5.	Cellulose acetate	20	1.1–1.3
6.	Nylon	10–20	1.5–2.00

Table 7.5 Per cent oversize of die orifice as compare to product thickness.

S.No.	Types of plastic	Profile width (%)	Profile height (%)
1.	Plasticised PVC	+ 20	+ 30
2.	Rigid PVC	+ 5	+ 5
3.	Polyethylene	+ 10	+ 15
4.	Cellulose acetate	+ 20	+ 20
5.	Polystyrene	+ 20	+ 20

Profile extrusion Profiles are the shapes that are other than circular or are not symmetrical. However, there are also symmetrical extruded products, such as capillary tubing, electrical cable casings, and window profiles. Profiles can be solid, hollow, or a combination of both and the design of the type of die depends on whether the product is hollow or compact/solid. Also whether the cross section is simple or complex.

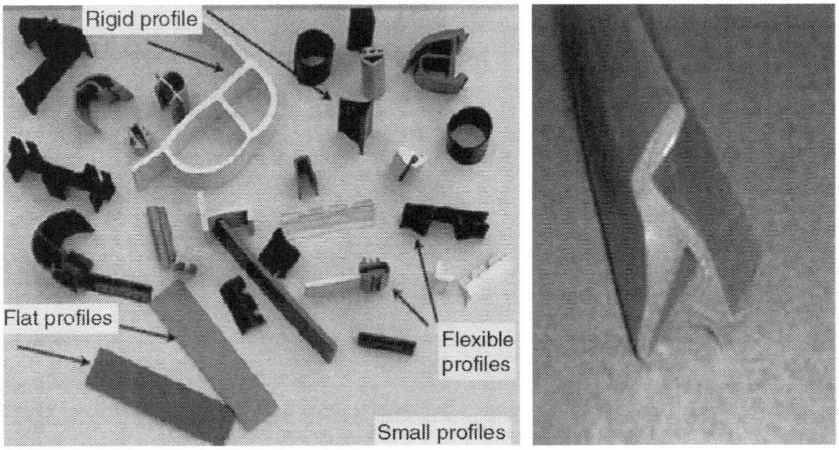

Fig. 7.15 Some products produced by profile extrusion.

Fig. 7.16 Modular designed window profile die.

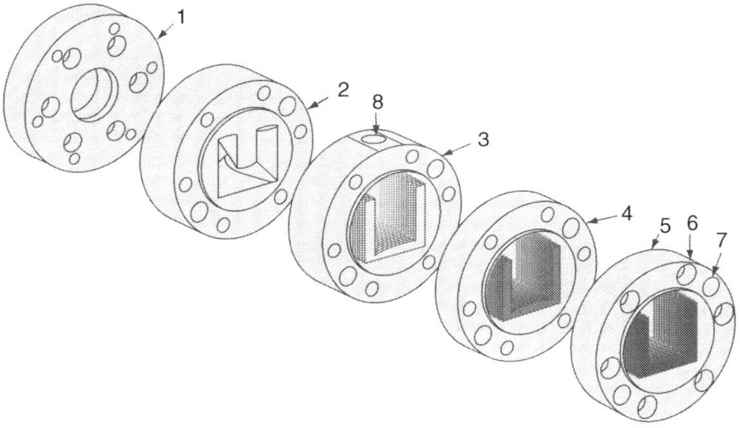

Fig. 7.17 U profile stack die: Exploded view.

The die has to be properly designed to have uniform flow and pressure drops in all the legs of the profile. Improperly designed dies can lead to severe warpage problems associated with the profiles.

The preferred geometry has uniform wall thickness around the entire part. Depending on the profile, constant wall thickness over the entire part is not always possible. The resistance to flow in a die channel is given by:

$$R \quad = \quad R = \frac{\Delta P}{n \times Q}$$

R = Resistance to flow

ΔP = Pressure drop

n = Melt viscosity

Q = Volumetric throughput rate

The volumetric throughput of the different channels has to be constant to prevent warpage. If the melt temperature is the same and the shear rates are similar, the melt viscosity in the different channels will be the same. This leaves the pressure drop to balance the resistance in each section or leg of the die.

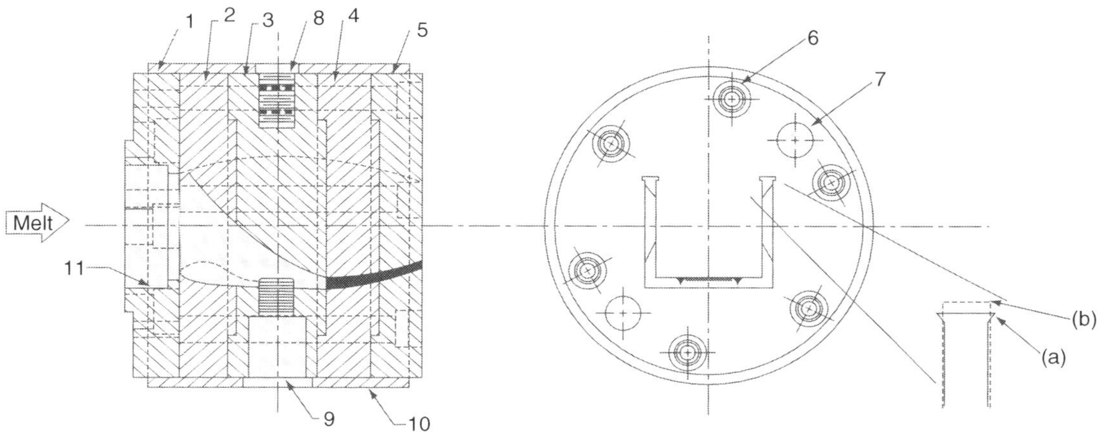

Fig. 7.18 U profile stack die: (1) Extruder mounting plate, (2) Die adapter plate, (3) Transition plate, (4) Preland plate, (5) Die land plate, (6) Die bolt hole, (7) Alignment dowel pin hole, (8) Thermocouple well, (9) Pressure transducer port, (10) Heater band, (11) Breaker plate recess, (a) Die exit profile, (b) Product profile.

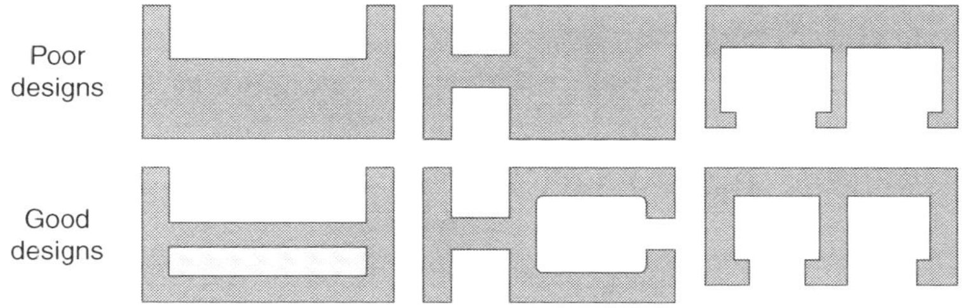

Fig. 7.19 Comparison of profiles with uniform versus non-uniform cross sections.

The pressure drop in the die is calculated from the following formula:

$$\Delta P = \frac{2 \times \tau \times l}{h}$$

where

τ = Shear stress = Force/area
l = channel length
h = channel height

The proper design of die is highly essential to produce uniform parts without warpage. The design criterion includes the following:

1. Thicker wall thickness components require more time for cooling, leading to lower production rate hence, proper calculation of wall thickness is to be done based on the product performance requirements. The ideal wall thickness is approximately 0.7 mm to 3.0 mm.
2. The wall thickness of the component should be reduced and the same stiffness can be achieved with the incorporation of ribs on the back side or the use of corrugated structures.

3. Uneven wall thicknesses lead to uneven flow in the die and differential profile cooling that can lead to warpage, bow, and twist in the final part. If more than one wall thickness is present in the part, make the transition from one wall thickness to the next gradual.
4. Ribs provided in the part leads to sink marks caused by the extra mass and the polymer shrinkage. The shrink marks can be eliminated by building the profile up opposite the rib or making the rib smaller so there is less material. The provision of radii in the part reduces shrink mark and the radii between the rib and the rest of the part are required to provide the necessary strength.
5. All inner and outer corners need to be designed with proper radii, eliminating sharp corners. Sharp corners generate high stress areas that cause premature part failure.
6. It is advisable to eliminate through-holes in the profile wherever possible; these are expensive from both a tooling perspective and the potential need for vacuum sizing. It is better to have through-holes in a profile than uneven wall thicknesses. Uneven walls can lead to more significant processing problems.

The control of die temperature is another key variable affecting the profile and uniformity. The die has to have uniform heat to provide a uniform melt viscosity across the entire profile. Large profiles may require insulation around the die to maintain uniform die temperature. The die face should be a large area with only a single heater band around the outside edge to maintain the temperature.

Changes in die temperature affect the melt temperature, with high temperatures producing lower melt viscosities.

The profile die construction is made out of high quality tool steel, hot die steel, P-20, H-11 steel and hardened. It should have gloss surface finish, the die land area is to be highly polished and chrome or nickel plated.

A profile die is attached to the extruder with an adapter that funnels the melt to a die cross section that is slightly larger than the profile cross section. Three profile die designs are used, depending on the production run length and the polymer being processed. The types of profile dies are flat back, semistreamlined and streamlined.

1. Flat Back Dies: These types of dies are used for low volume or prototype runs for duration of four to six hours long. A flat back die is a flat plate with the desired profile cut into its centre and mounted to a round tube that conveys the molten polymer from the extruder. The die plate is flat on the side facing the extruder and is bolted directly on to the tube comprising the die body. The disadvantage associated with a flat plate die are dead spaces in the die body around the front plate, where molten polymer can stagnate and degrade, and poor flow patterns within the die body.

2. A Semi-streamlined Die: The semistreamlined die, has a tapered channel in the die body that compresses the polymer flow as it approaches the die plate. It is designed to produce uniform flow velocity in all the channels with less extrudate swell and melt fracture. Streamlined dies are used in long production runs where polymer cannot be allowed to degrade. Semistreamlined dies are somewhere between the two die extremes. These types of dies are made with inserts allowing different shapes or slight modifications of a basic profile to be made more easily.

The main factors in semistreamlined die design are the land length and the draw down ratio.

The land length is defined as the uniform flow channel just prior to the die exit. The land lengths within the die determine the flow resistance for the different profile sections or channels. Thinner profile sections require less land length than thicker sections in the profile to produce the same flow resistance. Longer land lengths are used to produce profiles with higher gloss. The proper land length depends on the die gap, the orifice area, flow rates, melt viscosity, and the optimum pressure for the die and extruder. The die land length guidelines are detailed in Table 7.6.

Table 7.6 Recommended draw down ratios and land lengths for different materials.

Material	Die Land Length *	Draw Down, %
LDPE	15:1	30
MDPE	15:1	25
HDPE	15:1	20
PP	15:1	30
PPVC	15:1	5–25
UPVC	20:1	4–10
GPS	20:1	8–20
ABS	25:1	25
Nylon 6.6	20:1	15–20
Nylon 6	20:1	20
PC	10:1	15–25
Noryl®	20:1	25–40
* Land length to thickness ratio		

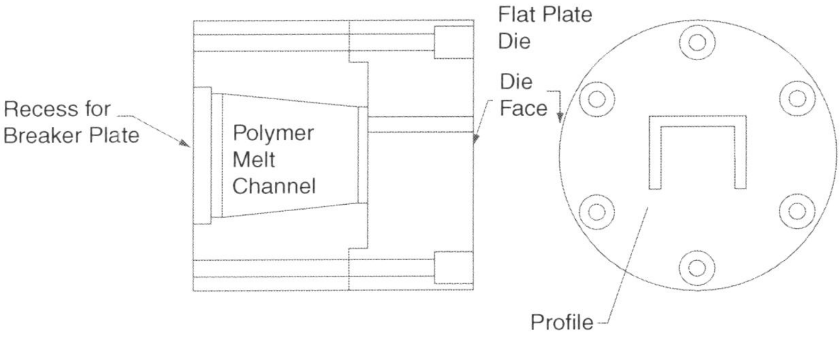

Fig. 7.20 A semi-streamlined die.

The land length is to provide sufficient land to generate 1000 to 1500 psi pressure across the land and the draw down ratio is the draw between the die and the final shape. As polymer exits the die, it is drawn by the puller and shaped by a sizing device to produce the desired shape. The draw determines the final molecular orientation in the part and the product performance. Higher draw ratios yield higher properties in the machine direction and lower properties in the transverse direction. With a specific cross section, the throughput rate is matched to one puller speed to produce the desired profile dimensions. The draw down ratio in a particular operation depends on the profile and the performance criteria.

7.10 BLOWN FILM DIES

Types of blown film dies Dies for tubular blown film can be classified as

1. Side fed dies or cross-head die
2. Centre fed dies, when the film is blown vertically upwards; this type is termed as *bottom fed*

7.10.1 Side Fed Dies

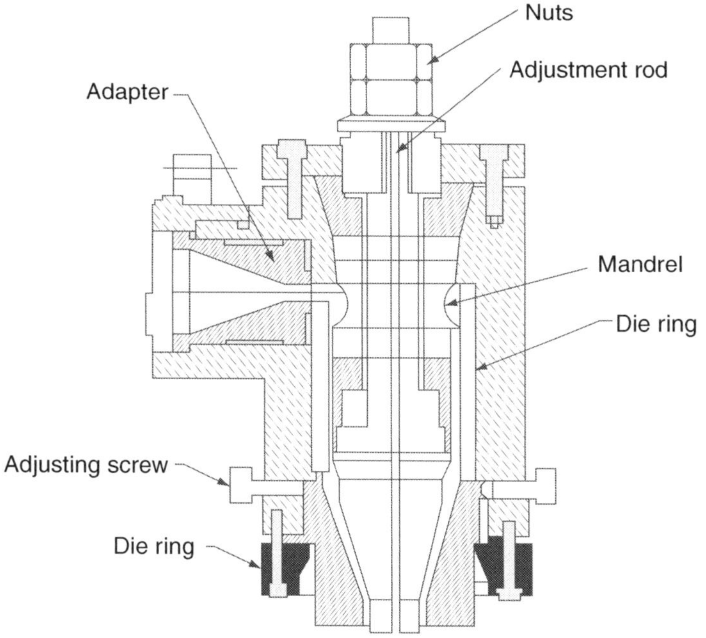

Fig. 7.21 Side fed dies.

The die body is kept vertical with the centrally located mandrel fixed to lower side. The adjustable die bush is located at the top and helps to adjust radially the die gap or orifice. The melt from the extruder flows through the adapter in a horizontal direction and turns through 90°, either in the upward or downward direction, and flows around the mandrel in the annular gap formed between the die body and the mandrel.

For large film sizes, the annular gap diameter is enlarged with a divergent taper. A land is provided at the end with radially adjustable die bush.

7.10.2 Centre Fed Dies (Bottom Fed Dies)

In this type of die the melt flow is turned through 90° before it enters the die. This is done by using an adapter located under the die. The axial flow path prior to the spider is sufficiently long. The melt then passes through the passages of the spider and further in an annular gap. The mandrel is either supported along the circumference by the spider or bolted down around the base.

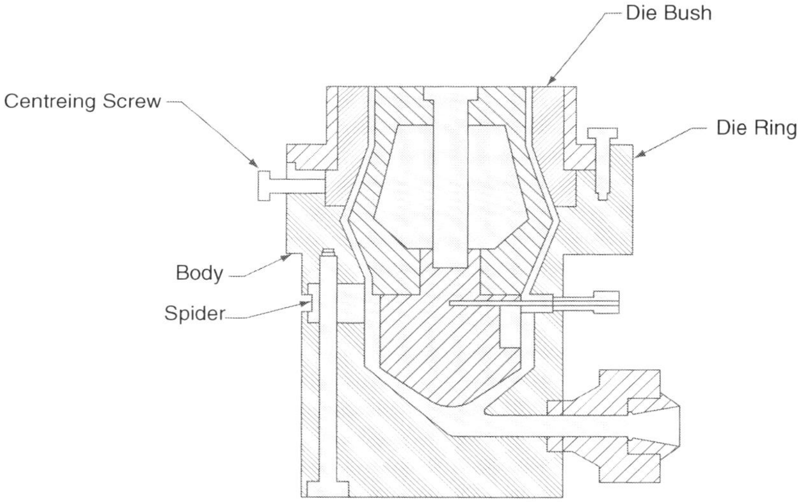

Fig. 7.22 Centre fed die.

7.10.3 Die Gap and Die Land

It is essential to obtain accurate concentricity of the die orifice or gap at the land portion. This is achieved either by a movable mandrel and/or by the die bush. In addition, a series of screw around the circumference of the bush may be positioned for adjustment effected by distorting of the die bush. The die gap is usually between 0.5–2 mm, depending on the thickness of the film required.

This can be determined by the following equation:

$$\text{Film thickness} = (\text{Die gap} \times \text{Swelling ratio}) / (\text{Blow ratio} \times \text{Draw-down ratio})$$

7.10.4 Rotating Tubular Die

This type of die is advancement of the bottom fed die. In a normal film die, localised variations in the film thickness are available, irrespective of accuracy of positioning the die ring in the centre around the mandrel. This means that any thin or thick spots are built at the same place on the finished film roll. In the rotating die the entire die body is slowly rotated around the vertical axis by an external drive. Any variation in thickness is therefore, evenly distributed. The rotation is provided through a spur gear system.

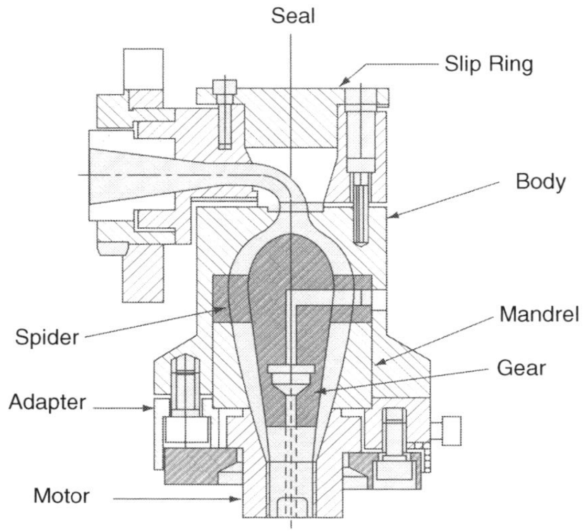

Fig. 7.23 Rotating tubular die.

7.10.5 Spiral Mandrel Die

In case of side fed or bottom fed dies, it is difficult to obtain uniformity of thickness of the extruded film. It is also difficult to recombine and knit the divided melt streams, which gives rise to weld or spider lines in the film. In addition, the flow in these dies occurs at low shear rates, resulting in poor purging performance and long melts residence times. The spiral-mandrel die is designed to eliminate these problems.

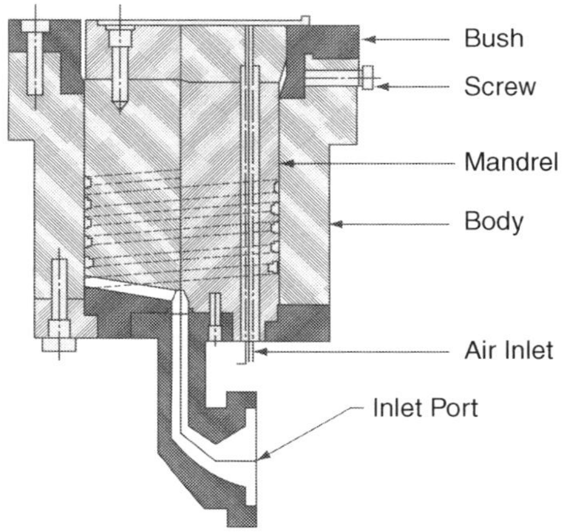

Fig. 7.24 Spiral mandrel die.

7.10.6 Parison Die

The parison dies are used to form parison in extrusion blow moulding process. The dies for parison are cross head dies and are classified as per flow of melt from extruder to the die. They are divided into two types. 1. Axial flow-head die 2. Radial flow cross-head die.

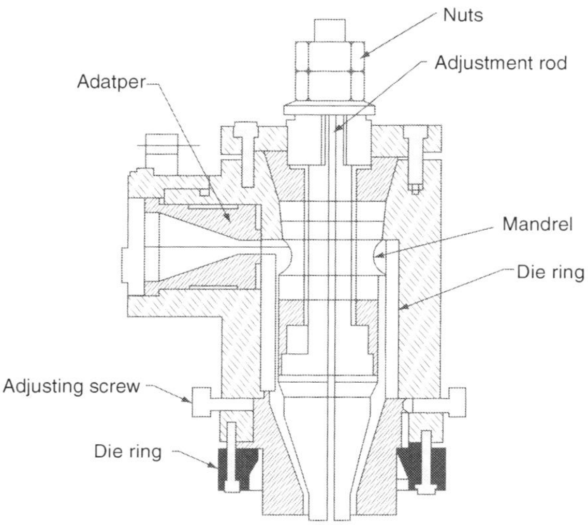

Fig. 7.25 Axial flow-head die.

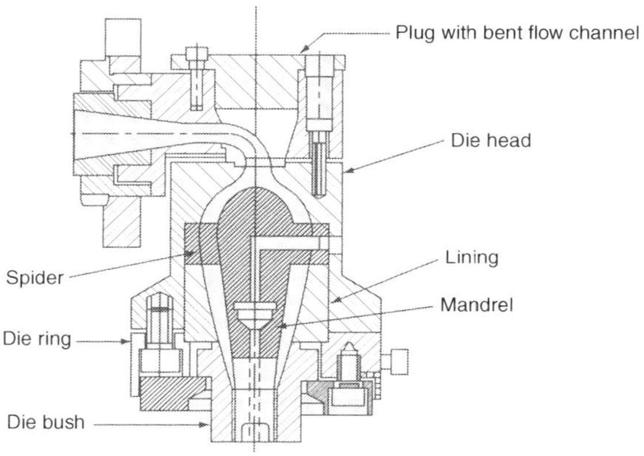

Fig. 7.26 Radial flow cross head die with through mandrel.

7.10.7 Control of Wall Thickness

Generally the wall thickness and the surface quality of the parison are influenced by the appropriate design of die and certain die components such as die body, mandrel or core, die bush, etc. When the melt enters the die, due to asymmetrical flow in the cross region, non-uniformities of stress are created. The stresses must be allowed to relax to provide a fully developed profile. Hence, a die relaxation zone is therefore, necessary. The convergent taper on the bush should be provided and the corresponding taper on the core should be greater. This will create a compression in this tapered annular gap. In some dies this taper is made divergent. These are sometimes known as expansion dies, and are used when the parison diameter has to be greater than the flow channel in the relaxation zone. The land section is either parallel, convergent or divergent. The latter two conical land sections are necessary for the regulation of wall thickness of the parison during extrusion.

Parallel land section The die length to gap ratio for parallel lands is usually between 5 to 20 mm, it depends on the blow moulding requirement. In blow moulded containers weight and wall thickness distribution can be controlled by adjusting the die gap in the tapered section. This method is particularly effective with melts having pronounced swell characteristics, and with short die land lengths.

When thin-walled parison with larger diameters are extruded, longitudinal folds may develop. These are formed with increasing parison length because swelling is obstructed below the die. This shortcoming can be overcome by changing the shape of the land.

Conical land sections The conical land sections are either convergent or divergent, and are suitable for manual gap adjustment, to regulate wall thickness and container weight. The die gap is adjusted by raising and lowering the die bush in axial flow cross heads, or by moving the core with a spindle in radial flow cross-heads with through mandrels.

1. The convergent land section should have an angle 20 to 30 degree on the core, and 30 to 50 degree on corresponding die bush portion. It is an advantage to incorporate a parallel land section, prior to the final convergent land section. This design is suitable for dies with small diameters only.

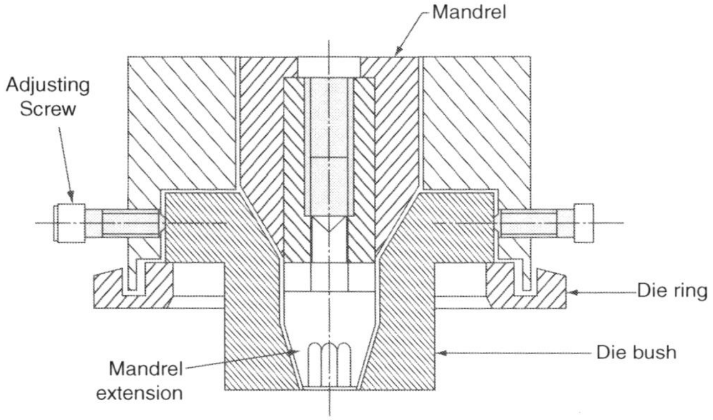

Fig. 7.27 Convergent land section.

2. The divergent land section should have an angle not exceeding 40 degree, with a parallel die gap prior to the divergent land section. This design is particularly suitable for the production of thin-walled parison, as it prevents longitudinal folds.

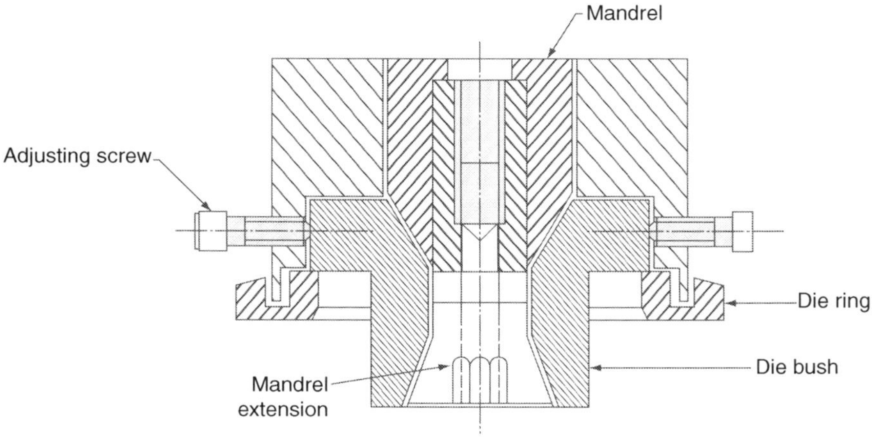

Fig. 7.28 Divergent land section.

Parison and die sizing The diameter of parison is related to the die annulus, the formula is

$D = SDm = 2L/\pi$

where D = Parison diameter ; Dm = Mean diameter

S = Die swell ratio for the polymer (see Table 7.7)

L = Pinch-off length ; $\pi - Pi$

The die gap in the land section can be obtained from the relationship between container weight and die dimensions. This is approximately given as

$t = W / C\, \pi \text{Dm LS}^2$

where t = Die gap (annulus gap) ; W = Weight of moulding including flash

D_m = Mean diameter ; L = Pinch-off length

S = Die swell ratio for the polymer (see table 7. 7) ; C = Constant (0.78)

Table 7.7 Die swell ratio for polymer.

Die swell ratio	Approximate shear rate (s^{-1})			
Polymer	10	100	400	700
Acrylic, high impact	1.17	1.27	1.35	1.42
Polyethylene low density	1.45	1.58	-	-
Polyethylene high density	1.49	1.92	2.15	2.35
Polypropylene copolymer	1.52	1.84	2.1	-
Polypropylene homopolymer	1.61	1.9	2.05	-
Polystyrene (G.P)	1.37	1.7	1.88	-
Polystyrene (toughtened)	1.22	1.4	-	-
Polyvinyl chloride (rigid)	1.35	1.5	1.52	1.53

7.11 FLAT FILM AND SHEET DIES

Dies used to make film less then 0.25 mm thickness uses flat, slit shaped dies called T-dies and it is similar to coat-hanger type dies. A variety of dies are used for extrusion of flat films and sheets are designed and fabricated. These are based on various features such as, manifold shape and layout, arrangements for control of flow and pressure, methods of die lip adjustment, configuration of the die face, etc. Based on the manifold system, dies can be classified as:

1. Manifold T- die
2. Fishtail die
3. Coat-hanger die

In the above said dies, the design features are added to facilitate extrusion of sheets, films and coated products.

7.11.1 Manifold T-Die

In manifold T-die, the melt from the extruder is fed at the centre of the die by an adapter attached to the extruder. The melt flows in a manifold of circular cross section, perpendicular to the extruder axis but in the same plane. The manifold extends over the entire die width. The melt is, therefore, distributed towards both the sides of the die. Through a slit in the manifold, the melt flows in the next portion of the die, known as damper section, which is in the form of a rectangular section, of suitable gap. The manifold can be streamlined by adopting cross sections of the shape of a tear-drop or a flat sided tear-drop. At the end of the damper section a special groove is provided extending over the entire die width. A metal bar matching the groove shape is placed in the groove. This is known as a *choker bar*. A nut is placed on the end of the stud which can be tightened to effect deflection of the choke bar, which acts like a valve with restriction in the flow passage. Choker bar acts as a restriction, so that a uniform pressure can be set up along the whole length of the die lips. The choker bar is therefore, used as a means of regulating the thickness of the sheet. The last portion of the die forms the lip section, with a gap smaller than the gap of the damper section. The lip section may be in the form of two jaws extending over the die width, the lower jaw being adjustable. The sides of the die are closed by end plates, which are easily provided with blender holes, to prevent stagnation of melt at the manifold ends. The die block is made into two halves, which are clamped by a series of cap screws, placed just behind the manifold region.

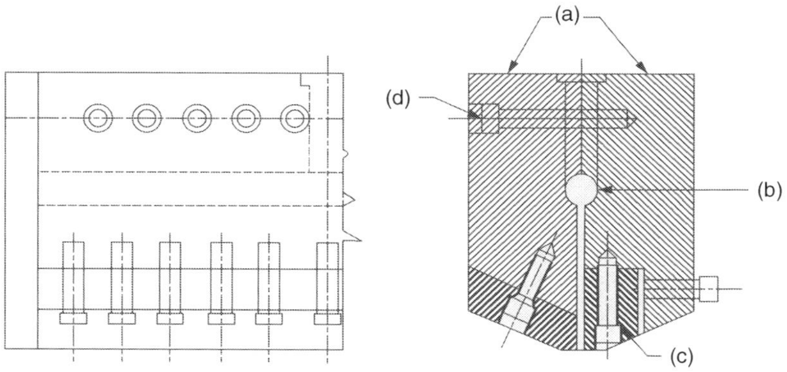

Fig. 7.29 Manifold T-die: (a) Die body, (b) Manifold, (c) Adjusting lip, (d) Clamping screw.

7.11.2 Fish tail Die

Fish tail type of die is used for extrusion of sheets of limited width or strips up to 150 mm width. It is a straight through die, in which the melt is fed from a circular adapter, also named as gate, which is attached to the extruder. The melt supplied at the centre of die flows radially over an arc subtended by an appropriate angle. The flow path cross section can be made slightly curved, with decreasing section towards the two sides. This can be considered as a fish tail section, and is of short length. The remaining length of the die consists of damper and lip sections in the form of slits, each having a different gap width. The die is formed of two plates which are held firmly by a series of bolts. The gaps of the damper section and fish tail section are made progressively greater, so that the melt flow accelerates from the die entrance to the lip section. All thermoplastic melts neck-in somewhat on exit from the die lips. Extruded sheets have therefore, to be trimmed on the width. An appropriate allowance is made in the width of the die opening. This die is more streamlined than other sheeting dies, and can be used for heat sensitive materials.

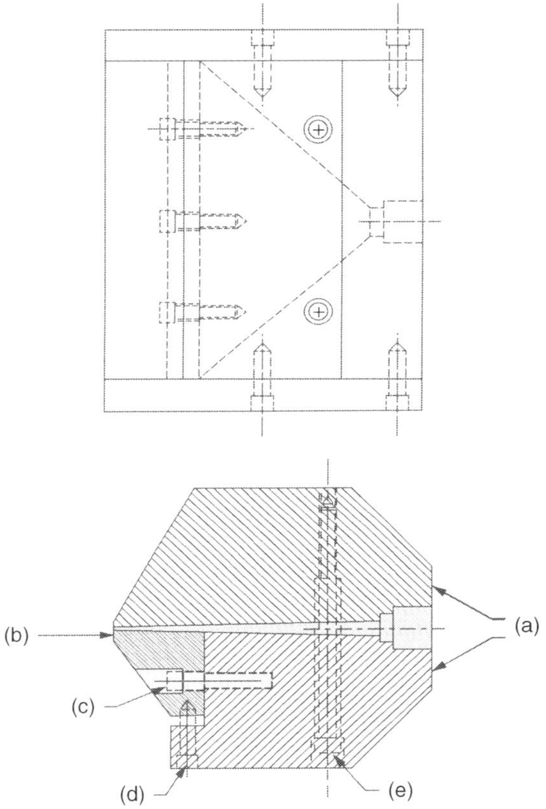

Fig. 7.30 Fish tail die: (a) Die body (b) Adjustable lip (c) Lip clamping screw (d) Lip adjusting screw (e) Clamping screw.

7.11.3 Coat-Hanger Die

The coat-hanger die is similar to the T-die, except for the cross section and layout of the manifold. In this case, the manifold, instead of being perpendicular to the flow direction of the feed, is sloping at an angle, and has, therefore-the shape of a coat-hanger. The manifold cross section in most cases gradually decreases from the central feed to the side of the die. Through a gap in the manifold, the melt flows in the coat-hanger, damper, choker bar and lip sections with acceleration due to progressively smaller gaps in each section. The gap of the choker bar at fully open position should not exceed the damper section gap. For extrusion of films, the flow path from the manifold gap to the die lips is a continuous taper and no sections or choker bar is provided. Circular manifold profiles are difficult and expensive to fabricate. It is easier to cut circular segments in each half of the die.

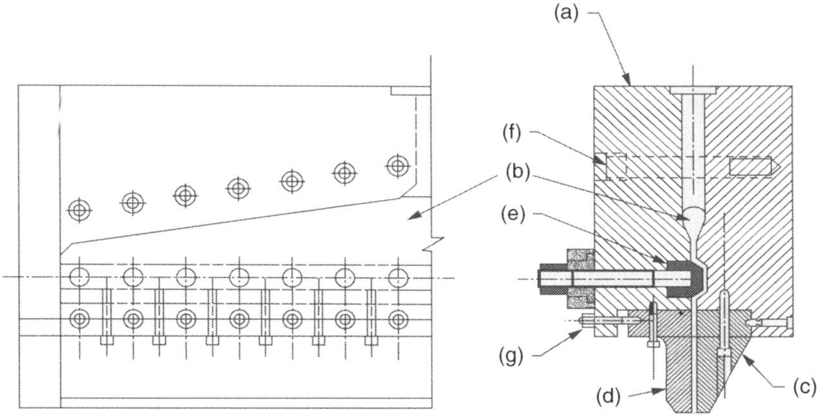

Fig. 7.31 Coat-hanger die: (a) Die body (b) Manifold (c) Fixed jaw (d) Movable jaw (e) Choker bar (f) Clamping screw (g) Jaw adjustment screw.

7.12 WIRE AND CABLE COATING DIES

The wire and cable coating dies are used for coating metallic wires or cables with a concentric layer of plastic material formed by the process of extrusion the wire or the cable is continuously fed to the die along with the plastic melt from the extruder. To facilitate continuous coating of the wire or cable the melt flow channel in the die is located at an angle of 30 °, 45 ° and 90 ° to the screw extruder axis, making it similar to a cross-head die. In order to make the design easier, construction of die head with right angle diversion of channel in the horizontal plane of the extruder axis is preferred.

Basic features This type of die is similar to a cross-head die. The melt from the extruder flows through a channel in an adapter and further around the core in the die head toward the die orifice. The core is in the form of a hollow cylinder through which the wire is fed to the die and is therefore, known as *wire or cable guide*. Two methods commonly used are known as a pressure die and tubing die.

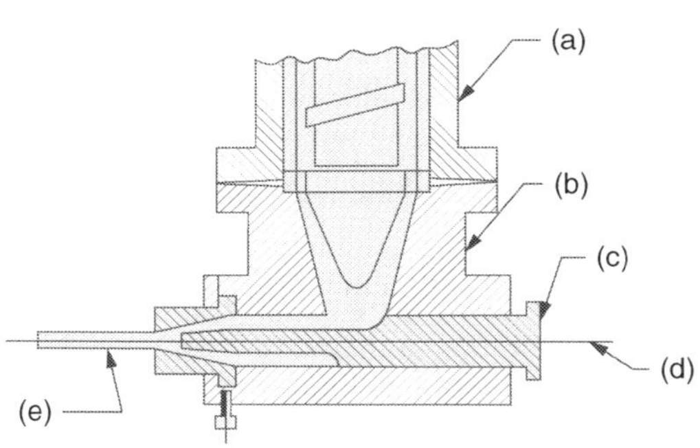

Fig. 7.32 Schematic arrangement of wire coating die: (a) Screw extruder (b) Adapter (c) Guide bush (d) Wire (e) Coated cable.

7.12.1 Pressure Die

In case of pressure die, the coating on wire takes place within the die with the melt under pressure, the melt flow surrounding the wire which leaves the die fully coated. The melt pressure produces a tight insulation which fills any space between conductors or those between insulated cores and produces a uniform circular surface.

The internal diameter of the die bush should be as close as possible to require coated wire diameter and the die land usually short which causes swelling of the extrudate. The radial clearance between the wire and the guide bush must be small, i.e., between 0.025 and 0.05 mm, to prevent the plastic melt entering the guide bush and obstructing the easy movement of wire.

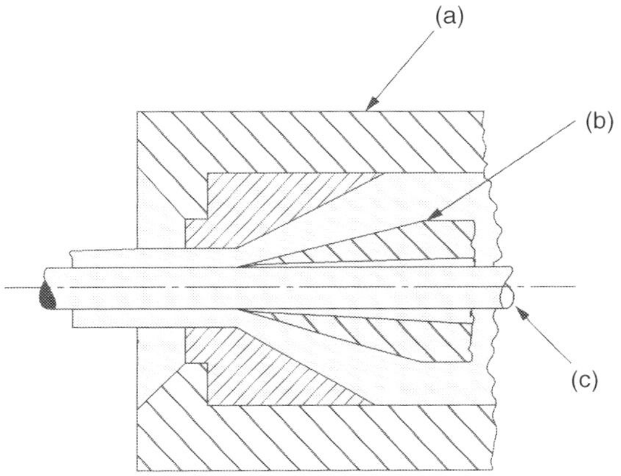

Fig. 7.33 Pressure die: (a) Die body (b) Guide bush (c) Wire.

7.12.2 Tubing Die

In case of tubing die, the thin walled tube of plastic material is extruded around the guide bush, without any contact with the wire of cable. This tube is drawn or shrunk on the wire

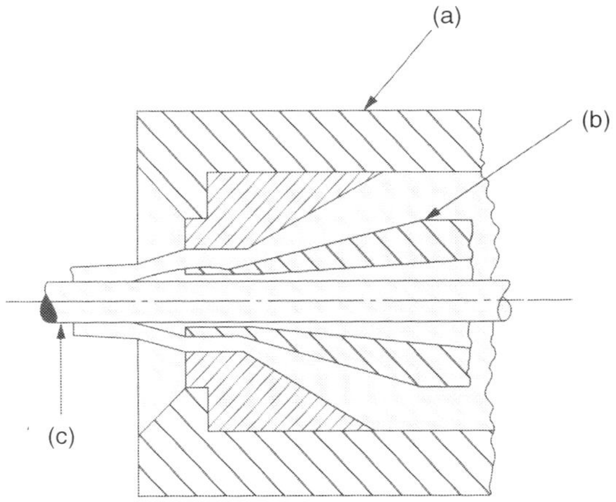

Fig. 7.34 Tubing die: (a) Die body (b) Guide bush (c) Wire.

immediately outside the die. A partial vacuum is sometimes produced in the tube to assist the shrinkage operation. Sufficient clearance must be maintained between the wire and the inside of the tube to draw vacuum. Tubing dies are commonly used for jacketing cables or coating very thin wires with viscous melt.

The ratio of area of the melt flow annulus channel forming the tube to the cross-sectional area of the final plastic coating is known as *draw down ratio*. This ratio is critical and has an effect on the surface roughness and internal stress in the coating.

Typical draw down ratios are:

Plasticised PVC and L.D.P.E	–	1. 5
HDPE	–	1. 2
Nylon	–	4. 0

To reduce the draw down to a minimum, the thickness of the guide bush should be as small as possible, consistent with its mechanical strength, so that the inside diameter of the annulus can be sufficiently small.

7.12.3 Core Deflector

The melt from the extruder flows through an adapter and enters into a channel formed between the inner and the outer hollow and cores. Since the melt tends to take the shortest path nearer the entrance side, this path is blocked by a heart shaped area, which fills the radial gap and forces the melt to follow along longer flow paths of more uniform length. This method is termed as die with core deflector.

7.12.4 Spiral Channel Die

It is a die where the melt entry is in a plane tangential to the core, so that the melt flow is not divided into two paths. The flow takes place in a spiral channel surrounding the core. The flight depth is reduced to increase the gap between the die body and the flight toward the end of the core. As the melt flows down the annular channel, the original helical flow changes gradually to an axial flow. This construction is similar to the spiral mandrel die used in film production. It is termed as tangential entry spiral channel die.

7.13 DIE MAKING

7.13.1 Steel Materials for Construction

The material commonly used for production of dies is alloy steel. Usually nickel-chromium (4–4$^1/_2$% Ni-Cr), nickel-chromium-molybdenum and high carbon high chromium steels (1.25–2.5% C, 12 to 14% Cr), H-11 steel, hot die steel and some varieties of stainless steels, with high carbon and high nickel-chromium content which have excellent corrosion resistance properties are used for manufacturing of dies. The stainless steels for dies must possess high strength, good hardenability, resistance to corrosion and must accept mirror polish. Easy machinability is a basic necessity in all die steels. Steel blocks in die making are vacuum degassed and are ultrasonically inspected to ensure that there are no pores, holes or voids. Die bodies are heat treated in order to obtain hardness between 42–48 HRC. A hard chromium plating or nickel plating of the flow path surface gives a high degree of surface finish.

7.13.2 Strength and Rigidity

As the dies are subjected to high pressures, it must be designed to withstand such high temperature and pressure. Also the variation in pressure and temperature inside the die must be taken into account. An analysis of the stresses and deformations can be made and values can be ascertained by application of strength of materials. In general, internal pressures produce either tensile stress of flexural stress. It is also necessary to determine accurately the extent of deformation or deflection of certain parts or components of the die, in particular die lips, which control the extrudate size. Similarly die walls and flanges must be strong and rigid, so that any joining made between such components will not develop any clearance, so as to avoid leakage of the melt. Screwed or welded joints must be rigid and perfectly aligned. Usually all types of dies overhang from the end of extruder barrel which causes deflection in the die components. Thick walls and sections will help to reduce deflection to a minimum values.

7.14 HEATING SYSTEM TEMPERATURE CONTROL

The extrusion dies require heating system to maintain a constant temperature of the melt during its passage through the die. This involves heating from normal temperature and subsequent heating to make up for the losses from the die surface. Heat is required to bring the dies to operating temperatures in a reasonable period, which may be about two to three hours and the losses are due to convection and radiation. Losses due to conduction can be eliminated if the joint or adopter between the extruder and die is properly heated.

Different electrical heating systems used for heating the die are detailed below:

7.14.1 Standard Voltage Resistance Heating

This method of heating is based on the main voltage electric supply and the resistance element is formed by embedding a resistance wire or tape in a refractory material enclosed in a metallic sheath. The elements can be formed to obtain suitable shapes such as strip, band, sleeve, horse-shoe, and cartridge. The type of the heaters depends on the shape and construction of the die.

Band heaters are the least expensive type and can be obtained in all shapes and sizes. It must be tight against the die surface otherwise that will burn out. Cartridge heater must also remain tight in the holes. Cast-in aluminium heaters are available in higher watt densities, but they are more expensive than band heaters. They have been used extensively for heating of large film and sheet dies, but the trend is toward the use of cartridge heaters. Since cartridge heaters are placed within the die, heat to atmosphere from the heater surface is negligible.

7.14.2 Low Voltage Resistance Heating

It requires a low voltage electric supply and the heating element made from the relatively thick wire which has a high specific resistance. The wire is formed into a shape such as a coil or a zigzag, which can be either located within the die or surrounding it.

7.14.3 Induction Heating

Induction heating is the process in which the heat generated in eddy currents induced by an alternating current which generates an alternating magnetic field around the conductor. The heating element consists of a heavy gauge copper wire wound into a former and bound with tape to form a coil of convenient size. The coil is embedded in or located around the die. When an alternating current is passed through the coil, eddy currents are included in the die, resulting in heat generation.

7.14.4 Design of Electric Heaters

The electric heaters used in dies are in the shape of a strip, clamp or cartridge; their capacities are based on watt density, which is the wattage per unit area. To assess the capacity of heaters for a die, the amount of heat for bringing the die to the operating temperature in the required time is primarily determined. Secondly, the amount of heat necessary to maintain the operating temperature is also determined. Whichever of the two is large will decide the heater size.

1. Required capacity for bringing the die up to the operating temperature (H_a):

H_A = Weight of the die × Specific heat of die material × Rise in temperature

= $W_D C_P (t_1 - t_2)$ kilocalories (Kcal)

W_D = Weight in kg

C_P = Specific heat in Kcal/°C/kg where, t_1 and t_2 in °C.

If the time required to attain the operating temperature (t_1) is T (hours), then, heater capacity (K.W) (for heating of die) = $H_A / T \times 860$

2. Heat losses (H_L): Losses due to convection and radiation. If the surface area of the die is calculated, kilowatts lost through convection and radiation can be calculated. Since the die is attached to the extruder, losses due to conduction may be neglected.

Average loss during heating up of the die = $H_L / 2$
Heater capacity (K.W.) = $(H_L / 2) / 860$

So, heater capacity for bringing the die up to the operating temperature, including losses (K.W.)

$$[A] = (H_A + H_L/2) / 860$$

$$\text{Total losses during operation (K.W.) } [B] = H_L / 860$$

The higher of the two values (*A* or *B*) is to be accepted for determining heater capacity. As a safety factor, an addition of 15 to 20 per cent may be made. Heater is made with densities ranging from 5 watts per sq. cm to 25 watts per sq. cm. For long life and best performance, heaters with lower densities are preferred.

7.15 MAINTENANCE AND CLEANING OF DIE

During the working process, the dies accumulate plastic material in certain channel, plate and flash occur. Accumulation of material can be avoided if the flow path of the melt is free and smooth. So any obstruction in the flow in the flow path should be streamlined and all corners should have a radius polyvinyl chloride component. They have a tendency to plate out, which forms deposits on the die channels. Suitable temperature at decompression zones help to reduce plate out. Flashing is the leakage of melt due to clearance between components. The die has to be cleaned often to remove such material, so as to avoid material degradation and blemishes on the extrudate surface. It should be possible to remove material deposits without dismantling the entire die. The material has to be scraped off and this should not damage the surface finish of the die channel. Cleaning should be carried out without creating any scratches on the surface.

After continuous use the die channels may wear out. The land section of the die may need adjustment or replacement after wear out. This part of the die being the most critical has to be checked for accurate dimensions and smooth surface finish. Dies should be designed with a view to facilitate easy adjustment and minimum wear.

7.16 DIE FABRICATION

Dies are fabricated from standard metals, such as rods, sheets, plate, section, billets, etc. The techniques adopted to machine the final precise dimension and to ensure the desired surface finish-such methods are planning, shaping, turning, milling, drilling, boring, reaming, broaching, grinding, honing, polishing, electroplating, coating, etc. The first group of fabrication techniques is mainly used for fabrication of these products which are available in all metals, both ferrous and non-ferrous. The die blocks cut to appropriate sizes are, therefore, used in die fabrication work. The initial machining operations of giving a suitable shape are carried on a lathe, planer or shaper. More complex shapes such as slots, recesses, grooves, special contours, etc., are carried out on milling machines. Drilling machines are used for drilling holes and tapping; for production precision cylindrical and flat surfaces grinding machines are used. In present days the dies are manufactured using advanced machining techniques using CNC machines to an accuracy of 5 microns.

Questions

1. What role does feed section play in extrusion process?
2. Describe transition section in extrusion process.
3. What is metreing section?
4. Describe the characteristics of single screw extruder.
5. Explain devolatising type of extruder.
6. Write the significance of mixing device extrusion machine.
7. Explain twin screw extruder.
8. Write a note on the following:
   a) Contra-rotating screw **b)** Corotating screw
9. Briefly explain the characteristics of twin screw extruder with output Vs head pressure graph.
10. How does a die play a vital role in extrusion process?
11. Write down the uses of following:
    a) Solid extrudate dies **b)** Hollow extrudate dies
12. How are the dies classified according to direction of flow of melt?
13. What are the types of dies according to the construction? Explain any one.
14. Briefly explain the die design procedure?
15. Explain operation of die with die and extruder characteristics curve.
16. What is the role of approach section in die? Explain with a neat sketch.
17. Explain die land design with a neat sketch.
18. Explain offset-dies with a neat sketch.
19. Describe co-extrusion in-Line pipe die with a neat sketch.
20. Draw a U profile stack die and name the various parts.
21. Write a short note on the following:
    a) Flat back plate **b)** Semistreamlined die
22. What is die gap and die land? How can we determine film thickness?
23. Explain wire and cable coating die with a neat sketch.
24. Explain pressure die with a neat sketch.
25. Describe tubing die with a neat sketch.
26. Explain core deflector.
27. Write down the different types of steel materials used for die making.
28. Write brief note on maintenance and cleaning of dies.
29. Write a short note on die fabrication.
30. Explain spiral channel die.
31. Describe the principle of extrusion process with a neat sketch.
32. Explain twin screw extrusion process.
33. Classify various types of dies and explain any two.

34. What are the types of dies according to the construction? Explain any one with a neat sketch.
35. What is die swell? Explain the factors which depend by die Swell with a neat sketch.
36. Explain rod die with its land and approach section calculation.
37. What is pipe die? Describe its function.
38. Explain in-line or straight-through dies with design calculation and sketch.
39. Classify blown film dies. Explain with a neat sketch
40. Explain rotating tubular die with a neat sketch.
41. Explain the spiral mandrel die with a neat sketch.
42. What are the various techniques to control the wall thickness die? Explain with a neat sketch.
43. Classify dies based on manifold system. Explain with a neat sketch.
44. How can we control heating system in die?
45. Explain the parison Die with a neat sketch.

References

1. Rosato D. V., *Extruding Plastics: A Practical Processing Handbook.*
2. Herold F. Giles, John R. Wagner and Eldridge M. Mount, *Extrusion: The Definitive Processing Guide and Handbook*, William Andrew Publishing, New York.
3. Joshi M. V., *Dies for Plastics Extrusion*, Macmillan India Ltd., Chennai.
4. Milivoje M Kostic, Dekalb and Louis G. Reifshneider, *Design of Extrusion Dies*, Paper submitted to Illinois State University, Illinois.
5. CIPET, *Technical Manual*, Chennai.

CHAPTER 8

CAD/CAM Applications in Mould Design

8.1 INTRODUCTION TO COMPUTERS

8.1.1 Introduction

Computer is an electronic machine that can perform arithmetic and logical calculations and data processing functions in accordance with a predetermined programme of instructions. It processes information based upon the instructions provided, and generates the desire output. Like any other system, a computer system also requires an input which is processed to get the desired output. In the case of computers, two kinds of inputs are required. One, the basic of raw data and two, a set of instructions containing the methodology to process this data. This set of instructions is called programme or software.

8.1.2 Characteristics of Computers

Every computer performs four basic functions. They are input, storage, processing and output of data. The power and usefulness of the computer are due to its following characteristics:

1. Computers are automatic machines, because once started on a job, they carry on, until the job is finished, normally without any human assistance.
2. A computer is a very fast device. It can perform in a few seconds, the amount of work that a human being can do in an entire year.
3. Computers are very accurate.
4. A computer can work for any longer time without getting tired like human beings.
5. Computers can carry out different functions or different operations simultaneously.
6. Any amount of data can be stored in a computer and these may be retrieved at any time we require.
7. A computer works according to the instructions given by the user and it can neither take decision on its own nor work by itself.
8. If an error occurs in the working of computer it is due to the faulty operation of a human being only.

9. There is a possibility of storing more data in a computer by using secondary storage devices such as pen drives, disk drives, etc. Also data can be easily transferred from one computer to other computers.

8.1.3 Types of Computers

Computers can be classified into five major types. They are:

1. Micro computers (portable computers)
2. Mini computers (personal computers)
3. Workstations and servers
4. Mainframe computers
5. Super computers.

Fig. 8.1 Palmtop.

Fig. 8.2 Laptop.

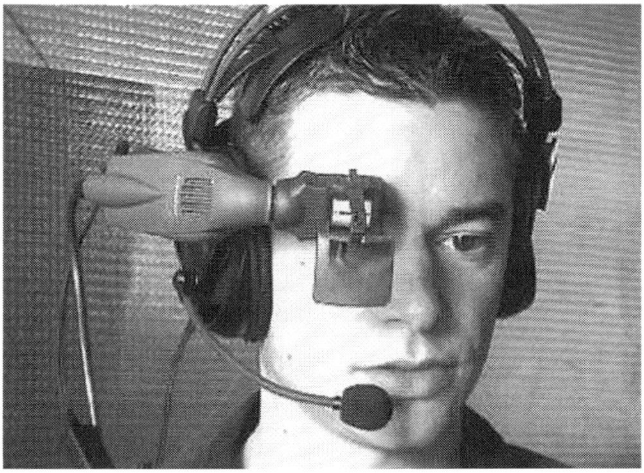

Fig. 8.3 Wearable computer.

1. Micro computers: (Portable computers): Portable computers, such as palmtop, wearable computers, and laptops come under this category. These computers have lower capabilities than that of desktop computers. These can be used for performing simple tasks such as word processing, spreadsheet calculations, presentations, simple multimedia tasks such as image and audio editing, browsing and playing small games, etc. These computers come under personal computers of portable type.

(a) Palmtop: With miniaturisation and high density packing of transistors on a chip, computers with capabilities nearly that of desktops which can be held in a palm, have emerged. These are called palmtops. It is also known as Personal Digital Assistant (PDA). Palmtop accepts hand written inputs using an electronic pen which can be used to write on a palm screen. It also has facilities to use it as mobile phone, fax and e-mail machine. At present these computers have 1 to 8 GB of storage capacities; 512 MB of memory; screen size varies from 3 to 7 inches. Operating system such as windows mobile is used in palmtops. Some of the company's manufacturing palmtops in today's market are Blackberry, Samsung, HTC, Acer, LG, Sony Ericsson, Nokia, etc.

Simputer is a mobile hand held computer with input through icons on a touch sensitive overlay on the LCD panel. It also has same facilities like palmtops. But, unlike palmtop, simputer uses free open source OS, hence the cost of a simputer is less than that of a palmtop. Simputer is an Indian initiative to meet the needs of rural population of developing countries.

(b) Laptops: Laptops are designed to run with batteries for few hours and are thus, designed to conserve energy by using power efficient chips. Many laptops can be connected to a network through wireless connectivity to read files from large stationary computers. Laptop uses miniature components which consume low power and have to be packaged in small volume. So they cost 2 to 3 times the cost of same capacity desktops. Laptops weigh around 1 to 5 kgs. It may be either AC powered, or battery powered or both. At present major laptop manufacturers in the Indian market are IBM, HP, Dell, HCL, Acer, Zenith, Wipro, Sony, etc. Laptops can be further classified into two categories are detailed below:

(i) Netbooks: They consume less power than that of desktop computers. But they are smaller in size. They are little more expensive than desktops. Netbooks are used for basic and common processing tasks such as word processing, spreadsheet calculations, presentations, audio and image editing, playing songs and CD quality movies, small games and browsing, etc. These netbooks are more compact and weigh less than 'notebooks'. The processing speed varies from 1 GHz to 1.5 GHz; screen size varies from 7 inches to 10 inches; memory capacity varies from 1 GB to 2 GB; and storage capacities vary from 160 GB to 320 GB. Intel atom processor based laptops are good examples for this category. Normally these laptops use Windows XP or UNIX or Linux operating systems. Recent netbooks have built in graphics processor, and main processor in the motherboard to reduce the space, power consumption, and to enhance the multimedia and graphics capabilities.

(ii) Notebooks: They are more powerful than netbooks and have emerged as desktop replacements. They are costlier than netbooks. They look similar to netbooks but are slightly bigger in size. Its processing speed varies from 2 GHz to 2.5 GHz; memory capacity varies from 2 GB to 4 GB; storage capacity varies from 250 GB to 500 GB; screen size varies from 10 inches to 17 inches. These notebooks can be used in moderate to high performance applications such as multimedia, home entertainment, running CAD/CAM/CAE software applications, etc., in addition to the common basic processing tasks. Recently Intel has launched core i3 processor for these laptops, which use Microsoft Windows 7 as OS.

(c) Wearable Computers: The size of the wearable computer is very small so that it can be worn on the body itself. It requires smaller processing power. It finds its application in the field of medicine. For example, a pacemaker is used to correct the heart beats; insulin metre is used to find the levels of insulin in the blood.

2. Mini Computers: (Personal Computers) Unlike laptops these computers are designed to use in a place and use standard size components hence the power consumption is more when compared to laptops. The size of the computer and its performance are more than that of laptops. This falls into two categories. They are:

Fig. 8.4 Net tops.

Fig. 8.5 Desktops.

(a) Nettops: A new category of desktop that came into the market recently is nettops. It has the same performance with very compact size CPU. Connections to the input and output devices are done through USB ports. Miniature storage and cabinets are used in this type of computer for compact CPU size. These net tops can do all normal processing such as browsing, word processing, spreadsheet calculations, presentations, gaming, etc. It can also be used as a multimedia PC. It can even support playing full HD quality videos because of the built-in graphics. Memory varies from 1 to 4 GB. Processor with built-in graphics is integrated in the main board to achieve compactness. The storage varies from 250 GB to 500 GB. The power consumption is less compared to desktops.

(b) Desktops: It is a normal personal computer used at home. It can perform the same normal processing tasks of nettops and notebooks with little high performance. The size of the desktop is bigger than nettops and the power consumption also is more. It can also perform higher multimedia tasks, and high end CAD/CAM/CAE software applications. The power and performance of the desktops now equal the workstations. Desktops range from 2 to 3 GHz processing speed with 1 to 6 GB of memory and 250 GB to 1 TB of storage capacity. Screen size varies from 15 inches to 24 inches. Intel's recent core i5 processor based systems are good example for this category.

3. Workstations and Servers: Workstations are also desktop computers, but it is for engineering applications such as CAD/CAM/CAE, multimedia software usage, and other types of applications that require moderate to high amount of computing power and relatively high quality graphics capabilities. Workstations generally come with a large, high resolution graphics screen, built-in network support and graphical user interface. The most common operating systems for workstations are UNIX, Windows server, etc. Workstations provide 2 to 5 times of processing speed than that of PCs, with high memory ranges from 4 GB to 16 GB, and the data are stored in a network with several times higher capacities normally in TB. Workstations are used for executing numeric and graphics intensive applications such as those which arise in computer aided design, simulation of complex systems and visualising the simulation results. Some manufacturers of workstations are, Silicon Graphics, IBM, Sun Micro systems, HP, HCL, etc.

Servers are high performance workstations intended to control the workstations under their network, to distribute the licenses among workstations, to manage the database, high performance numerical computing (called compute server), Web hosting, network printing, etc. Computer servers have high performance processors with large main memory; database servers have big on-line disk storage (terra bytes); and print servers support several high speed printers.

4. Mainframe Computers: Organisations such as banks, insurance companies, airlines and railways process large number of transactions on-line. They require computer with very large amount of disks to store several terra bytes of data and transfer data from disk to main memory at several hundred megabytes per second. The processing power needed from such computers is hundred million transactions per second. These computers are much bigger and faster than workstations and several times more expensive. They normally use proprietary operating systems which usually provide extensive services such as user accounting, file security and control. They are normally much more reliable when compared to operating systems on PCs. These types of computers are called *mainframe computers*. The most popular mainframe computer manufacturer is IBM.

5. Super Computers: Super computers are designed for ultra high performance tasks such as weather analysis, designing supersonic aircrafts, modelling complex molecules, study on the effect of collision of planets on space, very complex engineering analysis calculations, etc. Super computers are large, very expensive, usually owned by government agencies and large corporations. These super computers can carry out several trillion floating point operations per second. Super computers are also used to analyse large commercial databases, to produce animated movies, etc. Super computers also have parallel processing capabilities, (i.e., various workstations in the network can run different analysis by using processing capabilities of super computer at the same time). Examples for super computers manufacturers are IBM and Silicon Graphics.

Fig. 8.6 Mainframe computer.

Fig. 8.7 Super computer.

8.1.4 Merits and Demerits of Computer

Computer technology has revolutionised the way things were done before. They have entered our lives and have made things easier for us. It is one of the best inventions of this world. They have helped simplify complex task and have reduced the human efforts and work load. They have numerous benefits; however, they have certain drawbacks as well depending on how they are used.

Following are the merits of computer:

1. They have become a source of income for numerous individuals and companies.
2. It has made processing of difficult task easier.
3. It has made huge database management easier.
4. It saves time and money.
5. It has helped in convergence of several technologies like audio, video, mobile, etc.
6. Communication has been made easier by use of computers.
7. Source of information.
8. It helps in instant communication across different parts of the world.
9. It reduces or eliminates redundant activities.
10. It reduces the work load in offices and it has eliminated paper work.

Following are the demerits of computer:

1. It affects the concept of society in the world.
2. It isolates people from one another.
3. Some people waste a lot of time in meaningless activities.
4. It promotes unhealthy activities in youth.
5. It affects the health of the individuals.

8.1.5 Programmable Controllers

Programmable controller as defined by the National Electrical Manufacturers 'is a digitally operated electronic apparatus which uses a programmable memory for the internal storage of instructions for implementing specific functions such as logic sequencing, timing, counting, and arithmetic to control through digital or analog input / output modules, for various types of machines or processes.' Programmable controllers are used in many industrial applications including transfer machines, flow line conveyor systems, injection moulding, grinding, welding, cement processing, food processing, energy management, testing equipment, etc.

Basic components of programmable controller (PC):

The basic components of PC are:

1. Input/Output Interface: The PC is designed to be connected to industrial equipments. This connection is accomplished by means of the input/output interface. The input interface receives process and machine signals and converts them into usable form for the PC. The output interface converts the PC control signals into usable form for process equipment.

2. Processor: The processor, also called Central Processing Unit (CPU) is the central component of the PC. It performs arithmetic and logic operations on inputs, and determines the appropriate outputs.

3. Memory: The memory is used to store the programme. Its capacity ranges from 1K to more than 48 K.

4. Programming Device: The programme is loaded into the PC memory by a programming device. Two types of programming devices used are CRT terminal and keyboard device.

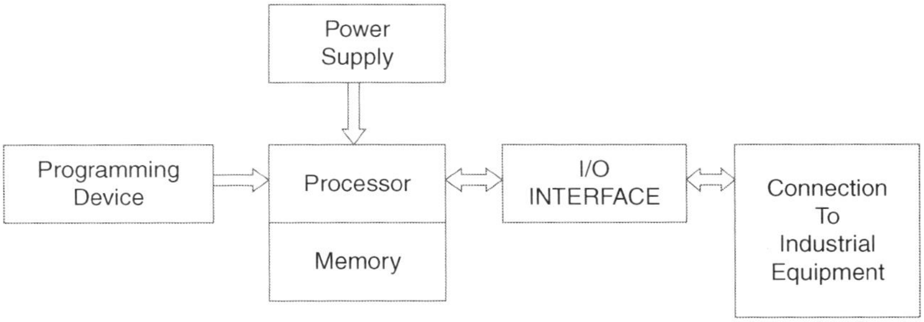

Fig. 8.8 Programmable controller.

5. Power Supply: The power supply runs the PC and provides a source power for the output signals.

Programming the programmable controller (PC): Ease of programming is one of the attractive features of a PC. Various PC manufacturers come out with different language formats but there are three basic types of PC programming languages. They are:

1. Relay Ladder Diagrams: It is currently the most popular type, as electricians, control engineers, and maintenance personnel prefer them.

2. Boolean Based Languages: It makes use of logic statements to establish relationships among PC inputs and outputs.

3. Mnemonics Languages: It is similar to computer assembly languages.

Functions of programmable controllers:

1. Control relay functions involve generation of output signal from the inputs contained in the PC memory.

2. Timing functions are used to generate an output signal of a specified delay time after an input signal has been received.

3. Counting functions in which the counter adds up the number of input contacts and generate a programmed output when the sum reaches a particular count.

4. Arithmetic functions are used in arithmetic and logic operations.

5. Analog control functions are used to accomplish proportional, integral, and derivative control functions.

Advantages of programmable controllers:

(a) Programming PC is much easier than the conventional.

(b) PC can be reprogrammed.

(c) It occupies less floor space.

(d) Maintenance is easier

(e) Better reliability

(f) It can be interfaced with plant computer systems more easily.

8.2 COMPUTER FUNDAMENTALS

There are two components in computer system, namely:

1. Computer hardware
2. Computer software

1. Computer Hardware: Computer hardware is termed as devices or parts of computer to perform a specific task.

2. Computer software: Computer software is a set of programmes used to manage the operations and perform the specific task of a computer.

8.2.1 Computer Hardware

A computer is designed using four basic units. They are:

1. Input unit
2. Central Processing Unit (CPU)
   a. Control unit
   b. Arithmetic Logic Unit (ALU)
   c. Registers
3. Memory unit
4. Output unit

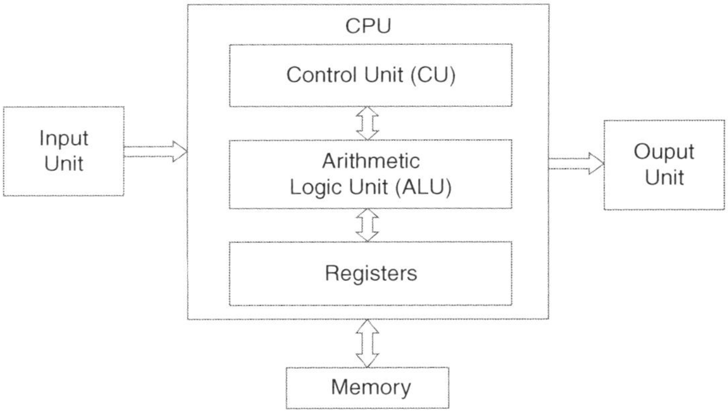

Fig. 8.9 Block diagram of computer.

1. Input Unit: Input unit performs the operation of receiving data and instructions from external world to the computer system. The input unit consists of one or more input devices such as keyboard and mouse.

2. Central Processing Unit: Central processing unit (CPU) is the main part of the computer system. It regulates the operation of all system components and performs the arithmetic and logical operations on the data. CPU consists of three operating units. They are:

(a) Control unit
(b) Arithmetic logic unit (ALU)
(c) Registers

(a) Control Unit: The control unit coordinates the various operations specified by the programme of instructions. These operations include receiving data and deciding how and when the data should be processed. The control unit directs the operation of the arithmetic logic unit (ALU). It sends data to the ALU and tells it what functions are to be performed on the data and where to store the results. It controls the input and output of information between the computer and the outside world through I/O section.

(b) Arithmetic Logic Unit (ALU): The arithmetic and logic unit performs operations such as addition, subtraction, multiplication, division, and comparisons according to the programmed instructions. These operations are carried out on data in binary form.

(c) Registers: Registers are small memory devices that can receive, hold, and transfer data. Both control unit and arithmetic logic unit perform their functions by utilising registers. Each register consists of binary cells to hold bits of data. The number of bits in the register establishes the word length that the computer is capable of handling. The number of bits per word can vary from 4 to 64.

Functions of registers:

(i) Programme Counter: The programme counter holds the location or address of the next instruction. An instruction word contains an operator and an operand. The operator defines the type of operation to be carried out and the operand specifies the data on which the operation is to be performed.

(ii) Memory Address Register: The location of data contained in the computer's memory unit is identified for an instruction by the memory address register. This unit is used to hold the address of data held in memory.

(iii) Instruction Register: The instruction register is used to hold the instruction for decoding. (Decoding refers to the interpretation of the coded instruction word so that the desired operation is carried out by the CPU.)

(iv) Accumulator: An accumulator is a temporary storage register which is used during an arithmetic or logic operation.

(v) Status Register: Status registers indicate the internal condition of the CPU.

(vi) Arithmetic Logic Unit: Provides the circuitry required to perform the various calculations and manipulations of data.

The digital information from one unit to another is carried by multiple wires called **buses**. There are three principal buses in a computer:

1. Address bus
2. Data bus
3. Control bus

3. Memory Unit: The data and the instructions required for processing have to be stored in the memory unit before the actual processing starts. The results generated from processing are also to be preserved before it is displayed. The memory unit thus, provides space to store input data, intermediate results and the final output generated. Additional memories are used to store huge information for future use.

4. Output Unit: It is used to print or display the results, which are stored in the memory unit. Output unit links the computer to the outside world. It consists of monitors, printers, speakers, etc.

8.2.2 Computer Software

Computer software is divided into two types. They are:

1. Operating system
2. Application software

1. Operating System: Operating system is a group of programmes which manage the operations of the computer. These programmes control the flow of information among the many units of the computer, namely monitor, keyboard, memory disk drives, printer, plotter, etc. The important functions of an operating system are:

(a) Transferring data between computer and peripheral devices for input and output.

(b) Managing computer files, directories and folders, etc.

(c) Loading computer programs into memory and controlling programs execution.

(d) The commands of operating system can be used to set the date and time, display the content of a file, display the contents of a directory, check disk status, clear display screen, setting resolution of monitor, etc.

Examples for operating systems are, Windows (98, 2000, NT, XP, Vista and 7), UNIX, Linux, MS DOS, etc.

2. Application Software: Application software is a set of programs used to perform a specific task. It is written in high level languages by specialist companies. These packages may be specifically designed for one type of computer, and ideally will be compatible for a range of different hardware suppliers. Most CAD application softwares are written in compiled languages such as FORTRAN, PASCAL, C, C++, etc., eg., AutoCAD, ProEngineer, Unigraphics, CATIA, MasterCAM, Moldflow, etc.

8.2.3 Integrated Circuits

In electronics, an **integrated circuit** is a miniaturised electronic circuit that has been manufactured in the surface of a thin substrate of semiconductor material. They are used in almost all electronic equipment in use today and have revolutionised the world of electronics. Computers, cellular phones, and other digital appliances are now inextricable parts of the structure of modern societies, made possible by the low cost production of integrated circuits. Integrated circuits were made possible by experimental discoveries which showed that semiconductor devices could perform the functions of vacuum tubes. The integration of large numbers of tiny transistors into a small chip was an enormous improvement over the manual assembly of circuits using electronic components. The integrated circuits mass production capability, reliability, and building-block approach to circuit design ensured the rapid adoption of standardised ICs in place of designs using discrete transistors.

There are two main advantages of ICs over discrete circuits: cost and performance. Cost is low because the chips, with all their components, are printed as a unit by photolithography rather than being constructed one transistor at a time. Furthermore, much less material is used to construct a packaged IC die than a discrete circuit. Performance is high since the components switch quickly and consumes little power compared to their discrete counterparts, because the components are small and positioned close together.

The first integrated circuits contained only a few transistors, called '**Small-Scale Integration**' (**SSI**), digital circuits containing transistors numbering in the tens provided a few logic gates.

The next step in the development of integrated circuits, which contained hundreds of transistors on each chip, called '**Medium-Scale Integration**' (**MSI**). They were attractive economically because while they cost little more to produce than SSI devices, they allowed more complex systems to be produced using smaller circuit boards, less assembly work because of fewer separate components, etc.

Further development, driven by the same economic factors, led to '**Large-Scale Integration**'(**LSI**), with tens of thousands of transistors per chip. Integrated circuits such as 1K-bit RAMs, calculator chips, and the first microprocessors, that began to be manufactured in moderate quantities in the early 1970s, had under 4000 transistors.

The final step in the development process, starting in the 1980s and continuing through the present, was '**Very Large-Scale Integration**' **(VLSI).** The development started with hundreds of thousands of transistors in the early 1980s, and continues beyond several billion transistors as of 2009.

8.2.4 Microprocessor

A microprocessor is a computer electronic component made from miniaturised transistors and other circuit elements on a single semiconductor integrated circuit. CPU, ALU, and control functions are combined in microprocessor.

The first microprocessors emerged in the early 1970s and were used for electronic calculators, using Binary-Coded Decimal (BCD) arithmetic on 4-bit words. The integration of a whole CPU onto a single chip greatly reduced the cost of processing power. From these humble beginnings, continued increases in microprocessor capacity have rendered other forms of computers almost completely obsolete. With one or more microprocessors used in everything from the smallest embedded systems and handheld devices to the largest mainframes and supercomputers.

Fig. 8.10 Intel 4004 - The first commercial general-purpose microprocessor.

8.2.5 Printed Circuit Board

A **Printed Circuit Board, (PCB)**, is used to mechanically support and electrically connect electronic components using conductive pathways, tracks or signal traces etched from copper sheets laminated onto a non-conductive substrate. It is also referred to as **Printed Wiring Board (PWB)** or **etched wiring board**. A PCB populated with electronic components is a **Printed Circuit Assembly (PCA)**, also known as a **Printed Circuit Board Assembly (PCBA).**

PCBs are inexpensive, and can be highly reliable. They require much more layout effort and higher initial cost than either wire wrap or point-to-point construction, but are much cheaper and faster for high-volume production. Much of the electronics industry's PCB design, assembly, and quality control needs are set by standards that are published by the **Institute for Printed Circuits (IPC)** organisation.

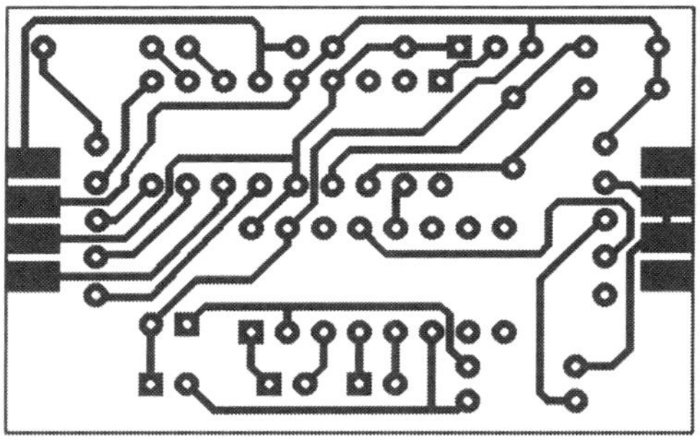

Fig. 8.11 Printed circuit board conductive pathways.

Fig. 8.12 Printed circuit board without components.

8.3 MEMORY

Memory in a computer is a device in which string of symbols (binary codes) representing either an instruction or a number can be stored. The individual symbol of a binary number is known as **bits** (unit of memory). Strings of bits make up a word. Word length or size of the word is the number of bits present in the word. The memory section consists of binary storage units, which are organised into bytes, usually having 8 bits per byte. Words are 4 to 64 bits long. The following table illustrates the details of the memory:

Table 8.1 Memory details.

Name	Shorthand	Value
Byte	1 B	8 bits
Kilobyte	1 KB	1024 B
Megabyte	1 MB	1024 KB

(Contd.)

Table 8.1 *(Contd.)*

Name	Shorthand	Value
Gigabyte	1 GB	1024 MB
Terabyte	1 TB	1024 GB
Petabyte	1 PB	1024 TB

The time required to find the correct address and fetch the contents of that memory location is called the **access time**. It determines the speed of the computer. It ranges from 100 ns to several microseconds.

8.3.1 Types of Memory

Computer memory can be divided into three categories:

1. Main memory (primary storage),
2. Auxiliary memory (secondary storage),
3. Cache memory

1. Main Memory (Primary Storage): The main memory (primary memory storage) refers to storage areas that are physically a part of the computer and connected directly to the CPU. It consists of the working registers and memory devices closely configured to the CPU. CPU communicates directly only with main memory. Main memory is faster, compact and consumes less power. The size of the main memory is much smaller than that of the secondary memory because of its high cost. The primary memory is a static device as there are no rotating parts in it.

Functions of the Primary Memory

(a) To contain a copy of the main software programme (operating system). This programme is loaded into the primary memory when the computer is turned on.

(b) Temporarily stores a copy of the application programme that is currently being executed.

(c) Temporarily stores the data input from the keyboard, which is required for processing.

(d) Temporarily stores the result, which is generated from processing until it is transferred to an output device.

Main memory or primary storage can be divided into two main categories:

(a) Read Only Memory (ROM): This is permanently engraved on a chip and can only be read on the computer. This consists of only a few kilobytes, enough to check the functioning of all hardware components and read the operating system, etc. This type of storage is non-volatile, which means that it can retain its data when power is interrupted. Read only memory is always there whenever the computer is switched on.

(b) Random Access Memory (RAM): RAM is accessible to users. The users can write information into RAM and read it. It has random access property, i.e., the access time for each memory location is same. The amount of RAM is important because it determines what software can be run on the machine. Some programme requires only a few kilobytes of RAM; the more sophisticated integrated packages take up much more (about several megabytes). RAM is volatile. It is erased when computer is switched off.

Fig. 8.13 ROM.

Fig. 8.14 RAM.

Types of computer storage technology used for main data storage are:

(a) Magnetic Core Storage: Where each data bits are represented by the magnetic state of ferromagnetic doughnut. This type of storage is non-volatile, which means that the data stored in it is not lost even when the power supply goes off.

(b) Semiconductor Storage: It consists of memory cells made up of transistor circuits. This type of storage is volatile, which means that the data is erased when the computer is turned off.

(c) Semiconductor Monolithic Storage: It is also called large scale integrated (LSI) memory circuitry which contains the equivalent of thousands of microminiaturised transistor memory cells. LSI memory uses less power than conventional transistor memory. It is volatile. It has slower rate of data transfer than semiconductor storage.

2. Auxiliary Memory (Secondary Storage): The programmes and data files are not directly available to the CPU, but through auxiliary devices that constitute a secondary storage and are physically external to computer. It is less expensive than main memory. It has large capacity of data storage. It is slow when compared to primary memory. It can be loaded into main memory as and when required.

There are two basic types of secondary storage:

(a) Sequential Access Storage: To read one particular data in the file, all records preceding it must also be read. It has lower access rate than direct access storage. It is suitable for applications that do not require a high level of file activity.

(b) Direct Access Storage: In this storage method, individual records can be located and read immediately without reading any other records. It has higher access rate. Cost per bit of data stored is higher when compared to sequential access method. It is more complicated and best suited to files where a high level of activity is involved, e.g., airline/railway reservation system.

3. Cache Memory: The cache memory is placed in between the CPU and the main memory. It is much faster than main memory. Its access time is much less compared to that of the main memory. The cache memory is an intermediate memory and is not accessible to users. It stores instructions and data, which are to be immediately executed. It is used to reduce the access time, thus, increases the operating speed of the system. It is more economical than use of fast memory devices to implement the entire main memory. But it is costlier than main memory.

8.3.2 Hardware Devices used in Computer for Storage Technology

1. Magnetic Tape Storage: It is a good example of sequential access storage. The data stored on magnetically coated mylar tape, which is similar to the magnetic tape used in the audio system. Recording and reading data are non-destructive, i.e., the tape can be erased and re-used. Access time is slow due to sequential storage. Low cost per bit and high capacity of magnetic tape make it ideal for system back up.

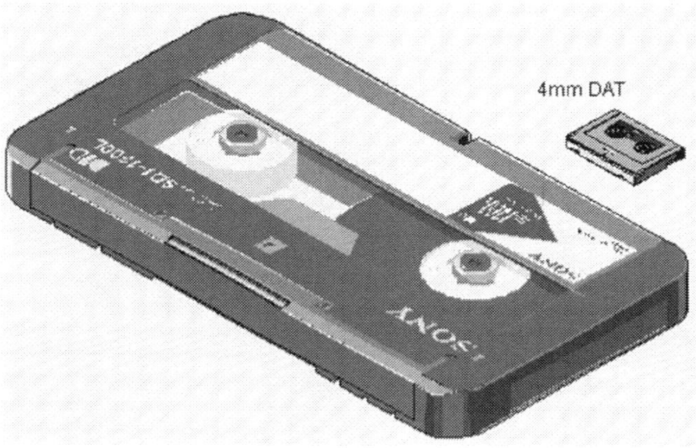

Fig. 8.15 Magnetic tape.

Fig. 8.16 Magnetic drum.

2. Magnetic Drum Storage: It is a random access storage device with high capacity and high access rates. It consists of a magnetically coated cylinder. During operation the drum is rotated at a constant speed and data are recorded in the form of magnetised spots. The drum can be read repeatedly without causing data loss. Read / write heads are used to read data to and from the drum as it rotates. The drum surface is divided into tracks, each with its own read / write head.

3. **Magnetic Disk Storage:** It is a direct access storage device in which the storage medium is a magnetically coated disk. Several types and sizes of disks are available to suit particular sets of applications.

Fig. 8.17 Floppy disk.

Fig. 8.18 Hard disk.

(a) The Flexible Diskette (Floppy Disk): 3.5, 5.25, 8 inch of several standard sizes and its packaged in a square plastic envelope to protect the magnetic surfaces. Reading and writing are accomplished through access holes in the envelope. Floppy disks are available with either one or both sides used for storing data. They can store 1.44 MB of data. This storage technology has become obsolete due to less storage capacity.

(b) The Hard Disk: It is a thin metal disk, which is coated on both sides with magnetic ferric oxide. Data are recorded in the form of magnetised spots on tracks on the disk surface. The hard disk contains several rotating disk plates which are encased in a hot disk drive. The disk rotation ranges from 5400 RPM to 10000 RPM. Data are transferred by moving a set of read / write heads (one per recording surface) to the appropriate track. Only one of the heads is used to transfer data at a time. The access time is dependant on the rotational speed of the disks and the capacity of the head to read from the disk surface. This rate is usually several megabytes per second. Recent hard disks are able to store from 80 GB to 2 TB of data. In five years from now, hard disks with Heat-Assisted Magnetic Recording (HAMR) technology will be common in the market, which is capable of storing data up to 30 TB.

4. Bubble Memory: It consists of microscopic magnetic bubbles on a thin crystalline magnetic film. It is impressive for its high storage density and random address capabilities.

Fig. 8.19 Bubble memory.

Fig. 8.20 Compact disk.

5. Optical Storage: Optical storage is large capacity data storage medium for computers on which information is stored at extremely high density in the form of tiny pits. The presence or absence of pits is read by a tightly focused laser beam. Optical disk consists of a rotating disk which is coated with a reflective material, e.g., CD, DVD, HD DVD, blue ray disk, etc. In the near future a new disk called Holographic Versatile Disk (HVD), will be in the market, which can store about 100 times more data compared to blue ray disk. All the existing optical storage such as CD, DVD, HD DVD, will become obsolete due to low storage capacity for future Ultra High Definition Video (UHDV) contents. This UHDV / super Hi vision has 16 times higher resolution than the current full HD videos (1920 × 1080). The following table explains various optical storage media and their storage capacity:

Table 8.2 Optical storage capacity details.

S. No.	Optical storage medium	Data storage capacity
1.	Compact Disk (CD)	700 MB.
2.	Digital Versatile/Video Disk (DVD)	4.7 GB – Single layer 8.5 GB – Dual layer
3.	High Definition/Density Digital Versatile Disk (HD DVD)	15 GB – Single layer 30 GB – Dual layer
4.	Blue ray Disk (BD)	25 GB – Single layer 50 GB – Dual layer

6. Electron Beam Accessible Memory Systems: EBAM relies on information storage in Metal Oxide Semiconductor (MOS). LSI memory chips are placed inside and near the face of a cathode ray tube. An electron beam is used to read and write on each chip. EBAM storage has advantages of fast access time, high data transfer rates and long-term data integrity.

8.4 COMMUNICATION DEVICES

The devices which allow the operator to create or modify an image on the screen or to enter alphanumeric data into the system through various programmed input functions are called **input devices**. The devices which allow the operator to bring the image or drawing from inside the computer to the outside world are called **output devices**.

8.4.1 Input Devices

The input devices can be divided into the following general categories. They are:

1. Cursor control devices
2. Digitisers
3. Alphanumeric and other keyboard terminals
4. Automated entry or scanners
5. Voice and video data entry

Cursor control devices and digitisers are both used for graphical interaction with the system. Keyboard terminals are used as input devices for commands and numerical data. There are two basic types of graphical interaction accomplished by means of cursor control and digitising. 1. Drawing and positioning new items on the screen. 2. Pointing at or identifying locations on the screen, usually associated with existing images.

1. Cursor Control Devices: The cursor appears in the form of a bright spot on the CRT screen that indicates where lettering or drawing will occur. The computer reads the position of the cursor. Hence the user is able to control the cursor position and we can also enter the locational data into the CAD system database, e.g., identifying a starting point of a line by locating cursor, select an item from menu of functions displayed on the screen. Some of the cursor control devices include:

(a) Thumb wheels
(b) Direction keys on a keyboard terminal
(c) Joysticks
(d) Tracker ball
(e) Light pen
(f) Electronic tablet / pen
(g) Mouse
(h) Touch screen

The first four items in the list provide control over the cursor without any direct physical contact of the screen by the user. Light pen, electronic tablet and touch screen require the user to control the cursor by touching the screen or some other flat surface which is related to the screen with a pen type device. Thumb wheels, direction keys, joysticks, and tracker balls are generally limited to cursor control. The light pen and tablet / pen are typically used for other input functions as well as cursor control. Some of the functions are:

(i) Selecting from a function menu

(ii) Drawing on the screen

(iii) Selecting a portion of the screen for enlargement of an existing image.

(a) Thumb wheel devices make use of two thumb wheels, one to control the horizontal position of the cursor, and the other to control the vertical position.

(b) Direction keys on the keyboard are used to control the cursor without graphics capabilities. There are four keys each used for the four directions in which the cursor can be moved.

(c) Joystick consists of a box with a vertical toggle stick that can be pushed in any direction which causes the cursor to move in that direction.

(d) Tracker ball is similar to that of the joystick except that an operator controlled ball is rotated to move the cursor in the desired direction on the screen.

(e) Light pen is a pointing device used to identify the cursor position where the light pen is in contact with the screen. It does not project light but detects it on the screen by using light sensor.

(f) Tablet and pen is an electronically sensitive tablet used with an electronic stylus. The tablet is a flat surface separate from the screen on which the user draws with the pen like stylus to input instructions or to control the cursor.

(g) Mouse operation initially appears to be similar to that of the command tablet / pen arrangement but the main difference is that the mouse is mounted on a roller ball or wheels. It does not require an electronic tablet surface. It may be moved on any flat surface to achieve the desired cursor location. The mouse is the most suited device for CAD systems. It is used not only for cursor control but also to select an item from the menu display on the screen.

Fig. 8.21 Thumb wheel.

Fig. 8.22 Direction keys.

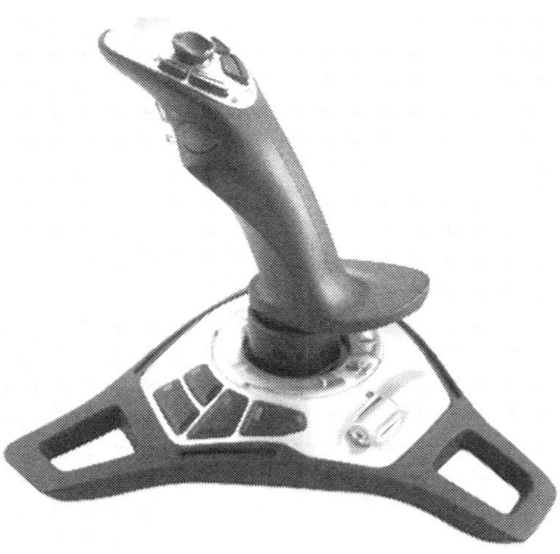

Fig. 8.23 Joy stick.

Fig. 8.24 Tracker ball.

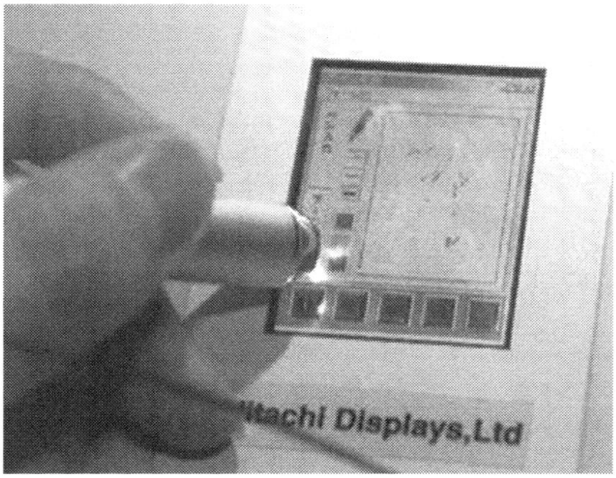

Fig. 8.25 Light pen.

Fig. 8.26 Tablet pen.

(h) Touch screens are used by simply touching the display with one's finger or a pointing device. Two types of touch screens are available in CAD systems – mechanical and optical. Mechanical type is a transparent screen overlay which detects the location of the touch. Optical touch screen systems use rows of light emitters and receptors mounted just in front of the screen with the touched location determined by broken beams. Mechanical systems have a position detection resolution of 0.25 mm and hence they have emerged as the most popular cursor control system in the recent period. Optical systems have low resolution and are mainly used for menu selection.

Fig. 8.27 Mouse.

Fig. 8.28 Touch screen.

2. Digitisers: The digitiser is an operator input device consisting of a large, smooth board like drawing board and an electronic tracking device which can be moved over the surface to follow existing lines. It is a common technique in CAD system for taking x, y, coordinates from a drawing. The electronic tracking device has a switch for the user to record the required x and y coordinate positions. The coordinates can then be entered into computer memory by a storage medium such as magnetic tape. The digitiser can be used to digitise line drawings.

3. Keyboard Terminals: Keyboard terminals are the most important input device used to enter alphanumeric data such as letters, numbers, symbols, etc. The keyboard contains a

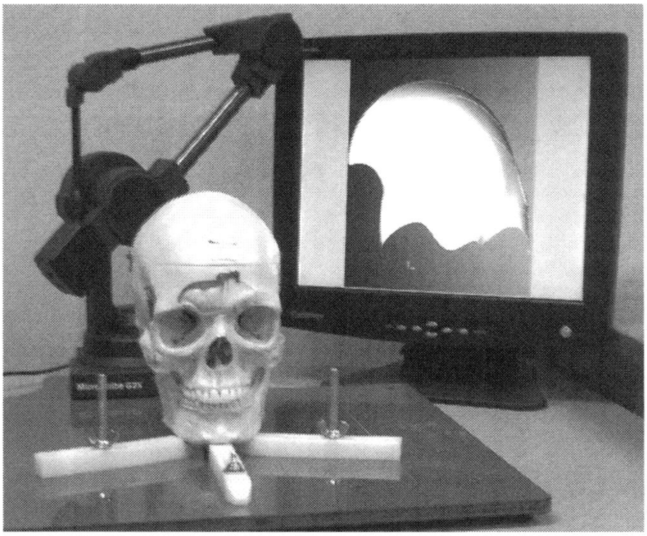

Fig. 8.29 Digitiser.

Fig. 8.30 Keyboard.

keyboard controller to check if any key is pressed or released. It has limited diagnostic and error checking capabilities. A buffer is normally available to store a certain number of key actions if the computer is busy. Some of the keyboard functions include:

(a) Selecting drawing command from screen menus.

(b) Entering drawing element size, components, symbols, etc.

(c) Entering text to the screen.

(d) Selecting CAD software and programming new software.

(e) To enter commands, functions, and supplemental data to the CAD system.

This information is displayed on CRT screen for verification. Some CAD systems make use of special function keyboards. These function keyboards are provided to eliminate extensive typing of commands, or calculate coordinate position, and other functions.

4. Scanners: A typical automated drawing entry device contains a scanner and a workstation for viewing and editing the drawing. The major requirement of design is the necessity to convert existing paper drawings to computer files. These units can scan a drawing and convert it into a form useful for CAD with no manual intervention. Software is required to control the scanned image. The pixel data may be stored and manipulated with the aid of sophisticated software.

5. Voice Data Entry: Voice data entry increases operator productivity in selecting menu items. The most important feature of voice data entry is that it gives the designer more mobility due to unrestricted hand and eye use. This is done with the help of voice recorders such as microphone.

6. Video Entry: This is the combination of both voice data entry as well as continuous image entry calculated in terms of frames per second (fps). Operators in complex manufacturing plants are trained with the working principles of machines in live video classes to absorb the techniques quickly. The devices used for video entry are digital cameras, Web cameras, camcoders or video cameras, etc.

Fig. 8.31 Scanner.

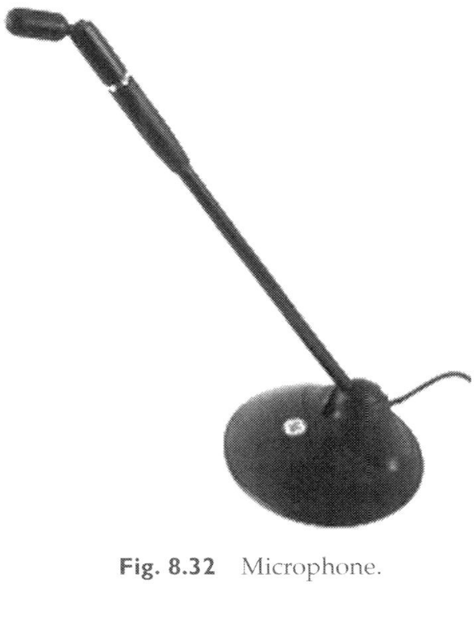

Fig. 8.32 Microphone.

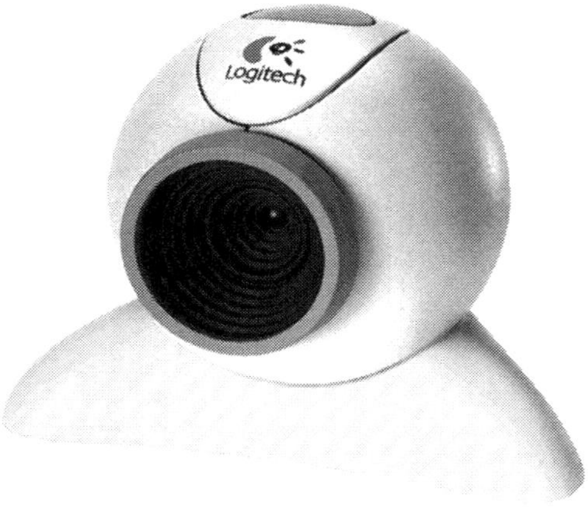

Fig. 8.33 Web camera.

Fig. 8.34 Camcoder.

8.4.2 Output Devices

The output devices can be divided into the following categories. They are:

1. Monitors
2. Projectors
3. Speakers
4. Printers
5. Plotters

Fig. 8.35 CRT monitor.

Fig. 8.36 LCD monitor.

1. Monitors: Monitors come under Visual Display Unit (VDU). It gives the video output to the user based on the input given by the user, so that the user can check the input data as well as the output data. Different types of monitors are available based on different computer graphics terminal, e.g., CRT monitors, flat screen monitors, plasma screen monitors, Liquid Crystal Display (LCD) monitors and LED monitors.

2. Projectors: Projectors also come under visual display unit. They also act like a monitor except that monitor can be restricted to a particular screen size, whereas the projectors can be zoomed to a large size of screen with the help of powerful focal lenses and display adjustments. It is mainly used in presentations in conference halls, film in theatres, etc., which require larger display.

3. Speakers: Speakers are voice output devices, based on the voice input data from microphone or voice recorders so that the user can check the input data.

Fig. 8.37 Projector.

Fig. 8.38 Speaker.

4. Printers: The workstations are equipped with a printer to get a hard copy as a permanent record of alpha-numeric commands and drawings on paper. A printer with graphics screen dump is sometimes referred to as a mini plotter. Five basic types of printers are available in CAD system, which are:

(a) Impact Printers: They use small hammers or print heads containing small pins to strike a ribbon to form dot matrix images. Colours are introduced through the use of multiple ribbons or single ribbons with different colour bands. Colour intensity is fixed. Copy quality is poor because of low resolution. Impact printers are suitable for works involving high speed, low cost, high volume hard copies and printing of text materials like bill of material, e.g., dot matrix printer.

(b) Electrostatic Printers: Electrostatic printers use electrodes that make a programmed set of point charges on to dielectric paper, which is then passed through a toner applicator. This is attracted to and fused into the charged areas thus, producing an image. Electrostatic plotters can produce black and white copies producing half tones. It is also possible to make colour copies. Electrostatic plotters are costly demanding frequent maintenance. Operating cost is high. They give fast output with high resolution.

(c) Inkjet Printers: Inkjet printer forms an image by spraying ink from a matrix of tiny jets on to the medium to be printed. Some of the droplets from the matrix get charged and are returned to the reservoir, while uncharged droplets are attached to the printing surface to form graphics.
(d) Thermal Printers: It is a non-impact printer that forms an image by moving heated stylus / nib over a specially treated paper. For colour production, the paper is passed through the thermal head three times, one for each primary colour (red, blue, green). Thermal printers are fast and noiseless but their resolution is low. The cost of consumables is high and therefore, the operating cost is high.
(e) Laser Printers: It is a high resolution printer that uses a version of the electrostatic reproduction technology of copying machines, to fuse text and graphic images to paper. It combines high speed with high resolution and the quality of output is very good. Laser printers are mostly preferred for faster and good quality printing.

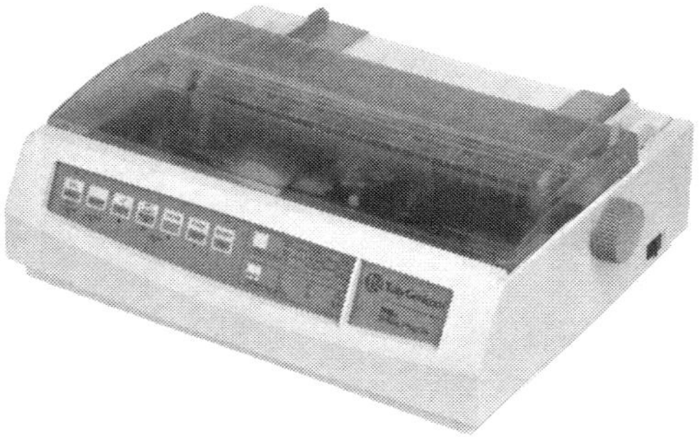

Fig. 8.39 Dot matrix printer.

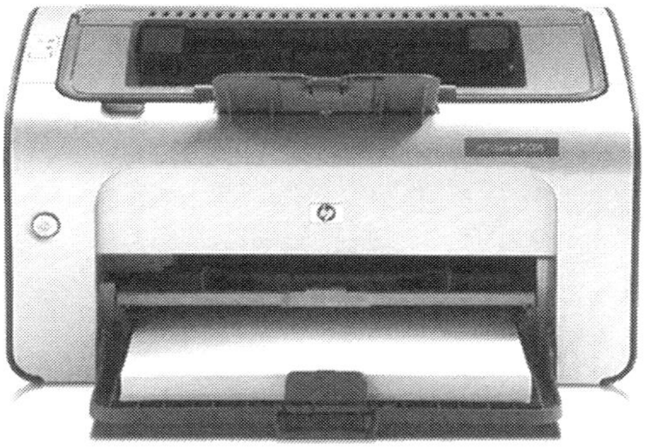

Fig. 8.40 Laser printer.

5. Plotters: A plotter is a printer that produces high quality graphical output by moving ink pens over the surface of the paper. The printer moves the pens under the instruction of the computer so that the printing is automatic. Plotters are commonly used for computer aided design (CAD) and presentation graphics. Common types of plotters used in CAD systems are of four types. They are:

(a) Vertical Drum Type Plotter: It has a roll of paper continuously wound and unwound on a rotating drum, in sequence with the moving pen. The paper may be a single sheet or continuous one. The vertical plotter is faster and cheaper. This may be used for lengthy production runs. They are commonly designed for A0 size papers.

(b) Flat Bed Plotter: This plots drawings or designs on a sheet of paper spread and fixed over a rectangular flat bed table. In this plotter the paper does not move but the pen moves to draw complex designs and drawings under computer control. Size ranges commonly from A3 to A0. The flat bed is more accurate than the vertical type and is thus, useful for applications such as template drawings. Also, flat bed plotters can accommodate sheets of inflexible materials such as plastic or steel.

(c) Electrostatic Plotter: It is a device which creates drawings from horizontal sweeps as a dot matrix printer does. It has a continuous roll of a special paper which passes across a writing head containing thousands of tiny electrodes arranged in the form of a thin stripe. When passed through a bath of toner a drawing is formed from a vast pattern of tiny dots created by the electrodes. Its main advantage is its ability to produce drawings unattended. Another advantage is its 'artistic' effects of extensive colour and variation of tone on hard copy. Like the drum plotter this also supports continuous printing on a role of paper. It may be utilised as a high speed line printer capable of printing 1200 lines of text per minute. Hence, they are being adopted in CAD system with complicated 3D modelling techniques. Its disadvantages include high cost and slight inaccuracy of line resolution.

(d) Inkjet Plotter: It can create hard copy images from numerous tiny dots. In the inkjet plotter, the writing head consists of a carriage containing three jets, which spill a controlled volume of ink at regular time intervals. The carriage slowly travels across a continuously rotating drum to which the paper is attached. Each jet confirms to a primary colour and can create a vast range of coloured images. It also can perform all the artistic effects like the electrostatic plotter.

Fig. 8.41 Inkjet plotter.

Fig. 8.42 Flatbed plotter.

Generally the printers and plotters are rated based on their print quality and print speed. Print quality is measured in terms of number of dots present in a linear inch. It is commonly known as **dpi** (dots per inch). Print speed is measured in terms of number of characters that can be printed in a second. It is commonly known as **cps** (characters per second). Print quality can also be measured in terms of number of pages that can be printed in a minute. It is commonly known as **ppm** (pages per minute).

8.4.3 Common Peripheral Devices Used for Computer Input / Output

1. Card Readers: The card reader transfers data from the punched card to the computer systems. The brush reader and the photoelectric reader are the two types of card readers currently in use.

2. Card Punches: A card punch transfers the output from the computer to the punched cards. The cards are reread to verify correct punching. Card punching speeds range from 100 to 300 cards per minute. Card readers and punches are often combined into a single unit.

3. Magnetic Tape Units: Magnetic tape units are used for programme and data storage, and they can be interfaced to the computer as both input and output units. The magnetic tape is used as a medium. During operation, the tape is passed through a read / write head, usually at a constant speed (25 to 200 in/s).

4. Punched Tape Readers: A punched tape reader reads data from punched holes on a strip of paper tape (punched tape) having 5 to 8 channels. As the tape is moved through the reading head, the presence or absence of holes is sensed. Paper tape data entry is usually slower than magnetic tape. In numerical control programming punched tapes and punched tape readers are mostly used.

Fig. 8.43 Punch card reader from IBM.

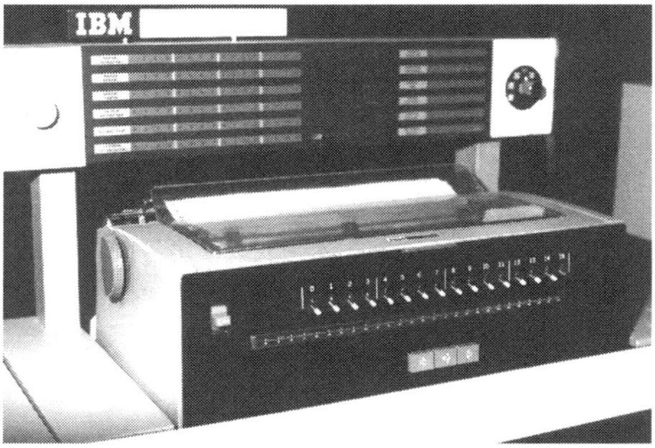

Fig. 8.44 Punch tape reader from IBM.

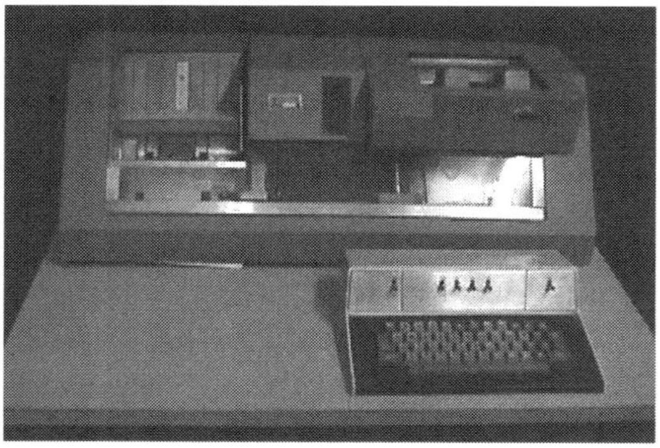

Fig. 8.45 Card punch from IBM.

Fig. 8.46 Magnetic tape unit from IBM.

5. Paper Tape Punches: Data from a computer system can be outputted onto a punched paper tape. Paper tape readers and punches are often combined into a single unit.

6. Keyboard Input Devices: Keyboard input devices are a typewriter like keyboard which can be used by a typist. Some of these devices feed the data and programmes directly to the computer. Others produce data on a special medium for subsequent input to the computer system.

7. **The Key Punch:** It is an electromechanical key board device which converts operator key strokes into machine readable holes on cards. The cards are then submitted through a card reader through the computer.

8. **Key to Tape Unit:** It is an electronic type writer device that converts operator key strokes into machine readable codes on a magnetic tape.

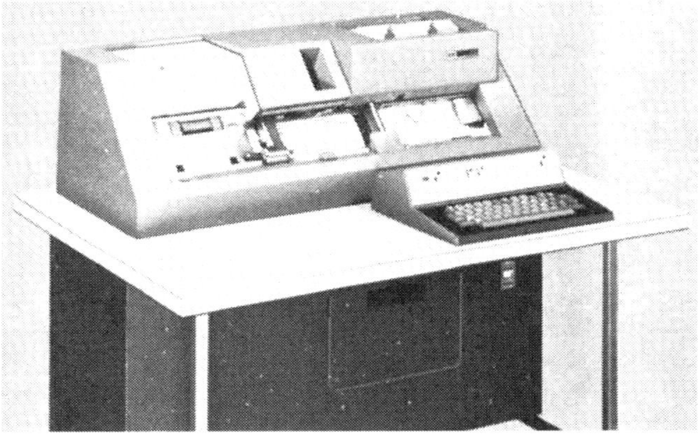

Fig. 8.47 Key punch from IBM.

Fig. 8.48 Teleprinter from Puma.

9. **Alphanumeric Displays:** This display consists of a typewriter like keyboard and a display screen (CRT), which can be used to display data. CRT terminals include screen, keyboard, communication interface, buffer memory, and sometimes a local microprocessor used for editing. It can be connected directly to a computer for online operation or it can be used with independent devices for offline operation. Transmission speeds are usually selectable from 110 to 9600 baud. A **baud** is a unit representing the number of discrete signal changes per second. For a binary system it is equal to the number of bits per second. Communication line quality

limits the speed of data transmission. Programming and data input on a CRT terminal are faster than for other keyboard entry devices.

10. Teleprinter: A teleprinter consists of an electromechanical or electronic typewriter keyboard and a hard copy printing device. It can function both as a remote data entry terminal and as an online output terminal. During input, data are usually transmitted, character by character as keys are pressed, although some units have buffer memory available to permit batched continuous transmission. Older electromechanical units print data at a rate between 110 and 300 baud. The newer electronic units operate at speeds of up to 9600 baud.

11. Magnetic Ink Character Recognition (MICR): These are electronic devices that operate by interpreting the sensed wave forms of the individual magnetic ink characters. This technology is used in the banking industry to facilitate mass handling of checks and deposit slips. These devices are capable of reading up to 1600 documents per minute.

12. Optical Character Recognition (OCR): This device is mainly used to recognise alphabetic and numeric characters printed on a paper. In this a mechanical drum is used to rotate documents past an optical scanning station. A light source and lens system can distinguish the patterns of the characters. These patterns are converted into electrical pulses, which are interpreted as individual characters. OCR is used in credit card billing, reading of pin code number, etc. It can be programmed to read a variety of character sets and even hand writing. Entire pages can be scanned and read very quickly. Data transfer occurs at a maximum rate of 3600 baud.

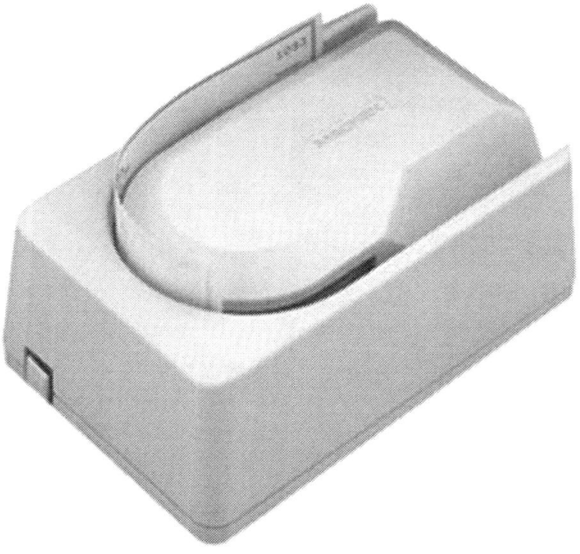

Fig. 8.49 MICR cheque reader.

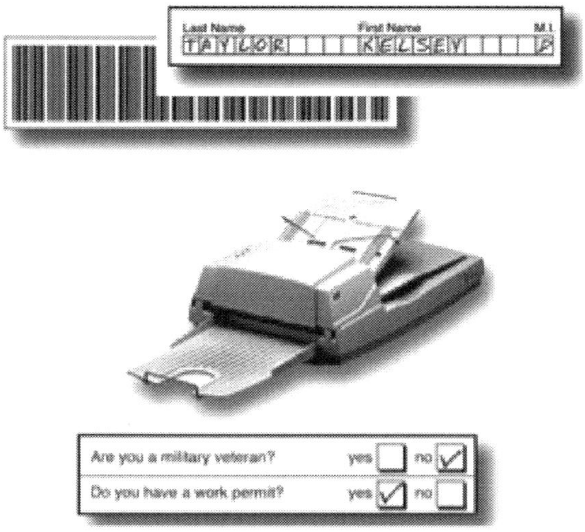

Fig. 8.50 OCR document reader.

13. Optical Mark Readers (OMR): This is mainly used in objective type tests. OMR recognises certain prespecified position of marks made by pencil or pen. The position is correlated to a previously defined character. The marks can be read from 80-column cards or full page documents. Data transmission speed depends on the speed of the feeding device. It is able to verify at the rate of 1500 forms per minute.

14. Optical Barcode Readers (OBR): This senses the configuration of shaded bars of a different width and correlates them to previously defined characters. An OBR can read characters at 50 to 400 per second. OBR is used widely in shopping malls for billing commodities.

15. Line Printers: The line printers print entire line at one cycle (80 to 132 characters per line) at rates that may exceed 1000 lines per minute. These units are expensive. One of the recent innovations in high speed printers combines laser and xerography technologies to achieve print speeds of about 10000 lines per minute.

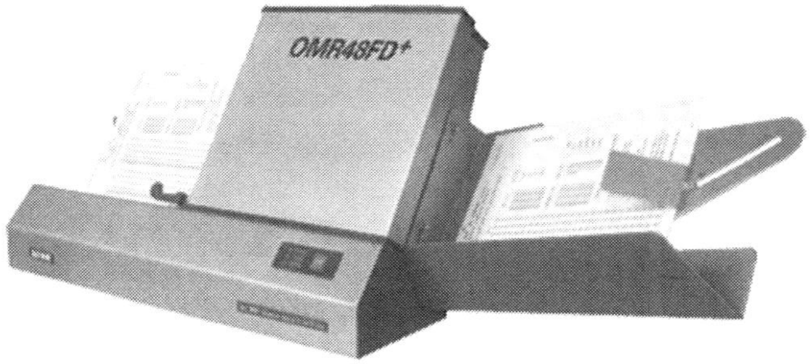

Fig. 8.51 OMR sheet scanner.

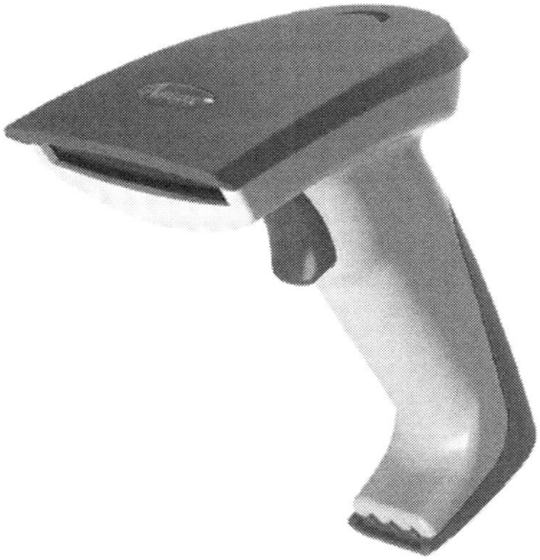

Fig. 8.52 Optical barcode reader.

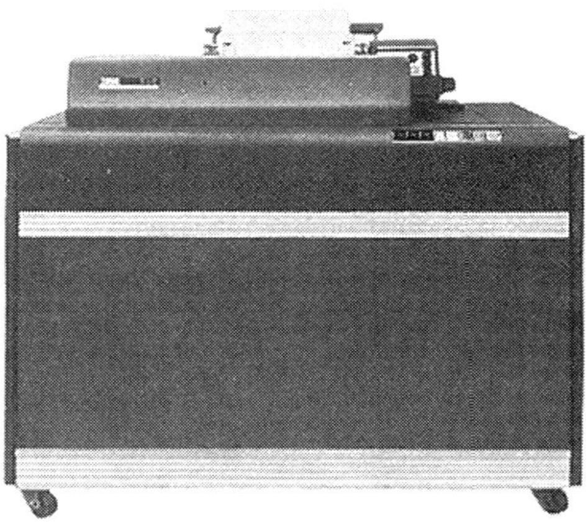

Fig. 8.53 Line printer from IBM.

8.5 PRINCIPLES OF PROGRAMMING

Language is defined as a means of communication. If we have to communicate with a computer we require a language which is called a programming language. The programme tells the computer what the user expects it to do for him.

8.5.1 Types of Programming Language

Computer programming language can be divided into three types:

1. Machine language
2. Assembly language
3. Procedure oriented (high level) language

1. Machine Language: The binary coded instructions that computers can understand are called machine language. Each machine language instruction contains an operation code and an operand. Operand might be a memory address or a device address or data. In machine language, programme storage locations are designated for the programme and data. These are used throughout the programme to refer to specific data or programme steps. The programmer must be familiar with the specific computer systems. It is machine dependant and is difficult to programme. Since machine language instructions are different for each computer, programming in machine language is tiresome, complicated, and time consuming.

2. Assembly Language: To reduce difficulties in writing programmes in binary coded instructions, symbolic languages have been developed, which substitute English like mnemonics for each binary instruction. Mnemonics are easier to remember. It helps speed up the programming process. A language consisting of mnemonic instructions is called an assembly language. Assembly languages are considered to be low level languages. They are the most efficient in terms of fast execution on the computer. A programmer has to write programmes for different applications for different computers. Assembly language programmes are be converted into machine language so that the computer can understand and execute the programme. The conversion is carried out by a programme called an assembler. An **assembler** is a programme which converts assembly language programmes into machine language programmes in such a way that a computer can understand.

3. High Level Language: Assembly languages are machine oriented whereas high level languages are procedure oriented. They are independent of the computer on which they are used. This means that a programme written on one computer can be run on a different computer without significant changes in the programme. The advantage of high level language is that it is not necessary for the programmer to be familiar with machine language. The programme is written as an algorithm to solve a problem. High level languages are also to be converted into machine languages. This is done by a special programme called a translator. **Translators** convert high level languages into binary messages which may be understood by the hardware. Translators may be of either interpreter type or compiler type. **Interpreters** translate high level languages into machine language in single statements (line by line execution) which is time consuming. One such interpreter language is BASIC. Interpreters are less expensive and more time consuming. **Compiler** translates the high level programme into machine language as a whole programme. Compilers are more expensive, but give a much faster computer response. They quickly translate the entire programme into binary language, e.g., for high level compiler languages are FORTRAN, COBOL, Pascal and C.

Some of the high level languages are:

1. FORTRAN (Formula Translation): It is one of the oldest and most popular of the high level languages. It is a compiler type high level language. It is quite efficient for mathematical computations but is not very efficient for applications such as file processing or document production.

2. COBOL (Common Business Oriented Language): It is a major computer language for business data processing applications. It is also a high level language of compiler type. COBOL is suitable for file environment but it is not capable for mathematical calculations.

3. BASIC (Beginners All Purpose Symbolic Instruction Code): It was developed in 1960s as an interactive language, where the user appears to be interacting directly with the computer. It is a high level language of interpreter type. BASIC is heavily used in the field of microcomputer and as an interactive teaching tool in schools. BASIC is inexpensive and versatile but slow computer response restricts it to the simpler types of programmes.

4. AutoLISP: AutoLISP is a dialect of LISP programming language built specifically for use with the full version of AutoCAD and its derivatives.

8.5.2 Number Systems

Numbers are symbols which hold certain values and have predefined meaning. They are basically classified into two types:

1. Non-positional number systems
2. Positional number systems

1. Non-Positional Number Systems: When counting started, human beings used their fingers to count. When ten fingers were not enough, they started using stones, pebbles and sticks which indicated values. For example, five objects can be represented as ***** or XXXXX. The overall value was the sum of all the individual values of the elements used. Such an approach is called the 'additive approach' or the non-positional number system. This type of system is very inconvenient when attempting to represent large quantities.

2. Positional Number Systems: Positional number systems were created in order to avoid the problem of creating and remembering a large number of symbols. These number systems use certain well defined symbols called **digits**. The value of the digit depends on the position in which it appears in the number. For example, let us take the decimal number 576, in which digit 5 represents 5 hundreds, 7 represents 7 tens, and 6 represents 6 ones. The 5 carries the highest value of the three digits; it is referred as **Most Significant Digit (MSD)**. The 6 carries the least weight and is called the **Least Significant Digit (LSD)**. Let us consider another example, 53.64. This number is equal to $5 \times 10 + 3 \times 1 + 6 \times 0.1 + 4 \times 0.01$. The decimal point is used to separate the integer and fractional parts of the number.

The various positional number systems used include;

(a) Decimal number system
(b) Binary number system
(c) Octal number system
(d) Hexadecimal number system.

(a) Decimal Number System: This is the most commonly used number system. The base of this number system is ten. The digits used by this system are 0,1,2,3,4,5,6,7,8,9. The successive positions from right to left represent units, tens, hundreds, thousands and so on, i.e., Each position represents a power of ten. The decimal point separates the positive powers of 10 from the negative powers. Example, decimal number 2453.26 can be written as $2453.26_{(10)}$.
Thus the number may be written as:

$$= 2 \times 10^3 + 4 \times 10^2 + 5 \times 10^1 + 3 \times 10^0 + 2 \times 10^{-1} + 6 \times 10^{-2}$$

$$= 2 \times 1000 + 4 \times 100 + 5 \times 10 + 3 \times 1 + 2 \times 0.1 + 6 \times 0.01$$

$$= 2453.26$$

The decimal number system does not permit itself to convenient implementation in digital systems. For example, it is very difficult to design electronic equipment so that it can work with 8 different voltage levels.

(b) Binary Number System: It is very easy to design simple accurate electronic circuits that operate with only two voltage levels. So every digital system uses the binary system as the basic number system for its operations. The base of this system is thus, two. The digits used by this system are 0 and 1. The principles of the decimal number system apply to the binary system also. Here, places to the left of the binary point are positive powers of 2 and places to the right are negative powers of 2. Example, the number 1011.101 can be written as:

$$= 1 \times 2^3 + 0 \times 2^2 + 1 \times 2^1 + 1 \times 2^0 + 1 \times 2^{-1} + 0 \times 2^{-2} + 1 \times 2^{-3}$$

$$= 1 \times 8 + 0 \times 4 + 1 \times 2 + 1 \times 1 + 1 \times 0.5 + 0 \times 0.25 + 1 \times 0.125$$

$$= 8 + 0 + 2 + 1 + 0.5 + 0 + 0.125$$

$$= 11.625_{(10)}$$

(c) Octal Number System: Octal number system operates with the base of 8. In this a total of eight symbols are used to represent the numbers. The digits used by this system are 0,1,2,3,4,5,6,7. The decimal equivalent of the octal number 2146 written as $2146_{(8)}$ can be evaluated by multiplying each octal digit by its positional weight. For example:

$$2146_{(8)} = 2 \times 8^3 + 1 \times 8^2 + 4 \times 8^1 + 6 \times 8^0$$

$$= 2 \times 512 + 1 \times 64 + 4 \times 8 + 6 \times 1$$

$$= 1024 + 64 + 32 + 6$$

$$= 1126_{(10)}$$

(d) Hexadecimal Number System: The base of the hexadecimal number system is 16. In this number system a total of sixteen symbols are used to represent the numbers. The first ten digits are the digits of the decimal system 0,1,2,3,4,5,6,7,8,9. The remaining six digits are denoted as A,B,C,D,E,F representing the decimal values 10,11,12,13,14,15 respectively. The decimal equivalent of a hexadecimal number ACE written as $ACE_{(16)}$ can be evaluated as given below:

$$= A \times 16^2 + C \times 16^1 + E \times 16^0$$

$$= 10 \times 256 + 12 \times 16 + 14 \times 1$$

$$= 2766_{(10)}$$

The following table shows the decimal numbers and their binary equivalent:

Table 8.3 Positional number systems.

Decimal	Binary	Octal	Hexadecimal
0	0000	0	0
1	0001	1	1
2	0010	2	2
3	0011	3	3
4	0100	4	4
5	0101	5	5
6	0110	6	6
7	0111	7	7
8	1000	10	8
9	1001	11	9
10	1010	12	A
11	1011	13	B
12	1100	14	C
13	1101	15	D
14	1110	16	E
15	1111	17	F

8.5.3 Algorithms

An algorithm is a step-by-step procedure to solve a given problem. The essential characteristics of an algorithm are:

1. Every step of an algorithm should perform a single task.
2. Confusion should not be there at any stage in an algorithm.
3. An algorithm should involve a finite number of steps to arrive at a solution.
4. Every algorithm should lead to a unique solution of the problem.
5. Each statement in the algorithm must be workable in finite time.
6. In case of repetition of steps, the number of repetitions must be finite.

Example of an algorithm to find the sum of two numbers:

Step1: Input number 1 and number 2

Step2: Compute sum ← number 1 + number 2

Step3: Output sum

Example of an algorithm to find the average of four numbers:

Step1: Start

Step2: Input values for *A*,*B*,*C* and *D*

Step3: Calculate average ← (A+B+C+D) / 4

Step4: Output average

Step5: Stop

Example of an algorithm to find the larger of two numbers:

Step1: Start

Step2: Input *A* and *B*

Step3: If (*A*>*B*) then

Output *A*

Else

Output *B*

End of if structure

Step4: Stop

8.5.4 Flowcharts

A flowchart is a pictorial or graphical representation of a solution to any problem. Flowcharts are constructed by using special geometrical symbols. Each symbol represents an activity, which could be either input or output of data, computation or processing of data, taking a decision or terminating the solution, and so on. The symbols used in a flowchart are joined by arrows.

Flowcharts are classified into two categories. They are programme flowcharts and system flowcharts. Programme flowcharts present a diagrammatic representation of a sequence of instructions for solving a problem whereas system flowchart indicates the flow of data throughout a data processing system, as well as the flow into and out of the system.

Symbols used in flowcharts and its purposes are given in the following table:

Table 8.4 Flowcharts: Symbols and purposes.

SYMBOLS	PURPOSE
	**Terminal:** The beginning, end, or a point of interruption in a programme.
	**Input / Output:** Input or output data or information.
	**Processing:** Represents calculations, data manipulations or information processing.
	**Preparation:** An instruction or group of instruction which changes the programme.
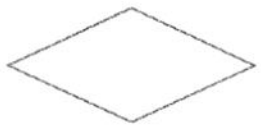	**Decision:** Represents a comparison, a question or a decision that determines alternative paths to be followed.
	**Connector:** An entry from, or an exit to, another part of the programme flowchart.
	**Flow Direction:** The direction of processing or data flow.

Flowchart provides an easy means of communication. It is easy to convert it into a programme. It is also independent of programming language.

Examples of flowcharts to find area of a square and to find the larger of two numbers:

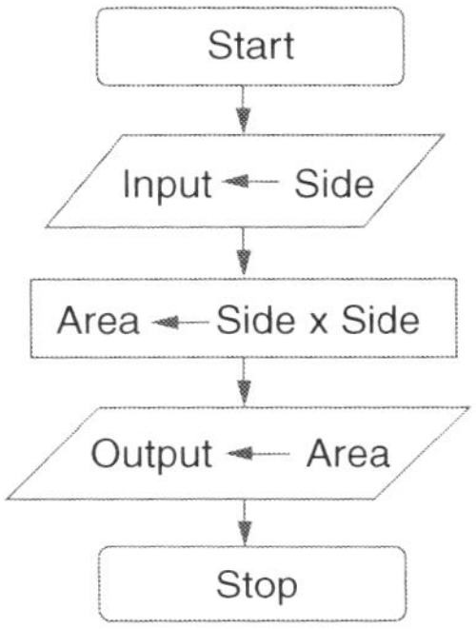

Fig. 8.54 Flowchart for area of square.

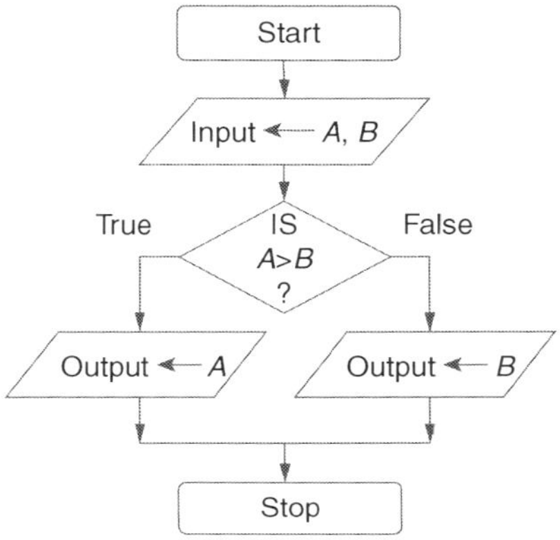

Fig. 8.55 Flowchart for larger number.

8.5.5 Constants and Variables

Constants are quantities whose values do not change during the execution of a programme. They are classified into numeric constants and string constants. All the numbers used in a programme are called **numeric constants**. A set of characters enclosed within quotations is called **string constants**. A string includes alphabets, numbers, symbols and blank spaces. It can also contain names, dates, addresses or any other information.

Variables are quantities whose values change during the execution of a programme. A variable always starts with an alphabet. Then it can include a combination of alphabets or digits. Blank spaces are not allowed. Variables are classified into two types. They are numeric variables and string variables. **Numeric variables** hold only numeric values. **String variables** are used to handle string constants.

8.5.6 Operators and Control Statements

Certain special symbols are used to indicate certain operations like addition, multiplications, etc. The symbols used for these are called **operators**. Operators are classified into arithmetic operators, relational operators, logical operators and assignment operators. **arithmetic operators** represent arithmetic operations such as addition, subtraction, multiplication, division. **Relational operators** are used to establish the relation between different components such as greater than (>), less than (<), greater or equal (> =), less or equal (< =) and not equal (<>). **Logical operators** are used to establish a total relation when the total relation depends on one or many subconditions such as AND, OR, etc. **Assignment operators** are used to assign the value of a constant or another variable or an expression to a variable.

Control statements are used in a programming language to perform some action. The statement indicates some change in the sequence of activities. When programming language was started sequence was the only facility available. The execution of a programme started at the first statement and continued with the execution of each line until the last statement was executed. Sometimes we jump to other parts of the programme or repeat a particular part of a programme. In this case normal sequence of programme is altered. A jump from one part to another causes transfer of control called **branching**. Repetition of the execution of a programme segment is called **looping**.

8.6 INTERACTIVE COMPUTER GRAPHICS

In CAD system there are two major components. One is interactive computer graphics (ICG), and the other is human designer. **Interactive Computer Graphics (ICG)** is a user-oriented system in which the computer is employed to create, transform, and display data in the form of pictures or symbols. ICG system is a combination of hardware and software. Hardware includes CPU, one or more workstations and peripheral devices such as printers, plotters and drafting equipment. Software consists of the computer programmes needed to implement graphics processing on the system:

CAD system would include one or more design workstations that include the following hardware components:

1. Graphics terminal or visual display unit (VDU)
2. Operator input devices
3. Operator output devices
4. Central processing unit (CPU)
5. Secondary storage

Layout of CAD system is shown below:

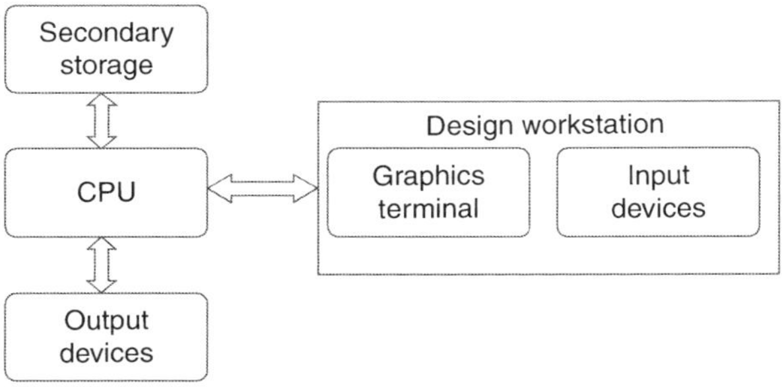

Fig. 8.56 Layout of hardware components in CAD system.

Design workstation must accomplish five functions. Those are:

1. It must interface with CPU.
2. It must generate a steady graphic image for the user.
3. It must provide digital descriptions of the graphic image.

4. It must translate computer commands into operating functions.
5. It must facilitate communication between the user and the system.

A typical interactive computer graphics workstation would consist of a graphics terminal and operator input devices.

8.6.1 Graphics Terminal

Till recently all the computer graphics terminals have been using the Cathode Ray Tube (CRT) as the display device.

The Operation of a Cathode Ray Tube (CRT):

A heated cathode emits a high speed electron beam onto a phosphor coated glass screen, causing it to glow at the points where the beam makes contact. The beam can be made to generate a picture on the CRT screen by focusing the electron beam, changing its intensity and controlling its points of contact against the phosphor coating through the use of a deflector system.

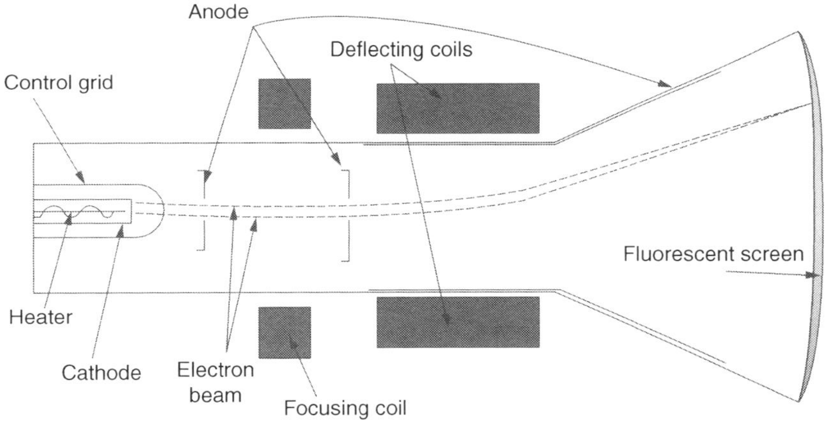

Fig. 8.57 Operation of CRT.

There are two basic techniques used in computer graphics terminal for generating the image on the CRT screen. They are:

1. Stroke writing
2. Raster scan

1. Stroke Writing: Stroke writing is otherwise known as line drawing, random position, vector writing and directed beam. It uses an electron beam which operates like a pencil to create a line image on the CRT screen. The image is constructed out of a sequence of straight line segment drawn on the screen by directing the beam to move from one point on the screen to the next, where each point is defined by its x and y coordinates. Smooth curves can be approximated by connecting short line segments.

2. Raster Scan: It is also called digital TV. In this approach the viewing screen is divided into a large number of discrete phosphor picture elements called **pixels**. The matrix of pixels constitutes the raster. The number of separate pixels in the raster display might range from 256 × 256 (a total of over 65,000) to 1920 × 1080 (a total of over 20, 00,000 points/pixels). Each pixel on the screen can be made to glow with a different brightness. During operation, an electron beam creates the image by sweeping along a horizontal line on the screen from left to right. This energises the pixels in that line during the sweep. When the sweep of one line is completed, the electron beam moves to the next line below and proceeds in a fixed pattern. After sweeping the entire screen the process is repeated at a rate of 30 to 60 entire scans of the screen per second.

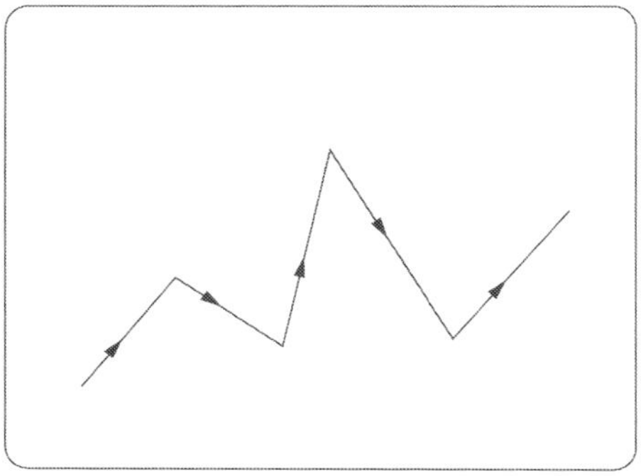

Fig. 8.58 Stroke writing.

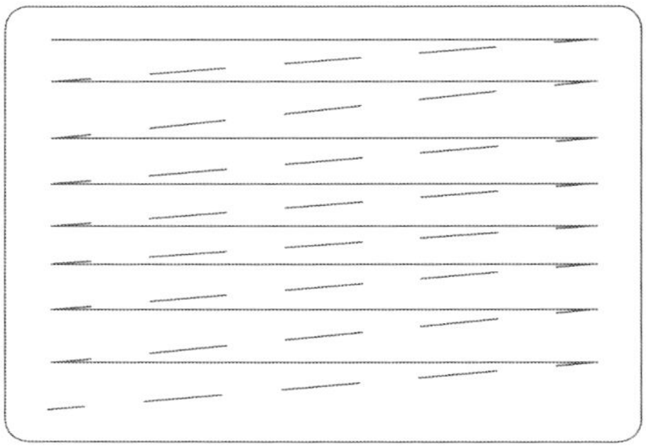

Fig. 8.59 Raster scan.

8.6.2 Factors to be Considered in Different Types of Graphics Terminals

The factors to be considered in different types of graphic terminals are:

1. The type of phosphor coating on the screen
2. Whether colour is required
3. The pixel density
4. The amount of computer memory required to generate picture

8.6.3 Types of Graphics Terminals

Three types of graphics terminals are used in CRT terminals for CAD system:

1. Directed Beam Refresh (DBR)
2. Direct View Storage Tube (DVST)
3. Raster scan (Digital TV)

1. Directed Beam Refresh (DBR): This terminal uses the stroke writing approach to generate the image on the CRT screen. The term 'refresh' means that the image must be regenerated many times per second in order to avoid noticeable flicker of the image. The phosphor elements on the screen surface are capable of maintaining their brightness only for short time (in micro seconds). If the images are to be continued, these picture tubes must be refreshed to retrace the image repeatedly by a directed beam. On densely filled screens, it is difficult to avoid flickering of the image with this process. There are several advantages associated with these systems. Because the image is being continually refreshed, selective erasure and alteration of the image is easily done. It is also possible to provide animation of the image. This system is the oldest of the modern graphics display technologies.

2. Direct View Storage Tube (DVST): DVST terminals also use the stroke writing approach to generate the image on the CRT screen. The term 'storage tube' refers to the ability of the screen to retain the image projected on it, thus, avoiding the need to rewrite the image constantly. In this approach an electron flood gun is directed at the phosphor coated screen, which keeps the phosphor elements illuminated once they have been energised by the stroke writing electron beam. The resulting image on the CRT screen is flicker free. The disadvantage in this approach is that individual lines cannot be selectively removed from the image. Storage tubes have been the lowest cost terminals, which are capable of displaying large amounts of data, either graphical or textual. Because of these features, there are probably more storage tube terminals in service. The major disadvantage of a storage CRT is that selective erasure is not possible. Other disadvantages include its lack of colour capability, the inability to use a light pen as a data entry device and its lack of animation capability.

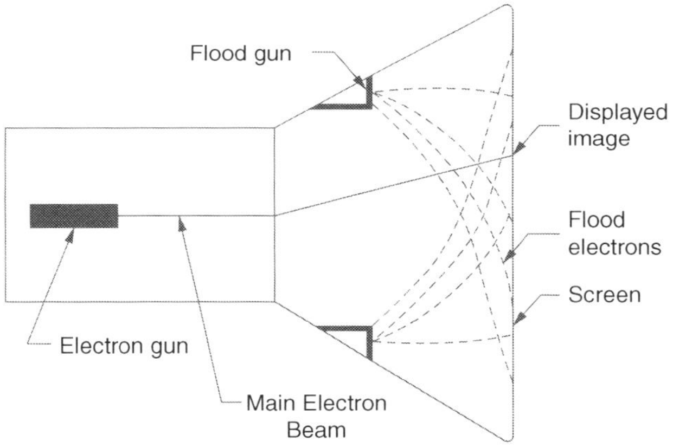

Fig. 8.60 DVST operating principle.

Operating principle of DVST: The DVST has the standard CRT electron gun and deflection system to locate the beam onto the screen. It also incorporates flood guns located between the main electron gun and the screen. The flood guns continuously emit flood electrons on the screen. This electrons cannot create image by themselves because of its less intensity. When the main electron beam hits the screen the flood electrons are attracted to the area of the image resulting in a switched on display. This is retained after the main beam is moved to another location.

3. Raster Scan Terminals (Digital TV): Raster scan terminals use raster scan technique to generate image on the CRT screen. It uses digital signals generated by a computer to construct an image. This raster scan graphics terminal using a refresh tube is based on the availability of the computer memory. For example, the simplest and lowest cost terminal in this category uses only two beam intensity levels, on or off. This means that each pixel in the viewing screen is either illuminated or dark. Each bit of memory contains the on / off status of the corresponding pixel on the CRT screen. This memory is called the **frame buffer** or **refresh buffer**. A 1920 × 1080 raster screen would require more than two million bits of storage in the frame buffer. The picture quality can be improved in two ways; by increasing the pixel density or adding a grey scale or colour. Increasing pixel density for the same size screen means adding more lines of resolution and more addressable points per line. For a colour display three times as many bits are required to get various intensity levels for each of the three primary colours; red, blue and green. A raster scan graphics terminal with high resolution and grey scale can require a very large capacity refresh buffer. Prices are competitive with the other two types. The advantages of the present raster scan terminals include the feasibility to use low cost TV monitors, colour capability, and the capability for animation of the image. These features, plus the continuing improvements being made in raster scan technology make it the fastest growing segment of the graphics display market.

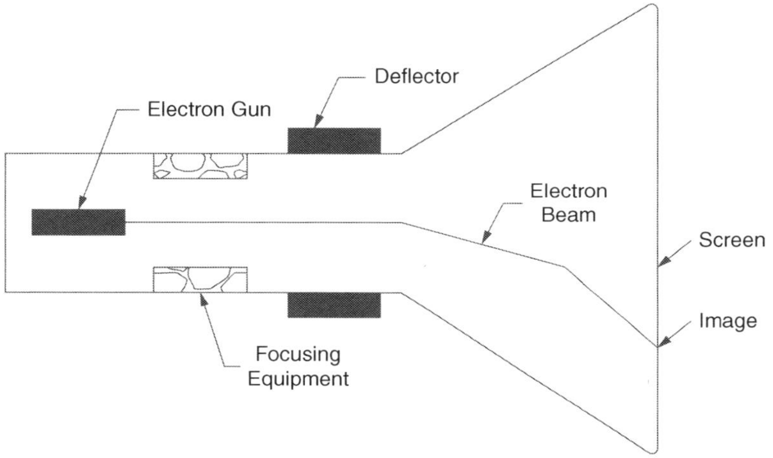

Fig. 8.61 Raster scan CRT principle.

Table 8.5 Comparison of the three graphics terminal features.

Features	Directed beam refresh (DBR)	Direct view storage tube (DVST)	Raster scan
1. Image generation	Stroke writing	Stroke writing	Raster scan
2. Picture quality	Excellent	Excellent	Moderate to good
3. Data content	Limited	High	High
4. Selective erase	Yes	No	Yes
5. Grey scale	Yes	No	Yes
6. Colour capability	Moderate	No	Yes
7. Animation capability	Yes	No	Moderate

Generally the display devices are rated in terms of their resolution, aspect ratio, brightness, contrast and response time. The number of pixels on a screen is its **resolution**. It is expressed in the number of pixels on each row of the display and the number of rows on the display. **Aspect ratio** in graphics is the ratio of the width of an image to its height. Two types of aspect ratio are available for computer monitors – 4:3 (standard ratio) and 16:9 (widescreen ratio). **Brightness** refers to the intensity of light source. **Contrast** refers to the degree of distinction between dark and light pixels. **Response time** refers to the time a monitor takes to carry out request. It is usually measured in terms of milliseconds (ms). LCD and LED monitors available today have response time that varies from 2 ms to 8 ms. Table 8.6 shows the resolution and aspect ratio of today's monitors with their pixel requirements.

Table 8.6 Resolution, aspect ratio and pixel requirements of today's monitors.

Resolution	No. of pixels used	Aspect ratio
640 × 480	307200	4:3
800 × 600	480000	4:3
1280 × 960	1228800	4:3
1400 × 1050	1470000	4:3
1600 × 1200	1920000	4:3
1280 × 720	921600	16:9
1366 × 768	1049088	16:9
1920 × 1080	2073600	16:9

8.6.4 Display Devices Being Used Recently in Graphics Terminals

CRT has the disadvantage that it is extremely bulky. Moreover, it consumes considerable power with increased heat dissipation requirements. Portability is reduced because of the size and the material of the CRT. These disadvantages have prompted the manufacturers to try different types of flat screens as output devices for computers.

1. Plasma Screen: Plasma screens are made with a fine mesh of wires between two glass panels. The space between the panels is filled with an inert gas, usually argon. When current passes through an intersection of wires, the pixel lights up. As in storage tube, pixel remains lit until it is erased. This type of screen is widely used in laptop or portable computers. The plasma screen ensures good colour capability though it is costly and consumes more power.

2. Liquid Crystal Display (LCD): A liquid crystal display (LCD) is a thin, flat electronic visual display that uses the light modulating properties of liquid crystals (LCs). LCs do not emit light directly. LCDs therefore, need a light source. They are used in a wide range of applications including: computer monitors, television, instrument panels, aircraft cockpit displays, etc. They are common in consumer devices such as video players, gaming devices, clocks, watches, calculators, and telephones. LCDs have replaced cathode ray tube (CRT) displays in most applications. They are usually more compact, lightweight, portable, and lower cost. They are available in a wider range of screen sizes than CRT and other flat panel displays. LCDs are more energy efficient, and offer safer disposal, than CRTs. Its low electrical power consumption enables it to be used in battery-powered electronic equipment.

3. Light Emitting Diode (LED) Display: This new technology uses light emitting diodes to light up pixels. This type of screen consumes very less power compared to LCD and plasma displays. This technology has all the advantages of LCD and in addition it gives better viewing angle, brightness, excellent colour capabilities, and longer life. LED technology has been occupying graphics terminals market from current LCD technology. Almost all the graphics terminal manufacturers have stopped production of CRT display.

8.7 DATABASE MANAGEMENT

Computer file is defined as a collection of related information stored in one place. Computer files are stored in disks or magnetic tapes. Each file has a file name and a file extension. File names are provided by the user for the identification of the file. Extensions are used to indicate the type of file, e.g., Brake.dwg, axle.dxf, planning.doc, flower.jpg, etc. A computer file can be created and edited by using utilities provided in operating systems such as notepad, wordpad, paintbrush, etc., or with the help of installed software such as Microsoft word, autoCAD, pro/engineer, etc. There are two types of data files, sequential and random files. A sequential file is a series of characters or values with no intrinsic format. Data is written, stored and accessed sequentially and therefore, the accessing is slow. Random files can be accessed faster.

8.7.1 Database

Database can be defined as a collection of related information about a subject recorded and organised in a useful format that establishes a relationship in form that can be handled by a computer. When data items are updated or modified, the new data are made available to all users at the same time. Several problems which arise due to the use of independent data files can be avoided if a single database is accessed by different users. Since the data is integrated in a single location, the compatibility improves. When same data is available in different files, there is a possibility that it may be updated in one file and not in the other. This results in conflicting information and reduced reliability of data. This problem is totally avoided in databases. Any change to the database should be the responsibility of a single database manager rather than the individual application programme.

8.7.2 Objectives of Database

1. Elimination of data repetition.
2. Integration of existing files.
3. Sharing of data among users.
4. Easy incorporation of changes.
5. Simplification of file use.
6. Lower cost of storage and retrieval.

7. Improved accuracy and integrity of data.
8. Prevention of unauthorised use of data.
9. Control over standards.

8.7.3 Disadvantages of Database

(a) Power database software is expensive and usually requires larger and faster hardware. This also leads to higher costs for the CPU and input/output devices apart from the software.

(b) Programmers and analysts have to be trained to become competent in database software.

(c) Because of the greater complexity of a database system a failure can be more difficult to solve.

(d) Many departments may be affected by a programme error instead of one department.

8.7.4 Design of Database

The database of a company will include data required for various departments like design, purchase, manufacturing, marketing, finance, inventory, etc.

Design data may include the following:

1. Standards: International, national and company standards
2. Material specifications and properties
3. Design theories and rules
4. Data based on experience
5. Test results of products and prototypes
6. Service feedback
7. Results of analysis of the product
8. Data on competitor's products

8.7.5 Principles of Database

Database contains the following subsections:

1. Files: Files of information each relating to a different category of data.

2. Records: Records inside each file. Every record of the file will be of identical configuration.

3. Fields: Fields of specific data on each record.

Example: In university database files are in different college names. Each college file has details regarding branch, P.G., U.G., courses called records. Each record has details regarding student's name, age, sex, etc., called fields.

8.7.6 Database Operation

The database operation includes the following steps:

1. **Mask Creation:** In this the design of the records according to company requirements is created. The number of fields per record, the maximum number of characters per field and the type of field are to be specified.
2. **Data Input:** This enables new records to be added to the file.
3. **Sorting:** Sorting the records in any order as per requirement.
4. **Interrogating:** Accessing the records in the order sorted.
5. **Editing:** Modifying or updating the field contents of an accessed and sorted record.
6. **Searching:** Individual records or group records may be quickly retrieved by quoting field numbers and field contents.

8.7.7 Database Structure or Database Models

It is defined as a set of data or element that is related to each other by a set of relations. In CAD/CAM a structure is a scheme, logic or sequence of steps developed to achieve a certain graphics or non-graphics application. With the help of data structure, database can be created. Database model defines the manner in which various files of a database are link together.

Types of data structures:

There are four types of data structure in design environment. They are as follows:

1. **Hierarchical Database:** Database of hierarchical type is based on a file structure called inverted tree (tree is represented with root at the top). Each circle shown in the tree is called a node. Each node can be related to another at a higher level, e.g., the data structure of a lathe. It is used in several database applications. However, its main limitation is that it does not support flexible data access because data elements can be accessed only by following paths formed by branches of the tree structure.

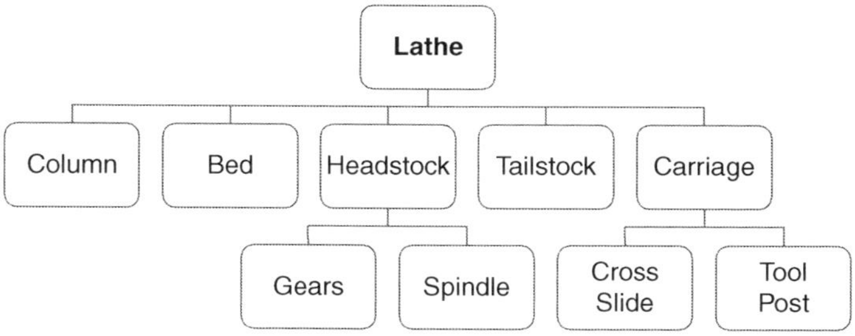

Fig. 8.62 Hierarchical data structure of a centre lathe.

2. **Relational Database:** In a relational database data elements are organised as multiple tables with rows and columns, each table is stored as a data field and each row a data record. Data in one file is related to that in another file with a common field. The relations are stored in files which can be accessed sequentially or in random access mode. Sequential access files are widely used. One

of the disadvantages of the relational database is that it requires substantial sorting, e.g., library database comprising of tables about members, book list and books borrowed / returned, etc.

3. Network Database: This allows the representation of arbitrary relationships between entities. Each entry is represented as a record, and it may be owned by more than one record, leading to a network structure. This approach permits modelling of many to many relations more directly than hierarchical approaches. The main disadvantage of network database approach is its undue complexity both in the structure and in the associated programming of it.

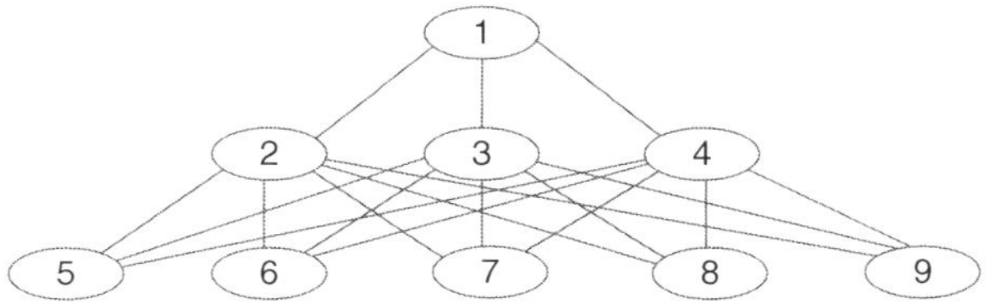

Fig. 8.63 Network database structure.

4. Object Oriented Database: An object oriented database is a collection of object whose behaviour, state and relationships are defined in accordance with object oriented concepts, such as objects, class, class hierarchy, etc. This database structure was introduced to overcome the shortcomings of conventional database models such as incorporation of model data of CAD/CAM/CAE and incorporation of multimedia contents (images, audio and video) in the documents.

The database may be applied to various fields. Some of the known applications are:

1. Computerised library system
2. Automated Teller Machine (ATM)
3. Flight / railway reservation system
4. Computerised inventory system.

8.8 COMPUTER NETWORKING

Many organisations have multiple users of computers. Some of these users are geographically remote from the offices of the organisations of the headquarters. Even within an office building there may be hundreds of employees to use a particular computer. Data communication is the electronic transfer of information between computers. Networking is a sharing of information and services. It is the collection of interconnected autonomous computers. Computer networking provides us with the communication tools to allow computers to share information.

8.8.1 Types of Network

Computer networks are often classified by their size, distance covered or structure. Accordingly network is classified into three types:

1. Local Area Network (LAN)
2. Wide Area Network (WAN)
3. Metropolitan Area Network (MAN)

1. Local Area Network (LAN): LAN is a combination of computer hardware and transmission media that is relatively small, i.e., LANs do not exceed tens of kilometre in size and use only one type of transmission medium. Usually a LAN is contained within a building or campus. The data transmission rate is higher when compared to WAN. Error rate is less. LAN is owned by a single organisation. Moreover, the cost of transmission is negligible as the transmission medium is owned by the user organisation.

2. Wide Area Network (WAN): WAN includes networks larger than LAN. They interconnect LANs which may be located around the world. The data rate is less but error rate is more. It is not owned by a single organisation. Different types of transmission medium can be used. Communication cost is more because it operates on telephone lines or satellite communications.

3. Metropolitan Area Network (MAN): MAN is used to refer to networks which connect systems or LANs within a city. It can support both data and voice and can include one or more LANs as well as telecommunication equipment. Data transmission rate is much higher when compared to WAN.

8.8.2 Internet and Intranet

The Internet is a worldwide interconnection of many different computers and networks. It allows the users to share information, programmes and equipment. An intranet is like a miniature Internet. It uses the same technologies as the Internet. It has Web servers for sharing information among computers. Internet is accessible to everyone, whereas intranet is a private network. Intranets are used by large corporations. Also intranets can be connected to the Internet so that intranet users can get access to Internet.

8.9 BASIC CONCEPTS OF CAD

Computer Aided Design can be defined as the use of computer systems to assist in the creation, modification, analysis or optimisation of a design.

8.9.1 Steps in General Design Process

The steps involved in general design process are:

1. Recognition of the Need: Involves the realisation that there is a need for a new product for which design should be thought off or some corrective action for an existing product.

2. Definition of Problem: Involves a thorough specification of the item to be designed such as physical and functional characteristics, cost, quality and performance.

3. Synthesis: Modelling the specified item was designed.

4. **Analysis and Optimisation:** Analysis involves checking the model. Synthesis and analysis are closely related and highly iterative in design process. Iterative design process is repeated until the design has been optimised.

5. **Evaluation:** Involves measuring the design against the specifications established in the problem during the definition phase.

6. **Presentation:** Presentation or documentation of the design by means of drawings, materials specifications, etc.

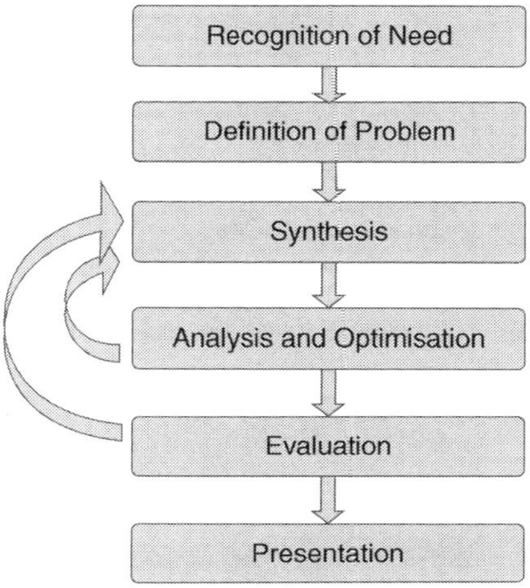

Fig. 8.64 General design process.

8.9.2 Steps in CAD Process

The steps involved in CAD process are:

1. **Geometric Modelling:** It corresponds to the synthesis phase in the general design process. It involves the development of a mathematical description of the geometry of an object. The mathematical description allows the image of the object to be displayed and manipulated on a graphics terminal using the CAD system.

2. **Engineering Analysis:** It corresponds to analysis phase of the general design process. This involves stress-strain calculations, heat transfer computations or the use of differential equations to describe the dynamic behaviour of the system. Computer can be used to aid in this analysis work. Two important examples are analysis of mass properties and finite element analysis.

3. **Design Review and Evaluation:** It corresponds to the evaluation process in the general design process. This involves checking the accuracy of the design on the graphics terminal. It helps to reduce the possibility of dimensioning errors. Examples for design review and evaluation are, layering, interference checking and simulation (or) kinematics.

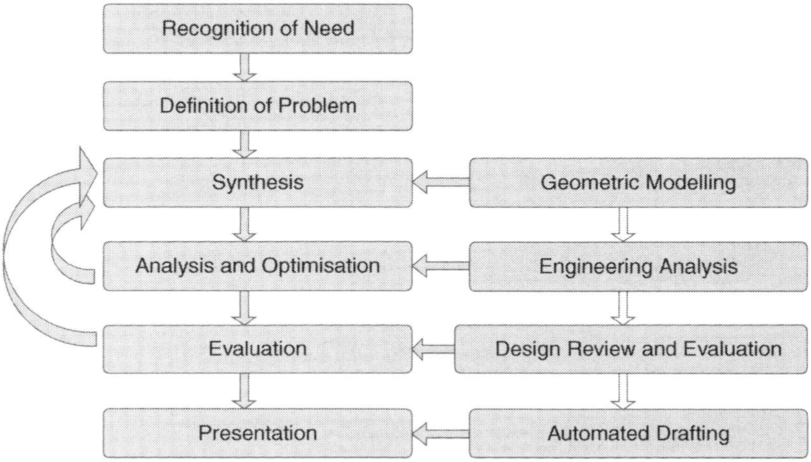

Fig. 8.65 Steps in CAD process.

4. Automated Drafting: It corresponds to the presentation phase of the general design process. It involves the creation of hard copy engineering drawings directly from the CAD database. It is also called Computer Aided Design and Drafting (CADD). CAD systems can increase productivity in the drafting function by five times approximately over manual drafting. Automated drafting features include automatic dimensioning, generation of cross-hatched areas, scaling of the drawing, and the capability to develop sectional views and enlarged views of a particular part. Engineering drawings can be made to adhere to company drafting standards by programming the standards into the CAD system.

8.9.3 Benefits of Implementing CAD

The following benefits can be achieved by implementing CAD:

1. Reduced engineering manpower requirements.
2. Easy to make modifications as per customer expectations.
3. Faster response to requests for quotations.
4. Goods may be delivered in time thereby avoiding subcontracting.
5. Accurate designs possible.
6. Transcription errors are reduced.
7. Recognition of component interaction analysis is easy.
8. Improvements in the engineering productivity.
9. Shorter time required for delivery of the product.
10. Better functional analysis facilities which reduce prototype testing.
11. Assistance in preparation of documentation.
12. Better designs provided.
13. Improved productivity in tool design.
14. Better knowledge of costs provided.
15. More standardised design.
16. Reduction in training time for drafting personnel.
17. Fewer errors in NC part programming.
18. Provision for using the existing parts or tools effectively.
19. Ensures designs appropriate as per existing manufacturing techniques.
20. Saves material and machining time by optimisation.
21. Operational status of work in progress provided.
22. Effective management of design personnel.
23. Assistance in inspection of complicated parts.
24. Better communication interfaces and understanding among the personnel involved.

8.9.4 Description of Shape and Size

Shape is defined as the geometry and topology of a feature, which are simple solid models, e.g., hole, boss, pocket, pad, slot, groove, block, cone, cylinder, sphere, etc. All the parameters and their values that fully define a feature are called **size** or size of a feature.

8.9.5 Parametric Programming

Parametric is a powerful characteristic of feature, which allows the designer to edit or re-use the model by changing its dimensions, thus, it saves the time and money instead of building the model again. Parametric programming or parametric modelling or parametric representation provides unlimited scaling capabilities. It creates families of parts. It supports the notion of changing and editing solid features as design changes occur in the product life cycle. It also supports the notion of model re-use. It allows designers to change dimensions in existing model to use it again. Parametric can save time and money by making a few changes to modify the existing model instead of building it again entirely.

8.9.6 Steps in Parametric Feature Creation

The steps involved in parametric feature creation are:

1. Creating the sketch
2. Defining the parameters
3. Assigning dimensions or values to parameters
4. Applying relations to parameters.
5. Applying constraints
6. Creating the feature

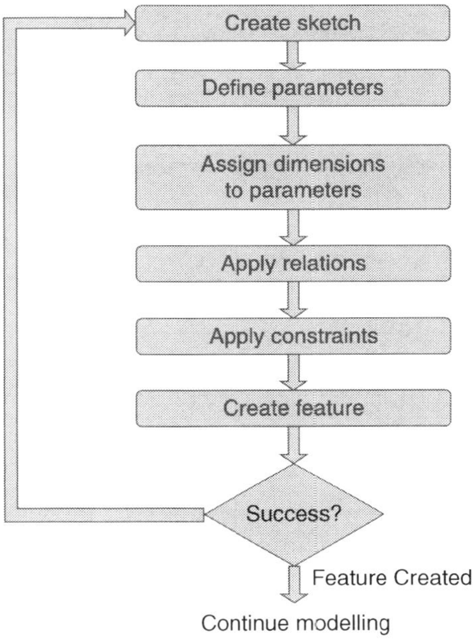

Fig. 8.66 Parametric feature creation.

8.9.7 Construction of Engineering Drawing

An engineering drawing serves as a tool to document a design and communicate it to the entire engineering enterprises, from production, inspection, and assembly to sales and marketing. After designers have finalised the design of the part or a product, they document their ideas of design and specifications in a drawing accurately and thoroughly. The drawing becomes a reference that is used during the remainder of the product life cycle after design.

1. Drawing Structure: The drawing structure consists of views, a title block, a bill of material, labels and notes. The views are laid out in an ordered fashion according to the rules of orthographic views. Isometric views and sectional views can be added to the drawing. Each view displays dimensions and tolerances to fully define the part. The title block is usually located in the bottom right corner of the drawing. It contains information such as company name, part number, drawing number, revision number, sheet number, materials and finish general tolerances, drawing scale, sheet size, revision block, angle of projection, unit, drawn by, checked by and approved by. The bill of material is a table that is usually located in the top right corner or above the title block of the drawing. It is used in assembly drawings to list item number, quantity, part number, material and description. The labels and notes provide additional information about the design that cannot be conveyed by dimensions. It includes instructions on manufacturing, machining, assembly, surface finish, etc.

2. Model and Drawing Associativity: CAD system provides a two-way associativity between model and drawing modes. If a designer makes a geometric change in the model mode, the change is reflected in the drawing mode. Alternatively if the designer changes a model dimension in one of the views of a drawing the model is automatically updated.

3. Drawing Content: A drawing has model views, dimensions, tolerances, annotations, bill of materials, assembly instructions, machining instructions, surface finish and roughness symbols. The generation of the drawing begins by defining and placing views in the drawing. After that they can be moved around to adjust their locations. We can scale them up or down. After ascertaining the locations and sizes of the views, dimensions and tolerances are added. After that we add labels and notes, fill in the title block and bill of material. The design is saved and printed.

4. Methods of Angle of Projection: The location of the projection plane relative to the observer and the model, defines the method of angle of projection also known as type of projection. The projection plane can either be between the observer and the model (3rd angle projection), or it can be behind the model and away from the observer (1st angle projection). In the first angle method the projected view is the opposite view of what the view name indicates. In the third angle method the projected view is the same as what the view name indicates. Some CAD systems allow users to set the type of projections.

5. Types of Views: CAD system provides us with various types of views that we can use in a drawing depending on the complexity of the model. Various types of views that can be included in drawing are:

(a) Projected View: It is a view that results from projecting an existing view in a given direction.

(b) Named View: It is a standard predefined view which includes the six standard 2D views (front, back, left, right, top and bottom) and the 3D views (isometric and trimetric).

(c) Auxiliary View: An auxiliary view is a custom view that is created using a custom viewing angle.

(d) Sectional View: This is the view that is obtained after it is cut and open the model to reveal important hidden details of its geometry (such as holes, pockets, etc.). It is required to define a viewing direction for the sectional view.

(e) Detailed View: It is to magnify a small portion of a given view, in order to show the details of the small portion only.

6. **Types of Dimensions:** Different types of dimensions are: Cartesian (along X and Y axes of the drawing), radial (a circle radius or diameter), angular (to dimension an angle), true length (for dimensions that are not along X or Y axis, e.g., true length of an arc.), and ordinate (uses same reference for all dimensions that are in one direction) dimensions.

7. **Annotations:** Annotating a drawing means adding notes and labels to it. These annotations are required to add information to a drawing above and beyond dimensions and tolerances. Notes and labels may be with leader or without the leader.

8. **Tolerances:** Tolerances allow for variability during manufacturing because there is no perfection in all manufacturing conditions. Designers are responsible for assigning tolerances to dimensions based on the functional requirements of the design. If the tolerances are higher the more expensive it will be to make the part.

9. **Dimensioning Rules:** The most important rules in dimensioning are:
   - (a) Provide the size and location of each feature in the drawing.
   - (b) Dimension features in the view should show their true size and shape.
   - (c) Use diameter dimension for circles and radial dimensions for arcs.
   - (d) Omit unnecessary dimensions.
   - (e) Ensure that dimensions are large enough to see, spaced out from each other, and are placed away from the profile.
   - (f) Provide a gap between profile lines and dimension extension lines to eliminate any ambiguity about the profile entities.
   - (g) Use a consistent size and style of leader lines, text and arrows throughout the drawing.
   - (h) Avoid over or under dimensioning.
   - (j) Use maximum material condition to display tolerance dimensions on a drawing.

10. **Editing Drawings:** Editing a drawing includes manipulating views, dimensions, and annotations. Manipulating views includes replacing, positioning, and scaling them as well as setting hidden line removal, tangent edge display, etc. Manipulating dimensions include positioning them, changing their text size, changing arrowheads, and changing their font, etc.

8.10 TWO-DIMENSIONAL DRAFTING

Creating a drawing in the X and Y coordinates on a plane or paper is termed as two-dimensional drafting. In 2D drawing distances are measured from the origin of two axes namely X and Y. The axes could be fixed during the creation of elements (called *absolute mode*) or there may be an option for the automatic shifting of axes, so that distances are measured from the end of each successive element (called *incremental mode*). All the existing CAD software divides options into three main categories – creation tools, editing / modifying tools and aid / support tools.

8.10.1 Creation Tools

All existing CAD systems provide users with curve entities, which can be divided into analytic and synthetic entities. **Analytic** entities are points, lines, arcs, circles, conics, rectangle and polygon. **Synthetic** entities include splines and bezier curves.

1. Points: Points have two coordinates namely X and Y, measured from origin either by absolute mode or by incremental mode.

2. Line: Straight lines connecting two points.

3. Arc: Arcs have three points. It can be made in different ways, e.g., start point, end point and point on arc or centre, start and end point.

4. Circle: Circles have either two or three points. It can be made in different ways, e.g., centre and radius or centre and diameter for two point method. Circle by three tangent lines for three-point method.

5. Rectangle: Rectangle can be created by either two-point method or three-point method.

6. Polygon: Polygons have two points. Polygon can be created in three ways (centre and side of polygon, centre and inscribed radius method and centre and circumscribed radius method). It requires number of sides for all the three types.

7. Conics: Lines, circles, ellipses, parabolas and hyperbolas are all special forms of conic curves. They all can be generated when a right circular cone of revolution is cut by planes at different angles relative to the cone axis, thus, the derivation of the name conics. Circles result when a cone is sectioned by a plane perpendicular to its axes. Ellipses, parabolas and hyperbolas are generated when the plane is inclined to the axis by various angles.

8. Splines: Splines are synthetic entities. They are of two types – cubic spline and b-spline. They are formed by two methods.

(a) Spline by points or curve interpolates

(b) Spline by poles or curve extrapolates.

In spline by points method the spline touches all the defining points. In spline by poles method the spline touches only start and end points, and it forms a curvature between intermediate points.

9. Bezier Curves: It is a synthetic entity defined by a set of data points. It may interpolate or extrapolate the data points which are used to control the shape of the resulting curves.

8.10.2 Editing Tools

Tools or options used to edit or modify the curves are called editing tools. Some of the commonly used editing tools are move, copy, cut, paste, offset, scale, mirror, array, fillet, chamfer, rotate, trim, extend, stretch, delete, divide, etc.

1. Move: This option moves an item from one location to another location. It is also called translate.

2. Copy: This option is otherwise called duplicate, because it duplicates the original one at another location. This copy function is similar to the move function except that it preserves a copy of the item at its original location.

3. Cut: This option is used to remove the drawing or drawing entity.

4. Paste: This option is used to paste or add the object which was already cut or copied.

5. Offset: This option offsets an item to the required distance. It differs from 'move', because a move translates an item from one point to another point, but offset transfers the curve in the normal direction as per the defined distance.

6. Scale: This option is used to increase or decrease the item to the required value. Scaling is of two types.

(a) Uniform scaling

(b) Non-uniform scaling. In uniform scaling an item is scaled in all directions equally. In non-uniform scaling an item is scaled to different values.

7. Mirror: This option creates a mirror image of the item about a specified plane.

8. Array: This option is also called 'instancing' or 'pattern'. Using this option you can make an item into a number of instances or arrays. There are two types of arrays, rectangular array and circular array.

9. Fillet: This option is blending the corners of the curves. It is also called 'blend'.

10. Chamfer: This option is beveling the corners of the curves. It is also called 'bevel'.

11. Rotate: This option is used to rotate an item through a specified angle from its original orientation.

12. Divide: This option is used to divide the entities into:

(a) number of elements

(b) between two entities

(c) a percentage of entities, etc.

13. Trim: This option is also called 'break'. It trims a line or other component by removing the unwanted portions which extend beyond the required limit.

14. Extend: This option extends the line or curve to the required limit.

15. Stretch: This option is also called 'dragging'. This is used to move the particular geometry along with adjacent geometries also.

16. Delete: It is also called 'remove' or 'erase'. This function causes the selected segment of the drawing to be removed from the screen and from the database.

8.10.3 Support Tools

Tools which support the geometric modelling or curve creation are called aid or support tools. Some of the common aid tools are Undo, Redo, Hatching or area filling, Grids, Layers, Limits, Snap, Boolean, Grouping, User Defined Features, Macros, Selection methods, Colours, Relations or Expressions and drawing tree. Some of the common tools which support graphical display are, Visualisation, Boundary setting or windowing, Fit, Zoom, Pan, and rotating the screen. Some tools support the 2D geometry in adding dimension, text, annotations, symbols, notes and labels, and bill of material or part list to the geometry.

1. **Undo:** It helps the user to get back to the previous step that was made by mistake.

2. **Redo:** It helps the user to get back to the 'undo' step.

3. **Hatching:** This option is used to fill the closed areas of a sectional portions or geometry. It is also called 'area filling'.

4. **Grids:** This forms points at regular intervals in X and Y coordinates, so that the user can easily understand the positions of the geometry. Grid points will not affect the geometry output.

5. **Layers:** This option is used to organise the objects in different layers so that the user can easily activate the required layer. The unwanted or other geometries will not be shown. Thus, it reduces the complexity in the display.

6. **Limits:** This option is used to set the size for drawing sheet.

7. **Snap:** This option helps to pick the required point of the existing geometry. Some of the snap options are, end point, midpoint, tangent point, intersection point, quadrant point, circle centre point, etc. This option is also called 'geometric modifiers'.

8. **Boolean:** This helps us to use the boolean operations such as unite, subtract, and intersect in the geometry.

9. **Grouping:** This option is used to organise the similar type of geometry and to name it. This helps the user to identify and group the similar type geometry, e.g., dimensions, text, curves, etc.

10. **User Defined Features (UDF):** This option is useful when making a same geometry with different values or data. In this the user has to make a UDF with the value that ranges from lower limit to upper limit and stores it in a UDF library. When the geometry requires a specific value, it can be taken from UDF library and set the required value for the geometry. Thus, UDF reduces the time in creating the same parts with different values, e.g., parts like nuts, bolts, etc.

11. **Visualisation:** This option is used to set background and foreground for the geometry.

12. **Colour:** It helps the user to differentiate geometry for a set of entities according to its purpose or use, e.g., to differentiate dimensions, texts, drawings, etc.

13. **Drawing Tree:** It is used to quickly view the options used to make the drawing. Thus, it helps in modifying or editing the drawing entities.

14. **Pan:** This option is used to move the drawing in the screen or display without changing its original location.

15. **Zoom:** This option is used to see the drawing closer by enlarging the portion of the drawing in display without changing its original size.

16. **Rotate:** This option is used to rotate the drawing in display without rotating the original position.

17. **Fit:** This option is used to bring the entire geometry into the display window.

18. Macros: This option is used to repeat the steps of geometry creation to the parts which are similar in shape and size. Instead of drawing it again we can record the geometry creation of the first entity and store it as macros. We can play the recorded macro which will do the same operations done in first one. Thus, it reduces design time for similar geometry creation using similar operations.

19. Entity Selection Methods: Entity selection methods help the user to pick the entity as per the user's wish. The user can select:

(a) Individual entity

(b) Group of entity

(c) Select the entity which comes inside a closed boundary

(d) Selecting the entity by its colour, etc.

20. Relations or Expressions: This option is used to make a drawing with relationship to other entities so that each can be adjusted when the user updates a particular value. This option helps to make geometries like gears.

8.10.4 Advantages of 2D Drawing in CAD

(a) 2D drawing requires less computer memory.

(b) It reduces the drawing time by approximately 1/5 times compared to manual drawing.

(c) Drawings have better accuracy than manual one.

(d) Modification and editing of the drawing are easier.

(e) Drawing presentation very good and neat.

8.10.5 Disadvantages of 2D Drawing

(a) It is very difficult to get the properties such as volume of the part, mass properties, centre of gravity, moment of inertia, etc.

(b) The understanding of the 2D drawing creates difficulties in case of complex parts.

(c) Workers should have a sound knowledge in engineering drawing.

8.11 THREE-DIMENSIONAL MODELLING

Creating geometry on x, y and z coordinates is termed as three-dimensional modelling. It is classified into three categories:

1. Wireframe modelling
2. Surface modelling
3. Solid modelling

8.11.1 Wireframe Modelling

A wireframe model is described in terms of points and lines. This is low level of modelling. This can be adequate for tasks involving simple shapes. One of the very common applications of wireframe modelling is the use of 3D tool path simulation displays for simple machining operations, such as 2 ½ axes and 3 axis milling. This is also called 2½ D geometry. It is less demanding on computer memory compared to the other two 3D modelling techniques.

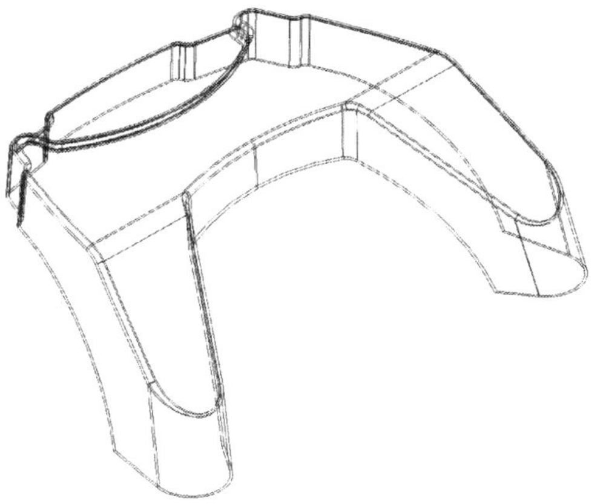

Fig. 8.67 Wireframe model of a game control.

Limitations of Wireframe Modelling

(a) Confusion is caused by the ambiguity of orientation and viewing plane.
(b) It cannot distinguish between visible and hidden edges.
(c) Inability to recognise curved profiles.
(d) Inability to detect interference between components, e.g., in machining, tool collisions cannot be automatically detected.
(e) Difficulty in calculating physical properties such as mass, surface area, volume, centres of gravity, etc.
(f) No facility for automatic shading.

8.11.2 Surface Modelling

A surface model is defined in terms of points, lines and faces. It is a higher level model than wireframe and is consequently far more versatile and advantageous. Surface modelling at present is the most suitable choice in some applications, especially those which involve the design and manufacture of complex curved surfaces, such as car body design and aerospace applications.

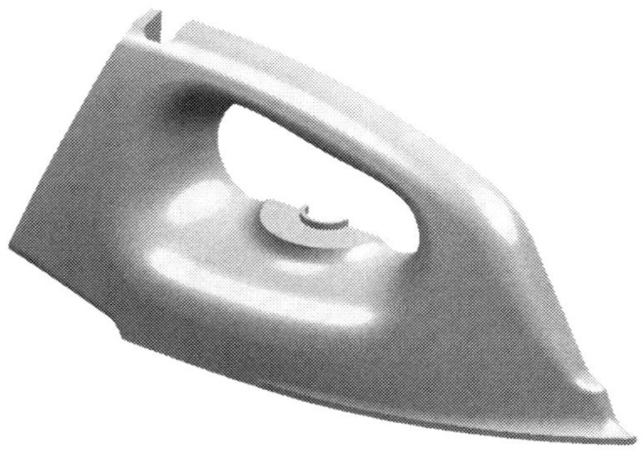

Fig. 8.68 Surface model of a casing of an iron box.

Advantages of surface modelling over wireframe modelling

(a) Ability to recognise and display complex curved profiles.
(b) Ability to recognise faces and thus, provide the facility of shaded surfaces in 3D.
(c) Ability to recognise facial features such as holes, bosses, etc.
(d) Ability to display superior tool-path simulations in 3D for multi-axis machining operations and complex shapes.
(e) Improved facility for simulation of robot handling.
(f) Ability to calculate cross sectional areas, surface areas, and volumes.
(g) Ability to interface with a finite stress analysis package.

Types of surfaces There are various types of surfaces which may be grouped under two heads⊠analytical surface and synthetic surface. They are explained as follows:

1. Analytical Surface: Analytical surfaces are classified into four types:

(a) Plane Surface: It is the simplest surface. It requires three non-coincident points to define plane. The plane surface can be used to generate cross sections by intersecting a solid with it (Fig. 8.69).

(b) Ruled (lofted) Surface: It is a linear surface. It interpolates linearly between two boundary curves that define the surface (rails or cross section). Rails can be any curves. This surface is ideal for representing surfaces that do not have any twists (Fig. 8.70).

(c) Surface of Revolution: It is an axisymmetric surface that can model axisymmetric (cylindrical) objects. It is generated by rotating a planer curve in space about the axis of symmetry to a certain angle (Fig. 8.71).

(d) Tabulated surface: It is a surface generated by translating a planer curve a certain distance along a specified direction (axis of the cylinder or directrix). The plane of the curve is perpendicular to the directrix (Fig. 8.72).

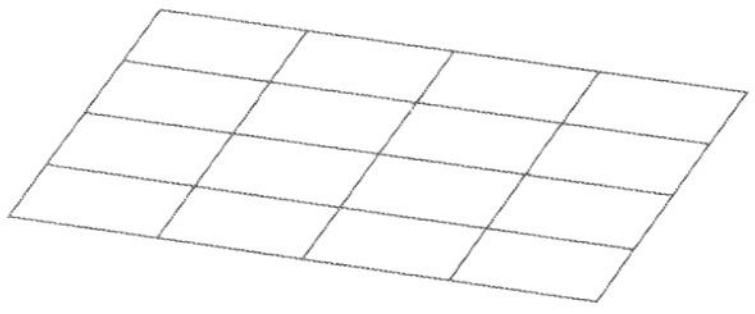

Fig. 8.69 Planer surface.

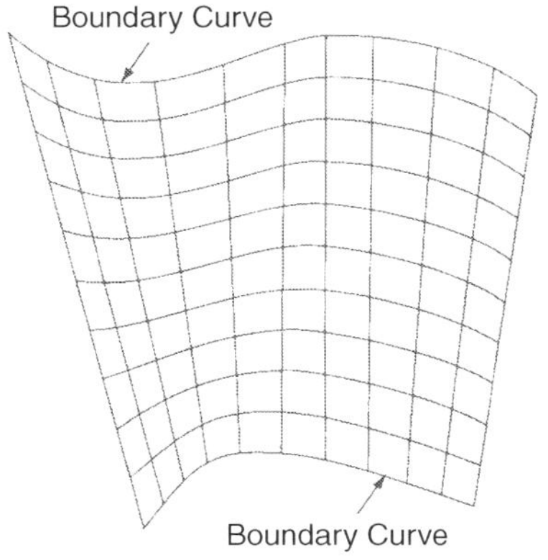

Fig. 8.70 Ruled surface.

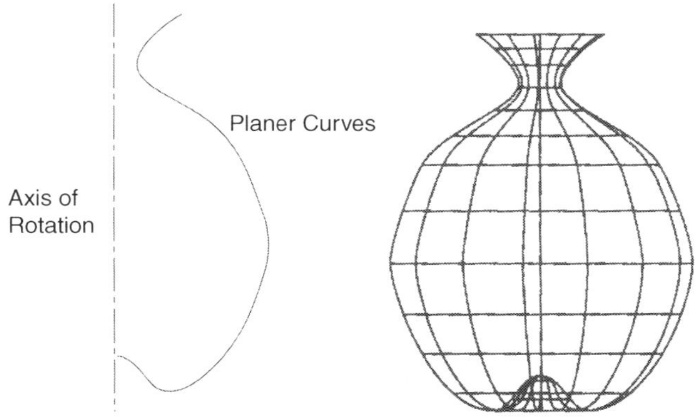

Fig. 8.71 Surface of revolution.

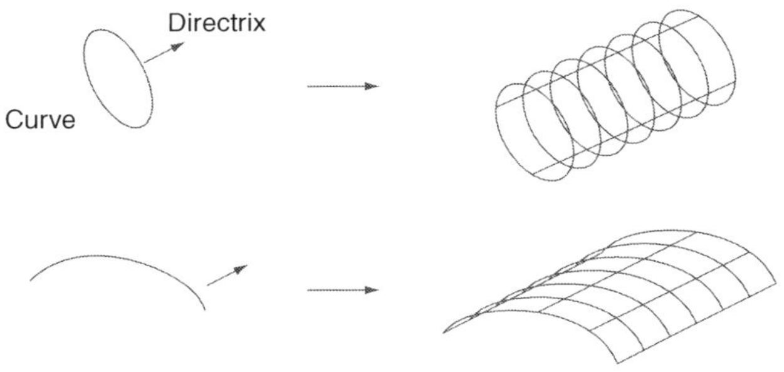

Fig. 8.72 Tabulated surfaces.

2. Synthetic Surface: Synthetic surfaces are of five types:

(a) Bezier Surface: It is a surface that approximates or interpolates the given input data. It extends the Bezier curve to surfaces. It is a general surface that permits twists. Bezier surface allows only global control of the surface (Fig. 8.73).

(b) B-spline Surface: It is a surface that can approximate or interpolate given input data. It is a general surface like the Bezier surface but with the advantage of permitting local control of the surface (Fig. 8.74).

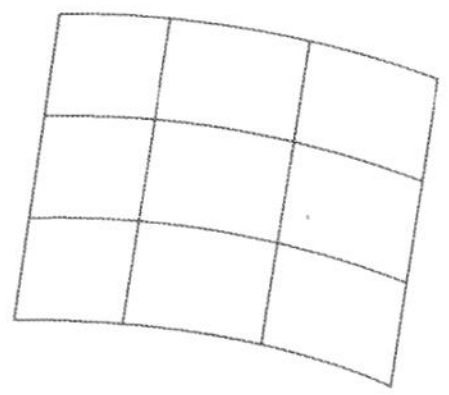

Fig. 8.73 Bezier surface.

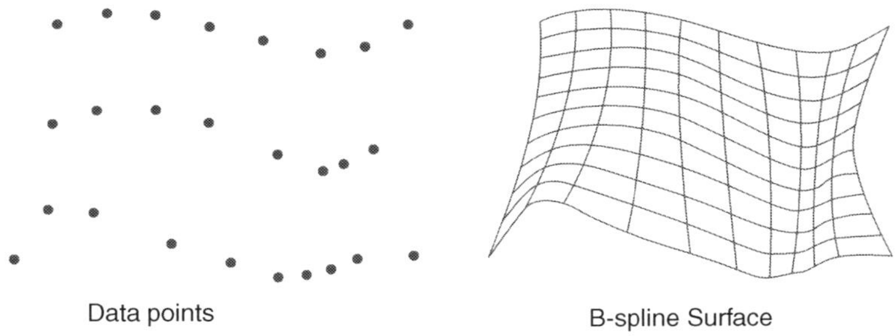

Fig. 8.74 B-Spline surface.

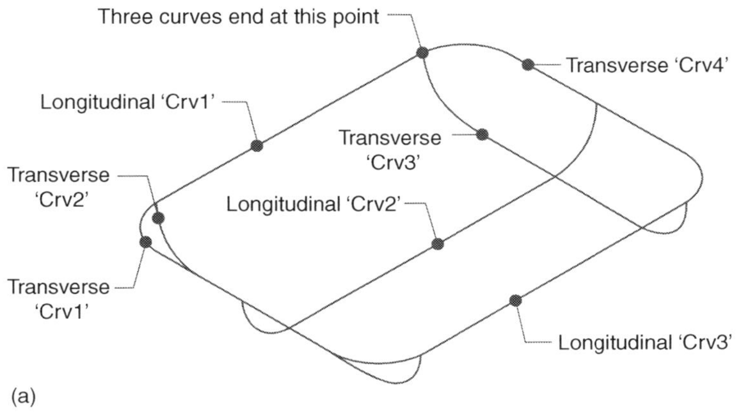

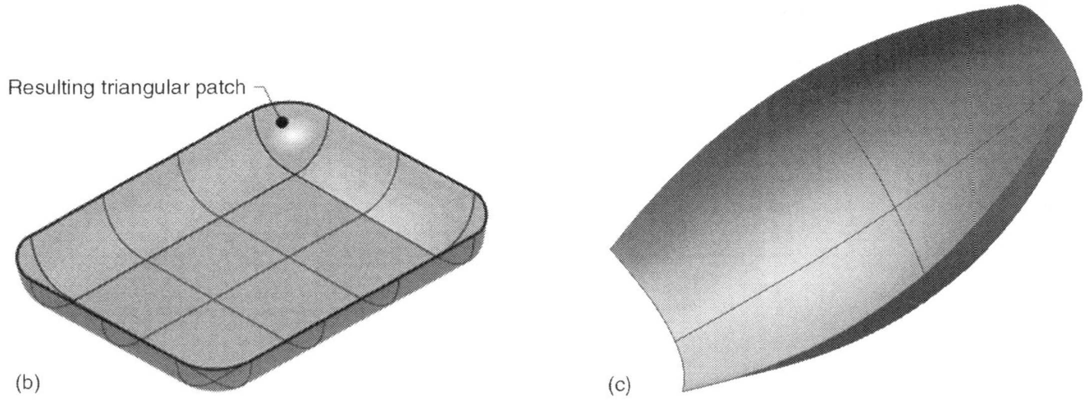

Fig. 8.75 (a), (b) and (c) Coon surfaces.

(c) Coons Surface: The previously described surfaces are used with either open boundaries or given data points. A coons patch is used to create a surface using curves that form closed boundaries (Fig. 8.75).

(d) Fillet Surface: It is a b-spline surface that blends two surfaces together. The two original surfaces may or may not be trimmed (Fig. 8.76).

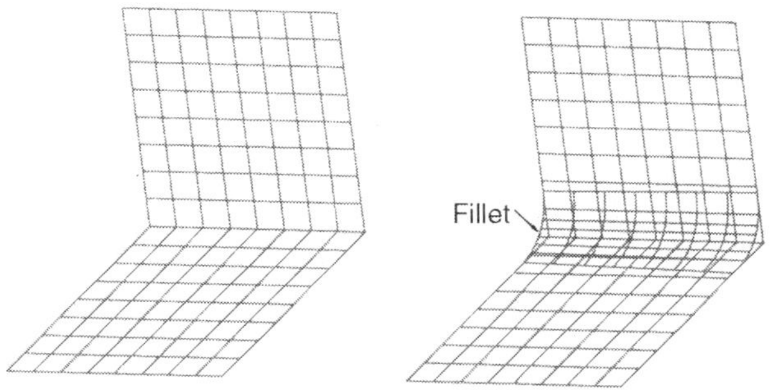

Fig. 8.76 Fillet surface.

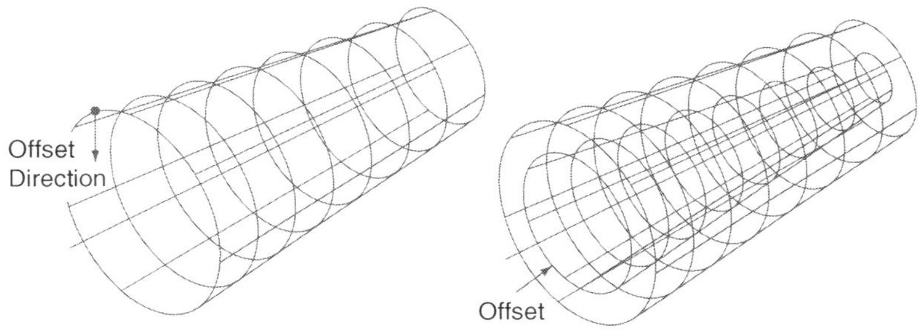

Fig. 8.77 Offset surface.

(e) Offset Surface: Existing surfaces can be offset to create new ones identical in shape but with different dimensions. It is a useful surface to speed up surface creation (Fig. 8.77).

Limitations of surface modelling

(a) Mass properties such as volume cannot be easily found.
(b) Hidden lines cannot be removed easily.
(c) Some complex surface models give unreliable volume data.

8.11.3 Solid Modelling

A solid model is described in terms of the volumetric shape which it occupies. Solid modelling is thus, the only technique that provides a full, unambiguous description of a 3D shape. This type of modelling is recent and the most advanced of the three models.

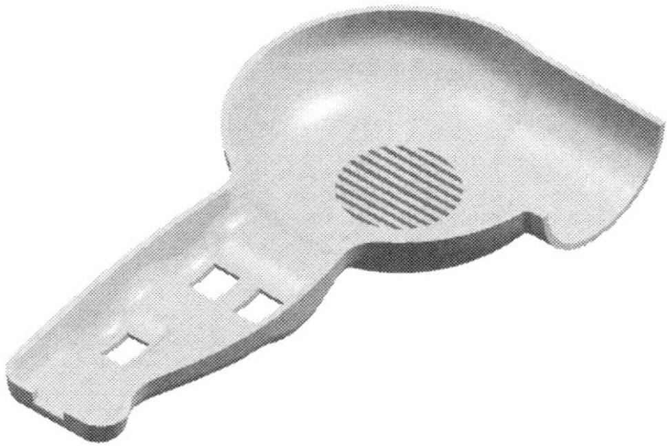

Fig. 8.78 Solid model of a hair dryer case.

Advantages of solid modelling

(a) Complete definition of volumetric shape, i.e., easy to calculate area, volume, mass, centre of gravity, and moment of inertia, etc.
(b) Ability to distinguish between the inside and outside of an object.
(c) Ability to detect unwanted interference between components.
(d) Ability to provide automatic removal of hidden lines.
(e) Ability to assemble 3D models and to see the exploded views.
(f) Solid modelling greatly improves efficiency in many design and manufacturing applications.
(g) Automatic linkage of data between all views.
(h) Large amount of editing is possible.
(i) Ability to provide clear, automatic 3D sectional views through components, with particular advantages on complex assemblies.
(j) Analytical advantages, including efficient construction of finite elements for analysis (such as structural, thermal, etc).

(k) Improved simulation of mechanism dynamics, tool path verification, and robot handling.

(l) Ability to incorporate extensive colour choice and tone control. It is also possible to manipulate the light source and produce shadow effects and background settings.

Solid modelling primitives Solid models are built from basic building blocks called solid modelling primitives. These are defined in terms of their solid shape, size, position and orientation. The primitives are combined by a mathematical set of Boolean operations to create solid models. Basic solid modelling primitives are:

1. Block
2. Cylinder
3. Cone
4. Sphere

Boolean operators These are the essential tools in building up the solid model. In this they define the relationship between neighbouring primitives. Boolean operators are based on algebraic set theory. The three Boolean operators are:

1. Union ($A \cup B$)
2. Difference (or) subtract ($A - B$)
3. Intersection ($A \cap B$)

Types of Solid Modelling

Solid modelling can be classified into three categories. They are:

1. Constructive Solid Geometry (CSG) Modelling: In CSG a complex model can be constructed in a building block manner using primitives by combining shapes or subtracting or intersecting them from each other. This is the first method developed for solid modelling. It is relatively easy to construct. It is compact in storage requirement, but slow in producing pictures. It is not possible to make complex shapes.

2. Constructive Representation (C-Rep) Modelling: Constructing the C-Rep model, solid primitives may be created by sweeping 2D areas in 3D space as with surface modelling, but with the essential difference that solid volume is now generated. Then by doing Boolean operations solid models are constructed. This modelling is also called sketcher modelling. This can support making complex shapes but still not is able to make contour shapes involving variable cross sections which sweeps along variable guides.

3. Boundary Representation (B-Rep) Modelling: To create models using B-Rep, the same techniques may be used, i.e., primitives may be formed with linear or rotational sweeps and composite shapes built up using Boolean operations. C-Rep modellers recognise a composite body in terms of the primitive solids from which it was built, where as a B-Rep modeller recognise that body in terms of the edges and faces. The main advantage of B-Rep modelling is the boundary profiles which may be more easily modified. B-Rep systems are more demanding on memory capacity. CAD/CAM software's employ C-Rep, CSG and B-Rep modelling concepts. B-Rep modelling gives the designer more freedom in building complex and contour shape models. It is more expensive on memory requirement. This modelling concept is the only one used in analysis software.

8.11.4 Assembly

An assembly is a collection of independent parts. It is important to understand the nature and the structure of dependencies between parts in an assembly to be able to model the assembly properly. The assembly model must include the spatial positions and hierarchical relationship among the parts, and the assembly relationships (mating conditions) between parts.

Assembly tree The most natural way to represent the hierarchical relationships between the various parts of an assembly is an assembly tree. An assembly is divided into several subassemblies at different levels. Each subassembly is composed of various parts. The leaves of the tree represent individual parts or subassemblies.

Assembly planning Assembly planning is a key to create successful assemblies, especially the large ones that are typically encountered in practise. The important issue is not only creating the assembly, but also updating it in the future when design changes are made to the individual parts. These updates should be done automatically and correctly. When a designer changes some of the assembly parameters, others should update accordingly.

Factors to be considered in assembly

1. Identifying the dependencies between the components of an assembly.
2. Identifying the dependencies between the features of each part.
3. Analysing the order of assembling the parts.

Mating Conditions Locating and orienting parts in their assembly is achieved by specifying mating conditions among them to specify relationship among the parts. When using mating conditions to assemble two parts, there must be six degrees of freedom for an object - three translations and three rotations. A part should be fully constrained to allow the creation of the correct assembly. Constraining the parts in assembly is achieved by options such as coincident, concentric, tangent, parallel, perpendicular, distance, etc.

Types of Assembly The assembly is divided into two types. They are:

1. Top down assembly
2. Bottom up assembly

1. Top down Assembly: It is a type of assembly approach in which the assembly layout communicates design criteria to subsystem developers including suppliers. It is ideal for large assemblies consisting of tens of thousands of components. It provides an effective tool and a well-organised approach to managing the design or large assemblies. It allows a project leader to break up product specifications, assign work teams, and enforce downstream design changes at a high level. The top down assembly approach fosters a systems engineering approach to product design, in which the assembly layout communicates design criteria to subsystem developers, including suppliers. This tight control allows distributed design teams to work concurrently within a common product framework. The major advantage of top down assembly is that if we change the layout sketch, the assembly and its parts are automatically updated.

2. **Bottom up Assembly:** It is a type of assembly approach in which the individual parts have been created and inserted into an assembly using the mating conditions to locate and orient them as required by the assembly design. In this assembly we make the parts individually. Assembly modelling process begins with creating a blank assembly model or file. Then import the assembly parts into this file one by one. The first part we insert is known as the base part of assembly. When inserting a part into an assembly, we insert copies of the parts. The copies are known as instances. The CAD software maintains a link between each instance and its original parts. If we change any original part the part in assembly will be automatically updated. We can modify the original part from the assembly also because assembly link is bidirectional. This is the preferred technique if the parts have already been constructed. It also allows designers to focus on the individual parts. It also makes it easier and simpler to maintain the relationships and regeneration behaviour of parts than in the top down approach. Bottom up approach is best suited for small assemblies.

8.12 CONCEPTS OF ENGINEERING DATABASE

Modelling data is generally of four types. These are shape, non-shape, design and manufacturing data. Shape data consists of both geometrical and topological information. Non-shape data includes images and measuring units of the database. Design data includes FEM/FEA. Manufacturing data includes tolerance and bill of materials. The transfer of modelling data between dissimilar CAD/CAM systems is achieved with the help of translators. There are two types of translators available namely direct and indirect.

8.12.1 Direct Translators

It translates the modelling data directly from one CAD/CAM system format to another, usually in one step. It requires the knowledge of both the native formats. These translators are suitable only for few systems normally between two or three. Direct translators run more quickly than the indirect translators. The data files produced by direct translators are smaller than the neutral files created by the indirect translators.

8.12.2 Indirect Translators

It translates the modelling data of one system to a neutral file, which is independent of any CAD/CAM system and from that neutral file it again converts it into modelling data of another system. The translator that converts data from the native format of the given CAD/CAM system to the neutral format is called a **preprocessor**. The translator that converts data from the neutral format to the native format is called **Postprocessor**. Indirect translator provides stable communication between CAD/CAM systems and eliminates the dependence on single system supplier. Most CAD/CAM systems use indirect translators.

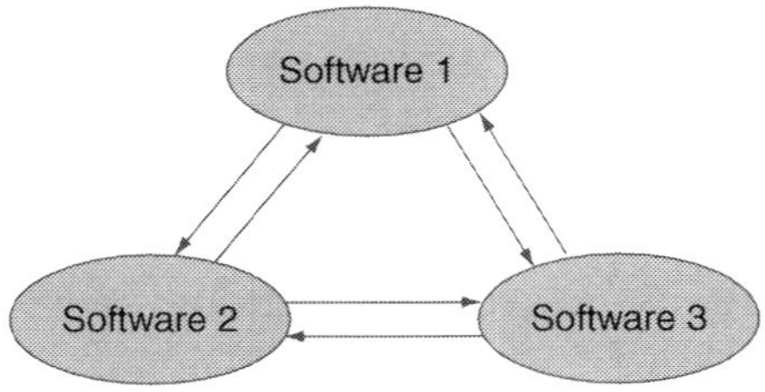

Fig. 8.79 Direct data translation.

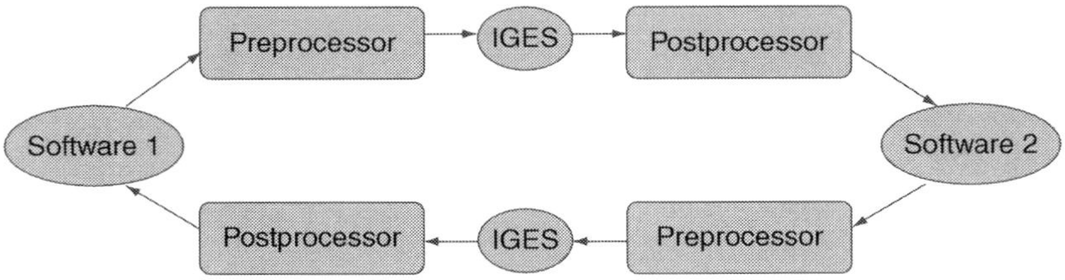

Fig. 8.80 Indirect data translation.

8.12.3 Translators used in CAD/CAM Software

The commonly used translators in CAD/CAM are of indirect type are:

1. DXF (Data Exchange Format): This is a popular data exchange format adopted by many CAD system vendors. DXF format is easy to interpret though it is a long file. It is a format that translates drawing files. It cannot translate part files effectively. DXF file comes in two formats ASCII and binary. ASCII (American Standard Code for Information Interchange) version is the most widely used in industry. DXF file consists of four sections. (a) Header (b) Tables (c) Block (d) Entities. Header section includes the autoCAD system settings such as dimension style and layers. Table section includes line styles and user defined coordinate systems. The block section includes drawing blocks. The entities section includes entity definition and data. Almost all the CAD/CAM software has this translator.

2. IGES (Initial Graphics Exchange Specification): This is the first exchange format developed to address the concept of communicating product data among dissimilar CAD/CAM systems. IGES is the ANSI standard. It can support solid modelling, including B-representation and CSG scheme. IGES has three data types; geometric, annotation and structure. Geometric entities define the product shape and include curves, surfaces and solids. Annotation entities include various types of dimensions, centrelines, nodes, labels, symbols and cross-hatching. Structure entities include views, drawings, attributes, properties, and macros. Almost all the CAD/CAM software has this translator also.

3. STEP (Standard for Exchange of Product data): This is an ISO standard. It has absorbed PDS (Product Data exchange Standard), which was an ANSI standard. The goal of STEP is to represent all product information in a common data format throughout a product's entire life cycle. The data being transferred is geometry, analysis, manufacturing, implementation, and testing procedures. STEP is a common structure, operating as a template for sharing the data among multiple users, across all functional areas. This translator available in mid and high end CAD/CAM software such as solid edge, solid works, IDEAS, CATIA, pro-E, unigraphics, etc.

4. Parasolid: It can support almost all the data of the model. Recent CAD/CAM software has this type of file format. It is a recently developed format among the translators. Some of the high end CAD/CAM software, which has this translator, are unigraphics, proengineer, etc.

8.13 ADVANCED CAD/CAM TECHNOLOGIES

Several advances have been made in CAD/CAM technology in recent years. This has made CAD/CAM more purposeful. Some of the advanced techniques which have become very popular are discussed below:

1. Rapid prototyping
2. Rapid tooling
3. Reverse engineering
4. Concurrent engineering
5. Quality function deployment

8.13.1 Rapid Prototyping

Rapid Prototyping (RP) is the automated fabrication technologies of seamless and rapidly creating accurate representative physical models of mechanical parts directly from 3-dimensional computer aided design (CAD) data without the use of tooling and with minimal human intervention.

RP uses state of the art laser technology, positioning systems, materials and computer technologies in the various processes. There are many RP processes that are widely used, each one using different methods and materials to produce the final part.

Benefits or advantages of rapid prototyping

(a) Reduces design phase cycle time and costs.
(b) Reduces the potential for expensive design errors.
(c) Reduces tooling costs on short-run parts.
(d) Impresses the customer with quick response.
(e) Reduces time of production and market.

Basic process steps in rapid prototyping

1. Create CAD model of the design.
2. Convert the CAD model to STL format.
3. Slice the STL file into thin cross-sectional layers.
4. Layer by layer construction.
5. Clean and finish the model.

Limitations of rapid prototyping

(a) The part produced by rapid prototyping is built to near net shape.
(b) Additional machining, polishing or other processes may be needed.
(c) Materials should have excellent machining properties.
(d) There may be quantity, size and use limitations.

Rapid prototyping materials

1. Metals: Steel alloys, aluminium, titanium, super alloys.
2. Non-metals: Plastics, ceramics, composites, blended materials in a single part.

Rapid prototyping applications

1. A design verification and optimisation tool to qualify the form/fit/function of individual parts and assemblies.
2. Concept visualisation tools to verify design details and gain internal design acceptance and justification.
3. Communication tools for internal design reviews, for design reviews with the customer and for dry fit installation checks.
4. As a three-dimensional fixture for bending or routing tubing or cables.
5. As an inspection fixture for parts with complex or compound surfaces and fixtures.
6. As a model to test airflow, ducting, diverters and channels.
7. To create light duty plastic parts for light duty use, such as ducting.
8. As a mould for gaskets, keypads, etc.
9. Most rapid prototyping parts may be machined, drilled and tapped, sanded, painted, baked, plated, bonded and coated with EMI protection.

Feature development in rapid prototyping

1. The first important development is accuracy and surface finish (~0.08 mm).
2. Another important development is increased size capacity.
3. Increase in the speed.

Rapid prototyping techniques

There are six techniques used in rapid prototyping technology. They are:

1. Stereolithography
2. Laminated object manufacturing
3. Selective laser sintering
4. Fused deposition modelling
5. Solid ground curing
6. 3D inkjet printing

1. Stereolithography (SLA): Stereolithography (SLA) is the first rapid prototyping process. It is the most widely used technology.

Highlights of stereolithography

1. Inexpensive compared to other techniques.
2. Uses a light-sensitive liquid polymer.
3. Requires post-curing since laser is not of high power to completely cure.
4. Long-term curing can lead to warping.
5. Parts are quite brittle and have a tacky surface.
6. No milling step so accuracy in Z can suffer.
7. Support structures are required.
8. Process is simple. No masking steps required.
9. Uncured material can be toxic. Ventilation is a must.

Working of SLA The implementation shown in Fig. 8.81 is used by 3D systems and some foreign manufacturers. A movable table, or elevator (a), initially is placed at a position just below the surface of a vat (b) filled with liquid photopolymer resin (c). This material has the property that when light of the correct colour strikes it, it turns from liquid to solid. The most common photopolymer materials used require an ultraviolet light, but resins that work with visible light are also utillised. The system is sealed to prevent the escape of fumes from the resin.

A laser beam is moved over the surface of the liquid photopolymer to trace the geometry of the cross section of the object. This causes the liquid to harden in areas where the laser strikes. The laser beam is moved in the X-Y directions by a scanner system (d). These are fast and highly controllable motors which drive mirrors and are guided by information from the CAD data.

The exact pattern that the laser traces is a combination of the information contained in the CAD system that describes the geometry of the object, and information from the rapid prototyping application software that optimises the faithfulness of the fabricated object.

After the layer is completely traced and for the most part hardened by the laser beam, the table is lowered into the vat a distance equal to the thickness of a layer. Early stereolithography systems drew a knife edge (e) over the surface to smooth it. More recently pump-driven recoating systems have been utilised. The tracing and recoating steps are repeated until the object is completely fabricated and sits on the table within the vat.

The geometry of some objects has overhangs or undercuts. These must be supported during the fabrication process. The support structures are either manually or automatically designed.

Upon completion of the fabrication process, the object is elevated from the vat and allowed to drain. Excess resin is swabbed manually from the surfaces. The object is often given a final cure by bathing it in intense light in a box resembling an oven called a Post Curing Apparatus (PCA). After final cure, supports are cut off the object and surfaces are sanded or otherwise finished.

Stereolithography provides the greatest accuracy and best surface finish of any rapid prototyping technology. Work continues to provide materials that have wider and more directly useable mechanical properties.

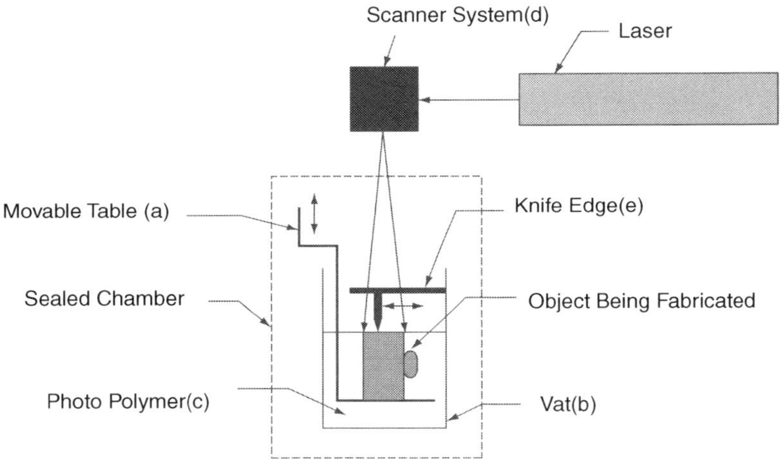

Fig. 8.81 Stereolithography.

2. Fused Deposition Modelling (FDM): FDM is the second most widely used rapid prototyping technology, after stereolithography.

Highlights of fused deposition modelling

1. Standard engineering thermoplastics, such as ABS, can be used to produce structurally functional models.
2. Parts up to 600 × 600 × 500 mm (24 × 24 × 20 inches) can be produced.
3. Filament of heated thermoplastic polymer is squeezed out like toothpaste from a tube.
4. Thermoplastic is cooled rapidly since the platform is maintained at a lower temperature.
5. Milling step is not included and layer deposition is sometimes non-uniform so 'plane' can become skewed.

Working of FDM A plastic filament, approximately 1/16 inch in diameter, is unwound from a coil (a) and supplies material to an extrusion nozzle (b). Some configurations of the machinery have used plastic pellets fed from a hopper rather than a filament. The nozzle is heated to melt the plastic and has a mechanism which allows the flow of the melted plastic to be controlled. The nozzle is mounted to a mechanical stage (c) which can be moved in horizontal and vertical directions.

As the nozzle is moved over the table (d) in the required geometry, it deposits a thin bead of extruded plastic to form each layer. The plastic hardens immediately after being squirted from the nozzle and bonds to the layer below. The entire system is contained within an oven chamber which is held at a temperature just below the melting point of the plastic. Thus, only a small amount of additional thermal energy needs to be supplied by the extrusion nozzle to cause the plastic to melt. This provides much better control of the process.

Support structures must be designed and fabricated for any overhanging geometries and are later removed in secondary operations. Several materials are available for the process including a nylon-like polymer and both machinable and investment casting waxes. The introduction of ABS plastic material led to much greater commercial acceptance of the method. It provides better layer to layer bonding than previous materials and consequently much more robust fabricated objects. The recent introduction of polycarbonate and polysulfone modelling materials has further extended the capabilities of the method in terms of strength and temperature range. Several other polymer systems as well as ceramic and metallic materials are under development.

The method is office-friendly and quiet. FDM is fairly fast for small parts on the order of a few cubic inches, or those that have tall, thin form-factors. It can be very slow for parts with wide cross sections. The finish of parts produced with the method has been greatly improved over the years, but are not quite on par with stereolithography. The closest competitor to the FDM process is probably three dimensional printing. However, FDM offers greater strength and wider range of materials than 3DP.

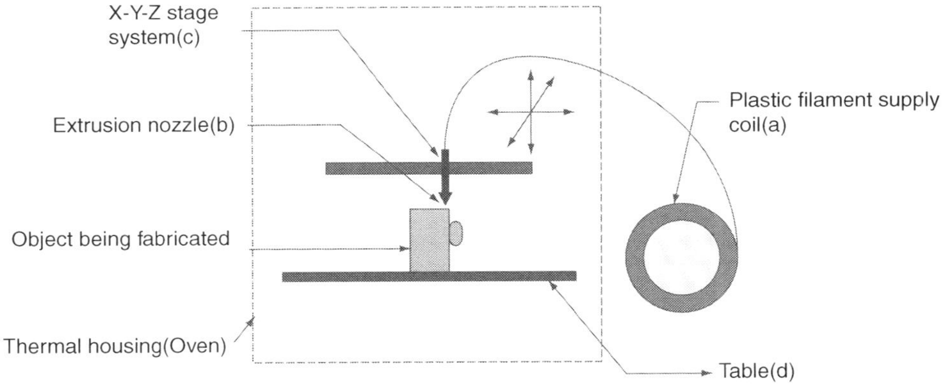

Fig. 8.82 Fused deposition modelling.

3. Laminated Object Manufacturing (LOM):

Highlights of laminated object manufacturing

1. Layers of glue-backed paper form the model.
2. Low cost: Raw material is readily available.

3. Large parts: Because there is no chemical reaction involved, parts can be made quite large.
4. No milling step. Accuracy in Z is less than that for SLA and SLS.
5. Outside of model, cross-hatching removes material.
6. Models should be sealed in order to prohibit moisture.
7. Before sealing, models have a wood-like texture.
8. Not as prevalent as SLA and SLS.

Working of LOM Profiles of object cross sections are cut from a paper using a CO_2 laser as shown in Fig. 8.83. The paper is unwound from a feed roll (a) onto the stack and bonded to the previous layer using a heated roller (b). The roller melts plastic coating on the bottom side of the paper to create the bond. The profiles are traced by an optics system that is mounted to an X-Y stage (c). The process generates considerable smoke. Either a chimney or a charcoal filtration system is required (e) and the build chamber must be sealed.

After cutting the geometric features of a layer is completed, the excess paper is cut away to separate the layer from the Web. The extra paper of the Web is wound on a take-up roll (d). The method is self-supporting for overhangs and undercuts. Areas of cross sections which are to be removed in the final object are heavily cross-hatched with the laser to facilitate removal. It can be time consuming to remove extra material for some geometry. The finish and accuracy are not good when compared to other methods.

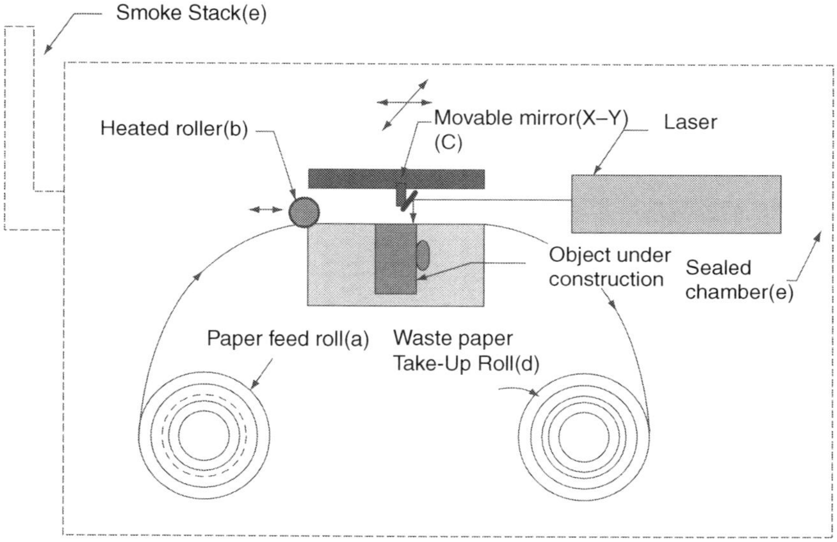

Fig. 8.83 Laminated object manufacturing.

4. Selective Laser Sintering (SLS):

Highlights of selective laser sintering

1. Considerably stronger than SLA; sometimes structurally functional parts are possible.
2. Laser beam selectively fuses powder materials such as nylon and elastomer.

3. Advantage over SLA: Variety of materials can be used.
4. No milling step so accuracy in Z can suffer.
5. Process is simple: No masking steps required.
6. Living hinges are possible with the thermoplastic-like materials.
7. Powdery, porous surface unless sealant is used. Sealant also strengthens part.
8. Uncured material is easily removed by brushing or blowing it off.

Working of SLS The process is somewhat similar to stereolithography in principle as can be seen from Fig. 8.84. In this case, laser beam is traced over the surface of a tightly compacted powder made of thermoplastic material (a). The powder is spread by a roller (b) over the surface of a build cylinder (c). A piston (d) moves down one object layer thickness to accommodate the layer of powder. The powder supply system (e) is similar in function to the build cylinder. It also comprises a cylinder and piston. In this case the piston moves upward incrementally to supply powder for the process.

Heat from the laser melts the powder where it strikes under guidance of the scanner system (f). The CO_2 laser used, provides a concentrated infrared heating beam. The entire fabrication chamber is sealed and maintained at a temperature just below the melting point of the plastic powder. Thus, heat from the laser need only elevate the temperature slightly to cause sintering, greatly speeding up the process. A nitrogen atmosphere is also maintained in the fabrication chamber which prevents the possibility of explosion in the handling of large quantities of powder.

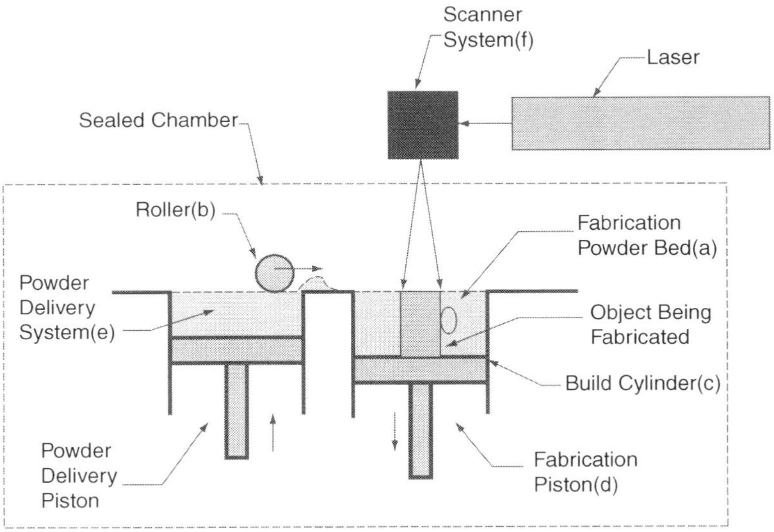

Fig. 8.84 Selective laser sintering.

After the object is fully formed, the piston is raised to elevate the object. Excess powder is simply brushed away and final manual finishing may be carried out. It may take a considerable time before the part cools down enough to be removed from the machine. Large parts with thin sections may require as much as two days of cooling time.

No supports are required with this method since overhangs and undercuts are supported by the solid powder bed. This saves some finishing time compared to stereolithography. No final curing is required as in stereolithography, but since the objects are sintered they are porous. Much progress has been made over the years in improving surface finish and porosity. The method has also been extended to provide direct fabrication of metal and ceramic objects and tools.

5. Solid Ground Curing (SGC):

Highlights of Solid Ground Curing

1. Large parts, 500 × 500 × 350 mm (20 × 20 × 14 in), can be fabricated quickly.
2. High speed allows production-like fabrication of many parts or large parts.
3. Masks are created with laser printing-like process, then full layer exposed at once.
4. No post cure required.
5. Milling step ensures flatness for subsequent layer.
6. Wax supports model: No extra supports needed.
7. Creates a lot of waste.
8. Not as prevalent as SLA and SLS.
9. Good accuracy and very high fabrication rate.
10. High operating cost and system complexity so poor market acceptance.

Working of SGC Instead of using a laser to expose and harden photopolymer element by element within a layer as is done in stereolithography, SGC uses a mask to expose the entire object layer at once with a burst of intense UV light. The method of generating the masks is based on electrophotography (xerography).

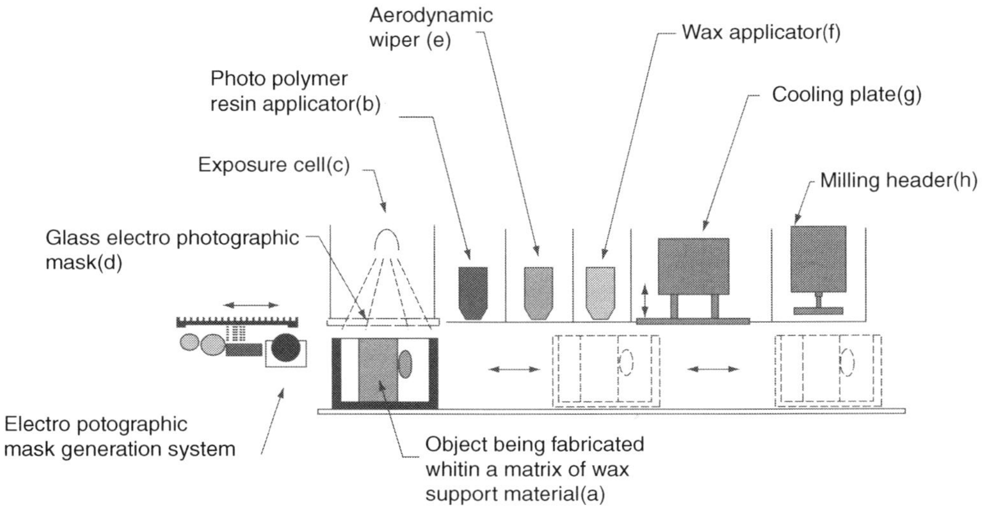

Fig. 8.85 Solid ground curing.

This is a two-cycle process having a mask generation cycle and a layer fabrication cycle. It takes about two minutes to complete all operations to make a layer:

(a) First the object under construction (a) is given a coating of photopolymer resin as it passes the resin applicator station (b) on its way to the exposure cell (c).

(b) A mask is generated by electrostatically transferring toner in the required object cross sectional image pattern to a glass plate (d). An electron gun writes a charge pattern on the plate which is developed with toner. The glass plate then moves to the exposure cell where it is positioned above the object under construction.

(c) A shutter is opened allowing the exposure light to pass through the mask and quickly cure the photopolymer layer in the required pattern. Because the light is so intense the layer is fully cured and no secondary curing operation is necessary as is the case with stereolithography.

(d) The mask and object under fabrication then part company. The glass mask is cleaned of toner and discharged. A new mask is electrophotographically generated on the plate to repeat the cycle.

(e) The object moves to the aerodynamic wiper (e) where any resin that wasn't hardened is vacuumed off and discarded.

(f) It then passes under a wax applicator (f) where the voids created by the removal of the unhardened resin are filled with wax. The wax is hardened by moving the object to the cooling station (g) where a cold plate is pressed against it.

(g) The final step involves running the object under the milling head (H). Both the wax and photopolymer are milled to a uniform thickness and the cycle is repeated until the object is completely formed within a wax matrix.

Secondary operations are required to remove the wax. It can either be melted away or dissolved using a dish-washing-like machine. The object is then sanded or otherwise finished as is done in stereolithography.

6. Three Dimensional Printing (3DP):

Highlights of 3DP

1. It offers multicolour possibilities
2. It emerges as a leading RP technique
3. No milling and masking steps required
4. Low cost

Working of 3DP The system is shown schematically in Fig. 8.86. The method resembles selective laser sintering, except that the laser is replaced by an inkjet head. The multichannel jetting head (a) deposits a liquid adhesive compound onto the top layer of a bed of powder object material (b). The particles of the powder become bonded in the areas where the adhesive is deposited.

Once a layer is completed, the piston (c) moves down by the thickness of a layer. As in selective laser sintering, the powder supply system (e) is similar in function to the build cylinder. In this case the piston moves upward incrementally to supply powder for the process and the roller (d) spreads and compresses the powder on the top of the build cylinder. The process is repeated until the entire object is completed within the powder bed. After completion, the object is elevated and the extra powder brushed away.

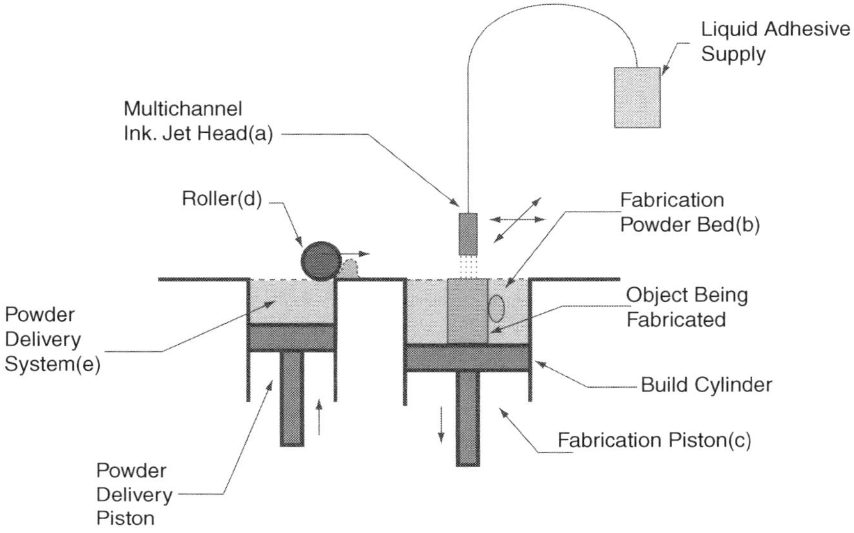

Fig. 8.86 Three-Dimensional printing.

8.13.2 Rapid Tooling

The term **Rapid Tooling (RT)** is typically used to describe a process which either uses a rapid prototyping (RP) model as a pattern to create a mould quickly or uses the rapid prototyping process directly to fabricate a tool for a limited volume of prototypes.

The difference between RT and conventional tooling

1. Tooling time is much shorter than for a conventional tool.
2. Tooling cost is much less than for a conventional tool. Cost can be below five percent of conventional tooling cost.
3. Tool life is considerably less than for a conventional tool.
4. Tolerances are wider than for a conventional tool.

8.13.3 Reverse Engineering

Reverse Engineering (RE) is a process that is used to create 3D CAD models directly from physical parts with little or no additional design documentation. It is simply the act of figuring out the real parts using software that you have.

Where reverse engineering is useful

1. In many cases, it is extremely difficult to clearly define physical part geometry using traditional measurement techniques.

2. Reverse engineering quickly and accurately bridges the gap between poorly or undocumented tooling and fully modifiable 3D CAD models useful for modern manufacturing methods.

Stages in reverse engineering process

1. Sample parts or objects provided are digitised using digitiser / 3D scanner / coordinate measuring machine.
2. The data is refined and used as a foundation for creating one of the various 3D file types available (STL, STEP, IGES, etc.).
3. Prototype models are made for design confirmation and market evaluation purposes.

Machines used in reverse engineering to digitise the product

1. Digitiser
2. Coordinate measuring machine

1. Digitiser: Product scanning also known as digitising or 3D digitising is a process that uses light to capture the shape of three-dimensional objects and recreate them in a virtual workspace. The data is collected as points and the resultant file is called a *point cloud*.

Steps in 3D digitising process

1. Light is projected onto the object.
2. The object reflects the light, which is then collected by a digital sensor.
3. Using algebraic equations, the 3D coordinates of the surface point are calculated.
4. The point's coordinate location is stored as part of a point cloud representing the physical part.
5. Millions of points are collected this way until the entire surface of the part or object has been digitised.
6. The digital data (point cloud) is used for reverse engineering, rapid prototyping and product inspection.

Use of 3D digitising

It captures high-density geometry, compound surface curvature, and draft that are difficult to measure using traditional measurement techniques.

Materials for digitising

1. Stone
2. Ceramic
3. Glass
4. Metal
5. Wood

6. Bone
7. Plastics
8. Rubber
9. Wax
10. Clay

Uses of point cloud

1. For making 3D model using reverse engineering.
2. Making prototype from the 3D model by converting it to STL format.
3. Product inspection purposes (comparing RP model with an existing CAD model and is used to create a colour error map).

2. Coordinate Measuring Machine (CMM): Coordinate measuring machines are mechanical systems designed to move a measuring probe to determine coordinates of points on a workpiece surface.

Main Components of CMM

1. Measuring probe
2. Computing unit
3. Measuring software
4. Machine itself

Specification for CMM

1. Measuring lengths along x, y and z-axes.
2. Resolution (accuracy)
3. Work piece weight and size.

Applications of CMM

1. Dimensional measurement
2. Profile measurement
3. Angularity or orientation measurement
4. Depth mapping
5. Digitising or imaging
6. Shaft measurement

Features common to CMM

1. Crash protection
2. Offline programming
3. Reverse engineering of part
4. Shop floor suitability
5. Software along with machine
6. Temperature compensation

8.13.4 Concurrent Engineering

Present scenario in manufacturing industry is that any new product raises customer expectations. Globalisation and developments in new technology complicate the matter further. Also the customers want access to information about the new product specification. People are aware of the fact whether the new product is pollution free. Finally there is intense competition among the producers to bring out the product to the market in a short time.

Product Development Cycle Generally the product development cycle starts with developing the product concept, evolving the design, engineering the product, planning, manufacturing the part, marketing and servicing.

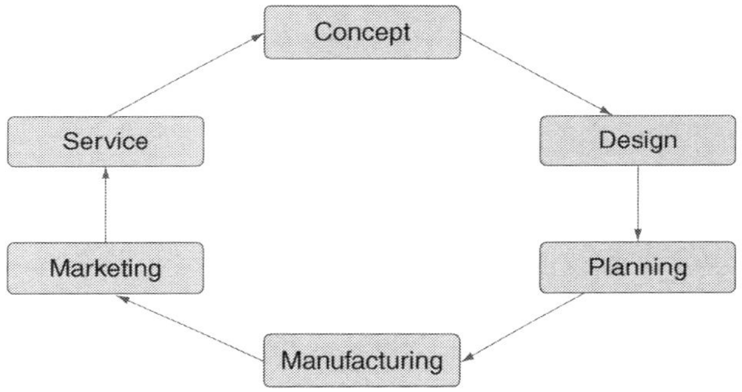

Fig. 8.87 Product development cycle

Sequential Engineering approach The traditional product development cycle is called sequential engineering. In sequential engineering product design, development of manufacturing process and supporting quality and testing activities, all carried out one after another. In this old product development cycle time taken for product development is long and the response to the market requirements is slow. In the age of reduced product life cycles, the gap between market demand and the introduction of new products in the markets has to be narrow.

Concurrent Engineering approach Concurrent Engineering (CE) or simultaneous engineering is a methodology of restructuring the product development activity in an organisation using a cross functional team. It is a technique adopted to improve the efficiency of product design and reduce the product design cycle time. This is also sometimes referred to as parallel engineering. Concurrent engineering brings together a wide spectrum of people from several functional areas in the design and manufacture of a product. Representatives from research and development, engineering, manufacturing, materials management, quality assurance, marketing, etc., develop the product as a team. Everyone interacts with other from the start, and they perform their tasks in parallel.

Concurrent engineering gives marketing and other groups the opportunity to review the design during the modelling, prototyping and soft tooling phases of development. Intensive teamwork between product developments, production planning and manufacturing is essential for satisfactory implementation of concurrent engineering. Hewlett Packard (HP) in Japan, Neon car in USA and Scooty moped from TVS-Suzuki Ltd. in India are the best examples for the companies using concurrent engineering approach.

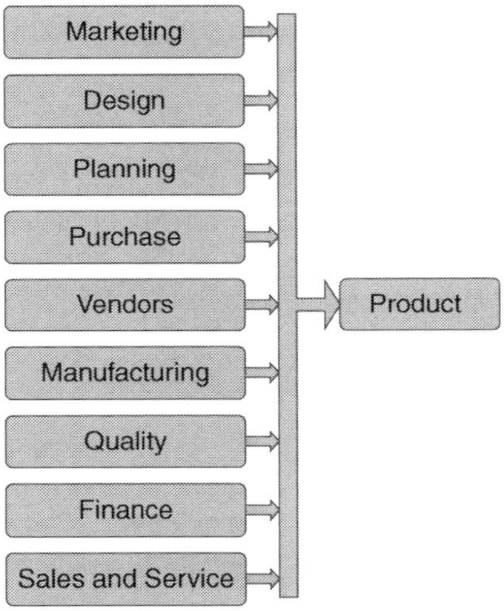

Fig. 8.88 Concurrent engineering.

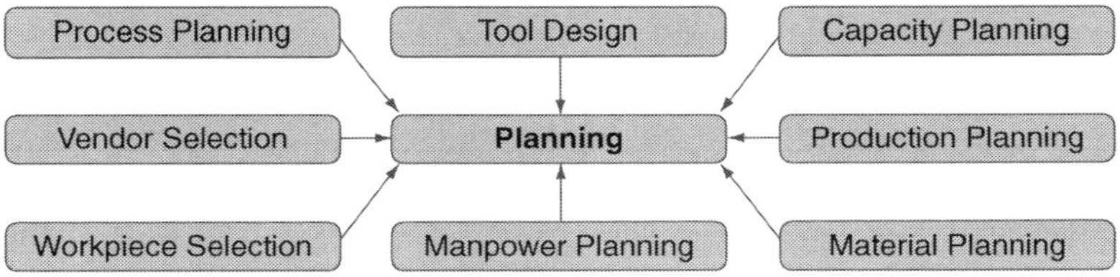

Fig. 8.89 Concurrent workflow within an activity.

Characteristics of concurrent engineering The concurrent engineering approach can be characterised by the following factors:

1. Integration of product development, process development and logistic support.
2. Closer attention of the needs of customers and new technologies.
3. Continuous review of design and development process.
4. Risk analysis at important stages of product development.
5. Rapid and automated information exchange through LANs, Internet, itranet.
6. Design team consisting of members drawn from different disciplines.
7. Rapid prototyping.

Advantages of concurrent engineering

(a) The cooperation between various specialists.
(b) Systematic application of special methods.
(c) Ensures quick optimisation of design.

(d) Early detection of possible faults in products and production planning. This additionally leads to reduction in lead-time, which reduces cost of production, and guarantees better quality.

8.13.5 Quality Function Deployment

Quality Function Deployment (QFD) is the latest approach to product design. It essentially consists of converting customer's need statement (which is usually qualitative) into technical specifications. For example, a user of automobile insists upon "easy closure" of the door. This voice of the customer enables the design task force to derive the specifications of door closing mechanism in terms of kilograms of force required for the mechanism. QFD enables organisations to be proactive rather than reactive in QC. QFD involves (a) the customer, (b) what are the customer requirements, and (c) how to fulfill his requirements.

Steps in quality function deployment

1. To objectively determine what groups constitute the customer base and who will benefit from the successful production or implementation of this product.
2. To determine the wants of each customer or customer group. (WHAT)
3. To determine a means by which to satisfy the customers. (HOW)
4. To determine the correlation between HOW and WHAT in the matrix.
5. An objective evaluation of the current performance of the organisation about what should be done and how it should be carried out.
6. The ratio of improvements (after execution) should be calculated.

Implementation of QFD in production A four-fold strategy for production engineering division can be emphasised as follows:

1. Establish a single centre for the integration and dissemination of tools to support the concurrent engineering process needed for QFD implementation on the shop floor.
2. Develop and validate analytical tools which increase the quality and quantity of information available to support the development of systems needed for incorporating the change in product or process.
3. Reduce the time and effort required to develop and transform systems into production by eliminating non-value adding activities.
4. Broaden the technology base by providing the ability to rapidly produce critical items for test, evaluation, and field trials.

The aim is to either locate or develop knowledge base or other tools related to the production function. Evaluate its effectiveness and utility to other groups. Then promote the transfer of this technology throughout the departments. The desired result is a more efficient and cost effective means for all organisations involved in the production process to share their collective knowledge and resources between various industries.

8.14 COMPUTER AIDED ENGINEERING

Computer Aided Engineering (CAE) is defined as the use of computer system to find the analytical solution for various engineering problems using different numerical methods. For many of the practical problems an engineer uses numerical methods to solve the problem and provide approximate but acceptable solution. For problems involving complex shapes, material properties and complicated boundary conditions, it is difficult to obtain analytical solutions. There are three numerical methods available in CAE to obtain analytical solution for engineering problems.

1. Functional approximation
2. Functional difference method
3. Finite Element Method (FEM) / Finite Element Analysis (FEA)

8.14.1 Finite Element Method

Finite element method / finite element analysis is one of the very popular mechanical engineering applications offered by existing CAD/CAM systems. Finite element method is the most popular numerical analysis technique for obtaining approximate solutions to a wide variety of engineering problems. It has grown out of the matrix analysis method used in aircraft design. This method is based on dividing a complex shape into small elements, solving the equilibrium equations at hand for each element, and then assembling the element's results to obtain the solution to the original problem. The shape divisions, the choice of the element and the analysis types are among the important decisions for the success of the method. The interpretation of the results of FEA requires a good understanding of the principles of engineering such as linear/non-linear mechanics, static and dynamic, heat transfer, fluid mechanics, etc.

8.14.2 Advantages of FEM

(a) This method is general enough to handle any complex shape or geometry, any material property, any boundary conditions and any loading conditions.

(b) FEM fits the analysis requirements of today's complex engineering systems and designs.

(c) It is an efficient design tool by which designers can perform parametric design studies by considering various design cases, analysing them and choosing the optimum design.

(d) It can solve a wide variety of engineering problems.

8.14.3 Steps in Finite Element Method

FEM and FEA follow a step by step process:

1. Creating the Finite Elements: The first step is to divide a continuum into a finite number of non-overlapping elements. This is achieved by replacing the continuum by a set of key points, called **nodes**, which when connected properly, produce **elements**. The collection of nodes and elements forms the finite element **mesh**. A variety of element shapes and types are available. The larger the number of nodes and elements, the more accurate the finite element solution, but also more expensive. More memory space is also needed to store the finite element model and result, and more computer time is needed to obtain the solution.

2. Approximating the Solution Within an Element: The variation of the unknown (called field variable) in the problem is approximated within each element by a polynomial. The field variable may be scalar (e.g., temperature) or a vector (e.g., displacement). Polynomials are used to approximate the solution over an element domain because they are easy to integrate and differentiate. The degree of polynomial depends on the number of nodes per element, number of unknowns at each node.

3. Developing Element Matrices and Equations: Once the nodes and material properties of a given element are defined, its corresponding matrices (stiffness matrix, mass matrix, etc.) and equations can be derived. Four methods are available to derive element matrices and equations.

(a) The direct method

(b) The variational method (suitable for solid mechanics problem)

(c) The weighted residual method (suitable for thermal fluids problems)

(d) The energy method.

4. Generating the Global System Matrix Equation: The individual element matrices are added together by summing the equilibrium equations of the elements to obtain the global matrices and the system of algebraic equations. Before solving this system, it must be modified by applying the boundary conditions. If boundary conditions are not applied, wrong results will be obtained.

5. Solving the Unknown at the Nodes: The global system of algebraic equations is solved via Gauss elimination methods to provide the values of the field variables at the nodes of the finite element mesh. Values of the field variables and their derivatives at the nodes form the complete finite element solution of the original continuum problem.

6. Interpreting the Results: The final step is to analyse the solution and results obtained from the previous step to make design decisions. The correct interpretation of these results requires a sound background in both engineering and FEA.

8.14.4 Types and Shapes of Elements

Elements are made from a number of nodes, and are geometrically simple in shape to form a continuum or full model. Different types of elements available are:

1. One-dimensional Elements: It is made from a minimum of 2 nodes to a maximum of 4 nodes. Line element is an example for one-dimensional elements. The popular one-dimensional elements in solid mechanics are the truss and beam elements.

2. Two-dimensional Elements: It is made from a minimum of 3 or 4 nodes to a maximum of 10 or 12 nodes. Triangular elements and quadrilateral elements are good examples for two-dimensional elements. Triangular elements are made of 3 nodes and quadrilateral elements are made of 4 nodes. Two-dimensional elements can be modeled to solve stress and strain problems.

3. Three-dimensional Elements: It is made from a minimum of 4 or 8 nodes to a maximum of 20 or 32 nodes. Tetrahedral elements and hexahedral elements are good examples for 3 dimensional elements. Tetrahedral elements are made of 4 nodes and hexahedral elements are made of 8 nodes. Three-dimensional elements can solve large number of problems.

All the elements are divided into three types according to their shape:

1. Linear Elements: These elements have nodes at the ends. These elements are made of straight lines which connect all the nodes of an element. It has minimum number of nodes in 1D, 2D and 3D elements. It requires less memory and time in modelling the continuum and analysing it. It is less expensive. It cannot support curved elements when compared to other two types.

2. Quadratic Elements: These elements have nodes at the ends as well as one in the middle of all line segments, which form elements by connecting all nodes. The node in the middle defines the curvature. It is somewhat more complex than linear type. But it can support curved elements. It requires more memory and time and is more expensive than linear type.

3. Cubic Elements: These elements have nodes at the end as well as two in between end nodes. It has a maximum number of nodes in 1D, 2D and 3D elements. It is the most complex one when compared to other. It can support all complex elements, but requires more memory and time. It is the most expensive one.

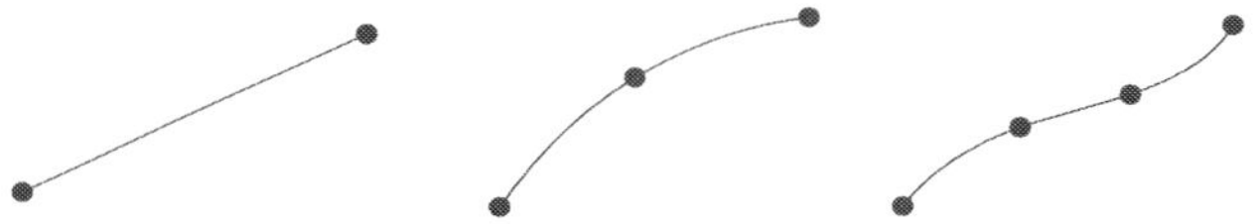

2 Nodes (Linear) 3 Nodes (Quadratic) 4 Nodes (Cubic)

One-Dimensional Elements (Line)

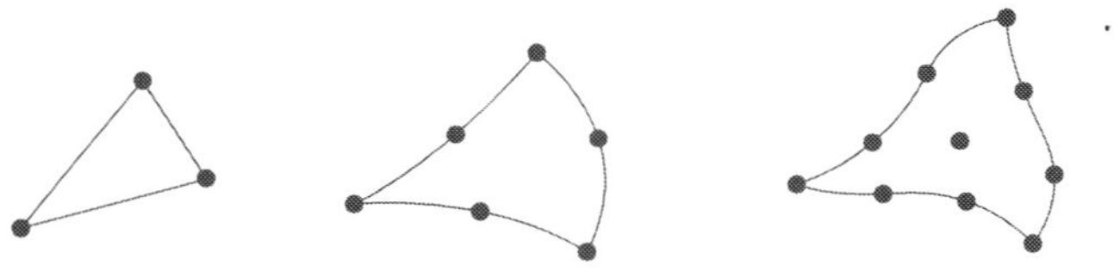

3 Nodes (Linear) 6 Nodes (Quadratic) 10 Nodes (Cubic)

2D triangular elements

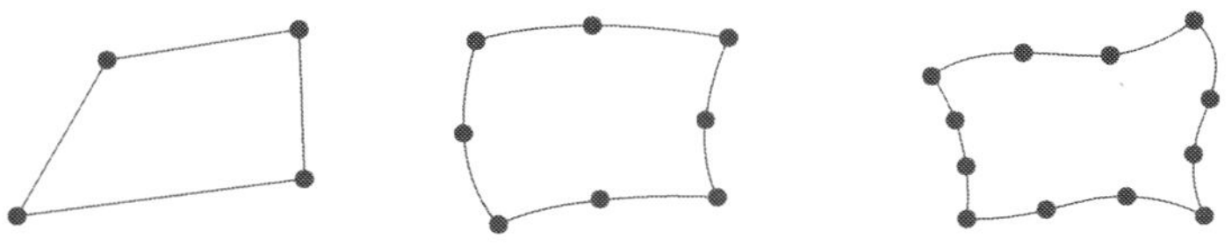

4 Nodes (Linear) 8 Nodes (Quadratic) 12 Nodes (Cubic)

2D Quadrilateral Elements

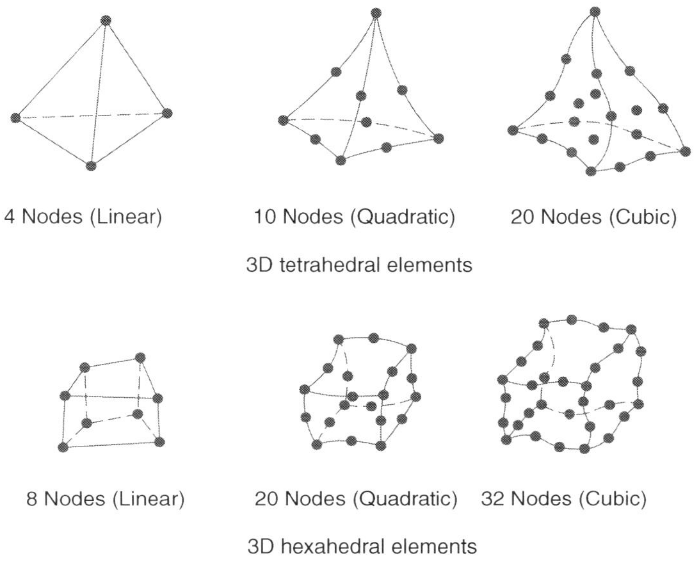

Fig. 8.90 Types of elements.

8.14.5 Mesh Generation

Mesh generation refers to the generation of nodal coordinates and elements. It is of two types:

1. Manual Meshing: In manual mesh generation the analyst discretises the simplified geometry of the object to be studied that is the geometric model of the object, into nodes and elements. Nodes are defined by specifying their coordinates, while element connectivity defines the elements. Manual meshing is inefficient and error prone; meshing data can grow rapidly and become confusing for complex objects especially for three-dimensional ones.

2. Automatic Meshing: It refers to the automatic creation and numbering of nodes and elements based on a minimal amount of user supplied data. Automatic mesh generation reduces errors, and saves a great deal of user time, therefore, reducing the FEA cost.

3. Requirements of a Good Mesh: The following are the requirements of a good mesh, which produces the correct FEA results:

1. Nodal locations should be precise and should not go beyond the boundary.
2. Various element types and shapes should be available to provide the user with more flexibility to meet the compatibility and requirements.
3. Mesh gradation or mesh smoothing and density control should be possible for users to control the mesh size.
4. To convert from one element type to another and it should be possible for the user.
5. Element aspect ratio (can be defined as the ratio between the element's maximum length to the elements minimum length) should be close to one for better results.

6. Mesh geometry and topology or mesh orientation should be uniform, e.g., for a cup model, the orientation of all cavity elements and all core elements should be uniform.
7. It should be compatible with different mesh generation types, e.g., mid plane meshing, fusion meshing and solid meshing.
8. The time taken to generate mesh and the time taken to perform FEA should be less. Mesh generation method should optimise the mesh and minimises the number of nodes and elements to reduce time and memory space requirements.

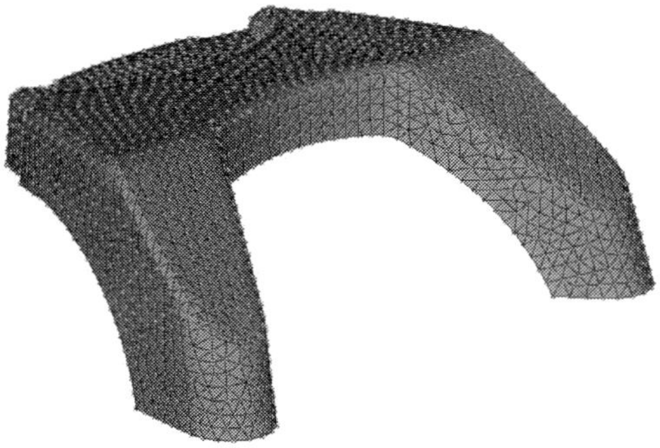

Fig. 8.91 Meshed game control.

8.14.6 Stages in FEA Program

The procedure in using an FEA program consists of three essential stages:

1. Preprocessing: This involves the preparation of the model data. The preprocessor is a programme, which enables the engineer to build a geometric model of a component design. From the model, the required mesh of finite elements may be generated. Required inputs to the preprocessor include:

(a) Geometric Parameters: For example, type of elements, nodal coordinates, variation of mesh density, etc.

(b) Loading Characteristics: For example, magnitudes, positions and directions of point loads, pressures; thermal loads, centrifugal loads, frequency and dependent forces, etc.

(c) Boundary Conditions: For example, positions and directions of nodal fixities, rotational axes and frictional resistances, etc.

(d) Material Properties: For example, Young's modulus, Poisson ratio, density, coefficient of friction, coefficient of expansion, shrinkage, melt temperature, etc.

2. Analysing the Model: After the preparation of the model data, the analysis is executed to get the result.

3. Post Processing: The postprocessor is a programme, which provides the engineer with tools to access the results of the model analysis. Analysed results output from the post-processor may be in either data form or graphical display. Typical post-processor output include:

(a) Printed nodal displacement values

(b) Printed element stress values

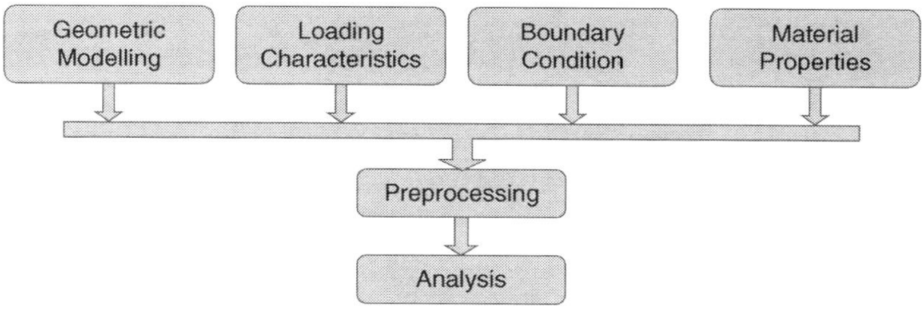

Fig. 8.92 Preprocessor diagram.

(c) Graphical display by a stated multiplication factor
(d) Graphical display of stress contours
(e) Colour or tone variation of element display according to stress range values or temperature range values
(f) Animated displays of moving displacements and varying element patterns for dynamic loading analysis

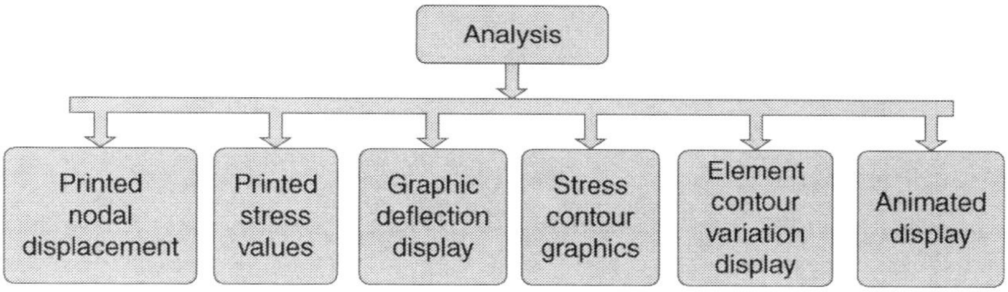

Fig. 8.93 Postprocessor diagram.

8.14.7 Factors to be Considered in FEA

The factors to be considered in FEA while making decisions are:

1. Type of analysis
2. The number of nodes
3. The degrees of freedom at each node
4. The element shape and type
5. The material type
6. The external loads
7. The boundary conditions
8. Interpretation of the results

8.14.8 Types of Analysis

There are wide varieties of analysis available. The major analysis types are:

1. Static and Dynamic Analysis: A structural model, thus, created can be used to predict the behaviour of the real structure, under the action of external forces. The response is usually measured in terms of deflection and stress. The response is static if the loads are static. This analysis is called static analysis. When the loads vary with time the analysis is called dynamic analysis.

2. Linear and Non-Linear Analysis: If the properties of the structure such as stiffness remain constant during the entire analysis, the analysis is called linear analysis. If these properties vary, the analysis is called non-linear analysis.

3. Thermal Analysis: FEA can be used for several design and analysis problems involving thermal stresses, thermal displacements, heat flow, temperature distribution, etc.

4. Fluid Flow Analysis: Finite element analysis can solve several types of fluid flow problems such as calculating velocity, discharge, reynolds number of fluid flowing in a pipe.

5. Field Analysis: Problems in magnetic and acoustics can be solved by FEA.

8.14.9 Disadvantages of FEA

(a) Likelihood of errors occurring in system codes due to their increasing size and complexity.
(b) Lack of standardisation between system codes of many software packages now in use.
(c) Deficiencies in the individual elements.
(d) Selection of unsuitable element types.
(e) Poor assessment of output data.
(f) Masking of important features by the output postprocessor.
(g) Inadequate understanding of the assumptions and limitations of the FEA technique.
(h) Poor representation of the component by the FEA model.

8.14.10 Introduction to Moldflow

Moldflow, formerly known as C-mould, is one of the leading software used in process wide plastics solutions. It is used by designers and manufacturers to produce optimal plastic parts. Many reputed companies use moldflow technology to produce billions of injection moulded parts each year.

Steps involved in using moldflow

1. Importing a model into Moldflow software
2. Creating a mesh to represent the model geometry
3. Repairing the mesh if it contains errors
4. Checking the mesh for aspect ratio, element overlapping, intersecting, thickness, etc.
5. Modelling the mould such as cavity layout, feed system and cooling
6. Selecting the analysis type
7. Selecting the gate location
8. Selecting the material

9. Specifying the moulding operation, moulding machine details, processing parameters, etc.
10. Analysing the study
11. Viewing and interpreting the results of the analysis
12. Adjusting the moulding parameters based on the analysis results

Types of mesh used in moldflow There are three types of mesh used in moldflow, which are:

1. Mid plane
2. Dual domain or fusion
3. Solid mesh

Mid-plane mesh is preferred for parts with uniform wall thickness. It uses triangular elements. Dual domain mesh is preferred for thin parts with variable wall thickness. It also uses triangular elements. Solid mesh is preferred for thick parts with variable wall thickness. It uses tetrahedral elements.

Various types of analyses used in moldflow

1. Fill Analysis: Fill analysis predicts the thermoplastic polymer flow inside the mould in the filling phase. It calculates a flow front that grows through the part incrementally from the injection location. The analysis continues until velocity / pressure switch over point has been reached. It gives results relating to time taken to fill the part, maximum injection pressure involved, maximum clamping force required, maximum and minimum temperatures of the melt flow, percentage / volume of fill, areas where air trapped, weld / meld lines, frozen layer fraction, etc.

2. Pack Analysis: Pack analysis predicts the thermoplastic polymer flow inside the mould in the packing phase. It is run as the second part of a fill added to pack analysis sequence. It is used to determine whether a cavity will be completely filled. The analysis continues until the flow front has expanded to fill the last location in the model.

3. Cool Analysis: Cool analysis products are used to analyse the flow of heat in plastic injection moulds. The outcome of this analysis is to analyse temperature in the plastic filled cavity and the cooling time. It interfaces with fill added to pack analyses. It also gives results related to velocity of coolant flow, temperature of the coolant at inlet / outlet, cooling circuit efficiency (in terms of heat obsorption and transfer), temperatures of the mould, time required for the component to reach the ejection temperature, percentage of frozen / melt layer of the part, temperatures of the core and cavity, etc.

1. Warp Analysis: Warp considers the cooling effects carried on to the fill plus pack analysis in order to compute the impact of differential temperature distribution on part warpage.

2. Shrinkage Analysis: Shrink analysis enables us to determine an appropriate shrinkage allowance.

3. Gate Location Analysis: This analysis is used to locate the possible spot for providing the gate through which the material can be injected.

4. Runner Balance Analysis: This analysis is used to determine the optimum volume for the sections of the runner system. It ensures that the parts fill evenly balanced.

5. Stress Analysis: Stress analysis predicts the actual moulding stiffness. It analyses a product for possible structural defects when the product is exposed to a load. It is used to identify the structural related problems such as the strength, stiffness and life expectancy of plastic products.

6. Moulding Window Analysis: The moulding window analysis is used to calculate the best preliminary process settings. It provides us with recommendations for the injection time, mould temperature and melt temperature values.

8.15 INTRODUCTION TO CAM

Computer Aided Manufacturing (CAM) refers to any automatic manufacturing process which is controlled by computers. Computer aided manufacturing may be defined as the use of computer systems to plan, manage, and control the operations of a manufacturing plant through either direct or indirect computer interface with plants production resources. Therefore, the applications of CAM can be broadly classified into two groups.

1. Computer Monitoring and Control: The computer is directly interfaced with the manufacturing machines for monitoring and control functions in the manufacturing process.

2. Manufacturing Support Applications: The computer is used offline to provide plans, schedules, forecasts, instructions and information by which the production resources of industries can be managed more effectively.

8.15.1 CAM Process

The CAM process on CAD/CAM systems is shown in Fig. 8.94. The geometric model developed during the CAD process forms the basis of the CAM activities. Interface algorithms are utilised to extract various information from CAD databases. In process planning, features that are utilised in manufacturing (such as holes, slots, etc.) must be recognised to enable efficient planning of manufacture. NC programs are prepared according to the process planning. When the parts are produced, CAD software can be used to inspect them. After the inspection, CAM software can be utilised to instruct the robots to assemble the parts, to produce the final products.

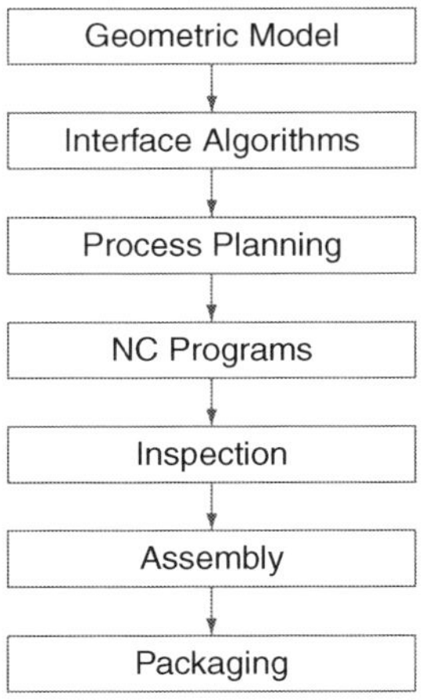

Fig. 8.94 CAM process on CAD/CAM system.

8.15.2 Applications of CAM

Computer numerical control now offers different automatic manufacturing processes, which include milling, turning, flame cutting, laser cutting, punching and spot-welding etc. Parallel developments in computer controlled robots and automated factories led to the development of computer manufacturing units, which are controlled by central computer systems organised under the concept known as flexible manufacturing systems.

The term CAM has come to be applied to many emerging computer controlled manufacturing technology. The important elements of CAM are:

1. Computer Numerical Control (CNC)
2. Direct Numerical Control (DNC)
3. Adaptive Control (AC)
4. Computer controlled robotics manufacture and assembly
5. Flexible Manufacturing Systems (FMS)
6. Computer Aided Inspection (CAI) techniques
7. Computer Aided Testing (CAT) techniques
8. Computer Aided Process Planning (CAPP) techniques
9. Computer Integrated Production Management Systems (CIPMS)
10. Computer Aided Quality Control (CAQC) techniques
11. Computer Integrated Manufacturing systems (CIM)
12. Variable Mission Manufacturing (VMM)

CAQC includes CAI and CAT. CIM is otherwise known as FMS or VMM or CMS (Computerised Manufacturing Systems)

8.15.3 Benefits and Advantages of CAM

Advantages of CAM are concerned with its accomplishment of the following objectives:

(a) Higher production rates with low work forces.
(b) Less likelihood of human error.
(c) Greater versatility of manufactured form.
(d) Cost savings due to increased production efficiency (e.g., less material wastage).
(e) Repeatability of production processes via storage of data.
(f) Superior products.

8.15.4 Reasons for Implementing CAD/CAM

1. Increased productivity due to less time consumed calculations, data storage and retrieval.
2. Better quality as the designer can examine a wide range of design alternatives and analyse each, more thoroughly before selecting one.
3. Better communication by design documents such as drawings, part list, bill of material and specifications as used to communicate to those who will manufacture it.

4. Common database is one of the most important benefits. This will help to eliminate the age old wall separating functions.

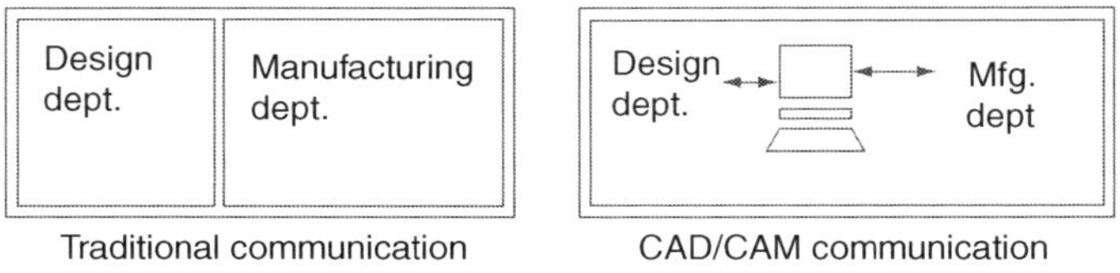

Fig. 8.95 Comparison between traditional and CAD/CAM communication.

8.16 NUMERICAL CONTROL

'**Numerical Control** can be defined as a form of programmable automation in which the process is controlled by numbers, letters, and symbols.' These numbers define the required position of each machine slide, feed, cutting speed and depth of cut. In addition, codes are used to control other functions like coolant ON/OFF, tool change, etc.

The data for preparing the coded instructions are called part programme. Instructions to the NC machines are fed through an external medium, (i.e.,) paper tape or magnetic tape. The information read from the tape, stored into the memory of the control system is called buffer storage and is processed by the machine step by step.

NC machines are also called tape controlled machines because the part cannot be produced without a tape being run through the machine. The tape has to be run repeatedly depending on the number of components to be produced. Also if there is even a minor change in the design of the component, the tape has to be discarded and a new tape with changed programme has to be produced.

8.16.1 Basic components of an NC system

An operational numerical control system consists of the following three basic components.

1. Programme of instructions
2. Controller unit also called Machine Control Unit (MCU)
3. Machine tool or other controlled process

1. Programme of Instructions: The programme of instructions is a detailed step by step set of directions which tell the machine tool what to do and in what sequence. The part programme is written in coded form and contains all the information needed for machining the component. It is fed to the machine control unit through some input medium. Various types of input media are:

(a) Punched cards

(b) Magnetic tapes and floppy disks

(c) Paper tape

The programme of instructions is prepared by a part programmer. The programmer's job is to provide a set of detailed instructions by which the sequence of processing step is to be performed.

2. Controller Unit: This consists of electronics and hardware that read and interpret the programme of instructions and convert it into mechanical actions of the machine tool.

The typical elements of a conventional NC controller unit include:

(a) Tape reader or programme reader

(b) A data buffer

(c) Signal output channel to the machine tool

(d) Feedback channel from the machine tool

The **tape reader** is an electromechanical device for winding and reading the punched tape containing the programme of instructions. The data contained on the tape are read into the **data buffer**. The purpose of this device is to store the input instructions in logical blocks of information. The **signal output channels** are connected to the servomotors and other controls in the machine tool through these channels the instructions are sent to the machine tool from the controller unit. To assure that the instructions have been properly executed by the machine, feedback data are sent back to the controller via the **feedback channels**. The most important function of this return loop is to assure that the table and workpart have been properly located with respect to the tool. Sequence controls coordinate the activities of the other elements of the controller unit. Another element of the NC system, is the control panel which contains the dials and switches by which the machine operator runs the NC system.

3. Machine Tool: It is part of the NC system which performs useful work. The machine tool consists of worktable and spindle as well as the motors and controls necessary to drive them. It also includes the cutting tools, work fixtures, and other auxiliary equipment needed in the machining operation.

The following figure shows the three basic components of an NC system - programme of instructions, controller unit and machine tool.

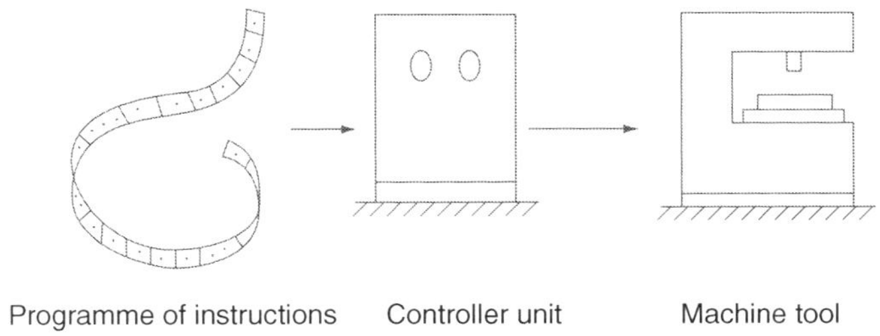

Fig. 8.96 Three basic components of NC system.

8.16.2 Steps in NC

1. Process planning
2. Part programming
3. Tape preparation
4. Tape verification
5. Production

8.16.3 Classification of NC System

The classification of NC machine tool system can be done in four ways. They are:

1. According to the type of machining or motion control system:

(a) Point–to–point

(b) Straight cut system

(c) Contouring system

(a) Point-to-point: Point-to-point systems represent the lowest level of motion control between the tool and workpiece. Point-to-point (PTP) is also sometimes called a positioning system. The objective of the machine tool control system is to move the cutting tool to a predefined location. Once the tool reaches the desired location, the machining operation is performed. NC drill presses are a good example of PTP systems. Point-to-point systems are simple machine control systems and are therefore, the least expensive of the three types. Figure 8.97 shows the point-to-point NC system.

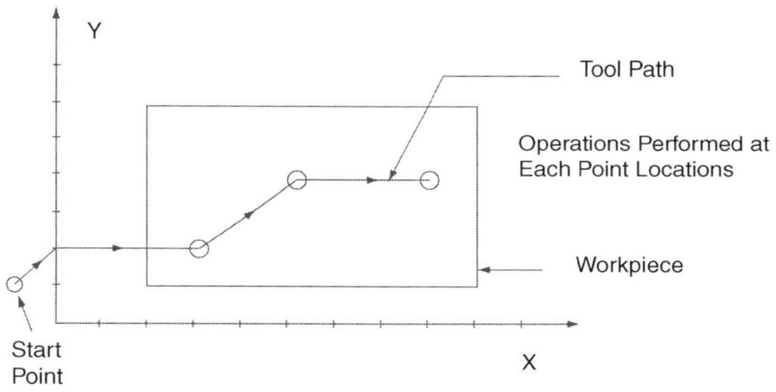

Fig. 8.97 Point-to-point system.

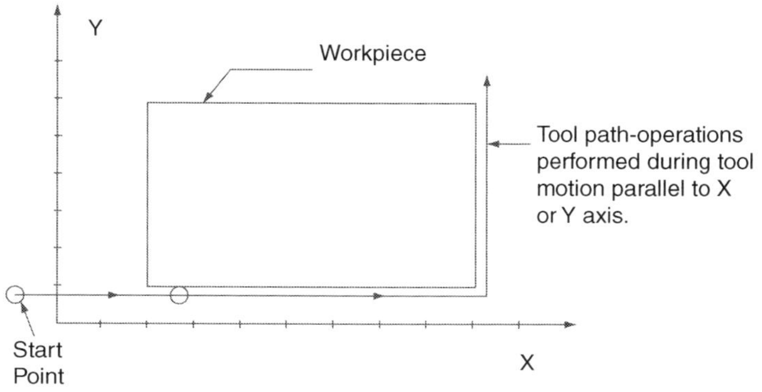

Fig. 8.98 Straight cut NC system.

(b) Straight cut NC system: Straight cut control systems are capable of moving the cutting tool parallel to one of the major axes at a controlled rate suitable for machining. It is therefore, appropriate for performing milling operations to fabricate workpieces of rectangular configurations. With this type of NC system it is not possible to combine movements in more than a single axis direction. Therefore, angular cuts on the workpiece would not be possible. An NC machine capable of straight cut movements is also capable of PTP movements. Figure 8.98 shows straight cut operation.

(c) Contouring NC system: Contouring is the most complex, the most flexible and the most expensive type of machine tool control. It is capable of performing both PTP and straight cut operations. The distinguishing feature of contouring NC system is their capacity for simultaneous control of more than one axis movement of the machine tool. The path of the cutter is continuously controlled to generate the desired geometry of the workpiece. For this reason, contouring systems are also called continuous path NC systems. Circular, conic and all mathematical forms are possible in contour system. Milling and turning operations are common examples of the use of contour control. Figure 8.99 shows the contour NC system.

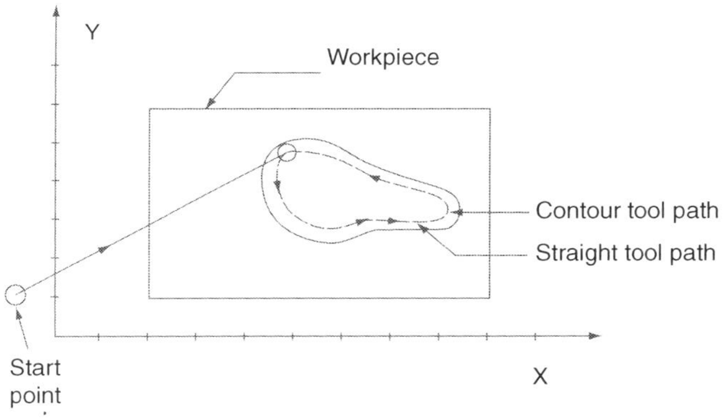

Fig. 8.99 Contour NC system.

2. **According to the structure of control circuit:**
   (a) Analogue control
   (b) Digital control

Analogue and Digital Control: In an analogue control system the quantities may vary continuously, while in digital systems they are varied discreetly such as the presence or absence of a quantity. The shortest cycle of presents and absence is the resolution of the digital systems of machine tools.

3. **According to the programming system:**
   (a) Absolute system
   (b) Incremental system

(a) Absolute System: An absolute system is one in which all moving commands are referred to as one reference point, which is the origin and will be called **zero point**. All position commands are given as absolute distance from that zero point.

(b) Incremental System: An incremental system is one in which the reference point to the next instruction is the end point of the preceding operation. Each dimensional data is applied to the system as a distance increment, measured from the preceding point at which the axis of motion was present.

4. According to the type of control loops:

(a) Open loop systems
(b) Closed loop systems

(a) Open Loop System: The term open loop control means that the one way communication between the controller unit and the machine tool. Controller unit sends instructions to the machine tool and the machine tool runs according to these instructions. Since it is an open loop there is no feedback from machine tool to controller unit. Controller has no reference to the result that machine tool produces.

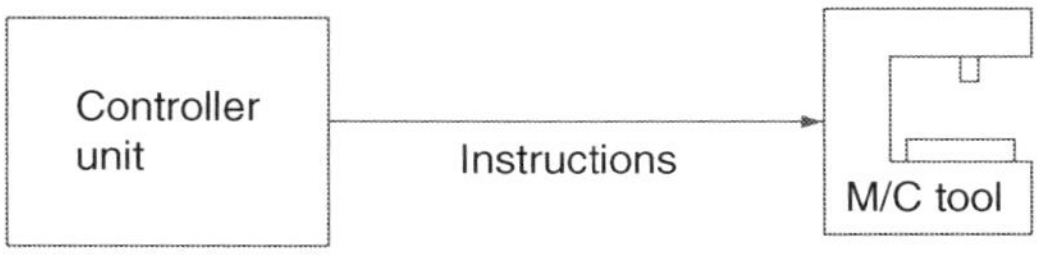

Fig. 8.100 Open loop system.

(b) Close Loop System: In the closed loop system two way communications occurs between controller and machine tool. Controller sends instructions to machine tool and the machine tool sends feedback data to the controller. The close loop control can compare the actual position of the axis and compares it with the desired reference position. The difference between the actual and the desired values is the error and it is designed in such away as to eliminate or to reduce the error to a minimum.

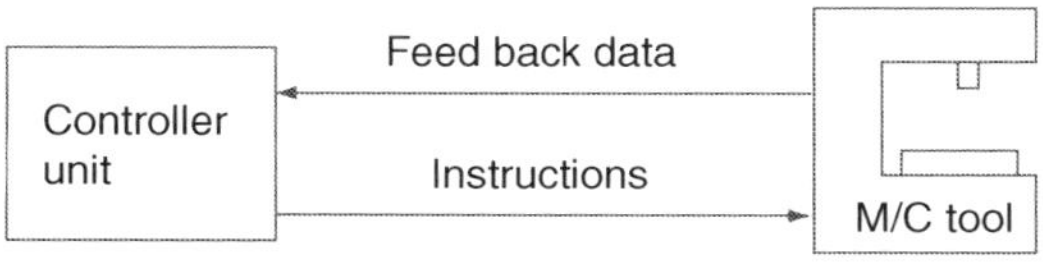

Fig. 8.101 Closed loop system.

8.16.4 Applications of NC

1. Numerical control systems are widely used in industry today, especially in the metal working industry. Various material removing processes include:

(a) Milling
(b) Drilling
(c) Boring
(d) Turning
(e) Grinding
(f) Sawing

2. In addition to metal machining NC has been applied to a lot of other operations, which include:

(a) Press working machine tool
(b) Welding machines
(c) Inspection machines
(d) Automatic drafting
(e) Assembly machines
(f) Tube bending
(g) Flame cutting
(h) Plasma arc cutting
(i) Laser beam processes
(j) Automated knitting machines
(k) Cloth cutting
(l) Automatic revetting
(m) Wire-wrap machines
(n) Injection moulding machines
(o) Blow moulding machines
(p) Extrusion machines

8.16.5 Suitability of NC Machines

NC machines are most suited for the following cases:

1. Parts are processed frequently and in small lot sizes.
2. The part geometry is complex.
3. Many operations must be performed on the part in its processing.
4. Much metal needs to be removed.
5. Engineering design changes are likely.
6. Close tolerances must be held on the work part.
7. It is an expensive part where mistakes in processing would be costly.
8. The parts require 100% inspection.

8.16.6 Advantages of NC

(a) NC reduces non-productive time by means of fewer set ups, less time in setting up, reduced work piece handling time, automatic tool changes on some machines, and so on.

(b) NC requires simpler fixtures and it is less costly to fabricate because the positioning is done by the NC tape rather than the jig or fixture.

(c) Jobs can be setup more quickly as fewer set ups are required.

(d) The time to deliver a job to the customer is reduced, thus NC reduces manufacturing lead time.

(e) NC is less difficult to adopt to engineering design changes. It has greater manufacturing flexibility.

(f) NC produces parts with greater accuracy, reduced scrap, lower inspection requirements. Thus, NC improves quality control.

(g) Owing to fewer steps and shorten lead times with NC the amount of inventory carried by the company is reduced.

(h) NC machines require less floor space than conventional ones. So it reduces floor space requirements.

8.16.7 Disadvantages of NC

(a) Higher investment cost.

(b) Higher maintenance cost.

(c) A highly skilled and properly trained programmer is needed.

8.16.8 Problems Faced in Conventional NC

1. Part programming mistakes are common.
2. Non-optimal speeds and feeds lower the optimum productivity.
3. Unreliable punched tape due to wear and tear because of frequent use.
4. Unreliable tape reader because it breaks down due to frequent use.
5. The control features cannot be easily altered to incorporate improvements into the unit.
6. It is not equipped to provide timely information on operational performance to managements.

8.17 PART PROGRAMMING

Numerical control **part programming** is the procedure by which the sequence of processing steps to be performed on the NC machine is planned and documented. It involves the preparation of a punched tape (or other input medium) used to transmit the processing instruction to the machine tool. The most common input medium is punched tape. There are two basic methods of preparing a punched tape.

1. Manual part programming
2. Computer assisted part programming

8.17.1 Manual Part Programming

To prepare a part programme using the manual method, the programmer writes the machine instructions on a special form called a part programming manuscript. The instructions must be prepared in a very precise manner because the typist prepares the NC tape directly from the manuscript. **Manuscript** is a listing of the relative tool and workpiece locations. It also includes other data such as preparatory commands, miscellaneous instructions, and speed/feed specifications, all of which are needed to operate the machine under tape control. Manual programming jobs can be divided into two categories; point-to-point jobs and contouring jobs. Manual programming is ideally suited for point-to-point applications.

8.17.2 Computer Assisted Part Programming

In the more complicated point-to-point jobs and in contouring applications, manual part programming becomes an extremely tedious task and subject to errors. In these instances it is much more appropriate to employ the high speed digital computer to assist in the part programming process. It saves time and results in a more accurate and more efficient part programme. In computer assisted part programming, punched tape preparation is divided into two parts – part programmer's job and computers job.

The part programmer's responsibility in computer assisted part programming consists of two basic steps.

1. Defining part geometry
2. Specifying the operation sequence and tool path. The part programmers job to enumerate the elements out of which the part is composed (such as raw material, finished parts, portions to be machined, etc.). Each geometric element must be identified and the dimensions and locations of the element explicitly defined (such as selection of machine coordinate system, tool, specific operation, etc.). After defining part geometry, the programmer must next construct the path that the cutter will follow to machine the part (such as step over distance, depth per cut, etc.). This tool path specification involves detailed step-by-step sequences of cutter moves along the geometry element, which has previously been defined.

The computer's job in computer assisted part programming consists of the following steps:

1. Input translation
2. Arithmetic calculation
3. Cutter offset computation
4. Post processor

The part programmer enters the programme written in the APT (Automatically Programmed Tools) or other language. The input translation component converts the coded instructions contained in the programme into computer usable form preparatory to further processing. The arithmetic calculation unit of the system consists of a set of subroutines for solving the mathematics required to generate the part surface. These subroutines are called by the various part programming language statements. The second task of the part programmer is to construct the tool path. The actual tool path is different from the part outline because the tool path is defined as the path taken by the centre of the cutter. The purpose of the cutter offset computation is to offset the tool path from the desired part surface. The post processor is a separate computer programme that has been written to prepare punched tape for a specific machine tool. The input of the post processor is the output from other three components and the output of the post processor is the NC tape.

8.17.3 NC Coordinate System

To plan the sequence of positions and movements of the cutting tool relative to the work piece by the part programmer, it is necessary to establish a standard axes system by which the relative positions can be specified. In an NC drill press, the drill spindle is in a fixed vertical position, and the table is moved and controlled relative to the spindle. In order to make things easier for the programmer, we adopt the view point that the work piece is stationary while the drill bit is moved relative to it. Accordingly, the coordinate system of axes is established with respect to the machine table.

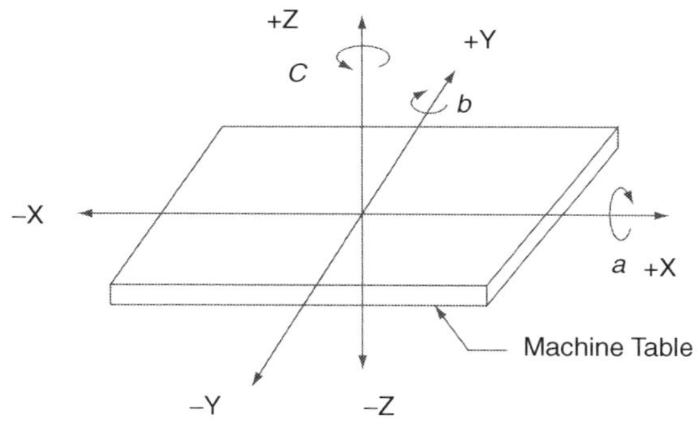

Fig. 8.102 NC machine tool axis system for milling and drilling operations.

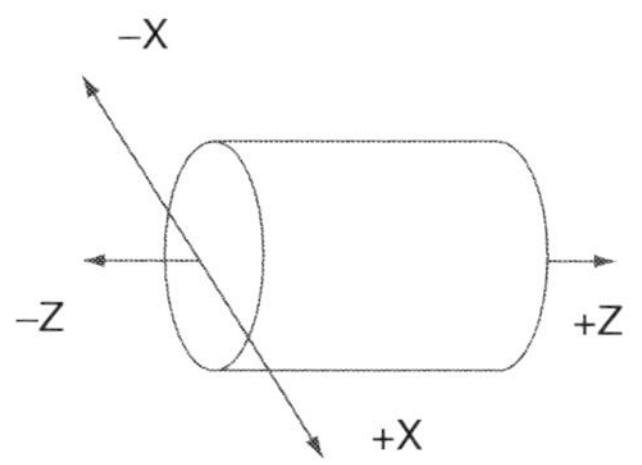

Fig. 8.103 NC machine tool axis system for lathe.

Two axes, X and Y are defined in the plane of the table. The Z axis is perpendicular to this plane and movement in the Z direction is controlled by the vertical motion of the spindle. The positive and negative directions of motion of tool relative to table along these axes are shown in Fig. 8.102. NC drill presses are classified as either two axis or three axis machines depending on whether or not they have the capability to control the Z axis.

A numerical control milling machine and similar machine tools posses the capacity to control one or more rotational axes. Three rotational axes are defined in NC, the a, b, c axes. These axes specify the angles about X, Y, and Z axes, respectively. Generally, X axis movement is called longitudinal and y axis movement called traverse.

For turning operations, two axes are normally used to control the movement of the tool relative to the rotating workpiece. The Z axis is the axis of rotation of the work part, and X axis defines the radial location of the cutting tool. This arrangement is shown below. The purpose of the coordinate system is to provide a means of locating the tool in relation to the work piece.

8.17.4 The Punched Tape

The part programme is converted into a sequence of machine tool motions by means of the input medium. The most common input medium is punched tape. The punched tape can be prepared by either manual part programming method or computer assisted part programming method. The punched tape used for NC is 1 inch wide. It is standardised by Electronics Industries Association (EIA). During production on a conventional NC machine the tape is fed through the tape reader once for work piece. While the machine tool is performing one instruction, the next instruction is being read into the controller unit's data buffer. After the last instruction has been read into the controller, the tape is rewound back to the start of the programme to be ready for the next work part.

There are eight regular columns of holes running in the length wise direction of the tape. There is also a ninth column of holes between the third and fourth regular column. These are smaller and are used as sprocket holes for feeding the tape. The coding of the tape is provided by either the presence or absence of the hole in various positions. This coding system is called **binary coding**. The NC tape coding system is used to code not only for number, but also alphabetical letters and other symbols.

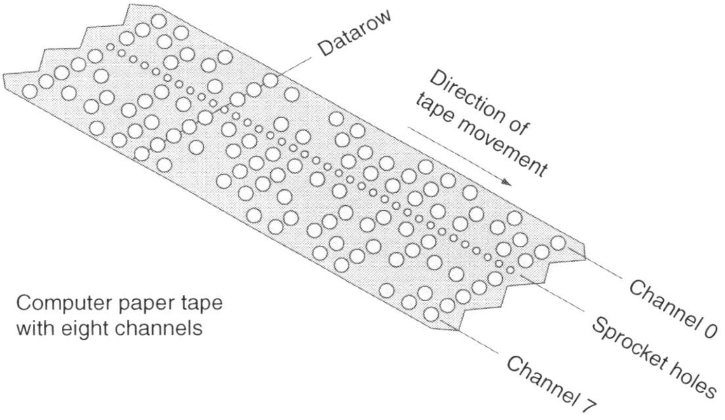

Fig. 8.104 The punched tape.

8.17.5 NC Words

NC words are a collection of characters used to form an instruction. A collection of NC words form a block and the block of words form a complete NC programme. Different types of words used in the formation of a block are:

1. **Sequence Number (N-Word)**: This is used to identify the block.

2. **Preparatory Function (G-Words):** This is used to prepare the controller for instructions that are to follow. The codes used in preparatory functions are given in Table 8.7.

3. **Coordinate (X, Y, and Z Words):** These give the coordinate positions to the tool.

4. **Feed Rate (F-Word):** This specifies the feed in a machine operation.

5. Cutting Speed (S-Word): This specifies the cutting speed of the process, the rate at which the spindle rotates.

6. Tool Selection (T-Word): This specifies which tool is to be used in the operation.

7. Miscellaneous Function (M-Word): This is used to specify certain miscellaneous or auxiliary functions which may be available on the machine tool. The codes used in miscellaneous functions are given in Table 8.8

8. End of Block (EOB): To identify the end of instruction an end of block symbol is punched on the tape.

Table 8.7 Preparatory function (G – Codes).

Code	Function
G00	Rapid traverse
G01	Linear interpolation
G02	Circular interpolation clockwise
G03	Circular interpolation counter clockwise
G04	Dwell
G05	Hold / delay
G17	XY plane designation
G18	ZX plane designation
G19	YZ plane designation
G33	Thread cutting
G40	Cutter compensation - cancel
G41	Cutter compensation - left
G42	Cutter compensation – right
G63	Thread cutting cycle
G70	Dimensioning in inch unit
G71	Dimensioning in metric unit
G80	Canned cycle cancel
G81–G89	Canned cycles
G90	Absolute dimensioning
G91	Incremental dimensioning
G92	Zero preset
G94	Feed rate mm/min.
G95	Feed rate mm/rev.

Table 8.8 Miscellaneous function (M – Codes).

Code	Function
M02	Programme stop
M03	Spindle start clockwise
M04	Spindle start counter clockwise
M05	Spindle stop
M06	Tool change
M08	Coolant on
M09	Coolant off
M30	Programme stop and tape rewind

8.17.6 Writing a Part Programme

In any part programme first instruction is to tell the control system about the various set up conditions for the machining tasks. The instructions should specify the following:

Block number (N-word)
Coordinate value -absolute or incremental (G90 or G91)
Dimensional units – inch or metric (G70 or G71)
Tool number (T-word)
Spindle speed (S-Word)
Feed function (G94 or G95)
End of Block (EOB)

Point-to-point machining In a point-to-point CNC system, the workpiece or the cutting tool moves from one point to another point and machining is done at a specific point. No machining is done when the spindle is moving between two points. To illustrate the point-to-point machining, let us consider the workpiece shown in Fig. 8.105 where three holes are to be drilled at different places. The depth of hole given is 20 mm; Z = 0 at the surface of the workpiece and the cutting tool is positioned above the workpiece.

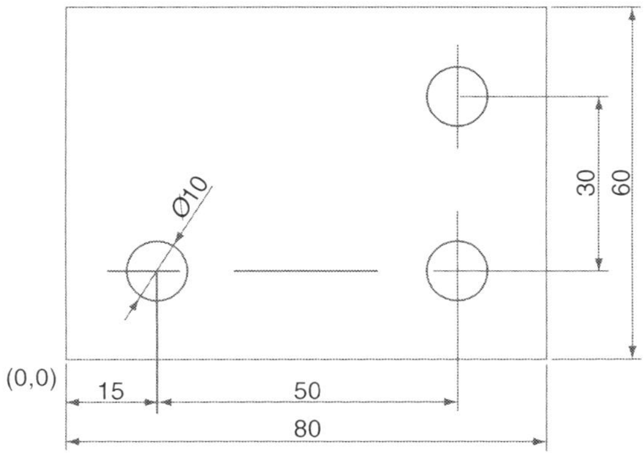

Fig. 8.105 Point-to-point operation.

Part program for the above drawing is given below:

N010 G71 G90 G94 EOB	-	Metric mode, absolute system, feed in mm/min.
N020 M03 F250 S1200 EOB	-	Spindle start CW, feed rate 250 mm/min at 1200 rpm.
N030 G00 X15.00 Y15.00 EOB	-	Move in rapid to point (15, 15).
N040 G00 Z2.00 EOB	-	Move in rapid to a point 2 mm above workpiece.
N050 G01 Z-20.00 EOB	-	Drill hole.
N060 G00 Z2.00 EOB	-	Move in rapid to point 2 mm above workpiece.
N070 G00 X65.00 EOB	-	Move in rapid to x = 65.
N080 G01 Z–20.00 EOB	-	Drill a hole.
N090 G00 Z2.00 EOB	-	Move in rapid to point 2 mm above workpiece.
N100 G00 Y45.00 EOB	-	Move in rapid to Y = 45.
N110 G01 Z–20.00 EOB	-	Drill a hole.
N120 G00 Z20.00 EOB	-	Move in rapid to a point 20 mm above workpiece.
N130 G00 X00 Y00 EOB	-	Move in rapid to (0, 0).
N140 M02	-	Programme end.

Machining along straight line Machining along straight line is done using linear interpolation. In case of straight line machining with a milling machine, machining can be started in either of the two ways. Firstly the tool is taken to required depth of cut outside the workpiece and then the tool is programmed to machine the component along straight line. Secondly the tool may be plunged to required depth of cut in to the workpiece and then machining along the straight line is started. Let us consider the part program for straight line milling on the component shown in Fig. 8.106. Machining is to be done at AB and BC. The part programme for the job is given below. In this programme Z = 0 is 10 mm above the surface of the workpiece and depth of cut is 10 mm.

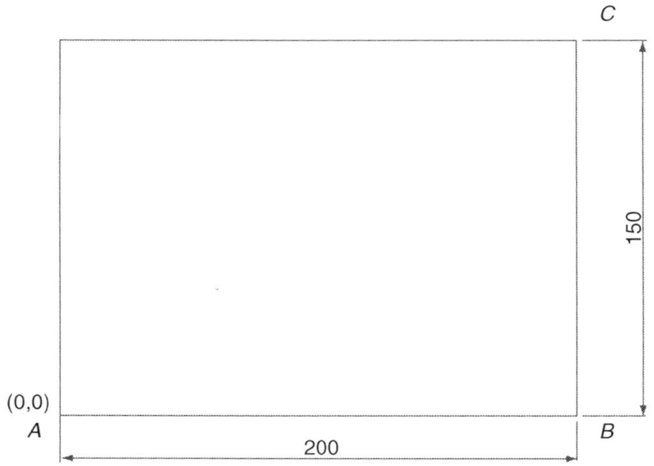

Fig. 8.106 Machining along straight line.

N0010 G71 G90 G94 EOB - Setting mode metric, absolute, feed rate mm/min.
N0020 F240 S1800 EOB - Set feed 240 mm/min and speed 1800 rpm.
N0030 M03 M08 EOB - Spindle on CW, coolant on.
N0040 G00 Z2.00 EOB - Spindle moves to Z = 2 in rapid mode.
N0050 X00 Y0.00 EOB - Move to X = 0, Y = 0.
N0060 Z–20.00 EOB - Spindle down to required depth of cut.
N0070 G01 X200.00 EOB - Linear interpolation to X = 200.
N0080 G01 Y150.00 EOB - Move to Y = 150.
N0090 G00 Z12 M09 EOB - Rapid spindle retract to Z = 12 and coolant off.
N0100 G00 X–10.00 Y0.00 EOB - Rapid to X = –10 and Y = 0.
N0110 M02 EOB - Programme end.

Facing in lathe In lathe operations only two axes are involved (X and Z axis). The Z axis is the axis of the spindle and the X axis is the direction of transverse motion of the tool post. To illustrate the facing operation let us consider the component shown in Fig. 8.107. The operations to be done on the workpiece are facing and to reduce the diameter of a bar from 24 mm to 20 mm. The part programme for the above workpiece is given below:

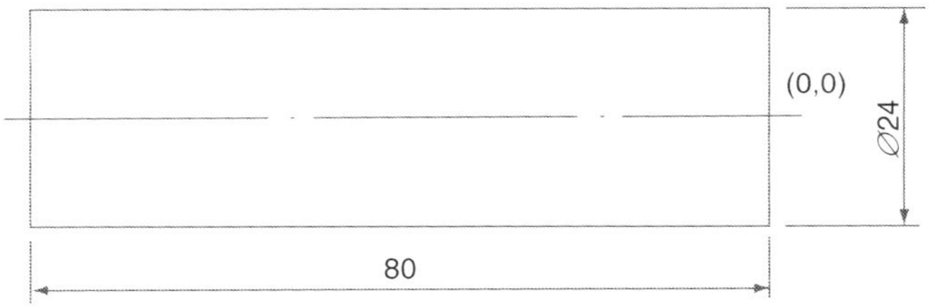

Fig. 8.107 Facing operation.

N0010 G71 G90 G94 EOB - Metric, absolute, feed rate mm / min.
N0020 T01 F250 M03 S600 EOB - Tool no.1, feed rate 250 mm/min, speed 600 rpm.
N0030 G00 X22.00 Z1.00 EOB - Rapid mode, move to point X = 22, Z = 1.
N0040 G00 X0 EOB - Move to X = 0, Z remaining constant.
N0050 G01 Z0 EOB - Go to Z = 0 in linear interpolation mode.
N0060 X24.00 EOB - Move to X = 24 in G01 mode, (facing).
N0070 Z–80.00 EOB - Move to Z = –80 in G01 mode (dummy cut).
N0080 G00 X26.00 EOB - Withdraw tool by 1 mm to X = 26.
N0090 G00 Z0 EOB - Move in rapid to Z = 0.
N0100 G01 X20.00 EOB - Move to X = 20 in G01 mode (Depth of cut of 2 mm).
N0110 G01 Z–80.00 EOB - Move to Z = –80 to turn the job.
N0120 G00 X26.00 EOB - Withdraw the tool to X = 26 in rapid movement.
N0130 Z20.00 EOB - Go to Z = 20.
N0140 M02 EOB - End of programme.

Taper turning in linear interpolation The part programme for taper turning operation is similar to the simple turning operation except that in this case both the coordinates, i.e., X and Z values of the final point are to be given in the programme. Consider the taper turning job given in Fig. 8.108. The raw material available is 20 mm diameter bar. The operations involved are, facing, turn to 15 mm diameter over 15 mm length and taper turning. The part programme for this job is given below in absolute mode.

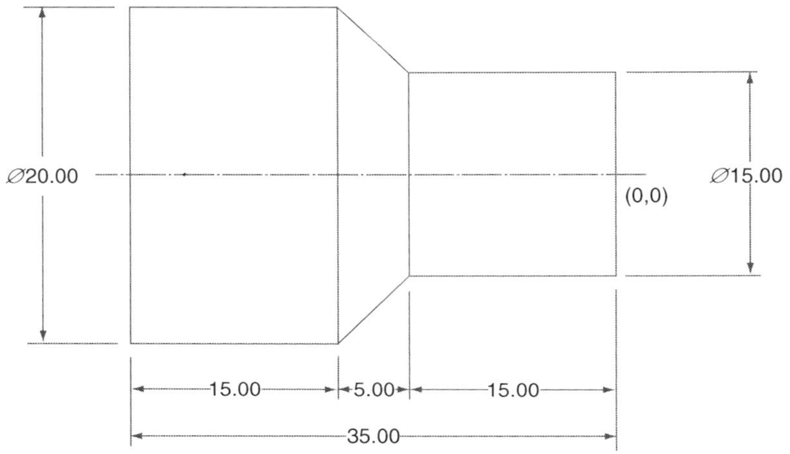

Fig. 8.108 Taper turning operation.

N0010 G71 G90 G94 EOB	-	Metric unit, absolute mode, feed rate mm/min.
N0020 T01 S1000 M03 EOB	-	Tool no. 1, speed 1000 rpm, spindle start CW.
N0030 G00 X22 Z0.5 EOB	-	Move rapid to X = 22, Z = 0.5.
N0040 G01 X0.00 F200 EOB	-	Move to X = 0 in G01 mode, feed 200 mm/min.
N0050 Z0.00 EOB	-	Move to Z = 0.
N0060 X20.00 EOB	-	Move to X = 20.
N0070 X15.00 EOB	-	Move to X = 15.
N0080 Z–15.00 EOB	-	Move to Z = –15.
N0090 X20.00 Z–20.00 EOB	-	Move to X = 20, Z = –20.
N0100 Z–35.00 EOB	-	Move to Z = –35 (X is constant).
N0110 G00 X25.00 Z20.00 EOB	-	Move to point X = 25, Z = 20 rapidly.
N0120 M02 EOB	-	End of programme.

8.18 COMPUTER NUMERICAL CONTROL

Computer Numerical Control (CNC) is an NC system that utilises a dedicated, shared programme computer to perform some or all of the basic numerical control functions. The external appearance of a CNC machine is similar to that of a conventional NC machine. Part programmes are initially entered in a similar manner. In NC the punched tape is cycled through the reader for every workpiece. In CNC the programme is entered once and then stored in the computer memory. Thus, the tape reader is used only for the original loading of the part programme and data. Compared to NC, CNC offers additional flexibility and computational capability. Simply by reprogramming the unit, new system options can be incorporated into the CNC controller.

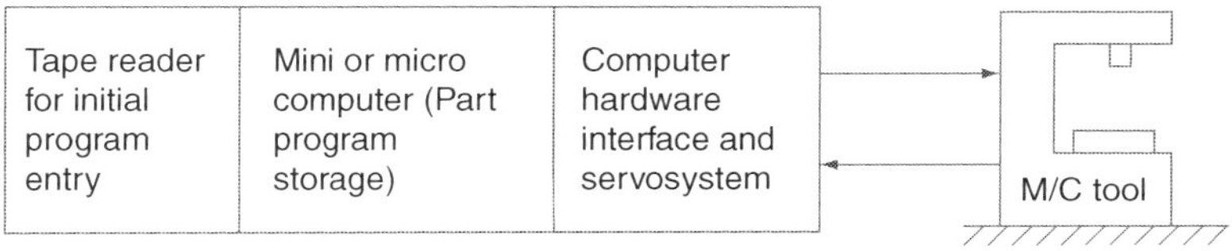

Fig. 8.109 General configuration of CNC system.

8.18.1 Functions of CNC

The principal functions of CNC are:

1. Machine tool control
2. In-process compensation
3. Improved programming and operating features
4. Diagnostics

1. Machine Tool Control: The primary function of the CNC system is control of the machine tool. This involves conversion of the part programme instructions into machine tool motions through the computer interface and servo systems.

2. In-Process Compensation: A function closely related to machine tool control is in process compensation. This involves the dynamic correction of the machine tool motions for changes or errors which occur during processing.

3. Improved Programming and Operating Features: CNC allows the introduction of many convenient programming and operating features including editing of part programmes at the machine, graphics display of the tool path to verify the tape, various types of interpolation, use of specially written subroutine, manual data input, and local storage of more than one part programme.

4. Diagnostics: CNC machines are often equipped with a diagnostics capability to assist in maintaining and repairing the system. The diagnostics subsystem would accomplish several functions. First the subsystem would be able to identify the reason for a down time occurrence so that the maintenance personnel could make repairs more quickly. Second the diagnostic subsystem would be alert to signs that indicate the imminent failure of a certain component. Hence maintenance personal could replace the faulty component in time. Third possible function is when one of these components fails; the diagnostics subsystem would automatically disconnect the faulty component and activate the redundant component. Repairs could thus, be accomplished without any break in normal operation.

8.18.2 Advantages of CNC

(a) The part programme tape and tape reader are used only once to enter the programme into computer memory. This results in improved reliability.
(b) The NC tape can be corrected and even optimised in the computer.
(c) CNC can accommodate conversion of tapes prepared in units of inches into the international system of units.
(d) New control options could be tried at low cost because of greater flexibility.
(e) Generation of specialised programs by the user (user written programs).

8.19 DIRECT NUMERICAL CONTROL

Direct Numerical Control (DNC) can be defined as a manufacturing system in which a number of machines are controlled by a computer through direct connection and in real time. Because of the omission of the tape reader, the DNC system has become more reliable than CNC. Instead of using the tape reader, the part program is transmitted to the machine tool directly from the computer memory. In principle, one large computer can be used to control more separate machines.

8.19.1 Components of DNC

A DNC consists of four basic components:

1. Central computer
2. Bulk memory (which stores the NC part programme)
3. Telecommunication lines
4. Machine tools

The computer calls the part programme instructions from bulk storage, and sends them to the individual machines as and when the need arises. It also receives data back from the machines. This two way information flow occurs in real time, each machine's requests for instructions are fulfilled instantaneously. The computer must always be ready to receive information from the machines and to respond accordingly. Thus, computer is servicing a large number of separate machine tools, all in real time.

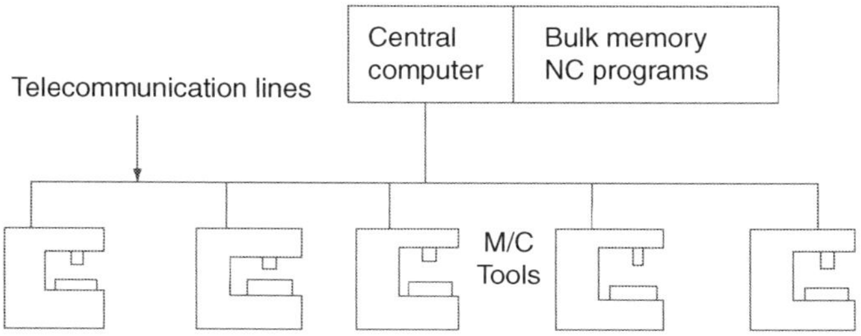

Fig. 8.110 General configuration of DNC system.

8.19.2 Functions of DNC

The principal functions of DNC are:

1. NC without punched tape
2. NC part programme storage
3. Data collection, processing and reporting
4. Communications

1. NC without Punched Tape: One of the main objectives in DNC was to eliminate the use of punched tape. All the costs and inconveniences of the punched tape are eliminated with DNC approach.

2. NC Part Programme Storage: A second important function of the DNC system is concerned with storing the programmes. The programme storage subsystem must be structured to satisfy several purposes. First, the programme must be made available for downloading from the central computer to the NC machine tools. Second, the subsystem must allow for new programmes to be entered, old programmes to be deleted and existing programmes to be edited. Third, the DNC software must accomplish the post processing function. The part program in a DNC system would typically be stored as the CLFILE, which must be converted into instructions for a particular machine tool. This conversion is performed by the post processor. Fourth, the storage subsystem must be structured to perform certain data processing and management functions such as file security, display of programme, manipulation of data, etc. The DNC program storage subsystem usually consists of an active storage and a secondary storage. The active storage would be used to store NC programmes which are frequently used. The secondary storage would be used for NC programmes which are not frequently used.

3. Data Collection, Processing and Reporting: Another important function of DNC involves the data collection, processing and reporting. DNC involves a two way transfer of data. The basic purpose behind the data collection, processing, and reporting function of DNC is to monitor production in the factory. The data are collected and processed by the DNC computer and reports are prepared to provide management with information necessary for running the plant.

4. Communications: A communication network is required to accomplish the previous three functions of DNC. The essential communication links in DNC are between the following components of the system.

(a) Central computer and machine tools
(b) Central computer and NC part programmer terminals
(c) Central computer and bulk memory which store the NC programmes.

The optional communication links may also be established between DNC systems such as shop floor control systems, corporate data processing, CAD system, remote maintenance diagnostics system, etc.

8.19.3 Advantages of DNC

(a) Elimination of punched tapes and tape readers.

(b) Greater computational capability and flexibility.

(c) Convenient storage of NC part programmes in computer files.

(d) Programmes stored as cutter path data (CLFILE) rather than post processed programmes for specific machine tools. Storing of the programmes in this general format affords the flexibility in production scheduling to process a job on any of several different machine tools.

(e) Reporting the shop performance by collecting, processing and reporting of data from the NC machine tools.

8.20 ADAPTIVE CONTROL

Adaptive Control (AC) denotes a control system that measures certain output process variables and uses these to control speed and feed. Some of the process variables that have been used in adaptive control machining systems include spindle deflection or force, torque, cutting temperature, vibration amplitude, and horse power.

Figure. 8.112 shows the schematic diagram illustrating the operation of the AC system, during the machining process. When the force increases due to increased work piece hardness or depth of cut, the feed rate is reduced to compensate. When air gaps are in the part, feed rate is increased to maximise the rate of metal removal.

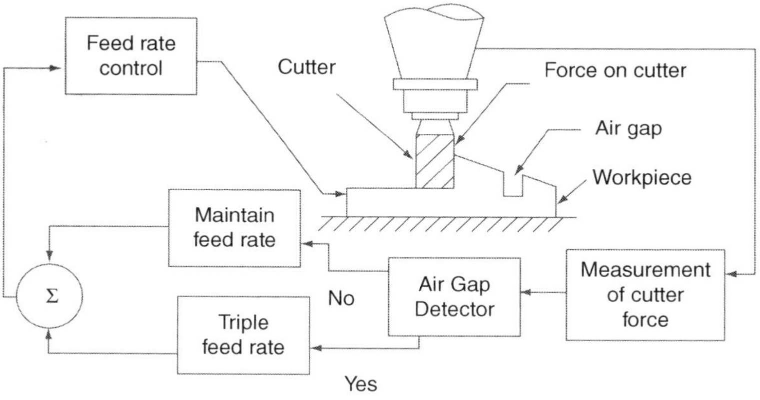

Fig. 8.112 General configuration of AC system.

8.20.1 Situations Where AC can be Suitably Applied

The adaptive control can be applied in situations where:

1. The in process time consumes a significant portion of the machine cycle time.
2. Significant sources of variability are found in the job.
3. The cost of operating the machine tool is high.
4. The typical jobs once involved steel, titanium, and high strength alloys.

5. There is variable geometry of cut in the form of changing depth of cut or width of cut (step over distance).
6. There is variable work piece hardness and variable machinability.
7. There is variable workpiece rigidity.
8. There is tool wear (AC will typically respond to tool wear by reducing the feed rate).
9. There are air gaps during cutting (AC automatically increases the feed rate where air gaps are found in the material).

8.20.2 Advantages of AC

(a) Increased production rate
(b) Increased tool life
(c) Greater part protection
(d) Less operator intervention

8.21 COMPUTER INTEGRATED MANUFACTURING

Computer Integrated Manufacturing (CIM) is defined as a production system which consists of a group of machines connected together by an automated material handling system and operating under computer control.

CIM is also called Flexible Manufacturing System (FMS), Variable Mission Manufacturing (VMM) and Computerised Manufacturing system. CIM incorporates the CAD/CAM and concepts which include CNC, DNC, computer process control, computer integrated production management, automated inspection methods and industrial robotics.

Computer integrated manufacturing systems are designed to fulfil the gap between high production transfer lines and low production NC machines. Transfer lines are very efficient when producing parts in large volumes at high output rates. The limitations on this mode of production are that the parts must be identical. These highly mechanised lines are inflexible and cannot tolerate variations in part design. Stand alone NC machines are ideally suited for variations in work piece. NC machine tools are appropriate for job shop and small batch manufacturing, because they can be conveniently reprogrammed to deal with product changeovers and part design changes.

A gap exists between the high production rate transfer machines and the highly flexible NC machines. This gap includes parts produced in midrange volumes. These parts are of fairly complex geometry, and the production equipment must be flexible enough to handle a variety of part designs. Transfer lines are not suited to this application because they are inflexible. NC machines are not suited to this application because their production rates are too slow. The solution to this mid volume production problem is the computer integrated manufacturing system. Figure 8.113 shows the general application guidelines for the CIM system.

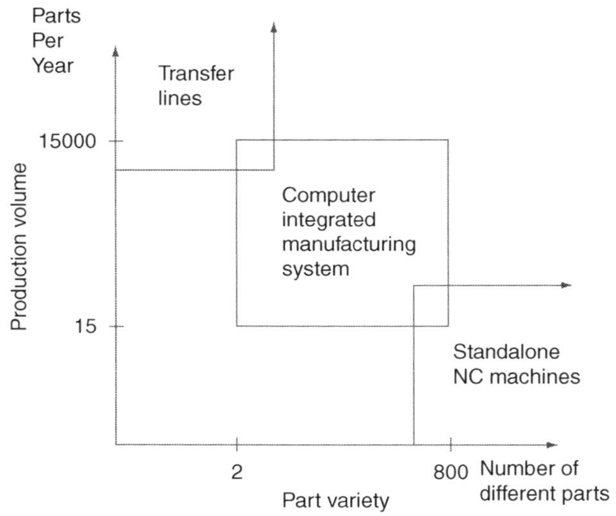

Fig. 8.113 General application guideline for CIM.

8.21.1 Types of CIM

The middle range, covering the medium part variety and medium production volume, can be further divided into three categories:

1. Special manufacturing system
2. Manufacturing cell
3. Flexible manufacturing system

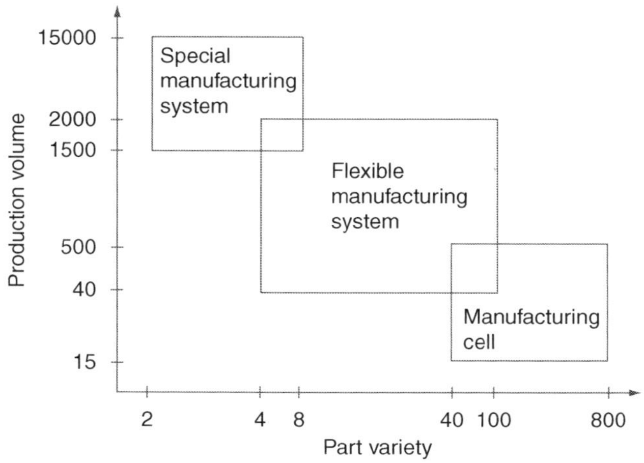

Fig. 8.114 Application guidelines for the three types of CIM.

1. Special Manufacturing System: It is the least flexible CIM system. It is designed to produce a very limited number of different parts. The number of different parts manufactured in this system is between 2 and 8. The annual production rate per part would lie between 1500 and 15000 pieces.

2. Manufacturing Cell: It is the most flexible but generally has the lowest production rate of the three types. The number of different parts manufactured in this system is between 40 and 800. The annual production rate per part would lie between 15 and 500.

3. Flexible Manufacturing System (FMS): FMS covers a wide middle territory within the mid volume, mid variety production range. The number of different parts manufactured in this system is between 4 and 100. The annual production rate per part would lie between 40 and 2000.

8.21.2 Advantages of CIM

(a) Increased machine utilisation
(b) Reduced direct and indirect labour
(c) Reduced manufacturing lead time
(d) Lower in process inventory
(e) Scheduling flexibility
(f) Remarkable manufacturing flexibility
(g) Higher rate of production with consistent high quality
(h) Uninterrupted production with negligible supervision
(i) Integrating of all factory functions, e.g., Material handling, tooling, metal cutting, inspection, etc.

8.21.3 Basic Components of CIM

1. Machine tools and related equipment
2. Material handling system
3. Computer system
4. Human labour

1. Machine Tool and Related Equipment: The machine tool and other equipment that comprise a CIM system include standard CNC machine tools, special purpose machine tools, tooling for these machines and inspection stations or special inspection probe used with the machine tools. The selection of the particular machine depends on the processing requirements to be accomplished by the system. Factors that define the processing requirements are:

(a) Part size
(b) Part shape
(c) Part variety
(d) Product life cycle
(e) Definition of future parts
(f) Operations other than machining

2. Material Handling System: The material handling system in CIM must be designed to serve two functions. The first function is to move work parts between machines. The second function is to orient and locate the work parts for processing at the machines. It can be divided into primary handling system and the secondary handling system. The primary work handling system is used to move parts between machine tools in the CIM. The secondary parts handling system must present parts to the individual machine tools in the CIM. Secondary system generally consists of one transport mechanism for each machine. The basic requirements of a material handling system are:

1. It must be compatible with computer control.
2. It must permit temporary storage or banking of work parts.
3. It must be interfaced with other material handling systems.
4. It should allow access to the machine tools for maintenance, tool changing, etc.
5. It must provide random, independent movement of palletised work parts between machine tools in the system.
6. It must provide for parts orientation and location at each workstation for processing.

The handling system could either be a separate unit like industrial robot, or a part changer which is an integral part of the machine itself. For separate unit the cost is more, and for integral unit the cost is less. The automatic changing of the workpiece in a machine tool is best satisfied by utilising handling equipment which is built integral with the machine tool, called computerised part changer.

Tooling System: Machine tool must be equipped either with turret or tool changer for supplying desired tools for cutting / machining. For larger components with longer cycle times, an automatic tool changer is required.

Transport System: Material handling system requirements include the transportation of part, raw material, final product, fixtures, tools, pallets and auxiliary materials. The material handling components for this are Automated Guided Vehicle (AGV), stacker cranes, conveyers, etc.

3. Computer System: The computer system monitors the entire manufacturing system. Computer performs large functions and sends reports. Functions of computer in CIM are:

(a) Machine control
(b) Direct numerical control
(c) Production control
(d) Traffic control
(e) Shuttle control
(f) Work handling system monitoring
(g) Tool control
(h) System performance monitoring and reporting

To control the operation of the manufacturing system the computer relies on data files. The principal data files required are: part program file, routing file, part production file, pallet reference file, station tool file, tool life file. The data collected during monitoring can be summarised for preparation of performance reports. The different categories of reports are: utilisation reports, production reports, status reports, tool reports.

4. Human Labour in CIM: The computer integrated manufacturing system has a highly automated production facility. However, human resources are required to operate, manage, maintain, and service the CIM. The personnel required for CIM are:

(a) System manager
(b) Electrical technician
(c) Mechanical / hydraulic technician
(d) Tool setter
(e) Fixture set up and lead man
(f) Load / unload man
(g) Rover operator or repairer

8.22 ADVANCED CAM TECHNIQUES

Several advances have been made in recent times in the field of computer aided manufacturing. Expert systems, artificial intelligence and smart manufacturing have revolutionised the modern world in such a way that the entire activities in a manufacturing plant can be handled by machines only.

8.22.1 Expert Systems

Expert systems are high performance computer programmes that deal with specialised areas of different problems and use knowledge and inference procedures developed from the knowledge of experts, to solve them. Expert system is a sort of database of knowledge, together with rules that reflect experience and decision making. Expert systems can be applied very usefully in solving practical problems in the manufacturing environments including material handling systems. The language used in computer programmes should be simple, concise and easy to understand.

A typical expert system includes a knowledge base, an inference engine, and a user interface. The knowledge base of an expert system consists of descriptions, relationships, and procedures pertaining to particular domain. Descriptions identify and differentiate objects and classes and generally include rules regarding the application of the descriptions in specific applications. Relationship describes the dependencies, and associations among the information in a knowledge base. Procedures specify the reasoning operation to be performed. The interface engine is the control system that directs the implementation of knowledge. It essentially decides which heuristic search techniques are used to determine, how the rules in the knowledge base are to be applied to the problem. The user interface of an expert system is intended for bidirectional communications.

8.22.2 Artificial Intelligence

Artificial intelligence is the subfield of computer science concerned with designing of intelligent computer systems that exhibit the characteristics associated with intelligence in human behaviour, understanding language, learning, reasoning, solving problems, etc. Artificial intelligence enables unattended manufacturing by capturing the craftsman's knowledge and using it automatically. It utilises an expert system (computer programme) developed from the knowledge of experts. The machine tools and factory systems can be controlled automatically when these are designed to have ability to detect and diagnose errors in manufacturing system.

Artificial intelligence represents a technique for solving problems in a better way than is available with conventional computer programmes. A conventional computer programme typically relies of algorithmic solutions in which a finite number of explicit steps produce the solution to a specific problem. Artificial intelligence uses heuristic search. Heuristic search is a rule of thumb approach that suggests a procedure for attempting to solve problem. Artificial intelligence can be used in almost any step of manufacturing life cycle, including design, engineering, production planning, and the actual management and its scheduling. It offers opportunities for diagnosis of machines and process and can be applied to sensor based monitoring and control. Artificial intelligence forms the basic tool for smart manufacturing.

8.22.3 Smart Manufacturing

Smart manufacturing is a manufacturing technique which uses artificial intelligence as a tool to make existing computer program (expert system) and the machines and systems they serve smarter. This is achieved by capturing the knowledge of their masters, the experts who make the routine decision about those machines and systems. Once captured, the resulting expert system makes it possible to free the human experts from making those routine decisions and thereby, makes them available to use their intelligence to deal with new challenges.

Questions

1. What is a computer?
2. What are the characteristics of computer?
3. Explain the types of computers.
4. What are micro computers?
5. What do you mean by a palmtop?
6. What do you mean by simputer?
7. What is the difference between notebooks and netbooks?
8. What is meant by wearable computer?
9. What is a personal computer?
10. Differentiate between nettops and desktops.
11. What is workstation? Explain its purpose.
12. What are the uses of mainframe computers?
13. What do you mean by super computer?
14. What are the merits of computer?
15. What are the demerits of computer?

16. Define a programmable controller. What are its basic components?
17. What are the functions of programmable controller? Mention its advantages.
18. What do you mean by computer hardware? Mention its four units.
19. Define registers. Mention their functions.
20. Define buses.
21. Define software. Explain the types.
22. Define integrated circuits.
23. Define microprocessor.
24. What do you mean by printed circuit board?
25. Define memory. Mention its unit.
26. What is meant by access time?
27. Explain the types of memory.
28. Differentiate RAM and ROM.
29. What do you understand by cache memory?
30. Explain the hardware devices used in computer for secondary storage technology.
31. What are input devices? Explain their types.
32. What are output devices? Explain their types.
33. Explain the term dpi, cps and ppm.
34. Explain common peripheral devices used for computer input / output.
35. Define baud.
36. Write a note on teleprinter.
37. Explain MICR, OCR, OMR and OBR.
38. Explain the types of computer programming language.
39. Define machine language.
40. Define assembly language.
41. What is assembler?
42. Explain high level language.
43. Define translators in programming language.
44. Differentiate compilers and interpreter.
45. Expand the term FORTRAN, COBOL and BASIC.
46. What do you mean by digit in number system?
47. What do you mean by MSD and LSD?
48. Explain various positional number systems.
49. Define algorithm.
50. Define flowchart.
51. Write an algorithm for finding the average of four numbers.
52. Write a flowchart for finding the larger of two numbers.
53. Differentiate constants and variables.

54. Define operators. Name their types.
55. What is the purpose of control statements? Mention their types.
56. Define ICG.
57. Explain the working of CRT with a neat diagram.
58. Differentiate stroke writing and raster scan.
59. Define pixels.
60. Explain the types of graphics terminals.
61. Expand the term DBR and DVST.
62. Define the term frame buffer.
63. How are monitors rated?
64. Explain the display devices used recently in graphics terminals.
65. Expand the term LCD and LED.
66. Define computer file.
67. Define data structure.
68. Define database.
69. What are the objectives of database?
70. What are the disadvantages of database?
71. Distinguish files, records and fields of database.
72. Explain the database operations.
73. Explain the types of data structure.
74. Expand the term LAN, WAN and MAN. Explain them.
75. Differentiate between Internet and intranet.
76. Define CAD.
77. Explain the steps involved in general design process.
78. Explain the steps involved in CAD process.
79. What are the benefits of implementing CAD?
80. Define shape and size.
81. Define parametric programming. Write down the steps involved in it.
82. What are the rules to be followed while dimensioning?
83. Define two dimensional drawing.
84. Explain various creation tools used in CAD software.
85. Explain various editing tools used in CAD software.
86. Explain various support tools used in CAD software.
87. Differentiate fillet and chamfer.
88. Differentiate trim and extend.
89. What are the advantages of CAD drawings over manual drawings?
90. Define wireframe modelling.
91. What are the limitations of wireframe modelling?

92. Define surface modelling.
93. What are the advantages of surface modelling over wireframe modelling?
94. Explain types of surface.
95. Define solid modelling.
96. What are the advantages of solid modelling?
97. Define primitives and give examples.
98. What do you mean by Boolean operators? Give examples.
99. Explain the term CSG, C-Rep and B-Rep.
100. Define assembly. What are the factors to be considered in assembly?
101. Differentiate top down and bottom up assemblies.
102. What are the four types of modelling data?
103. Name the commonly used translators in CAD/CAM software.
104. Expand DXF, IGES and STEP.
105. Define rapid prototyping.
106. Mention six techniques used in rapid prototyping.
107. What are the steps involved in rapid prototyping?
108. What are the advantages and limitations of rapid prototyping?
109. Write down the applications of rapid prototyping.
110. Explain stereolithography technique with the use of a neat sketch.
111. Explain fused deposition modelling technique with the use of a neat sketch.
112. Explain laminate object manufacturing technique with the use of a neat sketch.
113. Explain selective laser sintering technique with the use of a neat sketch.
114. Explain three dimensional printing technique with the use of a neat sketch.
115. Define rapid tooling.
116. Define reverse engineering.
117. Write down the stages involved in reverse engineering.
118. Compare digitisers and coordinate measuring machines.
119. Define concurrent engineering.
120. Define CAE.
121. Define FEM. Write down the steps involved in it.
122. What are the advantages of FEM?
123. Write down the types of elements.
124. Define mesh. What are the two types of meshing?
125. What are the requirements of a good mesh?
126. What are the stages in FEA program?
127. What are the factors to be considered in FEA?
128. Write down various types of analysis in FEA.
129. What are the disadvantages of FEA?

130. What is the purpose of moldflow software?
131. What are the steps involved in moldflow?
132. Name and explain various analyses done in moldflow.
133. Define CAM.
134. Explain the CAM process.
135. What are the applications of CAM?
136. What are the benefits and advantages of CAM?
137. Write down the reasons for implementing CAD/CAM.
138. Define numerical control. Name its basic components.
139. Define program of instructions.
140. What are the elements of controller unit?
141. Define tape reader.
142. What is the purpose of data buffer?
143. What is the purpose of signal output channel?
144. What is the purpose of feed back channel?
145. What are the steps involved in NC?
146. Explain the classification of NC system.
147. Differentiate open loop and closed loop system.
148. What are the applications of NC?
149. Write down where NC machines are most suited?
150. What are the advantages and disadvantages of NC?
151. Write down the problems faced while using conventional NC.
152. Define part programming. Name two types of it.
153. Define manuscript.
154. Explain the computer's job in computer assisted part programming.
155. What is the purpose of input translation in CAPP?
156. What is the purpose of cutter offset computation in CAPP?
157. Explain NC words.
158. What do you mean by binary coding?
159. Explain the punched tape.
160. What is CNC? Write down its functions.
161. What are the advantages of CNC over NC?
162. Define DNC systems. Name their components.
163. Write down the functions of DNC.
164. What are the advantages of DNC?
165. Define adaptive control. Explain its working.
166. Write down the situations where AC can be suitably applied?
167. What are the advantages of AC?

168. Define CIM. Name three types of it.
169. What is flexible manufacturing system?
170. What are the advantages of CIM?
171. Explain the basic components of CIM.
172. Define expert systems.
173. Define artificial intelligence.
174. Define smart manufacturing.

References

1. Adithan. M and Pabla B.S., *CNC Machines*, New Age International Publishers, New Delhi.
2. Besant C. B. and Lui C.W.K., *Computer Aided Design and Manufacture*, Affiliated East-West Press, New Delhi.
3. Chockalingam P., *Computer Aided Design and Manufacturing*, Palani Paramount Publications, Palani.
4. Groover P. Mikell and Zimmers, Jr. Emery W., *CAD/CAM Computer-Aided Design and Manufacturing*, Prentice Hall of India, New Delhi.
5. Gundu Rao H.K., Manjunath N.S., Nachappa M.N., *A Text Book of Computer Fundamentals*, Sumukha Publications, Bangalore.
6. Pfaffenberger, Bryan, *Webster's New World Computer Dictionary*, IDG Books, New Delhi.
7. Rajaraman V., *Fundamentals of Computers*, Prentice Hall of India, New Delhi.
8. Zeid, Ibrahim, *Mastering CAD/CAM*, Tata McGraw-Hill, New Delhi.

Made in the USA
Monee, IL
25 March 2026

46885630R00369